MIMICRY, CRYPSIS, MASQUERADE AND OTHER ADAPTIVE RESEMBLANCES

MIMICRY, CRYPSIS, MASQUERADE AND OTHER ADAPTIVE RESEMBLANCES

Donald L. J. Quicke FRES, PhD
Professor, Chulalongkorn University, Bangkok, Thailand

WILEY Blackwell

Registered Office(s)
John Wiley & Sons, Inc., 111 River Street, Hoboken, NJ 07030, USA
John Wiley & Sons Ltd, The Atrium, Southern Gate, Chichester, West Sussex, PO19 8SQ, UK

Editorial Office
9600 Garsington Road, Oxford, OX4 2DQ, UK

For details of our global editorial offices, customer services, and more information about Wiley products visit us at www.wiley.com.

Wiley also publishes its books in a variety of electronic formats and by print-on-demand. Some content that appears in standard print versions of this book may not be available in other formats.

A catalogue record for this book is available from the Library of Congress and the British Library.

ISBN 9781118931530

Cover images: courtesy of Donald L.J. Quicke (except top left on back cover, courtesy of Linda Pitkin)

Set in 9/11pt Photina by SPi Global, Pondicherry, India

Printed and bound in Singapore by Markono Print Media Pte Ltd

10 9 8 7 6 5 4 3 2 1

*For all the dogs who live on streets everywhere with no-one to love them,
and to those wonderful organisations such as Soi Dog Foundation in Thailand,
that work hard to care for them.*

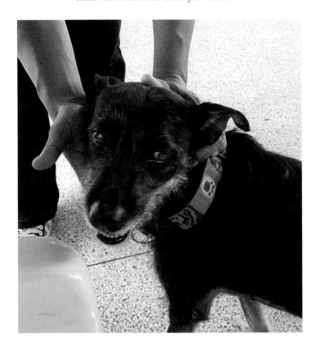

This is wonderful Puii. She lives at Saphan Taksin Pier.

CONTENTS

Preface, xiii
  A comment on statistics, xv
  A comment on scientific names, xvi

Acknowledgements, xvii

1 INTRODUCTION AND CLASSIFICATION OF MIMICRY SYSTEMS, 1

A brief history, 2
On definitions of 'mimicry' and adaptive resemblance, 3
  The concept of 'adaptive resemblance', 8
The classification of mimicry systems, 9
  Wickler's system, 9
  Vane-Wright's system, 10
  Georges Pasteur (1930–2015), 11
  Other approaches, 13
    Endler, 13
    Zabka & Tembrock, 13
    Maran, 14
Mimicry as demonstration of evolution, 14

2 CAMOUFLAGE: CRYPSIS AND DISRUPTIVE COLOURATION IN ANIMALS, 19

Introduction, 20
Distinguishing crypsis from masquerade, 20
Crypsis examples, 24
Countershading, 24
  Experimental tests of concealment by countershading, 27
  Bioluminescent counter-illumination, 28
Background matching, 29
  Visual sensitivity of predators, 30
  To make a perfect match or compromise, 31
Colour polymorphism, 32
  Seasonal colour polymorphism, 32

Butterfly pupal colour polymorphism, 32
  Winter pelage: pelts and plumage, 35
Melanism, 37
  Industrial melanism, 37
  Fire melanism, 40
Background selection, 41
  Orientation and positioning, 43
Transparency, 45
Reflectance and silvering, 47
Adaptive colour change, 49
  Caterpillars and food plant colouration, 50
  Daily and medium-paced changes, 54
  Rapid colour change, 56
    Chameleons, 56
    Cephalopod chromatophores and dermal papillae, 57
Bird eggs and their backgrounds, 58
Disguising your eyes, 61
Disruptive and distractive markings, 61
  Edge-intercepting patches, 61
  Distractive markings, 63
  Zebra stripes and tsetse flies, 66
Stripes and motion dazzle – more zebras, kraits and tigers, 69
  Computer graphics experiments with human subjects, 69
  Observations on real animals, 69
  Comparative analysis, 71
Dual signals, 72
Protective crypsis in non-visual modalities, 73
Apostatic and antiapostatic selection, 73
  Search images, 74
  Experimental tests of search image, 76
  Gestalt perception, 76
Effect of cryptic prey variability, 77
  Reflexive selection and aspect diversity, 77
  Searching for cryptic prey – mathematical models, 80

Ontogenetic changes and crypsis, 81
Hiding the evidence, 82
    Petiole clipping by caterpillars, 82
    Exogenous crypsis, 82
Military camouflage and masquerade, 85

3   CAMOUFLAGE: MASQUERADE, 87

Introduction, 88
Classic examples, 88
    Twigs as models, 88
    Leaves (alive or dead) as models, 88
    Bird dropping resemblances, 89
    Spider web stabilimenta, 93
    Tubeworms, etc., 94
Experimental tests of survival value of masquerade, 94
Ontogenetic changes and masquerade, 97
Thanatosis (death feigning), 97
    Feign or flee? The trade-offs of thanatosis, 100
    Other aspects of death mimicry, 100
Seedless seeds and seedless fruit, 100

4   APOSEMATISM AND ITS EVOLUTION, 103

Introduction, 104
Initial evolution of aposematism, 108
    Associations of unpalatable experience
      with place, 109
Mathematical models and ideas of warning colouration
    evolution, 112
    Kin selection models, 112
    Green beard selection, 112
    Family selection models, 113
    Individual selection models, 113
    Spatial models and metapopulations, 116
Handicap and signal honesty, 117
    Early warnings – reflex bleeding, vomiting
      and other noxious secretions, 120
Longevity of aposematic protected taxa, 121
Macroevolutionary consequences, 121
Experimental studies, 121
    Tough aposematic prey and individual selection, 121
    Pyrazine and other early warnings, 123
Learning and memorability, 124
    Strength of obnoxiousness, 126
    Is the nature of the protective compound
      important?, 126
    Neophobia and the role of novelty, 127
Innate responses of predators, 130
Aposematism and gregariousness, 132
    Phylogenetic analysis of aposematism
      and gregariousness, 134

Behaviour of protected aposematic animals, 135
    Of birds and butterflies, 135
    Evolution of sluggishness, 139
Origins of protective compounds, 140
    Plant-derived toxins, 140
    Cardiac glycosides, 141
    Pyrrolizidine alkaloids, 144
    De novo synthesis of protective compounds, 145
    Obtaining toxins from animal sources, 147
    Costs of chemical defence, 149
Aposematism with non-chemical defence, 150
    Escape speed and low profitability, 150
Parasitoids and aposematic insects, 152
Diversity of aposematic forms, 152
    Egg load assessment, 154
Proof of aposematism, 154
    Bioluminescence as a warning signal, 155
    Warning sounds, 155
    Warning colouration in mammals, 157
    Weapon advertisement, 158
    Mutualistic aposematism, 160
    Aposematism induced by a parasite, 161
    Aposematic commensalism, 161
Polymorphism and geographic variation in
    aposematic species, 161
Aposematism in plants, 163
    Synergistic selection of unpalatability
      in plants, 165
Aposematism in fungi, 166
Why are some unpalatable organisms aposematic
    and others not?, 167

5   ANTI-PREDATOR MIMICRY. I. MATHEMATICAL MODELS, 171

Introduction, 172
Properties of models, rewards, learning rates and
    numerical relationships, 172
Simple models and their limitations, 173
    Müller's original model, 173
    Simple models of Batesian and
      Müllerian mimicry, 173
    Are Batesian and Müllerian mimicry
      different?, 174
    An information theory model, 176
    Monte-Carlo simulations, 177
More refined models – time, learning, forgetting
    and sampling, 180
    Importance of alternative prey, 181
    Signal detection theory, 181
    Genetic and evolutionary models, 182
    Coevolutionary chases, 185

Models involving population dynamics, 185
Neural networks and evolution
of Batesian mimicry, 188
Automimicry in Batesian/Müllerian mimicry, 188
Predator's dilemma with potentially harmful prey, 190

6   ANTI-PREDATOR MIMICRY. II. EXPERIMENTAL TESTS, 191

Introduction, 192
Experimental tests of mimetic advantage, 192
How similar do mimics need to be?, 194
Is a two-step process necessary?, 198
Relative abundances of models and mimics in nature, 198
Sex-limited mimicries and mimetic load, 198
Mimetic load, 203
Apostatic selection and Batesian mimicry, 204
Müllerian mimicry and unequal defence, 204
Imperfect (satyric) mimicry, 206

7   ANTI-PREDATOR MIMICRY. III. BATESIAN AND MÜLLERIAN EXAMPLES, 213

Introduction, 214
Types of model, 214
Mimicry of slow flight in butterflies, 214
The Batesian/Müllerian spectrum, 215
Famous butterflies: ecology, genetics and supergenes, 216
Heliconius, 216
Hybrid zones, 217
Wing pattern genetics, 219
Modelling polymorphism, 220
Danaus and *Hypolimnas*, 220
Papilio dardanus, 221
Papilio glaucus, 223
Papilio memnon, 223
Supergenes and their origins, 223
Mimicry between caterpillars, 224
Some specific types of model among insects, 225
Wasp (and bee) mimicry, 225
How to look like a wasp, 228
Time of appearance of aculeate
mimics, 228
Pseudostings and pseudostinging
behaviour, 230
Wasmannian (or ant) mimicry, 231
Ant mimicry as defence against
predation, 231
Ant mimicry by spiders, 234
Spiders that feed on ants, 236
How to look like an ant or an ant carrying
something?, 236

Myrmecomorphy by caterpillars, 237
Ant chemical mimicry by parasitoid
wasps, 237
Protective mimicries among vertebrates, 239
Fish, 239
Batesian mimicry among fish, 239
Müllerian mimicry among fish, 239
Batesian and Müllerian mimicry among
terrestrial vertebrates, 239
The coral snake problem – Emsleyan
(or Mertensian) mimicry, 240
Other snakes, zig-zag markings and head
shape, 244
Mimicry of invertebrates by terrestrial
vertebrates, 246
Inaccurate (satyric) mimics, 248
Mimicry of model behaviour, 249
Aide mémoire mimicry, 250
Batesian–Poultonian (predator) mimicry, 251
Mimicry within predator–prey and host–parasite
systems, 253
Bluff and appearing larger than you are, 253
Collective mimicry including an aggressive mimicry, 255
Jamming, 255
Man as model – the case of the samurai crab, 258

8   ANTI-PREDATOR MIMICRY. ATTACK DEFLECTION, SCHOOLING, ETC., 259

Introduction, 260
Attack deflection devices, 260
Eyespots, 260
Experimental tests of importance of eyespot
features, 262
Eyespots in butterflies, 266
Wing marginal eyespots, 267
Eyes with sparkles, 267
Eyespots on caterpillars, 269
Importance of eyespot
conspicuousness, 269
Eyespots and fish, 269
Not just an eyespot but a whole head,
winking and other
enhancements, 271
Reverse mimicry, 271
Insects, 271
Reverse mimicry in flight, 275
Reverse mimicry in terrestrial
vertebrates, 275
Other deflectors, 277
Injury feigning in nesting birds, 277
Tail-shedding (urotomy) in lizards and snakes, 277

Flash and startle colouration, 280
    Intimidating displays and bizarre mimicries, 283
Schooling, flocking and predator confusion, 284
    'Social' mimicry in birds and fish, 286
    Alarm call mimicry for protection, 287

9 ANTI-HERBIVORY DECEPTIONS, 289

Introduction, 290
Crypsis as protection in plants, 290
    Leaf mottling and variegation for crypsis, 291
    Mistletoes and lianas, 293
    Fruit masquerade by leaves, 294
Protective Batesian and Müllerian mimicry in plants, 295
    False indicators of damage or likely
        future damage, 296
    Conspicuousness of leafmines, 297
    Dark central florets in some Apiaceae, 297
    Mimicry of silk or fungal hyphae, 299
    Insect egg mimics, 299
    Defensive aphid and caterpillar mimicry
        in plants, 300
    Aphid deterrence by alarm pheromone
        mimicry, 300
    Ant mimicry in plants, 301
    Of orchids and bees, 301
    Carrion mimicry as defence, 302
    Algae and corals, 302
    Plant galls, 302
    Experimental evidence for plant aposematism
        and Batesian mimetic potential in plants, 302

10 AGGRESSIVE DECEPTIONS, 305

Introduction, 306
    Cryptic versus alluring features, 307
Crypsis and masquerade by predators, 307
    Stealth, 307
    Shadowing, 308
    Seasonal polymorphisms in predators, 308
    Why seabirds are black and white (and grey), 309
    Chemical crypsis by a predatory fish, 309
Alluring mimicries, 310
    Flower mimicry, 312
    Rain mimicry, 315
    Physical lures, 315
    Angling fish, 315
    Caudal (and tongue) lures in reptiles, 317
    Caudal lure in a dragonfly, 318
    Death feigning as a lure, 318
    Other prey and food mimicry, 319
    The case of the German cockroach, 319
Wolves in sheeps' clothing, 319

Vulture-like hawks, 319
    Cleaner fish and their mimics, 320
    Mingling with an innocuous crowd, 322
Duping by mimicry of competitors, 323
Seeming to be conspecific, 324
    Getting close, 325
    Appearing to be a potential mate, 325
    Pheromone lures, 326
Mimicking danger as a flushing device, 328
    Human use of aggressive mimicry, 328
Cuckoldry, inquilines and brood parasitism, 329
    Cuckoldry in birds, 329
    Gentes and 'cuckoo' eggs, 332
        Cues for egg rejection, 335
    Mimicry by chicks – genetic and substantive
        differences, 338
        Cuckoo chick appearance, 338
        Begging calls, 339
    Cuckoo and host coevolution, 340
    Mimicry between adult cuckoos and their
        hosts, 340
    Hawk mimicry by adult cuckoos, 340
    Mimicry of harmless birds by adult cuckoos, 342
    Brood parasitism and inquilinism in
        social insects, 342
    Cuckoo bees and cuckoo wasps, 342
    Kleptoparasites of bees, 346
    Myrmecophily, 346
    Acquired chemical mimicry in social parasites
        and inquilines, 346
    Brood-parasitic and slave-making ants, 348
    Chemical mimicry and ant and termite
        inquilines, 349
        A brood-parasitic aphid, 349
    Ants and aphid trophallaxis, 349
    Aphidiine parasitoids of ant-attended
        aphids, 350
Does aggressive mimicry occur in plants?, 350

**11 SEXUAL MIMICRIES IN ANIMALS
(INCLUDING HUMANS), 353**

Introduction, 354
Mimicking the opposite sex, 354
    Female mimicry by males, 354
    Avoiding aggression from competing males, 357
    Mate guarding through distracting other
        males, 357
    Androchromatism and male mimicry by
        females, 358
    Egg dummies on fish, 360
    Food dummies and sex, 362

Mimicry by sperm-dependent all-female
lineages, 363
Female genital mimicry in a female, 363
Energy-saving cheating for sex, 364
Behavioural deceptions in higher vertebrates, 364
Polygynous birds, 364
Deceptive use of alarm calls and
paternity protection, 365
Female–female mounting behaviour in
mammals and birds, 365
Mimicry in humans, 367
Make-up, clothes and silicone, 367
Cryptic oestrus in humans, 368
Flirting in humans, 368

**12 REPRODUCTIVE MIMICRIES
IN PLANTS, 371**

Introduction, 372
Pollinator deception, 372
Pollinator sex pheromone mimicry, 376
Food deception, 382
Specific floral mimicry, 382
Generalised floral mimicry, 386
Mimicry of a fungus-infected plant, 388
Brood-site/oviposition-site deception, 388
Shelter mimicry, 392
Flower similarity over time, 392
Flower automimicry – intraspecific food deception
(bakerian mimicry), 393
Mathematical modelling of sexual deception by
plants, 394
Pollinator guild syndromes, 394
Bird-pollinated systems, 394

**13 INTRA- AND INTERSPECIFIC
COOPERATION, COMPETITION
AND HIERARCHIES, 399**

Introduction, 400
Remaining looking young, 400
Delayed plumage maturation, 400
Interspecific social dominance mimicry, 401
Bird song and alarm call mimicry – deceptive
acquisition of resources, 401
Wicklerian mimicry – mimicry of opposite sex to reduce
aggression, 403
Female resemblance in male primates, 403
Social appeasement by female mimicry
in an insect, 404
Hyperfemininity in prereproductive adolescent
primates, 404

Mimicry of male genitalia by females, 404
The case of the spotted hyaena, 404
Mimicry of male genitalia in other mammals, 404
Phallic mimicry by males, 405
Appetitive (foraging) mimicry, 406
Appetitive mimicry and deceptive
use of alarm calls, 406
Beau Geste and seeming to be more
than you are, 408
Appearing older than you are, 408
Weapon automimicry, 408

**14 ADAPTIVE RESEMBLANCES
AND DISPERSAL: SEEDS, SPORES
AND EGGS, 409**

Introduction, 410
Fruit and seed dispersal by birds, 410
Warningly coloured fruit, 414
Fruit mimicry by seeds, 414
Seed dispersal by humans, arable weeds and
Vavilovian mimicry, 414
Seed elaiosomes and their insect mimics, 415
Mimicry by parasites to facilitate host finding, 415
The trematode and the snail, 415
The trematode and the fish, 416
Pocketbook clams and fish, 416
'Termite balls', 417
Pseudoflowers, pseudo-anthers and
pseudo-pollen, 417
Truffles, 418
Mimicry of dead flesh by fungi and mosses, 419
Deception of dung beetles by fruit, 419

**15 MOLECULAR MIMICRY: PARASITES,
PATHOGENS AND PLANTS, 421**

Introduction, 422
Macro-animal systems, 422
Anemone fish, 422
Parasitic helminthes, 422
Platyhelminthes (Trematoda), 422
Tapeworms (Platyhelminthes: Cestoda), 423
Parasitic nematodes, 423
Parasitoid wasp eggs, 424
Pathogenic fungi, 424
Protista, 424
Chagas' disease, 424
Microbial systems, 424
Bacterial chemical mimicry and
autoimmune responses, 424

Helicobacter pylori, 425
Campylobacter jejuni, 425
Mimicry by plant-pathogenic bacteria, 425
Viruses, 425
Plants, 425
Sugar, toxin and satiation mimicry, 425
Phytoecdysteroids – plant chemicals that mimic
insect moulting hormone, 427
Plant oestrogens – phyto-contraceptives, 427

Extended glossary, 429

References, 445

Author index, 515

General index, 533

Taxonomic index, 539

PREFACE

The ever expanding field of mimicry requires a clear, but very elastic, definition which avoids hair splitting but allows for the constant stream of new examples and concepts.

Miriam Rothschild 1981

This book started almost 40 years ago with discussions with my old friend and fellow undergraduate Peter Kirby at Oxford University and subsequently in Derby and Nottingham, to whom I am greatly indebted for many a valuable discussion and pint of beer. Since then it has expanded due to discussions with many people. As things do, plans got shelved and occasionally revisited, and shelved again. This whole period has seen a remarkable increase in interest in various forms of mimicry and adaptive resemblance with a huge body of more experimental work in addition to theory supplementing the already vast number of casual and sometimes insightful descriptions of mimicry systems around the world.

Ever since Henry Bates (1825–92), an English naturalist working in tropical South America from 1848[1] to 1859, realised that some butterflies were not what they might at first appear to be, and interpreted this as being due to palatable species looking like unpalatable models (Bates 1862, 1864), mimetic phenomena have fascinated professional and amateur biologists alike, including Charles Darwin, whose theory of evolution had been part of Bates' inspiration (Moon 1976, Stearn 1981). As time passed the literature on the topic grew as more and more examples were described and as Holling (1963) said, "A small mountain of information about mimicry has been collected since Bates".

Unfortunately though, after a while this fascination became been tinged with rather negative views among some academic biologists who for some while tended to dismiss it as a quaint set of observations that are easily explained and not worth dwelling over in any great detail – though it was still used for its 'wow factor' in undergraduate lectures. Vane-Wright (1981) also lists some of the negative or simply dismissive views that were put forward, especially soon after Bates' publication, including that the resemblances described were just coincidental and therefore pointless to investigate. However, some researchers continued to investigate mimicry both theoretically and experimentally and, with clearer thought emerging about the detailed processes involved, mimicry (and camouflage) have had a resurgence of scientific interest with the result that many new examples have come to light in recent years and many new insights are continuing to emerge. Large-scale studies are becoming increasingly common; molecular techniques are allowing the evolution of mimicries and other adaptive resemblances to be viewed from a phylogenetic perspective; and increasingly sophisticated use of computer games allows testing of theories that are relatively new to the scene. Lichter-Marck et al.'s (2014) 4-year study of caterpillar predation in a temperate project is a nice example that combines all of these aspects and enabled comparison of the effectiveness of warning and camouflage strategies.

This book sets out to survey mimicry and camouflage (and the related topic of how aposematism evolved in the first place) and to place these in the context of results of the growing numbers of experimental tests that have been conducted. It further seeks to explain key and relevant models and experimental set-ups in a way intelligible to everyone and not just scientists. All of this draws on a wide range of examples from animals, plants, fungi and even protists, covering different modalities such as behaviour, colouration, bioluminescence, structure, chemistry and sound. Most examples are of the whole organism type but mimicry is also relevant to the success of various disease agents, so bacteria are also included. It also covers adaptive resemblances which have evolved for protection from predation or herbivory, to obtain prey (aggressive mimicries), to obtain

---

1. Initially he was accompanied by his friend Alfred Russel Wallace, but Wallace returned to England in 1852 and sadly his collection was lost at sea.

matings or, more precisely, fertilisations (sexual mimicry), to disperse seeds or spores, to avoid aggression from conspecifics and to protect from host immune systems, some of which lead to unfortunate autoimmunity consequences.

I have tried to combine as much interesting biology related to the topic as reasonable, and also to explain how mathematical models of various degrees of sophistication give new insights into how mimicry systems work and why warning signals and mimicry evolve under some circumstances and not others. I also touch on aspects such as the genetics underlying wing pattern polymorphisms in various insects, notably in the genus *Heliconius* and the sex-limited cases among swallowtails, which should at least give an inroad into the relevant booming genomics literature. Where enough data exist I have tried to separate out the more mathematical parts from descriptions of the mimetic systems themselves and also to some extent I have separated out some of the basic experimental tests of mimetic advantage when there are enough of them to make a separate coherent section. It is one thing to think that a harmless dronefly has evolved to resemble and mimic a stinging honey bee because of the potential protection that would afford it, but quite another to actually demonstrate that this is what has happened. Indeed, one could ask, why haven't honey bees evolved to look like wasps, or vice versa? Well, in some ways they do resemble one another, and as nearly all entomologists can vouch, a large number of non-biologists do confuse them. Indeed, in many languages there is no separate word for them – maybe they are just some sort of stinging insect or, maybe, Hymenoptera-like insect. And yet honey bees, despite many an illustration in children's books, are not normally boldly banded black and yellow; they may have orangey bands, but they are hardly highly conspicuous (see Fig. 7.24a). Not surprisingly, humans are also quite bad at distinguishing harmless hoverfly mimics from potentially stinging bees and wasps (Golding et al. 2005a). The fact that humans often do not distinguish wasps from bees does not mean that many insectivores do not, and indeed, the amazing similarities between some models and their mimics is testament to the amazing discriminatory powers of predators that have shaped and coloured them over evolutionary time. Wickler (1968), with rather fewer examples and far less experimental evidence, gives some lovely descriptions of many cases discussed here. His book was also beautifully and inspiringly illustrated so, although I have tried to obtain photographs to illustrate most systems, his work provides lots of informative pictures that I have cited where I could not find better.

For the vast majority of supposed cases of mimicry there have been no experimental tests, and for quite a few there is very little by way of field observations, perhaps just assumptions based upon museum specimens. Thus, I rather think that Vane-Wright's (1971) note with regard to his discussion of mimicry: "In this discussion such words as 'possibly', 'perhaps', 'presumably' are frequently omitted for convenience, where strictly they ought to be employed" could be applied to many examples herein. Certainly some suggested instances of mimicry could be considered as verging on the fanciful and some have been subsequently disproven. However, there is probably truth in the vast majority of cases. Applications of more modern techniques, such as visual modelling of potential predators, sometimes reveal things that human eyes miss, and may provide clearer explanation. As Grim (2013) points out, actually demonstrating that some feature has evolved due to mimicry is extremely difficult and proof positive can only be achieved by manipulatory experiments. Sometimes what human observers perceive as close resemblance may simply reflect our own limitations in discriminatory ability, and not necessarily those of the organisms involved. Nevertheless, although some suspected instances have turned out not to involve mimicry, I feel that the vast majority of described cases will be verified in due course. Further, as numbers of system types are subjected to experimental studies we might be justified in allowing some degree of inclusivity in terms of individual examples studied – after all, it is unrealistic to test experimentally, say, all cases of, for example, snakes using tails to lure prey – if it is found proven in a few species it seems likely that it will be true also of at least the majority of other species.

Where possible, I have included photographs to illustrate the main types of resemblances and adaptations that are discussed, though sadly space does not permit everything. In discussing individual examples, it also soon becomes clear that many adaptive resemblances serve dual functions (e.g. Gomez & Thery 2007). A bright yellow and black wasp once in the hand, so to speak, is clearly aposematic, but from a distance against a dry, yellow African savanna it may be hard to spot; similarly, bright red wasps against red lateritised tropical soils. The similarity of a flower mantis to a flower or cluster of flowers is both a device that helps it to avoid being detected by, and probably also positively attracts, the butterflies and bees upon which it feeds, but it also conceals it from avian or other predators; thus it is both a protective and an aggressive mimicry and it seems impossible to know which came first. Particularly convoluted is the case of 'cuckoos' and their hosts. Their eggs are typically very similar to those of the host bird, and that mimetic resemblance is determined by the mother cuckoo's genes, but once the cuckoo chick has hatched the young bird's gape, mimicking that of the host-bird's chick, is controlled by the chick's genes, as are the calls made by the cuckoo chick. But it doesn't stop there because in some species the adult birds may mimic their hosts' calls to distract them away from their nests or mimic prey with the same result and as an even more complicated twist in the case of some widow-birds, which also behave like cuckoos, the females will only mate if their male partner mimics the call of their brood host species. Given such complexity, I largely gave up in the case of cuckoos and discuss their various mimetic adaptations all together, despite their mimicries having multiple functions and involving different individuals, models and genes.

A comment on statistics

This book is meant to be of interest to both academic readers and lay people with an interest in natural history, mimicry, evolution, etc. Thus where I have presented equations I have tried to explain them, and I have also tried to explain the numerous graphs so that they are as comprehensible and interesting as possible. However, it must be understood that in the real world there is variation and 'noise' in the data that scientists collect in order to test hypotheses. Researchers therefore need to appraise their results against some 'yard-stick' – a measure of whether their particular findings are probably important or whether some observed trend might most likely just be due to chance. To make such assessments, researchers use a wide range of statistical tests which are essentially aimed at asking one (theoretically) simple question – how likely is it that the trend that has been observed is due to chance alone?

Just because, say, a bird in a study eats more red seeds than green ones when presented with equal numbers of each, it does not necessarily mean that this is a real effect. To assess whether the result reflects a true feature of that bird's behaviour rather than just being due to chance, scientists estimate, as accurately as possible, what the probability of obtaining that particular result purely by chance would have been. Then, completely arbitrarily but almost univer-sally accepted in biology, a result is deemed statistically significant if the chance of it having been found due just to noise in the system is less than one in 20. In other words, a result is deemed significant if it was likely to occur in less than 5% of observations/tests if there was no actual effect or interaction between the variables. In that case, the probability is expressed as a 'p-value' and it is written as '$p < 0.05$', i.e. less than 5%.

If the effect is strong, one might distinguish a p-value of <0.01 (one in a hundred) or <0.001 (one in a thousand). Of course, it does not necessarily mean that there is a real effect, perhaps a real causal relationship between observed measures, because if you do 20 experiments where you know that there cannot be any effect, on average one in 20 will generate results that have a p-value of about one in 20.

Finding the same effect in multiple *independent* experiments adds to the experimenter's confidence that they are dealing with a real phenomenon.

Many different tests are used by researchers depending upon the nature of the data, and the maths behind them can be quite complicated. One hopes that scientists have used appropriate tests and that referees during the peer-review process will have spotted any potential errors and had them corrected.

Let us go back to the bird and its choice of a particular number of red or green seeds. Even if the result seems extreme, say the bird eats nine green and only one red seed, when the null expectation given even numbers of each type were presented to start with, would be five red and five green, we want to know whether the 9:1 result can be considered a significant departure from random. If that result would be very unlikely to occur by chance alone, it might indicate a significant food colour bias by the bird. The appropriate sta-tistical test in such a case would be the chi-square (χ^2) test and in this case it is significant: $\chi^2 = 6.4$, d.f. $= 1$, $p = 0.01141$, i.e. the chance of the bird picking nine out of ten of all the same colour purely by chance if it had no particular bias is one in $1/0.01141$ or one in 87.6. A result of eight of one type and just two of the other is not significant (N.S.) at the 0.05 level ($\chi^2 = 3.6$, d.f. $= 1$, $p = 0.05778$). You can get a feel for this by thinking about tossing coins. If you toss ten coins, how likely is it that you will have nine or more out of ten heads, or nine or more out of ten tails? This sample size is small, but if you were gambling and there was a highly sig-nificant departure from 50:50 in the coin tosses, one might start to think that there was some skulduggery going on. Larger sample sizes provide better tests of the data.

The other thing that you will see in most of the results of statistical tests is something called 'degrees of freedom' or d.f. for short. This value is the number of values that went into the calculation of the statistic that are free to vary without *necessarily* changing the final statistic, and this is generally the number of classes or observations minus one, but for some tests there can be more than one degree of free-dom. For example, if there are ten numerical observations recorded then you can vary nine of them, and still obtain the same value of the test statistic by adjusting the 10th value, hence nine degrees of freedom.

The final important aspect to understand some of the experiments and results presented is independent. If the experimenter had got the N.S. 8:2 result the first time and then the next day ran the test again ran and got, say, 7:3 in the same direction, well then the total (15:5) is now signifi-cantly different from an even expectation ($p = 0.00254$), and may indeed reflect a true bias against eating red seeds. However, if the same individual bird were used, all it tells you is that one bird probably has a preference for green over red seeds. It does not necessarily mean that all members of its species have such a preference, nor that all birds do.

While mammalian and human physiologists have their laboratory rats (preferred, easy to work with, experimental animals, often rats but sometimes mice, cats, dogs, monkeys, etc., whatever is appropriate for a particular study) so too have experimenters on mimicry, and birds in particular. American blue jays, European blue tits and great tits have been the source of a very large proportion of the results referred to here. The hope, of course, is that the decisions they make are representative of those that many insectivorous birds would make, and while that might be true, there are bound to be exceptions.

For those wishing to get to grips with statistical analysis of data or making nice graphical outputs, I highly recommend Mick Crawley's 'R book' (Crawley 2007).

A comment on scientific names

Every formally named species is given a unique scientific name that comprises two parts, a species name and a genus name. A genus is a group of one or more species that taxonomists recognise as probably being a monophyletic unit, i.e. they have a single common ancestor and all the known descendants are included in the same genus. Genus and species names are distinguished from the rest of a text by being set differently, usually in italics. Genus names always start with a capital letter; specific names never do. No two different groups of animals are allowed to share the same generic name, and the same applies to plants, although the same species name may be used for different species so long as they are in different genera.

The taxonomic level of animal names between subtribe and superfamily is indicated by the suffix. Thus in increasing hierarchical order: -ina = subtribe, -ini = tribe, -inae = subfamily, -idae = family and -oidea = superfamily. For plants and fungi, family names end in -aceae. Animal names up to superfamily level are typified by a type genus, but above superfamily, such as order, class and phylum, the names are not standardised. In plants algae and fungi, all levels are typified, and phyla are traditionally called divisions. Fungal phylum names bear the suffix -mycota, and plant and algal phyla names have the suffix -phyta.

When higher group names are used as nouns they start with a capital letter, but when they are used as adjectives they do not (unless at the beginning of a sentence).

ACKNOWLEDGEMENTS

Many people have helped in the production of this book through discussion, drawing my attention to relevant articles, sending reprints, especially Lars Chittka, Peter Kirby, Armand Leroi, Shen-Horn Yen and Dick Vane-Wright. Obviously everyone who has supplied images that are so important to any book on mimicry is given credit for them in the captions, but I want to pay particular thanks to several who either waived normal fees out of friendship, or particularly sought out relevant things to photograph for me: Phil DeVries, Conrad Paulus G.T. Gillett, Dan Janzen, Simcha Lev-Yadun, James Mallet, Kenji Nishida, Linda Pitkin, Denis Reid and Claire Spottiswoode. I also wish to give special thanks to Yukiko Kayano for going to enormous trouble to obtain permission for me to include the photograph of the head of the Bunraku puppet Ki-Ichi, which is part of Japanese cultural heritage.

All graphs were produced using the statistical computing language R (R Development Core Team 2009). Image manipulations were carried out using GIMP (© 2001–2015 The GIMP Team).

Ward Cooper (my commissioning editor) at Wiley, Kelvin Matthews and, more recently, Sarah Keegan, Nick Morgan and freelance project manager Dr Nik Prowse, have helped enormously in overcoming many hurdles. Finally, I would like to say a special thank you to Dr Amoret Whitaker and Harriet Stewart-Jones for their very thorough copy-editing work, so any errors remaining are entirely down to me.

Chapter 1

INTRODUCTION AND CLASSIFICATION OF MIMICRY SYSTEMS

It is hardly an exaggeration to say, that whilst reading and reflecting on the various facts given in this Memoir, we feel to be as near witnesses, as we can ever hope to be, of the creation of a new species on this earth.

Charles Darwin (1863) referring to Henry Bates' 1862
account of mimicry in Brazil

Mimicry, Crypsis, Masquerade and other Adaptive Resemblances, First Edition. Donald L. J. Quicke.
© 2017 John Wiley & Sons Ltd. Published 2017 by John Wiley & Sons Ltd.

A BRIEF HISTORY

The first clear definition of biological mimicry was that of Henry Walter Bates (1825–92), a British naturalist who spent some 11 years collecting and researching in the Amazonas region of Brazil (Bates 1862, 1864, 1981, G. Woodcock 1969). However, as pointed out by Stearn (1981), Bates' concept of the evolution of mimicry would quite possibly have gone unnoticed were it not for Darwin's review of his book in *The Natural History Review* of 1863. Bates' observations of remarkable similarity between butterflies belonging to different families led him to ponder what might be the reason for this. He concluded that there must be some advantage, for example, for a 'white butterfly', *Dismorphia theucharila* (Pieridae), to depart from the typical form and colouration of the family, and instead to resemble unpalatable *Heliconius* species.[1] He also noticed that in all the bright and conspicuous butterfly colour pattern complexes there was at least one species that was distasteful to predators of butterflies (Sheppard 1959). Bates was also ahead of his time in his estimation of the huge and largely undescribed diversity of the Neotropical insect fauna. During his time in Amazonia he estimated that he had collected some 14,712 species, of which approximately 8000 were new, a number that seemed utterly implausible to most entomologists working in the UK at that time (Stearn 1981).

Some groups of insects seem to have an enormous propensity for evolving mimicry, and within apparently closely related groups can have evolved to resemble models of a wide range of colour patterns, shapes and sizes, such as, for example, the day-flying, chalcosiine zygaenid moths, which are no doubt mostly or entirely Müllerian mimics (Yen et al. 2005), or the day-flying Epicipeiidae moths which, with only 20 or so species, collectively mimic various papillionid, pierid, geometrid, zygaenid and lymantriid butterfly and moth models. No wonder this astonishing potential for variation has fascinated entomologists for years.

A lot of early research involved the collection and publication of field observations and relatively simple experiments, such as feeding various insects to predators and observing reactions (fine examples include G.A.K. Marshall & Poulton 1902, Swynnerton 1915b, R.T. Young 1916, Carpenter 1942). A rather lovely, if quaint, example is that of G.D.H. Carpenter (1921), a medical doctor by profession who was based in Uganda for some time before becoming Hope Professor of Zoology (Entomology) at Oxford University. He describes the results of extensive experiments in which insects were presented to a captive monkey and its responses observed. The article is over 100 pages long and in the foreword he notes that a lot of the observations are tabulated rather than given *seriatim* because of the "great increase in the cost of printing". Nevertheless, such observations are essential first steps in understanding whether species are models or mimics or have unsuspected defences.

Around the middle of the nineteenth century, another Englishman, Alfred Russel Wallace (1823–1913), an intrepid traveller, natural historian and thinker, was coming up with important notions concerned with mimicry and aposematism (Wallace 1867). He had earlier travelled to Brazil and collected with Henry Bates and later went on to explore South-East Asia. Indeed, he came up with the idea of evolution by natural selection more or less contemporaneously with Charles Darwin, though unlike Darwin he had little formal education (H.W. Greene & McDiarmid 2005). His early appreciation of the nature of aposematism and thoughts on poisonous snake mimicry are particularly pertinent here.

Mimicry and adaptive colouration have long been popular topics that have grabbed the imagination of both the public and academic biologists due to the incredible detail in many resemblances. Good early treatments include those of Poulton (1890), G.D.H. Carpenter & Ford (1933) and Cott (1940), all of which document numerous natural history observations and interesting ideas. Wolfgang Wickler's (1968) popular book on mimicry in plants and animals with many fine illustrations by H. Kacher no doubt fired many people (including myself) with enthusiasm for the topic. Komárek (2003) provides an excellent and more biographic description of the arguments, ideas and personalities that shaped our understanding of crypsis and mimicry up until 1955 (with some comments on subsequent works up to 1990). Other good general books include Pasteur (1972), D.F. Owen (1980), Forbes (2011) and J. Diamond & Bond (2013), as well as more academic works such as Ruxton et al. (2004a), Stevens & Merilaita (2011) and Stevens (2016). The book by Ruxton et al. provides a critical review of many experiments, models and arguments to do with anti-predator adaptations in general, not just mimicry and camouflage, but there is a great deal of overlap.

Many arguments, often heated, were also involved in the early discussions of mimicry. Some of the examples show such perfect matching of detail that many scientists found it hard to believe that they could have resulted from natural selection for progressively more similar forms from disparate starting points. Some thought that only major mutations could be involved rather than Darwinian gradual accumulation of small changes. This led to hearty debate about how natural selection and genetics work, for example

---

1. Butterfly systematics has progressed since Bates' time and many of the species he collected and referred to as Heliconiidae are now placed in the tribe Ithomiini in the nymphalid subfamily Danainae, while his Heliconini are now classified as a tribe of Nymphalidae.

Punnett (1915) and R.B. Goldschmidt (1945) on the side of major mutational leaps versus R.A. Fisher (1927, 1930), L.P. Brower et al. (1971) and, more or less, de Ruiter (1958) leaning towards gradualism. The current consensus is a combination, with an initial mutation that causes a large phenotypic shift followed by subsequent evolutionary refinement, called the 'two-step hypothesis', most probably being the major route, though gradualism might be sufficient in some circumstances (see Chapters 4 and 5). As J.R.G. Turner (1983) notes, in the complicated *Heliconius* system some quite large jumps in phenotype can occur as a result of simple genetic changes.

When it comes to camouflage, much credit should be given not to a scientist, but instead to the American portrait, animal and landscape artist Abbott Handerson Thayer (1849–1921), who discovered the principle of concealment by countershading, discussed disruptive colouration, and dazzle markings and distractive features, and even tried to help the military in disguising troops and ships (J. Diamond & Bond 2013). Interestingly, many of his suggestions came under attack from many naturalists and even hunters. While not all of his suggestions might have been correct, and indeed he probably went over the top in trying to explain all animal colouration as having some concealing function, the argumentation employed on both sides is of interest. People such as United States president Theodore Roosevelt, who was an enthusiastic hunter,[2] dismissed Thayer's claim that a zebra's stripes acts to help conceal it (Roosevelt 1911). Thayer's counter-argument was that just because someone saw something, it did not mean that they saw everything, because they do not know what they failed to notice. In an amusing section, Thayer wrote:

> Forty years of daily meeting the poacher at the post office does not strengthen his credit. And forty years of Roosevelt's seeing zebras not hidden by their costume, and failing to guess what the animal's stripes are for, are just as little to the point.

Kingsland (1978) wrote a very nice discussion of Thayer's work and how it was received, and two of Thayer's oil paintings illustrating camouflage are reproduced in J. Diamond & Bond (2013, pp. 44 and 45); Ruxton et al. (2004b) reproduce Thayer's photographs of a dead grouse positioned as 'in nature' and with its underside dyed darker, illustrating the effectiveness of countershading (see Chapter 2).

---

2. The large game and some 11,000 other specimens that Roosevelt and other expedition members shot or collected on the 1909 Smithsonian–Roosevelt African Expedition became a major part of the collections of both the Smithsonian Institution (Washington D.C.) and the American Museum of Natural History (New York). Some of the diaramas in the latter are real masterpieces of natural history display.

The past 30 or so years have seen an enormous resurgence in research on adaptive colouration and mimicry (Guilford 1990b, Komárek 2003), both experimental and theoretical, as can be seen by a quick scan of the dates of the articles cited here. Computer-generated graphics, usually but not always in conjunction with human subjects, have played an increasingly large role in investigations. Nevertheless, much is still being achieved with low-tech solutions, such as pastry model caterpillars exposed to predation by garden birds, or baited triangular shapes that roughly resemble moths resting on tree trunks exposed to woodland birds. Increasing awareness of the visual capabilities of predators, or in some cases of potential mates, is leading to quite a lot of more carefully controlled work, but there is still room for greater awareness. It is all too easy to think that because a model looks life-like to the experimenter, it will also appear life-like to a bird. Some birds can see well into the UV part of the spectrum, and if the signal receiver is an insect, it is important to understand that although insects can see UV light, most cannot see much at the red end of the spectrum.

Sometimes biologists get it wrong. For example, for a long while the North American viceroy butterfly, *Limenitis archippus* (Nymphalidae), was thought to be a Batesian mimic of the monarch butterfly, *Danaus plexippus* (e.g. J.V.Z. Brower 1958a). Now it is known to be actually unpalatable itself (Ritland & Brower 1991) (see Chapter 4, section *Plant-derived toxins*), and more recently it has been shown most probably to be a Müllerian mimic (S.B. Malcolm 1990, Guilford 1991, Ritland 1991, Ritland & Brower 1991, Rothschild 1991) and to contain phenolic glucosides (salicortin and tremuloidin) (Prudic et al. 2007b) sequestered from its *Salix* food plant, though these are rather different from those of the monarch and have different physiological effects.

Another often neglected aspect is the need for correct identification. D.F. Owen et al. (1994), for example, discovered that anomalous findings in an African butterfly mimicry system were resolved once it was realised that one of the mimic species was actually a pair of different but closely similar (cryptic) species.

ON DEFINITIONS OF 'MIMICRY' AND ADAPTIVE RESEMBLANCE

Mimicry can be defined in many ways but in most of these there exists the concept that some subject forms a model for the resemblance of another, the mimic. In biological examples, this resemblance also carries with it the notion that it serves to deceive another organism, though this is not as clear cut as it may seem and certainly deception is not

Table 1.1 Some selected definitions of mimicry and adaptive resemblance. (Adapted from Endler 1981 with permission from John Wiley & Sons.)

Publication	Definition
Cott 1940	'In the former [protective resemblance or crypsis] an animal resembles some object which is of no interest to its enemy, and in so doing is concealed; in the latter [protective mimicry] an animal resembles an object which is well known and avoided by its enemy, and in so doing becomes conspicuous.'
Wickler 1968	'If a signal of interest to the signal receiver is imitated, then this is a case of mimicry, whereas if the general uninteresting background or substrate is imitated, then camouflage (or mimesis) is involved.'
Wiens 1978	'… the process whereby the sensory systems of one animal (operator) are unable to discriminate consistently a second organism or parts thereof (mimic) from either another organism or the physical environment (the models), thereby increasing the fitness of the mimic.'
Vane-Wright 1980	'Mimicry involves an organism (the mimic) which simulates signal properties of a second living organism (the model) which are perceived as signals of interest by a third living organism (the operator) such that the mimic gains in fitness as a result of the operator identifying it as an example of the model.'
M.H. Robinson 1981	'Mimicry involves an organism (the mimic) which simulates signal properties of another organism (the model) so that the two are confused by a third living organism and the mimic gains protection, food, a mating advantage (or whatever else we can think of that is testable) as a consequence of the confusion.'
Maran 2005	'Proceeding from semiotics the essence of mimicry is the presence of two living beings (object) who have different applicability (interpretant) to the receiver (interpreter) and who because of the similarity of their messages or cues (representamen) are at least partly undistinctable for the receiver.'
Grim 2013	'Mimicry refers to functional 'model–mimic–selecting agent' trinity (with varying number of species involved) when the selecting agent (i.e. signal receiver) responds similarly to mimic and model to the advantage of the mimic.'
Speed 2014	'In its most general form mimicry refers to phenotypes of an organism that are adaptively modified to resemble living or nonliving components of its environment."
Dalziell et al. 2015	'…a [signal] is mimetic if the behaviour of the receiver changes after perceiving the … resemblance between the mimic and the model, and the behavioural change confers a selective advantage on the mimic.'

necessary for a large set of adaptive resemblances. A number of influential definitions of mimicry were cited by Endler (1981) and his Table 1 is reproduced here with a few additions (Table 1.1). Von Beeren et al. (2012) provided a summary of how various authors have applied terms related to camouflage and mimicry in relation to how the operator (dupe, signal receiver) perceives them and I have extended it with further examples in Table 1.2. Pasteur (1982, Fig. 1) presents a neat timeline showing where various authors drew the distinction between mimicry (homotypy in his terminology) and crypsis. The range is considerable – some workers restricted use of the term mimicry to anti-predator resemblances with defined models (essentially the Batesian–Müllerian spectrum with unpalatable models), while at the other extreme (e.g. Bates 1862, Turner 1970) all forms of camouflage were included under the definition as well. A number of people drew the line between crypsis due to background matching, countershading, etc. and cases

where the model was definable. In present parlance, that includes masquerade (i.e. resemblance of an organism to a definite object of no interest to a predator or herbivore) (Endler 1981), along with all classical protective, reproductive, dispersal and social mimicry cases. In a lot of older literature, masquerade is called mimesis.

That arguments have raged for years over the precise meaning of the term mimicry indicates that there are categories of relationships between organisms, or between organisms and inanimate subjects, which some people see as hard and fast examples of mimicry while others do not. Such grey areas serve to highlight what for more than a hundred years has been a fermenting, and sometimes acrimonious, debate. Probably the main reasons why there has been such a long history of debate on this issue is precisely because of some people's notions that 'mimicry' must involve deceit and that it must involve a definite model. Thus Müllerian mimicry, which is the convergence in

Table 1.2 Relationships proposed between various terms used in anti-predator mimicry and camouflage in relation to the predator's response to the prey, as employed by various authors. (Adapted from von Beeren et al. 2012 under the terms of the Creative Commons Attribution Licence CC BY 3.0.)

Predator's reaction to potential prey			
Not detected as a discrete entity (causing no reaction)	Detected as an uninteresting entity (causing no reaction)	Detected as an interesting entity (causing a reaction beneficial to the mimic)	Reference(s)
Crypsis	Masquerade	Mimicry	Endler 1981, 1988
Eucrypsis	Mimesis	Homotypy	Pasteur 1982
Eucrypsis	Plant-part mimicry	Mimicry	M.H. Robinson 1981
Crypsis	Masquerade	Mimicry	Ruxton et al. 2004a
Cryptic resemblance	Cryptic resemblance	Sematic resemblance	Starrett 1993
Crypsis	Masquerade	—	Stevens & Merilaita 2009b
Crypsis	Crypsis	Mimicry	Vane-Wright 1976, 1980
Camouflage or mimesis	Camouflage or mimesis	Mimicry	Wickler 1968
Crypsis	Masquerade	Mimicry	Endler 1981, 1988
Eucrypsis	Mimesis	Homotypy	Pasteur 1982
Eucrypsis	Plant-part mimicry	Mimicry	M.H. Robinson 1981
Crypsis	Masquerade	Mimicry	Ruxton et al. 2004a, Ruxton 2009
Cryptic resemblance	Cryptic resemblance	Sematic resemblance	Starrett 1993
Crypsis	Masquerade	—	Stevens & Merilaita 2009b

appearance between members of a guild of unpalatable species to mutual benefit, has often been excluded because the signal displayed by members of the guild does not deceive a predator.

Similarly, the form of camouflage called 'crypsis', in which organisms show a general resemblance to background properties such as colour, luminance and texture but not to any definite model, fails to satisfy the definitions of mimicry provided by many authors (Wickler 1968, Wiens 1978), whereas a camouflaging resemblance of an organism to a defined model, maybe a caterpillar to a twig, does. Examples of camouflage with a definite (definable) model are now usually referred to as 'masquerade' (see Chapter 3) but there is an inevitable grey area – after all, masquerading animals have almost always evolved from cryptic ones by gradual increases in similarity to a more precisely defined model. At some stage it becomes apparent that something is masquerading, but there are many intermediate cases. The differences may also depend on the perceptions of a given observer.

Many people have tried to make a clear distinction between mimicry and camouflage, for example Pasteur (1982), but this is not always an easy matter and may in fact be impossible in some cases. A particularly widespread problem when considering mimicry systems, alluded to by M. Edmunds (1974a), is our frequent lack of detailed understanding of the systems of interest. Edmunds considers various sea-slugs (nudibranch molluscs) that closely resemble, both in colouration and in texture, the sponges and hydroids upon which they feed (see Fig. 2.2). Whether such cases are mimetic or cryptic depends crucially on whether a potential predator, say a fish, ignores the sponges or hydroids because they are not suitable food, or actively avoids them because they pose a threat. Both situations are quite possible and may be expected to occur. Thus in these and perhaps many more instances, an adaptive resemblance can be both mimetic and cryptic, depending on which particular type of predator one is considering. Edmunds (1991) concluded that although some nudibranch molluscs are very brightly coloured (see Fig. 4.32), evidence that they are warningly coloured is rather scant (see also Guilford & Cuthill 1991 and Chapter 4, section *Evidence for individual selection*). Nevertheless, it would appear that some nudibranchs may act as Müllerian models for young fish (Randall & Emery 1971).

Early discussions of mimicry were almost entirely centred on examples at the extreme end of the spectrum of what are now considered to be mimetic resemblances (Rothschild 1981). This had the effect of emphasising differences, whereas in nature there is often a continuum; indeed there must be because of the way organisms evolve. That narrow view also tended to draw attention away from how the species got there in the first place, and, as Rothschild points out, "mimicry may be fluid, and the category we assign a species to may be different according to time and place".

Vane-Wright (1980) created a flurry of interest, debate and even dissent as a result of various aspects of his attempt to define mimicry (Table 1.1) in which he distinguished between models that are of no interest to the predator, such as twigs, dead leaves, etc., and those that are. Mimics of the former would then be regarded as cryptic by his definition. More than half of an issue of *Biological Journal of the Linnean Society* (volume 16, part 1) was devoted to criticism and useful commentary on Vane-Wright's definition (Cloudsley-Thompson 1981, M. Edmunds 1981, Endler 1981, M.H. Robinson 1981, Rothschild 1981) and, of course, Vane-Wright's (1981) reply. The issue also includes a shortened version of Henry Bates' famous 1862 paper which set out the basis of his thought on the mimicry that now bears his name.

Cloudsley-Thompson (1981) felt that the term mimicry should be restricted to resemblances of one animal by another and the word 'disguise' used for when an animal is "like a stick, lichen, bark, faeces, a stone or some other inanimate object unattractive to potential predators", though he notes that he might have been unduly influenced in this respect by his supervisor Hugh Cott, who considered that all animal resemblances to plants should be referred to as crypsis. Cloudsley-Thompson quotes from a written response to him from Dick Vane-Wright concerning the nature of mimicry and it is worth repeating here:

> I'm afraid I stick to my contention that it is the 'reference frame' of the operator that determines whether or not crypsis or mimicry should be employed. Tiger stripes 'are' aggressive crypsis, leaf-insects are defensively cryptic. But mantids which look like flowers (e.g. *Idolomantis diabolicum*) are aggressive mimics, as their victims are (presumably) actively seeking out flowers and we imagine that most flowers have evolved their 'flashy' signals to attract pollinators. I don't see how you could define such a general word as disguise in an operational way different to mimicry and crypsis as I have defined them, or as others have tried to do so. It might be useful as a word for lumping both phenomena together.

M. Edmunds (1981) argued that Vane-Wright's definition of mimicry could be seen as too broad, and presented a number of examples emphasising the fine borderline between a 'mimic' being of no interest to a receiver and being of potential interest. For example, in reference to the blue jay/stick caterpillar experiments of de Ruiter (1952), Edmunds notes that:

> [the] jays could not distinguish stick-like caterpillars from twigs until they accidentally trod on one. The question then is, was the signal of the caterpillar of no interest at all (crypsis), or was it of possible interest as an object to grip with the claw, in which case it becomes mimicry.

This seems rather nit-picking to me though I agree that some borderlines are hard to define. Vane-Wright (1981) published a rejoinder to Edmunds and noted, among other things, that his "original attempt at a general definition was introduced in the context of trying to define 40 theoretically different types of mimicry and to separate them from all possible types of crypsis."

M. Edmunds (1974a) defined the occurrence of Batesian mimicry as being when "a predaceous animal, which avoids eating one animal producing a particular signal, is deceived into avoiding a second animal which produces a similar but counterfeit signal". He went on to conclude that animals that precisely mimic other types of inedible objects, such as leaves, twigs, thorns, etc. in order to avoid predation, are Batesian mimics. I do not think that such broad definitions are useful because there are distinctly different effects on the models; if models are living then the more palatable Batesian mimics there are, the more the model species suffers, but it is hard for twigs or pebbles to suffer in the same sense.

Similarly, Endler (1981) also found difficulty with Vane-Wright's system when it comes to determining whether some resemblances are mimetic or cryptic because of the animate/inanimate distinction. For example, resemblances involving sticks or twigs (i.e. parts of a plant) involve a potentially animate model, whereas resemblances to a stone or some dung do not. Neither of these types of model is of interest to the predator, but Vane-Wright used the term mimicry for when the model "is an organism or part of one", which therefore excludes cases with stones or dung as models.

M.H. Robinson (1981) criticised Vane-Wright's definition primarily because it requires interpretation of the phrase "of interest" and as a remedy he couched his definition only in terms of confusion. While Robinson's definition (Table 1.1) overcomes some difficulties, as with Vane-Wright's (1980) definition, it precludes resemblances to non-living entities such as pebbles, bubbles, etc. Vane-Wright (1981) points out that Robinson's definition also ignores both evolution by natural selection and the receiver's perception, as well as being in itself a gross over-simplification of Bates' original notion. M.H. Robinson (1969), on the other hand, made an important contribution to classifying different types of crypsis, and particularly distinguishing eucrypsis (homochromy, countershading and disruptive colouration) from 'plant part mimicry' (*contra* Vane-Wright 1980) (i.e. Cott's (1940) 'special protective resemblance' and Turner's (1961) 'disguise').

Crypsis and hiding from predators are probably the basic (plesiomorphic) defensive strategies in most major groups (M. Edmunds 1990), and masquerade, mimicry and aposematism may evolve when circumstances are right

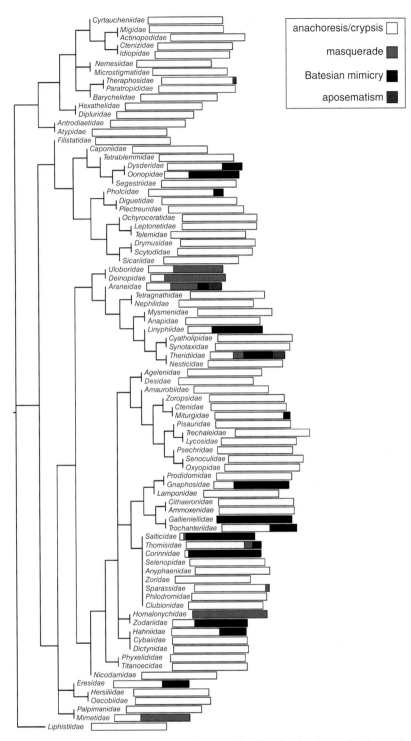

Fig. 1.1 Modes of passive defence mapped onto independent phylogeny of families of spiders (pruned to show only those families for which hypotheses about modes of anti-predator defence could be assessed). (Source: Adapted from Pekár 2014. Reproduced with permission from John Wiley & Sons.)

(i.e. probably when the background is not uniform but complex and contains many discrete or nearly discrete prey-sized items). Pekár (2014) has examined defensive strategies across the whole of the Arachnida (spiders) and shown that masquerade and mimicry have evolved in a restricted number of families in a very non-random way (i.e. the strategies are phylogenetically clustered) (Fig. 1.1). The most basal extant spiders, the liphistiids[3] and mygalomorphs, are generally large bodied, ground dwelling and nocturnal or spend most of their time concealed in silk-lined holes (anachoresis), so it is not really surprising that they have not evolved bright colours (with rare exceptions in the Theraphosidae, such as the red-headed mouse spider, *Missulena occatoria*) or mimetic resemblances to particular objects – the evolutionary advantage was not there. Orb-web weaving spiders such as aranaeids, tetragnathids and nephilids, however, are often fully visible in their webs in the daytime, and among these bright patterns are much more prevalent, and may, in fact, be attractive to certain prey (see Chapter 10, section *Alluring mimicries*). A similar phylogenetic pattern of ancestral crypsis has recently been shown for North American darkling beetles (Tenebrionidae: Asidini) by A.D. Smith et al. (2015), some of which are Batesian mimics of species of the related and well-protected genus *Eleodes* (see Chaper 6, section *Experimental tests of mimetic advantage*).

Schaefer & Ruxton (2009) coined the term 'exploitation of perceptual biases' (EPB) to differentiate what they regarded as true mimicry from simple exploitation of the sensory biases and loopholes of another organism. In their definition, the term mimicry only applies to cases where a receiver really misidentifies the mimic as a specific model.

Here I have decided first to discuss camouflage (crypsis and masquerade), as despite differences of opinion about what constitutes mimicry and what does not, these evolutionary strategies differ from what is generally nowadays called mimicry, in that the mistakes made by operator (dupe, receiver) have no effect on the model (Vane-Wright 1980, 1981, Endler 1981). Thereafter, I have largely followed a functional classification similar to that of S.B. Malcolm (1990) (Table 1.3), though with variations on the terminology and a greater number of divisions. Because of the many biological roles that mimicry affects there has long been a need to pigeon-hole examples to make it easier to compare and gain better understanding of different systems.

---

3. These South-East Asian spiders, which retain a partially segmented abdomen and lack venom glands, are the sister group to all other living spiders.

The concept of 'adaptive resemblance'

The term 'adaptive resemblance' (AR) was introduced by Starrett (1993) as a "broad inclusive term" for a wide range of mimetic and cryptic phenomena and was defined as: "any resemblance that has evolved or is maintained **as a result of selection for the resemblance.**" It specifically excludes "incidental resemblance or convergence, which is due to common adaptive responses to functional requirements."

AR incorporates everything normally regarded as mimicry as well as most things that are only occasionally regarded as mimicry. Starrett uses the term selective agent (SA) for what others have variously called the detectee, dupe, receiver, signal receiver or operator. AR includes both crypsis and masquerade (both crypsis in Starrett's definition) and makes no distinction about whether the model is alive, dead or inanimate. It [AR] encompasses Müllerian mimicry even though some authors have strongly objected on the grounds that no deception is involved, but it is clearly an adaptive resemblance. It also includes various topics rejected by Pasteur (1982), such as vocal mimicry (some bird song or alarm call mimicry) or behavioural mimicry by primates. Pasteur's reasoning was that vocal mimicry was a 'conscious imitation' and not something that had been a direct result of natural selection, though of course the cognitive ability or propensity to do so was selected. It is far from clear which instances of such mimicries are consciously done. When a female baboon goes off with a subordinate male for sex and they hide from the dominant male to avoid his aggression – that is probably conscious, but when a bird imitates alarm calls of conspecifics to deter a competing male, is that consciously thought out or a behavioural pattern that has been directly selected to be part of the species' repertoire? Starrett includes most if not all such cases under AR but also asks whether along the gradient of increasingly complex behavioural complexity (cognitive ability) there is a line "beyond which release from unconscious stereotyped behaviour allows discretionary situational and consciously deceptive imitation that might not be subject to natural selection". Surely there is, but beyond that we enter the realm of psychology and that goes largely beyond the scope of this book and AR. Nevertheless, I will discuss various human behaviours to highlight the similarities between what evolution has done and how people deceive, sometimes unconsciously.

This useful definition also makes for clearer thinking. Starrett pays particular attention to various other "troublesome" cases. For example, the widespread occurrence of bold black and white colouration in marine carnivores such as penguins, killer whales and various dolphins (see Fig. 10.3a–d) is clearly not due to the different species evolving to resemble one another. At least in *Spheniscus* penguins the colouration may be selected for because its

Table 1.3 Simple classification of mimicry based on kind of interaction, trophic level and selective agent as summarised by S.B. Malcolm. (Adapted from Malcolm 1990 with permission from Elsevier.)

Mimicry category	Interaction	Selective agent (operator)	Examples
Interspecific interactions			
Defensive	Prey–predator	Predator	Müllerian, Batesian, 'masquerade', 'predator mimicry'
Foraging mimicry	Prey–predator	Prey	Aggressive and 'masquerade' (lampyrid beetles, bolas spiders, cleaner wrasse)
	Flower–pollinator	Pollinator	Flower mimicry (orchid mimicry of bees)
Parasitic mimicry	Host–parasite	Host or vector	Parasites, dispersion and mating (cuckoos, trematodes, fruit dispersal, orchid mimicry of bees)
Intraspecific non-trophic interactions			
Sexual mimicry	Male–female	Same, or opposite, sex	Mating lures ('sneaky' mating, egg or prey dummies)
Social mimicry	Any sex, juveniles	Same, or opposite, sex	Signals of hierarchy, deceptive alarm calls

conspicuousness causes schools of prey fish to 'depolarise', making some prey individuals easier to see and capture (R.P. Wilson et al. 1987). Why exactly these conspicuous markings might have this effect is unclear; maybe they 'mimic' signals from other potential predators that might best be thwarted by a confused response. After all, over millions of years of evolutionary time, if the bold black and white pattern of piscivores leads to greater overall mortality one might have expected schooling fish not to respond to it any longer.

THE CLASSIFICATION OF MIMICRY SYSTEMS

While the concept of mimicry is generally understood by everyone, mimetic relationships have evolved in relation to a wide range of biological processes, which has made attempts to classify the different types complicated. Different workers have also highlighted different factors as being important, and there is considerable overlap. While some authors are clearly willing to accept the idea that some resemblances have elements of more than one type of mimicry, for example many cases of crypsis, and sometimes masquerade, serve to hide the aggressor simultaneously from potential predators and from potential prey (see Figs 2.29 and 10.7d,e). There may be other 'grey areas'. Perhaps some authors have concentrated heavily on pigeon-holing particular hard-to-place examples. While this can serve as a test of classificatory systems, it is hard to imagine that any one simple system can accommodate all the results of evolution.

In terms of creating frameworks for classifying different types of mimicry, three people in particularly have made great advances – Wolfgang Wickler, Richard [Dick] Vane-Wright and Georges Pasteur. I will discuss their contributions in some detail below.

Wickler's system

Wickler (1965) proposed a formal notation for mimicry as a communication network in which three players interact and on which the costs or benefits of the signals to the participants could be represented. Two of the parties he termed signal senders (S, called S1 and S2) and one the signal receiver (R), which responds to the signals emitted by the senders. By convention, the model is represented by S1 and the mimic by S2. Endler (1981) preferred to use the notation P, S and R, referring to primary and secondary signal generators and a receiver, but here I use Wickler's system. Wickler then provided the following formal definition of mimicry as a system:

1. comprising two signal senders (S1 + S2) which have one or more receiver(s) in common,
2. in which the receiver(s) respond similarly to the signals of the two senders,
3. in which it is advantageous for the receiver to respond to one signal sender in a given way but disadvantageous for it to respond to the other signal sender in the same way.

In terms of Wickler's notation this is

$$S1 + R - S2$$

from which the mimic in the system can be defined as the signal sender that elicits a response in the receiver that would be disadvantageous (negative) for the receiver, while the other sender is the model (S1). In other words, the mimic is the party that emits counterfeit signals that elicit maladapted responses in the receiver. As noted by Wickler, the mimic always has a selective advantage. This can be represented in terms of his notation as

$$S1 + R - + S2$$

Note that the mimic (S2) must always benefit because otherwise the mimicry would not have been selected for in the first place. The receiver's response, of course, does not have to be negative to the other sender and in the broader context of adaptive resemblance it is possible for all the interactions to be positive, as for example in classical 'Müllerian mimicry', which would be represented as

$$S1 + + R + + S2$$

Vane-Wright's system

Vane-Wright (1976) provided a broader basis for the classification of mimetic relationships, extending Wickler's analysis of interactions between the three components of the system, S1, S2 and R. Vane-Wright's (1976, 1981) definition differs from that of Wickler (1968) in that it does away with the need for deception – thus it comfortably incorporates Müllerian mimicry in which the receiver is not deceived, and single or mixed schools of fish in which one individual gains protection because of predator's difficulty in visually separating it from all the other school members (arithmetic mimicry).

The first part of Vane-Wright's analysis considered whether signals from the mimic (S2) are advantageous or disadvantageous to the receiver (R), and whether signals from the mimic (S2) are disadvantageous to the model (S1). Systems in which the mimic's signal are not harmful to the model are referred to as 'synergic' and in the opposite case, 'antergic'. He depicted the full range of possibilities as shown in Fig. 1.2.

The second part of Vane-Wright's system considered the actual embodiment(s) of these three components (i.e. the species they belong to). While for many familiar cases of mimicry, the model, mimic and receiver are each different species, termed 'disjunct' systems, there are cases in which all three are members of the same species (conjunct systems), or any two may belong to one species (bipolar systems).

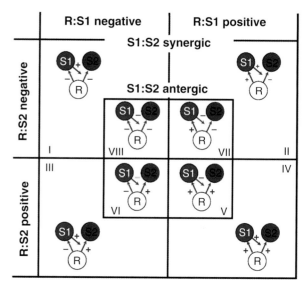

Fig. 1.2 The eight interactions that can occur in mimetic relationships with the signal receiver (R or dupe) and two signal transmitters (S1 and S2) showing how the signals and responses of each member of the triad are either beneficial (positive) or harmful (negative) to the others. The outer relationships are synergic, meaning that the existence of the mimic (S2) is beneficial to the model (S1), whereas in the inner set of four the mimic has a detrimental effect on the model. (Source: Adapted from Vane-Wright 1976 with permission from John Wiley & Sons.)

The three possible polar systems can be represented by $S1 + R/S2$, $S2 + R/S1$ and $S1 + S2/R$, as illustrated in Fig. 1.3.

Vane-Wright's classification was not intended to be a complete treatment of adaptive resemblances as it only deals with cases in which the model is definable and animate; his choice of criteria in fact prevents the inclusion in his classification of crypsis and masquerade. Nor does it embrace ill-defined features such as deflective markings and dazzle, as these cannot be classified as being either synergic or antergic and do not have a clearly defined model.

Out of the 40 conceivable types of mimetic relationship allowed for in his classification (Table 1.4), Vane-Wright was able to find non-human examples for 21 with some certainty, a couple rather more speculatively, and human examples (e.g. military decoys and spies) for a further three. Vane-Wright notes that the bipolar $S1/S2 + R$ system, in which the mimic and dupe belong to one species while the model belongs to another, seemed to be completely empty, though he did suggest that some courtship activities of spiders and empidid flies which incorporate simulation of prey organisms could belong to this category. To these possibilities might be added the food lures used by certain male fish that resemble invertebrates and

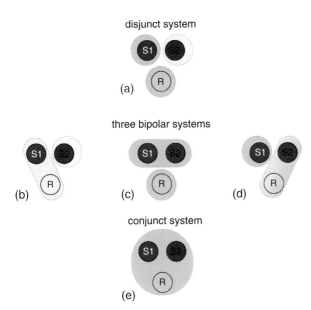

disjunct system

(a)

three bipolar systems

(b)   (c)   (d)

conjunct system

(e)

Fig. 1.3 The different relationships at species level between the signal receiver (R or dupe) and two signal transmitters (S1 and S2) with one disjunct tripolar system, three different disjunct bipolar systems and one conjunct system in which all three aspects of the mimetic relationship are confined to a single species. (Source: Adapted from Vane-Wright 1976 with permission from John Wiley & Sons.)

attract females (Kolm et al. 2012; see Chapter 11, section *Food dummies and sex*), as well as human Native North American use of bird call mimicry to deceive European settlers and troops about their intent.

Following Vane-Wright's explicit and objective classification of mimicry systems, several authors have pointed out that it will remain difficult to pigeon-hole many examples because we usually do not know exactly what an operator (e.g. a predator's) perceptions of a given situation are, nor is there any reason why all operators should share the same perceptions or why any one operator should have the same perception all of the time. Some interesting examples were cited by M. Edmunds (1981). For example, many sea-slugs (Nudibranchia) resemble the food organisms on which they sit and feed not only in general colouration and texture, but sometimes also in more specific detail. Further, in some instances, the sea-slug's prey/substrate may be unpalatable, as with many sea anemones and sponges. Whether, then, such examples should be classed as crypsis or as Batesian mimicry will depend upon whether the model is neutral to a potential predator, such as a fish, or avoided by it. In a similar vein M. Edmunds cites Fricke's (1970) description of the fish *Siphamia* sp., which aggregate so as to resemble sea urchins (*Astropyga* spp.). Should sea urchins be regarded as

neutral background elements to potential predators of *Siphamia* or objects to be avoided because of their spines? Probably both to different predators.

Vane-Wright thus accepts Miriam Rothschild's (1975) dual signals concept to explain how the nature of a signal from a deceptive organism can change with distance (see Chapter 2, section *Dual signals*). He also notes, in response to a criticism from Rothschild (1981) concerning how cuckoos change during development from rejecting many typically unpalatable insect food items to readily accepting them as adults, that "the terms model and mimic describe functional entities in a schematised biological interaction: they are not fixed properties of individuals" (Vane-Wright 1981). Some resemblances might therefore be able to represent more than one of Vane-Wright's 40 categories simultaneously, depending on the receiver. A given pair of butterfly species may be aposematic and unpalatable (i.e. Müllerian mimicry) to one predator, but one of them may be quite palatable to a different predator (Batesian mimicry).

Georges Pasteur (1930–2015)

Pasteur (1982) developed a largely three-dimensional classification of mimicry, combining the concepts of functionality with the disjunct, bipolar and conjunct systems of Wickler (1965) and whether the dupe is indifferent to the model or whether it is of interest to the dupe, and in the latter case, whether it is agreeable or forbidding to the dupe. Seven functional classes were recognised (Table 1.5), which together with the five possible species compositions of a mimicry system and three main classes of the dupe's interest in the model gives a maximum of 105 different combinations, though Pasteur only found examples for a far smaller number. Pasteur was also responsible for naming many classes of mimicry after their discoverers or co-discoverers, without due regard to the nature of the interactions involved, and this has tended to lead to confusion.

Unfortunately, many of Pasteur's categories still combine examples of mimicries with very different functions. For example, as discussed by Quicke et al. (1992), the term Kirbyan mimicry was employed to cover aggressive/reproductive S1 + R/S2 systems involved in brood parasitism. Within this seemingly tightly defined system, Pasteur included the egg mimicry of cuckoos and the mimicry of Hymenoptera by robberflies (Diptera: Asilidae) whose larvae are predators of their model's larvae (see Chapter 10, section *Cuckoldry, inquilines and brood parasitism*). He also suggested that this category might include the vocal mimicry of host young by cuckoo chicks which resemble those of their hosts (M.G. Anderson et al. 2009; see Chapter 10, section *Gentes and 'cuckoo' eggs*). It is clear that several very

Table 1.4 Classification of 'mimicry' systems according to Vane-Wright (1976) with a few extra examples added (note S1 = model, S2 = mimic, R = dupe). (Vane-Wright 1976. Reproduced with permission from John Wiley & Sons.)

Model and mimic	Synergic (i.e. mimicry is good or neutral for the model)				Antergic (i.e. mimicry is bad for the model)			
	I Warning	II Aggressive	III Defensive	IV Inviting	V Inviting	VI Defensive	VII Aggressive	VIII Warning
A (Disjunct, i.e. 3 separate species)	Müllerian, quasi-Müllerian	Angling with lures		Arithmetic	Useful weeds	Batesian, quasi-Batesian, hawk mimicry by cuckoos	Peckhamian, hawk mimicry by cuckoo	?
B S1 + R/S2				Trophobionts	?	Vine 'eggs', predator mimicry by prey, monkey call mimicry by cats, antigenic mimicry by parasites	Cuckoos, monkey calls of big cats, pollinator sexual deception	
C S1 + S2/R	Aposematism	Alarm calls of antshrike	Automimicry, thanatosis	Herding, schooling	Competition			
D S2 + R/S1				Food lure for courtship				
E Conjunct	Military uniform	Military decoys	Bluff	Egg dummies in fish	Sexual competition	Bluff, appetitive use of alarm calls	Spies	

Table 1.5 Pasteur's (1982) seven functional classes of mimicry. (Data from Pasteur 1982.)

Class	Biological function
1	Aggressive
2	Aggressive/reproductive
3	Reproductive
4	Reproductive/mutualistic
5	Mutualistic
6	Commensalist
7	Protective

different types of mimetic relationship are involved here. Cuckoo eggs usually have to be more or less accurate mimics of host eggs in order to avoid rejection (Brooke & Davies 1988, Lotem et al. 2009).

In another, perhaps even more extreme example, Pasteur's term Wicklerian–Barlowian mimicry was coined for conjunct reproductive mimicries. This includes cases where females mimic males (androchromatism; see Chapter 11, section *Mate guarding through distracting other males*) as well as the similarity between male and female flowers in monoecious plants (see Chapter 12, section *Flower automimicry – intraspecific food deception (Bakerian mimicry)*) and the mimicry of the erect male penis by the female's clitoris in spotted hyaenas (see Chapter 13, section *The case of the spotted hyaena*). Thus while Pasteur's system may allow most cases of mimicry to be classified, his choice of functional categories seems to be too general to lead to groupings with much internal consistency, and certainly should not be followed without due caution. Similarly, the term Wasmannian mimicry has been employed as a general term for mimicries of ants, irrespective of whether it is Batesian or aggressive in nature.

One distinction that Pasteur uses is between cases of mimicry when there is a single model (or small tightly defined set of models) whose species identity is obvious (for example, sex pheromone mimicry by bolas spiders or sexually deceptive orchids or cases of Batesian mimicry in which a particular species is the unpalatable one), referred to as 'concrete homotypies', in contrast to 'abstract homotypy' in which the model is only definable in a vague way, such as a snake or a pair of vertebrate eyes. Semi-abstract examples might include the vermiform lures of anglerfish and alligator snapping turtle. Even more extreme is when no particular type of model is represented, as in the case of some deimatic startle displays, but the display is interpreted as a possible source of danger. In a sexual (reproductive)

mimicry setting, Pasteur cites the case of some *Oncidium* orchids whose flowers are so loosely attached that they dance around in the slightest breeze and thus attract highly territorial male *Centris* bees, which attack almost anything flying by (Dodson & Frymire 1961).

Other approaches

Endler

Endler (1981) considered that the result of a predator making a mistake should be taken into account as a basic characteristic of mimetic and cryptic systems (Table 1.6). Two criteria are considered: first, whether the mistake confuses the signal sender (mimic) with the background or with a specific thing, and second, whether the mistake affects the population dynamics of the model (if relevant). Endler superimposed on his classification table the extents of the definitions of mimicry applied by various other workers, and shows clearly that different authors have included and excluded large categories from consideration.

Zabka & Tembrock

Zabka & Tembrock (1986) proposed an alternative approach to the classification of mimicry systems based on "the aims" of the mimics and the behavioural responses involved. They separate camouflage (crypsis in their terminology) from other mimicries on the basis that the model is irrelevant to the signal receiver, whereas it is relevant in true mimicries and in those cases may cause an adverse or an appetitive (i.e. feeding) reaction (Fig. 1.4, Table 1.7). To put it another way, "mimicry is a phenomenon of the relevant environment of the signal receiver". Irrelevant models may be either abiotic or biotic but relevant ones are almost certainly always going to be biotic. Zabka & Tembrock's system can be seen as a refinement of the broadly biological classification of Wickler (1968) and Table 1.8 shows how they classified various of Wickler's (1968) examples. It will be apparent that Zabka & Tembrock's classification of reproductive mimicry does not make allowance for bipartite situations in which the model and mimic belong to the same species (i.e. S1 + S2/R). However, such systems do exist, for example, in the eastern Mediterranean cucurbit *Ecballium elaterium*, the relatively rare female flowers, which do not produce nectar, effectively mimic the commoner, nectiferous male flowers on the same plant (Dukas 1987) (see Chapter 12, section *Flower automimicry – intraspecific food deception (Bakerian mimicry)*). However, their lumping under the category 'reproductive mimicry' of all systems in which some aspect of sexual behaviour of the dupe is mimicked seems to me

Table 1.6 Effects of predator (P) making an error in discriminating model from mimic (S) and how different authors relate this to the distinction between mimicry and crypsis. (Adapted from Endler 1981 with permission from John Wiley & Sons.)

	Mistake has no effect on population dynamics or evolution of P (model = inanimate or background)	Mistake affects population dynamics of single P species (model = 1 species)	Mistake affects population dynamics or evolution of more than 1 P species (model = more than 1 species)
Mistake depends on the relationship between S and background	Crypsis Crypsis (=Eucrypsis) Aggressive mimicry (a)	Polymorphism Polymorphism and apostacy	Convergence Convergent evolution
Mistake depends on similarity to specific object or species, not background	Masquerade Plant-part mimicry Dung and stone mimicry Aggressive mimicry (b)	Batesian mimicry Batesian mimicry Dispersal mimicry Reproductive mimicry Group mimicry Aggressive mimicry (c)	Müllerism Müllerian mimicry Mertensian mimicry

☐ versus ☐ & ■ criterion of Cott (1940) and Vane-Wright (1980)

☐ & ☐ versus ■ criterion of Wickler (1968)

unhelpful (Table 1.9), and I prefer to use the term 'sexual mimicry' specifically for cases that have evolved to increase fertilisations/parenthood directly (see Chapter 11).

Maran

Maran (2007, 2010) applies semiotic theory and notation, essentially to Vane-Wright's (1981) classificatory system, that allows for easy depiction of how different receivers (or perhaps the same receiver at different times or distances, or different parts of the mimic, etc.) perceives a mimetic system (Fig. 1.5). In Fig. 1.5a, for example, the mimic is an arbitrary anglerfish (as in Fig. 10.7), but it uses two separate models for different purposes: the bulk of the fish's body is cryptic, i.e. resembling the background encrusting organisms, while the lure is mimicking a worm-like object. The same receiver, the potential prey, responds to both.

MIMICRY AS DEMONSTRATION OF EVOLUTION

Mimicry has long been seen as important for understanding the process of natural selection, and indeed Batesian and Müllerian mimicry are probably the best studied examples when it comes to natural populations (Orr & Coyne 1992). Indeed the degree of perfection in many mimicry systems as we perceive them led it, for a while, to be used as

an argument against evolution. For example, the butterfly expert and geneticist Reginald Punnett (1915) was fervently against the idea that gradualism could lead to the high-fidelity mimicry he saw between many unrelated species of butterflies during his visit to Sri Lanka (formerly Ceylon).

Mimicry has been particularly important when it comes to the genetic variation upon which evolution acts; does evolution proceed mostly in small incremental steps in a neo-Darwinian framework, or do large leaps occur due to mutations in genes that have a large effect? The micro-evolutionary paradigm has been largely accepted in most fields, yet it is clear that genes of large effect are involved in many cases of aposematism and mimicry (though with other modifier genes having lesser effect). As Orr & Coyne (1992) stated, "[it] might be objected … that mimicry is not a 'typical' adaptation" because it so often seems to involve large phenotypic jumps. This question is still largely open to debate, despite modern genomics approaches shedding a lot of new light on the subject, but the general consensus seems to be that a two-step process, with an initial mutation causing a large shift in phenotype followed by gradual fine-tuning, is probably involved in many cases, although a gradual shift is possible under certain circumstances (Lindström et al. 1999a, see Chapter 4).

Nadeau & Jiggins (2010) suggest that the genetic variation observed in natural populations that is responsible for many quantitative traits does not represent the

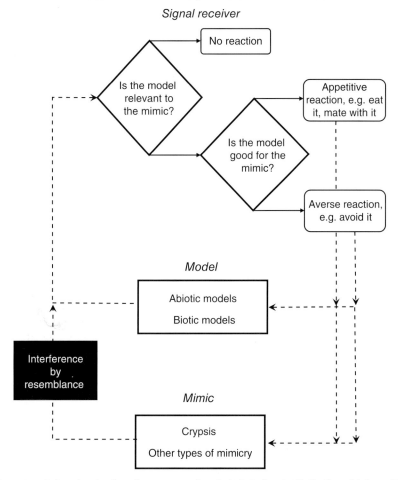

Classification of mimicry based on aims of mimics

Fig. 1.4 Diagrammatic representation showing how the presence of a mimic 'interferes' with the flow of information between the model and signal receiver and showing how the information can be interpreted by the receiver (dupe) as either relevant or irrelevant. (Source: Adapted from Zabka & Tembrock 1986.)

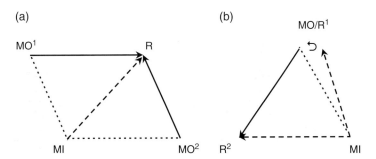

Fig. 1.5 Depictions of two mimetic examples using semiotic notation. (a) This describes the relationship of an anglerfish (such as those in Fig. 10.7) which is the mimic (MI) but has two links to the receiver (R^1 and R^2), one via its worm-like lure with an unspecified worm as model 1 (MO^1) and the other through its cryptic resemblance to the background (MO^2); (b) a depiction of relationships of a cuckoo bumblebee, *Bombus* (*'Psithyrus'*) sp. (MI), that is a brood parasite and mimic of a normal bumblebee, *Bombus* sp. (MO) which is also (possibly: see Chapter 10: section *Cuckoo bees and cuckoo wasps*) one of the dupes of the resemblance (R^1), the other receiver (R^2) being a potential predator of both such as a bird. (Source: Adapted from Maran 2010. Reproduced with permission from Springer.)

Table 1.7 Classification of mimetic and cryptic relationships involved in predator–prey interactions. (Adapted from Zabka & Tembrock 1986 with permission from Elsevier.)

| State of the mimic | Imitation of signals of the: | |
	relevant environment	irrelevant environment
Aversion (Protection)	PROTECTIVE MIMICRY Batesian and Müllerian mimicry; eyespots; butterfly egg mimicry in plants; jamming	PROTECTIVE CRYPSIS Crypsis of prey; 'fly mimicry' of beetles; thanatosis by prey; stone mimicry in plants (e.g. *Lithops*)
Appetence (Feeding)	AGGRESSIVE MIMICRY Angling; cleaner-fish mimicry; perhaps some insectivorous plants e.g. *Nepenthes* (see Chapter 10, section *Does aggressive mimicry occur in plants?*)	AGGRESSIVE CRYPSIS Crypsis of predators; death-feigning by predators; 'vulture mimicry' by zone-tailed hawk

Table 1.8 Zabka & Tembrock's (1986) interpretation of the classification of mimicry systems suggested by Wickler (1968). (Adapted from Zabka & Tembrock 1986 with permission from Elsevier.)

Type of system	Examples
I. CRYPSIS (mimesis & camouflage)	
II. MIMICRY	
1.1. Batesian mimicry	Wasp mimicry by hoverflies
1.2. Emsleyan mimicry	Coral snake mimicry
2. Aggressive mimicry	
2.1. Disjunct, i.e. with three or more participating species:	Anglerfish; cleanerfish mimicry; carrion flowers; sporocyst of *Leucochloridium macrostomum*
2.2. Bipartite	
2.2 a) Conspecific	Non-stinging male wasps mimicking stinging females
2.2 b) Conspecific	Males and females of *Corynopoma riisei*
2.2 c) Conspecific	Cuckoos; *Ophrys* orchids; female *Photuris* fireflies
2.3. Conjunct	Egg mimicry by mouth-breeding fish

Table 1.9 Overview of 'reproductive mimicry' as interpreted by Zabka & Tembrock (1986) but which contains such diverse functions as increasing fertilisations, obtaining food, and obtaining dispersal. (Adapted from Zabka & Tembrock 1986 with permission from Elsevier.)

| Signal-receiver | Imitations of: | |
	non-communicative releaser	communicative releasers
Interspecific receiver	DISJUNCT, e.g. carrion flowers, sporocysts of *Leucochloridium*, male *Photuris* fireflies	S1 + R/S2, cuckoos, brood parasitism in insects, orchid pollination by sexual deception
Intraspecific receiver	S1/S2 + R, e.g. prey mimicry by *Corynopoma riisei* male fish	CONJUNCT, e.g. egg-dummies in mouth-breeding cichlids, female mimicry by male scorpionflies to obtain food

spectrum of variation that leads to major adaptive traits, and that the latter includes mutations that are rare and perhaps not seen under normal circumstances in, say, laboratory selection experiments. For example, almost all examples of resistance to insecticides observed in the field (such as mosquitoes to DDT) involve a single mutation that confers virtually complete protection. Artificial selection experiments in the laboratory almost inevitably lead to polygenic changes, i.e. several genes, each contributing a little to the improved insecticide resistance. The difference is one of scale – the populations under selection in the lab are likely to be a few thousands or tens of thousands of insects, whereas in the field, blanket pesticide spraying is aimed at killing many millions of individuals and the selection is much stronger. With each spray application the pest controllers are trying to kill all the insects, whereas in the lab the aim is to leave a few alive after each round of selection, from which to breed the next generation.

Chapter 2

CAMOUFLAGE: CRYPSIS AND DISRUPTIVE COLOURATION IN ANIMALS

'It's like this,' [Pooh] said. 'When you go after honey with a balloon, the great thing is not to let the bees know you're coming. Now, if you have a green balloon, they might think you were only part of the tree, and not notice you, and if you have a blue balloon, they might think you were only part of the sky, and not notice you, and the question is: Which is most likely?'

'Wouldn't they notice you underneath the balloon?' [Christopher Robin] asked.

'They might or they might not,' said Winnie-the-Pooh. 'You never can tell with bees.' He thought for a moment and said: 'I shall try to look like a small black cloud. That will deceive them'

'Then you had better have the blue balloon' [Christopher Robin] said: and so it was decided...

...Winnie-the-Pooh went to a very muddy place that he knew of, and rolled and rolled until he was black all over; and then, when the balloon was blown up as big as big, and [Christopher Robin] and Pooh were both holding on to the string, [Christopher Robin] let go suddenly, and Pooh Bear floated gracefully up into the sky'

Milne (1926)[1]

---

1. Thanks initially to a blog by Karen Summers that drew my attention to this perceptive excerpt.

Mimicry, Crypsis, Masquerade and other Adaptive Resemblances, First Edition. Donald L. J. Quicke.
© 2017 John Wiley & Sons Ltd. Published 2017 by John Wiley & Sons Ltd.

INTRODUCTION

The vast majority of animals defend themselves from predation by making it hard for predators to detect them. In evolutionary terms, natural selection has favoured those traits that reduce the probability of detection by predators, and camouflage is undoubtedly the most commonly evolved system (Waldbauer 1988a, Ruxton et al. 2004a). An alternative evolutionary result – quite the opposite – is where selection has favoured conspicuous signals such as bright, contrasting colours to let predators know that they are not profitable to attack, either honestly (see Chapter 4) or dishonestly (see Chapters 5–7). Here, and in the next chapter, I deal with the adaptations that have evolved to reduce the probability of a prey organism being detected by its predators.

The various types of adaptation that are recognised under the heading 'camouflage' have not changed much since Cott (1940), such as aspects of colour, pattern, texture, outline modification or obliteration, shadow reduction and disruption.

DISTINGUISHING CRYPSIS FROM MASQUERADE

Camouflage is broadly defined here and includes very simple features, such as: countershading, which helps to make an organism seem less three-dimensional or make its shadows less conspicuous; crypsis, which makes an organism inconspicuous against its background by matching such features as colour and texture (Fig. 2.1) or even by being transparent; and masquerade,[2] in which the organism remains clearly distinguishable as an object, but not an object that a predator would consider to be prey (Endler 1981, J.A. Allen & Cooper 1985, Skelhorn et al. 2010c). Endler (1981) also distinguishes masquerade from Batesian mimicry because the predator's action or inaction has no effect on the population dynamics of the model (Skelhorn et al. 2010c). The distinction between crypsis and masquerade was often not made in the past and even now can be quite blurred. For example, in Fig. 2.2, are the sea-slugs all being cryptic on their sponge, hydroid or algal backgrounds

_____

2. The term masquerade was introduced for this class of resemblance to provide a clearer definition of what had generally been referred to as 'special resemblance' (Cott 1940), 'procryptic resemblance' (Carrick 1936), 'concealing imitation' (Hailman 1977) and 'mimesis' (Pasteur 1982).

or are some masquerading as fronds, stalks or lumps of sponge? Skelhorn et al. (2010c) try to clarify the distinction as follows:

> The visual appearance of a cryptic species hinders its detection, whereas the visual appearance of a masquerading species hinders its correct identification.

and

> masquerade is not thought to hinder the viewers' ability to detect that there is an entity in the spatial position of the focal organism but, instead, the viewer is thought to misclassify the identity of the focal organism as something in which it has no interest (i.e. something which is not a profitable food source or a threat).

In the quote at the beginning of this chapter from Winnie-the-Pooh, the character Pooh is making the distinction between a blue balloon which the target bees might not notice against the blue sky, which is an example of crypsis, or looking like a small black cloud which the bees might well notice but would not consider to be a honey-thief and therefore would ignore, i.e. this would be masquerade.

Skelhorn et al. (2010b) identified the following three criteria that collectively determine whether a resemblance is masquerade:

1. The species is misidentified or misclassified by either its predators, its prey or both;
2. Being misidentified confers fitness benefits on the masquerading species;
3. The presence of the masquerader does not change the behaviour of the signal receiver in a way that influences the population or evolutionary dynamics of the model (if the object is not a living species then this is self evident).

The critical thing is whether the predator fails to detect the organism or not – but I think even that is open to confusion. As a result of the above criteria and further argumentation, Skelhorn et al. (2010c) came up with the following definition of masquerade:

> [an animal whose] appearance causes its predators or prey to misclassify it as a specific object found in the environment, causing the observer to change its behaviour in a way that enhances the survival of the masquerader. Any change in the population/evolutionary dynamics of the model caused by the presence of the masquerader will not be as a result of the signal receiver changing its behaviour towards the model.

The objects that are the models for masquerade are very varied, including fairly well-studied examples such

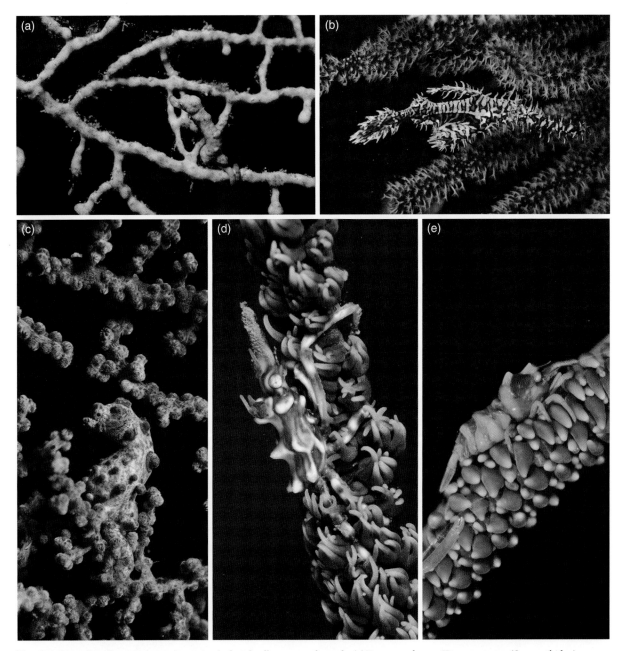

Fig. 2.1 Examples of crypsis in marine animals that dwell among soft corals. (a) Pygmy seahorse, *Hippocampus* sp. (Syngnathidae); (b) ornate ghost pipefish, *Solenostomus paradoxus* (Solenostomidae), among gorgonian coral, Indonesia; (c) pygmy seahorse, *Hippocampus bargibanki,* Komodo National Park, Indonesia; (d) whip coral spider crab, *Xenocarcinus* sp. (Epialtidae) on a gorgonian, Ambon Island, Indonesia; (e) whip coral shrimp, *?Dasycaris* sp. (Pontoniidae), on gorgonian, Ambon Island, Indonesia. (Source: © Linda Pitkin Underwater Photography. Reproduced with permission from Linda Pitkin.)

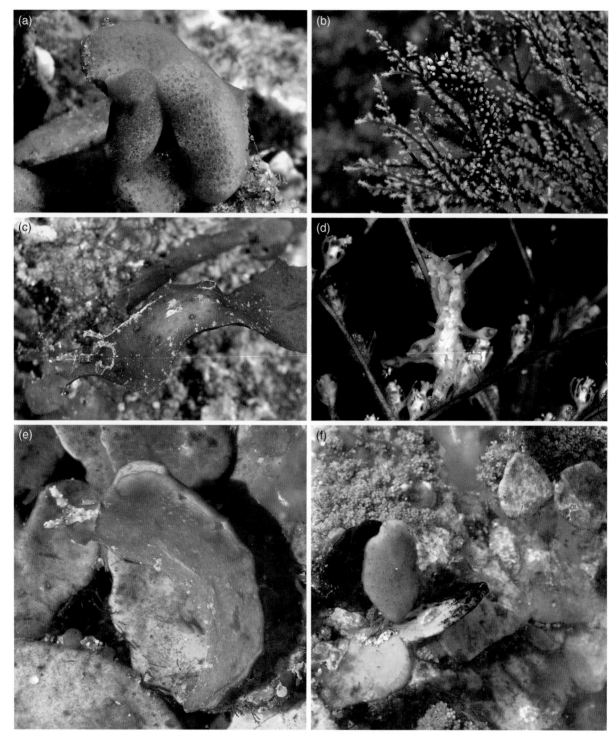

Fig. 2.2 Examples of crypsis/masquerade in specialist nudibranchs that feed on unprotected or weakly protected foods. (a) Two sea-slugs of the genus *Jorunna* (Discodorididae) feeding on unidentified sponge; (b) *Protaeolidiella juliae* on a hydrozoan *Solanderia* species (possibly *fusca*); (c) *Dolabrifera dolabrifera* on a brown alga (*Sargassum* or *Cystophora* sp.); (d) *Myja longicornis* on the stinging hydroid *Pennaria disticha*; (e) *Bosellia* sp. on the alga *Halimeda discoidea*; (f) *Elysiella pusilla* on *H. discoidea*. (Source: Denis Riek. Reproduced with permission from Denis Riek.)

Fig. 2.3 A nocturnal potoo, *Nyctibius jamaicensis*, resting in the daytime and masquerading as a broken branch. (Source: © Professor Phil deVries. Reproduced with permission from Professor deVries.)

as branches (Tate 1994) (Fig. 2.3), twigs (de Ruiter 1952; see Fig. 2.27a,b), thorns (L. Roy et al. 2007), leaves or dead leaves (Nickle & Castner 1995, Solano-Ugalde 2011, Suzuki et al. 2014) (see Figs 3.1c–f and 8.21c,d), bird droppings (Nentwig 1985a, Minno & Emmel 1992, Liu et al. 2014) (see Figs 3.2, 3.3 and 3.5), flowers (Koptur 1989) (see Fig. 2.30) or pebbles (Wiens 1978). Some beetles are thought to masquerade as seeds (Cloudsley-Thompson 1977), some wasp cocoons as flowers (Koptur 1989) and, famously, leafy sea dragons, *Phyllopteryx eques* (Syngnathidae), as fronds of seaweed. In the latter case, as with the charaxine butterfly caterpillar in Fig. 2.4 and many similar instances, I vacillate between considering them cryptic or masquerading. It depends whether the sea dragon, for example, resembles a distinct frond of seaweed or just has the texture, shape and colour to make a predator overlook it against a similarly coloured and textured seaweed background. Starrett (1993) comes to our aid here in that he recognises gradients in the degree of abstractness, such that there may be no clear distinction between crypsis and masquerade even though the majority of cases may be clearly categorisable.

Further, an individual may be masquerading in one situation, but cryptic in another and almost all masquerading animals are cryptic from a distance when a predator's visual acuity is no longer sufficient to resolve their exact form. These do not mar the distinction because, in most given situations, it will be clear whether the animal (or plant) is seen but not recognised for what it really is (masquerade) or is not recognised as an entity from the background (crypsis) (Stevens & Merilaita 2009a). (See the Introduction to Chapter 3 for further discussion.)

Fig. 2.4 Larva (a) and host plant (probably *Parkia speciosa*, Fabaceae) frond (b) of the plain nawab butterfly, *Polyura hebe* (Nymphalidae, Charaxinae; identified by John Horstman).

CRYPSIS EXAMPLES

There are many wonderful examples of camouflage among arthropods, especially insects (Figs 2.5b–e and 2.6) as well as among marine and terrestrial vertebrates (see Figs 2.1a–c and 7.20) and invertebrates (see Fig. 2.2). Evolution has led to similarities in colour and luminance, patchiness of patterns, overall body shape and, in quite a lot of cases, texture and ornamentation. In almost all likely situations there will also be conflicting selectional forces. Colour and pattern may be important in background matching and thus protection from predators, but any changes in overall light absorbance will also cause changes in body temperature. As exemplified by John Endler's large body of research on guppies, there will also be conflicts in other dimensions, such as the selection of males to be as brightly coloured as possible to attract females (Endler 1978). Thus, at any one place and time, the degree of crypsis of a prey species is likely to be balanced against the need to attract a mate. Being essentially visual creatures ourselves, most research has focused on visual appearance, though there can be little doubt that similar balances are being struck by species in other sensory modalities.

Different types of patterning appear to be associated with whether or not an organism avoids capture primarily by remaining motionless or by fleeing. Animals that freeze in the presence of predators often have irregular camouflage patterns, such as stippling, whereas those that flee often have striped patterns. Some have a combination of the two on different parts of their bodies. This is referred to as bimodal. Irregular patterns are thus associated with 'static camouflage' and stripiness with 'motion camouflage' (see section *Stripes and motion dazzle – more zebras, kraits and tigers* later). K.L.A. Marshall & Gluckman (2015) illustrate this by comparing the patterns of two groups of birds: aquatic waterfowl (Anseriformes: 118 spp.) and terrestrial game birds (Galliformes: 170 spp.). The former normally take flight and the latter freeze. Their analysis showed that regular patterns did indeed predominate among waterfowl and irregular or bimodal patterns among game birds. Interestingly they also found that irregular patterns could evolve into regular ones, but not the reverse. In the case of birds, and probably other animals, with bimodal pattern types, the regular parts are probably associated with intraspecific communication rather than camouflage; and in some instances barred patterns also serve a communication role (Gluckman & Cardoso 2010).

COUNTERSHADING

Probably the most widespread form of crypsis is countershading, which for most animals means that their dorsums are more darkly pigmented than their venters (Rowland 2009).

Understanding of this phenomenon is most widely attributed to Abbott H. Thayer (1896) and a lot of detail of his work is reported in a paper by his son, G.H. Thayer (1909), though Poulton (1888, 1902) had published essentially the same idea in relation to caterpillars. The way countershading was suggested to work by Abbott Thayer is through balancing out the shadow that light from overhead causes on a three-dimensional body, such that instead of appearing paler above and darker below, as three-dimensonal objects do, the body appears more or less uniformly dark from top to bottom. This has been termed self-shadow concealment (SSC), and certainly when one experimentally manipulates a countershaded animal such that it is upside down, it appears far more conspicuous with greatly enhanced contrasts (see Fig. 2.7). Therefore, to a predator that is cued into using apparent three-dimensionality as an attribute of potential prey, an animal with good SSC ought to be less conspicuous.

Ruxton et al. (2004a, 2004b) reviewed the evidence for countershading's function and found that, despite its ubiquity, there was very little actual evidence that countershading, and especially SSC, actually works in nature. Instead they suggest most of the evidence indicates that the dorsum and ventrum are often simply responding to different selection pressures and that this is often to do with background matching. Very little experimental work has been done (see later) and there are caveats with all of them such that indirect evidence plays quite a large role in interpreting the results.

Nevertheless, the idea of SSC is very intuitive, and I suspect that it generally plays a more important role in larger bodied creatures. SSC is further supported by various observations, such as the fact that in some animals that routinely live upside-down, the darkness gradient is reversed. Classic examples are the upside-down catfish, *Synodontus nigriventris* (Mochokidae), which, as its specific epiphet suggests, has a darker ventral surface, and this corresponds to their normal habit of swimming upside down foraging at the surface of rivers, or underneath submerged logs. There are in fact a number of other, similarly coloured and behaving congeners, such as *Mystus leucophasis* (Bagridae), a catfish from South-East Asia. Another indication that SSC is important in concealment is that it is primarily associated with animals that are cryptic in other ways and not brightly, warningly coloured – as the latter animals have not evolved to hide from predators.

Kamilar & Bradley (2011) examined how countershading patterns are related to normal positioning behaviour in primates. Unlike most mammals, among primates there is considerable variation in countershading, with larger bodied species displaying it less prominently than smaller bodied ones. Using colour-corrected photographs of museum skins from 113 primate species, Kamilar & Bradley compared

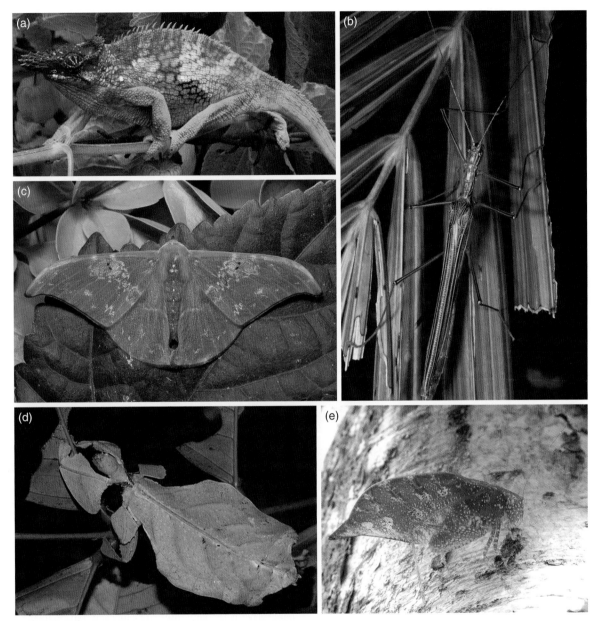

Fig. 2.5 Examples of animals cryptic on, or masquerading as, green leaves. (a) A chameleon, *Kinyongia* sp., from East Africa; (b) unidentified stick insect (or phasmid) on typical background leaves; (c) green geometrid moth, *Tanaorhinus viridiluteatus*, from China also showing mimicry of fungus-infected green leaf tissue; (d) giant leaf insect, *Phyllium giganteum* (Phyllidae); (e) unidentified bush cricket (katydid) masquerading as fungus-infected leaf, Costa Rica. (Source: a, Ales.kocourek at the English language Wikipedia. Reproduced under the terms of the Creative Commons Attribution Share-Alike Licence CC BY-SA-3.0, via Wikimedia Commons; b, © Professor Phil deVries. Reproduced with permission from Professor deVries; c, Itchydogimages. Reproduced with permission from John Horstman; d, Bernard DUPONT 2013. Reproduced under the terms of the Creative Commons Attribution Share-Alike Licence CC BY-SA 2.0, via Flickr; e, Kenji Nishida. Reproduced with permission from Kenji Nishida.)

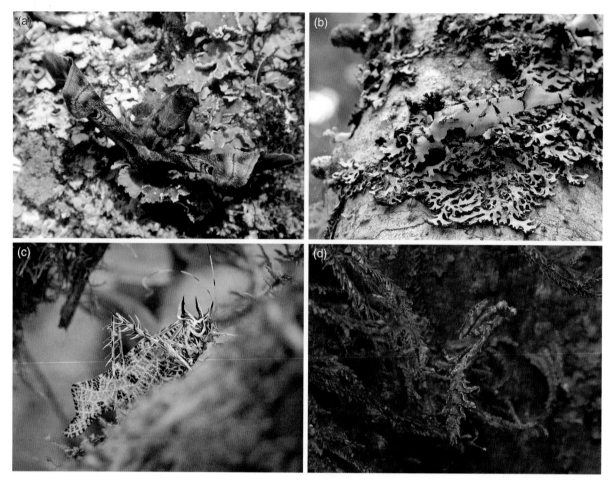

Fig. 2.6 Examples of crypsis with lichen or moss as models. (a) A uraniid moth from China; (b,c) *Lichenomorphus ?carlosmendesi* and *Markia hystrix* (Tettigoniidae) katydids from Costa Rica; (d) unidentified moss-mimicking praying mantid from Australia, possibly a *Pogonogaster* sp. (Thespidae). (Source: a, Itchydogimages. Reproduced with permission from John Horstman; b,c, Kenji Nishida. Reproduced with permission from Kenji Nishida.; d, Sam Wilson. Reproduced with permission from Sam Wilson.)

dorsal and ventral luminance values and correlated them in a phylogenetically independent way with the frequency that the species adopted a vertical posture. The strength of countershading was significantly negatively correlated with frequency of adopting an upright posture, independent of body size. This further seems to support its protective role, since having the countershading in the wrong orientation makes an animal more, not less, conspicuous.

To illustrate the likely concealing function of countershading, a classic example among insects involves the large, usually predominantly green, caterpillars of various hawkmoths (Sphingidae) or giant silk moths (Saturniidae). The former are often illustrated in their sphinx-like posture, sitting on a twig or branch. However, probably due to their weight, they normally rest hanging beneath the plant and typically have their ventral surface somewhat darker than

their dorsal surface, another example of reverse countershading. Figure 2.7 shows two examples of hawkmoth caterpillar in their normal daytime resting position underneath a leaf or stalk of its food plant where its countershading almost perfectly obliterates its cylindrical nature, its uppermost and lowermost parts appearing approximately equally luminous. In the views where I have rotated the caterpillar through 180° (Fig. 2.7a,c) the light from above makes its paler dorsal surface highly contrasted with its darker underparts. Add to that the shadow it creates and it is clearly a three-dimensional potential prey.

As Ruxton et al. (2004a) discuss, countershading is largely associated with cryptic animals and they therefore suggest that only their upper sides are darker because if pigment synthesis is expensive it ought to be placed only in the parts of the body where it will be most effective against

Fig. 2.7 Effectiveness of countershading demonstrated with two species of hawkmoth (Sphingidae) caterpillars. (a,b) Fourth instar larva of convolvulus hawkmoth, *Herse convolvuli* (Sphingidae), on *Ipomoea rosea* leaf, (a) with the leaf manually rotated so the caterpillar is sitting above it in an unnatural position for daytime, its pale dorsal surface contrasts markedly against its darker sides and shadowed, darker ventral surface with real shadow cast beneath it, and (b) in normal daytime resting position, still illuminated by light from sky, note that it has a far more even reflectance; (c,d) eastern death's-head hawkmoth, *Acherontia styx* (identified by Ian J. Kitching, The Natural History Museum, London), again in unnatural and natural orientations, respectively – note also the distinctive defensive 'sphinx posture'.

predators. That argument would also seem to apply to aposematic prey, because there would be little point in putting expensive pigments where a predator would be

unlikely to see them. However, the majority of unpalatable aposematic animals have their venters as strongly pigmented and patterned as their upper sides (see Figs 4.3c,d and 4.34) though, of course, if the prey were to end up on its back during an attack it might be evolutionarily favourable to continue displaying the warning.

Another way that some insects have evolved to eliminate shadow is through possession of lateral flanges that can be pressed close to the substrate and a generally wider body form than some close relatives (see Fig. 2.23a).

Experimental tests of concealment by countershading

Although Abbott H. Thayer demonstrated how countershading worked using painted wooden blocks on wire legs to convince his audience, it was not until the second half of the twentieth century that experimentation really started.

E.R.A. Turner (1961) employed pastry models of insect larvae as bait for wild birds on a suburban lawn to investigate the type of features that would provide the baits with a significant degree of camouflage. The findings showed that small changes in background resemblance, disruptive patterning, countershading and special resemblance could all provide a significant degree of protection from predation. Unfortunately, some problems with experimental design, pseudoreplication and possibility of local predator community learning casted some doubt on the validity and generalisability of his results (Kiltie 1988), but further work has supported his conclusions (see later).

Despite some design issues with E.R.A. Turner's (1961) earlier experiments, his general experimental procedure has been widely employed in subsequent studies. Edmunds & Dewhirst (1994), using the same system, set out to perform a slightly more controlled test, setting their prey out on various lawns in the south of England. They employed four sorts of model caterpillar baits which were coloured, albeit rather more crudely than most natural examples. The 'caterpillars' were either evenly coloured with light or darker green food colourant that was tasteless (at least to humans), or made darker above (countershaded) or darker below (reverse shaded, i.e. the same as countershaded but laid out upside down), and nine tests were carried out.[3] The results (Fig. 2.8) showed that all treatments differed

---

3. In each test 25 copies of each type of model caterpillar were laid randomly in a 10×10 grid, each bait 1 m from the next. The predatory birds were quite varied, mainly house sparrow (*Passer domesticus*), chaffinch (*Fringilla coelebs*), starling (*Sturnus vulgaris*), blackbird (*Turdus merula*), song thrush (*Turdus philomelos*), robin (*Erithacus rubecula*), dunnock (*Prunella modularis*), blue tit (*Cyanistes coeruleus*) and great tit (*Parus major*).

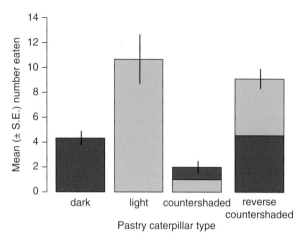

Fig. 2.8 Protective value of countershading detected using coloured pastry 'caterpillar' prey with wild garden birds as predators in a lawn setting. (Source: Adapted from Edmunds & Dewhirst 1994. Reproduced with permission from John Wiley & Sons.)

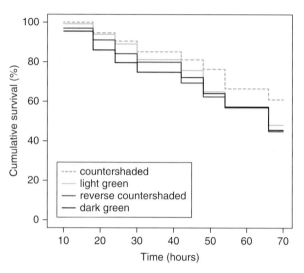

Fig. 2.9 Cumulative survival plots demonstrating protective value of countershading detected using coloured pastry 'caterpillar' prey in a woodland setting. (Source: Adapted from Rowland et al. 2008. Reproduced with permission from The Royal Society.)

significantly from one another apart from reverse countershading versus light green with, as probably expected, countershading providing significantly greater protection than uniform dark or light green (paired t-tests, $p = 0.007$ and $p = 0.002$, respectively).

The effectiveness of countershading as a camouflage device depends on many things, one of which is the predator species. Speed et al. (2005), in a series of tests with wild garden birds in the UK, failed to find the protective effect that Edmunds & Dewhirst (1994) found. Examination of their data showed that whereas countershading was at least partially effective as a defence against blackbirds, *Turdus merula*, it did not provide any significant protection against predation by blue tits, *Cyanistes caeruleus*, or robins, *Erithacus rubecula*. This led Rowland et al. (2007a) to carry out a very similar experiment to Edmund & Dewhirst's (1994), presenting four types of pastry model[4] to wild birds feeding from lawns but with a larger sample size and some further refinements. Their results were in agreement with those of Edmund & Dewhirst (1994) rather than Speed et al. (2005) in that countershaded 'prey' experienced significantly lower predation than each of the other three types, though reverse countershaded ones experienced an intermediate level between those of uniformly dark and light types.

Rowland et al. (2008) extended the study of 'caterpillars' to a woodland situation with the pastry models pinned to the upper surfaces of tree branches – a much more natural

setting than the lawn experiments. The results demonstrate a significant benefit of countershading (Fig. 2.9) (Wilcoxon tests showed that countershaded prey had significantly higher survival rates than dark; $z = -4.306$, $p < 0.001$, pale, $z = -3.298$, $p = 0.001$, and reverse-shaded ones, $z = -3.141$, $p = 0.002$). In addition, they also carried out an experiment with the pastry 'caterpillars' attached to the undersides of branches and found that in this case the reverse countershaded ones (see Fig. 2.7) had significantly higher survival rates. See also Chapter 8, section *Eyespots on caterpillars*.

Bioluminescent counter-illumination

In the deep ocean the vast majority of larger organisms employ bioluminescence for some purpose (Dahlgren 1916, O'Day 1973, R.E. Young 1983). A number of fish and squid have a high density of photophores on their ventral surface, and the idea that they might function to render them less visible to predators below was first put forward by Dahlgren (1916). He suggested that the ventral photophores of some squid that emit a bluish light would help them blend with the sky. Thus the use of bioluminescence helps to conceal prey from predators approaching from below who would otherwise be able to see their silhouette against a somewhat paler sky. Counter-illumination will obviously only be successful at appropriate depths where there is still some sky light visible and the water is clear. Close to the surface, the illumination from above is likely to be more spatially

---

4. Light, dark, countershaded and reverse countershaded, coloured with different concentrations of green SuperCook™ food dye.

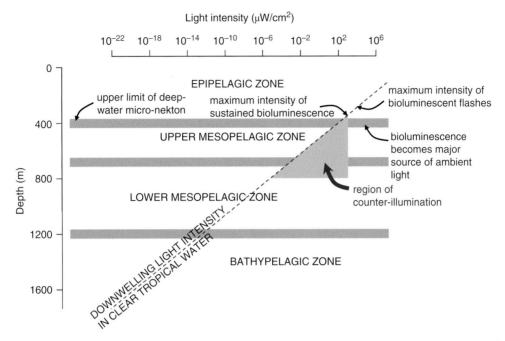

Fig. 2.10 Diagram showing the depth and light conditions that allow marine organisms to use bioluminescence for counter-illumination. Too close to the surface and organisms cannot produce enough light to match the downwelling daylight, and too deep there is no daylight to counter. (Source: Adapted from Young 1983 with permission from Bulletin of Marine Science.)

heterogeneous, thus making accurate counter-illumination more challenging, and even closer, too bright for a bioluminescent animal to be able to match. Off Hawaii, it has been estimated that the lowest level where this might be useful is at a depth of between 750 and 800 meters (R.E. Young 1983). It is important to realise that deep-sea fish have extremely light-sensitive eyes, estimated to be between 10 and 100 times more sensitive than human eyes (G. L. Clarke & Denton 1962). Figure 2.10 shows the depth and light intensity zone where bioluminescence can be used to achieve counter-illumination in daytime (R.E. Young 1983). At shallow depths, it becomes physiologically/energetically unfeasible for organisms to sustain continuous light emission, though here short bright flashes may be employed as warning signals.

Successful counter-illumination depends on regulating the amount of light given out, and its angular distribution, to match that of the background illumination from the sky. Animals vary in their ability to do this, but in some it can be very accurate. A number of laboratory studies have been carried out to test this for various species both of fish and squid and also some prawns (Case et al. 1977, R.E. Young & Roper 1977, Warner et al. 1979, R.E. Young & Mencher 1980, R.D. Harper & Case 1999, Stevens et al. 2014a).

BACKGROUND MATCHING

Possibly the commonest form of camouflage is background matching, in which a prey animal (or possibly plant) has the same overall colouration and texture as the environment or microhabitat where it lives (e.g. Cott 1940, Stoner et al. 2003, Stevens & Merilaita 2009a, Caro 2014). Background matching has evolved to enable prey to conceal themselves effectively against backgrounds as simple as smooth green leaves, sand, gravel, soil, various tree bark textures, mosses, lichens, etc. Using a model run as a neural network, Merilaita (2003) came up with several insights into the evolution of camouflage, some obvious, some requiring further investigation:

1. It is harder to detect camouflaged prey against visually more complex backgrounds, which is fairly intuitive;
2. It is easier to reduce detection through camouflage against visually more complex backgrounds;
3. Selection on camouflage can exploit limitations in predators' information processing;
4. There are shortcomings in using the degree of background matching as the measure of camouflage.

The first two of these are because complex backgrounds require a lower degree of background matching for successful camouflage than do uniform or more even backgrounds, and the last point follows from that.

The majority of mobile cryptic species show behavioural preferences for resting (and orientating) on appropriately coloured/textured backgrounds and a good deal of work has been carried out with various Lepidoptera (see especially sections *Melanism* and *Orientation and positioning* later). Others are capable of changing their colours either quickly or slowly in relation to the environment. Sometimes, selecting an appropriate background has to be done quickly, as in the case of grasshoppers that jump to safety (Eterovick & Côrtes Figuera 1997). Eterovick et al. (2010) found a similar, crypsis-enhancing background selection in threatened tadpoles.

Visual sensitivity of predators

What humans often forget is that many potential predators and other visual signal receivers do not have the same range of light sensitivity as us and may, even within the region of the human visual spectrum, have more or less discriminatory ability. Even among humans, there is some age-dependent variation in sensitivity at the violet–ultraviolet transition, and, probably through cultural influences, some populations subdivide the visible spectrum up far more finely and discriminate shades of, for example, green, that most western readers would find very hard or impossible to distinguish on a colour chart. Figure 2.11 compares the sensitivities of the photoreceptor pigments in a snake and a passerine bird in relation to what a human can see (i.e. not UV). It is not surprising, therefore, that some things that humans cannot distinguish may be perfectly distinguishable to other taxa. For example, M.D. Eaton (2005) shows that several birds that humans have previously considered to be sexually monomorphic are, in fact, dimorphic to their conspecifics, and Garcia et al. (2013) found the same trade-off between sexual signaling and camouflage in Australian Mallee dragon lizards, *Ctenophorus fordi* (Squamata: Agamidae).

It is widely known among European lepidopterists that certain cryptic caterpillars, such as those of the emperor

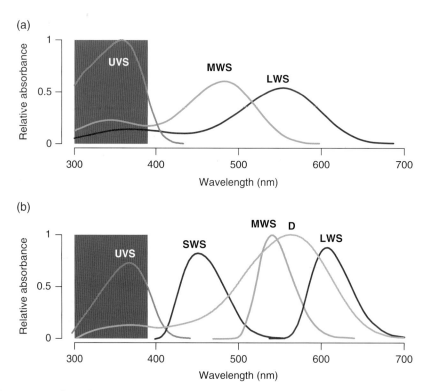

Fig. 2.11 Spectral sensitivities of visual pigments: (a) in the colubrid snake *Thamnophis sirtalis*; (b) in birds with UV sensitivity. The dark grey areas at short wavelengths indicate the part of the spectrum that humans typically cannot see, though young children tend to have sensitivity extending slightly further into the UV. (Source: Adapted from Stuart-Fox 2008. Reproduced under the terms of the Creative Commons Attribution Licence CC BY, via PLoS ONE.)

moths, *Saturnia pavonia* and *S. pavoniella* (see Fig. 4.35d), which often feed on *Erica* and *Calluna* and are normally very difficult to spot, are nevertheless easily visible to dichromatic (red–green colour-blind) people (E.B. Ford 1967, M.J. Morgan et al. 1992). However, Cournoyer & Cohen (2011) found no differences in the potential abilities of a dichromat and a trichromat predator to detect the cryptic arrowshrimp, *Tozeuma carolinense*, which lives among seagrass (Alismatales) and is under-represented in the diet of predatory fish. However, their visual chromatic contrast models for two of the shrimp's potential predators, *Cynoscion nebulosus* (a dichromat) and *Sciaenops ocellatus* (a trichomat), which are otherwise quite closely related to one another, showed that the green and brown shrimp colour morphs were in each case a good match to the background of dead and living seagrasses, and that out of three species of seagrass in the area, green-morph shrimp were only found on the two whose colour they best matched. Similarly, using hyperspectral imaging, Chiao et al. (2011) investigated how cuttlefish would appear to the eyes of potential predatory fish and found they would indeed also be highly cryptic to these.

Differences between the vision of a predator and that of a prey organism can be exploited because often a prey species needs to communicate visually with conspecifics, for example to locate mates. Håstad et al. (2005) found that the intraspecific sexually dimorphic plumages of Swedish songbirds (Passeriformes) were more conspicuous to songbirds in general than they were to predatory raptors or corvids (crows and allies) because the predators largely lack the UV visual sensitivity that passerines have (Hart & Hunt 2007). Schaeffer et al. (2007) came to the same conclusion based on their analysis of spectral sensitivities of fruit/seed-dispersing passerines (see Chapter 14, section *Fruit and seed dispersal by birds*).

To make a perfect match or compromise

Despite the impressive resemblances, in terms of colour, reflectance, pattern and even texture, that many animals achieve to the backgrounds against which they conceal themselves, natural environments are seldom uniform or ideal. Being an extremely close match, say, to the bark of one particular kind of tree will give a moth good protection on that tree species, but will inevitably limit the places where it can rest most safely. This might not be a short-term problem if, say, the moth was a specialist feeder on that type of tree, but for a potential generalist it might mean that some places where the larvae could find food are effectively out of bounds. One evolutionary option might be to produce various colour morphs such that they behaviourally select the best matching backgrounds (e.g. Reimchen

1979, Tsurui et al. 2010), or locally adapted forms. Another option might be to compromise so that rather than precisely resembling one substrate type, the moth instead has a general resemblance to a wide range of bark-type backgrounds (D.L. Evans 1985). Merilaita et al. (2001) and Merilaita & Dimitrova (2014) approached this problem experimentally with avian predators, and Merilaita et al. (1999) and Houston et al. (2007) using a mathematical approach. In Houston et al.'s (2007) system, several factors were found to be important in determining whether a prey would evolve a compromise pattern between two types of background or specialise on just one. These included relative frequencies of the patch types, the travel time of predators between patches, the mean prey number in each patch type, and the trade-off function between the levels of crypsis in the patch types. If one patch type (background type) is rare, it is easy to appreciate that there would be limited opportunities for species to evolve specialisation, and J.M. Cooper & Allen (1994) showed in their experiments with wild birds that prey that matched rarer backgrounds suffered higher relative levels of predation. The easier it is for predators to move between patch types (shorter travel time) the more a compromise background-matching strategy becomes optimal. Such might be the case in a woodland or forest with many tree species, largely randomly distributed.

In the bird experiments, Merilaita et al. (2001) used great tits, *Parus major*, foraging for artificial prey against two backgrounds, one with a large-grained pattern and one with a small one. Prey also had large or small patterns, but in addition there was an intermediate, medium-sized pattern. The results were asymmetric, with a clear trend in predation (or search time) on pattern size against the small-grained pattern, but no difference between medium and large patterns against the large pattern background, indicating that a compromise pattern might be selectively advantageous under some circumstances. In Merilaita & Dimitrova's (2014) experiment, wild-caught blue tits, *Cyanistes caeruleus*, were trained to find triangular pieces of patterned paper (artificial moths) with peanut rewards. The experimental backgrounds comprised a jumble of shapes of varying complexities (similar to those in Fig. 2.40d,e but black on white) and the patterns on the artificial moths resembled the backgrounds to various degrees. When only one background type was used, the birds exerted strong selection for close matching of the prey to the background, but when birds experienced the same range of prey against different backgrounds in a series of tests, the compromise pattern prey were as hard for them to locate as the best matching ones. Again, this supports the theory that when there is heterogeneity of background types, compromise might be a good strategy.

COLOUR POLYMORPHISM

Seasonal colour polymorphism

In many parts of the tropics where the temperatures are high enough to permit species to have two or more generations throughout a year with little break in between, and also where there are distinctly different seasons (usually wet and dry with distinctly different predominant background colours), a lot of plurivoltine species show seasonal colour polymorphisms. These typically involve one predominantly green phenotype (see Fig. 8.22c) that is present in the wet season when there is an abundance of fresh green vegetation, and a brown phenotype (see Fig. 8.22d) that is expressed when the habitat dries out and much of the local vegetation turns yellow/brown until the next rains. As food is normally limited during the dry, brown season, the brown phenotype is also frequently a dormant phase of development, such as a pupa, and one that enters diapause until a combination of signals trigger further development. Many examples of seasonal polymorphism exist among insects and these have been particularly well reported from various parts of Africa, but it is also known in some frogs (Toledo & Haddad 2009).

The European map butterfly, *Araschnia levana* (Nymphalidae), has distinct seasonal patterns, with the spring form having a more or less fritillary-like orange and black pattern and the summer form being black with white lines. Such differences might suggest that selective predation has played a role in this, perhaps with one form being more unpalatable, but Ihalainen & Lindstedt (2012) found that both phenotypes were palatable to birds (blue tits, *Cyanistes caeruleus*) and, using models, that both forms were equally subject to predation irrespective of season.

It has been suggested that darker seasonal forms of some tropical pierid butterflies, a group that in general are anything but cryptic, could be a form of seasonal mimicry (Canfield & Pierce 2010). The darker wet season forms of *Appias lyncida*, *Prioneris thestylis* and *Cepora nerissa* make them all resemble more closely various aposematic, and presumably unpalatable species, given that their larval host plants, *Delias* species, are predominant at that time of year.

Butterfly pupal colour polymorphism

Background selection is important in many other situations. Many Lepidoptera, especially various butterflies of the family Papilionidae, develop their pupae in exposed and visible locations, and often exhibit a colour polymorphism such that the pupa matches the general colour of its pupation background (e.g. D.F. Owen 1971a) (Fig. 2.12). It is interesting that exposed butterfly pupae are almost always cryptic, even if chemically protected, and the adults and/or larvae aposematic, one exception being the highly chemically defended lycaenid *Eumaeus atala*, which has bright orange pupae that contrast strongly with the green vegetation they are found on (Bowers & Larin 1989). A probable explanation for pupae generally being cryptic was provided by Wiklund & Sillén-Tullberg (1985), who showed that pupae were almost invariably killed when attacked by Japanese quail, *Coturnix coturnix*, whereas larvae and adults often deterred a predator before they were seriously harmed (Fig. 2.13). In contrast to those of butterflies, moth pupae are nearly always formed in concealed places and/or in tough cocoons, even when the moths themselves are highly protected. The survival advantage of cryptic butterfly pupae was tested experimentally using the two pupal colour forms of the European swallowtail, *Papilio machaon*, by Wiklund (1974). One of each pupal morph were set out in the field in pairs against either green or brown backgrounds. During the summer generation, cryptic pupae had approximately 50% greater survival than non-cryptic ones, though there was no difference during the overwintering period.

Poulton (1887a, 1892) carried out experiments on the 'special colour relation' between pupae and their backgrounds and showed the importance of environmental factors. While most research has focused on the swallowtails, some species of other butterfly groups also exhibit pupal colour polymorphisms, including the Pieridae (A.G. Smith 1980) and Danainae (D.A.S. Smith et al. 1988).

Normally, environmental factors act as triggers for developmental switches and there have been numerous experimental investigations of what the environmental triggers might be (e.g. Hazel & West 1979, 1982, West & Hazel 1985) and I will only describe a few here. West & Hazel (1979) marked prepupal caterpillars of three American swallowtail butterfly species with UV-fluorescent paint, released them in their natural habitat to go and pupate, and located them later with the aid of a blacklight. Their experimental species differed in their habitat and pupation sites and, in agreement with C.A. Clarke & Sheppard (1972b), they found that those that inhabit stable habitat such as shaded forest, e.g. *Papilio glaucus*, tend to have monomorphic pupal colouration, whereas those that live in more varied, open, grassy habitat, e.g. *P. polyxenes*, are polymorphic, forming green or yellow pupae depending upon the background they are formed on (i.e. grass or wood). The pipevine swallowtail, *Battus philenor*, however, while a forest species and predominantly forming brown pupae, did occasionally form green ones when pupation occurred on the "slenderest twigs" as also noted by C.A. Clarke & Sheppard (1972b) (Fig. 2.14). These latter authors had investigated the genetic component of pupal colour in *B. philenor* and in *P. polytes*.

Fig. 2.12 Seasonal pupal forms of citrus swallowtail butterfly, *Papilio demodocus* (Papilionidae), that are determined by combined environmental and genetic factors. (Source: Chris Jeffs. Reproduced with permission from Chris Jeffs.)

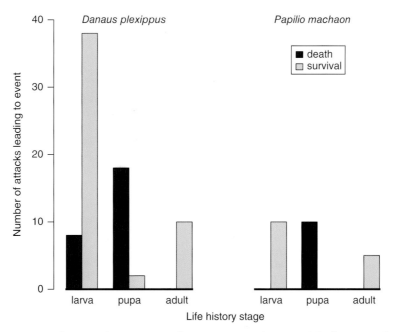

Fig. 2.13 Number of times seizure of an insect by Japanese quail, *Coturnix japonica* (as *coturnix*), leads to survival or death of different life history stages of two unpalatable butterflies – the monarch, *Danaus plexippus*, and the European swallowtail, *Papilio machaon* – demonstrating that pupal stages are much more vulnerable than either larvae or adults. In all cases survival of pupae differed highly significantly (Fisher exact tests, $p < 0.001$) from both larvae and adults, but adult and larval survival numbers did not differ significantly. (Source: Wiklund & Sillén-Tullberg 1985. Reproduced with permission from John Wiley & Sons.)

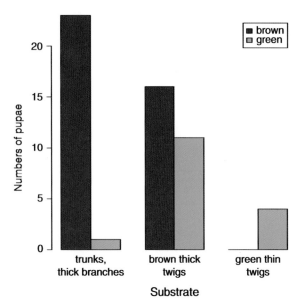

Fig. 2.14 Pupal colour of the swallowtail butterfly, *Battus philenor*, in relation to pupation site. (Source: Clarke & Sheppard 1972b. Reproduced with permission from John Wiley & Sons.)

The environmental factor(s) that actually trigger pupal colour development pathway appear not to be the background colour *per se* but some other correlated variable, and C.A. Clarke & Sheppard (1972b) suggest humidity or CO_2 concentration might be involved in the case of *P. polytes* because of their observation that they nearly always form green pupae if they pupate in a small enclosed plastic box. They performed a short selection experiment with *P. polytes* covering approximately five generations with two selection lines, one selected for those whose pupal colour matched the background (green or brown) and the other for those that did not. A significant difference between the final populations of the selection lines ($\chi^2 = 18.85$, d.f. = 2, $p < 0.001$) confirmed the existence of a genetic component. That there was no significant difference in the proportion of green and brown pupae on a brown background ($\chi^2 = 0.471$, d.f. = 1, $p > 0.07$) but a highly significant difference in the pupation site chosen between populations ($\chi^2 = 17.95$, d.f. = 1, $p < 0.001$) provides strong evidence that selection was acting on the pupation site chosen by the caterpillars. Comparison of results for various swallowtail species, some of which are associated with fairly uniform pupation sites and are monomorphic, and some that are associated with heterogeneous pupation backgrounds and are polymorphic, suggests that the genetic component generally relates to the effectiveness of environmental factors in determining cryptic pupal colouration. The rarity of brown pupae being formed on a green background may be because such pupae

are more conspicuous than green ones on a brown background and therefore selection would have acted on the sensitivity of the environmental switch so it is biased in that direction. C.A. Clarke & Sheppard (1972b) also conclude that the wide distribution of pupal polymorphism in the Papilionidae suggests that the underlying mechanisms probably evolved long ago and have been inherited across speciation events.

Sims (1983) studied the genetics of the colour polymorphism in *Papilio zelicaon*. In this species both univoltine and diapausing multivoltine races are known, and the brown pupal morph is positively correlated with the diapausing race, but it is a propensity rather than a fixed feature, with the proportion of pupal forms coming from broods of each race being normally distributed and reciprocal crosses showing intermediate proportions. The proportion of brown morph pupae produced was easily selected for, and their production was influenced mostly by short day length and low temperatures, irrespective of the substrate colour, whereas with long day length and warm temperatures, pupal colour was affected by the colour of the pupation substrate. Hazel (1977) similarly found that the pupal colour polymorphism in *Papilio polyxenes* involved quantitative genetic variation regulating the threshold for a switch, and suggested that the polymorphism was subject to weak stabilising selection.

Studies on several species of swallowtail with polymorphic pupal colouration suggest a possible relationship between the cues used and the time of day that the fully fed butterfly larvae evacuate their guts and start wandering from their feeding site to find a pupation site, a process that might take up to 12 hours (West & Hazel 1985).

Wohlfahrt (1954, 1957) showed that in the multivoltine European scarce swallowtail[5] butterfly, *Iphiclides podalirius* (Papilionidae), larvae growing under long day conditions, which correspond to the first generation, produce mostly green pupae, whereas short days give rise to a brown phenotype. Further, this is strongly correlated with whether the pupa will diapause, the first generation not diapausing but the second and occasionally third generations entering diapause to overwinter. Stefanescu (2004) monitored caterpillars of the swallowtail butterfly, *Iphiclides podalirius*, and confirmed that season had a strong influence on pupal colour (Fig. 2.15), but the latter also correlated strongly with whether the pupae were formed on the host plant or elsewhere, often among leaf litter on the ground, showing that pupal site selection is also influenced by day length. It is believed that in these insects the plesiomorphic pupal colour is brown, but these suffer considerable predation by small

---

5. Despite its common English name, this is not a scarce species.

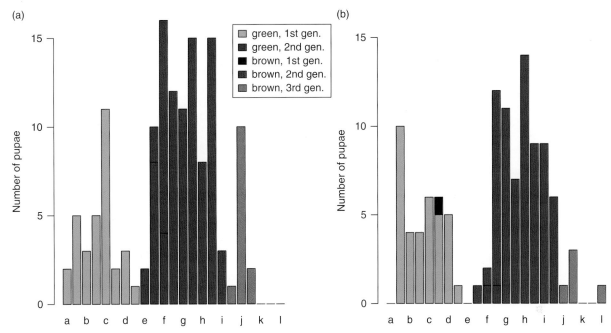

Fig. 2.15 Effect of season and day length on pupal colouration in the multivoltine swallowtail butterfly, *Iphiclides podalirius*. (a) Numbers of pupae of different colours formed by 137 larvae found in the wild and subsequently monitored; (b) data from 112 experimental larvae reared outdoors. Letters indicate weeks: a, 1–7.vi; b, 15–21.vi; c, 29–5.vii; d, 13–19.vii; e, 27–2.viii; f, 10–16.viii; g, 24–30.viii; h, 7–13. ix; i, 21–27.ix; j, 5–11.x; k, 19–25.x; l, 2–8.xi. (Source: Stefanescu 2004. Reproduced with permission from C. Stefanescu.)

mammals, thus evolution favoured cryptic green morphs formed among green vegetation in members of the non-diapausing first generation. Interestingly, whereas all the second and third generation brown pupae in Stefanescu's (2004) study entered diapause, the one brown pupa resulting from a first generation caterpillar (Fig. 2.15b) did not.

Mellencamp et al. (2007), working on the same four North American papilionids, used cauterisation to show that the larval eyes (stemmata) were the organs responsible for detecting background colouration and are also responsible for the larva's selection of a suitable background against which to pupate. Two types of stemmatal photoreceptor cells, those most sensitive to blue and to green light, appear to be those directly involved in determining pupal colour.

As in the Papilionidae, pupal colouration in other butterflies showing polymorphism is controlled by a combination of genetic and environmental factors, e.g. the development of either green or pink pupae by the plain tiger butterfly, *Danaus chrysippus* (D.A.S. Smith et al. 1988). The majority of such studies, however, simply examine environmental influences, for example, the work of pierid butterfly pupal colouration (Bowden 1952, Kayser & Angersbach 1974, Rothschild et al. 1975, A.G. Smith 1980), or there are just casual observations of colour polymorphism (e.g. Nymphalidae: Morphinae – Heredia & Alvarez-Lopez 2004).

At a physiological level, colour is determined by a neuropeptide hormone called pupal melanisation-reducing factor (PMRF), which causes both a reduction in the amount of melanin and an increase in the amount of the yellow carotenoid lutein in the cuticle at metamorphosis (Bückmann & Maisch 1987, Maisch & Bückmann 1987). PMRF is present throughout the nervous system in both the nymphalid *Inachis io* and the pierid *Pieris brassicae*. At the time of moulting into the pupal stage, its release is controlled by the effect of light-sensory neurons in the brain: when pupation is taking place on a pale background the neuropeptide is released. In contrast to those two butterflies, Starnecker & Hazel (1999) have shown that in the papilionid *Papilio polyxenes* the same neuropeptide has exactly the opposite effect, i.e. it causes production of brown pupae, and therefore the same facultative pupal colour polymorphism must have evolved independently in the two groups of butterfly.

Winter pelage: pelts and plumage

Similar seasonal changes are apparent among a number of birds and mammals whose normal environment is reliably snow-covered during winter months. Well-known mammal examples include various hares and rabbits, e.g. *Lepus arcticus*

Fig. 2.16 Summer and winter pelage and plumage, respectively: (a,b) stoat (*Mustela erminea*), the white winter form is sometimes called an ermine, also note the conspicuous black tip to the tail; (c,d) rock ptarmigan (*Lagopus muta*). (Source: a, soumyajit nandy 2013. Reproduced under the terms of the Creative Commons Attribution Share-Alike Licence CC BY-SA 2.0, via Flickr.; b, © iStock.com/ mihailzhukov; c, H. Zell 2009. Reproduced under the terms of the Creative Commons Attribution Share-Alike Licence CC BY-SA-3.0, via Wikimedia Commons; d, Alpsdake 2002. Reproduced under the terms of the Creative Commons Attribution Share-Alike Licence CC BY-SA 3.0, via Wikimedia Commons.)

(Stoner et al. 2003), arctic fox (*Vulpes lagopus*), various ermines (*Mustela* spp.), i.e. species of stoat and weasel that have white winter coats (Fig. 2.16a,b), and the barren-ground caribou (*Rangifer tarandus groenlandicus*). Among birds there are only various grouse (*Lagopus* spp.), including the famous ptarmigan,[6] that moult to white feathers in winter (Fig. 2.16c,d). This only occurs in the northern subspecies or races of some of the species, because only in the northern parts of their ranges is snow highly predictable and long-lasting. Obviously being pure white would render them highly susceptible to detection by predators if there was no snow. Several workers have wondered whether,

since dark objects both absorb and lose radiant heat more quickly than pale ones, these winter pelages have thermal consequences. Ward et al. (2007) examined this experimentally in two subspecies of grouse, the rufous-coloured Scottish and the white-coloured Scandinavian subspecies of *L. lagopus*, using simulated solar radiation. No difference was found in windy conditions but in still air the Scottish form gained heat faster than the white form, as expected, thus suggesting that there may be a fitness trade-off between heat gain and predator detection in these birds.

Some species have their moults to winter pelage triggered by short day length whereas others have endogenous rhythms. In mammals, two hormones have been shown to play important roles and probably directly control the moult

---

6. In some countries also called the rock ptarmigan.

type: prolactin triggers moult to summer pelage in hamsters, lemmings, minks and voles (M.J. Duncan & Goldman 1984, Martinet et al. 1984, Smale et al. 1990, B.A. Gower et al. 1993), while melatonin causes moult to winter coats in these species, and probably also inhibits prolactin secretion (Rust & Meyer 1969, Martinet et al. 1983, Lamberts & Macleod 1990, Badura & Goldman 1992, B.A. Gower et al. 1993).

Of course, global climate change is likely to reduce the extent of winter snowfall, and whether those species that obtain protection from predation can adapt may be critical to their local survival. If climate change leads to a delay in the onset of winter snowfall, and birds' and mammals' moult to winter pelage is triggered by day length, then there will be a mismatch, most likely with the animals turning white at an inappropriate time and thus suffering increased predation (Imperio et al. 2013). Evidence from recent observations of North American snowshoe hares, *Lepus americanus*, suggests that they lack sufficient phenotypic plasticity to cope with current levels of change (Zimova et al. 2014).

Analogous to the seasonally white pelages of various arctic animals is the ontogenetic shift from white to grey of harp seals and grey seals that are born on the arctic snow. Although the newborns of both are initially yellowish, stained with amniotic fluid, the yellow fades or wears off to white after a couple of days so they are very well camouflaged, then within approximately two weeks the grey coat suitable for swimming grows through and they become 'greycoats' (Komárek 2003).

While winter coat-colour change is a highly conspicuous feature of various northern rabbits and hares, Stoner et al.'s (2003) study also showed that pelage colour in lagomorphs generally matches their backgrounds even outside of winter. Belk & Smith (1996) also demonstrated some regional background reflectance matching in Oldfield mice, *Peromyscus polionotus*, either in dull daylight or moonlight conditions, but there was no relationship in terms of hue or chroma. Hoekstra et al. (2004) investigated the genetics of similar background matching in rock pocket mice, *Chaetodipus intermedius*, and found that coat colour was controlled by two alleles of the melanocortin-1 receptor gene (*Mc1r*), a dominant allele causing melanism. There was a strong correlation between *Mc1r* allele frequency and habitat but no correlation for two neutral mitochondrial genes, showing that there was strong selection acting on the coat-colour genes.

MELANISM

Many species of animals, including many insects, birds, mammals and reptiles, occasionally produce dark-coloured, melanic individuals. There are trends with latitude, such that in poikilothermic animals, melanism may be more frequent in colder climates and may be related to thermoregulation, darker bodied individuals being able to gain heat from sunlight more quickly, and there is evidence that this may give a significant advantage (e.g. Currey & Cain 1968, Stiles 1979, Andrén & Nilson 1981, Brakefield 1985, Fields & McNeil 1988, P.H. Williams 2007). Depending on the environment, melanic individuals may, however, suffer increased risk of being detected by visual predators, so there is likely to be selective trade-off. Sometimes either natural phenomena (fires, volcanism) or man-made events (environmental pollution) can lead to transient changes in general background darkness, and these situations may afford an advantage to melanic genotypes of species that perhaps, more normally, live in undarkened habitats.

Industrial melanism

Perhaps the best known demonstration of camouflage, not to mention natural selection in action, concerns the phenomenon of industrial melanism and in particular the European peppered moth, *Biston betularia* (Geometridae), a moderately large geometrid that often rests on tree trunks during the daytime (Fig. 2.17). By the latter half of the nineteenth century, the cumulative effects of unmoderated pollution from heavy industry had resulted in a thick black soot deposit on trees and buildings all around many of Europe's large cities (Berry 1990). Tutt (1896) postulated that the increase in the proportion of the melanic form of the peppered moth was due to industrial pollution killing off pale-coloured lichens and replacing them with soot. In the northern English cities of Manchester, Newcastle and Liverpool the situation was perhaps as bad as anywhere (Fig. 2.18) (Kettlewell 1958), though other large industrial cities, including some in North America, also went through a period of greatly increased pollution. It would be interesting to know what is happening to the frequencies of melanic moths in China's large industrial hubs right now.

The case of the peppered moth became a major example of natural selection in action, with visible evolution occurring well within the time frame of a person's life. A lot of research was carried out by Bernard Kettlewell (Kettlewell 1955a, 1955b, 1956, 1959) and other members of E.B. Ford's influential ecological genetics group at Oxford University, including Niko Tinbergen. Kettlewell's work and his popularisation of it has been subject to much scrutiny and disagreement (Majerus 1998, L.M. Cook 2000, 2003, Hooper 2002, Rudge 2003, 2005). Some of the attacks on his experiments were from so-called 'creation

Fig. 2.17 The famous illustrations of the likely different protective values of pale (form *typica*) and melanic (form *carbonaria*) morphs of the peppered moth (*Biston betularia*): (a) resting on an unpolluted, lichen-encrusted tree trunk in Dorset, UK; (b) on a soot-darkened tree trunk in Birmingham, UK. (Source: Kettlewell 1956. Reproduced with permission from Macmillan.)

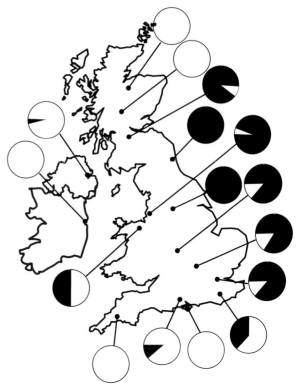

Fig. 2.18 Frequencies of melanic peppered moths (forms *carbonaria* + *insularia*) at sites across the UK during a period of severe industrial pollution, 1952 65. (Source: Adapted from Kettlewell 1958. Reproduced with permission from Macmillan.)

science' advocates, with their own anti-scientific agendas (L.M. Cook 2000), whereas other criticisms against Kettlewell suggest he was scientifically fraudulent. Some of this was initiated by comments of the famous Oxford biologist J.B.S. Haldane, who thought that Kettlewell was capitalising on his own observations of some 30 years earlier and that the data Kettlewell presented too neatly fitted a bird predation selection hypothesis, whereas other factors,

such as thermoregulation and/or physiological effects of pollutants, could also be playing a role (True 2003 summarises the numerous non-visual selective forces that may operate on melanic species). Kettlewell himself noted that the darker form will be less conspicuous in flight during twilight periods as compared to pale moths, a phenomenon he termed aerial crypsis (Kettlewell 1961). Sargent (1969) claimed that his own experiments on a North American moth, *Phigalia titea*, which is very similar in appearance to the peppered moth, failed to show differential bird predation of colour morphs on pale and dark backgrounds and suggested that Kettlewell may effectively have trained his birds to attack moths on tree trunks whereas that is not the most common daytime resting place. Hooper's (2002) book was even more damning and claimed that many of Kettlewell's published data were fabricated, noting that his field notebooks could not be found. Jerry Coyne (1998), in his review of Majerus's (1998) book, while being objectively scientific, fanned the flames against Kettlewell's research and the use of the peppered moth as a classic example of natural selection. He concluded:

> It is clear that, as with most other work in evolutionary biology, understanding selection in *Biston* will require much more information about the animal's habits. Evolutionists may bridle at such a conclusion, because ecological data are very hard to gather. Nevertheless, there is no other way to unravel the forces changing a character.

Michael Majerus and others carried out thorough re-examinations of the Kettlewell studies and replicated some experiments on a grand scale – indeed the largest predation experiment ever performed (a total of 4864 peppered moths were released). His results, which were published posthumously (L.M. Cook et al. 2012), vindicated Kettlewell. That is not to say that there were not any problems with the studies – some experiments involved dead (pinned) moths placed out in a way that might not truly resemble normal resting attitude; some experiments involved releasing moths at abnormally high densities; the behaviours of reared insects from distant localities might not have been the same as local ones; and so on (L.M. Cook et al. 2012). The other criticism made by many Kettlewell critics and evolution-sceptics was that peppered moths do not normally rest on tree trunks. However, Majerus showed that out of 135 individuals surveyed 48 (35%) were found resting on trunks and a further 70 (52%) were on branches which might afford similar polluted or unpolluted backgrounds (Table 1 in L.M. Cook et al. 2012). Further, and importantly, Majerus found that melanic forms released in unpolluted habitat suffered a 9% reduction in daily survival compared to typical form moths, and as adult moths live for several days, the level of selection against melanics throughout all their adult lifespan is likely to be considerable. Of course, Majerus was not able to conduct the reverse experiment comparing survival of forms in a highly polluted habitat, as such habitats have since been cleaned up. The evidence now available, in my view, suggests that the hyperbole against Kettlewell and the Oxford school were effectively unfounded, vindictive and, in some cases, motivated by anti-evolutionary views. The following therefore assumes that vision-mediated predation is the major selective force driving the evolution of melanism. In addition to selective predation on (industrial) melanic morphs in the peppered moth, there is also similar evidence for several other moth species (see Steward 1977a).

With the introduction of clean air acts throughout much of Europe over the past couple of decades, tree trunks in general have become far less discoloured, and in consequence the frequencies of the melanic morphs of the peppered and other moths have gradually declined (B.S. Grant et al. 1996, L.M. Cook 2003, L.M. Cook & Turner 2008). General 'cleaning up' of and/or decline of some large

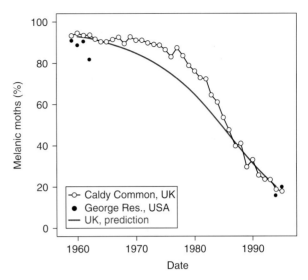

Fig. 2.19 Recovery of non-melanic forms of peppered moth, *Biston betularia*, in the UK and in the USA. The UK site is located near Liverpool, in the industrial north, and was sampled continuously from 1959 to 1965, whereas the US site, Edwin S. George Reserve, was only sampled at the beginning and end of this period. The latter is in quite heavily industrialised southern Michigan though quite far from any large industrial cities – Detroit being 65 km away towards the east. The blue continuous line is what would have been expected to happen at the UK site if there had been a constant level of selection against the dominant melanic allele. (Source: Adapted from Grant et al. 1996. Reproduced with permission from Oxford University Press.)

industries also occurred in parts of North America, though there is far less documentation of the incidence of melanism – insect collecting is just not so common a hobby there. Nevertheless, in 1994 and 1995, B.S. Grant et al. (1996) returned to a site in southern Michigan state where peppered moths had been sampled in 1959–62. Although there are no intervening data points, melanism frequencies at the beginning and ends of the time series are remarkably similar between the British and North American cases (Fig. 2.19). The observed melanic moth frequencies and those that would have been expected if there had been a constant selection against the dominant allele with a selection coefficient, s, of 0.153 (blue continuous line in Fig. 2.19) do not differ significantly. Grant et al. predicted that the melanic form would have declined to virtually zero by the year 2010. With more up to date data Saccheri et al. (2008) examined the spatial pattern of melanic frequencies along a transect from the almost pollution-free town of Abersoch on the Llyn peninsula in North Wales through to industrialised Leeds in northern England at three time intervals (Fig. 2.20). Their updated data gave a new

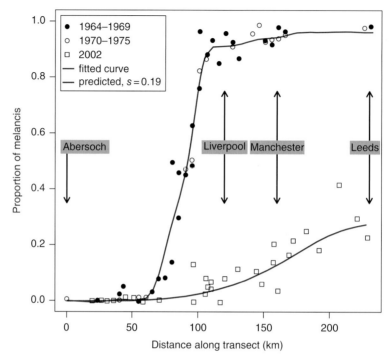

Fig. 2.20 Frequencies of melanic and normal (including intermediate f. *insularia*) morphs of the peppered moth, *Biston beltularia*, along a transect from unpolluted Abersoch (towards the west edge of Wales) to Leeds (in England). The red line is a Bezier fit to the 1967–75 data, the blue line is the prediction made based on the estimated selection coefficient against the melanic form and the mean parent–offspring dispersal based on 30 generations after the 1972–75 data when the cost of *carbonaria* and dispersal are set to their maximum likelihood values ($s = 0.19$ and $\sigma^2 = 184$ km^2). (Source: Adapted from Saccheri et al. 2008 with permission from National Academy of Sciences, USA.)

estimate of 0.19 for the selection pressure against the melanic allele. Furthermore, they were able to estimate how much of the observed melanic frequency was due to dispersal of individual moths between generations, and they obtained a surprisingly high value for the mean square parent–offspring distance of 184 km.

In most instances of industrial melanism in insects, a dominant allele at a single locus is involved (e.g. Kettlewell 1973, Steward 1977b, Lorimer 1979), but in the noctuid moth, *Simyra albovenosa*, Vakkari (2009) has shown that two loci are involved, one controlling an intermediate form which is partially dominant to the pale form, the other controlling the completely melanic morph which is dominant to the pale form but epistatic to (partially dependent on) the intermediate form, giving rise to three phenotypes in the progeny when individuals that are heterozygous for the melanic allele are crossed, depending on the alleles they have at the intermediate locus. The various degrees of melanism exhibited by the peppered moth have been shown by extensive genetic analysis to result from various combinations of multiple alleles at a single locus. Van't Hof et al. (2011) actually identified the core gene sequence (200 kilobases) responsible for the melanic *carbonaria* form to a region homologous to the

silkworm chromosome 17, and found that the sequence variant shows a strong signal for recent selection as evidenced by synonymous to non-synonymous mutation ratios. Industrial melanism, although most often considered in relation to moths, also affects other insects; for example, Creed (1971) reported it in the aposematic ladybird beetle, *Adalia bipunctata*, Popescu et al. (1978) described its occurrence in a tree trunk inhabiting member of the bark-lice (Psocoptera), and Stewart & Lees (1987) found it in the highly polymorphic spittlebug, *Philaenus spumarius*.

Fire melanism

Although darkened tree trunks caused by industrial pollution have been evident for the past couple of hundred years or so, natural events such as extensive forest fires or, less commonly, volcanic eruptions, have provided millions of years for natural selection to favour dark forms of prey organisms when appropriate backgrounds become a common feature of the landscape. Hocking (1964) notes that over much of sub-Saharan Africa grassland vegetation is burnt off annually or every second or third year, and that although more recent fires have been caused by humans, in the past this probably happened naturally as a result of thunderstorms. Thus on a frequent and perhaps regular basis, the background against which surviving insects will find themselves can change very dramatically and very quickly.

Evolution of fire melanism can be very rapid, as evidenced in pygmy grasshoppers, *Tetrix subulata*, by Forsman et al. (2011). The authors showed significant increases in the proportions of melanistic forms in areas that had been ravaged by fire during the previous year, compared to non-fire-affected places nearby. Of course such changes could only occur if there were already some melanic forms in the population. Karpestam et al. (2012) used images of melanistic pygmy grasshoppers against different backgrounds on computer screens, with humans as virtual predators, to demonstrate that they become harder to spot the more burnt the background was, thus supporting the notion that predation pressure was the driving force in the evolution of the melanism. Survival of morphs against different backgrounds was also size-, and hence sex-, dependent in this species (Karpestam et al. 2014b).

Even among mammals there are instances, or at least suggestions, of the occurrence fire melanism. Unlike albinism, melanic individuals often reach quite high population densities, and they are often associated with areas where fires occur quite frequently. Guthrie (1967) discusses the case of the ground squirrel, *Citellus osgoodi*, in which melanics can constitute up to 20% of populations in Alaska. Populations of the fox squirrel, *Sciurus niger*, in the lower Mississippi River drainage in the USA similarly frequently include melanistic individuals, though the genetics of the melanism appears to differ from that in other parts of the species' range (Kiltie 1989). Examination of possible causative factors revealed a significant positive correlation with the frequency of fires caused by lightning strike, but interestingly not with those caused by humans, suggesting that melanic gene frequency has been driven over a longer evolutionary period of time than the more recent period of human farming and recreation.

Although most work on crypsis against fire-affected backgrounds concerns animals, Lev-Yadun & Ne'eman (2013) suggest that the colouration of seeds of some arid-land plants, in their case, *Pinus halepensis*, may be adaptive to reduce predation by seed-eating passerine birds. The winged, wind-dispersed seeds of *P. halepensis* have one side pale coloured and one side nearly black. Each side provides the seed with good crypsis but against different soils, the pale side being particularly good against light-grey, ash-covered soil. The authors suggest that such intra-seed colour variation might actually be more widespread than previously thought and that it is particularly found in plants whose habitats are either frequently ravaged by fire or are otherwise heterogeneous.

BACKGROUND SELECTION

It is obvious in all of the above examples that crypsis will not normally be effective if the animal rests against some sort of conspicuously different background, and therefore the animal's behaviour in choosing an appropriate background is often paramount, especially if the animal is a medium to small-sized invertebrate, i.e. prone to predation. Thus the effectiveness of camouflage often depends heavily on the selection of a suitable background, and also on assuming the correct posture, 'or positional aspect'. W.M. Malcolm & Hanks (1973) demonstrated that *Agonopterix pulvipennella* (Oecophoridae), a cryptic 'micro-lepidopteran', spends considerable time searching for a background with the same degree of reflectance as its wings before settling to rest for the day. Background matching might be fairly general, such as in the case of a dark-winged cryptic moth choosing to spend the day on a dark background, and a pale one on a paler background.

Adult nocturnal Lepidoptera have provided a very suitable subject for the investigation of site selection (Sargent 1966, Boardman *et al.* 1974), and one of the first species to be specifically investigated was the peppered moth. In many areas, not all trees or surfaces were evenly darkened by pollution, and similarly, the peppered moths in these areas were often polymorphic for colour morph. This posed the question as to whether the different colour morphs of the moths select similarly coloured backgrounds to rest on during the daytime. This problem was examined by Kettlewell (1955a), who carried out a simple experiment using a large, empty cider barrel, the inner surface of which was covered with alternate broad stripes of black and white paper. Each evening, three black (*carbonaria* form) and three pale (typical form) peppered moths were released into the barrel, and the following morning their final resting positions were scored. In all, 108 individuals were tested, and the results (Table 2.1) show very clearly that the *carbonaria* forms had a preference for resting on the dark background while the converse was true for the typical, pale form.

Depending upon the time of year, the tree trunk resting places for many moths appear different (due to rainfall, algal growth, etc.) and tree trunk resting moths also show a progressive change in colouration through the year such that species more closely resemble the appearance of bark at the time of year during which they normally fly (Endler 1984). Z. Wang & Schaefer (2012) reanalysed Endler's data and found that moths with horizontal bands (see e.g. Fig. 2.22 top right) normally rested on small herbs or on bark rather than on dead leaves, larger herbs or shrubs.

Sargent (1968) tried to determine whether background selection by a cryptic moth was achieved by comparing the light coming from substrates with the moth's own circumocular scales by painting them a different colour.

Table 2.1 Background preferences of pale and melanic forms of peppered moths (*Biston betularia*), pale brindled beauty moths (*Phigalia pilosaria*), and green brindled crescent moths (*Allophyes oxyacanthae*) in choice experiments using a cider barrel or other cylinder lined with alternate strips of black and white paper. Peppered moths ($\chi^2 = 9.799$, d.f.=1, $p=0.00175$) and green brindled crescents ($\chi^2 = 4.913$, d.f. = 1, $p=0.0266$), showed a significant preferences for settling on matching backgrounds, but *Phigalia* ($\chi^2 = 2.02$, d.f. = 1, $p=0.1552$) did not. (Data from Kettlewell 1955a, Lees 1975 and from Steward 1985.)

Background selected	Moth	
	Peppered moth (*B. betularia*)	
	Pale form (f. *typical*)	Black form (f. *carbonaria*)
White background	39	21
Black background	20	38
	Pale brindled beauty moth (*P. pilosaria*)	
	Pale form	Black form (f. *monacharia*)
White background	11	32
Black background	19	26
	Green brindled crescent moth (*A. oxyacanthae*): <2 or ≥2 out of 4 trials	
	Typical form	Melanic form
Pale bark	63	103
Dark bark	16	56

His experiments involved the dark-coloured *Catocala antinympha* and the pale-coloured *Campaea perlata*, which habitually rest on backgrounds of varying darkness, and he painted the scales around their eyes either black or white. The treatments had no effect on the choice of settling backgrounds and therefore he concluded that choice must be exercised by absolute measures of luminosity. He further concluded that the choice in both species was genetically determined and related this to the potential problem faced by species in which melanics occur sporadically, suggesting that for the melanism to be advantageous the moths must also evolve independent genetic control of resting site choice, because if it relied on matching the background luminescence to something the moth would also be able to see (i.e. its circumocular scales) his treatments ought to have changed their behaviour. In further support of this Sargent cites the case of the geometrid moth, *Cosymbia pendulineata*, in which the occasional melanics that appeared in 1967 following a particularly cold winter, still selected to rest on pale backgrounds. Steward (1985) studied background selection in families of the European green-brindled crescent moth, *Allophyes oxyacanthae* (Noctuidae), and similarly found that the gene(s) controlling background selection were not closely associated with the main melanic gene locus, though they might be linked to other loci involved in darkening.

It should be noted, however, that while no moths have been found to select backgrounds by matching the colour of their own visible (circumocular) scales, Gillis (1982) was able to reverse the background matching behaviour of the North American wrangler grasshopper, *Circotettix rabula rabula*, which has greyish-green and red morphs, by painting red around the eyes of individuals of the green variety.

Heiling et al. (2003) investigated the anthophilic crab spider, *Thomisus spectabilis* from Australia, that has white and yellow colour forms. Unlike that of *Misumena vatia* which can change colour (see *Daily and medium-paced changes*), individuals sit on flowers of their preferred colour. Surveys of spiders on flowers in the field as well as in laboratory choice tests (Heiling et al. 2005) showed that yellow spiders were only found on yellow flowers but white spiders were equally likely to be found on yellow or white daisies. Surprisingly, they found that while appearing well camouflaged to human vision against white flowers of the daisy, *Chrysanthemum frutescens*, the white form of the spider is highly reflective in the UV whereas the flowers it rests on are not, and thus the spider would be easily seen by pollinating insects such as bees (see Cloudsley-Thompson 1981). This appears to be an adaptive trait as its contrast in the UV renders the flowers more attractive to potential pollinators. The only colour combination that deterred bees from landing was yellow spiders on a white flower, something the spiders do not do. It seems likely, given the bees' visual spectrum, that the colour contrast formed by the spider/flower combinations is deceiving the bees by

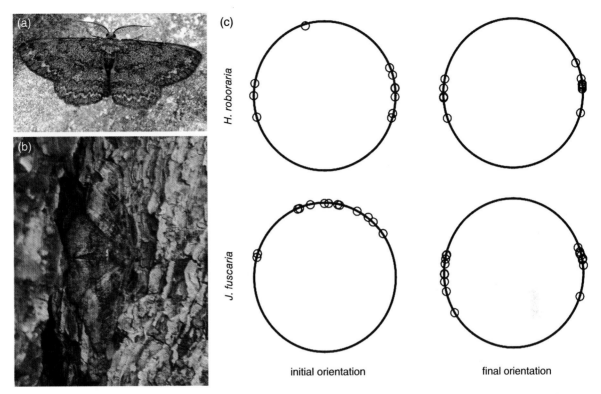

Fig. 2.21 Settling and resting behaviour on tree bark of two species of geometrid moths, (a) *Hypomecis roboraria* and (b) *Jankowskia fuscaria*, that (c) adjust their final resting sites and orientations after their initial landing to afford themselves better crypsis. (Source: a, Jeff de Longe 2005; b, Dr Changku Kang. Reproduced with permission from Dr Kang; c, adapted from Kang et al. 2012. Reproduced with permission from John Wiley & Sons.)

rendering the flowers more attractive, perhaps by mimicking some more generalised rewarding flower visual model in the bee's brain.

Orientation and positioning

Depending on shape or pattern, it is not always sufficient for an animal to choose to rest on a given background. For example, for moths with a striped pattern on their wings, the orientation that they rest in may be crucial for effective crypsis (Pietrewicz & Kamil 1977).

Kang et al. (2012) released and then observed individuals of two moth species, *Hypomecis roboraria* (Fig. 2.21a) and *Jankowskia fuscaria* (Fig. 2.21b) (both Geometridae), as they settled onto tree bark, their normal resting place. The *Hypomecis* settled facing sideways and remained that way (Fig. 2.21c, top), whereas *Jankowskia* almost always settled with the head pointing upwards (Fig. 2.21c, bottom left) but then reorientated itself so that it was facing to the side (Fig. 2.21c, bottom right). Further, neither species stayed still at their initial landing site, but instead walked slowly,

raising and lowering their wings, until they chose a final spot where their crypsis would be more effective. In addition, Kang et al. (2013, 2014) show that *Jankowskia fuscaria* use several sorts of information to choose the orientation direction, including the pattern of the bark.

Webster et al. (2009) used human subjects to spot and 'attack' computer images or real specimens of two North American bark-resting moths, *Catocala cerogama* and *Euphyia intermediata*, against real bark image backgrounds orientated either vertically or horizontally. Field surveys had shown that *Catocala* moths had a strong preference for resting in the head-up or head-down position, though species differ in which way up they rest (Sargent & Keiper 1969). Although *Catocala* moths do not have conspicuous stripes, the human 'predators' most frequently failed to attack them when their images were in the head-down position against vertically orientated bark (Fig. 2.22a), whereas they survived least well in that position when the bark was horizontal (Fig. 2.22c). *Euphyia*, on the other hand, with its quite strong transverse striping, fared badly on vertical bark when facing up or down (Fig. 2.22b) and generally survived less well at almost any orientation other than when it was

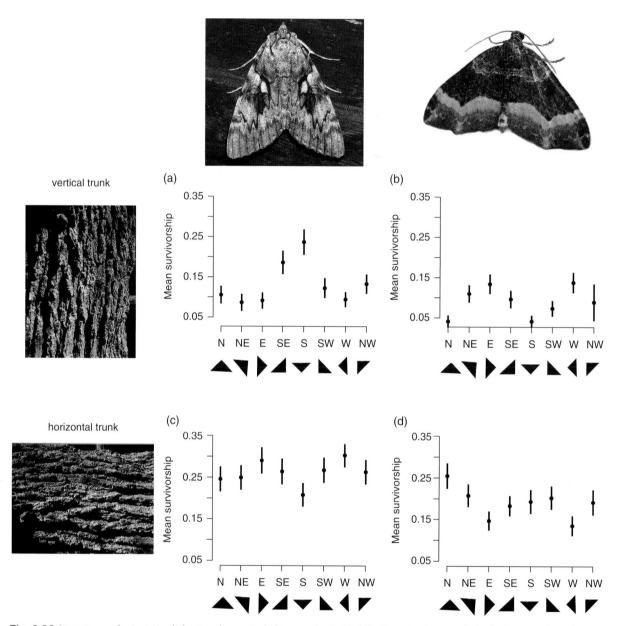

Fig. 2.22 Importance of orientation behaviour for survival of two moths that habitually rest on tree trunks in the daytime: *Catocala cerogama* (Noctuidae) and *Euphyia intermediata* (Geometridae) on sugar maple trees, *Acer saccharum*. Orientations of models are indicated by points of the compass and pictograms, background bark illustrated on left. (Source: Graphs adapted from Webster et al. 2009. Reproduced with permission from The Royal Society; image of *C. cerogama*, D. Gordon E. Robertson 2011. Reproduced under the terms of the Creative Commons Attribution Share-Alike Licence CC BY-SA 3.0, via Wikimedia Commons; image of *E. unangulata*, which is visually virtually identical to *E. intermedia*, Dieder Plu 2011. Reproduced under the terms of the Creative Commons Attributions Share-Alike Licence CC BY-SA 3.)

against a horizontal bark background (Fig. 2.22d). In an experiment using model moths with either horizontal or vertical stripes and giving a mealworm[7] reward to wild birds, Z. Wang & Schaefer (2012) demonstrated highly significant effects of orientation on survival. Moths with horizontal stripes had higher survivorships when orientated facing to the side, whereas model moths with a vertical (i.e. longitudinal) stripe had higher survivorship when facing upwards.

Behavioural selection of appropriate backgrounds for effective crypsis are widespread not just in moths, but also among tropical marine fish (N.J. Marshall 2000) (see Fig. 10.7d,e).

TRANSPARENCY

When thinking of camouflage one mostly thinks of organisms having the same colouration and pattern as their background. However in some cases, organisms may achieve inconspicuousness through simply being transparent, at least in part. Among terrestrial organisms transparency is rather uncommon, but some butterflies, especially some South American ithomiine and other nymphalids (Fig. 2.23b,c) and pierids, have wings that are largely transparent, and transparent patches are present in the wings of a number of moths such as the clearwings (Sessiidae) (Duckworth & Eichlin 1974), bee hawk-moths (Sphingidae) (Fig. 2.23d) and some diurnal wasp mimic, or simply aposematic, e.g. tiger moths (Erebidae[8]: Syntomini). In many of the diurnal moths with transparent wings, such as those in Fig. 4.2 and others such as the neotropical *Dinia eagrus* (Erebidae: Arctiinae), the body is boldly marked bright red on black, making it distinctly aposematic if the transparency has failed to prevent its detection (see section *Dual signals*). Some tortoise beetles (Cassidinae) have large transparent patches that probably have multiple functions, including crypsis, shape disguise and possible emphasis of distinctive, non-transparent, markings (Fig. 2.23a). The wings of some bee hawk-moths are almost entirely transparent. Well-known to aquarists is the freshwater glass

catfish, *Kryptopterus bicirrhis* and *K. vitreolus* (Siluridae), from South-East Asia and Indonesia. In the sea there are numerous transparent organisms belonging to a wide range of phyla. Johnsen (2001) surveyed transparency in marine taxa and found that it has evolved multiple times, and although it is particularly prevalent among Ctenophora (comb jellies) and Cnidaria (jellyfish, siphonophores, etc.) it is also a feature of some annelids, molluscs, crustaceans and urochordates (sea squirts), the larvae of various fish including eels (which are called leptocephali) (Meyer-Rochow 1974) and the amazing sea sapphires (Crustacea: Copepodidae: *Sapphirina* spp.), whose transparency can be replaced in a fraction of a second by spectacular iridescence caused by regular arrays of guanine crystals depending on the viewing angle. Most notably, transparency is strongly associated with planktonic pelagic species. Although many jellyfish are coloured, most members of the Cubomedusae and Hydromedusae as well as most comb jellies (Ctenophora) are virtually transparent, as well as many planktonic salps (Urochordata: Larvacea), planktonic arrowworms (Chaetognatha), some crustaceans, notably amphipods, and pteropod molluscs, and the near-invisible planktonic polychaete worm *Tomopteris*, to name but a few. Transparent planktonic animals are not completely immune from protection though, due to their necessarily higher refractive index (see later), and some may be chemically defended (McClintock et al. 1996).

Cnidarians of the neustonic zone, i.e. living at the water's surface, such as the Portuguese man o' war (*Physalia*) and the by-the-wind sailor (*Velella*), tend to be bluish. In deep water, however, where little or no visible light penetrates, it is interesting to note that many fish, jellyfish and comb jellies are red, the colour having no visual significance, since in these deep zones bioluminescence is the predominant source of light, so red-coloured animals are effectively black. Zylinski & Johnsen (2011) found that transparency was facultative in some mesopelagic (600–1000 m deep) cephalopods and that they can switch rapidly between transparency and pigmentation. When *Japetella heathi* is camouflaged against sky light it is transparent, but when illuminated by bioluminescent light from potential predators it reflects a good deal of light, making it visible, so by rapidly expanding its red-brown chromatophores it significantly reduces the amount of light it reflects back to the predator.

Achieving transparency in a Lepidoptera wing is not always just a matter of not incorporating pigments. The areas are largely devoid of scales that would scatter light, but there is still the problem of reflectance. Yoshida et al. (1997) showed that in the Japanese bee hawk-moth, *Cephonodes hylas*, the wing surface is highly specialised to minimise reflectance, being covered in a regular array of hexagonal pegs (nipples) approximately 250 nm high and

---

7. Mealworms are the larvae of the tenebrionid beetle *Tenebrio mollitor* and are widely used in experiments as a 'standard palatable insect' as well as for feeding various pet animals.

8. Up until quite recently the tiger moths (and some other mostly day-flying aposematic noctuid moths) were treated as separate families (Arctiidae, Ctenuchidae, Syntomidae), but recent molecular phylogenetic work has shown these (a) to form a monophyletic group and (b) to be nested within the family Erebidae which was separated off from the Noctuidae (Zahiri et al. 2011, 2012).

Fig. 2.23 Transparency. (a) Tortoise beetle, *Aspidomorpha* sp. (Chrysomelidae), from Thailand, with transparent edge-intercepting parts of the pronotum and elytra and dark and strongly reflecting bold markings that collectively disguise its beetle-like shape; the species are also probably toxic and take flight readily, as well as having a broad and flattened shape that makes them hard to handle; (b) *Cithaerias menander* (Nymphalidae: Satyrinae) from Costa Rica; (c) glasswinged butterfly, *Greta oto*; (d) narrow-bordered bee hawkmoth, *Hemaris tityus* (Sphingidae). (Source: a, Bernard DUPONT 2013. Reproduced under the terms of the Creative Commons Attribution Share-Alike Licence CC BY-SA 2.0, via Flickr. b, Kenji Nishida. Reproduced with permission from Kenji Nishida; c, Liz West 2008. Reproduced under the terms of the Creative Commons Attribution Licence CC BY 2.0, via Flickr.; d, Marcin Kutera 2010. Reproduced under the terms of the Creative Commons Attribution Share-Alike Licence CC BY-SA 4.0, via Wikimedia Commons.)

200 nm apart, which reduce reflectance across a broad range of wavelengths from red to UV by impedance matching between the wing cuticle and the air, the individual nipples being smaller than the wavelength of light.

One of the most difficult things for a transparent organism to do is hide their eyes, because photo-receptors must necessarily absorb light (Feller & Cronin 2014).

Having coloured prey is also a problem for the transparent animal. Using juvenile coho salmon (*Oncorhynchus kisutch*) as predators, Giguère & Northcote (1987) showed that glassworms, the predatory aquatic larvae of a small nematoceran fly, *Chaoborus*, which are normally almost completely transparent except for two pairs of small airsacs, suffer a considerably increased risk of predation if they have

food in their guts, and predation risk increased in proportion to meal size. Larvae with full guts suffered three times higher risk of predation than those with empty or near-empty guts. In a slightly more controlled experiment Bohl (1982) tested whether predatory fish would preferentially attack egg-bearing *Daphnia* that are more visible since their eggs are opaque, rather than non-pregnant ones which are essentially transparent. This was the case in lit conditions, but not when there was no light, so it was their visibility rather than other signals that led to the greater mortality of the pregnant, and therefore more visible, *Daphnia*.

It is not only food in the gut that can affect the visibility of otherwise cryptic transparent prey, but also parasites. Infection with the parasitic marine isopod crustacean

Probopyrus pandalicola significantly increases the risk of predation by fish of its normally largely transparent host shrimp, *Palaemonetes pugio* (Brinton & Curran 2015). This results from a combination of visual and behavioural changes.

Another problem with achieving transparency is that cells rapidly become opaque after death. As all cells die eventually, and many do while the animal is still alive, maintaining transparency ought to be quite physiologically costly (Hamner 1995, Ruxton et al. 2004a).

In theory, the necessarily higher refractive indices of prey compared to the surrounding water should mean transparent planktonic animals would become detectable to predators from below by virtue of them bending light more (see Ruxton et al. 2004a, p. 41). As described in Chapter 10, section *Why seabirds are black and white (and grey)* (see also Fig. 10.2), an underwater animal's view of the sky is limited to a circular area of the surface called Snell's window. This optical effect has another consequence as described by Ruxton et al. (2004a, p. 41), in that transparent zooplankton just below the surface and just outside of the edge of the window, and especially close to the surface, might become detectable to predators from below because of the way their higher refractive indices will distort the passage of light entering from above. This means that those individuals just outside of the Snell's window cone where no light from above should be invisible, may bend the light rays such that they appear as bright entities against an otherwise dark background. However, the inevitable natural turbulence of the water surface probably means that cases where this might render the plankton easily distinguishable are most likely rather uncommon, except on very still days in freshwater ponds and lakes.

Greer et al. (2016) have made an alternative suggestion about some largely transparent fish larvae, which is that they might actually be Batesian mimics of transparent, but protected, models. For example, leptocephalus eel larvae might be mimics of gelatinous, cestid ctenophores.

REFLECTANCE AND SILVERING

Another manner in which some organisms achieve crypsis is by reflecting the colours of their surroundings. Of course, all surfaces apart from black do this to some extent, but if the surface is uncoloured (i.e. white) and has a high reflectance, then a good degree of background matching can be achieved. The wing undersides of the angled sunbeam butterfly, *Curetis acuta*, is particularly impressive in this respect (Wilts et al. 2013). The upper surfaces are sexually dimorphic, dark brown and white in the female and bark brown and orange in the male (Fig. 2.24a,b, respectively), but the underside of the wings in both sexes is brilliant white (Fig. 2.24c). A scanning electron micrograph (SEM) of the surface of a single brown scale (Fig. 2.24d) shows the general structural arrangement, with longitudinal ridges spaced at approximately 2 μm intervals and numerous cross-ribs. Figures 2.24e,f show a transverse section and an SEM of a reflecting scale from the undersurface of the wing which has the surface between longitudinal ridged and cross-ribs enclosed by a nearly intact, thin transparent chitinous plate separated from the undersurface of the scale by an approximately 0.3-μm-wide air-filled gap. The butterflies often choose to rest on the leaves of local sclerophyllous, wax-covered plants such as Japanese oak, *Quercus acuta*, and flee to such places when under threat. In such semi-shaded places the undersides of the wings reflect the colour of the substrate they are resting on, and so become virtually indistinguishable from the background to a predator (Fig. 2.25).

Several beetles are extremely reflective, such as some cassidine leafbeetles (e.g. *Aspidimorpha* spp. (Chrysomelidae) which combine silvery reflectance with transparency; see Fig. 2.23a) and predominantly Meso-American jewel scarabs, *Chrysina*[9] spp. (Scarabaeidae), which are very popular among collectors. The function or functions of such a shiny cuticle are uncertain. For some of the very shiny tropical species, Max Barclay (pers. comm.) has observed that when resting by the mid-rib of some large leaf, they can easily be mistaken for a large droplet of rain water. There is also some evidence that shininess and iridescence may be aposematic (Fabricant et al. 2014), but in the case *Chrysina gloriosa*, its striped reflective green and black pattern is most likely cryptic against the *Juniper* trees it sits on in the daytime (F.N. Young 1957).

The silvery scales of many fish also reflect ambient colours (as well as contributing to countershading (q.v.). To function properly, however, the fish needs to remain at an appropriate angle to the water surface and this can be in conflict with the need to eat, because planktonivorous fish often have to attack prey that are above them (Dare & Montgomery in press). Silvery fish scales differ both structurally and functionally from those of *Curetis* described above in that they possess multiple (at least 11) protein platelet layers (Denton 1971, Denton et al. 1972, Jordan et al. 2012; see also Sun et al. 2013 or Vukusic & Chittka 2013 for nice descriptions of the physics of structural colours). This structural difference enables them to reflect incident light from any (or at least a very wide range of) angles,

---

9. Many species were formerly placed in the genus *Plusiotis*.

Fig. 2.24 The Japanese angled sunbeam butterfly, *Curetis acuta*, and micrographs showing difference between pigmented and reflecting wing scales. (a,b) Female and male upper sides, respectively; (c) highly reflective underside of a female specimen, which in the field reflects a large amount of the incident light, preserving its colour; (d) scanning electron micrograph (SEM) of an orange-red scale of a male showing typical structural arrangement of butterfly wing scales, with regularly spaced parallel ridges connected by cross-ribs; (e) transmission electron microscope section through reflecting scale showing strong ridges on outer surface and irregularly spaced air-filled gaps between trabeculae connecting upper and lower surfaces; (f) SEM of a reflecting scale showing the thin, largely intact windows between more widely spaced ridges with barely visible cross-ribs. (Source: Dr Bodo D. Wilts. Reproduced with permission from Dr Wilts.)

Fig. 2.25 *Curetis* (Lycaenidae) butterfly wing undersides reflect surrounding colours. (a) Diagram showing how the highly reflective Japanese sunbeam butterfly, *Curetis acuta*, can appear green to a predator or human if observed resting on a leaf that reflects significant green specular light. (b) *Curetis bulis* in Thailand, reflecting colour of the concrete upon which it is settled. (Source: a, Adapted from Wilts et al. 2013 with permission from John Wiley & Sons.)

without polarisation, whereas the *Curetis* scales strongly polarise light at oblique angles. Many animals are capable of detecting polarisation (even humans have a miniscule ability), but perhaps significantly, it is widespread in both arthropods and fish (Hawryshyn 1992), which could mean that any fish reflecting polarised light would be easily detected by other fish or by cephalopods, which also use polarised reflected light in intraspecific signalling (Cronin et al. 2003).

To animals with the ability to detect polarised light, the open ocean is, in fact, a far from a static and uniform visual environment, and provides a particularly difficult situation for camouflage to work in. Brady et al. (2015) used video-polarometry to examine a wide range of open-ocean fish viewed from different angles and from different directions, to assess the effectiveness of their camouflage to polarisation-sensitive predators. Open-ocean fish were found to have significantly better polarisation camouflage acting in the direction from which predators might attack compared to nearshore fish; this supports the idea that natural selection in the open ocean has resulted in effective polarocrypsis in many species.

ADAPTIVE COLOUR CHANGE

Some organisms are able to dramatically change their colouration in a short period of time, chameleons (Stuart-Fox et al. 2006) and various cephalopod molluscs being examples *par excellence*. Pigments in the skin of colour-changing animals are contained in special cells called chromatophores, and different coloured chromatophores are given different names: melanophores (dark brown–black), xanthophores (yellow), erythrophores (red), leucophores (white), iridophores (iridescent/reflecting) and rare cyanophores (blue). Colour change in various organisms occurs over a wide range of time scales and involves several different physiological mechanisms, which can be divided into two major types: morphological and physiological (Stuart-Fox & Moussalli 2009), which have different implications for their role in camouflage. Morphological colour change results from changes in the number and quality of chromatophore cells and typically occurs over a period of days or even months (Bagnara & Hadley 1973). In contrast, physiological colour change occurs due to movement (dispersion or concentration) of pigment granules within chromatophores or changing shape of chromatophores or migration of amoeboid chromatophores (R. Fujii 2000, Umbers et al. 2014), and can happen in milliseconds up to an hour or so. Most examples of physiological colour change are the results of hormonal (neuroendocrine) actions but the situation in cephalopods is unique and involves specialised muscular units under direct neuronal control. This is discussed separately later. I think it is worth noting a comment by Stuart-Fox & Moussalli (2008):

> Ironically, the best camouflaged animals are often the hardest to study because they are difficult to find in the wild – and this is particularly true of many colour changing animals.

As a consequence almost all we know is either anecdotal or based on laboratory systems. Also, while chameleons are probably the colour-changing animals that most people would immediately think of, and indeed the animals that Stuart-Fox and collaborators mainly work on, a lot of their colour change is to do with intraspecific signalling (Stuart-Fox & Moussalli 2008), though camouflage is also an important aspect of it.

An important distinction also needs to be made between changes that are reversible and those that are not. Many changes, even environmentally triggered changes that occur during an organism's development or those triggered by the parent, are irreversible. Other changes are plastic within the lifetime of an individual, but physiological constraints usually severely limit the speed at which colour changes can take place.

Caterpillars and food plant colouration

Cryptic caterpillars of some moths that feed on a range of plants varying in general colouration have sometimes evolved developmental switches such that their exposure and/or feeding on a given host species results in the development of different colouration. This sort of flexible colour matching may be a response to the visual background as seen by the insect itself or may be induced by diet. Poulton (1892) distinguished colour effects due to what the caterpillar could see, which he called 'phytoscopic effects', and those caused by what the caterpillar was eating, which he called 'phytophagous effects'. He showed, for example, that peppered moth caterpillars could be made to develop in brown or green forms depending upon what twigs (or pieces of paper) they first experienced as their microhabitat. In the case of the caterpillars of the North American bivoltine geometrid moth *Nemoria arizonica*, both generations feed on oaks (*Quercus* spp.), but the spring generation grows to become mimics of oak catkins whereas the summer generation ones mimic oak twigs, the catkins all having vanished

by that time of year. E. Greene (1989) showed that caterpillar morphotype was determined by diet; if they fed on catkins, they became catkin mimics and if they fed on oak leaves, they became twig mimics. This effect was not influenced either by temperature or photoperiod, but the development of higher levels of dietary tannins in the *Quercus* food plant induces a developmental switch from an early season catkin-mimicking caterpillar form to the later season twig-mimicking form. However, the 1989 results did not control for the light experienced by the caterpillars, i.e. more yellow-green in the catkin-feeding stage and darker green in the mature leaf-feeding stage. In follow-up experiments Greene confirmed that it was diet alone that induced the appropriate developmental pathways (E. Greene 1996).

Caterpillars of the European poplar hawkmoth, *Laothoe populi*, and eyed hawkmoth, *Smerinthus ocellatus*, which feed on either *Populus* or *Salix* species, resemble their particular host plants in colour, typically being either yellow-green or grey-green, the latter if they are feeding on white poplar, *Populus alba* or *Salix alba* (Poulton 1885, 1886). Transfer experiments show that the larva can switch between green and whitish forms depending on the background colour of their environment (Grayson & Edmunds 1989, M. Edmunds & Grayson 1991) (Fig. 2.26). Grayson et al. (1991) went on to demonstrate that at least in the case of the poplar hawkmoth, it is the amounts of dietary carotenoids, lutein and *cis*-lutein, that are translocated from the gut to the integument that causes the colour change, though the mechanism by which the perceived reflected light causes this remains unknown.

Similarly, Noor et al. (2008) also showed that caterpillars of the polyphagous American subspecies of the peppered moth, *Biston betularia cognataria* (Geometridae), can change their colours reversibly so as to match their background, and they do so by detecting the spectra of the light around them, with their food plant species having little or no influence (Fig. 2.27a,b). Skelhorn & Ruxton (2011) interpret this in a slightly more complicated way and show that when appropriate twigs are present the caterpillars are protected by masquerading them, however, even when twigs have been removed from the host plant branches, caterpillars fared better on branches of their host plant, thus showing that they are also cryptic. In this case, given the diversity of its food plants,[10] as well as the possibility that a larva might move between host plant species during its development, it is clear that being able to respond to direct visual stimuli will confer a considerable survival advantage, though it should

---

10. Food plants of *B. betularia* in the USA include: various species of *Acer*, *Alnus*, *Amelanchier*, *Aster*, *Betula*, *Juglans*, *Larix*, *Malus*, *Prunus*, *Quercus*, *Rhus*, *Ribes* and *Salix*.

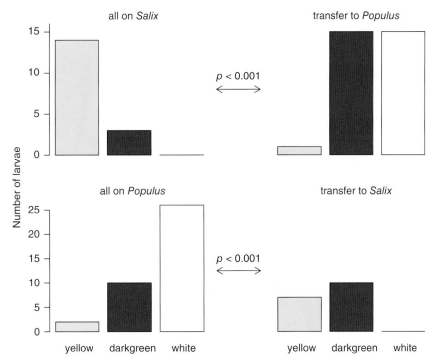

Fig. 2.26 Effects of larval food plant on the colour morphs of poplar hawkmoth, *Laothoe populi*, caterpillars. The left hand pair of plots show the colours they form when maintained on their natal food plant, the corresponding right hand pair after they have been transferred to the alternate host at the end of the second instar. (Source: Adapted from Grayson & Edmunds 1989. Reproduced with permission from John Wiley & Sons.)

be noted that colour change requires a moult and cannot occur within the duration of a larval instar. Skelhorn & Ruxton (2011) suggest that the ability to change phenotype may have actually facilitated the moth evolving to be so polyphagous. These more modern studies thus confirmed the far earlier one by Poulton (1892). *Biston*'s masquerading caterpillars might be doing something more to protect themselves because their potential predators are not limited to visually influenced birds. The Oriental species *Biston robustum* is also polyphagous and has distinct morphotypes associated with three of its principal food plants – cherry, *Prunus yedoensis*, a chinquapin, *Castanopsis cuspidata*, and a camellia, *Camellia japonica*. In this species, Akino et al. (2004) found that ants (*Lasius* and *Formica* species) frequently walked over its twig-like caterpillars and even antennated them but do not seem to recognise them as anything of interest, suggesting that chemical crypsis may also be involved. Indeed analyses of cuticular hydrocarbons from caterpillars reared on cherry, chinquapin and camellia showed that they closely resembled the food plant. By feeding caterpillars that were resting on cherry, leaves of the other food plant species, they found that cuticular chemistry also changed after the next moult, but not the other way around. The cuticular mimicry is triggered by their diet, and the hydrocarbons were subsequently shown to be sequestered via feeding on the food plant rather than through body contact (Akino 2005), and although some processing and chemical selection may be involved, it might be more a passive than an active crypsis. Further, the caterpillars also showed a preference to rest upon the plant upon which they had been feeding in a three-way choice experiment (Fig. 2.28), which may be an important adaptation as the different food plants could commonly overlap in the field, and birds would find them easier to detect if they wandered onto a tree species where their masquerade was less perfect. In contrast to the experiments on the American *Biston*, in *B. robustum* Akino et al. (2004) did not notice any diet-induced effect of morphological phenotype when caterpillars were moved onto other host plants, but they do point out that their experiment was not designed to test that. Nevertheless, it would be interesting to know how its morphotypes are formed. Incorporation of dietary components to effect what

Fig. 2.27 Caterpillar colour polymorphism in two species of moth. (a,b) Reversible colour polyphenism in the North American subspecies of the peppered moth, *Biston betularia cognataria* (Geometridae) that detect and respond to the wavelengths of the ambient light up to their final instar; (c,d) fixed colour polymorphism in final instar larvae of the hawkmoth *Theretra oldenlandiae oldenlandiae* (Sphingidae; identified by Ian J. Kitching, The Natural History Museum, London) in Vietnam, rarer green morph and commoner brown forms. (Source: a,b, Noor et al. 2008. Reproduced under the terms of the Creative Commons Attribution Licence CC BY, via PLoS ONE.)

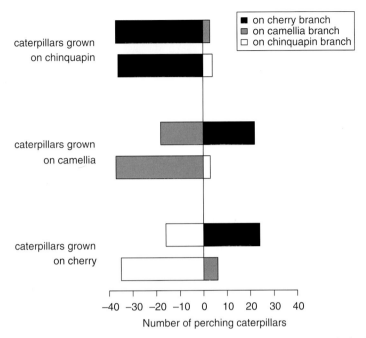

Fig. 2.28 Resting plant preferences of geometrid moth caterpillars (*Biston robustum*) depending upon which of three plant species they have been reared upon. (Source: Adapted from Akino et al. 2004. Reproduced with permission from Springer.)

is probably chemical crypsis is also known in a marine mollusc (Fishlyn & Phillips 1980) but see section *Protective crypsis in non-visual modalities* for further discussion of this.

The caterpillars of some other hawkmoths show even greater amounts of colour polymorphism. For example, those of *Eumorpha fasciata* can be green, pink or pink and yellow in earlier instars and either green or multicoloured in later ones (Fink 1995). The mechanism in this species is only partially understood, with all morphs occurring on all food plant species in the wild, but there also being food plant-induced effects on relative morph frequencies. It is quite common that earlier instars may be uniformly coloured, e.g. green in the Eurasian elephant hawkmoth, *Deilephila elpenor*, but with some individuals retaining this colour in the final instar, whereas most change to a brown background colour. An example of this in another hawkmoth species is shown in Fig. 2.27c,d. Upon moult to the final instar the green larvae of the nymphalid butterfly *Parage xiphia* become either green or brown. Experiments showed in this case that environmental factors were largely responsible, the most interesting of which was crowding, which induced a higher proportion to become brown (Gotthard et al. 2009). Apart from camouflage, another aspect of melanin production in insects is that it is

associated with increased immune defensive function, because compounds produced along the melanin biosynthesis pathways are also important in processes such as encapsulation. Thus the alternative brown form seems to be partly an evolutionary response to the likely increased risk of disease transmission when local densities are high. Nokelainen et al. (2013b) similarly found that the colour morphs of the European wood tiger moth, *Parasemia plantaginis* (Erebidae), were associated with differences in immune capability and this affected their differential survival in crowded and dispersed rearing conditions, though the effects were more complicated.

Direct utilisation of food pigments is known in a number of animals. The northern kelp crab, *Pugettia producta* (Epialtidae) (see Fig. 2.53a), in addition to doing some decorating of its carapace (see section *Exogenous crypsis*), sequesters pigments from the algae upon which it feeds and deposits them in its cuticle, thereby achieving background matching by direct use of background pigments (Hultgren & Stachowicz 2008). Young crabs inhabit intertidal and shallow subtidal red algae and are correspondingly red in colour, whereas larger individuals inhabit the amber-coloured kelp, *Macrocystis pyrifera*, and are brown in colour. Their means of sequestering algal pigments means that their crypsis parallels their ontogenetic shift in location and diet, though it is a little more complicated because older crabs have higher growth rates when feeding on kelp and red crabs, whether small or large, suffer higher mortality when feeding on or living among kelp, presumably from predation.

Daily and medium-paced changes

Among fish, amphibians, crabs and some other groups, integument colour varies between day and night conditions or depending on ambient light and on their background (Kats & van Dragt 1986, Stevens et al. 2014a, 2014b). These changes are the result of changes in the distribution of either light-absorbing pigment granules (e.g. in melanocytes) or broadly white, light-reflecting ones (e.g. in iridophores) (McNamara & Taylor 1987, Umbers et al. 2014). These changes are nearly always under hormonal control (e.g. Carlson 1935, Abramowitz 1937, Sandeen 1950, Rao et al. 1967). In Crustaceae, or at least in Malacostraca, the major lightening (due to contraction of melanophores) is caused by hormones released from neuroendocrine glands in the eye-stalks (or below the eyes if they are sessile) (C.L. Thurman 1988). The main organs are the medulla terminalis X-organ, many of whose cell bodies have axons terminating in the sinus gland located between the two optic ganglia and which is the site of release of colour change hormones. Injection of eye-stalk homogenate into most species causes paling as a result of aggregation of melanin granules in the melanocytes. Thus amputation or ligaturing of the eye-stalks leads to darkening. Conversely, eye-stalk gland secretions cause dispersion of the white granules in the leucophores, but instead of the sinus gland, it seems that aggregation of the white pigment is due to hormones released by the circumoesophageal neural connectives. The principal hormones involved are appropriately termed lightening hormone and darkening hormone, and they have been subject to many physiological studies as well as biology student practical classes (see Fingerman 2013). Additionally, the chromatophores themselves may also detect and respond directly to ambient light in some species and often display circadian rhythms (D. Atkins 1926, Abramowitz 1937, Darnell 2012) though Detto et al. (2008) could not detect any endogenous rhythms in the fiddler crab, *Uca capricornis*, in contrast to what has been reported for some other crabs, including congeners. Stevens et al. (2013a) investigated colour change in the Asian ghost crab, *Ocypode ceratophthalmus*, which, typical of many crustaceans, shows a circadian rhythm in its brightness that helps it remain camouflaged, but this is also fine-tuned by responding to light reflected from the background (Keeble & Gamble 1900, Kleinholz & Welsh 1937).

Mäthger & Hanlon (2007) have shown that in the squid *Loligo pealeii*, light reflected by various iridophores gets filtered by overlying chromatophores, and the iridophores themselves can polarise reflective light, all adding to the complexity of signals and level of camouflage that they can achieve.

In various fish, such as the neon tetra, *Paracheirodon innesi* (Characidae), spacing between the reflective guanine layers of the iridophores changes according to lighting conditions even in decapitated fish (Lythgoe & Shand 1982, Kasai & Oshima 2006). Although in intact fish they are under the control of the sympathetic nervous system, the iridophores themselves contain visual pigments and respond directly to light, as well as showing a circadian rhythm (Lythgoe et al. 1984, Oshima et al. 1998, Oshima 2001). Dermal light sensitivity is also involved in light–dark matching in some lizards; for example, Fulgione et al. (2014) found that the common Moorish gecko, *Tarentola mauritanica*, still responds to illumination and background when blindfolded, and even when their eyes were not covered their colour change was blocked when their flanks were covered. This was the first study demonstrating crypsis mediated by dermal light sensitivity in amniotes.

In crustaceans, and in many other groups, colour change is brought about by intracellular migration of pigments within chromatophore cells. Each cell is large, uninucleate and asymmetric and contains many pigment granules that can be of just one or of several types. Cell filaments contain many microtubules and it seems likely that these are largely responsible for moving pigment granules into and out of the cell filaments. The individual chromatophore cells often form clusters of between 2 and 15 cells and these complexes are called chromatosomes (McNamara & Taylor 1987).

Several crab spiders (Thomisidae) that are ambush predators, usually waiting for prey such as bees or butterflies on flowers, are able to change the colour of their cephalothorax and opisthosoma (abdomen) to match the colour of the flower. The best studied example is the Holarctic spider *Misumena vatia* (Théry & Casas 2002, Théry et al. 2005), females of which are most frequently yellow and sit on yellow flowers (Fig. 2.29a), though their basic colour is white due to reflectance from guanine crystals in their ramified guts seen through their translucent cuticle, and white is the colour they become or remain when resting on predominantly white flowers (Fig. 2.29b). Colour change to yellow is a result of the background colour seen by the spider and can be blocked by painting over their eyes. Complete change takes between 10 and 25 days and is reversible, involving transportation of diet-derived, yellow pigments from internal reserves to cells in the hypodermis, though if the spiders

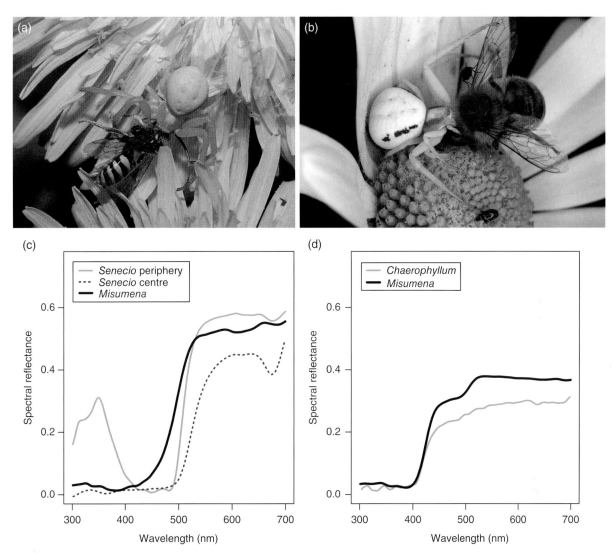

Fig. 2.29 The spider, *Misumena vatia*, which can adapt its colour over a period of a week or so to match white or yellow flowers where it is an ambush predator, and comparison of its and its substrates' reflectance spectra. (a) Spider on yellow Asteraceae flower having caught a *Nomada* bee; (b) on an Asteraceae flower with white petals having caught a honey bee; (c) reflectance spectra of yellow spider and yellow outer and inner parts of flower of *Senecio vernalis* showing mismatch in the UV region of the spectrum for the periphery of the flower; (d) reflectance spectra for white spider and white *Cheirophyllum temulum* flower. (Source: a, Olaf Leillinger 1998. Reproduced under the terms of the Creative Commons Attribution Share-Alike Licence CC BY-SA-2.0, via Wikimedia Commons; b, © H. Krisp 2011. Reproduced under the terms of the Creative Commons Attribution Licence CC BY 3.0, via Wikimedia Commons; c,d, adapted from Chittka 2001. Reproduced with permission from Schweizerbart.)

are maintained for a long period on a white background, they gradually excrete the pigment and so further change to yellow is retarded until they have had time to synthesise more. Colour matching is very convincing to humans but typical prey such as honey bees (*Apis mellifera*) have a different spectral sensitivity to ours, being unable to see red, but seeing into the UV spectrum. Chittka (2001) measured the reflectance spectra from white and yellow forms of *M. vatia* and flowers upon which they are often found, and used a model for bee colour-vision. White spiders were found to be generally good matches to white background flowers (e.g. *Cheirophyllum temulum*; Fig. 2.29d), but their crypsis was less perfect against some yellow flowers, including the eastern groundsel, *Senecio vernalis*, upon which they often sit in Europe, because the spider's cuticle is UV-absorbing whereas many yellow flowers are UV-reflecting (Fig. 2.29c, note the curve for the flower periphery). Nevertheless, Chittka (2001) argues that the spiders might still be well camouflaged as bees might only pay attention to their green light receptor input at the distances where the spider might be detectable, and indeed honey bees form a large part of the spider's prey even when they are sitting on yellow flowers. Colour changes are also known in some other thomisids (*Misumenoides formosipes*, *Misumenops asperatus* and *Thomisus onustus*) and, as in *M. vatia*, colouration is also partially affected by ommochromes in their diet and can be modified experimentally by feeding them, for example, normal red-eyed *Drosophila* or white-eyed mutants which differ in their ommochrome pigment load (Schmalhofer 2000, Théry 2007).

Slow (i.e. several days to a few weeks) background-matching colour change has also been shown in the chameleon prawn (*Hippolyte varians*), which is popular with marine aquarium hobbyists because it can change colour to red, yellow, brown, green or blue (Keeble & Gamble 1900), and *Crangon* shrimps are almost as adept (Koller 1927).

Many fish, amphibians and reptiles also show a combination of diurnal lightening and darkening as well as some ability to match background colours and textures (K.S. Norris & Lowe 1964). Larvae of two closely related species of salamander, *Ambystoma barbouri* and *A. texanum*, differ in their normal colour, background selection and response to predation threat (Garcia & Sih 2003). *A. texanum* generally selects backgrounds that match its current colour when given a choice and did not change its behaviour in response to threat, whereas *A. barbouri* larvae immediately sought out dark substrates when threatened and then changed their colour to match it.

A very interesting, multifunctional instance of colour change occurs in the dusky dottyback, *Pseudochromis fuscus*, a predatory coral reef fish (Cortesi et al. 2015). This change serves both cryptic and aggressive functions, and, remarkably, the dottyback changes its colour to match that of the other fish with which it co-schools.

Rapid colour change

In contrast to all the above, there are a few organisms that can change their colours and pattern to match their backgrounds exceedingly quickly. The best studied by far are the cephalopod molluscs, whose colour changes can be amazingly dynamic and used for intraspecific signalling as well as crypsis. It can be remarkably hard for humans to spot cuttlefish (*Sepia* spp.) against a wide variety of natural backgrounds (see Fig. 2.32) and importantly, their crypsis works well even under the different visual systems of potential fish predators as shown by hyperspectral imaging (Chiao et al. 2011). J.J. Allen et al. (2010) have shown that cuttlefish generally do not have a preference for any particular type of substrate, except for soft sand into which they can bury themselves, and match all substrates with the same ease, suggesting that there is no significant energetic difference in producing different background-matching patterns. When presented with two very different backgrounds they responded to the average of the two rather than showing different responses on different parts of their body.

Rapid colour change has also been demonstrated in at least some fish. Marked changes in appearance to match the degree of background splotchiness was achieved in just 2–8 seconds by the tropical left-eye flounder, *Bothus ocellatus* (Ramachandran et al. 1996). Ari (2014), Stevens et al. (2014a) and Watson et al. (2014) have reported rapid colour change in a manta ray, in a rockpool goby (*Gobius paganellus*) and a grouper, respectively. However, the general view that all flatfish are capable of rapid adaptive colour change, or even slower change, is not supported. Early work by Sumner (1911) with *Paralichthys albigutta*, a cold water flounder, which reported marked change over a few days was not found to be repeatable by W.M. Saidel (1978) (cited in Ramachandran et al. 1996) nor by Ramachandran et al. (1996) themselves. Another flounder, *B. lunatus*, has limited colour change capability but compensates by selecting to rest only on those backgrounds that are within its mimetic capability (Tyrie et al. 2015). Somewhat slower colour change, over a period of 20 minutes or so, is more widespread and predominantly exhibited by benthic species (e.g. S. Cox et al. 2009, JM Clarke & Schluter 2011). Some aggressive mimicry by fish is also achieved using quite rapid colour change capabilities (see Chapter 10, sections *Getting close* and *Cleaner fish and their mimics*).

Chameleons

The mechanism by which chameleons change colour has only recently been resolved and may be unique. For a long time it was thought that it was brought about by migration of pigment granules within cells, but this is not the case, or at least not a major factor in the rapid colour changes. Teyssier et al. (2015) have shown that above a layer of pigment-containing iridophores (the deep or D-iridophores) there is another layer of special iridophores (superficial or S-iridophores) that contain photonic crystals, specifically regularly spaced guanine nanocrystals which produce a large range of structural colours. Stimulation of the S-iridophores changes the spacing of the nanocrystals and therefore the colours produced and the wavelengths that can be reflected through from the deep layer. So far only the one species has been studied and therefore it is not known whether there are variants on this theme, nor is it known exactly how the S-iridophores are stimulated.

Although, as mentioned earlier, much of the well-known rapid colour change ability of chameleons is to do with intraspecific signalling, they also use it for crypsis and indeed some species use it in a facultative way, that is they show a different colour change response in relation to different perceived predator threats (Stuart-Fox & Moussalli 2009). The authors presented dwarf chameleons, *Bradypodion* spp., with either a stuffed fiscal shrike, *Lanius collaris*, or a resin cast of a fresh specimen of the diurnal, visually hunting, boomslang snake, *Dispholidus typus*, both of which are common local predators of chameleons. The colour changes of facultative species are such that they make the animal optimally cryptic to the visual spectrum of the particular perceived predator. Plotting ability to show

facultative colour change on a molecular phylogeny of *Bradypodion* species and populations (Fig. 2.30) shows that facultative crypsis ability has been lost on several occasions and it is hypothesised that this might reflect local variation in the presence or abundance of different types of chameleon predator, since there is probably a cost associated with having the facultative response which could be saved if one of the response types is no longer needed.

Cephalopod chromatophores and dermal papillae

In some cephalopods, complete colour and pattern changes can be achieved in under a second, based on a muscular system. In addition to their ability to change their colours, some octopuses can also change their texture, which, in combination, allows the animals to achieve remarkable matches to a wide range of backgrounds, also through muscle activity. Colour change in cephalopods is based upon chromatophore organs in the skin, each formed of special

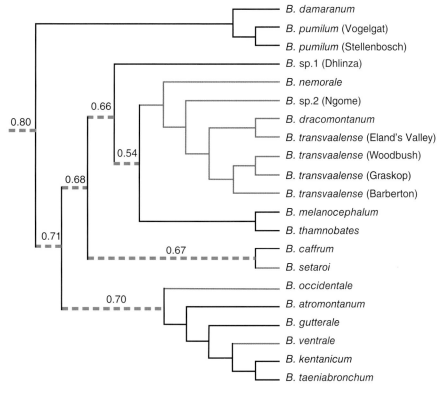

Fig. 2.30 Estimated phylogeny of 21 lineages of African dwarf chameleons, *Bradypodion* spp., with their ability to show facultative crypsis indicated in black, and four clades showing apparent losses of the ability indicated in orange (tree based on maximum likelihood analysis of two mitochondrial gene fragments; thick, broken grey lines are branched on which trait reconstruction was equivocal and the numbers above them are maximum likelihood estimates of the ancestor along that branch having facultative crypsis). (Source: Adapted from Stuart-Fox & Moussalli 2009 with permission from The Royal Society.)

combinations of cells. A central and very elastic pigment-containing cell (cytoelastic sacculus) is surrounded by a ring of 10–30 radially arranged muscle cells which, when they contract, cause the pigment cell to stretch to many times its contracted size, making a large area of colour (Florey 1969, Messenger 2001) (Fig. 2.31c–e). The chromatophore muscle cells are innervated directly from the central nervous system (CNS) and in particular a part of the brain called the posterior chromatophore lobe. Release of the neurotransmitter L-glutamate from neuromuscular junctions along the length of the muscle cell causes fast contractions of the muscles, leading to expansion of the pigment-containing cell (Fig. 2.31f) (Messenger et al. 1997). The serotonergic neurotransmitter 5-HT (5-hydroxytryptamine) causes muscle relaxation, allowing the elastic pigment-containing cells to quickly contract back so they present just small coloured dots, as can be seen in Fig. 2.31d. In addition, the neuropeptide Phe-Met-Arg-Phe-NH$_2$ (commonly called FMRFamide – pronounced 'fermerfamide') causes slower muscle contraction; acetylcholine has some presynaptic action, and recently, nitric oxide has been shown to cause slow muscle contraction via phosphorylation pathways of receptors on the muscles' internal membrane-bound calcium stores (Mattiello et al. 2010). The pigments, which range from yellow through orange and red to black, are probably all ommochromes and the different types can all be changed, expanded or contracted independently (Froesch & Messenger 1978).

Surprisingly, octopuses and cuttlefish are colour blind (Messenger 1977), so the effectiveness of their background matching depends upon evolution having moulded their response appropriately to the texture of the background that they see. That no visual feedback (i.e. comparing its own colouration with that of the background) is involved is nicely demonstrated by Messenger (2001) who placed a collar around young cuttlefish behind their eyes to prevent them seeing their own mantles – these animals responded only to the texture that they saw in front of them. Hanlon et al. (1999), Barbosa et al. (2008) and Zylinski et al. (2009b, 2009c) have made detailed studies of how the common cuttlefish, *Sepia officinalis*, classifies backgrounds to achieve the best camouflage. They found that the animals needed to see the whole outline of objects in order to determine the graininess of the background and to decide whether or not to employ the pattern known as 'Disruptive', in which there are a number of discrete components including a white square, white head bar and white mantle bar (Fig. 2.32) (Hanlon & Messenger 1988). Cuttlefish even deploy their camouflaging 'Disruptive', 'Mottle' and 'Uniform' patterns appropriately at night, indicating that they have very good night vision (Hanlon et al. 2011); from analysis of thousands of photographs it turns out that these are the only three camouflaging patterns that common cuttlefish have (Hanlon et al. 2009).

In several cephalopods, countershading is also a dynamic reflex with the statocysts supplying information about orientation, and the nerve-controlled chromatophores responding on a quadrant by quadrant basis with those dark pigment chromatophore units that the animal interprets as facing upwards expanding, thus darkening the uppermost surface (Ferguson & Messenger 1991, Ferguson et al. 1994).

While chromatophores are the best-studied aspect of cephalopod camouflage, they can also rapidly change the texture of their skin through numerous neuronally controlled papillae. The biomechanics of skin papillae in the cuttlefish, *Sepia officinalis*, was investigated by J.J. Allen et al. (2013) who found that they had concentric circular muscles responsible for the basic erectile response, and horizontal muscles that compressed the papillae laterally, so changing their shape. Retraction of the papillae appeared likely to be partially passive through the release of tension stored in the connective tissue during erection. J.J. Allen et al. (2014) compared the skins and textural capabilities of six species of cephalopod and found a diverse range of erectile papillae. Most of the papillae described had two sets of muscles, as in the cuttlefish.

BIRD EGGS AND THEIR BACKGROUNDS

Bird eggshells exhibit wide variation in their colour and pattern. Many ground-nesting birds have mottled, broadly camouflaged eggs that match the colour and texture of their typical nesting sites (Fig. 2.33), and this can be refined by the birds adding particular, carefully selected camouflaging material (J.C. Solis & de Lope 1995). Despite the obvious similarity to the background of many ground-nesting bird eggs, it has only recently been demonstrated that the level of camouflage, as would be perceived by their predators, actually affects clutch survival (Troscianko et al. 2016).

Female Japanese quail, *Coturnix japonica*, show considerable variation between individuals in the size of the blotches on their eggs and have been shown to select a substrate to lay on that best matches their individual egg patterns (Lovell et al. 2013). What makes this parental background matching interesting is that, unlike the case of polymorphic melanic moths (see section *Melanism* earlier), the variation between females is continuous rather than discrete. It is not known whether there is a genetic link between behaviour and egg patterning or whether birds learn their own egg features on laying their first clutch and modify their background selection correspondingly (see Stevens 2013a,

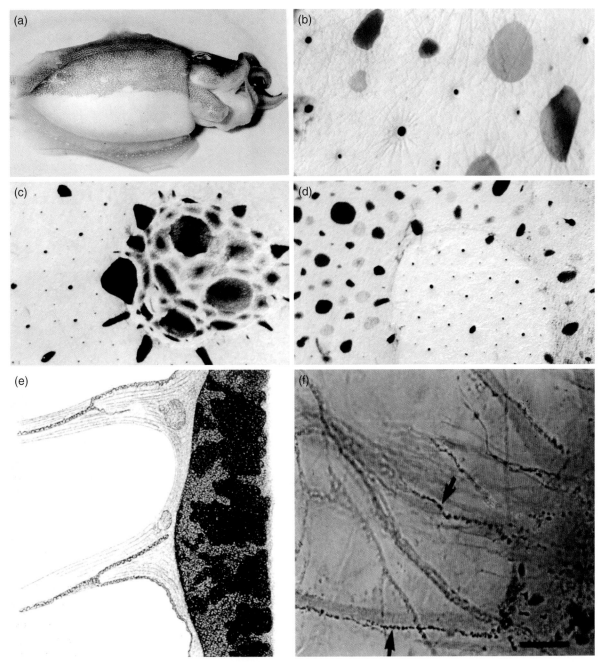

Fig. 2.31 Colour change and chromatophores of cephalopod molluscs. (a) Lightly anaesthetised *Sepia officinalis* rotated through 90 degrees showing countershading reflex affecting only half the body; (b–d) chromatophores of *Alloteuthis subulata*: (b) piece of skin with melanophores contracted naturally in response to pale surroundings; (c) piece of skin with area exposed to L-glutamate (5×10^{-4} M) resulting in the contraction of radial muscles and consequent expansion of the elastic chromatophore sac cells to form large coloured blobs; (d) piece of skin with exposed area flooded with 5-HT (1×10^{-5} M) resulting in the relaxation of radial muscles and consequent contraction of the elastic chromatophore sac cells to just small dense spots; (e,f) details of chromatophores of common squid, *Loligo vulgaris*: (e) part of a single chromatophore stained with methylene blue showing two radial muscles attached to pigment cell (right), each muscle having associated excitatory and inhibitory neurons running along its length; (f) chromatophore stained using peroxidase–antiperoxidase/diaminobenzidine (PAP/DAB) following incubation in L-glutamate antiserum and showing the reactivity along the length of the chromatophore radial muscles. (Source: Professor John B. Messenger. Reproduced with permission from Professor Messenger.)

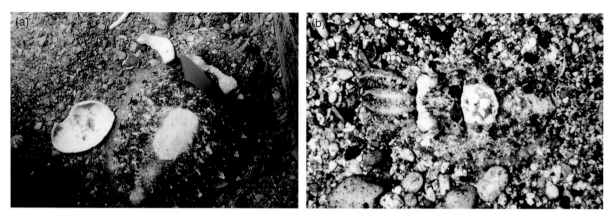

Fig. 2.32 Two individuals of the common cuttlefish, *Sepia officinalis*, employing the pattern known as 'disruptive' which is providing effective crypsis against two quite different large-grained backgrounds. (Source: Professor John B. Messenger. Reproduced with permission from Professor Messenger.)

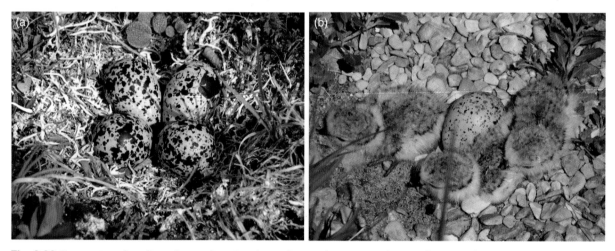

Fig. 2.33 Eggs, nests and nestlings of ground-nesting plovers showing disruptive markings and cryptic colouration. (a) American golden plover, *Pluvialis dominica*, eggs in nest, Alaska; (b) piping plover, *Charadrius melodus*, chicks and one unhatched egg in nest scrape. (Source: a, C. Meegs 2003. Reproduced under the terms of the Creative Commons Attribution Share-Alike Licence CC BY-SA 3.0, via Wikimedia Commons; b, USFWS Mountain-Prairie 2007. Reproduced under the terms of the Creative Commons Attribution Licence CC BY 2.0, via Flickr.)

for a commentary on Lovell et al.'s 2013 paper). If it is the former case, it is quite hard to imagine what the nature of the underlying genetics might be.

P. Mason & Rothstein (1987) note that birds that lay eggs in open cup-nests often lay speckled eggs, as do many non-nest-making species (N. Tinbergen et al. 1962, Lack 1968, Montevecchi 1976, Oniki 1979) and colouration in these is primarily to do with crypsis (Westmoreland & Kiltie 1996, Skrade & Dinsmore 2013). In the case of most plovers, their eggs and chicks are blotched (see Fig. 2.35), and, importantly, the pattern of markings is irregular and can vary considerably, even within the same clutch. Whether it is possible to generalise their result is not clear, but Westmoreland & Kiltie (1996) found that among the small group of species they studied (three species of North American Icteridae) mean clutch pattern disparity was highly correlated with mean background disparity ($r = 0.999$) which suggests that disparity matching has been the main driving force in evolution rather than crypsis *per se*. In other words, they have evolved such that their clutches of eggs do stand out from the nest background through being overly uniform or perhaps overly variable. Similar intra-clutch variation was found in the African black oystercatcher, *Haematopus moquini*, by Hockey (1982).

Background matching of eggs is also a reason behind the high tenacity that some individual ground-nesting birds display (Sánchez et al. 2004) effectively reducing the potential heterogeneity of background types to find a match with.

The maculation of ground-nesting bird eggs is due to the biliverdin and protoporphyrins, both of which play other physiological roles in the adult bird, and are therefore likely to be under constraints and possibly involved in fitness trade-offs. Duval et al. (2016) explored this by measuring the extent of eggshell maculation and the concentrations of the pigments in Japanese quail, *Coturnix coturnix japonica*, that were either maintained on a normal, ad libitum diet, or on a restricted one. Restricting the female diet led to an increase in the concentration of protoporphyrin, a decrease in that of biliverdin, and an increase in spot coverage. Protoporphyrins also play a role in strengthening eggshells and may help to compensate for inadequate dietary calcium (Gosler et al. 2011).

Once a chick has hatched, the white or pale inner surface of the eggshell may make the nesting site more conspicuous and vulnerable to predator detection. N. Tinbergen et al. (1962) thus suggested the shell removal behaviour of the parent birds is actually a component in the nest/nestling camouflage strategy.

DISGUISING YOUR EYES

Vertebrate eyes are a feature that most vertebrates, at least, are very cued into as they can signify danger. Cott (1940) pointed out a number of examples whereby the location or presence of an animal's eye is obscured, for example, in the western Atlantic and Caribbean jack-knife fish, *Equetus lanceolatus* (Sciaenidae), in which one of three dark brown stripes passes right around the side and top of the head through the eye. In fact when a vertebrate has striped markings one of the stripes nearly always passes across the eyes (Neudecker 1989). Often the only stripe is associated with the eye (Fig. 2.34c). Many geckos have intricately patterned irises so the only true eye-like feature is the rather small (in daylight), slit-like pupil (Fig. 2.34d).

G.W. Barlow (1972) examined the orientation of eye-crossing lines ('eye-lines') in an unbiased sample of South African fish and found that stripe direction was related to the fishes' overall length-to-depth ratio, but perhaps in a quite complicated way. Essentially, relatively deeper bodied fish tended to have more vertically orientated eye-crossing lines (Fig. 2.34a) compared with more elongate species that have more shallowly sloped stripes (Fig. 2.34b), but there were many exceptions, and line orientation is clearly not independent of other features of fish shape or their resting attitudes.

DISRUPTIVE AND DISTRACTIVE MARKINGS

Perhaps especially when hunting for a known prey type, a predator is likely to have some general mental model of the shape of the prey, for example, moths resting on tree trunks are likely to have a roughly triangular outline (e.g. see Figs 2.17 and 2.22), toads hunting for worms are stimulated by horizontally moving linear objects (Ewert 1974), etc. Natural selection is therefore likely to lead to features that confound the detection of such basic attributes. Disruptive colouration is the term generally used to indicate features of an animal's pattern that, instead of presenting a single recognisable prey-like entity such as a caterpillar, visually break up the body into two (or more separate parts) or blend different parts with parts of the background. A classic example is the dark saddle on the larva of the European puss moth, *Cerura vinula* (Notodontidae) (Cott 1940, Komárek 2003), which seems to make it appear perhaps as two separate green, uncaterpillar-like shapes, possibly small leaves. In general, disruptive markings have high contrast with the rest of the animal as well as with the background, such as the narrow white border around *Cerura*'s saddle (Thayer 1909, Cott 1940, I.C. Cuthill et al. 2005, Stobbe & Schaefer 2008).

Edge-intercepting patches

The number of colour or dark patches that are 'truncated' by the border of an organism, so called edge-intercepting or disruptive patches, is important in reducing prey detectability against complex backgrounds such as many tree barks, seaweeds, pebbles, etc. Merilaita (1998) showed that the markings on the common and highly variable marine isopod crustacean *Idotea balthica*[11] intercepted the animal edge significantly more often than would be expected by chance. Edge-intercepting patches are found in many cryptic animals; for example, the cross-banding patterns of many cryptic snakes have been suggested to function not only to match background texture but to disguise the elongate snake-like form (Thayer 1909, Cott 1940, Pough 1976, W.L. Allen et al. 2013).

It is not easy to identify what individual roles disruptive colouration and background matching play in real camouflage. Two independent experimental investigations were carried out more or less at the same time: Schaefer & Stobbe (2006) and I.C. Cuthill et al. (2005, 2006). Schaefer & Stobbe (2006) used manipulated, baited digital images of the European peach blossom moth, *Thyatira batis*, which

---

11. Frequently misspelt as '*baltica*'.

Fig. 2.34 Markings that make true eyes less conspicuous. (a) Beaked coralfish, *Chelmon rostratus* (Chaetodontidae), with vertical stripe across eye, and (b) immaculate damsel *Mecaenichthys immaculatus* (Pomacentridae), both with large eyespots posteriorly; (c) Asian pit viper, *Trimeresurus* (*Popeia*) *popeiorum*, an arboreal snake, showing eye stripe; (d) unidentified lizard, probably Gekkonidae, from Costa Rica, with slit pupil and intricately patterned iris.[12] (Source: a,b, Denis Riek. Reproduced with permission from Denis Riek.; c, Itchydogimages. Reproduced with permission from John Horstman.; d, Kenji Nishida. Reproduced with permission from Kenji Nishida.)

has a quite unusual colour pattern (Fig. 2.35). The real moth has a brownish background with some pink edge-intercepting patches, but they also made versions with colours switched and in each case with the patches either in their natural position or moved inwards so they no longer intercepted the moth image's outline. Tests were carried out against both a birch tree trunk and a moss background and results showed that having markings that intercepted the edge significantly improved survivorship (Fig. 2.36).

I.C. Cuthill et al. (2006) carried out an experiment on the effects of pattern features on wild bird predation levels using patterned, moth-like, baited, cardboard triangles (Fig. 2.37) pinned out against oak tree (*Quercus*) trunks. Survivorship of the 'moths' was enhanced by having an asymmetric pattern, and having edge-intercepting patches (i.e. disruptive patterns), and symmetry reduced survival when patterns were either edge-intercepting or non-disruptive (Fig. 2.38). Their investigation of symmetry is rather misleading in that insects,[12] and indeed arthropods in general, are highly constrained to bilateral symmetry,[13] and all Lepidoptera have very symmetric wing patterns, even down to intricate detail. This is quite remarkable and implies deep genetic inertia because asymmetric patterning would clearly have a

---

12. Slit pupils are always associated with multifocal lenses and are associated with eyes that have to function in low light intensities, but probably secondarily, this enables their eyes to be rather camouflaged in bright light conditions.

13. Asymmetry is largely restricted to some internal organs (such as a looped gut) and genitalia, there being one (in this context well-known) Mexican water bug that can have either left- or right-handed genitalia.

Fig. 2.35 Set specimen of the unusually patterned peach blossom moth, *Thyatira batis*, which has been suggested to have disruptive or even distractive markings. Note also that the edges of the peach-coloured blotches are enhanced by narrow, high-contrast white and black borders. (Source: Adapted from Didier Descouens 2014. Reproduced under the terms of the Creative Commons Attribution Share-Alike Licence CC BY-SA 4.0, via Wikimedia Commons.)

large impact on survival were it possible to evolve it. In a twist to the crypsis and asymmetry situation, Forsman & Herrström (2004) showed that when it comes to aposematic warning colouration, asymmetry has a quite negative effect on its efficacy.

Using humans as model predators, Webster et al. (2013) explored the way in which the number of edge-intercepting patches on the cryptic target affected the searcher's ability to detect it. Their results show that the time required for subjects to locate 'moth' images increased broadly linearly with the number of edge-intercepting patches, (Fig. 2.39a), and the same was true for mean 'survival' of the patterns (Fig. 2.39b).

The effect of edge-intercepting and other disruptive patches in many animals may be increased by enhancing the contrast at their borders, such that the edges of the markings are easier to detect (Osorio & Srinivasan 1991). This is likely an important factor as most vertebrate visual processing involves edge-detecting neuronal wiring. However, Stevens et al. (2009b), using wild birds as predators and baited moth-shaped targets with various patterns, found that having high-contrast (but asymmetric) irregular markings away from the target's outline and lower contrast edge-intercepting markings significantly enhanced survival compared to having low-contrast markings away from the edge and high-contrast ones intercepting the edge. This effect, which they call 'surface disruption' is, I think, rather due to that fact that high-contrast edge-intercepting patches

with perfectly triangular targets draws attention to the edge more, hence revealing the 'moth' shape.

In addition to colour markings, many cryptic animals also have irregular cuticular protuberances that make their basic body outlines hard to discern, for example, various frogfish and filefish (J.J. Allen et al. 2015) (see also section *Cephalopod chromatophores and dermal papillae* earlier).

Distractive markings

Distractive markings were proposed by Thayer (1909) as devices that would draw a potential predator's attention away from other more diagnostic features, and as such might render the prey even more conspicuous but less recognisable as potential prey. Possible examples among north temperate Lepidoptera have been suggested to include the underside wing markings of the comma butterfly, *Polygonia c-album* (Olofsson et al. 2013), the conspicuous white 'Y' shape of the silver-Y moth, *Autographa gamma*, and the gold mirror patch on the gold spangle, *A. bactrea* (Stevens et al. 2013b), while Caro (2011) suggested that the white tail tips of various small mammalian predators might have this function and Stevens et al. (2013a) suggest a number of possibilities among the world of fish. However, not surprisingly, the idea that conspicuous marks could help conceal prey is rather controversial (Stevens & Merilaita 2009b, Merilaita et al. 2013). Stevens et al. (2013c) note 'distractive markings are not automatically "by definition a camouflage strategy," but rather a theory for how one type of concealment might potentially work.' They suggest that if distractive markings can act to reduce predation then they ought to fall under the definition of camouflage and thus the word camouflage should be defined by function rather than appearance.

There is considerable difficulty in trying to design experiments to test Thayer's theory and three important experiments have not all given the same results. The first, using wild birds and triangular moth-like shapes baited with mealworms, pinned to tree trunks was by Stevens et al. (2008d). Their results showed clearly that the more contrasting the distractive markings, the lower the prey's survival, thus apparently clearly disproving Thayer's notion, albeit only in this context. This was followed by Dimitrova et al. (2009), who tried to test Thayer's proposition by predation experiments with captive blue tits, *Cyanistes caeruleus*, again with baited triangular moth-shaped models, with or without a bright distractive marking (Fig. 2.40a–c), on artificial complex backgrounds. This artificial set-up not only allowed for the prey targets to be modified by either increasing or reducing the contrast of their distractive marks but also meant the backgrounds could be

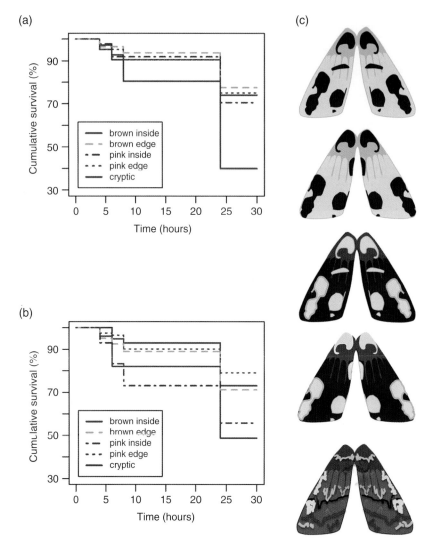

Fig. 2.36 Experiment with wild birds to distinguish effects of disruptive colouration and background matching in providing camouflage, using manipulated digital photographs of peach blossom moth, *Thyatira batis*, baited with mealworm. (a) Targets placed on birch trunks; (b) targets placed on moss; (c) stylised images of the moth targets in order from top to bottom: with brown marking located in from the wing margin, brown markings intercepting wing margin, pink markings set in from wing margin, pink markings intercepting wing margin, and a cryptic control. (Source: Schaefer & Stobbe 2006. Reproduced with permission from The Royal Society.)

modified (Fig. 2.40d,e). Their results were equally clear but in the opposite direction: the birds took longer to locate prey with more conspicuous marks irrespective of what background they were placed against and, also, all prey types were harder for the blue tits to locate when they were on the background with high-contrast distractive marks (see figure legend for more details).

To try to resolve this, Stevens et al. (2013b) carried out further experiments using baited moth-shaped models in the field and computer-screen experiments with naïve human predators. This rather more sophisticated set of experiments, with rather more realistic moth targets, found that neither in the field nor in the computer screen cases did distractive markings aid survival; their impact was either neutral or more often led to increased detection. Stevens et al. (2013c) discuss the possible reasons why Dimitrova et al.'s (2009) results were so different, and conclude that Dimitrova et al.'s study had elements of pseudoreplication

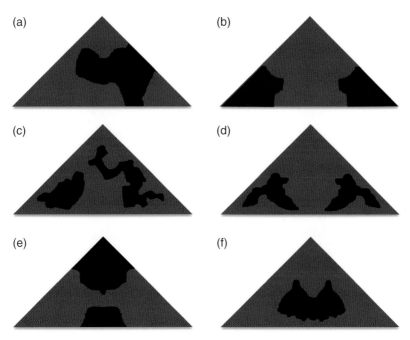

Fig. 2.37 Examples of baited moth shapes placed out on tree trunks to test the effects of symmetry (b,d,e,f) versus asymmetry (a,c), midline marks (e,f) and edge disruption (a,b,e) on avian predation; results shown in Fig. 2.40. (Source: Adapted from Cuthill et al. 2006. Reproduced with permission from Oxford University Press.)

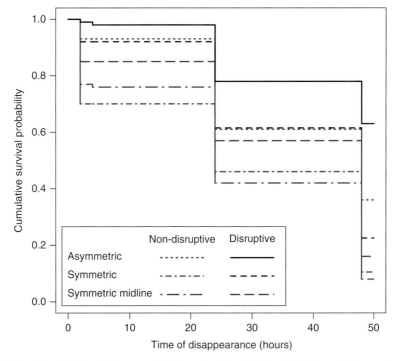

Fig. 2.38 Survivorship of baited artificial moth shapes and patterns (see Fig. 2.39) showing that predation was increased by symmetric patterned models and ones that did not have the dark markings disrupting their outlines (edge-intercepting patches). (Source: Adapted from Cuthill et al. 2006. Reproduced with permission from Oxford University Press.)

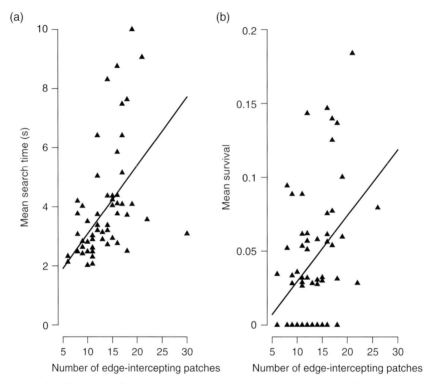

Fig. 2.39 Experimental results of human 'predators' searching for cryptic, computer-generated 'moth' images against a similar complex background, and showing that the number of patches on the 'moth' that intersect with the edge of its outline significantly affects the mean time taken to detect them. Higher numbers increase the probability of the 'moth' surviving the predator's search (i.e. not being detected), even at the expense of reducing the targets' overall similarity to the background. (Source: Adapted from Webster et al. 2013.)

and that the pre-training they gave to the birds may have caused them to overlook the high-contrast prey in the experimental trials. Another factor might have been the highly artificial patterning of the backgrounds and targets, whereas Stevens et al. (2013b) used natural bark backgrounds and bark-patterned artificial prey with wild birds.

It is possible, however, that the experimenters' attempts to avoid particular biases by using abstract shapes as the distractive marks does not fully test the hypothesis; evolution may have honed particular cases rather more precisely than the simple addition of a white mark. The above three experiments did not use real insects. Olofsson et al. (2013) carried out a test with pairs of real comma butterflies, *Polygonia c-album*, one of which had had the conspicuous hind wing comma mark obliterated, using blue tits, *Cyanistes caeruleus*, as predators. Separate series of trials were carried out for male and female butterflies as the 'comma' of the males has a markedly greater reflectance across a broad range of wavelengths. While the apparently well thought out experiments of Stevens et al. (2008d, 2013c) failed to find an

advantage of distractive marks, Olofsson et al. (2013) found that for both male and female comma butterflies, the presence of an intact comma significantly reduced bird attacks, i.e. the birds attacked the butterfly without the comma mark first (binomial test, $n = 62$, $p = 0.03$) (Fig. 2.41).

Zebra stripes and tsetse flies

Nearly all large herbivorous mammals are classically countershaded and are uniformly variously shades of grey, fawn, tan, brown or blackish dorsally. There are a few exceptions though, notably giraffes and zebras. Indeed the zebra's stripes are the sources of many myths and fables, both in Africa and the west; an internet search for "How the zebra got its stripes" scored more than 16,000 hits. More scientifically, however, this is perhaps an even more interesting question, and like so many adaptive resemblances there may be more than one causal factor; even Charles Darwin and Alfred Russel Wallace disagreed on the subject.

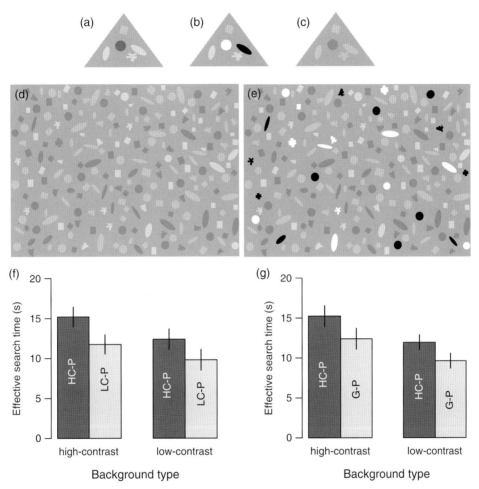

Fig. 2.40 Experimental background and baited 'moth' targets used in experiment to test the effectiveness of distractive markings, and results of two experiments. The models differed in the relative conspicuousness of some markings and the intensities chosen so that the luminance range was appropriate to the blue tit predators. The results seem to indicate that models with conspicuous (distractive) markings take significantly longer for the blue tit predators to locate. (a–c) Targets of normal (G-P), high contrast (HC-P) and low contrast (LC-P), respectively; (d,e) backgrounds of low and high contrast, respectively; (f) predators (blue tits) took significantly longer to locate the high-contrast prey (HC-P) than low-contrast ones on both background types ($F_{1,32} = 7.72$; $p = 0.009$), and search times were higher against the high-contrast background ($F_{1,32} = 9.43$; $p = 0.004$); (g) as in (f) but instead of low-contrast prey the comparison was made with targets that matched the background. (Source: Adapted from Dimitrova et al. 2009.)

Ruxton (2002) provides a review of theories, which fall into four main categories: predator avoidance, social signalling, thermoregulation and protection from tsetse flies. A fifth theory, that perhaps they appear to be taller than they really are through v. Helmholtz's (1867) illusion, could also play a role (see Chapter 11, section *Make-up, clothes and silicone*). The predation protection theme also breaks down into a number of subcategories: stripes may make a prey look bigger (taller); moving stripes dazzle predators; striped camouflage against tall grasses; and striped shapes in general are harder to interpret in poor light.

Zebras (*Equus burchelli*, *E. grevyi* and *E. zebra*) are entirely African and, like most large mammals there, they are prone to being bitten by blood-sucking flies, including members of the tsetse fly genus, *Glossina*, which are vectors of trypanosomiasis (sleeping sickness), especially when they visit water holes. The idea that zebra stripes might be involved in their apparently low attraction to tsetse flies was first proposed by R.H.T.P. Harris (1930) and subsequent experimental tests of the hypothesis have been carried out by Waage (1981), G. Gibson (1992) and Brady & Shereni (1988). The general conclusion is that tsetse are less attracted to striped

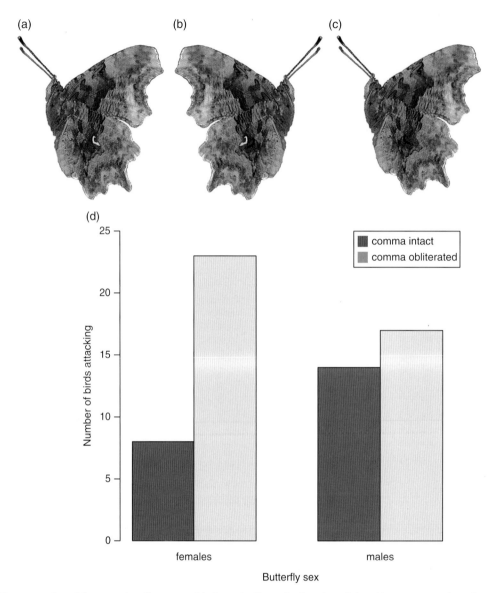

Fig. 2.41 Demonstration of the protective effect, ostensibly due to its distractive function, of the white comma mark on the underside of the hind wing of a comma butterfly, *Polygonia c-album* (Nymphalidae). (a) Unmodified comma male; (b) with additional sham-painted black comma to control for the presence of a comma-shaped mark; (c) sham-painted but with white comma mark obliterated; (d) results of pairwise presentation of butterflies as in (b) and (c) to wild-caught blue tits, *Cyanistes caeruleus*. (Source: a–c, Adapted from Zeynel Cebeci 2013. Reproduced under the terms of the Creative Commons Attribution Share-Alike Licence CC BY-SA 3.0, via Wikimedia Commons; d, adapted from Olofsson et al. 2013. Reproduced with permission from Elsevier.)

targets than to solid dark shapes, thus supporting the idea that the zebra's stripes might make the zebra a less likely feeding target to the flies.

While most work has concentrated on the behaviours of tsetse flies because of their disease transmission, Egri et al. (2012) showed that tabanid flies are also far less strongly attracted to striped horse models than to uniform black, brown or white horse models, and that lack of polarisation of reflected light also plays an important role since tabanids are positively polarotactic. Tabanid flies can have a considerable negative fitness impact on their hosts, both through feeding intensity and transmission of parasites. Indeed dark-coloured models are far more attractive than light ones to both tsetse flies and tabanids, and Horváth et al. (2010) note that this means that white horses, despite being prone to malignant skin cancer and visual deficiencies owing to their high sensitivity, may experience considerably reduced fly-transmitted parasite loads. Since female horseflies (Tabanidae) are strongly attracted to both CO_2 emanating from their host's breath, and ammonia formed by traces of old urine, they can theoretically locate hosts from a distance easily, but despite local concentrations of the attractant chemicals, striped targets (and quite probably spotted ones also) provide such a weak acceptance signal that the flies do not land on them (Blahó et al. 2013).

STRIPES AND MOTION DAZZLE – MORE ZEBRAS, KRAITS AND TIGERS

While the stripiness of zebras may well afford them some protection from insect-borne diseases, other functions might also be important. Apart from zebras and tigers, stripes are widespread among fish and in various snakes (such as bands as in some of the kraits, *Bungarus* spp. and coral snakes) and have a profound effect on the perception of movement and speed (Godfrey et al. 1987).

Computer graphics experiments with human subjects

Stevens et al. (2008c) found that targets with a range of high-contrast banded and zig-zag patterns as well as uniformly camouflaged ones were hard to capture, especially against heterogeneous (leafy) background images. Stevens et al. (2011) used a large number of human subjects playing a special computer game to unravel the possible mechanisms by which stripes reduce a predator's ability to capture the 'prey', and carried out tests in both stationary situations and when the striped target was moving against a complex (i.e. busy) background of the same mean luminosity. The subjects had to hit the targets using a touch screen. The six patterns and the background are shown in Fig. 2.42a,b, respectively. In experiments to locate the stationary targets, after a target was located the background was replaced briefly by uniform grey, followed by a new background on which a fresh target was already present, so that the participant would not spot the appearance of the new target. When targets were moving, striped and uniform grey patterns were hit least often (Fig. 2.42c,d), whereas when they were stationary, the two types of blotched patterns did best, followed by uniform grey (Fig. 2.42e,f). The results seem to

demonstrate that stripiness would be a particularly effective protection for moving animals.

Using computer images with either uniformly shaded, longitudinal or transverse stripes moving on a complex background (Fig. 2.43), and human volunteers, von Helversen et al. (2013) found that targets with both types of stripe were perceived as moving faster than unstriped ones and had a small but significant effect on hit rate. A similar effect of stripes on speed perception was found by Scott-Samuel et al. (2011). However, in their trials they found that longitudinally striped targets were hit more often than either uniform or transversely striped ones, which was contrary to expectation. I suspect this was due, at least in part, to the artificiality of the system as real animals seldom move in perfectly straight lines and snakes slither along convoluted paths.

How & Zanker (2014) go on to show, using a visual model, that the motion of stripes of a zebra can "flood" an observer's visual system with erroneous motion signals, and that these closely correspond to two well-known visual illusions: (i) the wagon-wheel effect (a perceived reverse motion as often seen in 'cowboy and Indian' films due to the wrong alignment of wheel spokes in consecutive images), and (ii) the barber-pole illusion (misperceived direction of motion due to the aperture effect). They conclude that these illusions could be effective in confusing both biting insects and mammalian predators, especially when two or more zebras are moving together in a herd.

Observations on real animals

Lindell & Forsman (1996) studied a polymorphic population of adders, *Vipera berus*, which included both zig-zag striped and monochromatic melanistic forms. Male adders are far more active than females, spending a lot of their time moving around in search of mates, and are consequently far more likely to be noticed by predators. Mark–release–recapture results showed that zig-zag patterned males survived better than monomorphic males, whereas melanistic females survived better than zig-zag patterned ones, thus supporting the 'flicker-fusion' hypothesis, i.e. the zig-zag pattern makes moving snakes harder to follow and catch. Flicker-fusion may also play a significant role in the case of aposematically coloured coral snake mimics (see Fig. 7.16d–f), that in many areas occur outside of their apparent model's range, especially under low light level conditions such as at dusk (Titcomb et al. 2014). Indeed Brodie III (1992) had found a complex correlation between striped patterns of juvenile garter snakes, *Thamnophis ordinoides*, and their tendency to perform reversals during their escape behaviour. Striped snakes had higher fitness if they did not perform

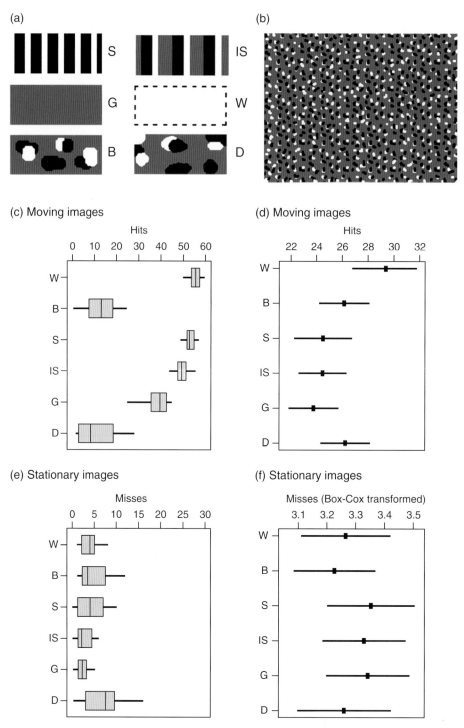

Fig. 2.42 Testing whether stripes help to camouflage and to confuse in motion using computer-generated images and observers whose aim were to successfully target them. (a,b) Sorts of patterns and background used to test the effects of the patterns on the detectability of targets; (c) hits by observer, the two striped patterns (S and IS) and the uniform grey target (G) were the least easily targeted when moving, and correspondingly (d), they were the ones most often missed; (e,f) experiments with stationary targets showing highest protection for the two blotched target types. (Means ± 95% confidence intervals). (Source: a,b, © M. Stevens. Reproduced with permission from M. Stevens and under the terms of the Creative Commons Attribution Licence CC BY 2.0; c–f, Adapted from Stevens et al. 2011. Reproduced under the terms of the Creative Commons Attribution Licence CC BY 2.0.)

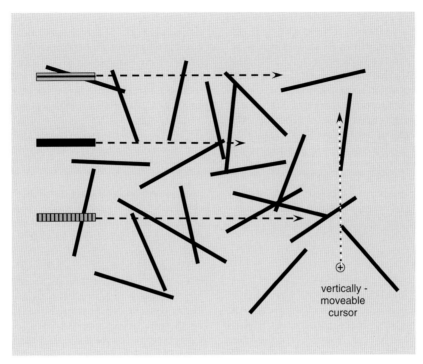

vertically -
moveable
cursor

Fig. 2.43 Computer game set-up designed to investigate whether longitudinal or vertical stripes affect the ability of humans to target and hit moving targets with a joystick-controlled cursor with a 'fire' button that can only be moved up and down as the target shapes move left to right across the screen. (Source: Adapted from von Helversen et al. 2013 under the terms of the Creative Commons Attribution Licence CC BY, via PLoS ONE.)

reversals, whereas plainer snakes were fittest if they did, giving rise to a saddle-shaped fitness surface. Given that the traits are under genetic control, the correlational selection would be expected to lead to linkage disequilibrium, i.e. non-random associations between alleles at the separate loci.

Pough (1976) describes almost the opposite effect in young North American water snakes, *Natrix sipedon* (Colubridae), which have conspicuous dorsal cross-bands. When disturbed, they move very rapidly, such that to the human observer, rather than displaying camouflaging stripes, the pattern blurs, making it appear a uniformly coloured snake. However, after a relatively short distance (10–50 cm) the snakes stop so suddenly that to the observer there is no apparent slowing down, the moving grey uniform image disappears and the banding pattern again renders it cryptic.

Cephalopods that have the ability to change their appearance rapidly (see section *Cephalopod chromatophores and dermal papillae*) ought, in principle, to be able to employ motion dazzle at will. However Zylinski et al. (2009a) found that, at least in the case of the common cuttlefish, *Sepia officinalis*, this was not the case. Instead of changing to a high-contrast pattern (see Fig. 2.32) during movement they adopted a

low-contrast, fine-grained one. A possible explanation for this is that they are simply too conspicuous while moving for any motion dazzle to be effective.

Comparative analysis

J.F. Jackson et al. (1976) used multiple discriminant analysis to investigate the ecological and behavioural correlates of colour pattern for all Mexican and North American snake species based on the authors' own classification of pattern classes. W.L. Allen et al. (2013) carried out a similar comparative analysis of snake patterns and behavioural features including a total of 171 species from Australia and North America, but used computer algorithms to classify patterns. Both studies produced similar findings, with behavioural rather than habitat variables generally being more important determinants of colour pattern. Longitudinal stripes were found to be associated with rapid escape speed, plain colours with active hunting, blotched patterns with ambush predators and spotted patterns with living close to cover. J.F. Jackson et al. (1976) had found a negative association between the number of regular transverse bands and escape speed (based on morphological proxy-variables) and a positive correlation with body size, which would be consistent with a flicker-fusion role. No direct evidence was found by W.W. Allen et al. (2013) that transverse striped patterns were associated with either highly venomous species (but given the existence of many mimics this is perhaps not surprising) or rapid escape speed, thus apparently not supporting the flicker-fusion hypothesis as the sole explanation, but the authors argue that it might be a compromise strategy, providing disruptive pattern benefits when the snakes are stationary but invoking flicker-fusion when escaping. Their study was only concerned with pattern and not with colour, so aposematic coral snake type patterns were not distinguished from other less conspicuous cross-banded ones.

DUAL SIGNALS

It is easy to imagine that many brightly coloured animals sporting bold patterns of reds, yellows, black, etc. must be the most conspicuous things to see in the natural world, which is dominated by greens and browns. However, that this is not necessarily the case was first proposed by Miriam Rothschild (1975), who drew attention to organisms such as the black and yellow banded larvae of the European cinnabar moth, *Tyria jacobaeae* (Erebidae: Arctiinae), which feed on ragwort, *Senecio jacobaea* (Asteraceae), and sequester toxins from their food plant which render these insects highly unpalatable both as larvae and adults. Their black bands effectively break up their outline and their yellow colouring closely resembles that of the food plant's flowers; often the give-away to their presence in years when they are very abundant are the almost completely denuded plants they leave in their wake. Quicke (1986a) presented an informal survey of the distributions of colour patterns in large braconine braconid wasps in the Afrotropics and showed, albeit without statistics or phylogenetic control, that bright red species were far more abundant in the north, east and south where there is a preponderance of red lateritised soils. Predominantly yellow species, typically with dark wing markings and a black tip to the metasoma, had a similar distribution and are hard to see against dry grasses; and black species with bright yellow heads and largely black species with dark red head and mesosoma were largely restricted to the equatorial forested zones from western Uganda to the Atlantic. All these patterns occur in multiple genera and most genera include species with at least two of them, and, in some cases, with all of them. When any one of these insects is seen close-up, it is very obviously aposematic, and the detailed resemblances between the patterns in different genera strongly indicates that they belong to mimicry complexes.

Papageorgis (1975) and K.S. Brown (1988) extended Rothschild's (1975) dual signal concept to the levels in the forest in which warningly coloured butterflies fly, and hence the level of illumination and colour composition of the background. At any one site in South American forests, there are usually butterflies belonging to several mimicry complexes (Müllerian assemblages), some black with cream and blue markings, and some orange with brown patterning, and some transparent. These collectively belong to several taxonomic groups, especially the unpalatable and brightly coloured Heliconiinae and Ithomiini (Danainae). Papageorgis (1975) found that these homeochromatic groups were not uniformly distributed throughout the forest layers. Transparent-winged species fly near to the ground, tiger pattern orange and black (see Fig. 7.2e,j) slightly higher, followed by the red (see Fig. 7.2k–t), then blue complexes higher still. The brightest orange species avoid the darker, shadier internal zone where they would be especially conspicuous, and fly above the canopy. Overall, darker patterns were concentrated at mid levels within the forest. Thus, despite their bright colours when seen close-up, these butterflies may be hard to detect or follow as they fly through the dappled shade of the forest (Endler 1981, Burd 1994). Hespenheide (1996) also noted that various beetle mimicry complexes were substrate-specific or associated with different parts of the local vegetation, and that this was true of many insect mimicry systems. Locally, the community structure of unpalatable, mimetic butterflies is non-random, with communities at a given altitude having similar higher taxon (phylogenetic) composition, and the higher the altitude the greater the phylogenetic diversity in the local community (Chazot et al. 2014). Also, the geographic ranges of species involved in Müllerian complexes were significantly larger than those of species not involved in such mutualistic relationships. The first experimental proof of this came from Tullberg et al. (2005), who showed that the striped larvae of the European swallowtail butterfly, *Papilio machaon*, functions in two different ways depending on the distance from which they are viewed. Entomologists have long known that these swallowtail caterpillars are quite hard to spot.

Dual signals might also frequently involve different classes of predator. Insects in general cannot see the colour red (it would 'appear' black to them). So an animal that is conspicuous to one class of predators, such as birds, which can see reds, might be cryptic to others that cannot. Since not all classes of predator respond equally to particular toxic compounds, it can be the case that a prey is simultaneously cryptic to those predators that might find them quite palatable, and aposematic (see Chapter 4) to those that find them distasteful (see Fabricant & Herberstein 2015 for an example).

Schultz (1986) examined the ability of polymorphic North American tiger beetles (Cicindelidae) to escape from predators and showed that while they are generally somewhat camouflaged, those individuals that are most conspicuous against the backgrounds on which they were collected, also showed the greatest escape flight distances. That is, they were more wary of potential predators that might spot them more easily.

PROTECTIVE CRYPSIS IN NON-VISUAL MODALITIES

Ruxton (2009) reviews the evidence that animals may be cryptic to senses other than vision. The strongest evidence concerns adaptive silence both for protection from predators and when hunting (see also Chapter 10, section *Stealth*). The situation with chemical crypsis is often confusing in that many cases are most likely mimetic (for example of social insects by inquilines (see Chapter 10, section *Acquired chemical mimicry in social parasites and inqualines*) but there appear to be some cases that are better regarded as crypsis, such as the adopted chemical signature of the caterpillars of *Biston robustum* (Akino et al. 2004; see section *Caterpillars and food plant colouration*). A similar case was described by Portugal & Trigo (2005), in which larvae of the nymphalid butterfly *Mechanitis polymnia* on their natural food plant are ignored by ants but are attacked if transplanted onto other plants whose odour they do not resemble.

It is often difficult to determine whether an animal is not attacked because it is unrecognised as potential food, or whether a predator is actively ignoring it because of prior bad experience. A typical difficult example, discussed by Ruxton (2009), is that of the limpet *Notoacmea paleacea*, which feeds on surfgrass, *Phyllospadix* sp., described by Fishlyn & Phillips (1980). These limpets would be suitable prey to predatory *Leptasterias* starfish, but the starfish simply walk over the limpets when they encounter them on the surfgrass. Fishlyn & Phillips' (1980) interpretation is that this is because the limpet incorporates chemicals (flavonoids) from the food into its shell's outer layer but not into its flesh, and that the starfish misinterpret them as surfgrass and walk over them without attacking. However, it is possible that the flavonoids are toxic and that the starfish are therefore rejecting the limpets as apparently inedible.

There is also some evidence, though not completely conclusive, that various fish employ forms of crypsis against detection by electrosensory systems such that they become less detectable through a sort of cloaking effect (P.K. Stoddard & Markham 2008).

APOSTATIC AND ANTIAPOSTATIC SELECTION

While not universal, it is commonly observed that when predators have access to two distinguishable, palatable food types, the commoner one will be attacked disproportionately more often than the rarer one, a phenomenon referred to as apostatic selection (B. Clarke 1962, J.A. Allen & Clarke 1968, J.A. Allen 1972, J.J.G. Greenwood 1984, 1985, Raymond 1984, J.A. Allen et al. 1988, Tucker & Allen 1988). This phenomenon has been postulated as being important in maintaining polymorphism in situations where both prey types are palatable. Apostatic selection might also have a significant role to play in Batesian mimicry. In contrast, antiapostatic selection, in which predators preferentially attack the rarer prey type, will have a stabilising effect within a species, eliminating new phenotypes (J.A. Allen 1972, Lindström et al. 2001a). These two types of predator behaviour ought to be expected under different conditions. In single or mixed species schools of fish, atypical individuals are easily targeted by predators (see Chapter 8, section *Schooling, flocking and predator confusion*) whereas searching images will always lead to balancing selection that maintains polymorphism.

Apostatic selection, and, for that matter, antiapostatic selection, involves some sort of perceptual switching. A significant amount of work has been carried out on apostatic selection of cryptic prey, some with captive predators, some with wild ones such as garden birds, and some with human subjects, typically trying to locate cryptic prey in computer displays. Bond (1983) used three domestic pigeons (*Columba livia*) to explore switching between two artificial prey types (black beans and red crumbs) on a visually complex background, and found that all three birds showed very similar switching to the commoner form when the frequency differed a small amount from 50:50 (Fig. 2.44). J.M. Cooper (1984) investigated how apostatic selection might act on cryptic prey types by presenting wild birds with various frequencies of orange and grey pastry prey, presented against each of three types of background: a hessian sheet scattered with either orange and grey stones (the 'matching' background), lilac and yellow stones, or green stones (two sorts of 'control' background). In all three cases the birds exerted apostatic selection, but rather surprisingly, the effect was greatest with the matching background. Apostatic selection does not necessarily have to be mediated visually, however, and Soane & Clarke (1973) demonstrated it with laboratory mice foraging on vanilla- or peppermint-scented baits.

Bond & Kamil (1998) carried out a novel experiment with blue jays, *Cyanocitta cristata*, by using digital images of cryptic *Catocala* moths (see Fig. 8.21a,b) drawn from a virtual population. The relative numbers of each type that

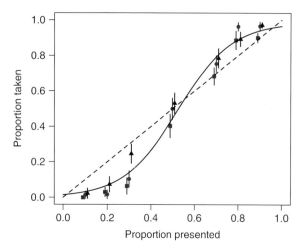

Fig. 2.44 Apostatic selection and the perceptual switching of three different individual pigeons that were presented with different proportions of two food items (black beans and red wheat) presented on a visually complex gravel background – the proportions shown are for the beans (means ±2 S.E.; curve is a three-parameter logistic regression fitted to the means only). (Source: Adapted from Bond 1983 with permission from APA.)

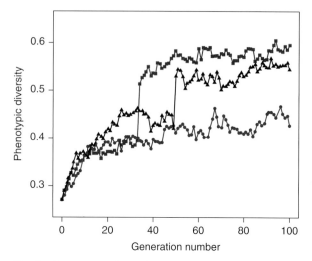

Fig. 2.45 Experimental test of the effect of apostatic selection by blue jays on phenotypic diversity of digital images of cryptic *Catocala* (Noctuidae) moths. The three lines are results of three separate selection series, and each shows that 'predation' by the birds led to an increase in moth image phenotypic diversity. (Source: Adapted from Bond & Kamil 2002 with permission from Macmillan.)

escaped detection by the predator were then used to adjust the relative frequencies with which images of the moth types were subsequently presented to the jays. The results showed that the birds did indeed operate apostatic selection such that the asymptotic virtual prey population had each prey type represented at some characteristic abundance, presumably based on the detectability of each individual prey pattern. Using the same set-up, Bond & Kamil (2002) showed that blue jays searching for evolving digital moth images often failed to detect atypically patterned moth images, and this selection process led, as expected, to phenotypic variation or polymorphism in cryptic prey species (Fig. 2.45). Such behaviour would also be expected to increase aspect diversity between cryptic prey species (Rand 1967, Bond 2007) (see also section *Effect of cryptic prey variability* later).

Observation of feeding behaviour and food selection in the wild is usually rather difficult and detailed studies are scarce. Wulff (1994) reports a long-term study of feeding by various grazing tropical Caribbean reef fish and observations of over 4000 bites at sponges by three species. Although sponges were attacked in order of their abundance, rare species were attacked disproportionately more, i.e. the fish collectively showed antiapostatic selection. More interesting is that in 92% of cases that a fish fed on two sponges consecutively, the sponges belonged to different species. In this situation it seems likely that the fish were minimising their exposure to any given sponge toxin,

as sponge-feeding seems to be quite a specialist occupation (Randall & Hartman 1968). Further evidence that apostatic selection is involved in the maintenance of polymorphism in the wild is hard to come by, but it has been suggested as likely to be involved in a number of cases, such as colour polymorphisms in intertidal periwinkles (W.D. Atkinson & Warwick 1983), *Cepaea* land snails (Cain & Sheppard 1950, 1954), salamanders (Fitzpatrick et al. 2009) and fish (Olendorf et al. 2006) among others.

Search images

It is a common human experience that after a little practice at trying to find inconspicuous items, such as small beads dropped on a patterned carpet, the searcher's efficiency becomes markedly improved. The original concept based on introspection was first proposed by von Uexküll (1934, translated 1992), who thought of it as 'learning to see'. Such anthropocentric observations gave rise to the concept of the search image, equivalent to the more common colloquialism, 'getting one's eye in'. In a biological context, Luuk Tinbergen proposed the same notion for the way birds can locate generally well-camouflaged prey such as caterpillars on woodland trees (L. Tinbergen 1960, published posthumously), and the notion is generally referred to as forming a 'search image' or 'searching image'. Tinbergen (p. 332) wrote "this implies that the birds perform a highly selective sieving operation on the visual signal that reached their retinas leading to selective attention for a particular type of signal" (Lawrence & Allen 1983, Bond 1983, Gendron 1986, Endler 1988). The way many subsequent workers sought to test whether a predator forms a search image was by seeing whether, after progressively encountering one or more randomly distributed cryptic prey, they detected them at an increasing rate, and this was found commonly to be the case. Plaisted & MacKintosh (1995), using pigeons and computer-simulated targets, found that their discovery of cryptic targets of one type improved when the same target was presented consecutively, but not when intermixed with a second type, and that the improvement plateaued and was only temporary.

As J.A. Allen (1989) pointed out, because the basic way experimenters tell that a predator has detected a prey is whether or not they eat it, a better operational definition would be that a predator shows a perceptual change in its ability to detect a *familiar* cryptic prey, rather than being perhaps more willing to consume it.

A corollary of search images is that any improvement in finding the particular object or pattern does not come without a trade-off; specifically, the process of getting 'one's eye in' results not just in an improvement in detecting the particular item, but also in a reduction in the likelihood of noticing other potentially interesting items that would otherwise have come to the person's attention. Another way of looking at this is to view it as the temporary acquisition and utilisation of selective attention. That an organism can temporarily improve its ability to detect some things to the detriment of noticing others has been widely applied to prey location behaviour, particularly cryptic prey. However, despite many experiments such as those of Marian Dawkins (1971) with domestic chicks searching for cryptic or conspicuously coloured grains, whose results are perfectly consistent with the perceptual changes envisioned by L. Tinbergen (1960), there is still a great deal of debate as to whether or not animals actually do form search images, or whether their prey detection behaviours can be explained adequately by other means.

It turns out that an increase in the rate at which a predator locates cryptic prey with experience does not necessarily mean that they are using a 'search image' at all. Guilford & Dawkins (1987) showed that the same improvement would be achieved if the predator slowed the rate at which it scanned the environment, giving itself more time to interpret the complicated visual image they have acquired. There followed various attempts to distinguish between true search image and the effects of changes in scan rate. Lawrence (1985a, 1985b), for example, video-taped wild European blackbirds, *Turdus merula*, foraging for dyed

cryptic or conspicuous artificial prey, and found what appeared to be evidence for an improved ability to detect the cryptic prey after some experience. However, even after reanalysis of the video recordings of the blackbirds (Lawrence 1989), Guilford & Dawkins (1989a) pointed out that it was still not possible to be certain exactly what was happening with the blackbirds' scanning – perhaps they were moving their eyes and scanning quickly rather than moving their heads, which is what Lawrence (1989) had interpreted as indicating scans. Again using video recordings, Lawrence (1986) compared how young great tits' abilities to locate and capture conspicuous and cryptic artificial prey on a bark background changed with experience and mode of hunting. The birds could either spot a prey from a perch 0.5 m away from the bark, or land on the bark and search. Initially, conspicuous prey were detected easily but cryptic prey were often located accidentally. However, they improved their search ability remarkably through the trials, especially when it came to spotting prey from a distance, again consistent with search image acquisition.

Although an improvement in prey detection rate is explicable both through modification of search rate and through search image acquisition, and the two are not mutually exclusive, search rate slowing ought to improve detection of all cryptic prey, whereas search image formation ought only to work for one prey type corresponding to the search image (Guilford & Dawkins 1987, 1989a, 1989b). A prediction from this is that a predator using a search image should initially improve its rate of detection of one prey type, then, after having depleted them, start to use another, again improving its rate. Other aspects of search image acquisition may also reflect other behavioural processes. L. Tinbergen's (1960) original idea stemmed from the observation that predation by tits on a certain type of caterpillar was lower than expected by chance when their prey density was low, but Gibb (1962) points out that this could be because the birds did discover them but rejected them as unprofitable.

If search images really do exist (see later), which I believe they do, defining what they are exactly has proved surprisingly hard. Croze (1970), for example, stated:

> We have as yet no vocabulary to give a clear and distinct meaning to the basic mechanics of Searching Image.

Staddon & Gendron (1983) showed using signal detection theory (see Chapter 5, section *Signal detection theory*) that even in the absence of search image, optimal predator behaviour may lead to prey switching and that dispersion would therefore be favoured in cryptic prey. This results from the fact that a predator foraging optimally should take cryptic prey disproportionately more often when their density is high (as it would be if they were locally clustered), whereas for conspicuous prey an optimally foraging predator should take them only in proportion to their abundance. J.J.D. Greenwood (1986) showed that the Staddon & Gendron (1983) model also applied to situations of searching for masquerading prey or for Batesian mimics by, for example, setting the cost of rejecting a prey to zero, and it turned out that Geeenwood's modified equations were simplified forms of those used by Getty (1985) (see also Getty 1987).

R.A. Morgan & Brown (1996) distinguish two types of search image: passive, which means that a predator responds to an encounter with a prey type, and active, when the predator can choose between different image-related search modes. Theoretically these could be distinguished by the giving-up density of the food items in either monotypic or mixed patches, however, their experiment using wild squirrels, *Sciurus carolinensis*, actually provided no evidence of search image use at all.

Because humans are predominantly visual animals, it is sometimes hard to imagine how other animals perceive their environment, but there is no reason to suppose that the search image concept could not apply to other sensory modalities. Nams (1991) describes how the olfactory reaction distance of striped skunks, *Mephitis mephitis*, changes in response to experience with a given food in a way that would be perfectly consistent with a search image-type sensory filtering.

Experimental tests of search image

Guilford & Dawkins (1987) have reviewed several experimental studies purporting to provide evidence in support of the specific search image hypothesis. In experiments with pigeons, *Columba livia*, foraging for variously wheat dyed various colours, Reid & Shettleworth (1992) found that when birds were presented with two cryptic types of grain, they preferentially took the type they had most recently experienced, which would be consistent with them using a search image. However, they also found that experience of finding one of the two cryptic types of wheat also improved the bird's ability to locate the second type, which would not be expected for a search image, and instead proposed that "Search image may better be thought of as priming of attention to those features of the prey type that best distinguish the prey from the background".

Karpestam et al. (2014a) tested whether the natural colour variation of the grasshopper *Tetrix subulata*, which occurs in black, striped or grey, used colour morphs, adding to their protection. Human subjects were given the task of locating images of the morphs against typical photographic backgrounds on a computer screen. When the three morphs

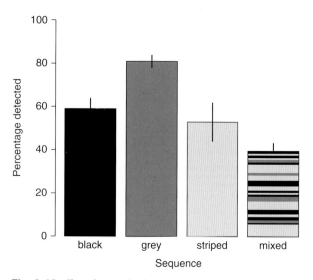

Fig. 2.46 Effect of natural colour polymorphism on human detection of images of a cryptic grasshopper, *Tetrix subulata*, against photographs of natural background showing that 'prey capture' was significantly reduced if they were presented in a mixed sequence of forms. (Source: Adapted from Karpestam et al. 2014a with permission from John Wiley & Sons.)

were presented in mixed sequences, significantly fewer were spotted than when they were presented in monomorphic sequences (paired *t*-test based on data for all three morphs: $t = 5.4$, d.f. $= 29$, $p < 0.0001$) (Fig. 2.46). However, there was no evidence that the subjects were forming search images because slopes of the regressions of proportion of grasshoppers spotted versus presentation number did not differ significantly from zero for any of the four presentation sequences (Fig. 2.47).

Gestalt perception

Trying to understand how disruptive selection leads to perfect polymorphic mimicry, Ikin & Turner (1972) explored whether a predator might be able to use Gestalt perception, i.e. recognition of a pattern in its entirety rather than recognising only individual parts of it, using wild birds and coloured and patterned targets. Their rationale was that if a predator, for example, were only to recognise parts of a mimic's pattern, say body colour or wing colour, and separately learn that each was associated with palatable prey, it would not matter what combinations of wing and body colour there were as all would be attacked if they had just one feature (wing OR body colour) that was associated with palatability. They placed pieces of either palatable or

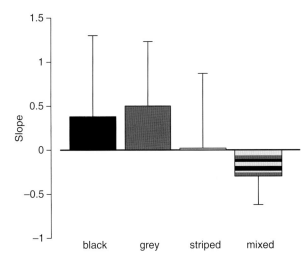

Fig. 2.47 Evidence that the human 'predators' whose 'prey' capture results are shown in Fig. 2.46, did not form search images as none of the slopes of capture rate against position of prey in sequence differed significantly from zero. (Source: Adapted from Karpestam et al. 2014a with permission from John Wiley & Sons.)

unpalatable[14] pastry on variously coloured and patterned cardboard squares (Fig. 2.48a) laid out on a lawn. Analysis of variance showed a significant interaction between prey colour and pattern which confirms that wild birds do perceive patterns as a whole rather than relying on a single element (Fig. 2.48b).

EFFECT OF CRYPTIC PREY VARIABILITY

There are surprisingly few experimental studies regarding how much protection aspect diversity affords, or even if it does afford any. Irrespective of whether predators form search images or not, recognising prey against a complex background will involve learning, and the fewer patterns the prey display the easier it ought to be for a predator to learn to distinguish them. Therefore it is predicted that polymorphism of the cryptic pattern ought to provide protection as it makes it harder for a predator to learn. Knill & Allen (1995) used a computer game involving 'zapping' cryptic targets with 80 schoolgirls aged 12–13 years old as predators and also found that predation (zap) rate was significantly negatively correlated with the number of prey morphs.

---

14. Soaked for 45 minutes in 70% solution of quinine dihydrochloride; controls were soaked in water.

Reflexive selection and aspect diversity

In sharp contrast to the convergence on a particular pattern or form shown by many mimetic and cryptic systems, Moment (1962a, 1962b) drew attention to a number of cases in which organisms seemed to display excessive levels of polymorphism. Examples are common among marine benthic organisms, including various bivalves such as the coquina, *Donax variabilis* and *Donacilla cornea* (D.F. Owen & Whiteley 1988, Whiteley et al. 1997), marine and brackish-water gastropods, sea anemones, tube-dwelling annelids (Fig. 2.49a) and echinoderms, notably brittle and feather stars (Fig. 2.49b), but also examples such as the marine gastropods *Nucella lamellosa* and *Neritina communis*, terrestrial snail *Cepaea nemoralis* (Cain & Sheppard 1950, 1954) and *C. hortorum*, *Amphidromus* tree snails, and the isopod crustacean *Idotea balthica* (Merilaita 1998). Sherzer (1896) noted the high level of variability in beans of an unidentified Philippine plant that locals apparently collected for the manufacture of soap. These resembled the diversity of pebbles on the shore where the beans wash up, and presumably some finally germinate and grow near the shore line. This intraspecific variability can be the result of true polymorphism, or be a pseudopolymorphism with all possible intermediates occurring (e.g. Grüneberg 1982). The genetics of the colour and banding patterns in *Cepaea* snails was investigated by Cain et al. (1960) and Wolda (1967); some features show complete dominance, others are controlled by multiple genes.

In some cases the extreme variation in appearance appears to be associated with highly heterogeneous backgrounds such as is the case of the chiton *Ischnochiton striolatus* (Gonçalves Rodrigues & Silva Absalão 2005), and indeed the same has been suggested in the case of *Cepaea* snails, i.e. that disruptive selection is caused by local presence of many particular types of background.

In general, the taxa involved are at a low trophic level, which may consequently be associated with high local abundance. To account for this, Moment (1962a, 1962b) suggested that this was an evolutionary strategy to reduce predation by reducing the possibility that a predator could form a search image (see section *Search images*) for a given prey and so increase its efficiency at finding them. He also coined the term 'reflexive selection' for the phenomenon. Moment's hypothesis received a good degree of criticism but Rand (1967) independently came to the same conclusion that what he referred to as 'aspect diversity' had evolved because it thwarted some of the advantage gained from a predator developing search images. Moment's reflexive selection is simply an extreme result of apostatic selection, with predators preferentially attacking the commonest morph until its numbers have dwindled, whereupon they switch to the next most common, and so on (J.A. Allen

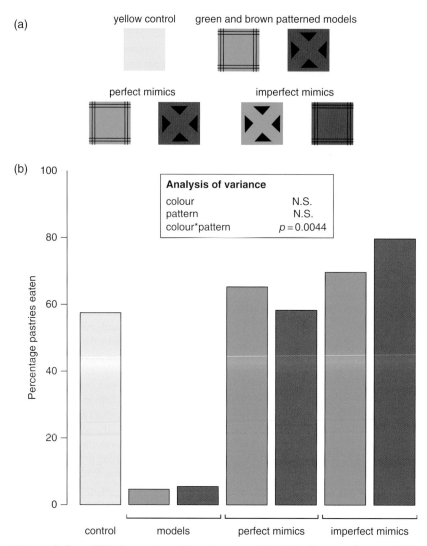

Fig. 2.48 Experiment to test whether wild birds recognise entire patterns or only individual pattern elements. (a) Patterns and colour combinations for card rectangles placed out with either unpalatable pastry food (model training) or subsequently with palatable food (perfect and imperfect mimicry tests); (b) summary of numbers of pastry foods eaten and results of statistical analysis comparing perfect and imperfect mimicry results. (Source: Adapted from Ikin & Turner 1972 with permission from Macmillan.)

1988, Ruxton et al. 2004a). Clarke (1962) had earlier introduced the term apostatic selection to describe this advantage to diversity *per se*, as it is selection in favour of those individuals that differ from the rest of the population. With the massive diversity that some of the proposed examples show, it is the diversity itself that is beneficial as it thwarts a predator's search use of search image (D.F. Owen & Whiteley 1986). Li & Moment (1962) discuss possible genetic mechanisms that could bring about such high levels of phenotypic polymorphism.

Even in situations where the colour polymorphism is not as extreme as in the above examples, polymorphism has been postulated as a mechanism that will reduce overall predation. However, Wennersten & Forsman (2009), who set out 2976 pastry 'caterpillars' either in monomorphic or four colour groups (red, yellow, brown, green), found no such evidence (Fig. 2.50), indeed the opposite. However, their prey were not protected and some were without doubt very conspicuous, such that in polymorphic groups the red 'caterpillars' always went extinct first, and green morphs

Fig. 2.49 Possible examples of reflexive selection, i.e. species that display huge amounts of intraspecific colour variation. (a) Christmas tree worms, *Spirobranchus giganteus*, Thailand; (b) several specimens of an unidentified feather star (Crinoidea), Komodo Island, Indonesia. (Source: © Linda Pitkin Underwater Photography. Reproduced with permission from Linda Pitkin.)

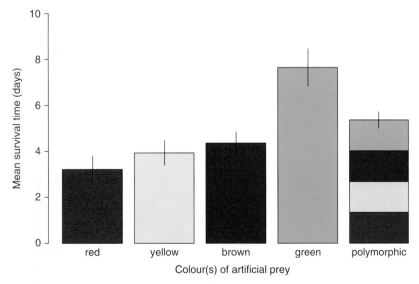

Fig. 2.50 Results of field predation experiment with four differently coloured pastry baits presented in tree trunks in groups of four identical ones or in 'polymorphic' groups showing that the polymorphism treatment did not confer any marked level of protection. (Source: Adapted from Wennersten & Forsman 2009.)

last, with yellow and brown models having intermediate survivorships. Thus the heavy predation on the undoubtedly aposematic red prey caused overall predation pressure to be higher, and the authors suggest that if one of the morphs of a polymorphic prey actually attracts predators, it will have a negative effect on the whole prey population, what they call the 'give-away cue' hypothesis. I also suspect that predation was high because the baits were set out in groups, which is rather unrealistic, and a fairer test would be to have larger scale blocks with larger numbers of baits set out over areas comprising several trees – though of course the experimental logistics then start to become quite difficult.

The mechanisms generating and perpetuating these massive levels of visual polymorphism have thus far only been investigated in depth in *Cepaea nemoralis* (Cain & Sheppard 1950, 1954), but as the second of their papers explains, different workers have come to quite different conclusions, some previous workers having concluded that it was random.

As a corollary about diverse appearance thwarting search images, Schall & Pianka (1980) presented a theoretical argument combined with data from *Cnemidophorus* lizards that selection ought to favour a diversity of different escape behaviours, as this will also thwart a predator's ability to learn how to deal with prey that are trying to escape – in the case of the lizards, some via tail autotomy.

Searching for cryptic prey – mathematical models

Gendron & Staddon (1983) developed a simple mathematical model based on Holling's famous disc equation (Holling 1959) to examine the effect of search rate and different degrees of crypticity and conspicuousness on the probability of a prey item being detected by a predator

$$N = aD(T - Nh)$$

where N is the number of prey captured, a is a constant referred to as the rate of successful search, D is prey density, T is the total time foraging, and h the handling time per prey discovered. Rearranging to put the Ns all on one side gives

$$N = \frac{aDT}{1 + aDh}$$

Then, dividing both sides by foraging time gives an equation for the rate of prey capture, R,

$$R = \frac{aD}{1 + aDh}$$

Gendron & Staddon (1983) then defined an equation that relates probability of a prey being detected (strictly the conditional probability of an encountered prey being detected), P_d, to the search rate, S, the maximum search rate at which the predator can detect any prey

$$R = \frac{DSP_d}{1 + DSP_d h}$$

P_d is therefore an inverse measure of how cryptic the prey is, and also must be inversely related to search rate, S. To determine the effect of different levels of prey crypticity or conspicuousness, Gendron & Staddon (1983) introduce a variable, K, which they term a 'conspicuousness index'. This index relates the probability of prey detection to relative search rate thus

$$P_d = \left[1 - (S/M)^K\right]^{\frac{1}{K}}$$

If the search rate at which recognising any cryptic prey is called M, which is therefore the search rate at which $P_d = 0$, then relative search rate can be defined as S/M. Figure 2.51 illustrates the relationships between a predator's probability of detecting prey and relative search rate for four different levels of prey conspicuousness, K. The curves are obtained by substituting for P_d in the equation for rate of prey capture, R, above. As would be expected intuitively, the most conspicuous prey are detected easily across almost all search rates and only fail to be detected when searching is

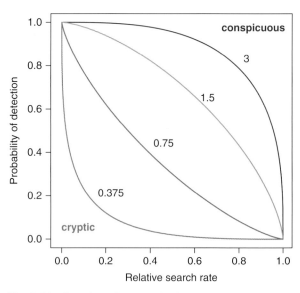

Fig. 2.51 Effect of search rate on the detection of prey of varying degrees of conspicuousness, parameter k. (Source: Adapted from Gendron 1986 with permission from Elsevier.)

very fast, whereas highly cryptic prey only have an high chance of being detected if the search rate is very low.

Gendron & Staddon (1983) then went on to examine the optimal search rate in situations with either one prey type of given crypticity or two prey types differing in crypticity. The result of the latter model is particularly interesting as it showed that optimal search rate increases with the relative proportion of the less cryptic prey type, and while this may seem intuitively obvious, it results in a dramatic drop in the detection of the more cryptic prey type with even slight differences in crypticity. So mutations making an organism only slightly more cryptic may nevertheless provide a strong selective advantage and so spread rapidly through the population. When the different individuals belong to different species, the effect will be that the slightly less cryptic one will suffer markedly greater levels of predation.

J.J.D. Greenwood (1986) notes that Staddon & Gendron's (1983) work pertains equally well to masquerade and to Batesian mimicry and he developed their model further to predict how optimally foraging predators should decide to switch between alternate prey types and incorporate imperfection.

While there is a tendency for aposematic prey to be clumped in the environment, cryptic ones are typically more dispersed (Ford 1945, N. Tinbergen et al. 1967, Benson 1971). This makes locating any particular type difficult, and indeed if only one type were to be eaten, searching would almost certainly mean passing by other types of suitable cryptic prey.

Staddon & Gendron (1983) noted that there are four possible outcomes of a predator searching for cryptic or masquerading prey. They may:

1. Correctly identify something as prey and catch it;
2. Correctly identify something as prey but miss it;
3. Incorrectly identify something (such as a twig) as prey and attack it;
4. Incorrectly identify something as prey but not attack it.

A bird, for example, might 'act rashly' and attack anything that looked vaguely like a twig caterpillar – this would mean that they had a high chance of getting a benefit (B) if it was a caterpillar, but would incur a cost, even if only the energy spent in making an attack, or the time missed during the attack that could have been used for more searching. Thus you can express the choices made as conditional probabilities.

$$p(A|P)$$

is the probability of an attack *if* the item detected is real prey, and

$$p(A|!P)$$

is the probability of attacking something that is *not* real prey (!P).

If v and u are costs associated with failing to attack a real prey and attacking a non-prey, respectively, the net benefit (H), given densities of prey and non-prey (D_p and $D_{!p}$, respectively) can be written thus

$$H = D_p \left\{ p\langle A|P \rangle B - \left[1 - p\langle A|P \rangle v \right] \right\} - D_{!p} p\langle A|!P \rangle u$$

The authors liken the relationship to the receiver operating characteristic (ROC) curves that radio users used to use, which typically have the form

$$p\langle A|P \rangle = \langle A|!P \rangle^s$$

where s is a measure of crypticity ($0 < s < 1$). Thus if s is close to unity, prey are very hard to distinguish from non-prey so the probabilities of attacking prey or non-prey are nearly the same, i.e.

$$p(A|P) \approx p(A|!P)$$

Substituting for this approximation we get

$$H = pD_p(B+v) - vD_p - D_{!p} u p^{1/s}$$

Predators are expected to maximise their benefit, i.e. they will have evolved to behave in such a way as to maximise H. Taking densities and costs to be constant, differentiating the above equation with respect to p will give the value of p that maximises H, hence

$$p = \left\{ S(D_P/D_N) \left[(B+v)/u \right] \right\}^r$$

where $r = s/(1-s)$.

However, if crypticity, the density of non-prey, and the costs and benefits are all constant this rather cumbersome equation simplifies greatly such that p is proportional to prey density to the power of r. Having now allowed for both search rate and crypticity, P_d was substituted into Holling's disc equation (Holling 1959) to give the actual feeding rate. When the prey are at least partially cryptic, i.e. $S > 0$, feeding rate was predicted to increase with prey density to the power $1/(1-s)$, and therefore since $S > 0$ the power is also greater than one. Thus as prey density increases, cryptic prey will be consumed disproportionately more often, which is exactly what a search image would lead to, without any of the model having search image consequences included.

ONTOGENETIC CHANGES AND CRYPSIS

Many animals show marked changes in colouration throughout development. Typically, among mammals and birds the young are generally rather drab and cryptic, and brighter colouration is developed only on sexual maturity, and often only in males (Booth 1990). The opportunity for crypsis may be negated as the animal grows for a variety of reasons – sheer physical size might make it impossible to hide effectively, body form may change, they may have to shift habitats, their food source might necessarily change thus potentially altering the compounds that are available to synthesise pigments, etc. Alternatively, an animal's ability to defend itself may increase markedly with its age/size; for example, Nylin et al. (2001) found that chicks initially attacked intact spiny 5th instar larvae of the comma butterfly, *Polygonia c-album*, which are palatable if their spines have been removed, but quickly learned to avoid them – interestingly, the larvae were not harmed during the attacks.

J.B. Grant (2007) examined the ontogenetic colour change of panic moth, *Saucrobotys futilalis*, caterpillars which are initially green and cryptic but which progressively become aposematic orange with black spots and show corresponding behavioural changes. These larvae feed on toxic hemp dogbane, *Apocynum cannabinum* (Apocynaceae), and defend themselves, as do many other caterpillars that feed on toxic plants, by regurgitating what they have recently eaten when threatened. The 'unpleasantness' of the regurgitate puts many predators off before the larvae are harmed, and as they rest together in clusters, even if one is killed, the siblings are likely to survive. Such changes can be explained if the initial cryptic phase imposes an opportunity cost (Merilaita & Tullberg 2005, Speed & Ruxton 2005a), for example limiting where the caterpillar can forage for food.

Many Lepidoptera larvae that were cryptic on their host plants when fully grown change to a brown colour before wandering away from the food plant across the ground to find a suitable pupation site.

Among mammals, too, some show a change in colour pattern from dappled/spotted young, as in wild boar, many deer, tapirs, etc., to their final, usually more sombre and solid colouration after they are no longer small enough to hide. Not only size but also microhabitat may change with development. Juveniles of the Australian python, *Morelia viridis* (Pythonidae), live on the ground among leaf-litter and are cryptically coloured yellow or red-brown to match, but as they grow they move to live in trees and become green; this developmental change helps protect all life stages from avian predators (D. Wilson et al. 2007).

HIDING THE EVIDENCE

Petiole clipping by caterpillars

In several instances in which the normal activities of an organism leave marks on the surroundings that could assist a potential predator in locating it, selection has led to behaviours that reduce the conspicuousness of the activity. Some of the best examples occur among cryptically coloured Lepidoptera larvae that feed openly on vegetation but behave so as to conceal the damage their feeding has inflicted on their host plants. Feeding on leaves will inevitably leave tell-tale signs (Fig. 2.52a,b) and these can be used as a cue to the potential presence of a prey insect by a predatory bird. These can be disguised by feeding on the leaves in such a way that the remaining leaf material is not so conspicuously ragged. Some caterpillars align their bodies along the eaten edge so that their bodies appear to replace the missing leaf tissue. Other, mostly larger species, will bite off any remaining leaf (Fig. 2.52c) or bite through the petiole of the leaf they have been consuming (Fig. 2.52d) so that its ragged outline will not be seen by diurnally foraging predators.

Heinrich (1979) hypothesised that foraging behaviour would be different in palatable and unpalatable caterpillars, with the former normally being cryptic, feeding or resting on the underside of leaves, feeding primarily at night and commuting between feeding and resting sites so that leaf damage visible during daylight hours does not give an accurate clue as to their whereabouts. Unpalatable aposematic species should not do any of these and in fact often rest exposed on the tops of leaves during the day, and many, such as tiger moth caterpillars (Erebidae: Arctiinae), are messy feeders, leaving conspicuous signs of their feeding activity. Lederhouse (1990) has shown that the type of feeding disguise displayed by unpalatable tiger swallowtail caterpillars, *Papilio glaucus*, depends on the leaf shape of the plant on which they are feeding. Thus, if they are on large-leaved plants where feeding holes would be very conspicuous they frequently clip the remaining leaf off at the petiole. However, when feeding on smaller leaved plants, depending on size, they either eat all the leaflets on one side of the midrib or eat the whole leaflet, neither of which leaves a very conspicuous clue.

Disguising leaf damage may have resulted from selection pressure imposed not only by the obvious visually hunting birds (Heinrich 1979, Heinrich & Collins 1983), but also some parasitoids, such as the braconid wasp *Cotesia glomerata*, which attacks larvae of the large white butterfly, *Pieris brassicae*, and which have been shown to be attracted to visually damaged plant leaves (Sato 1979), though otherwise they rely heavily on odours.

Exogenous crypsis

A number of animals, and indeed some plants, gain cryptic protection either through actively attaching common items from their surroundings onto the outsides of their bodies or

Fig. 2.52 Insect herbivore leaf damage and concealment of damage on plants from Thailand. (a,b) Two examples of unidentified plants showing conspicuous leaf damage caused by beetle and caterpillar respectively; (c,d) two examples of petiole clipping, stereotypic cutting across near base of leaf, and completely chewing through petioles respectively.

through passively having external objects attach to them. Among insects at least, this trait dates back to the Cretaceous (B. Wang et al. 2016) and is termed exogenous crypsis (Pasteur 1982). There are numerous examples but the best known are probably the 'decorator' crabs such as *Hyastenus elatus*, *Maja varrucosa*, *M. squinado* and *Pisa tetraodon*. 'Decorator' crabs are distributed among a number of families, most notably Epialtidae, Inachidae, Majidae, Oregoniidae and Pisidae, and occur in all parts of the world. Their behaviour, which has been known for a long time (see Bateson 1889, Poulton 1890), is primarily associated with crypsis (Fig. 2.53a–d), but in some species a preference for attaching stinging sessile cnidarians means that as well as probably functioning as a cryptic device, it has a secondary defensive function too (Fig. 2.53e), and similarly, the spider crab, *Libinia dubia* (Pisidae), decorates its carapace

with poisonous *Dictyota* algae (Stachowicz & Hay 1999). The surface of the carapace, and sometimes legs, are furnished with small hooked setae, often described as being Velcro-like, which aid the 'adhesion' of the camouflaging material. If the attached material is alive, such as pieces of sponge or some algae, tunicates, etc., these may subsequently grow over the crab's body almost completely (Fig. 2.53b–e). Hultgren & Stachowicz (2009) found, in a phylogenetic context, that the distribution of hooked setae on the body was a good predictor of decorating behaviour in the wild. In many of these majoid species, the extent of the hooked setae decreases with body size (ontogeny), probably indicating that either decorating is less effective as a camouflage in larger individuals, or they need it less because they are otherwise better able to defend themselves. As long ago as 1907, Minkiewicz found that experimentally denuded

Fig. 2.53 Examples of crabs that camouflage their carapaces and/or legs by attaching objects, seaweeds or creatures from their environment to them. (a) Northern kelp crab, *Pugettia producta* (Epialtidae), from the Pacific coast of North America with *Ulva* type seaweed attached, and colour partly derived from ingested seaweed pigments; (b,c) unidentified dresser crabs from East Timor, Indonesia; (d) graceful decorator crab, *Oregonia gracilis* (Oregoniidae), from California, dressed in sponges; (e) a spider crab *?Achaeus spinosus* (Inachidae) from the Pacific, covered in *Ianthella blasta* polyps. (Source: a, D. Gordon E. Robertson 2008. Reproduced under the terms of the Creative Commons Attribution Share-Alike Licence CC BY-SA 3.0, via Wikimedia Commons; b,c, Andrepiazza 2011. Reproduced under the terms of the Creative Commons Attribution Share-Alike Licence CC BY-SA 3.0, via Wikimedia Commons.; d, Ed Bierman 2011. Reproduced under the terms of the Creative Commons Attribution Licence, CC BY 2, via Flickr; e, Nick Hobgood 2006. Reproduced under the terms of the Creative Commons Attribution Share-Alike Licence CC BY-SA 3.0, via Wikimedia Commons.)

crabs would select items to redecorate themselves, ensuring that they matched the background they were placed on.

Hultgren & Stachowicz (2008) found evidence of a trade-off between decoration and colour change in three decorating species of kelp crabs, *Pugettia richii*, *P. producta* and *Mimulus foliata* (all Epialtidae). The first of these species decorates the most but displays the lowest colour change capability, whereas the second decorates the least and has the greatest colour change ability, with the *Mimulus* species being intermediate in both respects. Further, *P. producta* sequesters pigments from the algae upon which it feeds (see Chapter 4, section *Origins of protective compounds*). *Pugettia* kelp crabs also shift habitat as they grow larger and thus the types and colours of seaweed from which they sequester camouflaging pigments changes in parallel (Hultgren & Stachowicz 2010). The choice of decorating items of species may vary geographically so, for example, in some parts of its range, *L. dubia* utilises the toxic sponge *Hymeniacidon heliophila* instead of toxic algae for camouflage and protection (Stachowicz & Hay 2000).

A similar protective use of decoration is shown by the caterpillar of the North American wavy-lined emerald moth, *Synchlora aerata* (Geometridae), which attach petals of their food flowers on their backs using their silk, and replace them with fresher ones as the older petals fade (Treiber 1979). The related *S. frondaria* also employs decoration and additionally displays diet-induced phenotypic plasticity (Cranfield et al. 2009).

Eisner et al. (1978a) describe yet another system in a quite different arthropod. Larvae of the predatory lacewing *Chrysopa slossonae* (Neuroptera: Chrysopidae) develop among infestations of their prey, the woolly alder aphid, *Prociphilus tessellatus* (Hemiptera: Aphididae), and they disguise themselves from the ants that shepherd the aphids by plucking some of the prey's waxy wool and attaching it to their own backs. Experimental denudation of the lacewing larvae removed their protection from ant attack and the importance was demonstrated by the fact that hungry denuded lacewing larvae gave re-investiture of the wax approximately the same priority as feeding. Larvae of another lacewing, *Mallada desjardinsi*, attach the carcasses of the aphids they have preyed upon on their backs, thus providing them with a source of the aphids' own cuticular hydrocarbons, so protecting them from attack by *Tetramorium* ants attending the aphids. Hayashi et al. (2016) have shown that the effectiveness of this as a defence relies on the ants having had prior experience of the aphids, which enables them to learn the aphids' chemical signature.

Some birds, such as long-tailed tits, *Aegithalos caudatus*, and blue-grey gnatcatchers, *Polioptila caerulea*, decorate the outer surfaces of their nests with special types of items such as spider cocoons or flakes of lichen, both of which are conspicuously pale or white. Hansell (1996) tried to test whether the function was essentially to achieve crypsis by making the nests resemble the trees' branches or whether they reflect ambient light to effectively make the nest 'dissolve' into the background and not appear to be a solid object. Although the decoration applied by some species examined might have enabled the nests to resemble lichen-covered branches, the bulk of the evidence suggests that the application of white particles primarily serves the general light reflection function.

Even large animals may inadvertently gain advantage from general dusting. Zebras, for example, are always more or less black and white striped in zoos, give or take a bit of mud, but in wild areas with red, lateratised soils such as in large parts of Tsavo National Park, Kenya, their white stripes in particular become almost a dark orange, markedly reducing their conspicuousness at a distance. Some small succulent plants that live in sandy environments have sticky surfaces so sand grains will adhere to them. This will reduce their conspicuousness and may have other physical effects (Lev-Yadun 2006). Some aggressive mimics do the same and in these cases the crypsis no doubt has a dual function. Not all coverings made by animals are necessarily part of any crypsis – many case-bearing caddis fly (Trichoptera) larvae can appear quite conspicuous, but the toughness and inedibility of their cases probably renders them rather low profitability prey, and this might make their conspicuousness an honest signal (see Gohli & Högstedt 2010).

MILITARY CAMOUFLAGE AND MASQUERADE

Although really a form of crypsis, soldiers, and especially snipers, go through considerable training in order to learn how to remain unnoticed – failure to do so being as likely to have fatal consequences. Soldiers or indeed hunters use very much the same principles that animals have evolved – general colour matching, appropriately sized irregular and asymmetric blotches (or pure white if they are in the high arctic), attaching pieces of vegetation, and concealing the eyes with paint. Ghillie suits, which get their name from the Scottish gamekeepers who are often called ghillies, cover the entire body including hands and face and are covered in ruffles and strips of material that break up the outline and create confusing and changing shadows (Fig. 2.54a).

Early attempts to persuade the military (the British War Office) to employ camouflage for its troops and battleships

Fig. 2.54 Use of camouflage and disruptive and motion dazzle effects by the military. (a) A United States Marine Corps sniper dressed in a ghillie suit during a training exercise; (b) the U.S.S. West Mahomet in port (c. 1918), painted with a dazzle camouflage scheme that also distorts the outline of her bow. (Source: a, U.S. Marine Corps. 1988; b, Bureau of Ships Collection in the U.S. National Archives.)

involved a collaboration between Abbott Thayer and another British artist, John Singer Sargent, who had better social connections there. Unfortunately, one of his suggestions, to paint warships white, came undone when one of the painted ships was torpedoed and the navy abandoned the idea (Kingsland 1978). However the use of disruptive stripes and dazzle effects were more successful and widely applied (Fig. 2.54b) (D. Williams 2001). Nowadays, visual sighting of ship, aeroplane and tank targets has been superseded in many cases by radar detection, so the sharp angular designs of stealth versions are rendering them effectively cryptic by minimising the likelihood of an enemy's radar pulses being reflected back to the detectors.

During World War II, armies on both sides went to elaborate lengths to deceive the enemy (Reit 1970). There was even a special British army unit (the 602 Camouflage Engineers) dedicated to the business. Examples include inflatable rubber and, later on, plastic tanks (bluff and decoy) which were first deployed during World War I, but far more during World War II and also successfully during the Kosovo and Gulf Wars. A nice arms race in mimicry not only led to greater apparent authenticity, but also the incorporation of heating systems, so recent examples mimic the heat signatures of real tanks, thus deceiving enemies with infrared detectors. Decoys also included entire false towns in the countryside where bombing would do minimal damage (decoy), rubber dummy paratroopers called 'paradummies' to make parachute invasions appear larger than they were (bluff), headquarters disguised as rubbish dumps (defensive masquerade), and snipers in artificial tree trunks (aggressive masquerade), etc.

Chapter 3

CAMOUFLAGE: MASQUERADE

[C]an there be any edge in looking 5 per cent like a turd?

Steven J. Gould (1977)

Mimicry, Crypsis, Masquerade and other Adaptive Resemblances, First Edition. Donald L. J. Quicke.
© 2017 John Wiley & Sons Ltd. Published 2017 by John Wiley & Sons Ltd.

INTRODUCTION

In masquerade, the organism has not evolved to avoid detection; it can readily be detected but appears uninteresting to a potential predator, i.e. the object being mimicked is usually inanimate and/or inedible such as a twig, dead leaf or bird dropping. Cott (1940) used the term 'special resemblance' to include a number of examples of what we now call masquerade, and in a lot of the literature it is referred to as mimesis. Some workers make a distinction between cases of masquerade in which the resemblance is to a dominant or common element of the environment, such as leaves (dead or alive), twigs, clumps of moss, fronds of plants or seaweeds, sometimes pebbles (if pebbles are abundant), for which the terms 'cryptic mimesis' and 'element imitators' have often been used (Hailman 1977, pp. 174–76, Pasteur 1982, Toledo & Haddad 2009), and when the discrete item being mimicked is a rare and inanimate thing, such as a bird dropping or a pebble in a place where pebbles are not particularly abundant. Here the terms 'phaneric mimesis' and 'object imitators' have been used (Pasteur 1982, Skelhorn & Ruxton 2010). With the pebble example it is clear that the distinction between the two imitator types is not always going to be simple, and I do not think this distinction is especially helpful, though Skelhorn et al. (2010c) point out that object imitators (e.g. bird dropping mimics) are likely to be misclassified over a wider range of backgrounds compared with element imitators. A consequence of this is that element imitators are likely to obtain additional advantage from crypsis; a leaf-like insect on a leafy plant will often merge into the general background of foliage unless it draws attention to itself by moving in a strange way.

Skelhorn et al. (2010c) point out some small problems with Endler's (1981) definition of masquerade in that when the masquerader is phytophagous and feeds on its plant model, then it will actually have some demographic effect upon on the latter – an outbreak of stick insects can defoliate and kill many trees. Alternatively, if the masquerader is an entomophagous predator that, for example, resembles a flower, such as the mantids in Fig. 10.6, then depending upon whether they eat pollinators or herbivores they too will have effects on the model's fitness. Skelhorn et al. therefore propose a modification of Endler's definition to allow inclusion of stick/leaf insects, flower mantids, etc., within the category of masquerade, because, according to Endler's definition, they would be considered as Batesian mimics even though this is not the context in which that term is usually employed. Specifically, Skelhorn et al. (2010c) make the distinction that Batesian mimics influence their model's population dynamics/survival indirectly through the action of a mutual predator, whereas stick insects and mantids do so more directly, either by consuming the plant or by killing its pollinators or other herbivores.

CLASSIC EXAMPLES

Twigs as models

Many caterpillars, and, of course, stick insects, bear a close resemblance to twigs or narrow stems. Among the Lepidoptera, larvae of the family Geometridae are often exceptionally well adapted to resemble a twig and adopt a twig-like posture when at rest. The family is largely characterised by a reduction in the number of prolegs (the fleshy abdominal ones rather than the true legs on the three thoracic segments) with only two pairs remaining, one on the 6th abdominal segment and the anal claspers on the 10th (see Fig. 2.27a,b). This results in a distinctive gait which gives them their common names, loopers or inchworms, depending upon which part of the world you come from. Most twig-mimicking caterpillars are primarily nocturnal feeders, presumably because in the daytime any movement would reveal them for what they are. Stick-mimicking caterpillars often possess refinements in addition to colour, shape and texture, such as marks mimicking leaf-scars or lenticels (Heinrich 1993). Twig masquerade is also known in some snakes, notably the Madagascan *Mimophis*; as with geometrid caterpillars, *Mimophis* stick their head and neck out at a sharp angle and can remain motionless in this posture for a very long time (Mertens 1960).

Higginson et al. (2012) found that twig masquerade by temperate caterpillars was especially common among species that overwinter as larvae, whereas species with cryptic caterpillars mostly overwintered as pupae, which makes sense since typical cryptic species, i.e. ones with green caterpillars, might be too visible against a largely grey-brown background of deciduous trees in winter. Further, masquerading species were generally larger, which probably fits with the size of their models – there are very few diminutive woody twigs. Their findings also serve to emphasise the importance of distinguishing crypsis from masquerade, as they can have quite different ecological and evolutionary implications.

On a larger scale, some birds that rest during the daytime resemble small broken branches, notable the Neotropical potoos, *Nyctibius* spp. (Nyctibiidae) (see Fig. 2.3) (Solano-Ugalde 2011), the Australian tawny frogmouth, *Podargus strigoides* (Podargidae), and Eurasian wrynecks, *Jynx* spp. (Picidae).

Leaves (alive or dead) as models

When it comes to resembling leaves, the distinction between masquerade and crypsis may depend on the precise location of the animal and also upon the distance from which it is viewed. Reasonably good camouflage against a green leaf background may be achieved from a distance simply by being an appropriate shade of green, as in chameleon in Fig. 2.5a, and can be enhanced by having an approximately similar shape to the leaves around (see Fig. 2.5b), but natural selection has led many insects to resemble living leaves far more accurately, with convincing analogues to shape, such as midribs with diverging veins (see Fig. 2.5d). In the moist tropics, where leaves may persist for years and often display signs of fungal infection, infection-mimicking markings are also frequently present (see Figs 2.5c,e). A similar spectrum can be seen in the case of dead leaf mimicry. Figure 3.1a,b, for example, shows two forest toads from Panama that live among the leaf litter but that, apart from mid-dorsal markings which might partly give the illusion of the midrib of a leaf, show no remarkable leaf-like resemblance. On the other hand bush crickets (Tettigoniidae) (Fig. 3.1c,d) have clearly evolved a strong resemblance to particular dead leaf shapes, colours and markings and so are clearly masquerading (some noteworthy examples include, *Mimetica* spp., *Orophus tesselatus*, *Pterochroza ocellata*,[1] *Systella rafflesii* and *Typophyllum* spp.).[2] Often the shape of a dead leaf mimic (e.g. *Typophyllum bolivari*) mimics a leaf that has been partly consumed by another insect. Bush crickets that mimic green leaves often have markings resembling patches of surface fungus or 'windows' resembling areas that have been leaf-mined or been eaten by a beetle.

The moth shown in Fig. 3.1e, for example, has evolved a resting shape very like a dead leaf; its head and the distinction between left and right wings are effectively invisible from above unless studied very closely, and in addition to a conspicuous midrib there are some fine dark lines reminiscent of lateral leaf veins, though they run in the same direction on either side of the midrib. The remarkable Indian leaf butterfly, *Kallima inachus* (Fig. 3.1f) mimics a dead leaf with its wing undersides when at rest, and has evolved conspicuous leaf vein markings that converge on the midrib line. Suzuki (2013) showed that the way in which Lepidoptera wing patterns are largely based on a standard groundplan[3] (Nijhout 1990, 1991, Nijhout et al. 1990), such that different elements can evolve to some extent independently or be combined in different groups called modules, has facilitated evolution of accurate dead leaf mimicry in divergent groups. As Nijhout (1994) points out, the compartmentalisation of pattern elements in butterfly wings due to their

developmental pathway means that small genetic changes may have quite profound phenotypic effects. In the case of leaf butterflies, Suzuki et al. (2014), in a beautiful paper, explored how the remarkable leaf mimicry of the *Kallima* species and closely related genera evolved using a comparative phylogenetic and morphological analysis, and showed how the pattern evolved gradually from non-mimetic ancestral nymphalid ground state, with different elements evolving independently. That *Kallima*'s accurate masquerade was shown to have been selected for gradually indicates Darwin's and Wallace's views and refutes R.B. Goldschmidt's (1945) idea that its accurate leaf mimicry came about through salutatory evolution, i.e. in one large jump. *Kallima* is also represented in Africa, whereas, in the neotropics there are quite a few other nymphalid leaf wing butterflies in several tribes that have independently evolved very similar leaf shape and colour to *Kallima*, including *Fountainea nobilis*, *Zaretis itys* and species of *Anaea*, *Coenophlebia*, *Historis*, *Memphis* and *Siderone*, and in Australia there is the leafwing, *Doleschallia bisaltide*.

Bird dropping resemblances

A number of insects and arachnids are good masqueraders of bird droppings. Many caterpillars (Fig. 3.2d,e), quite a few adult moths (Fig. 3.3), some beetles, at least one orthopteran (*Sathrophyllia* sp., Tettigoniidae), a frog (Keeton & Gould 1986 cited in Starrett 1993) and the nest of a potter wasp (Auko et al. 2015) are also effective bird dropping mimics. The European Chinese character moth, *Cilix glaucata*, and garden carpet moth, *Xanthorhoe fluctuata* (Boardman et al. 1974), and the North American pearly wood nymph moth, *Eudryas unio*, and the Arizona bird-dropping moth, *Ponometia elegantula* (both Noctuidae), are often-cited examples, though there are numerous others. Similarly, the early instar larvae of various caterpillars, such as the European alder moth, *Acronicta alni*, and those of many *Papilio* butterfly species (Fig. 3.2f), notably those of *P. cresphontes*, *glaucus*, *polydamas*, *polymnestor*, *polytes*, *rutulus* and *Troilus* (Minno & Emmel 1992), are dropping-like, and the early larva of *P. palimedes* is dropping-like but also has a well-developed eyespot up close. It seems that for caterpillars, bird dropping mimicry only works for small species, so in larger species it is only the earliest instars that are dropping mimics and they undergo a marked transformation in later instars. The final instar of the alder moth quite closely resembles the caterpillar shown in Fig. 4.3d, which is not remotely dropping-like. Interestingly, alder moth caterpillars are unpalatable to birds at all developmental stages but the bird dropping mimicry would not be effective in the larger active caterpillars when they are searching for pupation sites – hence the

---

1. When disturbed, this species reveals its hind wings which have bold eyespots.
2. Quite a lot of these also have green colour morphs which resemble green leaves rather accurately too.
3. The best worked out version is the nymphalid ground plan (NGP) (Nijhout 1990, 1991) but it can also be applied to wing pattern elements on many other butterflies and moths.

Fig. 3.1 Dead leaf mimicry and leaf-litter camouflage in various animals. (a,b) Two species of leaf-litter toads from Panama showing background matching and disruptive colouration; (c,d) two unidentified species of dead leaf mimic bush crickets (katydid) from Costa Rica; (e) an unidentified leaf-mimicking moth from Costa Rica displaying Oudemans' phenomenon, in which marking on separate structures align to form a continuum; (f) Indian leaf butterfly, *Kallima inachus* (Nymphalidae), showing leaf mimetic wing undersides and brightly coloured uppersides. (Source: a,b, Chris Jeffs. Reproduced with permission from Chris Jeffs; c–e, Kenji Nishida. Reproduced with permission from Kenji Nishida; f, Sarefo 2007. Reproduced under the terms of the Creative Commons Attribution Share-Alike Licence CC BY-SA 3.0, via Wikimedia Commons.)

Fig. 3.2 Examples of bird dropping masquerade mimicry. (a) *Phrynarachne* cf *ceylonica* from Cat Tin N.P., Vietnam; (b) *Phrynarachne* sp. (Thomisidae); (c) *Eriovixia laglaisei* (Araneidae); (d) caterpillar of *Phauda* sp. (Zygaenidae) from China; (e) unidentified Chinese moth caterpillar; (f) unidentified papilionid butterfly caterpillar, probably *Papilio cresphontes*, Thailand. (Source: b–e, Itchydogimages. Reproduced with permission from John Horstman.)

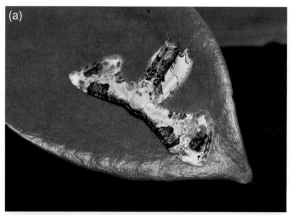

Fig. 3.3 Two adult moths masquerading as bird droppings. (a) *Oroplema plagifera* (Uraniidae); (b) the Asian geometrid moth, *Macrocilix maia*, apparently showing a whole bird dropping-related scene on its wings, the white and black hind wing markings and surrounding orange-brown representing a bird dropping, and each fore wing with what appears to be a fly feeding on the dropping. (Source: Itchydogimages. Reproduced with permission from John Horstman.)

Fig. 3.4 Female *Argiope* sp. (Araneidae) sitting in the middle of silk stabilimentum pattern in her web, Vietnam.

evolution of an ontogenetic switch in defensive strategy (Valkonen et al. 2014).

The Asian geometrid moth, *Macrocilix maia*, is an example of bird dropping mimicry *par excellence*, with its wings clearly appearing to show two flies feeding on a patch of excrement (Fig. 3.3b). A very similar double fly mimicry is present on the otherwise translucent wings of the fly *Goniurellia tridens* (Tephritidae) from the Middle East, though the markings might represent ants, spiders or flies and so it might be more appropriately a case of predator mimicry (see Chapter 7, section *Batesian–Poultonian (predator) mimicry*). Although bird droppings *per se* are not uncommon, when I first went to Africa I was particularly keen to discover some impressive bird dropping insect mimics and for many days I investigated every dropping-like entity that I saw – they were all the real thing.

Among spiders, notable examples are found in the genera *Pasilobus* and *Celaenia* (both Araneidae), and *Phrynarachne*[4] (Thomisidae). Some bird dropping mimicking *Phrynarachne* not only have bodies that have the texture and colour of a bird's seed-rich dropping (Fig. 3.2a,b), they can also smell very foul (Gray 1991). Some *Phrynarachne* species, e.g. *P. ceylonica*, *P. rothschildi* and relatives, have markedly contrasting forelegs (Fig. 3.2a,b) whose function is not known. It seems likely, especially given their smell, that they are combining a protective and an aggressive mimicry, and, supposing that a prey insect lands in front of them, their positioning towards the apex of a leaf will provide a great opportunity for prey capture (see also *Leistotrophus* in Chapter 10, section *Death feigning as a lure*).

---

4. An amusing generic synonym of *Phrynarachne* is *Ornithoscatoides*, which translates as 'like a bird dropping'.

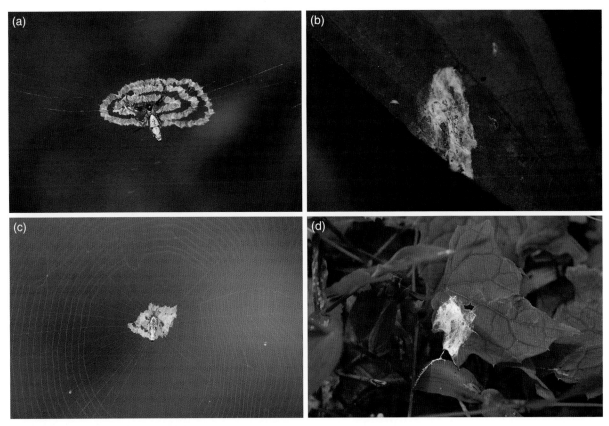

Fig. 3.5 Photographs of *Cyclosa* spiders in their webs with bird dropping mimicking silken weaving disguising the spider (a,c) and very similar bird poo (b,d). (Source: Professor I.-M. Tso. Reproduced with permission from Professor Tso.)

Spider web stabilimenta

The name stabilimentum is generally used in a broad sense to include various structures, sometimes of plant or animal origin, or spider egg sacs, usually placed near the centre of a spider's orb; Wickler (1968) called them 'pseudo-platforms'. Sometimes they are elaborate structures, the best known of these being the silk patterns made by the cross spiders, *Argiope* spp. (Araneidae) (Fig. 3.4) and members of the related genus *Neogea*. Egg sac stabilimenta and detritus stabilimenta are largely assumed to provide camouflage for the spider (Wickler 1968: p. 57, Rovner 1976, J. Edmunds 1986), although M.H. Robinson & Robinson (1970) argued, based on observations of 2500 *Argiope argentata* webs, that the elaborate structures of these have a purely mechanical function and are only created as a final adjustment when necessary. I think that this is unlikely to be the whole explanation, however, as most other spiders that make large orb webs do not seem to require such structures, and when stabilimenta are made, the female spider often rests against them.

W.G. Eberhard (2003) suggests that silk stabilimenta in the spider *Allocyclosa* act as camouflage devices for egg sacs and there seems to be no reason why they could not serve multiple functions. In the case of *Cyclosa mulmeinensis* from South-East Asia, the spider wraps the remains of previously consumed prey in bundles of similar size to the spider's own body and fixes them in the pseudo-platforms, adding to the deception (Tan & Li 2009). Nakata (2009) explored factors affecting stabilimenta construction by *C. argentoalba* and found that it was significantly increased when spiders were tricked by a tuning fork to perceive a higher risk of predation, but that it was unaffected by prey availability. Liu et al. (2014), however, provide evidence that *Cyclosa* spider stabilimenta are actually bird dropping mimics that serve to camouflage the spider (Fig. 3.5).

Tubeworms, etc.

While bird droppings, twigs and thorns may be typical uninteresting but readily definable items to predators and thus appropriate for masquerade on land, the sea and other aquatic systems have their own equivalents. Thus Reimchen (1989) demonstrated that some juveniles of the North Atlantic winkle, *Littorina mariae* (Littorinidae), possess a spiral white stripe that gives them a strong resemblance to the calcareous tubes of the polychaete worm genus *Spirorbis*, which occur in large numbers on the wrack seaweed, *Fucus serratus*, on which the winkles also live. The likely mimetic resemblance was demonstrated using blennies (*Blennius pholis*) as test predators. These fish exerted a lower level of predation on the white-striped (and hence tubeworm like) morphs of *L. mariae* than on the other morphs (uniform yellow or reticulate), apparently especially when *Spirorbis* were also present. As *Spirorbis* tubes might contain living worms and would therefore provide food for a predatory fish, one might wonder why the fish should prefer the snails. Reimchen (1989) suggested that one reason could be that the reward from a *Spirorbis* tube might be low compared with a snail because removal of the former from its foothold would require expenditure of considerable time and effort, and many *Spirorbis* tubes could be empty and therefore offer no reward at all.

EXPERIMENTAL TESTS OF SURVIVAL VALUE OF MASQUERADE

Superficially, the degree of similarity of many animals to the objects they mimic can be very impressive, and such masquerade can be highly effective. Regarding stick-like caterpillars, de Ruiter (1952) comments that "it is very unlikely that a jay or [sic] a chaffinch will peck at a resting stick caterpillar if chance brings it near one, because it has the experience that stick-like objects are inedible." Masquerade may have evolved to protect an animal from predators or to enable it to avoid being recognised as a danger by its prey.

As Skelhorn et al. (2010b) pointed out, masquerade has not been extensively investigated experimentally and doing so is not that straightforward: "It is difficult to determine whether a predator has detected and misidentified an individual (that is, masquerade) or whether it has simply failed to detect the prey item (which would be crypsis)".

De Ruiter (1952) was the first person to attempt an experimental test of the benefit of masquerade by presenting jays and chaffinches with twigs and stick-like caterpillars of three species of geometrid moth – the oak beauty, *Biston strataria*, *B. hirtaria* and the candy-shouldered thorn,

Ennomos alniaria. The birds initially ignored twigs with which they were familiar but, having accidentally discovered that some of the 'twigs' were in fact caterpillars, the birds then started pecking at all similar objects until they again became 'discouraged' by having encountered a succession of real twigs. De Ruiter (1956) went on to compare the nature of countershading in cryptic caterpillars of 17 British moths that rested either on top of or below a branch or twig and found that whether they showed normal (dorsal darker) or reverse (dorsal paler) countershading was perfectly correlated with their resting position. Their resting positions could even be manipulated by shining a bright light on them from below.

Neither of these studies is particularly easy to interpret because of an absence of controls. Therefore to test the hypothesis that twig-like caterpillars gain some protection from bird predation through their appearance, Skelhorn & Ruxton (2010) and Skelhorn et al. (2010c) carried out two neat experiments using domestic chicks as predators. As real prey Skelhorn et al. (2010c) used the twig-like caterpillars of two geometrid moths, the brimstone, *Opisthograptis luteolata*, and the early thorn, *Selenia dentaria*, both of which commonly feed on the shrub hawthorn (*Crataegus* spp.). The birds were presented either with caterpillars in isolation, caterpillars together with unmodified hawthorn sprigs, or caterpillars with hawthorn sprigs that had been disguised by finely binding them with purple cotton. The brimstone moth caterpillars were found to be slightly more protected than those of the early thorn, but the main result was that it took the birds significantly longer to make their first peck at the prey or target (a twig) when they were presented together with natural hawthorn branches than when presented with modified hawthorn, and the latter was significantly longer than when prey/targets were presented alone (Fig. 3.6). The treatments also significantly affected handling time of the target (Fig. 3.7), with birds taking longer to handle targets when presented in a more natural context. The birds look longer to attack twigs than caterpillars whether presented singly or in pairs, and longer to attack single caterpillars than a caterpillar when presented together with a twig (Fig. 3.8), showing that they could distinguish the stick caterpillars from twigs but did so easier if they could make a direct comparison between the two.

In a variant of the above experiment, Skelhorn & Ruxton (2010) tested whether the presence of twig models affected the abilities of chicks to mistake a twig-like caterpillar for a twig by presenting the birds, in the presence of an unmanipulated hawthorn branch or a manipulated one, with various combinations of one or two caterpillars, one or two twigs or a twig and a caterpillar. When manipulated branches were used there was little difference in the latency

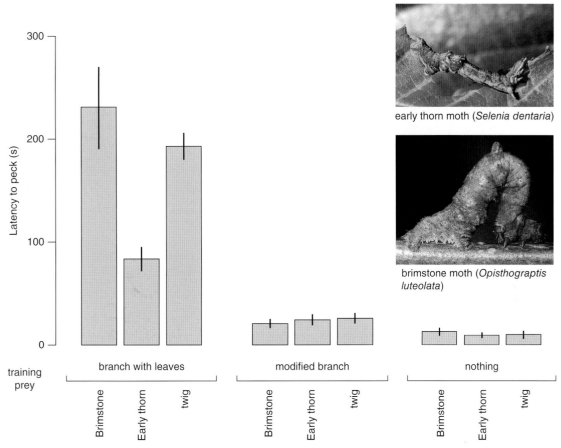

early thorn moth (*Selenia dentaria*)

brimstone moth (*Opisthograptis luteolata*)

Fig. 3.6 Experimental results demonstrating the protective benefit (latency to attack) of masquerade in the presence of the model, in this case by twig-like caterpillars of brimstone or early thorn moths in the presence of twigs on branches, manipulated branches which were wrapped in purple cotton, and presented alone (means ± S.E.). Predators (domestic chicks) took significantly longer to peck at the caterpillars or twig targets when presented with unmodified food plant (hawthorn) branches (Kruskal–Wallis test; $\chi^2 = 47.39$, $p < 0.001$, d.f. = 1). (Sources: Adapted from Skelhorn et al. 2010c. Reproduced with permission from AAAS. Image of early thorn larva, Gyorgy Csoka, Hungary Forest Research Institute 2008. Reproduced under the terms of the Creative Commons Attribution Licence, CC BY 3.0, via www.bugwood.org. Image of brimstone moth larva, Soebe, 2005. Reproduced under the terms of the Creative Commons Attribution Share-Alike Licence CC BY-SA 3.0, via Wikimedia Commons.)

to attack either twigs or caterpillars in any combination – the birds quickly identified the targets and pecked at them. However, with natural branches the trial results fell roughly into four groups; for unmanipulated branches, mean latencies to peck were greatest when chicks were presented only with twigs, but significantly lower when they were presented with just caterpillars and lower again when presented with one twig and one caterpillar. This showed that the birds were better able to recognise a caterpillar as being a suitable prey if they could compare it to its model. In contrast with manipulated branches, latencies to peck at all combinations of twigs and caterpillars were effectively the

same and significantly shorter than for twigs or caterpillars on the unmanipulated branches.

Not surprisingly, Skelhorn et al. (2011) found that the density of twig-like caterpillars affected the level of protection that the deception gave. The greater the relative abundance of model twigs, the less likely predators were to expend effort in trying to detect masquerading caterpillars, and the more difficult the predators found it to detect masquerading prey. Different types of masquerade are likely to be affected differently because, in some cases, such as with twig-like caterpillars among twigs, there will be numerous real twigs (models) nearby, and the predators will quickly

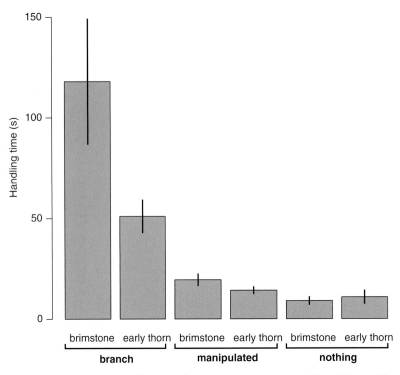

Fig. 3.7 Handling times (means ± S.E., $n = 8$) for twig-like caterpillars or target twigs were significantly longer (Kruskal–Wallis test: $\chi^2 = 31.16$, $p < 0.001$, d.f. = 1) when presented together with unmodified food plant branches than with modified food plant or with nothing (see legend to Fig. 3.6 for further details). (Source: Adapted from Skelhorn et al. 2010c. Reproduced with permission from AAAS.)

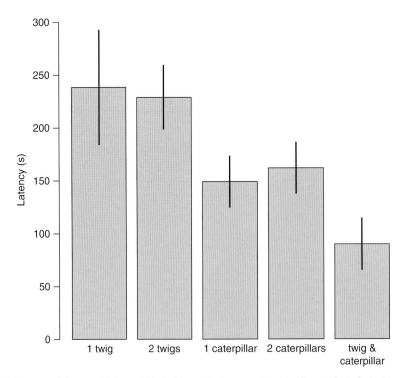

Fig. 3.8 Latency of chicks to attack (means ± S.E., $n = 8$) twig-like, early thorn moth caterpillars (*Selenia dentaria*) or target twigs when presented individually or in pairs. Birds took significantly longer to attack twigs than caterpillars (Kruskal–Wallis test: $\chi^2 = 5.29$, d.f. = 1, $p < 0.021$) when presented in pairs, and they took significantly longer to attack caterpillars when presented alone than when one caterpillar was presented together with one twig (Kruskal–Wallis test: $\chi^2 = 9.57$, d.f. = 1, $p < 0.002$) (see legend to Fig. 3.6 for further details). (Source: Adapted from Skelhorn & Ruxton 2010.)

learn that pecking at twig-like objects has a low success rate. However, for masqueraders such as bird dropping mimics, whose models will be relatively rare (unless they are on a tree that acts as a major roosting site for birds) predator attacks might have a higher success rate.

Skelhorn & Ruxton (2013) demonstrated that the effectiveness of twig masquerade by early thorn caterpillars depends not only on the appearance but also the size of the masquerader relative to the twigs being mimicked, and this assumes greater importance if predators (domestic chicks) have been trained to recognise that size is a useful cue to edibility. The caterpillars, on the other hand, were shown to minimise predation risk by selecting microhabitats where there was a lower level of size disparity between themselves and the twigs they masqueraded (Skelhorn et al. 2010a, Skelhorn & Ruxton 2013).

ONTOGENETIC CHANGES AND MASQUERADE

Of course, the *Spirorbis* tubes described above are too small to be models for older larger snails, so after reaching a certain size the snails develop different shell patterns. Such changes in the nature of what is being mimicked are common and often reflect the differences in the scale of the substrate. Hacker & Madin (1991) show, for example, that members of two genera of pelagic shrimp that live among *Sargassum* floating seaweed[5] change as they grow from being unicolourous and resembling pieces of the seaweed, such as frond or bladder pieces, to being disruptively coloured. This is effectively a shift from masquerade to crypsis. These ontogenetic shifts are often referred to as transformational mimicry. Pasteur (1982) illustrates the problem in reference to sturgeon fish (Acipenseridae), whose young look like drifting twigs, hardly something possible for the adults which are often between 2 and 5 m long.

THANATOSIS (DEATH FEIGNING)

Many animals feign death when threatened, probably because most predators only eat live prey; this is generally called thanatosis (from the Greek word for death) though the actual state is also called catalepsis or 'tonic immobility' by some. In many treatments of mimicry it is treated in

some *ad hoc* way, though as Pasteur (1982) points out, this is a sort of self-mimesis, making the animal appear uninteresting to the potential predator.

Thanatosis usually involves tonic immobility, with the animal remaining still and unresponsive to external stimuli. This type of automimicry is distinct from remaining still to avoid detection because it occurs after the prey has been detected and often after it has been attacked (Ruxton et al. 2004a). This mimicry could act in more than one way depending upon the exact biologies of the mimic and its predator. For example, sometimes the animal simply stays very still, and therefore may render itself difficult to detect for predators that rely to a large extent on movement. This sort of simple immobility is exhibited by many insects and other terrestrial arthropods. Often the potential prey animal may drop from its substrate and remain motionless where it falls; if the predator did not observe exactly where it fell, the prey may be exceedingly difficult to spot, as any field entomologist can testify.

In other, more sophisticated forms of thanatosis, the mimic may assume a posture that is much more obvious and will be clearly visible to a predator. Among vertebrates, classic examples include the European grass snake, *Natrix natrix*, and dice snake, *N. tesselata* (Colubridae) (Fig. 3.9f) and various other reptiles and amphibians (e.g. Holmes 1906, dos Santos et al. 2010, Mirza et al. 2011, Sannolo et al. 2014), which will lie still in a contorted, upside-down position that looks very much like a dead individual (Fig. 3.9a–c). Of course we get the expression 'playing possum' from the thanatosis exhibited by the North American possum (or opossum), *Didelphis virginiensis* (Francq 1969). Similar behaviour is also exhibited by at least some gundies (Ctenodactylidae) (Grenot 1973) and the African ground squirrel, *Xerus erythropus* (Ewer 1966), both of which can hold their breath for over a minute, thus hiding the slightest visible sign of being alive. Even some birds will play possum in the face of predators (Sargeant & Eberhardt 1975), and this has even been observed to occur sometimes during cock fights (Harold Herzog quoted in Milius 2006).

Toledo et al. (2010) surveyed information on a large number of anuran (frog and toad) species and found that more than a single type of behaviour was being lumped under the term thanatosis. Some of the frogs exhibited typical thanatosis (Fig. 3.9a–c), whereas others showed a behaviour he termed 'shrinking' (Fig. 3.9d,e) but which is called, more appropriately, 'puffing' by Escobar-Lasso & González-Duran (2012). Toledo et al. (2010) found that whichever one of these behaviours a species displayed, it was closely associated with the animal's toxicity. Species that displayed thanatosis were mostly non-toxic and therefore likely to be advertising that they are dead and unpalatable, but the toxic shrinkers/puffers might be using an *aide mémoire*

---

5. See also pp. 20–26 in J. Diamond & Bond (2013) for more information of the early discovery of crypsis by animals living among *Sargassum* by the eighteenth-century explorer Peter Osbeck; also photographs of some examples.

Fig. 3.9 Examples of thanatosis (a–c,f) and 'shrinking' behaviour (d,e) in frogs, salamander and snake. (a) *Scinax fuscomarginatus*; (b) *Ischnocnema guentheri*; (c) *Neurergus kaiseri*; (d) *Bokermannohyla luctuosa*; (e) *Rhinella icterica* (=*Chaunus ictericus*); (f) Iranian grass snake, *Natrix tesselata*. (Source: a–e, Luis Felipe Toledo. Reproduced with permission from Luis Felipe Toledo; c,f, Omid Mozaffari. Reproduced with permission from Omid Mozaffari.)

posture. Sometimes the postures reveal aposematic markings on the undersides, and when the tongue is extruded, it is often also aposematically marked. In some species the thanatotic posture is accompanied by the release of an unpleasant odour, for example, *Leptopelis rufus* (Arthroleptidae) displays a typical thanatosis posture with an open mouth from which a strong ammonia-like smell is emitted. As noted by Toledo et al. (2010), further study is needed to ascertain how effective each behaviour is as a defensive strategy. The effect of immobility is also enhanced spectacularly in some snakes: dwarf boas (Tropidophiidae) autohaemorrhage from the eyes and mouth and hognosed snakes, *Heterodon platyrhinos* (Colubridae), will release foul-smelling faecal matter and sometimes bleed from the cloaca (Burghardt & Greene 1988, D. Gower et al. 2012).

Thanatosis is well known among various groups of insect, including various beetles (Chemsak & Linsley 1970, Prohammer & Wade 1981, A.A. Allen 1990, Oliver 1996, Acheampong & Mitchell 1997, Miyatake 2001), especially ones that are not particularly good fliers, Odonata (Abbott 1926), Orthoptera (Nishino & Sakai 1996), Hemiptera (Holmes 1906), Hymenoptera (King & Leaich 2006) and even some Lepidoptera, notably unpalatable butterflies (Dudley 1989, Larsen 1991) but also quite a few moths (D.L. Evans 1983). It is perhaps particularly well known and studied in stick insects (Phasmatoda). Ulf Carlberg has made an extensive set of studies on thanatosis and other types of defence shown by stick insects (Carlberg 1980, 1981a, 1981b, 1983, 1985, 1985b, 1985c, 1986a, 1986b), though in experiments with caged rats as predators and a palatable species of phasmid as prey it was found only to provide weak protection (Carlberg 1986b).

In a two-way selection experiment, Miyatake et al. (2004) investigated the heritability of thanatosis duration in the red flour beetle, *Tribolium castaneum* (Tenebrionidae). The trait was easily selected for and there was also a correlated response in the frequency of death feigning in the artificially selected lines. They then went on to test whether the thanatotic behaviour provided a selective advantage by subjecting members of the selected lines to predation by females of the Adanson jumping spider, *Hasarius adansoni* (Salticidae), and found that survivorship was significantly higher for beetles from the lines selected for increased duration of death feigning.

In the spider *Pisaura mirabilis* (Pisauridae), only males display thanatosis and they do so as part of their courtship, which involves presenting a nuptial gift to a potentially cannibalistic mate. It might seem reasonable, then, to conclude that thanatosis serves to protect them from female aggression (Bilde et al. 2006). Spinner Hansen et al. (2008) tested the prediction that if male thanatosis has truly evolved as an anti-predation strategy then the males should exhibit thanatosis more when they are threatened by female aggression and therefore more vulnerable. In their experiments, they approximated the aggressiveness of individual females by observing their interactions with prey, and rendered males more vulnerable by removing one of their legs. However, they found no evidence that the probability of a male displaying thanatosis was influenced either by the aggressiveness of the female or by the male's vulnerability. Instead they found that the thanatosis increased both the probability of the male gaining a copulation and the duration of the copulation, and so the behaviour might have been selected for directly. This then raises the question as to why there is so much variation among males in their propensity to feign death, even though all were capable of doing so. The authors suggest that there is a potential cost associated with thanatosis and therefore some balancing selection is in operation.

Thanatosis, being a very defined behaviour pattern, lends itself to neurophysiological investigation, though there have been few studies to date. Neurophysiological correlates of thanatosis have been investigated in the common stick insect, *Carausius morosus* (Godden 1972, 1974). T.C. Jones et al. (2011) have investigated the actions of circulating biogenic amine levels (also referred to as neurohormones) on thanatosis in an orb-weaving spider *Larinioides cornutus* (Araneidae). They found that elevating their octopamine level significantly shortened the duration of thanatotic behaviour, whereas raised serotonin lengthened it.

An interesting exception to the normal neuronal thanatosis of most species is the 'severe' thanatosis exhibited by some burnet moths when injured. In these moths, tissue damage leads quickly to the enzymatic production of hydrocyanic acid, which not only acts as a strong anti-feedant stimulus to a potential predator, but also leads to anoxia and thus immobility in the moth (Balleto et al. 1987). If the damage incurred and cyanide production are not too great, however, physiological mechanisms will slowly remove the cyanide and the insect can recover.

It should be noted that some reported examples of thanatosis might be due to misinterpretation of a different sort of defence. The static and 'unwieldy' postures that some grasshoppers and beetles, e.g. *Sagra* spp., assume when threatened are more likely to make them difficult to swallow or handle (Honma et al. 2006, Ruxton 2006).

Some cichlid fish 'pretend' to be dead as a lure for scavengers but because they are alert to, and respond to, the potential prey I don't think that this really ought to be termed thanatosis (see Chapter 10, section *Death feigning as a lure*).

Feign or flee? The trade-offs of thanatosis

Among all animals, there is a trade-off between thanatosis and flight. Among snakes, it seems that death feigning is particularly prevalent among young individuals, which fortuitously makes it far easier to study. Gerald & Claussen (2006) and Gerald (2008) examined this behaviour in small, young brown snakes, *Storeria dekayi* (Colubridae), and found that they were more likely to exhibit thanatosis when in water and also, contrary to expectation, when the ambient temperature was higher, despite this meaning that they probably had a potentially greater escape speed. At suboptimal temperatures the snakes appeared to use other defence behaviours.

How long an animal remains in thanatosis is a trade-off between the perceived risk of coming out of it too soon and the benefit of having more time foraging, seeking mates, etc. Kuriwada et al. (2009) explored this in the sweet potato weevil, *Cylas formicarius*, and found that for females the duration of death feigning was reduced in those that had mated and been inseminated. Males also showed reduced thanatosis duration if they had already mated on multiple occasions. This suggests that if members of either sex had already had a chance to pass their genes into the next generation it was worth taking a bit of extra risk. It would be interesting to know whether for females there was any additional effect of having already oviposited.

The readiness of bird species to go into thanatosis has been shown to correlate with the local intensity of predation by gosshawks, *Accipiter gentilis*, and cats (Møller et al. 2011), but it is not clear whether this is due to local learning and assessment of predator type and density or whether it is genetic and reflects past selection. These particular predators are probably the ones most likely to attack and kill only living and mobile prey, and R.K.R Thompson et al. (1981) confirm this for quail and cats.

Other aspects of death mimicry

Kirkpatrick (1957) described an instance analogous to thanatosis in which a Caribbean slug caterpillar larva (Cochlidae) spins a cocoon with four small round holes that give it the appearance of having been previously parasitised by a parasitic wasp. This signal might be effective against birds and also possible parasitic wasps, though the latter might be expected to rely more on chemical cues.

A strange, and most likely erroneous, example referred to by Pasteur (1982) involves a behavioural modification of its host caterpillar by the parasitic braconid wasp *Aleiodes alternator* (as *Rhogas* [sic] *geniculator*). In France, the wasp and its host caterpillars *Arctia* often inhabit the same locations as various *Zygaena* caterpillars, and the latter sometimes get infected with an entomopathogenic fungus (*Isaria* sp.), resulting in them being ignored by bird predators. Before killing the *Arctia*, the braconid causes its host to climb up a grass stem and then kills it, mummifying it in the same posture as the fungus-infected *Zygaena*. This was interpreted by Giard (1894) as a case of protective mimicry and was one of the examples used by Pasteur (1982) to illustrate indirect mimicry. Indeed, as he wrote, such cases are always very specific and thus the concept of indirect mimicry is of little use. However, *A. alternator* attacks many low-feeding hairy caterpillars and, like several other *Aleiodes* that cause hosts to climb up stems before mummifying them, the host ends up facing downwards. Thus I suspect that the resemblance Giard noticed was just a coincidence, and the parasitoid had evolved to modify the caterpillar's behaviour so that its mummy would be formed in a generally favourable situation, maybe warmer or less prone to general predators. However, that does not necessarily mean that in places where there are also fungus-attacked *Zygaena* larvae, the *Aleiodes* doesn't gain some additional indirect mimetic benefit. I previously published a short note regarding some other *Aleiodes* in the UK that cause their hosts similarly to climb up grass stems before mummification, ending up resembling resting syrphid flies (Quicke 1984). At the time I thought that these were models that had the potential to escape birds easily because of the escape speed, but I now think that any mimetic benefit to this was probably incidental to other factors, the most important probably being temperature to enable these plurivoltine species to complete their current generation quickly.

SEEDLESS SEEDS AND SEEDLESS FRUIT

Myczko et al. (2015) describe a situation analogous to death mimicry in animals in the Scots pine tree, *Pinus sylvestris*. The trees produce three kinds of 'seed': dark and pale full (viable) ones, as well as empty (seedless) wings. Typical seeds have a dark seed contrasting with the pale buff-coloured wing attachment. These seeds are a rich food resource for birds and are highly predated. The empty seeds are produced in vast quantities and are entirely pale, and seed predators quickly learn that these are worthless. However, the tree also produces full seeds in which the true seed body and the wing attachment are both pale coloured. In experiments with chaffinches, *Fringilla coelebs*, Myczko et al. showed that dark-coloured seeds had the highest probability of predation, and the empty seeds the lowest. When 'seeds' of the three types were presented to the birds in equal numbers, dark ones were predated at almost twice the rate as both the full pale seeds and the empty ones (generalised linear mixed model: $F_{2,2337} = 69.12$, $p < 0.001$, $n = 2340$ seeds and 26 plots), but there was no significant difference in predation probability between the pale true seeds and the empty wings. The pale seed form is therefore mimicking the worthless, non-viable empty seeds which birds learn to attack at a lower rate. Many other plants produce large numbers of empty seeds that look like full ones and there is evidence in some other species that these also lead to reduced predation on true seeds (Fuentes & Schupp 1998). This makes evolutionary sense, especially if the pericarp (flesh) of a fruit is relatively inexpensive to make and of interest only to seed dispersers and not to seed predators.

Chapter 4

APOSEMATISM AND ITS EVOLUTION

When in doubt, wear red.

Bill Blass, American designer (1922–2002)

Mimicry, Crypsis, Masquerade and other Adaptive Resemblances, First Edition. Donald L. J. Quicke.
© 2017 John Wiley & Sons Ltd. Published 2017 by John Wiley & Sons Ltd.

INTRODUCTION

A substantial topic within mimicry, and the one that has attracted by far the most theoretical and experimental interest, concerns the resemblance of palatable species to less palatable ones that are brightly coloured or conspicuous in some other way. However, before this Batesian mimicry can be discussed, it is pertinent to consider how the less palatable models evolved to be conspicuous in the first place. Brightly coloured animals that contrast with their background and are unpalatable are termed aposematic or 'warningly coloured'. Harvey & Paxton (1981), following Järvi et al. (1981a), define an aposematic organism as one that satisfies two criteria:

1. the animal must be unpalatable and
2. it must be easily recognisable by potential predators so that they can learn to avoid it easily.

Järvi et al. (1981a) added to the second of these criteria that its appearance should enable a predator "to avoid [it] after a first encounter". The appearance should indeed be easily recognisable and Harvey & Paxton (1981) point out that this comprises two components:

3a. to be easily learned and
3b. remembered for a long period of time.

Detectability, discriminatory ease and memorability are key features of receiver psychology that help determine the efficiency of any animal signals (Guilford & Dawkins 1991). As such, bright colouration *per se* or contrast with the background are common features of aposematism, though they are not absolutely required (see Chapter 2, section *Dual signals*). As a corollary, one might consider that camouflage patterns should do the opposite, they should not only be hard to spot but also hard to learn and memorise (Troscianko et al. 2013). Guilford (1990c) points out that the two components of aposematism – unpalatability and signal – are likely to result from different types of selection, with the former resulting from either kin or individual selection, and the latter also through synergistic selection. That is, the benefits of being warningly coloured preferentially involve other individuals, not necessarily relatives, that possess the same feature (Leimar & Tuomi 1998).

The visual aposematism of animals may be considered as a primary defence in that it warns a predator from a distance of likely unpalatability (M. Edmunds 1974a, Endler 1988, Tullberg & Hunter 1996, Ruxton et al. 2004a). Guilford (1986) proposed that one of the functions of conspicuousness might be that it reduces the errors that predators might make. In Guilford's scenario, predators were assumed sometimes to make mistakes, failing to recognise that a particular prey was unpalatable, and thus in attacking/consuming it, suffering from exposure to its toxins or other defences. If aposematic prey are detected by predators further away than less aposematic ones, they also give the predator longer to recall past negative experiences, which may, therefore, reduce the risk of a predator making a recognition error in haste and attacking regardless (Gamberale-Stille 2000).

Field tests of aposematic advantage are few and far between. In a very large study of bird predation on a community of temperate forest caterpillars, including exclusion experiments, Lichter-Marck et al. (2014) demonstrated a clear advantage to species that display warning signals, as both aposematic unpalatable species and their mimics experienced significantly less mortality due to bird predation than did camouflaged species. Furthermore, among species that rely on camouflage (crypsis and masquerade), those showing the best camouflage according to human criteria also fared best.

As mentioned above, being aposematic usually, though not always, means displaying vivid colours that differ from the background, and the pattern often has repeated elements such as stripes or spots. These markings also often have high-contrast boundaries, e.g. with narrow pale lines around black spots, and the animals may also possess features that enhance their body outline in contrast to drawing attention away from it, as occurs in cryptic taxa (cf. features generally associated with crypsis; see also *Cerura*) (Stevens 2007). Black and red (Fig. 4.1), and black and yellow (Fig. 4.2, see also Fig. 7.8a–c), black and white or black and cream (Fig. 4.3) are particularly prevalent among insects and some other invertebrates, because green or brown backgrounds are the commonest backgrounds against which aposematic terrestrial animals are located. Physiological constraints and also a large element of nocturnality mean that the aposematic colours mainly used by mammals are black and white (Fig. 4.3e,f, see Fig. 4.36, and section *Warning colouration in mammals*, later).

These common red, orange, yellow, black and white combinations are important because they provide both strong internal colour contrast and colour contrast against the background (Aronsson & Gamberale-Stille 2009). However, contrasting patterns within the prey were found to be less important in avoidance learning by domestic chicks than contrast between the prey and its background. By examining the colours of various ladybirds (Coccinellidae) with an avian visual model, Arenas et al. (2014) found that red and yellow provide a higher contrast against green backgrounds than other colours such as blue, and also this contrast remains strong, and hence stable, throughout the day and under both full daylight and overcast conditions. Herrera (1985) suggests that some aposematic insects could

Fig. 4.1 Prominence of red and black colouration in various aposematic taxa. (a) Ladybird beetles, *Harmonia dimidiata* or possibly the very similar-looking *H. axyridis*, in hibernation aggregation, showing part of range of intraspecific pattern variation; (b) *Drosicha* sp. (Monophlebidae); (c) *Mylabris* sp. (Meloidae); (d) *Graphosoma italicum* (Pentatomidae); (e) *Nicrophorus investigator*; (f) unidentified tipulid fly from China; (g) pink and black millipedes, *Orthomorpha* sp. (Paradoxosomatidae) from Vietnam (identified by Professor Somsak Panha). (Source: a,b,e, Itchydogimages. Reproduced with permission from John Horstman; c,d,f,g, Conrad P.D.T. Gillett. Reproduced with permission from Conrad P.D.T. Gillett.)

Fig. 4.2 Warningly coloured and probably toxic day-flying, Asian moths that may, in some cases, also be wasp mimics, and have large transparent wing patches. (a) Unidentified erebid from China; (b) *Amata ?polymita*, from China; (c) *Amata* sp., from China; (d) *Syntomoides imaon* (Erebidae: Ctenuchini), from China. (Source: a–d, Itchydogimages. Reproduced with permission from John Horstman.)

potentially be mistaken for fruit by frugivorous birds, but notes that most aposematic insects are patterned whereas most fruit are not, and that when aposematic insects are unicolourous they are often metallic blue or green and thus are not fruit-like at all. Some bright-coloured seeds might, on the other hand, be mimicking aposematic insects (see Chapter 14, section *Fruit and seed dispersal by birds*).

In marine environments, aposematic colours may not always be so constrained (see Fig. 4.32) and encrusting marine organisms display a far wider range of colours. Furthermore, both red light and ultraviolet light are strongly absorbed by water so it is not surprising that UV vision is exhibited by only a few species of fish that live close to the surface and have corresponding colour receptors.

In a nice experiment making use of natural colour mutants of the typically red and black patterned, distasteful bug *Pyrrhocoris apterus*, Exnerová et al. (2009) showed that wild-caught individuals of several European insectivorous passerine birds that would probably have had

experience of the unpalatable red form in the wild, avoided wild-type red as well as orange mutant bugs but did not avoid white mutants or brown-painted cryptic ones. Response to the yellow and black mutants varied between bird species, perhaps because yellow and black is a frequently encountered warning colour combination in some, but not all, situations. With the same bug, Svádová et al. (2009) demonstrated that naïve great tits, *Parus major*, attacked the firebug mutants and wild-type red and black ones, irrespective of colour. However, the birds generalised their avoidance differentially, depending on what colour morph they had been trained upon. Thus birds trained on red firebugs "did not generalise their experience to yellow or white mutants whereas birds that learned to avoid yellow mutants generalised their experience to red firebugs".

In an elegant experiment with domestic chicks which involved training them with palatable and unpalatable food crumbs presented in blue or red paper cones on blue or red

Fig. 4.3 Common black and white-striped aposematism (and mimicry) in Vietnam and Thailand. (a) *Bipalium rauchi* (Platyhelminthes: Turbellaria); (b) caterpillar of *Tirumala* cf. *gautama* (Nymphalidae: Danainae) (identified by Keith Wolfe), Vietnam; (c) caterpillar of *Asota plana* (Noctuidae), China; (d) caterpillar of *Tinolius* sp. (Noctuidae) (identified by HM Yeshwanth); (e) highly venomous juvenile Malayan krait, *Bungarus candidus* (Elapidae) from Thailand; (f) non-venomous wolf snake, *Lycodon ophiophagus* (Colubridae) from same locality as previous figure. (Source: c, Itchydogimages. Reproduced with permission from John Horstman.)

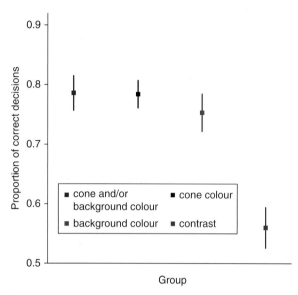

Fig. 4.4 Demonstration that prey colour is more important than contrast in learning to discriminate unpalatable prey. Palatable or unpalatable food crumbs were placed in red or blue paper cones attached to red or blue cardboard circles, and domestic chicks had to learn to avoid those colour combinations associated with unpalatability. The graph (mean number of correct decisions ± S.E.) shows that they could easily learn to recognise a given colour associated with the noxious food but were hardly able to recognise that colour contrast was the cue. (Source: Adapted from Gamberale-Stille & Guildford 2003. Reproduced with permission from Elsevier.)

backgrounds, Gamberale-Stille & Guilford (2003) showed that error rate in avoiding unpalatable food was far lower if the negative experience was associated with contrasting cone and background colours (Fig. 4.4). Among insects, we generally think that green colouration has evolved to render individuals cryptic, but there can be little doubt that metallic colouration can be aposematic, as in the case of many chemically well-defended metallic green beetles, such as the Spanish fly beetle, *Lytta vesicatoria* (Meloideae), as well as various oedemerids (false blister beetles). Jewel beetles (Buprestidae), although probably most being quite edible, are exceedingly difficult to capture (see section *Escape speed and low profitability*, later) and males of the tropical Old World frog-legged beetles, *Sagra* spp. (Chrysomelidae), aside from being very tough, can make themselves very difficult to handle or swallow by a bird by sticking out their much-enlarged hind legs.

In addition to bright colours, aposematic animals and plants often exhibit some secondary repellent defence that operates during an encounter with a predator, such as the emission of noxious compounds that provide further

evidence of unpalatability which may lead to a predator giving up an attack before the prey has received significant harm (Rothschild 1961), but this is not always the case. However, whereas it is obvious that noxious compounds might deter a predator before it has completed its attack, true olfactory aposematism would involve association of a non-noxious odour with adverse effects. Eisner & Grant (1981) proposed that this might be the main aposematic type of signalling by poisonous plants, many of whose actual dangerous toxins are non-volatile, though there is little evidence to support that idea. Camazine (1985) tested whether the harmless compound octen-3-ol that naturally occurs in some mushrooms could cause the American possum, *Didelphis virginiana*, to avoid foods that smelled of it after a learning trial with fungi food that caused the possum subsequently to become ill due to the presence of the toxin muscimol. Most humans know that after being very sick even a day after eating some foods, they can hardly ever face eating them again, at least not for many years (Palmerino et al. 1980) and the same is true for other mammals (e.g. Ralphs 1997). Camazine (1985) found this was true also of the possums.

INITIAL EVOLUTION OF APOSEMATISM

For years, ever since Darwin, aposematic colouration has proved a really tricky thing for evolutionists to explain, and it has even been used as evidence for group selection (Uyenoyama & Feldman 1980) though this is not a plausible evolutionary mechanism. The problem is simple: even if an organism is unpalatable, how can it benefit by making itself more apparent to predators? Of course, if there is a sufficiently high density of individuals displaying the aposematic signal and honestly signalling their unpalatability, then predators will learn to avoid like-looking individuals after they have attacked and perhaps killed one – though how initially high frequencies may come about is less clear (Ruxton et al. 2007). Furthermore, predators forget and re-sample, even after negative experiences, and thus, as would be the case when a new aposematic mutant first evolves, they may well have forgotten by the time they next encounter an individual with the aposematic mutation. As Yachi & Higashi (1998) put it:

> a rare and conspicuous mutant in a population of unpalatable cryptic prey must overcome a *double disadvantage*: a greater risk of being detected (as a result of being more conspicuous) and of being attacked (because its rarity results in a decreased association with aversion) by a predator. [my emphasis]

However, imagine how a mutant gene that reduces the crypticity of its unpalatable bearer can spread through the population in the first place since it must be rare initially. Sufficient evidence in various forms is now available to show that this apparent paradox does not really pose any problems (Mappes et al. 2005, Marples et al. 2005). Important factors that help a mutant allele for aposematic (or at least more conspicuous) appearance to survive include recessiveness rather than dominance, neophobia (dietary conservatism) of predators, associations of unpalatable experience with other prey, associations of unpalatable experience with place or a given food plant, and gregariousness and close proximity of kin. These are all discussed in more detail below.

J.R.G. Turner et al. (1984) suggested that the "usual evolutionary pathway to full aposematism is predicted to be distastefulness first and warning colour second." Therefore, another way of viewing the problem of evolving aposematism is to consider what benefit would be gained by what is initially a palatable cryptic species in evolving to develop costly secondary defences, such as sequestering or synthesising toxins (Lindstedt et al. 2011). As discussed in the section *Mathematical models and ideas of warning colouration evolution*, crypsis may entail opportunity costs in foraging which aposematism overcomes (Merilaita & Tullberg 2005). What is apparently critical from the point of view of the evolutionary stability of secondary defence if it has a non-zero cost, as revealed by mathematical models, is that if the secondary defence is on a continuum then stable strategies will always be for either all or none of the members of a population evolving it (Broom et al. 2005).

Associations of unpalatable experience with place

Rothschild (1964a, 1985a) put forward the idea that because many unpalatable aposematic insects sequester their toxins from their food plants, then "any highly coloured mutation would most likely be avoided by birds, reinforced by their experience of other aposematic species present in the immediate environment." Gittleman & Harvey (1980) asked why distasteful prey are not cryptic. In order to answer this, they presented 4-day-old chicks with blue- and green-dyed food items (chick food crumbs) against either a blue or a green background, the chick food crumbs having been rendered distasteful by inclusion of a mustard and quinine sulphate mixture. The chicks were therefore presented with two types of distasteful prey, one cryptic, the other not. The results were very clear cut (Fig. 4.5) (see also Harvey et al. 1983). Although more of the conspicuous crumb type were taken initially, the chicks learned that they were distasteful more quickly than they did the cryptic ones.

However, the experiment did not distinguish between two possible mechanisms for the difference:

- the conspicuous type may be easier to learn,
- more of the conspicuous type were eaten more quickly and therefore may have formed a stronger learning stimulus.

Several learning experiments conducted using domestic chicks have shown that the primary feature they attend to is the colour of the 'prey' (Aronsson & Gamberale-Stille 2008). There is a strong suggestion in the literature that protected aposematic taxa should be expected to be larger than equivalent non-aposematic taxa because their aposematic signal will be more prominent on a larger body (Forsman & Merilaita 1999, M. Nilsson & Forsman 2003). This may be the case in dendrobatid poison dart frogs (see Fig. 4.31) (Hagman & Forsman 2003). It is probably also relevant that many Lepidoptera that have aposematic late instar larvae have cryptic or masquerading earlier instars (Sandre et al. 2007). To test this, Remmel & Tammaru (2011) used artificial caterpillars of different sizes and aposematic signal strength, and wild great tits, *Parus major*, as predators. In their experiment they found that strength of the aposematic signal was more important in predator learning than the size of the prey but could not exclude that the size and conspicuousness might have a synergistic effect. Mänd et al. (2007) had earlier found that for artificial caterpillar prey against complex (naturalistic) backgrounds, great tit predation was not related to prey size, whereas it was for conspicuously coloured prey. Using various combinations of three colours, three patterns and three shapes of target with mealworm rewards (Fig. 4.6), Kazemi et al. (2014) found that wild-caught blue tits, *Cyanistes caeruleus* (Paridae), responded primarily to colour as do domestic chicks, but were also capable of learning shapes and patterns to discriminate unrewarding targets from rewarding ones which were associated with mealworm rewards (Fig. 4.7) albeit slower and with pattern perhaps being rather more easily learned than target shape. However, when the birds were presented with any of three different combinations of colour, shape and pattern associated with unrewarding prey targets, no significant difference was found in the birds' abilities to learn any particular pattern (Fig. 4.8).

Mappes et al. (1999) drew attention to an additional issue, that if some aposematic unpalatable individuals are less aposematic than others, they might benefit from proximity to better defended and aposematic neighbours. Using great tits and mealworms they found that the birds readily learned to avoid both groups of aposematic unpalatable prey and singletons, and also avoided mixed groups. Different species of predator may also respond differently – Exnerová et al. (2003) found that wild-caught insectivorous passerines

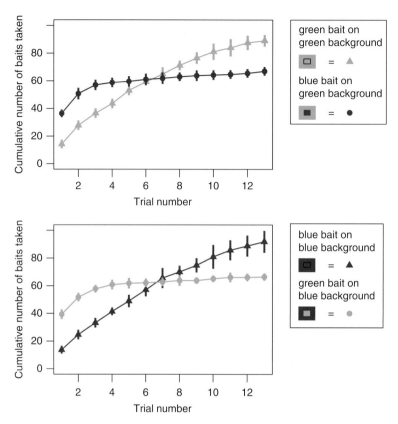

Fig. 4.5 Domestic chicks learning to avoid unpalatable food crumbs that were either colour contrasting or background matching showing that they learn far more quickly when the unpalatable (quinine-treated) crumbs contrast with the background. By performing reciprocal set-ups the authors discount possibility of innate biases. Means ± 95% confidence intervals shown. (Source: Adapted from Gittleman & Harvey 1980. Reproduced with permission from Macmillan.)

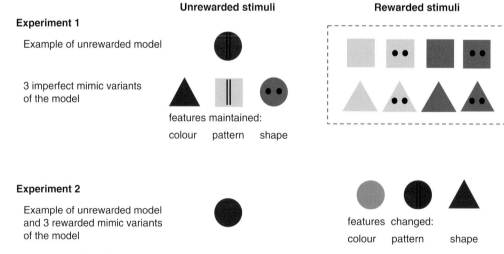

Fig. 4.6 Artificial prey used in two discriminant learning experiments with blue tits (*Cyanistes caeruleus*) as predators; mealworm rewards were placed in wells under the artificial prey targets. (Source: Adapted from Kazemi et al. 2014 with permission from Elsevier.)

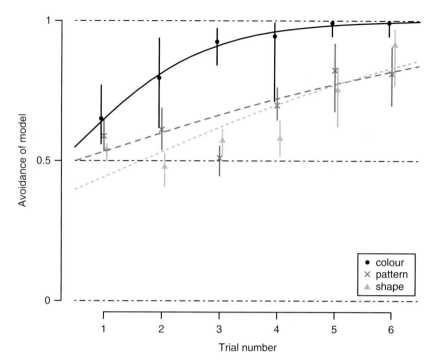

Fig. 4.7 Predator learning curves for avoiding distasteful models where the artificial 'aposematic' signal is based on based on colour, pattern or shape (means ± S.E.). (Source: Adapted from Kazemi et al. 2014 with permission from Elsevier.)

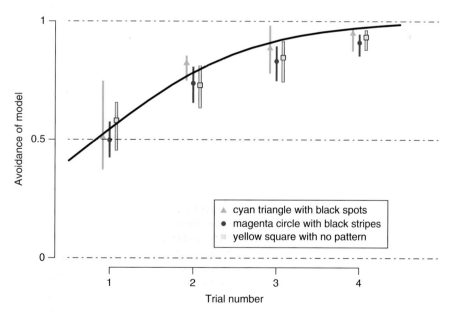

Fig. 4.8 Predator learning curves for avoiding distasteful models where the artificial 'aposematic' signal is based on based on three different colour, pattern and shape combinations (means ± S.E. and fitted logistic regression line). (Source: Adapted from Kazemi et al. 2014 with permission from Elsevier.)

readily distinguished natural aposematic and artificially-made brown forms of the unpalatable firebug, *Pyrrhocoris apterus*, but largely granivorous species did not.

Sword (1999) presented evidence that warning colouration can be density-dependent in the grasshopper/locust, *Schistocerca emarginata* (=*lineata*), with them being cryptic in situations where they are rare but aposematic when they are sufficiently common for predators to learn their pattern easily, such as when they are swarming.

MATHEMATICAL MODELS AND IDEAS OF WARNING COLOURATION EVOLUTION

Kin selection models

Since individuals of an unpalatable species that display mutations making them more conspicuous would seem to be placing themselves at greater risk, it has often been assumed that kin selection is involved, whereby the death of one prey individual leads to avoidance learning by the predator which then does not attack and kill close relatives nearby, such as siblings from the same brood. This mode of selection has been seen by some workers as absolutely necessary for the evolution of aposematism (e.g. R.A. Fisher 1958, Harvey & Greenwood 1978), whereas others (e.g. Järvi et al. 1981b, Sillén-Tullberg & Bryant 1983, Engen et al. 1986), while accepting that it can be important, think that other mechanisms can also work independent of kin selection. Guilford (1985) summarised the general consensus view of 30 years ago as being that both kin and individual selection play their roles and the remaining question was – what was their relative importance?

Some predators that forage in groups may also learn to avoid particular prey by observing the reactions of conspecifics that have attacked one (Harvey et al. 1982, Rothschild 1985b). For example, Rothschild & Ford (1968) found that Eurasian starlings, *Sturnus vulgaris*, often give out warning signals when rejecting an unpalatable food item, and Swynnerton (1915b) noted that caged birds on his veranda in in Chirinda Forest, Southern Rhodesia (now Zimbabwe) would simultaneously begin beak-wiping behaviour when he walked past them with an unpalatable *Danaus* specimen in his forceps. Such indirect learning will have the effect of amplifying the protection afforded to the aposematic organism, and Rothschild & Lane (1960) reported other cases. An additional aspect of some predators' behaviour is that they may avoid attacking other, non-aposematic potential prey that are in close proximity to a previously experienced, aposematic distasteful one. For example, Rothschild (1964b) described how a tame carrion crow, *Corvus corone*, that normally ate cryptic caterpillars refused them if they were offered to it close to an aposematic prey type that the bird had previously learned to reject. As nearby conspecific individuals, such as insects on a plant, will often be from the same brood, such behaviour, if widespread, may further aid the spread of alleles for aposematism that may be recessive.

S.B. Malcolm (1986) discusses the case of the yellow greenfly (*Aphis nerii*: Hemiptera) which commonly forms large colonies on its food plants, such as oleander. When first presented to two very different predators – great tits, *Parus major*, and a spider, *Zygiella x-notata* – many aphids are killed and consumed until the predator starts to recognise their unpalatability. Their conspicuousness thus evolved despite high predator-induced mortality, but being aphids their colonies comprise entirely parthenogenetically produced, genetically identical individuals and thus it seems highly likely that kin selection, through their clones, was involved.

Green beard selection

Guilford (1988) summarised a number of evolutionary scenarios for the evolution of conspicuous colouration in relation to the order of evolution of conspicuousness and unpalability. From his logical approach it can be seen that aposematism *would* be expected if unpalatability and conspicuousness were coevolving traits, and *could* be expected to evolve if the prey had previously evolved unpalatability and subsequently evolved conspicuousness. However, if conspicuousness had evolved first and later unpalatability, this would not be a case of aposematism. Guilford (1988) also discussed what he considered to be the misuse and overuse of the term kin selection in scenarios for the evolution of aposematism in unpalatable prey. He proposed a modification of green beard selection as an alternative. The idea of green beard selection is that an organism displays an honest signal that they are altruistic, for example a green beard. Individuals with green beards may therefore be expected to be likely to reciprocate favours. In the context of the evolution of aposematism, irrespective of how closely related they are to one another, the presence of individuals with the honest aposematic signal serves to help others with the same feature.

Since individuals that display a green beard character are not immune to cheating, such traits are likely to evolve more easily within family groups because the high level of relatedness makes them less likely to be cheats. As the aposematic allele becomes commoner, the importance of family grouping will dwindle (Guilford 1985). If independent mutations give rise to the same green beard (i.e. aposematic) features in other protected individuals, they will further enhance survival overall and benefit from alternative aposematic alleles. Eventually, genetic drift, as well perhaps as slight differences in costs or efficacy, will lead to fixation of one particular aposematic allele.

Family selection models

Harvey et al. (1982) developed a family selection model for the evolution of aposematic colouration in a distasteful organism which examined the survival of individuals and families within a given territory. A predator's territory is assumed to have within it several families of prey such that each aggregated family (brood) are genetically identical and either cryptic or aposematic, but both unpalatable. The predator, on discovering a family, kills some individuals before learning to avoid that type. In their model they consider the proportion of individuals of type i in a population that survive predation through a season, S_i, which is given by

$$S_i = \frac{(nf - (1 - Q_i^n)V_i)}{nf}$$
$$= 1 - (V_i / f)((1 - Q_i^n)/n)$$

where n is the number of families of type i, f is the family size of the prey, V_i is the number of prey of type i that are killed while the predator is learning to avoid similar prey, and Q_i is the probability that a predator does not detect a particular prey family.

If aposematic individuals (and hence their families) are initially rare, and if a territory is taken to contain k families, then in most territories all k families are cryptic and only a few territories will comprise a single aposematic family together with $k - 1$ cryptic ones. For aposematic individuals to increase in number within a territory, their survival in a territory with $k - 1$ cryptic families (i.e. just a single aposematic family) will have to be greater than the survival of cryptic individuals in all cryptic territories. From the above equation this will be if

$$(f - P_a V_a)/f > (kf - (1 - Q_c^k)V_c)/kf$$

On the left is the mean survival of aposematic individuals given the probability of detection and number killed if detected as a proportion of the number in the family to start with. On the right is the equivalent for the $k - 1$ families of cryptic prey. This reduces to

$$1 - Q_c^k > \frac{kP_a V_a}{V_c}$$

In reality, for this condition to be satisfied and aposematism to spread, the right-hand side of the equation must be small, which means that the number of families should be small and/or the number of cryptics killed in an attack (V_c) should be large, relative to aposematics killed (V_a), and the probability of detecting aposematic families (P_a) not much greater than detecting cryptic ones. Extending this to a more realistic case of sexually reproducing prey but with a dominant aposematic allele doubles the right-hand part of the inequality, restricting the conditions necessary for aposematism to spread. If predators learn to avoid aposematic prey more quickly than they do cryptic ones or remember to avoid them for longer (see section *Learning and memorability* later) then this will help a lot because it reduces the V_a/V_c ratio.

Sillén-Tullberg & Leimar (1988) followed up Sillén-Tullberg's earlier work by the development of a model designed to investigate the effects of group size, unpalatability and aposematism on individual survival. From this, death rate can be calculated as

$$\text{Death rate} = A(-e^{-n/n'})[1 - (1 - h^n)]/nh$$

where A is attack rate, n is the number of prey in a group, n' is effectively a threshold group size where the rate of detection begins to level off, and h is the probability that a predator, having sampled some unpalatable prey in the group, is inhibited from further attack. The left-hand part of the death rate equation represents detection rate and the right-hand part the proportion of a group that will be attacked (potentially killed) before the predator has discovered their unpalatability. The curves obtained all show an increase in mortality as group size initially increases from unity but this peaks (depending on the value of h) such that it then decreases again for larger aggregations (Fig. 4.9). To illustrate the likely validity of the model Sillén-Tullberg & Leimar presented plots of group size (or surrogates thereof) for caterpillars of two types of North American Lepidoptera, both of which show marked under-representation of intermediate sized aggregations (Tables 4.1 and 4.2).

Individual selection models

Although there is quite a lot of evidence that many, even rather soft-bodied insects are actually fairly resilient to the initial handling by a predator (see section *Evidence for individual selection* later) and as argued by D.L. Evans (1987) if their survival in an initial attack is high enough and if during or shortly after an attack predators detect their unpalatability, then they may avoid similar-looking individuals in the future. Thus mutations that make an initially cryptic pattern more

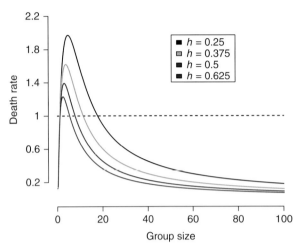

Fig. 4.9 Effect of aggregation size of an unpalatable prey and the probability (h) that a predator will stop attacking after sampling one or more unpalatable individuals. The initial peak shows that small groups of aposematic prey will suffer markedly, but that large groups will benefit. (Source: Adapted from Sillén-Tullberg & Leimar 1988.)

Table 4.1 Lepidoptera caterpillar group sizes for 61 species from Canada showing deficit of species with small groups (only species with reduced mouthparts included as these have a high representation of gregarious species). (Data from Sillén-Tullberg & Leimar 1988 and from P.D.N. Hebert 1983.)

Caterpillar group size	Percentage of species
1	42
2–3	25
4–7	0
8–15	5
16–31	16
32–63	10
64–128	2

memorable may be favoured in an animal that is already unpalatable but fairly resilient to attack.

Sillén-Tullberg & Bryant (1983), following on from Wiklund & Järvi (1982), used three probabilities to describe predator–prey interactions:

- P_d = the probability that a prey is detected,
- $P_{s|d}$ = the conditional probability that a prey is seized once it is detected, and
- $P_{k|s}$ = the conditional probability that a prey is killed once it has been seized.

Table 4.2 Egg clutch sizes for North American Nymphalinae butterflies. (Data from Sillén-Tullberg & Leimar 1988 and from Scott 1986.)

Egg cluster size	Percentage of species
1	75
2–3	3
4–7	0
8–15	3
16–31	0
32–63	5
64–127	4
128 or more	10

The probability of a prey being killed by a predator, Ø, is therefore,

$$Ø = P_d . P_{s|d} . P_{k|s}$$

If a prey is distasteful, then there is a non-zero chance of it being released unharmed after being seized because the predator may be able detect its unpalatable nature quickly, i.e. $P_{k|s} < 1$. In any given situation we can specify a conditional value of Ø, call it $Ø_k$, which depends only on the probability of discovery (P_d) and the probability of seizure ($P_{s|d}$), i.e.

$$Ø_k = P_d . P_{s|d}$$
$$P_{s|d} = Ø_k / P_d$$

This lets us produce the hyperbolic curve in Fig. 4.10, which shows that for any mutation in the prey that increases its aposematism, and therefore its probability of discovery (P_d) to be selected for, it must disproportionally reduce the probability of seizure ($P_{s|d}$).

Engen et al. (1986) produced a mathematical model of the evolution of aposematism through individual selection based on what they refer to as a 'life-span survival model', which takes into account that predators may be able to assess the unpalatability of a prey individual after seizing it but before killing or eating it. The model in this case predicts that aposematism ought to evolve more easily in highly unpalatable species with long pre-reproductive phases, especially if they occupy a habitat that affords little opportunity for concealment. Their model predicts that a two-step process must be involved with a threshold value of conspicuousness that has to be crossed before survival probability starts to increase. The learning ability of the predator then determines whether the prey will evolve to become highly conspicuous or be limited at some weaker level of conspicuousness.

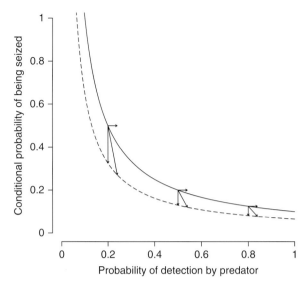

Fig. 4.10 Graphs showing how evolution of warning colouration has to proceed. In order to evolve greater conspicuousness (higher probability of detection by a predator) there must be a concomitant decrease in the probability of being killed or injured if the prey is discovered. (Source: Adapted from Sillén-Tullberg & Bryant 1983. Reproduced with permission from John Wiley & Sons.)

Leimar et al. (1986) constructed an evolutionary model based on two one-dimensional characters: prey 'colour' and prey unprofitability. The lowest values of 'colour' correspond to cryptic prey and higher ones to progressively more aposematic ones. With this construct they showed that an evolutionarily stable strategy (ESS) for prey colouration would exist if the predator showed either, or both, of two behaviours: (1) a reluctance to attack prey that are more conspicuous than those so far encountered, and (2) faster learning of more conspicuously coloured prey. Similarly, an ESS for prey unprofitability will exist if one or both of the following are satisfied: (1) increased survival of attacks, and (2) increased learning by the predator of more unprofitable prey. Their model assumes that whether or not a prey is attacked is solely dependent on its 'colour' value and whether or not it is killed depends only on its unprofitability. Their findings are illustrated in Figs 4.11 and 4.12, which show the effect, given a fixed level of unpalatability (y) in their equations, of increasing aposematism on detection rate and survival. The key points are that with naïve predators the best strategy is always to be maximally cryptic, but with well-experienced predators, attack is minimised at an intermediate level of conspicuousness (Fig. 4.11, lower lines), which implies that for aposematism to work the prey must be sufficiently abundant so that many predators will have multiple encounters with them during their lifetimes.

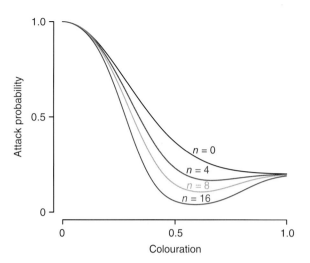

Fig. 4.11 Effect of increasing aposematism ('colour' axis) on the probability of attack for three levels of encounter number, i.e. $n=0$ means a naïve predator, $n=5$ means that the predator has had five previous experiences with the animal, $n=20$ the predator is very experienced and thus there is a well-developed optimal non-extreme level of aposematism. (Source: Adapted from Leimar et al. 1986.)

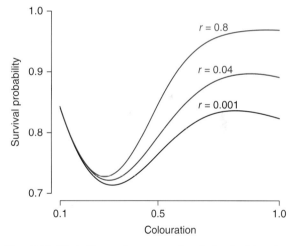

Fig. 4.12 Effect of increasing aposematism ('colour' axis) on the probability of survival from attack. (The lines presented are close approximations to those in figure 5 of Leimar et al. 1986, but as I could not find the precise parameters used in their models, the results should be considered purely qualitatively; here the values are $y_0 = q = 0.05$, $x_0 = y_1 = s = 0.1$, $x_1 = 0.9$, $T = 1$, $N = 100$, $N_p = 40$.) (Source: Adapted from Leimar et al. 1986.)

Figure 4.12 illustrates the effects of kin structure, their parameter r, on the probability of survival. The starting point of the curve along the 'colour' (phenotype) axis is $x_c = x_0 = 0.1$, and if the survival remains below this value for increasing aposematism, then there is no aposematic ESS, which is the case when there is no kin structure ($r=0$). By introducing some degree of relatedness between the individuals that a predator encounters, among individuals in the 'surrounding field' (as they term it), there comes a point when an aposematic strategy can do better; in the parameterised example, this is when $r=0.03$. A value of $r=0.5$ indicates that all individuals encountered are siblings, and therefore only a small proportion of the 'field' need to be kin for there to be no cryptic ESS.

Leimar et al.'s (1986) work led Broom et al. (2006) to develop an analytical ESS model based on there being a range of predator types with sequential replacement of experienced predators by naïve ones. This then formed the basis of an even more general model by Ruxton et al. (2007), which examined the coevolution of secondary defences and aposematism and did not require that the first conspicuous mutants were necessarily already protected. Not surprisingly, aposematism evolved easier if the prey were initially already protected, providing that there was "sufficient investment in toxicity to make the prey aversive". They also found that coevolving conspicuous signals should also be higher in species that are initially protected. Another possible result was that species could evolve high levels of protection while not evolving aposematism.

A completely different modelling approach was presented by Speed & Ruxton (2005a) where they introduce the notion of opportunity costs. Foraging widely to obtain more resources over varying backgrounds necessarily makes the prey more conspicuous and detectable. They show that if being cryptic does not affect an organism's ability to gain resources they should not expend energy in secondary defences, but if it does constrain them, then acquisition of expensive secondary defences would be favoured and with that the evolution of aposematism, as this would enable the species to forage more widely.

Servedio (2000), in a rather more complex model with the predator displaying variation in the classical conditioning features of learning and forgetting, and also making recognition errors, showed that recognition errors had a particularly strong influence on whether conspicuousness would spread. In the extreme case of one-trial learning with no forgetting (i.e. the obnoxiousness of the prey is very high) the effect of conspicuousness is essentially one of reducing the probability of recognition errors by the predator. In less extreme, perhaps more natural conditions, with gradual avoidance learning and gradual forgetting, Servedio (2000) found that it was very difficult for aposematism to become established "unless bright individuals cross an often high threshold frequency through chance factors". This is exactly what happened in my spatial simulations (see section *Spatial models and metapopulations* later).

Ruxton et al.'s (2007) models explored whether aposematism can evolve in prey that have a rather lower secondary defence ability than is normally assumed. As expected, warning colouration evolved more easily in prey that have good secondary defences, but also evolving expensive secondary defences is quite compatible with remaining maximally cryptic.

Puurtinen & Kaitala (2006) presented a deterministic numerical simulation model from which they could determine the strength of newly emerged warning colouration signals under a wide range of biological assumptions, including aspects of predator psychology (learning, errors, neophobia) (Guilford & Dawkins 1991, 1993a) and forgetting (Speed 2000). One of their important findings was that selection in favour of a novel aposematic form is dependent on the number of individuals rather than their frequency, with a threshold value being necessary for the warning colouration to spread (see also section *Spatial models and metapopulations* later). They also found that conspicuousness and distinctiveness were related, in that more conspicuous colouration can only evolve if it simultaneously makes the prey easier to distinguish from palatable ones. Easily learned warning signals and long-lasting neophobia both facilitate the evolution of warning signals, but a counterintuitive result, at least in my opinion, was that the threshold number of individuals for the spread of the warning signal was lower when predation pressure was higher. To understand this it is important to separate predation pressure from number of predators. If the predation pressure increases due to there being more predators, then there are more predators that have to learn the signal, but if it increases, say, because of a scarcity of alternative prey, then individual predators are more likely to encounter additional aposematic forms before they have forgotten the negative experience from sampling the first.

Franks & Noble (2004) developed the coevolutionary model first put forward by Sherratt (2002b). Both studies showed that predators would evolve a preference to attacking cryptic prey. However, in the particular simulations Franks & Noble were unable to get conspicuous colouration to evolve.

Spatial models and metapopulations

Virtually all of the empirical and theoretical work on the initial evolution and spread of a conspicuous form of an animal have considered a uniform field of prey, predator and genotypes in some, usually large or infinite population. Lee & Speed (2010) note that many organisms exist in fragmented populations with some degree of migration between patches, i.e. metapopulations. They present the first explicit model that takes this uneven patchiness into account, although Mallet & Singer (1987) and Endler & Rojas (2009) both realised the likely importance of uneven spatial distribution and the former postulated that if an aposematic allele came to fixation in one population, it would probably then be easy for it to spread to all others. Nonacs (1985) found that non-random distribution of model and mimic pastry prey and a wild population of chipmunks, *Eutamias quadrimaculatus*, increased the amount of foraging on the baits. Nonacs (1985) also noted that the chipmunks "seemed particularly adept at detecting a favourable change" in the local proportion of edible models.

In Lee & Speed's (2010) study, predator/prey populations were simulated in a 50 × 50 square grid of cells containing predator and prey populations, though some cells received NULL predators. Predators were initially naïve but learned to avoid noxious prey. I present in Fig. 4.13 some results from a very similar simulation, using a 40 × 40 grid with opposite edges wrapped so that, for example, prey in left-edge squares that migrate left are added to squares on the right-hand margin of the array. In my simulations, all cells were occupied by predators and thus there is no predator-free space. Nevertheless, depending on the initial random distribution of aposematic individuals, and the values of four user-defined variables (1) the proportion migrating each generation (equivalent to 10 attacks on individuals in a cell), (2) relative conspicuousness of aposematic and cryptic prey, (3) the length of avoidance memory that they induce in the predator, and (4) the rate of mutation or recombination leading to the replacement of a cryptic individual with an aposematic one, the aposematic forms will go extinct, climb or fall from initial abundance to a long-term persistent stochastic equilibrium (Fig. 4.13a) or form one or more large aggregations (Fig. 4.13b) that eventually take over the environment. When a long-term equilibrium is reached, cells with aposematic individuals tended to form drifting networks with large areas of space dominated by uneducated predators because the appearance of new aposematics was low relative to migration. In these models and those of Lee & Speed (2010), the continued appearance of new aposematic individuals is what would be expected if the aposematic allele was recessive, such that the population might have a reasonably large number of cryptic heterozygotes. If there is any small initial degree of heterozygous advantage (heterosis) then we might expect occasional aposematic individuals to appear relatively frequently over long periods of time and also perhaps to appear in a more clumped spatial pattern, which would facilitate them forming local clusters sufficient to educate predators and thus provide temporary havens from predation for new aposematics or kin to survive in. These models rapidly become too parameter-rich for easy investigation,

and also the values of parameters have little foundation in empirical biological data, and the modes of migration are highly simplified compared to reality. Nevertheless, I think they provide some insight into the sorts of factors that are likely to be important.

The results of these simulations indicate that the higher the level of inter-population migration, the less likely it is that a novel aposematic allele will survive and spread, and therefore the evolution of aposematism is likely to be favoured by high levels of population fragmentation with low dispersal rates. Lee & Speed's (2010) results also suggest that initially when an allele for aposematism is low in an isolated population, migration may reduce its numbers still further, but after a while, reverse migration of aposematic forms into the original population will reverse that trend and ultimately the allele may become locally fixed as the local population of predators become educated.

HANDICAP AND SIGNAL HONESTY

Zahavi (1977) introduced the idea of handicap as a selective trait, the notion being that if an individual, almost always a male, could be seen to have reached adulthood while sporting something that is evidently costly and only expressed in males, it indicates that the overall genetic make-up of the individual is likely to be good in order to compensate. Examples often cited include such things as the enormously large and spectacular tails of peacocks, which are heavy, conspicuous, liable to get damaged and almost certainly increase the possessor's risk of predation. A male that has survived these costs and increased risks while still being able to flaunt his magnificent tail in courtship is likely to be fit in other terms. Thus a female that chooses to mate with such an individual may gain doubly. Her sons may inherit a fine tail and be attractive, and both her sons and daughters are likely to carry good alleles of other fitness traits. Several people questioned whether this principle could actually function in practice, as cheating would seem to make such costly 'honest' signalling unviable. Grafen (1990) presented a game theory model which vindicated Zahavi's (1977) idea as a viable ESS under quite a broad range of conditions, however cheats may still occur. M.S. Dawkins & Guilford (1991) point out that costs are often not only borne by the signal sender, but that at least in intraspecific behavioural interactions, the receiver may also bear some because they may have to reply. The conclusion is that honest signalling systems are always at risk of being corrupted by mimics that have evolved less costly imitations of the true signal, and that true signalling (i.e. handicaps) will be relatively rare.

While it is well recognised that aposematic displays indicate unpalatability in a qualitative fashion, at least in some

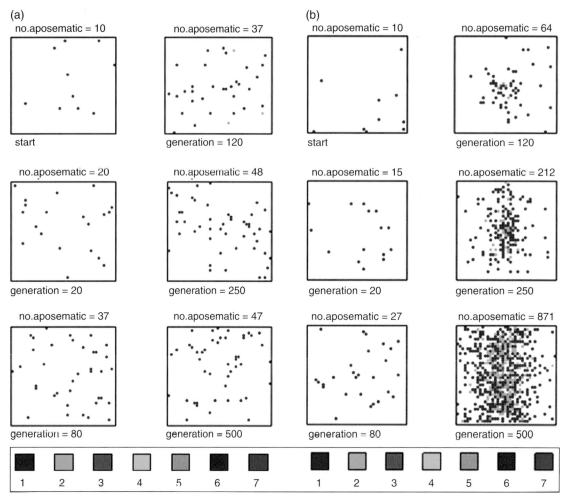

Fig. 4.13 Snapshots of results of two spatially explicit simulations of the occurrence of novel aposematic and unpalatable forms in a population of cryptic relatives over 500 generations subject to predation by a learning/forgetting predator.[1] (a) Long-term persistence of aposematic individuals at population density higher than the original seed, but small local clusters and interconnecting threads rapidly 'dissolve' due to high migration rate and low rate of appearance of new aposematic individuals, with large spaces full of naïve predators;[2] (b) rapid and continuous increase in proportion of aposematic individuals after a sufficiently dense local cluster is formed.[3]

---

1. Each cell contains 100 prey and 1 predator at the beginning of each generation and during one generation each predator encounters ten potential prey and attacks with the probabilities $P_{cryptic}$ and $P_{aposematic}$ accordingly. After attacking an aposematic prey the predator rejects them for its next $Pred_{memory}$ encounters with potential prey. At the end of each generation $P_{migrate}$ of each cell's population are migrated with equal probability to the eight adjacent squares. At the beginning of each generation the population of each cell is rounded to 100 with the same proportions of cryptic to aposematic as remained at the end of the previous generation as if other density dependence was acting primarily at a non-mimetic juvenile stage.

cases the degree of conspicuousness may also be an honest quantitative signal of toxicity or other defence (M. Edmunds 1974a, Blount et al. 2009, 2012). As Speed & Ruxton (2007) put it, "the nastiest prey might 'shout loudest' about their unprofitability". There would seem to be good evolutionary reason for this; for example, when chicks have

2. $P_{migrate} = 0.2$; $Pred_{memory} = 60$ encounters; $P_{cryptic} = 0.1$ and $P_{aposematic} = 0.2$; 1 new aposematic added per generation (out of 160,000 potential prey).

3. As above but $P_{migrate} = 0.15$.

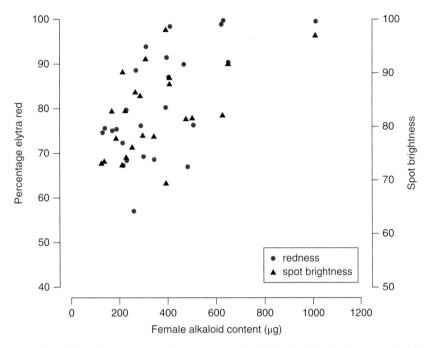

Fig. 4.14 Positive intraspecific relationship between conspicuousness and toxicity in the Asian harlequin ladybird, *Harmonia axyridis* (Coccinelidae) (see Fig. 4.1a). For the percentage of the elytra that was red, the correlation was $r_s = 0.61$ ($p < 0.01$), and for the brightness of the black spots, i.e. faintness, $r_s = 0.56$ ($p < 0.01$). (Source: Adapted from Bezzerides et al. 2007 with permission from Springer.)

been trained to avoid aposematic prey they subsequently show greater avoidance of more brightly coloured ones (Gamberale-Stille & Tullberg 1999).

Variation in toxicity and aposematism may be studied both between species, in a phylogenetic context, and within species. Vidal-Cordero et al. (2012) have recently found that in the European paper wasp, *Polistes dominula*, those individuals with brighter, more conspicuously aposematic colour patterns, on average, have larger poison glands. Thus the degree of aposematism in this intraspecific example is a fairly honest signal of potential noxiousness. Similarly, conspicuousness is also related to colour pattern in the ladybird beetle, *Harmonia axyridis* (Coccinellidae) (Fig. 4.14) (Bezzerides et al. 2007). Daly & Myers (1967) failed to find any correlation among populations of the Mesoamerican poison frog *Oophaga* (as *Dendrobates*) *pumilio* (see Fig. 4.31a,b) but Maan & Cummings' (2012) investigation of dendrobatid colouration did provide an honest signal, particularly to birds. In terms of interspecific studies, the toxicities of opisthobranch sea-slugs (Cortesi & Cheney 2010) and dendrobatid tree frogs (Summers & Clough 2001) have been shown to correlate with the degree of aposematism, and Hensel & Brodie (1976) presented evidence in the same direction in the North American salamander, *Plethodon*

jordani (see also Chapter 6, section *Experimental tests of mimetic advantage*). In contrast, Darst et al. (2006) and I.J. Wang (2011) both found inverse correlations between toxicity and conspicuousness in poison dart frogs – Darst et al. between species and I.J. Wang within a polymorphic species. Speed & Ruxton (2007) sought to explain these contradictory findings using an optimality model with the defensive trait having associated costs and benefits. They showed that a positive correlation could be expected with optimal prey toxicity, but the opposite could be the case if costs of the aposematic display varied a lot between populations.

However, when Guilford & Dawkins (1993b) explored handicap theory using realistic models of predator behaviour on the presence of warning colouration and Batesian mimicry, they found no support for the handicap model and that Batesian mimicry was predicted by conventional signalling not by handicap theory. As Ruxton et al. (2004a) put it, "the presence of some degree of signal reliability does not necessarily mean that warning displays are handicap signals." In particular, Guilford & Dawkins' (1993b) models showed that pattern similarity between aposematic species could be adequately explained purely on the basis of predator discrimination rather than the need for species to suffer similar degrees of handicap.

Sherratt & Franks (2005) asked whether the pressure provided by mimics would therefore select overall for models to evolve signalling systems that were hard to evolve. Their model, based on signal detection theory, showed that natural selection in unpalatable models should lead to the use of signals (colours and/or patterns) that a palatable potential mimic could only evolve if their mimicry was fairly exact. Their model actually has further implications; for example, it is one of the first investigations of the evolution of warning colouration to take into account the potential concomitant effect of mimics.

Early warnings – reflex bleeding, vomiting and other noxious secretions

Predators might be put off attacking a given prey type either because it is toxic, so causes distress such as heart palpitations or vomiting (Swynnerton 1915a, Carpenter 1942, L.P. Brower et al. 1968, Rothschild & Kellett 1972; see figure 7 in L.P. Brower 1971), or because it tastes bad. Further, if the risk from genuine toxicity is high, a predator would probably benefit if it could assess the taste of a potential prey before consuming all of it. This is what Holen (2013) terms 'go-slow behaviour'.

There is now growing evidence that insectivorous birds reject prey, often before killing them, on the basis of taste and/or smell (Skelhorn & Rowe 2006a, 2006b), therefore the sooner they can taste a distasteful prey during an encounter, the sooner they may decide to let it go. Monarch butterflies, for example, deposit the highest concentration of their sequestered cardiac glycosides in their wings, thus affording predatory birds that have missed the body the best opportunity to determine that they are chemically defended (see section *Cardiac glycosides* later).

The distinctive odours given off by many unpalatable insects when threatened provide the potential predator with a signal of their possible toxicity before they have a chance to kill the prey and/or experience its unpleasant effects. Thus they can be thought of as a sort of olfactory aposematism (Eisner & Grant 1981) and provide early warnings. These odours are typically only released when the insect is physically molested, as their continuous release would not only be costly but would also provide a signal to specialist predators or parasitoids that are not deterred by them. Caterpillars feeding on noxious plants often regurgitate their food (probably along with other compounds), while others have eversible 'scent'-releasing structures called osmeteria that are involved in repelling predators. These structures are primarily found in swallowtail butterfly caterpillars (Papilionidae) and in these they are Y-shaped and located between the head and the prothorax. They release a strong-smelling mixture of isobutyric and 2-methylbutyric acids (Eisner et al. 1970, Damman 1986). Caterpillars of some moths, e.g. some Lymantriidae, may have abdominal osmeteria, and notodontids of the genus *Cerura* have whip-like abdominal organs of similar function, and many other notodontids have eversible glands located ventrally between the head and front legs, which are called adenosma (Bowers 1993).

Well known to many people will be the odorous blood of ladybirds (Coleoptera: Coccinellidae), which is released from special structures at the basal joints of the legs – in this case the main compound is a pyrazine. The chemically well-defended lycid beetle *Metriorrhynchus rhipidius* gives off a distinctive scent dominated by the related compound 2-methoxy-3-isopropylpyrazine (B.P. Moore & Brown 1981), and *Lycus* species (Lycidae) have similarly been shown to contain pyrazines (Eisner et al. 2008). Pyrazine compounds may either be synthesised *de novo* by animals (isopropyl derivatives and secondary butyl derivatives) or sequestered from plant sources (only isobutyl homologues) (B.P. Moore et al. 1990, Dettner & Liepert 1994).

Many staphylinids have strong odours; on one occasion I was walking my dog in some woods in England when I noticed and picked up a *Creophilus maxillosus*, a large rove beetle that feeds on carrion-associated insects; its odour was unpleasant, nauseating but not strong. After releasing it I offered my dog a treat with the same hand but he would not come near it. The protective chemicals produced by rove beetles include various compounds used as defensive secretions in ants and lacewings, such as iridodial, actinidine, isopentyl acetate and dihydronepetalactone (Jefson et al. 1983). However, none of these is actually harmful; some are used in perfume manufacture, some attract cats and some are used in commercial insect repellants. Yet their combined effect seems to be a quite powerful deterrent to mammals with no prior experience, and rove beetles in general tend to be widely dispersed unless feeding at a particular source of food such as dung or a carcass.

Some arctiine moths secrete a defensive pyrrolizidine alkaloid-containing froth from pores between their first and second thoracic tergites (see figure 3 in Rothschild 1961, figure 13 in Blest 1964 or figure 1A in A. González et al. 1999). Such predator behaviour would be particularly significant for both Batesian and automimicry. If you are operating a light trap, especially in the tropics, you very soon become aware of how potent and unpleasant-smelling this yellow foam is, but having got considerable amounts on my hand I can testify that there were no other ill-effects.

Halpin et al. (2008) carried out an experiment with domestic chicks and coloured (cryptic and conspicuous) food crumbs, some of which were rendered unpalatable to taste, and found that chicks could readily learn to avoid the unpalatable crumbs if they were conspicuous. More interestingly, if instead of attack, actual consumption was used as a measure of protection, Halpin et al. found that quick-acting, external taste signals of unpalatibility can confer protection even in the first avoidance learning trial, because the prey is released rather than consumed. Bad taste may therefore be an alternative to true toxicity as well as an adjunct in the evolution of mimicry systems (see Speed 1993b).

Holen (2013) used a classic diet model to explore the separate roles of taste and toxicity in Batesian and quasi-Batesian[4] scenarios. In a way analogous to the proposed Emsleyan mimicry (Wickler 1968) of highly venomous snakes, Holen (2013) found that if a defensive toxin was sufficiently distasteful then the members of a Batesian mimicry complex might collectively be better defended against predation if the model is only moderately rather than very toxic. This also means that taste might also be profitably mimicked. If one goes through any old compendium of organic chemicals, there is often a comment on a compound's taste, and in very many cases it says "bitter".

LONGEVITY OF APOSEMATIC PROTECTED TAXA

Many aposematic animals are remarkably long-lived for their size and taxonomic group. Monarch butterflies, *Danaus plexippus*, famously migrate as far as between Canada and Mexico, overwintering in large aggregations and returning in the spring. *Heliconius* and *Ithomia* butterflies live for several months as adults (J.A. Scott 1973). In *Heliconius* this is no doubt aided by their pollen-feeding habit, which also means that they can carry on producing more eggs throughout their life (Dunlap-Pianka et al. 1977, K.S. Brown 1981). Blest (1963) made an interesting proposal about the potentially contrasting consequences of aposematism and crypsis on longevity. Although he couched his paper in terms of group selection he referred to kin, suggesting that as was common at the time he did not make a distinction between group and kin selection; it seems to me that there may well be truth in his assessment. Blest (1963) noted that if predators such as birds search for, and occasionally detect, cryptic prey and in doing so get better at hunting for them (see Chapter 1, section *Search images*), then once an individual prey has reached the end of its reproductive life, if it were to remain alive it could only improve the chances of a predator finding other similar, possibly related, prey nearby, some of which might still have considerable reproductive potential.

In contrast, the persistence of post-reproductive, aposematic individuals can potentially still lead to more predators learning to associate their signals with unpalatability, and thus increase the potential fitness of nearby, potentially related individuals that had not reached the end of their reproductive lives. However, Carroll et al. (2011) re-examined Blest's (1963) hypothesis and found the evidence to be "weak at best" and suggested instead that differential predation might be largely responsible for any observed differences in the field.

There is a considerable need for more research on this topic. Another possible example I came across is by Hetz & Slobodchikoff (1990) working with two tenebrionid beetles from North America, *Eleodes obscura* the model and its mimic *Stenomorpha marginata*. They note that adults of the models live for up to 4 years, whereas those of the mimic only live for approximately three months.

MACROEVOLUTIONARY CONSEQUENCES

An under-investigated area is whether aposematic unpalatable taxa evolve differently from others. Przeczek et al. (2008) performed a comparative analysis of 14 focal groups of aposematic taxa (collectively including various frogs and toads, salamanders, spiders, beetles, hymenopterans and butterflies) and their non-aposematic sister groups. In line with their prediction, that aposematic species should be better protected and therefore less likely to go extinct, they found that in 11 of the 14 cases the aposematic taxon had the higher number of species. Comparison of the differences in log-transformed numbers of species in sister taxa showed the difference to be significant (Wilcoxon signed rank test: $W = 36.5, p = 0.01$). If this interesting finding holds up with further taxon inclusion it will be a startling demonstration of the importance of aposematism in shaping the tree of life. Of course, it is not possible from these data to distinguish whether the difference is due to reduced extinction rates or greater speciation rates or both, though that ought to be possible by studying dated phylogenies, and comparing lineages through time plots.

EXPERIMENTAL STUDIES

Tough aposematic prey and individual selection

A major new wave of thought about the evolution of aposematism in unpalatable prey came about largely as a result of the experimental work of Järvi et al. (1981a) and Wiklund & Järvi (1982), who demonstrated that several aposematic and unpalatable insects could survive attack by avian predators.

---

4. Term used to indicate a Batesian mimic that is distinctly unpalatable but not so unpalatable as the model.

Indeed, Poulton (1887b) had noted this much earlier, observing that after tasting secretions from them, the birds desisted and went in search of other food. There is now ample evidence that many noxious prey are able to survive attack by predators due to a combination of innate toughness and the release of chemicals that deter the predator from continuing the attack. For example, Ohara et al. (1993) found that while cryptic green caterpillars of small white butterflies, *Pieris rapae*, feeding on cabbage were readily eaten by naïve domestic chicks, *Gallus gallus*, the black larvae of the sawfly *Athalia rosae*, were conspicuous and unpalatable, and were seldom injured by the chicks.

While it was not so surprising that hard-bodied aposematic insects such as ladybird beetles (Coccinellidae) could survive a bird attack, it had not previously been so obvious that small caterpillars, such as those of the European swallowtail butterfly, *Papilio machaon*, could also survive a bird attack. Nevertheless, the occurrence of so many aposematic soft-bodied organisms still seemed to pose a considerable problem. Several studies on the predation of soft-bodied aposematic animals have focused on marine invertebrates such as sea-slugs (Tullrot & Sundberg 1991, Tullrot 1994) and the apparently extremely fragile Nemertea (Sundberg 1979, 1987, Tullrot 1994). Tullrot (1994) tested a total of 108 individuals, collectively representing seven species of soft-bodied marine invertebrate (three nemerteans and four opisthobranchs), against wild-caught juvenile cod (*Gadus morhua*). Two of the nemerteans and two of the sea-slugs were thought to be cryptic, and one of the brightly coloured sea-slugs was also probably cryptic against its hydroid food. The other species were considered to be aposematic. A very high proportion, 104 out of the 108 invertebrate prey tested, survived attack by the cod, with just one aposematic and three cryptic nemerteans being eaten. These results suggest that two factors might be important in individual survival: toughness, or in the case of soft-bodied animals, flexibility, and bad taste, the latter leading to relatively short handling times and therefore lessened risk of serious injury. Many tropical marine polyclad flatworms (Platyhelminthes) are also very brightly coloured and despite swimming in the open are shunned by predatory fish (L. Newman & Cannon 2003). In their case, a likely explanation is that even if they are quite badly harmed by an encounter with a fish but then not actually killed or eaten, they have remarkable powers of regeneration.

Many readers will have at some point in their lives tried to pick up a ladybird beetle (Coccinellidae) and in many cases will have found that their typically smooth, nearly hemispherical dorsums make them hard to manipulate without dropping them, and that once handled, they usually secrete a yellow liquid with a somewhat unpleasant odour. Such odours are typically not harmful *per se*, but may indicate that the animal is also otherwise protected (see section *Early warnings – reflex bleeding, vomiting and other noxious secretions*). Even if a predator is not deterred by the secretion, the handling time and probable low capture rate may serve to reduce a predator's future attacks.

D.L. Evans et al. (1986) found that a high percentage (85%) of the normally unpalatable and aposematic bug *Caenocoris nerii* (Lygaeidae) survived attack by naïve common quail, *Coturnix coturnix*. They manipulated some bugs so as not to contain cardiac glucosides by feeding them on sunflower rather than oleander seeds; this group did suffer higher mortality (eaten or receiving a fatal peck) but nevertheless survivorship was still remarkably high.

There are some aposematic and unpalatable taxa that have undoubtedly evolved aposematism through individual selection. Rosenberg (1989) describes one such example, the marine gastropod *Cyphoma gibbosum*, also known as the flamingo tongue snail, which feeds on gorgonian soft corals in the tropical Atlantic. Since this species and all close relatives have planktonic larvae which become widely dispersed, there can never be any opportunity for kin selection. The same is true for various aposematic nudibranch sea-slugs. Guilford & Cuthill (1991) questioned whether there was sufficient evidence that any marine gastropods were aposematic, despite Rosenberg (1989) demonstrating that *Cyphoma* was unpalatable to some fish. Of course, demonstrating that the distinctive colours of *Cyphoma* or of various sea-slugs function aposematically is quite another thing. However, apparent Batesian mimicry of aeolid sea-slugs by an amphipod crustacean in the Pacific (Goddard 2015) would seem to provide some support for the hypothesis. Guilford & Cuthill (1991) also pointed out that pure individual selection seems unlikely because of what they term 'synergistic selection', that is there are always likely to be other, quite possibly unrelated, individuals nearby with a similar phenotype that have undoubtedly evolved aposematism through individual selection. In his reply to Guilford & Cuthill (1991), Rosenberg (1991) notes that as of that time no experiments had demonstrated two other essential features of aposematism (*vide* Edmunds 1987), that predators learn to avoid attacking them because of their colouration rather than their shape or odour, and crucially, that more aposematic individuals within a species accordingly gain better protection. Since *Cyphoma* have planktonic larvae there will be no phenotypic clustering within their populations, and the co-occurrence of multiple aposematic individuals within groups will be determined by the Poisson distribution. Rosenberg (1991) showed that this means that on average 50% of conspicuous individuals would occur in groups with at least one other conspicuous individual. Therefore synergistic selection would act on 50% of the population of aposematic individuals.

Gamberale-Stille & Tullberg (1996) demonstrated another aspect that could facilitate evolution of aposematism by counteracting the increased risk of more conspicuous signallers to predation, as was suggested as a result of Leimar et al.'s (1986) individual selection model (see earlier). They found that if predators experienced an unpalatable moderately conspicuous (subadult) prey, in their case domestic chicks preying on different instars of the aposematic bug *Tropidothorax leucopterus* (Lygaeidae), they not only developed an aversion to that stage but showed a greater aversion to the similarly defended but larger and more conspicuous adult bug. They call this a peak-shift in generalisation, the implication being that while individuals with the same conspicuousness as the subadult may gain protection, mutants that are even more conspicuous might gain an even greater degree of protection.

Pyrazine and other early warnings

Numerous aposematic insects use the compound pyrazine to warn predators against attack. Guilford et al. (1987) found that after giving domestic chicks experience of pyrazine-tainted water they developed a neophobic reaction and Lindström et al. (2001c) showed that pyrazine odour makes chicks significantly less likely to attack conspicuously coloured food crumbs (Fig. 4.15). However, C. Rowe & Guilford (1996) had shown that pyrazine odours are not innately aversive; they specifically induce aversion to red and or yellow (i.e. aposematic) prey, but they do not affect predation on cryptic food. In a similar vein, Marples & Roper (1996) compared responses of naïve chicks to water with familiar or novel colour in the presence or absence of each of five harmless odour compounds, including two different pyrazine derivatives (2-methoxy-3-sec-butyl pyrazine, 2-methoxy-3-isobutyl pyrazine, almond oil, vanilla oil and thiazole). Again, chicks showed greater neophobia to colours in the presence of the pyrazines and also almond oil, but there was hardly any effect of the other two compounds. Interpreting such experiments is complicated, however, because we do not know how sensitive the chicks or other predators are to the different compounds and without further checks it could always be the case that they simply cannot detect a particular compound rather than detecting it but not responding to it, just as some people can smell the perfume of *Freesia* flowers while others cannot. Also, it is unknown whether this suggests that evolution has led to an inherent association of aposematic colouration and pyrazine odours or whether evolution has favoured the use of pyrazine because of innately present neuronal circuitry that associates it with potentially toxic food.

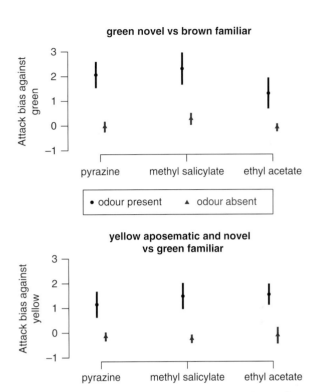

Fig. 4.15 Modulating effect of pyrazine odour on predation of aposematic prey by domestic chicks. The *y*-axis bias measure is the difference in the number of cryptic crumbs eaten compared to conspicuous ones, such that positive values indicate that the chicks ate more cryptic than conspicuous crumbs. (Source: Adapted from Lindström et al. 2001c. Reproduced with permission from The Royal Society.)

Jezt et al. (2001) carried out two neat experiments with domestic chicks to test the effect of chemical odours on food choice. In the first experiment they showed that chicks became biased against pecking at an unfamiliar coloured food crumb type when they were also exposed to a novel organic odour. No difference was found between the three odour treatments: pyrazine, the plant signalling compound methyl salicylate, and one effectively completely novel compound, ethyl acetate (i.e. of no relevant biological significance). In their second experiment they only used ethyl acetate as the test odour, and either allowed chicks to become accustomed to it before the trial or used naïve chicks. Their results were very clear: chicks show avoidance of novel coloured food when it is accompanied by a novel odour, but not when they are already familiar with the same odour; familiarity with the organic odour is just the same as if no odour was present. There was, however, an effect of whether one of the two food colours was familiar or whether both were novel. When both were novel, the chicks tended, in the presence of the novel odour, to avoid the aposematic yellow crumbs more than the unfamiliar green ones. Whether this is generalisable to other aposematic signals is unknown.

Instead of using domestic chicks, Kelly & Marples (2004) used zebra finches, *Taeniopygia guttata*, which not only extended findings to passerine birds but allowed a separation of time to contact the food of novel colour or novel colour plus odour (pyrazine). They found that whereas the latency to contact did not differ (both were significantly higher than familiar colour and odourless controls), the latency to food consumption was greatly increased if the prey had both novel colour and odour. This indicates that the neural processing leading to these two stages in predation are not the same.

In a series of feeding trials, Vasconcellos-Neto & Lewinsohn (1984) found that among various Brazilian butterflies well known to be unpalatable, the large orb web-weaving spider *Nephila clavipes* (Nephilidae[5]) almost always released members of the Ithomiini and some species of Danainae, but invariably consumed a range of taxa including heliconiines (e.g. Acraeini) (Fig. 4.16). In the case of this predator, experience with one prey type did not influence its behaviour in the next encounter, in contrast to the way that an unpalatable prey item may make a vertebrate predator generally more wary immediately afterwards. This difference is most likely due to the fact that the spider uses a web to trap prey, which gives it more time to assess them. It does not have to consume any part of them if it can detect unpalatability by other senses. L.P. Brower et al. (1963) carried out a similar comparative palatability test of various heliconiines in feeding trials with silver-beaked tanagers, *Ramphocelus carbo*, for which the tested species were all unpalatable, though *Dryas iulia*, *Agraulis vanillae* and *H. doris* were less unpalatable than *H. erato*, *H. melpomene*, *H. numata* and *H. sara*.

Very few mammals and birds contain toxins but some are more palatable than others. Cott (1947) reports on the edibility to humans, cats and hornets of various birds as food and revealed considerable variation between species. Dumbacher et al. (1992) report on the occurrence of high concentrations of an alkaloid called homobatrachotoxin in New Guinean birds of the genus *Pitohui*. Homobatrachotoxin is present in the birds' feathers and skin and, in *P. dichrous*, also in striated muscles. Contact with it causes numbness, 'burning' and sneezing in humans when it contacts buccal membranes, and convulsions when extracts are injected into mice. Its mode of action is similar to that of a poison dart from the poison dart frog (*Phyllobates*) in that it holds open

Na[+] (sodium) channels. The local people refer to these as 'rubbish birds' since they cannot be eaten. Mouritsen & Madsen (1994) proposed that although this toxin might help protect against predators such as birds, snakes and mammals, it also serves to reduce parasite load as *Pitohui* have very low infestation rates compared with other birds. The widespread publicity about *Pitohui* gave the impression that these were the only toxic birds and indeed until recently no birds were included in general lists of toxic animals, but Dumbacher & Pruett-Jones (1996) have shown this is quite wrong, and that many other bird species have either unpalatable or definitely toxic compounds in them. Although unpalatability in birds is taxonomically widespread, Cott (1947) and other data compiled by Dumbacher & Pruett-Jones (1996) suggest that there are some taxonomic clusters, notably various crows (Corvidae) and drongos (Dicruridae) which are often all black, and kingfishers which are brightly coloured. After 50 years, Cott's (1947) work became the focus of two further studies. Götmark (1994) reanalysed Cott's (1947) data, taking into account male plumage colouration and, although not phylogenetically controlled, showed fairly convincingly that unpalatability was generally associated with bright colouration. Weldon & Rappole (1997) revisited this topic by surveying the opinions and experiences of ornithologists and found evidence for unpalatability in members of some 30 genera as well as three whole families, distributed across 13 orders. Considerably more were reported to be odorous, some of them being scavengers and some sea birds, hence likely to be associated with their diet. The hoatzin, *Opisthocomus hoazin*, is well known for its foul odour, as indicated by some of its vernacular names, such as stink bird or stinking pheasant. This is clearly an area that requires a great deal of further research and experimental testing.

LEARNING AND MEMORABILITY

Two separate topics may be considered here, the memorability of the patterns or other aposematic signal produced by an organism, and the memorability of the noxious stimulus. The more conspicuous a prey, the longer a predator remembers it and avoids attacking it or similar-looking individuals in future encounters (Gittleman et al. 1980, Roper & Redston 1987, Roper 1994). In a number of experiments involving chicks drinking red- or blue-coloured, pure or quinine-tainted (unpalatable) water, Shettleworth (1972) and Roper (1993) both found that in general, novelty facilitated taste-avoidance learning. However, familiarity played a confounding role. Roper & Wistow (1986) similarly found that pecking rate by domestic chicks on contrasting, unpalatable food crumbs declined more quickly than against non-contrasting unpalatable ones.

---

5. Previously often considered to belong to the Araneidae.

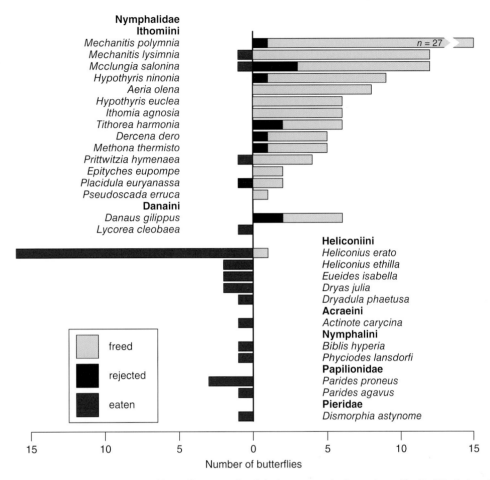

Fig. 4.16 Results of tossing various Neotropical butterflies into webs of the large tropical orb-weaving spider *Nephila clavipes*, showing large differences between taxa, with *Heliconius* and various others largely accepted and eaten but most Ithomiinae rejected and often actively freed from the web. (Source: Adapted from Vasconcellos-Neto & Lewinsohn 1984. Reproduced with permission from John Wiley & Sons.)

Terrick et al. (1995) investigated the role of colour associated with a noxious stimulus using laboratory-born garter snakes, *Thamnophis radix*, that will accept food such as fish and earthworms offered to them in forceps. This way the associated colour can be painted on the forceps (in this case black, neutral green, aposematic black and yellow or unpainted), leaving all prey effectively identical in appearance. Snakes fed fish were then injected after 5 minutes with a small dose of lithium chloride (LiCl), enough to give them an unpleasant experience and induce regurgitation, or in the case of controls, with physiological saline. Having determined that the snakes showed no *a priori* preference or aversion to any of the forceps types, though a slightly greater latency to attack fish rather than earthworm, Terrick et al. found a highly significant associated aversion effect of the LiCl treatment and fish prey that lasted some

three weeks, irrespective of whether fish were presented in aposematic or plain forceps. However, retention of the aversion to fish associated with LiCl was significantly higher for snakes whose bad experience was associated with the yellow and black aposematic forceps ($F = 9.37$, d.f. $= 1,10$, $p = 0.012$). A similar example is afforded by the use of the commercial goose repellent methyl anthranilate (Rejex-It AG-36); the number of geese visiting a field and damaging the treated crop can be reduced if the presence of Rejex-It on the crop is also signalled visually by spraying them with a white paint pigment (titanium dioxide) (J.R. Mason & Clark 1996).

Reiskind (1965) conducted an experiment on the learning rate and memory retention period of caged North American black-capped chickadees, *Parus atricapillus*, when trained on sunflower seeds (*Helianthus annuus*) that were

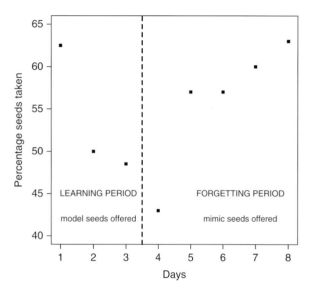

Fig. 4.17 Rapid avoidance learning for quinine-treated coloured seeds by black-capped chickadees, *Parus atricapillus*, and rapid loss of avoidance when the seeds were once again made palatable, showing the flexibility that the birds have as a result of continued sampling of potential prey. (Source: Adapted from Reiskind 1965. Reproduced with permission from Elsevier.)

either rendered noxious by dipping in quinine hydrochloride, thus simulating models, or controls that were simply dipped in distilled water, and stained either green or orange. Birds learned to avoid the distasteful model seeds during a training period of 4 days (Fig. 4.17), but when the seeds were replaced by palatable ones of the same colour, the birds sampled them and quickly learned that they were no longer unpalatable.

Strength of obnoxiousness

Intuitively, the more unpalatable a conspicuous model species, the less predation it will suffer and therefore the degree of protection afforded by aposematism would be expected to be positively correlated with the amount of distasteful compound present. Holling (1963) in his early modelling of the relationship between the proportion of models in the population and degree of protection from predation, showed that the level of unpalatability was very important. From his graph (figure 1 loc. cit.) weakly unpalatable models never afforded a very high level of protection, but moderately and strongly unpalatable ones did, indeed up to 100% protection.

To test this assumption, O'Donald & Pilecki (1970) presented data on predation of coloured (blue, green or yellow) and variously distasteful (untreated, dipped in 1% or 3% quinine solution) flour and lard baits put out in an array on a bird table and predated by house sparrows, *Passer domesticus*. Their results were reinterpreted by J. Edmunds & Edmunds (1974), who concluded that they showed that when a weakly distasteful model has many palatable mimics, it is at a disadvantage, but that highly unpalatable ones were not adversely affected. Conversely, a common mimic is probably relatively more disadvantaged if its model is only mildly distasteful rather than very distasteful (Lindström et al. 1997) (see also Speed et al. 2000; Chapter 6, section *Müllerian mimicry and unequal defence*).

In another experiment with free-ranging garden birds as predators, M.A. Goodale & Sneddon (1977) tested whether birds that had only experienced mildly distasteful model baits would reject fewer palatable mimic food items than if they had experienced very distasteful ones. Birds were first trained to take yellow or red food pellets from paving stones: red pellets at sites A and B and yellow pellets at sites C and D. In the next stage, 25% of the pellets were exchanged for unpalatable versions of the alternative colour; at two sites the 'models' were mildly unpalatable (20% quinine solution treated) and at the other two sites the 'models' were very unpalatable (treated with 80% quinine solution). After this training to recognise the unpalatable models, the third stage involved all palatable pellets of both colours plus three intermediate shades of orange. At sites where the birds had experienced only mildly unpalatable models they ate relatively more of the pellets closer in colour to the form that had previously been unpalatable than did birds trained to recognise highly unpalatable ones (Fig. 4.18). The higher proportions eaten when the unpalatable models were yellow suggests that these wild birds had a stronger basic aversion to red prey.

Is the nature of the protective compound important?

Skelhorn & Rowe (2005) investigated how predators, in their case domestic chicks, responded to Müllerian mimic crumbs when the ones presented contained all the same unpleasant compounds or when two different compounds (Bitrex® and quinine) were represented. It turned out that mimic crumbs with more than one unpalatable compound experienced lower predation (Fig. 4.19). Skelhorn & Rowe (2009) further investigated whether Bitrex, which has a very bitter taste but is completely non-toxic, could act as a defensive compound in its own right. European starlings, *Sturnus vulgaris*, were trained on mealworms, *Tenebrio molitor*, with associated colour signals that had either been injected with Bitrex or had the same amount painted on to

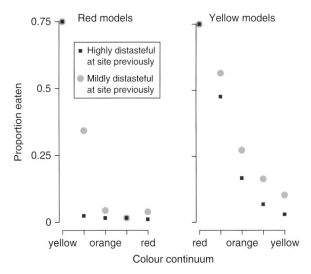

Fig. 4.18 Predation levels by wild suburban birds on mimetic pastries depends on whether the models they had previously experienced were highly unpalatable or only mildly so. (Source: Adapted from Goodale & Sneddon 1977. Reproduced with permission from Elsevier.)

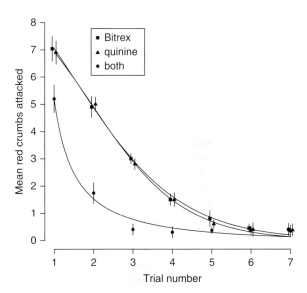

Fig. 4.19 Effect of predators (chicks) experiencing unpalatable aposematic 'prey' with just one distasteful agent (Bitrex or quinine) or a mix of prey representing both types of unpalatability. (Source: Adapted from Skelhorn & Rowe 2005. Reproduced with permission from The Royal Society.)

them so that the birds would taste it when eating them. The birds that tasted the compound on the outside of the mealworms displayed 'disgust responses', but those that ate injected prey did not. In subsequent paired trials, birds read-

ily ate injected mealworms but ate significantly fewer of the ones coated in Bitrex, indicating that an unpleasant taste might in itself provide protection from predation under some circumstances (see Chapter 6, section *Müllerian mimicry and unequal defence*).

This finding lends strong support to the probabilistic model of Gohli & Högstedt (2009), which postulates that the evolution of warning colours can be explained if secondary defence compounds are secreted on to the prey's exterior such that a predator will experience them before they actually kill or ingest them. This is likely to be enhanced if conspicuous colouration causes predators to spend longer examining a potential prey before actually consuming it. If the secreted compounds are also smells then the period of opportunity for the predator to decide not to consume it will be increased further. One animal, a harvestman, seems to have mimicked prior release of secondary defensive compounds successfully (see Chapter 7, section *Aide mémoire mimicry*).

Neophobia and the role of novelty

Coppinger (1969) investigated the responses of 171 wild-caught North American blue jays, *Cyanocitta cristata*, to a selection of Neotropical butterflies whose colour patterns would have been unfamiliar to the bird, but not all of these were attacked. The blue jays showed various degrees of preference for familiar food, generalisation from previous experience, innate avoidance of specific stimuli, and innate avoidance of novel stimuli. The experiments were quite complicated, involving a number of different training diet regimes which had significant effects on the birds' responses to novel insect prey, and the results showed complex interactions. One of Coppinger's (1969) conclusions though was that Blest's (1957b) suggestion that innate recognition of eye-spot markings in Lepidoptera by potential predators served as a defence might not be correct, because the blue jays did not reject novel prey due to any particular component of their pattern, but rather based on their past experience. Coppinger (1970) therefore extended the study to naïve birds,[6] which had been hand-reared, and later trained to accept mealworms to ensure that they were experienced with eating a palatable insect. He then presented them with frozen-thawed specimens of three species of similarly sized tropical nymphalid butterflies, *Anartia amalthea, A. jatrophae*

---

6. Seventeen blue jays (*Cyanocitta cristata*), seven grackles (*Quiscalus quiscula*) and six red-winged blackbirds (*Agelaius phoeniceus*).

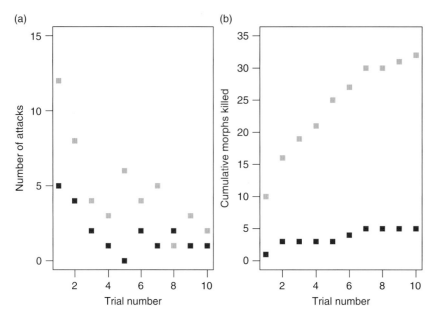

Fig. 4.20 Attacks and survivorship of cryptic (grey squares) and aposematic (red squares) colour morphs of the unpalatable bug *Lygaeus equestris* (Lygaeidae) by great tits, *Parus major*, showing that birds learned to avoid the aposematic morphs more quickly and hence killed fewer of them over the series of trials. (a) Attacks in each successive trial; (b) cumulative numbers of morphs killed. (Source: Adapted from Sillén-Tullberg 1985a. Reproduced with permission from Springer.)

and *Protogonius hippon*, in different sequences. One of the most startling observations he made was

> In most cases, the first butterflies, regardless of species, caused alarm. Blue jays raised their crests or gave alarm calls. These birds typically retreated to far corners of the cage, or often flew rapidly from side to side of the cage, banging loudly against the side. In short, rejection of the first butterfly or the first few was usually an active rejection.

Thus these birds were not innately able to recognise a butterfly as prey, quite the opposite. After the birds had got used to accepting butterflies as food, the results showed that rejection of novel prey depended on how different they were to the prey that the birds were familiar with, and that conspicuous colouration, in the absence of unpalatability, also provided a degree of protection. Ruxton et al. (2004a) pointed out a potential problem with Coppinger's (1969) work in that the effect of novelty might in fact have been confounded by an innate aversion to red prey.

The unpalatable bug *Lygaeus equestris* has both aposematic and cryptic colour forms and in experiments with hand-reared great tits, *Parus major*, Sillén-Tullberg (1985a) found that the aposematic form had higher survival because the birds showed greater initial reluctance to attack them, attacked them less lethally and learned to avoid them

quicker (Fig. 4.20). Sillén-Tullberg (1985b) used naïve rented pairs of zebra finches, *Taeniopygia guttata*, as predators with grey and red forms of *L. equestris* and again found that the red form was attacked less often but also when it was on a red rather than a grey background and hence not aposematic in its surroundings.

Lindström et al. (2001b) tested whether neophobia of predators might give initially aposematic prey an advantage by training wild-caught great tits, *Parus major*, on two novel prey types (novel world method) and found that prior experience of one novel type and learning that it was acceptable food did not influence the birds' avoidance of a second aposematic novel type. Thus, some predators appear to have been pre-programmed by evolution to be reluctant to attack novel prey which, when familiar prey are abundant, might mean that novel aposematic forms survive to reproduce. In contrast, Alcock (1973) found that hand-reared American red-winged blackbirds, *Agelaius phoeniceus*, showed no neophobia when presented with individuals of the distasteful stink bug *Euschistus consistus*. Neophobia might therefore be quite species and/or context dependent. *E. consistus*, despite its noxious taste, is a mottled dull brown rather than being aposematic.

Thomas et al. (2003) carried out a test using 10 wild-caught European robins, *Erithacus rubecula*, foraging on

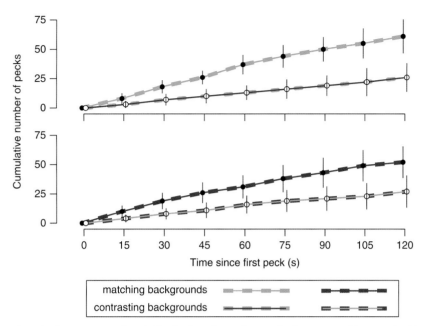

Fig. 4.21 Experimental results demonstrating that chicks that have learned to avoid either green or blue unpalatable food lose the avoidance memory when the food is now made palatable more slowly if the food is on a contrasting background: blue on green in the top figure and green on blue in the lower one. (Source: Adapted from Roper 1994. Reproduced with permission from John Wiley & Sons.)

palatable pastry prey of different colours in batches of 20. After each foraging session the proportion of prey types was adjusted, in an artificial evolution way, to reflect the survivorship of types in the previous test. The results were quite startling – the robins showed a strong avoidance of the novel prey type such that the novel form quickly increased to fixation in 14 out of the 40 prey 'populations', and also did so at least once in the case of eight of the experimental birds' trials. Attaining fixation occurred more quickly when the novel morph was of one of the classic aposematic colours, red or yellow, but also did so when the novel prey type was blue or green. Not only might such avoidance of novelty give initially rare and more conspicuous phenotypes of an unpalatable species an initial foothold, it also has some implications for the frequent occurrence of polymorphism within unpalatable taxa and within Müllerian complexes. Essentially, if fitness increases as the population density increases, then monomorphism should be favoured, but if it declines with abundance, polymorphism may be expected (Speed 2014).

C. Rowe et al. (2004), using the wild-caught great tits, *Parus major*, as predators and almond slices as artificial prey in a novel world set-up (prey paper envelopes marked with crosses, crossed squares, stars and squares), found that pattern similarity had no effect on the speed of avoidance learning, i.e. there was no evidence of predators

generalising between novel symbols. Veselý et al. (2013) explored whether European insectivorous birds, great tits and also blue tits, *Cyanistes caeruleus*, would recognise the novel aposematic south Asian red cotton bug, *Dysdercus cingulatus*, as unpalatable, having already been trained on the unpalatable European red firebug, *Pyrrhocoris apterus*, which is supposedly part of the same Müllerian complex. However, the birds initially could distinguish the Asian bug from the European one, showing that they did not include in the *Pyrrhocoris* mimicry ring, but they soon learned to avoid it, suggesting that the red cotton bugs had their own effective chemical protection.

In a novel experiment using stuffed European blackbirds, *Turdus merula*, which are important prey to sparrowhawks, *Accipiter nisus*, Götmark (1996) demonstrated that birds with wings painted bright red actually suffered fewer attacks than those with natural black wings, although they were more easily spotted by human observers. Thus, although individual sparrowhawks that did attack a redwing mutant would most likely learn that it was just as palatable as the typical form, an initial neophobia or perhaps assessment that it might be unpalatable, could permit the mutant allele to persist (Lindström et al. 2001b). In an earlier experiment, Götmark (1992) had similarly shown, by placing stuffed specimens out in the field, that highly contrastingly black and white male specimens of the pied flycatcher, *Ficedula hypoleuca*, suffered a lower level of preda-tion than the dull-coloured females. It therefore seems likely that conspicuous plumage of many male birds, especially during the breeding season, has functions other than simply signalling to conspecifics, and possibly support the unprofit-able prey/escape mimicry hypothesis (see section *Escape speed and low profitability* later).

Roper (1993) presented evidence that novelty affected the ability of domestic chicks to learn to discriminate palatable and unpalatable food. If they had been trained to drink pal-atable uncoloured fluid and were then required to discrimi-nate that from unpalatable (quinine-tainted) blue fluid, they learned to do so readily, but their avoidance of novel blue palatable fluid confounded the reverse experiment, i.e. they failed to discriminate between palatable and unpalata-ble water if both types had a novel colour. In further experi-ments it appeared that novelty facilitates visually mediated taste-avoidance learning in chicks. Roper (1994) followed this up again using domestic chicks as predators and showed that the conspicuousness of a model that a predator has learned to avoid reduces the probability that learned avoid-ance will be reversed by subsequent experience of similarly coloured palatable prey (Fig. 4.21).

INNATE RESPONSES OF PREDATORS

Earlier workers on warning colouration generally believed that avoidance of aposematic prey was inevitably the result of learning (e.g. Lloyd Morgan 1896, Cott 1940), but over the past 40 years or so evidence has been obtained that shows that predators may have innate (instinctive) aversions to par-ticular colours or patterns (e.g. Lindström et al. 1999b). While most predators do learn to avoid unpalatable prey, especially if the latter are aposematic, there is a strong body of evidence that many predators also have innate avoidance responses to particular patterns and/or colours, though as Guilford (1990b) in his mini review put it "there seems to be no simple rule relating colour pattern to innate aversion". One of the first studies on innate avoidance responses was presented by Rubinoff & Kropach (1970), who showed that various predatory fish avoided warningly coloured sea snakes, and Caldwell & Rubinoff (1983) confirmed that naïve herons and egrets avoid the incredibly venomous[7] and highly aposematic, black and yellow, yellow-bellied sea snake, *Pelamis platura*, which occasionally come to shore.

Humans have a predisposition to being able to detect snakes or snake-like things such as worms, spiders, centipedes, etc., even as young children (LoBue & DeLoache 2008), although fear *per se* appears to be learned. There are appropriate medical terms for many of these fears, for exam-ple, arachnophobia for fear of spiders. It is not hard to realise the potential evolutionary significance of this, especially if a person has grown up in a country where seriously venom-ous and dangerous spiders, snakes, etc. are common.

Such phobias are not restricted to humans but are also apparent in some primates. In Thailand where I now reside, in car parks where there are a lot of monkeys that are prone to damage aerials and windscreen wipers, etc., you can often hire large crocodile cuddly toys to drape over your vehicle (Fig. 4.22). This surprisingly simple and inexpensive exercise is an effective deterrent to all but the bravest indi-viduals, the monkeys clearly having a completely innate wariness of crocodilians, even though in nearly all these places not one of the living animals would ever have encountered a crocodile, and certainly not have witnessed another individual being attacked and consumed by one. These monkey troops are usually near temples in the hills and a long way away from any living crocodiles.

Little is known about how widespread such aversions are among visually hunting predators, nor how important they might be in shaping the evolution of aposematic colours and forms among their potential prey. It is also important to distinguish between aversions to particular colours or pat-terns and aversion to novelty – but how a predator perceives novelty is also a difficult question (see section *Neophobia and the role of novelty* earlier).

Schuler & Hesse (1985) found that naïve, young domestic chicks when presented with mealworms, *Tenebrio molitor*, painted either green, olive or with black and yellow stripes, although directing the same pecks at them at the same rate, consumed the black and yellow ones at a significantly lower rate. The olive-coloured prey were the result of painting the mealworms with the black and yellow paints mixed and therefore the aversion to eating them was not the result of some chemical in either the black or yellow paints but an in-borne behavioural response to the resulting colour. Roper & Cook's (1989) and Roper's (1990) experiments with domestic chicks revealed that certain coloured artifi-cial prey were avoided but there was no universal rule. Naïve chicks were strongly averse to novel black prey, mildly so to black and yellow striped prey (confirming Schuler & Hesse 1985) but only marginally so to red ones, and they were not averse to various other bright colours.

Naïve domestic chicks are omnivores and will eat grain, berries and insects, and this led Gamberale-Stille & Tullberg (2001) to test whether unlearned colour preferences dif-fered depending on whether chicks were presented with insect or fruit-like food items coloured red or green. In their first experiment they found that chicks had a significant

---

7. The venom of *Pelamis* is many times more toxic to vertebrates than that of rattlesnakes, cobras or coral snakes.

Fig. 4.22 Use of large crocodile toys to protect parked vehicles from monkey damage at the monkey temple, Wat Tham Khao Luang, Phetchabury, Thailand. (a) Crocodile toys available to hire; (b) 'crocodiles' doing their work.

preference for green compared with red insects ($\chi^2 = 6.9$, d.f. $= 1$, $p < 0.001$), but not if the food were the artificial fruit, and therefore the chicks' innate colour biases were context dependent. Their second experiment examined whether it was important that the insect prey were alive or not, and found no difference between treatments, and therefore the birds were able to recognise that the prey were animals rather than fruit just by their appearance; whether or not they were moving was unimportant.

The innate aversion to red (and yellow) shown by domestic chicks is readily enhanced by association with other features typical of warningly coloured prey such as bad-tasting or odoriferous chemical defence or sounds (e.g. G.D.H. Carpenter 1938). The aversion can even be increased if the red food is palatable but given in association with unpalatable brown standard food or simply colourless quinine-flavoured water (C. Rowe & Skelhorn 2005). In nearly all studies involving taste, experimenters used quinine solution as the aversive taste compound. Skelhorn et al. (2008) wondered whether this was part of a general phenomenon or something specific to quinine (which is a natural compound chemically fairly closely related to some of the protective chemicals used by insects). Comparison between quinine and Bitrex or quinine plus Bitrex showed that actually the chicks' avoidance learning effect was entirely limited to quinine and there was no additive or synergistic effect of including Bitrex.

Coral snake patterns are found in animals as far north as Minnesota, where coral snakes do not occur. Indeed, coral snake patterns occur in mimics in large parts of North America allopatric to the models. Pfennig & Mullen (2010) discussed their significance and note three possible explanations : (1) the supposed Batesian mimics might in fact be unpalatable, (2) predators may show innate avoidance, or (3) the patterns might be the result of convergence due to different reasons. Another possibility could be that some important predators are migratory and do learn from experience when in regions in which the model lives. S.M Smith (1975, 1977) found that hand-reared members of two Neotropical bird species, turquoise-browed motmots (*Eumomota superciliosa*) and great kiskadees (*Pitangus sulphuratus*), both of which are regular snake predators, have an innate aversion to a red and yellow ring pattern such as that found in coral snakes. In contrast, S.M. Smith (1980) found no such innate aversion to the pattern in three Nearctic bird species, house sparrows (*Passer domesticus*), blue jays (*Cyanocitta cristata*) and red-winged blackbirds (*Agelaius phoeniceus*), although two of these would not normally be expected to attack snakes.

It has been known for a long time that birds may show innate aversion to insects of different colours (F.M. Jones 1932), and as Miriam Rothschild noted (1986), for human observers, "red has a remarkable capacity to draw attention to itself". Dolenská et al. (2009) found that hand-reared great tits, *Parus major*, displayed innate aversion to the patterns of two ladybird beetles (*C. septempunctata* and *Scymnus frontalis*), but Exnerová et al. (2007) found that naïve great tits would readily attack chemically defended red and black firebugs, *Pyrrhocoris apterus* (Pyrrhocoridae), whereas wild-caught birds did not. It is not clear how to explain this difference, since Exnerová et al. (2007) did find evidence for innate aversion to the red and black bug in some other tit species. Forsman & Merilaita (1999) presented artificial 'butterfly' prey to domestic chicks with different visual signals and found that the protective value of visual warning displays is enhanced by increased size of the signal pattern elements and decreased by asymmetry.

Central to the classical cases of Batesian and Müllerian mimicry is that the models are not only unpalatable but also display warning colouration, i.e. they are boldly patterned and stand out from the surrounding environment, thus advertising themselves. This poses an immediate evolutionary conundrum, because although once evolved, such signals can be learned by predators which subsequently avoid similarly coloured conspecifics as well as heterospecific mimics, surely the first individuals to display conspicuous markings would put themselves at a selective disadvantage. Predators would be particularly likely to spot them, attack them and, quite possibly, kill them, thus selecting against the genetic combination or mutation causing their conspicuousness.

In Müllerian mimicry, locally at least, selection will show density dependence, that is, the level of predation they experience increases with the total population density of the members of the complex because of the proportional opportunity for local predators to learn to avoid them (Llaurens et al. 2013) (see also *Spatial models and metapopulations* earlier).

APOSEMATISM AND GREGARIOUSNESS

Many aposematic and unpalatable species live gregariously (e.g. Aldrich & Blum 1978) and it has often been assumed that living gregariously, and particularly in groups of related individuals, is an important factor in the evolution of warning colouration. Indeed, several models for the evolution of warning colouration take as an assumption the existence of family groups that share unpalatability and other genes though several other studies, both experimental and theoretical, have shown that kin selection need not be involved, even though there is evidence that in some species, kin members preferentially aggregate with one another (Waldman & Adler 1979, Waldman 1982).

Sillén-Tullberg (1990) tested whether aggregations of aposematic individuals are attacked less than solitary ones, because although it was a widely held view in the literature

that groups of aposematic individuals would be attacked less because the gregariousness served to enhance the aposematic signal, there had been no tests of this assumption. As she later admitted, she had expected to find that groups would suffer less but they did not. Her findings were criticised by W.E. Cooper Jr (1992) because Sillén-Tullberg's (1990) conclusion was based on a significant interaction term[8] for the two prey species tested; i.e. most attacks on the *Papilio machaon* caterpillars occurred when they were presented in groups and most attacks on the sawfly larvae (*Neodiprion sertifer*) occurred when they were presented as individuals. Interpreting interactions like this is potentially risky. Sillén-Tullberg (1992) presented further data on attacks by quail on *P. machaon* caterpillars, either presented as solitary larvae or in groups of 20, and found in three separate tests that groups suffered significantly more attacks. Alatalo & Mappes (1996) carried out a novel world experiment and obtained evidence that gregariousness favoured the initial evolution of aposematism and then once predators had learned the warning signals, gregariousness favoured the evolution of Müllerian mimics even among solitary individuals. This work was followed up by Tullberg et al. (2000), who first reanalysed Alatalo & Mappes' (1996) data by excluding attacks on palatable cryptic 'prey', and the redrawn graphs indicated no difference in relative mortality between solitary and gregarious treatments. Tullberg et al. (2000) then carried out fresh novel world experiments comparing attacks by great tits on aggregated or solitary cryptic unpalatable and aposematic unpalatable prey. Their results showed no difference between aposematic and cryptic prey in the aggregated presentations, but aposematic prey were attacked significantly less after the birds had first learned their unpalatability in the solitary situation, thereby providing no evidence of an advantage of gregariousness in terms of individuals killed. Similar lack of evidence for a direct benefit of aggregation was presented by Lindström et al. (1999b) from experiments with wild-caught and hand-reared great tits.

Gagliardo & Guilford (1993) reversed the standard question by asking why aposematic insects live gregariously. Using domestic chicks and yellow- and green-dyed crumbs, they manipulated whether prey were visible singly or in groups and whether birds seeing groups could only ingest single prey items. In the experimental phase, chicks were first trained on unpalatable crumbs (sprayed with a quinine and mustard mixture) under various treatments and the cumulative numbers attacked levelled off, as expected. They then switched the prey from being unpalatable back to palatable and followed

---

8. The way the response variable behaves to one explanatory variable is dependent on the value of a second explanatory variable.

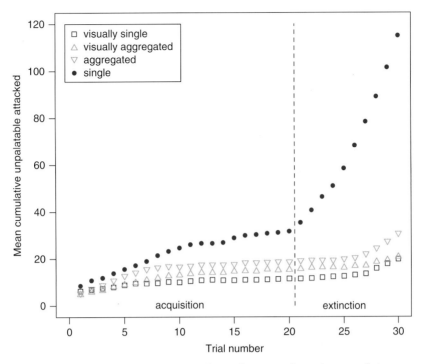

Fig. 4.23 Effect of whether artificial unpalatable prey are presented to predators visually singly or visually in groups on avoidance learning ('acquisition') and subsequent forgetting ('extinction') following rendering them palatable. Experiments were with domestic chicks and coloured food crumbs, and in the case of the treatments labeled "visually ..." birds were able to see all the crumbs presented but could only attack one. (Source: Adapted from Gagliardo & Guilford 1993 with permission from The Royal Society.)

the birds as they relearned that they are palatable. They found a startling and highly significant effect of whether prey were presented singly or in groups (Fig. 4.23). When 'prey' items were aggregated visually, the birds both learned to avoid them quicker and were slower at relearning that the yellow crumbs were once again palatable.

Alatalo & Mappes (1996) compared attacks by great tits, *Parus major*, on three tree types (palatable cryptic, unpalatable cryptic and unpalatable aposematic) presented either singly or in groups in a novel world environment to avoid problems with pre-existing biases, and concluded from the results that selection would favour the evolution of aposematism for prey living in groups but not for solitary prey. However, Tullberg et al. (2000) pointed out that the proper comparison to be made would only involve unpalatable prey. They reassessed Alatalo & Mappes' (1996) data and carried out new experiments with wild-caught great tits. Contrary to Alatalo & Mappes' (1996) conclusion they found no evidence that the birds differed in their attack rates on cryptic or aposematic prey presented in aggregations.

Gamberale-Stille (2000) suggested that gregariousness makes the aposematic signal stronger and therefore means that predators will spot the cluster from a greater distance

away than they would an isolated individual, and therefore have more time in which to make sure that they want to attack, i.e. remember a past negative experience (see Endler 1988 for discussion). To test this, Gamberale-Stille (2000) presented domestic chicks with live aposematic and non-aposematic prey either as single individuals or in groups of nine and found that latency to attack clustered aposematic prey was significantly longer than for solitary ones (Fig. 4.24), and that learning to recognise unpalatable aposematic prey was also quicker when prey were presented in groups. Furthermore, chicks were more ready to attack and eat defended prey when a competitor bird was present (Fig. 4.24, compare right-hand pairs of points with left-hand pairs), with significantly increased numbers of prey killed and slower latency to attack.

Riipi et al. (2001) presented great tits, *Parus major*, with a novel world environment in which there were various-sized groups of prey of various levels of conspicuousness. One experiment tested the effect of appearance and group size on detectability and involved entirely palatable prey, and a second experiment had unpalatable aposematic prey in varying group sizes as well as cryptic solitary palatable ones. They found that risk of detection increased only

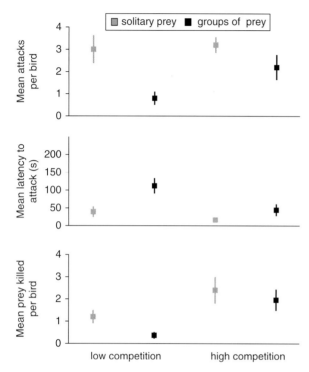

Fig. 4.24 Results of experiment with domestic chicks predating live aposematic defended prey presented either as single prey items or in groups of nine, with and without competing birds present, and showing that prey in groups, representing gregarious aposematic wild prey were attacked more slowly and had fewer eaten compared to solitary prey. (Source: Adapted from Gamberale-Stille 2000. Reproduced with permission from Elsevier.)

asymptotically with group size and that adding increased conspicuous colouration had only a small additional effect on group detectability. Because naïve great tits that encountered an unpalatable individual in a group abandoned the group, and also learned more rapidly when unpalatable prey were in a group, gregarious behaviour should be selected for in aposematic taxa. Riipi et al.'s (2001) results indicate that gregariousness should be expected to evolve in aposematic prey even without invoking kin selection.

Direct demonstrations that warning colour patterns experience positive frequency-dependent selection in the field are not common. Chouteau et al. (2016) were able to demonstrate it in the multiple mimicry complexes displayed by unpalatable butterflies in South America (see also Chapter 7, section *Heliconius*). They showed that in local prey assemblages, what potential distasteful prey predators learned to avoid depended on the relative local abundance of their aposematic phenotype. Therefore, as a corollary, life history traits that favour predator learning of warning patterns, such as enhancing their visibility, will be strongly selected for. Gregariousness is one obvious

trait that will do this, but others, such as convergence of niches (microhabitats, phenology) between co-mimics can also contribute to enhanced predator learning.

Phylogenetic analysis of aposematism and gregariousness

It would be easy to fall into the trap of assuming that because quite a lot of aposematic taxa live gregariously, it was gregariousness that facilitated the evolution of aposematism, but the other possibility, that initial aposematism favoured the evolution of gregariousness, also needs to be considered. One way of trying to tease these things apart is by the use of phylogenetic comparative methods, whereby life history traits such as aposematism and gregariousness are mapped onto independent phylogenies.

Sillén-Tullberg (1988) examined the incidence of both aposematism and larval gregariousness from a phylogenetic view point. Using independently derived cladograms for various selected groups of butterflies, she showed that gregariousness had evolved on some 23 occasions, while aposematism appeared to have evolved on some 12 separated occasions, including five times in cryptic species of unknown palatability. For those groups of butterflies in which both larval gregariousness and aposematism occur and in which the evolution of the two characters could be distinguished in the cladograms, five separate evolutions of warning colouration could be distinguished, which were followed by the evolution of gregariousness on 15 separate occasions. Importantly, on no occasion did it appear that the evolution of gregariousness had preceded the evolution of aposematism. Thus while published phylogenies cannot necessarily be taken to be accurate estimates of phylogenetic history, the figures that have emerged from Sillén-Tullberg's (1988) analysis seem to strongly support the notion that gregariousness is disadvantageous for palatable taxa and is therefore only likely to evolve in unpalatable organisms. As Sillén-Tullberg (1988) put it, "unpalatability is an important predisposing factor for the evolution of egg clustering and larval gregariousness in butterflies". Furthermore, her data also indicate that kin selection, at least in the butterflies investigated, probably plays only a minor role in the evolution of gregariousness in butterfly larvae.

Tullberg & Hunter (1996) carried out such an analysis for Lepidoptera using those phylogenies for various lineages that were available. Reconstructing the evolution of warning colouration showed that the great majority of transitions occurred in lineages that had solitary larvae (Table 4.3), thus supporting the idea that aposematism, at least in these moths, evolved predominantly through individual selection, though some role of kin and family

Table 4.3 Transitions from having solitary caterpillars to gregariously living ones are far more likely to occur if larvae are already warningly coloured (sign test, $p = 0.019$) and/or have repellent defence mechanisms (sign test, $p = 0.002$) as shown by phylogenetic independent contrasts analysis. A dash means no data. (Data from Tullberg & Hunter 1996. Reproduced with permission from Elsevier.)

| Taxon for which contrasts could be calculated | Contrasts with highest proportion of transitions to gregariousness in branches with: | | | |
	warningly coloured	cryptic	repellent defence	no (or unknown) repellency
Bombycoidea		1	1	
Sphingidae	1		1	
Lasiocampidae	1		—	—
Saturniidae	1		—	—
Geometridae	1		1*	
Notodontidae	3	1	2	
Noctuidae	3		2	
Pieridae	—	—	1	
Nymphalidae	—	—	1	
TOTALS	**10**	**2**	**9**	**0**

* Defensive feature involved is a leaf roll or tie.

selection cannot be specifically ruled out, since related caterpillars could still be living close-by.

BEHAVIOUR OF PROTECTED APOSEMATIC ANIMALS

Aposematic taxa, no doubt because on average predators have learned to avoid them, tend to behave differently from those species whose primary defence is to flee when detected.

Of birds and butterflies

Entomologists, and perhaps butterfly collectors in particular, have long noted that many well-known distasteful aposematic species tend to fly slowly and lazily, and some, when threatened, do not immediately flee to safety, something noticed by both Bates (1862) and Wallace (1865). Palatable species, on the other hand, generally fly quickly and erratically, though not those palatable species that are Batesian mimics (Srygley 2004). Srygley & Chai (1990b) summarise descriptions in the literature thus:

> [M]ost pierid and nymphaline butterflies have been described as "fast," "erratic," "strong," "swift," and "zigzagging," whereas others, such as those of many heliconiine, danaine, and ithomiine butterflies and their mimics, are said to be "slow," "heavy," "fluttery," "feeble," and "deliberate".

Miriam Rothschild (1963) suggested that probably all butterflies were in some way protected as only then would they be able to fly so freely in the daytime and generally sport such conspicuous colours. Anyone who has observed large numbers of several species of butterfly visiting a 'butterfly bush' such as *Buddleja*,[9] with not a single insectivorous bird paying attention to them, would not be surprised. I have only ever witnessed one bird chasing a butterfly (a house sparrow, *Passer domesticus*, chasing a *Pieris* sp., and failing to capture it), and entomological friends of mine also report seeing very few such incidents; many naturalists in the early part of the twentieth century thought that most butterflies were too agile to be caught on the wing by birds (see later). Therefore, several workers have questioned whether bird predation could be sufficient to lead to the evolution of some of the extremely accurate mimicries found among them (see G.A.K. Marshall 1909 for discussion of contemporary views at the time), or whether bird predation is really very important in regulating butterfly populations (Muyshondt & Muyshondt Jr 1976). Studies of temperate insectivorous bird stomach contents (e.g. Collinge 1937) showed that butterflies constitute only a small part of the birds' diets but this does not mean that they do not have an important selection effect.

---

9. Article 60 of the 2006 International Code of Botanical Nomenclature (Melbourne Code) requires that Linnaeus' original spelling should be used rather than *Buddleia*, which has been widely used for a long time.

Fig. 4.25 Nymphalid butterflies showing wing damage due to bird attack. See also Chapter 8, section *Eyespots*. (a) Lemon pansy, *Junonia lemonias*, Vietnam; (b) intact and attacked individuals of the common fivering, *Ypthima baldus*, Thailand; (c,d) banded treebrown (*Lethe confusa*) from China showing symmetric damage. (Source: c,d, Itchydogimages. Reproduced with permission from John Horstman.)

Collenette (1935) and G.D.H. Carpenter (1937, 1939) sum-marised a considerable number of observations by corre-spondents indicating that birds have been seen to attack butterflies (often missing them) as well as evidence from beak marks left on the wings of butterflies that managed to escape (Fig. 4.25) (D.A.S. Smith 1979, Wourms & Wasserman 1985a). While direct observations of birds eating butterflies, especially ones in flight, are rather rare (Pinheiro 2004, Kiritani et al. 2013), there are some birds that do seem to be more specialist butterfly eaters. The bulk of wild observations of wild-bird predation on butterflies come from the tropics (K.S. Brown & Vasconcellos-Neto 1976) and/or places where butterflies and some aposematic moths are aggregated either as overwintering roosts (Carpenter 1933), aestivating or while puddling[10] (see cover image) (Rothschild 1986, Burger

& Gochfeld 2001). The latter may give some clue as to the apparent rarity of casual observation because, in general, butterflies are sparsely distributed (Bowers et al. 1985). Since observation of individual chases/captures of prey in the wild is rare, direct observation of individual predators sampling potentially unpalatable prey and subsequently rejecting them in the wild is exceedingly unlikely (L.P. Brower 1963, Boyden 1976).

It should also be noted that birds are not the only preda-tors of free-foraging butterflies as there are a number of observations in warmer parts of the world of lizards taking them (Ehrlich & Ehrlich 1982, Odenaal et al. 1987), not to mention praying mantids and spiders (Vasconcellos-Neto & Lewinsohn 1984, Sourakov 2013a – see also Chapter 8, section *Reverse mimicry*). Ehrlich & Ehrlich (1982) com-ment that the butterflies they observed being eaten by igua-nid lizards at Igaçu Falls in southern Brazil were completely devoured, wings and all. In the case of the *Anolis* lizards observed by Odenaal et al. (1987) it is interesting that they consumed the aposematic and toxic swallowtail *Battus*

---

10. Males of many species of tropical butterfly visit damp soil/gravel patches and drink from them to gain nutrients (see cover photograph).

Table 4.4 Incidence of bird beak marks on museum specimens of three groups of tropical butterflies; *Euploea* from Asia, and other danaines from Africa (both Nymphalidae: Danaini) are generally unpalatable, whereas *Colotis** (Pieridae) are generally thought to be relatively palatable. The proportions with beak marks differ highly significantly between the groups with the pierine specimens having a very much lower incidence ($\chi^2 = 96.4434$, d.f. $= 2$, $p < 0.0001$) and are believed to indicate that many more of the palatable *Colotis* attacked by birds do not survive to be collected. (Data from Carpenter 1941. Reproduced with permission from John Wiley & Sons.)

Butterfly group	Number of specimens	Number showing beak marks	Percentage with beak marks
Euploea	10,661	123	1.15
African Danainae	6,322	117	1.85
African *Colotis*	7,163	11	0.15

* A common, principally African genus of white butterflies often with coloured wing tips, with approximately 50 described species.

philenor readily and apparently without ill-effect, and J.V.Z. Brower & Brower (1961) found the same was true of white-footed mice, *Peromyscus leucopus*.

Examination of butterfly specimens in collections and in the field shows that some have quite distinct damage due to attacks by birds' beaks (Lamborn 1920) – beak marks – and for a while this was quite a popular and fruitful area of study for those interested in mimicry. G.D.H. Carpenter (1941) summarised the incidence of beak marks on three groups of butterflies in the old Hope Department of Entomology collection in Oxford University Museum, and the results are shown in Table 4.4. The clearly lower proportion of beak marks on the more palatable *Colotis* led Carpenter (1941) to write:

> These facts are interpreted as evidence of the greater destruction of species not furnished with aposematic characters, which, when attacked, do not escape like the tougher, more distasteful species.

That, indeed, is the perceived wisdom. The other possibility, of course, is that the pierids successfully elude their predators until the latter give up, and numerous casual observations cited by Collenette (1935) and Carpenter (1941) include ones where the birds were unsuccessful in their chase. M. Edmunds (1974b) is, for the same reason, critical of how various authors have interpreted the meaning of different proportions of beak marks on different species. Even with mark–release–recapture data, without knowing predation figures accurately it may not be possible to guarantee interpreting the data correctly. He also suggests the possibility that eyespots on butterfly wings, while possibly acting as deflectors of attack (see Chapter 8, section *Attack deflection devices*) might actually encourage attack. Balgooyen (1997) interestingly describes mimicry of a butterfly by a grasshopper, which is attributed to the butterfly's

ability to evade capture. In this case the mimic outnumbers the model by between two- and four-fold but is less apparent to avian predators because of where it normally rests; birds probably detect many more of the butterflies.

It has long been proposed that unpalatable species of butterflies nearly always have tough wings that allow them to withstand some degree of handling by predators, although there was very little by way of supporting evidence. DeVries (2002, 2003), using a simple method of applying weights to wings, compared the toughness of the hind wings of various Afrotropical nymphalid butterflies. The first study compared *Junonia terea* (Nymphalinae), *Bicyclus safitza* (Satyrinae), *Amauris niavius* (Danainae), *Acraea insignis* and *A. johnstoni* (Acraeinae) of which the latter three are considered unpalatable. The second study compared *Amauris albimaculata*, *Pseudacraea lucretia* and *Cymothoe herminia*, which are, respectively, an unpalatable model, its mimic, and a close but non-mimetic relative of the mimic. In all cases the models had significantly tougher wings than their mimic or non-warningly coloured species, and in the second study, the mimic had significantly tougher wings than the non-mimic. Although only eight species in total were compared, his data support the suggestion that the unpalatable model is tough enough to withstand some attack, its mimic which presumably attracts attack by naïve predators also has some degree of physical protection, and more so than a less conspicuous relative. DeVries's (2002, 2003) findings provided some harder evidence for what was already widely appreciated by natural historians interested in butterfly mimicry such as Poulton, Marshall, and Carpenter (e.g. Poulton 1887b, Rothschild 1971). Kassarov (2004) levelled considerable criticism at DeVries's (2002, 2003) conclusions, stating "I fully disagree with these concepts. I consider them the result of

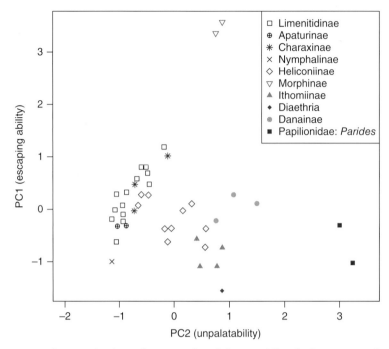

Fig. 4.26 Principal component ordination of 42 butterfly species released close to wild kingbirds, *Tyrannus melancholicus* (Tyrannidae) in Brazil showing the apparent trade-off between unpalatability and flight escape ability. (Source: Adapted from Pinheiro 1996. Reproduced with permission from Elsevier.)

conclusions made on the basis of an experimental design that does not mimic natural conditions". However, Kassarov (2004) misses the point – the model may well be attacked, especially by naïve predators, and thus they will need to have wings that can withstand such attacks until the predator detects their odour or a small amount of taste. Chai (1996) found that hand-reared naïve rufous-tailed jacamars, *Galbula ruficauda*, which are relatively specialised butterfly predators, readily attacked live butterflies when they were first offered as food and, importantly, quickly taste-rejected unpalatable individuals, most of which survived the encounter. In addition, unpalatable models often have higher longevities (Blest 1963, Blanco & Sherman 2005) as do species with defences such as eyespots (Beck & Fiedler 2009) and thus wing toughness might simply allow them to live longer. The fact that distasteful butterflies often have slow flight should not be taken as an indication that they are not tough. Indeed, reading a book about butterfly collecting as a child, and the 'collector's pinch'[11] that

---

11. A sharp, reasonably strong pinch of a butterfly's thorax that 'kills' it quickly, causing little damage, no doubt largely by damaging the three thoracic ganglia that control leg and wing movement.

butterfly collectors use to disable their specimens, I recall it saying that the pinch does not work for *Danaus plexippus*, which is an infrequent and distasteful vagrant across the Atlantic to the UK. In contrast, palatable butterflies, especially if they have distracting wing marks such as eyespots, ought to be expected to have more fragile wings, such that if a bird pecks at them the wing will tear easily, allowing to butterfly to escape largely unharmed (for example, the peacock pansy butterfly, *Junonia almana*, see Fig. 8.1d). Butterflies can still fly reasonably well with surprisingly large proportions of their wings missing (Kolyer 1968). Kassarov (1999) thought it unlikely that birds would be able to taste-reject butterflies on the basis of tasting their wings because their taste buds are located at the base of the tongue, but does not consider that while grabbing the potential prey they might still be able to smell it.

By releasing 668 individuals of 98 different butterfly species in Brazil close to wild kingbirds, *Tyrannus melancholicus* (Tyrannidae), which readily chase, attack and eat flying butterflies, Pinheiro (1996) found a clear trade-off between unpalatability and flight escape success (Fig. 4.26). He reported that:

> Only the troidine swallowtails (*Parides* and *Battus*; Papilionidae) were consistently rejected on taste and elicited aversive behaviours in birds. Most other aposematic and/or mimetic

species in the genera *Danaus* and *Lycorea* (Danainae), *Dione*, *Eueides* and *Heliconius* (Heliconiinae), *Hypothyris*, *Mechanitis* and *Melinaea* (Ithomiinae), *Biblis*, *Callicore* and *Diaethria* (Limenitidinae) were generally eaten.

This indicates that, like cuckoos that regularly feed on prey that are generally unpalatable to other birds, kingbirds are probably able to withstand at least moderate levels of several toxins. It might be argued that the diversity of tropical butterflies and other insect prey means that they ingest a variety of different types of defensive toxins but hardly ever sufficient of any one toxin to cause serious harm. Freeland & Janzen (1974) proposed the same idea about tropical forest seed-eating mammals, that perhaps they only sample small amounts of each of the various toxic seeds and so avoid overdosing on any single set of phytotoxins. Several studies have shown that generalists do better on mixed diets, but this has most often been assumed to be due to complementarity of nutrients rather than balancing toxin loads. Nevertheless, the latter certainly cannot be ignored – indeed toxins may be seen simply as negative nutrients (Westoby 1978).

Humans almost certainly do not have the same taste responses as many avian predators, but some butterflies (the usual suspects) were found by Larsen (1983) in a taste test, to be rather noxious. The most unpleasant he tried was one individual of *Danaus chrysippus*, which even left his lips blistered, although two others were less unpalatable. Second worst was an *Acraea eponina*, which left "a distinctly nauseous aftertaste". Larsen (1983) noted that he found nothing unpalatable about the three pierid butterflies that he sampled (*Mylothris ?phyleris*, *Anaphaeis creona* and *Catopsilia florella*). The question therefore arises as to why the vast majority of pierid butterflies are so conspicuous, predominantly largely white or yellow, usually with some red, orange, purple or black marks mainly at the apex of the forewing. Differences between predator species in taste response to, or induced sickness from, eating butterflies is clearly of great importance in understanding particular aposematic and mimicry systems (Boyden 1976). Nevertheless, even ants have been known to find some butterfly species far more unpalatable than others. Experiments with feeding ants drops of mashed Ugandan butterfly body tissue Molleman et al. (2010) revealed an eight-fold difference in ant feeding bout duration between butterfly species.

A surprisingly high proportion of the European observations of birds chasing or attacking butterflies in G.D.H. Carpenter's (1937) paper involved white *Pieris* butterflies or other pierid butterflies (14 cases out of 21), but whether this was because they are a favoured prey item or particularly common or just more conspicuous to observers than other species is impossible to know – probably a combination of all three. *Pieris* larvae feed almost exclusively on Brassicaceae, plants that are fairly well-defended by their glycosinolate precursors of 'mustard oils' that are produced as a result of the plant being chewed (Aplin et al. 1975). Lyytinen et al. (1999) tried to test whether white butterflies (Pieridae) were protected and thus their conspicuous colour, aposematic, using captive pied flycatchers, *Ficedula hypoleuca*, as predators. The results were not very conclusive but in some more naturalistic settings they found that the pied flycatchers often did not attack or rejected free-flying pierids, suggesting that maybe their flight agility or odours made them unprofitable prey.

Evolution of sluggishness

It is one thing to note that defended aposematic prey tend to fly slowly or be generally sluggish in their behaviour, another to understand how and why that behaviour might have evolved in the first place. Moving less quickly might be energy-saving and/or might allow a predator to recognise the aposematic colour pattern more easily, but that must necessarily mean the predator has a better chance of catching the prey. Hatle et al. (2002) hypothesised that sluggishness might evolve in some groups through motion-orientated predators culling the fastest or first-to-move individuals of an aposematic species. Many predators, such as frogs and toads and some mammals, rely for prey detection largely on seeing prey movement rather than 'carefully' matching prey shapes. Hatle et al. (2002) carried out experiments to test this using the gregarious noxious eastern lubber grasshopper, *Romalea guttata* (=*microptera*), and northern leopard frog, *Rana pipiens*, as predator. First they demonstrated consistent inter-individual variation in rate of movement of otherwise matched grasshopper individuals, therefore showing that there is statistically significant variation for selection to work on (although they did not investigate its heritability it would seem probable that there is some genetic component). They then showed in predation experiments that, indeed, the more sluggish individual grasshoppers were significantly less likely to be eaten by the frogs (two-tailed Wilcoxon test; $T+ = 61$, $p = 0.0098$) and that non-moving grasshoppers were significantly better protected than sluggishly moving ones ($T+ = 21$, $p = 0.05$). Further experiments corroborated this with aggregations of prey. Therefore, with motion-orientated predators, and possibly ambush predators more generally, it seems that selective predation operating against faster moving individuals could explain the evolution of sluggishness as a protection/defence against predation.

With more pattern-discriminating predators such as birds, and non-aggregating prey such as butterflies, it must still be true that the more slowly moving aposematic individuals must have been at a selective advantage, but the operating mechanism must be different.

ORIGINS OF PROTECTIVE COMPOUNDS

Many toxic herbivorous insects, such as many butterflies and moths (Lepidoptera) and beetles (Coleoptera), obtain their chemical toxins by sequestering them from their host plants (Nishida 2002), though particular groups are well known for synthesising their own toxins *de novo* (Eisner 1970). Plant-derived carotenoid pigments may also contribute to aposematic colouration (Rothschild & Mummery 1985, Rothschild et al. 1986). In contrast, vertebrates, in general, synthesise their toxins *de novo*, although there are exceptions (see poison dart frogs later).

Rothschild et al. (1970) referred to sequestered toxins as category one compounds, and *de novo* synthesised ones as category two. L.P. Brower (1984) introduced the idea of "class II chemicals", such as pyrazines, which are in themselves harmless but are involved in many of these cases as signals of likely toxicity (Rothschild et al. 1984b, Guilford et al. 1987).

Sometimes a chemical associated with a dangerous animal may be acquired in strange ways. For example, Clucas et al. (2008) showed that ground squirrels, *Spermophilus* spp., apply rattlesnake scent to themselves by vigorously licking their fur after chewing on shed rattlesnake skins. This then helps ward off other predators that might normally attack these relatively defenceless rodents. Similarly, Kobayashi & Watanabe (1981) describe how Siberian chipmunks, *Eutamias sibiricus asiaticus*, chew on carcasses of various snakes and then self-groom, sometimes even anointing themselves with snake urine.

Humans have also employed the same technique. During the Vietnam War, it was common, and still is, for farms and other properties to keep geese around them as they were excellent at raising the alarm if a stranger such as a Viet Cong fighter ('sapper') was trespassing. But geese react quite differently to the sight and smell of snakes, keeping quiet and – anthropomorphically speaking – hoping not to be detected. This behaviour was utilised by the Viet Cong as described by Col. Hoang Ngoc Lung, a former senior South Vietnamese intelligence officer, in Sorley's (2010) book on the war:

> The sapper threat was recognized and given high priority by security units. Small outposts took inexpensive measures for detecting infiltration which were nevertheless effective, such as raising dogs, geese and ducks on the outer perimeter of their positions.... To deal with geese and ducks, they [the sappers]

attached a stalk of blackened water potato plant to the end of a walking stick and dangled it upwind in front of the birds. Thinking they saw snakes, the birds did not dare make a sound. Another way they distracted the ducks and geese was to run green onion leaves on the sappers' bodies. The smell frightened the birds because they thought they smelled vipers.

It is relatively easy to conceive how herbivores and plants are in an arms race when it comes to protective compounds. Plant and animal physiologies have many differences, so plants may evolve pathways for synthesis of compounds that will poison herbivores and so gain protection, then herbivores that have mutations enabling greater tolerance, detoxification or often sequestration of the plant toxins will be able to utilise these plants again. For animals to evolve a mechanism for synthesising toxins that affect other animals with similar physiologies may be evolutionarily more difficult. However, having the ability to tolerate and sequester toxic food plant chemicals may also provide the basis for the evolution of biochemical pathways for *de novo* synthesis (K.S. Brown et al. 1991).

Plant-derived toxins

Plants synthesise a wide range of protective compounds (Nishida 2002) though particular classes of compound tend to be synthesised by different clades or subsets of plant families. This means that those specialist herbivores that can overcome particular classes of plant toxin are more or less restricted to feeding on those plant families. However, there are some generalist species that seem to be able to tolerate a fairly wide range of plant-derived toxins. For example, caterpillars of the European garden tiger moth, *Arctia caja* (Erebidae: Arctiinae), feed on a very wide range of plants. Tiger moths, in general, gain a lot of their protection from their dense, long setosity or bristles. Nevertheless, they also sequester a wide range of plant chemicals including cardiac glycosides, biogenic amines, iridoid glycosides and pyrrolizidine alkaloids (see later) (Rothschild 1972, 1985a). In addition, they can accumulate compounds such as cannabinoids when fed novel toxic host plants such as cannabis (Rothschild et al. 1977), and thus presumably have some more general mechanism for recognising, transporting and compartmentalising foreign molecules. How their physiologies cope with such a diversity of toxins is not fully understood.

There is evidence that tropical plants are, on average, more toxic than temperate ones (Levin 1976, Moody 1978) and certainly the caterpillar faunas of moist tropical forests seem to be dominated by highly aposematic forms, and these often share very similar colour patterns, presumably a mixture of Batesian and Müllerian

Fig. 4.27 An assortment of common, aposematic caterpillars from Area de Conservación Guanacaste, Costa Rica, and what birds feed their nestlings. (a) *Phoebis sennae* (Pieridae); (b) *Poliopastea laciades* (Erebidae: Arctiinae); (c) *Epithisanotia sanctijohannis* (Erebidae); (d) *Chrysoplectrum* sp. (species code Burns01) (Hesperiidae); (e) *Heterochroma sarepta* (Erebidae); (f) *Pseudosphinx tetrio* (Sphingidae); (g) *Euglyphis lankesteri* (Lasiocampidae); (h) *E. jessiehillae*. (Source: Professor Dan Janzen. Reproduced with permission from Professor Janzen.)

relationships (Fig. 4.27). It is interesting that the caterpillars that most insectivorous birds in the same places feed to their chicks do not belong to this complex, and are instead dominated by cryptic green species (Fig. 4.28). No doubt, some cryptic species are unpalatable in these forests too, and are rejected by the adult birds on contact, but certainly the majority will be more or less palatable whereas a large proportion of the aposematic ones are not.

Cardiac glycosides

Cardiac glycosides (CGs), also called cardenolides, are a family of chemicals comprising a sugar group attached to a steroid and, pharmacologically, they affect vertebrate heart muscle. Many are exceedingly toxic and some have been used medicinally since ancient times (Norn & Kruse 2004). Although they are produced by quite a few different

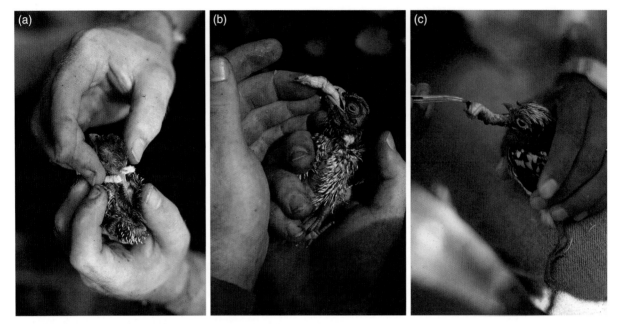

Fig. 4.28 Nestling of black-headed trogon, *Trogon melanocephalus*, from Area de Conservación Guanacaste, Costa Rica, and the diet it is fed. (a) A chick with a soft pipe cleaner put around its throat so that it cannot swallow big caterpillars; (b,c) the same chick with two different species of edible sphingid caterpillars (especially if their gut contents are removed), *Perigonia lusca* and *Manduca florestan*, respectively, fed to it by its parents. (Source: Professor Dan Janzen. Reproduced with permission from Professor Janzen.)

plant groups they are particularly well known in various Apocynaceae, which now includes the milkweeds (Asclepiadoideae[12,13]), but also occur in Scrophulariaceae, and some Asteraceae and Menthaceae. CGs and other alkaloids produced by these plants are sequestered by various danaine butterflies, such as the boldly orange and black marked African queen, *Danaus chrysippus* (see Fig. 4.30a) and monarch, *D. plexippus*, butterflies upon which a great deal of research has been conducted. Some of these CGs, depending on dosage, have a very unpleasant, emetic effect on birds that eat them in larvae, pupae or adults (L.P. Brower et al. 1968, 1975), though vertebrates vary enormously in their susceptibility to them. For example, to quote Rothschild & Kellett (1972), a quail they tested "was virtually unaffected by a dose of digitoxin (80 mg/kg) sufficient to kill 50 men". Different subspecies of *D. plexippus* rely on CGs for protection to different degrees (Rothschild et al. 1984a) and some may rely more on pyrrolizidine alkaloids (PAs) obtained during their adult lifespans (see later).

In the wild, *D. plexippus* butterfly larvae feed almost entirely on *Asclepias* species, but will sequester various chemically related toxins if these are painted onto food plant leaves, for example, ouabain from *Strophanthus* (Apocynaceae), digitoxin, digoxin and digoxigenin from *Digitalis* species (Plantaginaceae), and oleandrin from *Nerium* (Apocynaceae) (Frick & Wink 1995).

Hristov & Conner (2005), in one of the few studies of unpalatability of Lepidoptera to bats, compared how the palatability of different tiger moths (Erebidae: Arctiinae) was affected by the various classes of compounds in their larval diets. Arctiine or control palatable noctuid moths were presented in dishes to caged, wild-caught individuals of the insectivorous big brown bat, *Eptesicus fuscus*. Feeding on the arctiine moths followed a consistent sequence based upon what their larvae had fed upon and therefore contained. From least palatable to most palatable the sequence was:

cardiac glycosides < pyrrolizidine alkaloids
< (biogenic amines = iridoid glycosides)

CGs are not uniformly distributed within an individual butterfly nor between the sexes (Fig. 4.29) (L.P. Brower & Glazier 1975). Females have higher levels than males, and the highest concentrations in both sexes are in the wings,

---

12. Recent evidence (Endress & Bruyns 2000) indicates that the milkweeds (formerly Asclepiadaceae) are derived members of the Apocynaceae and given subfamily rank.

13. Often misspelt as Asclepiaceae.

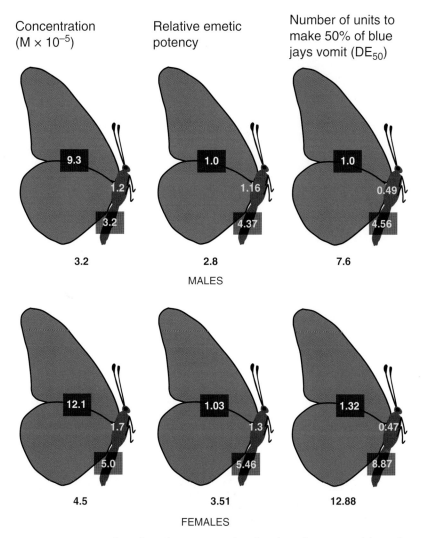

Concentration (M × 10⁻⁵)

Relative emetic potency

Number of units to make 50% of blue jays vomit (DE₅₀)

MALES

FEMALES

Fig. 4.29 Concentrations, emetic potency and numbers of emetic units of cardiac glycosides in monarch butterflies, *Danaus plexippus*, reared on the milkweed *Asclepias curassavica*. Total values of whole butterflies are given below each diagram, and values for wings, thorax and abdomen separately given on the relevant part of each diagram. (Source: Adapted from Brower & Glazier 1975 with permission from AAAS and L. Brower.)

although the abdomens are more potently emetic and contain the highest number of ED_{50} units (i.e. doses required to make 50% of blue jays vomit). Despite having the highest concentrations of CGs, the wings were found to be the least emetic. Females sequestered more powerfully emetic glycosides than did males, which suggests that the compounds are sequestered at a cost but deployed to body parts to provide maximum protection from predation. It is important to realise that predators grabbing wings will experience the highest concentrations. Vomiting takes a while to occur after consuming the toxins and birds do not consume butterfly wings when they eat them, so would only experience the emetic effect after consuming the body. The thorax is mainly occupied by flight muscles so there is not much room for tissues such as fat body, where large quantities of CGs could be stored. Interestingly, some of the interspecific differences in predation behaviour of North American birds on a Batesian mimicry system reported by Alcock (1971) may have been due to differences in their pre-consumption manipulation of the quinine-treated mealworm prey because some behaviours might lead to a higher chance of detecting the prey's noxiousness.

(a)

(b)

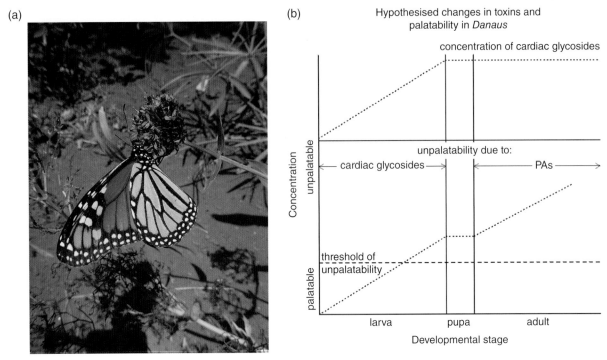

Fig. 4.30 Protected danaine butterflies and suspected changes in palatability as a result of toxin sequestration. (a) African queen butterfly, *Danaus chrysippus*; (b) levels of protective compounds through development due to larval consumption of cardiac glycosides from milkweed food plants and adult consumption of pyrrolizidine alkaloids. Even though the amount of protection from cardiac glycosides may be limited by the host plant's content, the adults can still build up chemical defences. (Source: a, Conrad P.D.T. Gillett. Reproduced with permission from Conrad P.D.T. Gillett; b, adapted from Gibson & Mani 1984. Reproduced with permission from The Royal Society.)

Variation in the dietary concentration of CGs in milkweeds and sequestered toxins in other plants, both inter- and intraspecifically, means that the adult butterflies will often show substantial intraspecific variation in palatability in a specific location. Thus the most, and possibly almost completely, palatable individuals are effectively Batesian automimics of the more unpalatable individuals (e.g. L.P. Brower et al. 1968, Bowers & Williams 1995). D.O. Gibson & Mani (1984) made a detailed study of glycoside concentrations in the African queen butterfly, *D. chrysippus*, which is summarised in Fig. 4.30b. The upper part of the figure shows that with regard to adult butterflies, the situation is complicated by the fact that male Danainae (including Ithomiini) and some other butterfly and moth species are known to feed as adults on sources of pyrrolizidine alkaloids, which are also protective (see later).

Although CGs are normally sequestered from food, at least some chrysomelid beetles have been found to be able to synthesise their own (Pasteels & Daloze 1977, Daloze & Pasteels 1979). In some cases the CGs are secreted by the larvae and adult beetles as part of their defensive behaviour, including, in some cases, reflex bleeding (Deroe & Pasteels 1977, Daloze & Pasteels 1979). Pasteels et al.

(1979) suggest that the bitterness of the cardenolides is the primary reason that vertebrates avoid eating the beetles.

Pyrrolizidine alkaloids

Pyrrolizidine alkaloids (PAs) are plant-derived toxic compounds sequestered and utilised for protection by a number of insects, notably danaine (including ithomiine) butterflies and tiger moths (Erebidae: Arctiinae) (Schneider et al. 1975, Edgar et al. 1976, 1979, K.S. Brown 1984a, 1984b, Eisner & Meinwald 1987, Cardoso 1997). In danaines and some arctiines they are the chemical precursors for the synthesis of male sex pheromones (Boppré 1978, 1990, C.A. Kelly et al. 2012).

PAs are produced by many members of more than 20 plant families, most notably Apocynaceae, Asteraceae (especially *Senecio* spp.), Boraginaceae (especially *Heliotropium*) and Fabaceae (notably monocrotaline in *Crotalaria*). However, these are not the principal host plants of some of the PA-sequestering insects, for example, danaine butterflies; it is the adult male butterflies rather than their caterpillars that seek out these PA sources to feed on.

Dead or rotting vegetation of PA-containing plants is one of the most commonly used sources (Hopkins & Buxton 1926), but they can also be obtained from the nectar of some plant species (Boppré 1978, 1990) or even from the bodies of injured, PA-sequestering grasshoppers (Bernays et al. 1977). The congregation of butterflies around living PA-containing plants suggests that they are also the plants' primary pollinators (Pliske 1975).

Males of both danaine butterflies and of the American arctiine moth *Utetheisa ornatrix* often transfer a considerable quantity of PAs to females during mating (Dussourd et al. 1989) even though the larvae of the latter feed on PA-containing *Crotalaria*. Mated females gain protection after consumption or after receiving PAs through mating and they can remain in the adult insect for a considerable period, at least 20 days (A. González et al. 1999). Females also transfer some of their PAs to their eggs, which therefore gain protection from predation as a consequence (Dussourd et al. 1989, Hare & Eisner 1993). Since the PA content of a male is directly related to the amount of sex pheromone he can produce (see earlier), females can choose to mate with those males that will provide them with the largest PA gift in their spermatophore (Iyengar et al. 2001). Alternatively, females may be highly polyandrous, each time receiving both PAs and nutrients from their spermatophores; in the case of *U. ornatrix*, each female may mate with up to 20 males (C.A. Kelly et al. 2012).

Rowell-Rahier et al. (1995) compared predation by wild-caught red-winged blackbirds, *Agelaius phoeniceus*, on larvae of two species of *Oreina* beetles (Chrysomelidae), one of which sequesters PAs, the other synthesising a CG *de novo*. Repeatedly stimulating the larvae to release their toxic secretions from their pronotal and elytral glands and washing can remove the glycosides and also the PAs on the surface, but the PA-feeding species still contains PAs after cleaning, whereas the glycosides are only on the surface and so effectively entirely removed by washing, thus leaving that species completely undefended. The birds still rejected the PA-containing larvae even after washing, and it seems possible that they can detect the PAs by their scent.

D.O. Gibson and Mani (1984), in considering the variable protection afforded to danaine butterflies according to the concentration of larval food plant CGs, pointed out that even in individuals that have low concentrations of these at emergence, chemical protection can be built up throughout the adult lifespan by PA consumption (Fig. 4.30b). In some contrast to Gibson & Mani's (1984) findings and illustration, Alonso-Mejía & Brower (1994) found that CG concentrations actually decline with age in adult *D. plexippus*, thus, in terms of this one class of compounds they become more palatable as they get older, but PAs may help compensate. The fact that danaids are reliant on obtaining both CGs from their Asclepiadoideae larval food plants and PAs

from other plants (since PAs do not occur in Asclepiadoideae) led Edgar et al. (1974) to speculate that the ancestral food plant of milkweed butterflies probably contained both classes of compound, but subsequent phylogenetic divergence, with the butterflies only tracking the CG-containing lineage, meant that they are now also reliant on finding other sources of PAs to enable successful courtship.

De novo synthesis of protective compounds

A number of herbivorous insects are also capable of synthesising toxins or their precursors themselves. Notable among these are cyanogenic species that make CGs such as lotaustralin, linamarin and samentosin (Heliconiini butterflies: Nahrstedt & Davis 1983; Acraeinae butterflies: K.S. Brown & Francini 1990; zygaenid moth eggs, caterpillars, pupae and adults: D.A. Jones et al. 1962, Holzkamp & Nahrstedt 1994, Zagrobelny et al. 2007a, 2007b, 2008; *Abraxas* (Geometridae): Nishida 1995).[14] Various other Lepidoptera are known to synthesise quantities of histamine, acetylcholine (and analogues) and proteinaceous toxins themselves (Rothschild et al. 1970, 1979). As with cantharophilic insects (see section *Obtaining toxins from animal sources* later), males of the cyanogenic burnet moth, *Zygaena filipendulae*, also transfer cyanogenic glucosides to females during mating, and females prefer males with a higher cyanogenic glucoside content (Zagrobelny et al. 2007a). The story here may be even more complicated physiologically because the greater general emission of hydrogen cyanide from adult females compared with males suggests that cyanide may be acting as a female sex pheromone, and larvae reared on acyanogenic plants do not develop as rapidly as ones feeding on cyanogenic plants (Zagrobelny et al. 2007b).

Cyanogenic capability in some insects is often associated with feeding on cyanogenic plants, as in the case with zygaenid moths, and the question arises as to which came first (K.S. Brown & Trigo 1995). If an insect has evolved *de novo* capability to synthesise cyanogenic compounds and concordant ability to withstand the inevitable occasional release of some cyanide into its tissues, it might well preadapt it to being able to colonise cyanogenic host plants (Rothschild et al. 1970, Nahrstedt 1988). Conversely, insects may have evolved to cope with cyanogenic host plants such as the bird's foot trefoil, *Lotus corniculatus* (Fabaceae), which is the food plant of *Zygaena* species, and maybe enzymatic mechanisms involved were then co-opted by evolution for defence.

---

14. Many other insects and myriapods are cyanogenic and the pathways have obviously evolved independently in many groups, though few have been studied in detail; see Zagrobelny et al. (2008) for a review.

Table 4.5 The nature of various animal-synthesised toxins that afford defence against predators.

Category of compounds	Specific names	Taxa that employ them for protection	References
Acetylenic acids	Lycidic acid	Lycid beetles (Lycidae)	B.P. Moore & Brown 1981, Eisner et al. 2008
Alkaloids	Coccinelline	Ladybirds (Coccinellidae)	Tursch et al. 1971, de Jong et al. 1991, G.J. Holloway et al. 1991, Daloze et al. 1994
	Precoccineline	Ladybirds (Coccinellidae), soldier beetles (Cantharidae)	B.P. Moore & Brown 1978
	Hypodamine	Ladybirds (Coccinellidae), soldier beetles (Cantharidae)	B.P. Moore & Brown 1978
Anthraquinones		Some leaf beetles (Chrysomelidae)	A. Kunze et al. 1996
Aristolochic acid		Some birdwing butterflies (Papilionidae: Troidini)	Nishida 1995, 2002, Klitzke & Brown 2000
Steroids		Firefly beetles (Lampyridae)	Blum & Sannasi 1974, Eisner et al. 1978b
Glycosides	Carminic acid (a hydroxylanthrapurin glucoside)	Cochineal insects (Hemiptera: *Dactylopius coccus*)	Eisner et al. 1994
Cyanogenic glycosides	Linamarin, lotaustralin	Burnet moths (Zygaenidae)	Zagrobelny et al. 2004
		Acreinae butterflies	K.S. Brown & Francini 1990
		Heliconiini butterflies	Nahrstedt & Davis 1985
Cardiac glycosides		Chrysomelid beetles	Pasteels & Daloze 1977, Daloze & Pasteels 1979, Pasteels et al. 1984, Rowell-Rahier et al. 1995
		A toad	Horiger et al. 1970, Y. Fujii et al. 1975
Peptides, proteins	—	Tiger moths (Erebidae: Arctiinae: *Arctia caja*)	Rothschild et al. 1979, Hsiao et al. 1980
	Leptinotarsin	Colorado potato beetles (Chrysomelidae: *Leptinotarsa*)	Daloze et al. 1986
Terpenoids	Cantharidine and palasonin (demethylcantharidine)	blister beetles (Meloidae: e.g. *Lytta vesicatoria* but widespread among others), false blister beetles (Oedemeridae)	Dettner 1997
(Z)-Dihydromatricaria acid		Beetles (Cantharidae)	Meinwald et al. 1968

Of course, animals collectively synthesise an enormous range of toxins that may deter predators and it would be impractical to attempt anything like a comprehensive treatment of them here. Nevertheless, it seems sensible to present some sort of summary about major toxin groups and their distributions in animals, and especially those creatures that have been investigated from the point of view of Batesian or Müllerian mimicry. Table 4.5 lists some of the protective toxins synthesised by various, reasonably well-studied, aposematic insects, and table 5.1 in Ruxton et al. (2004a) lists a number of examples of plant-derived toxins that various insects sequester for protection. Two indirect defences of this nature are shown by the marine amphipod crustacean *Pseudamphithoides incurvaria*, from the Caribbean and Bahamas, which constructs domicile pods from its toxic brown-alga food (*Dictyota* spp.) (Lewis & Kensley 1982), and the spider crab *Libinia dubia* (Pisidae), which decorates its carapace with *D. menstrualis* (Stachowicz & Hay 1999).

Coccinelid (ladybird) beetles often synthesise and secrete various alkaloids, notably coccinelline, precoccinelline and hippodamine. Surprisingly, the unrelated, aposematic cantharid beetle *Chauliognathus pulchellus* has been found to synthesise the exact same alkaloid, no doubt representing a case of chemical convergence (B.P. Moore & Brown 1978). Several leaf beetles (Chrysomelidae: Galerucinae) are known to contain anthraquinones, and although synthesis of these by the beetles has not been conclusively demonstrated, the evidence is very strong (A. Kunze et al. 1996). Interestingly, one of the compounds in the beetles, 9,10-anthraquinone, is a chemical that is used commercially by fruit-growers to deter frugivorous birds. Hilker & Köpf (1994) found that although not causing strong aversive behaviour in tits (*Parus* spp.), it nevertheless slowed the rate of consumption, and this might be a successful strategy if there are other sources of tit food available.

Obtaining toxins from animal sources

Whereas it is widely known that many herbivores sequester protective toxic compounds from their food plants, several carnivorous animals also obtain their toxic protection from their prey. Among insects, for example, several beetles, predominantly in the Tenebrionoidea, are strongly attracted to prey that contain the highly toxic terpenoid cantharidine (see Table 4.5) (Holz et al. 1994, Dettner 1997). Bright red pyrochroid beetles have a particularly interesting association with the chemical. Male pyrochroids are attracted to and feed on cantharidine-containing prey (Meloidae and Oedemeridae) and store the cantharidine both in a cephalic exocrine gland, which affords the males some protection from predators, and in their reproductive system, transferring it to their spermatophore. Females selectively mate with males that have higher loads of cantharidine and once they have received it from the male's reproductive system during copulation, they transfer it to their eggs (Nardi & Bologna 2000), thus providing them and young larvae with chemical protection too. In the pyrochroid *Schizotus pectinicornis*, approximately 45% of the cantharidine load of a mated female comes from the male and a substantial part of this is transferred by her to her eggs, thus affording her offspring chemical protection from the outset. In contrast, in the cantharidine-producing oedemerid beetle *Oedemera femorata* males transfer hardly any cantharidine during copulation (Holz et al. 1994).

Two predators of cochineal insects, the ladybird (*Hyperaspis trifurcata*) and a chamaemyiid fly (*Leucopis* sp.),

utilise protective carminic acid from cochineal insect prey, *Dactylopius coccus* (Dactylopiidae), but only the ladybird is aposematic and sequesters the carminic acid to be released in its reflex bleeding fluid (Eisner et al. 1994).

Cerambycid beetles (*Elytroleptus* spp.) are excellent mimics of toxic lycid beetles and also, as adults, are predators on them, leading Eisner et al. (1962) to postulate that they may sequester from their prey some of their toxins, thus also developing their own chemical protection. This turns out not to be the case, however, and the mimetic longhorn beetles remain relatively palatable, and hence are Batesian mimics (Rettenmeyer 1970, Eisner et al. 2008).

Some amphibians, including dendrobatid poison dart frogs (Dendrobatidae) (Fig. 4.31), obtain some of their skin toxins from their diet, specifically through myrmecophagy (Saporito et al. 2004, Mebs et al. 2010). However, not all species sequester toxins. For example, the microhylid investigated, which also feeds mainly on ants, did not. Other exogenous sources of poison frog toxins include oribatid mites, centipedes and bacteria (Saporito et al. 2007) and potentially some beetles (Dumbacher et al. 2004). Indeed, most of the 800+ toxic alkaloids discovered in the poison frog families Dendrobatidae, Mantellidae and Myobatrachidae are sequestered from dietary sources (Daly 1998), including those of the most toxic species of all, *Phyllobates terribilis*. Toxins in dendrobatids are stored in glands that are uniformly distributed in the skin (Prates et al. 2012) and because the toxins are highly potent, the skin does not actually have a very high density of poison glands compared to some other amphibian taxa.

Another interesting case concerns the acquisition and redeployment of undischarged nematocysts of their cnidarian food by certain aolid sea-slugs,[15] a process known as kleptocnidae, or the similar acquisition of sponge spicules by others. Just how much protection these afford to the sea-slugs is a topic that has spawned considerable debate over the years (M. Edmunds 2009). *Glaucus atlanticus*, for example, is a pelagic nudibranch that feeds on the Portuguese man o' war, *Physalia physalis*. Because it can store the latter's undischarged nematocysts, picking up a *Glaucus* that has washed up on the shore can deliver a quite painful and sustained sting (T.E. Thompson & Bennett 1969, 1970), reputedly causing a human death on at least one occasion. These still functional, 'stolen' nematocytes pass through the gut wall of the sea-slug undischarged and are transported to special muscular structures called cnidosacs at the ends of the cerata,[16] but the mechanism by which the sea-slug

---

15. One ctenophore, *Haeckelia rubra*, and some flatworms also do this.

16. Dorsal protuberances of many sea-slugs, usually arranged in rows.

Fig. 4.31 Various poison arrow frogs from Central and South America. (a,b) Strawberry poison dart frog, *Oophaga* (formerly *Dendrobates*) *pumilio* from Panama; (c) *D. tinctorius* f. *azureus*;[17] (d) *Dendrobates auratus*; (e) *D. tinctorius*; (f) *D. leucomelas*. (Source: a,b, Chris Jeffs. Reproduced with permission from Chris Jeffs; c, Michael Gäbler 2009, Reproduced under the terms of the Creative Commons Attribution Licence CC BY 3.0, via Wikimedia Commons; d, Wildfeuer 2009. Reproduced under the terms of the Creative Commons Attribution Share-Alike Licence CC BY-SA 3.0, via Wikimedia Commons; e, Olaf Leillinger 2006. Reproduced under the terms of the Creative Commons Attribution Share-Alike Licence CC BY-SA 2.5, via Wikimedia Commons; f, © H. Krisp 2011. Reproduced under the terms of the Creative Commons Attribution Licence CC BY 3.0, via Wikimedia Commons.)

---

17. See Wollenberg et al. 2006.

Fig. 4.32 Three brightly coloured nudibranch molluscs. (a) *Chromodoris magnifica* from Bangka; (b) *Goniobranchus kuniei* from Rinca; (c) pyjama nudibranch, *Chromodoris quadricolor*, from Jackson Reef. (Source: © Linda Pitkin Underwater Photography. Reproduced with permission from Linda Pitkin; identifications by Richard Willan.)

prevents their discharge is as yet unknown (see also *Amphiprion* in section *Aposematic commensalism* later). *Glaucus* is certainly easily recognisable but its transparency and blue colouration suggest that it is essentially cryptic, either within the surface of the sea when viewed from above or when feeding on its bluish prey. Interestingly, the *Glaucus* is not immune to being stung by the nematocytes. During copulation, pairs hold their cerata well away from their mate, and if one is touched accidentally, the discharging nematocysts cause the other to flinch (Ross & Quetin 1990). On the other hand, other nudibranch molluscs (family Aeolidae: Fig. 2.2a,b; family Tergepedidae: Fig. 2.2d) which also sequester nematocysts (R.G. Greenwood & Mariscal 1984) are harmless to human touch though no doubt can defend themselves to some extent against fish and possibly other predators. Their extreme flexibility probably means that unless they are actually swallowed they may suffer relatively little harm in an initial strike if the fish then spits them out. Some sponge-feeding nudibranchs (e.g. Chromodorididae, Discodorididae and Doridae) can sequester sponge toxins as well as utilising their spicules, and other

species are known to synthesise toxins *de novo* (though for most the details are not known: Fuhrman et al. 1979). Given the very bright colours of many (Fig. 4.32) and that they appear to have few predators (L.G. Harris 1973), it seems likely that all of these are protected in some way. Other sponge (Fig. 2.2a) and algal feeding species (Fig. 2.2e,f) are well camouflaged but may well be toxic also, given their diets. Cheney et al. (2014b) found, somewhat surprisingly, that the very aposematic patterns of sea-slugs did not evolve with increasing body size, but actually decreased, i.e. larger species are generally more cryptic. The study controlled for phylogeny so this is not simply due to a few very large species of sea-hares (Aplysiomorpha clade).

Costs of chemical defence

That many unpalatable insects sequester their protective toxins from host plants begs the question as to how they themselves are not poisoned by the chemicals (K.S. Brown & Trigo 1995). There are necessarily going to be a number of different answers depending upon the nature of the toxins. A number of studies have revealed fitness trade-offs associated with sequestering toxins from food plants (Cohen 1985, Rowell-Rahier & Pasteels 1986, Bowers 1988, Camara 1997). While most studies on trade-offs have centred on growth or fecundity, with insect larvae, immune defence against pathogens and parasitoids may also be involved. Smilanich et al. (2009) found that larvae of the nymphalid butterfly *Junonia coenia*, which sequesters iridoid glycosides from Plantaginaceae and Scrophulariaceae, were less able to encapsulate injected glass beads if they had eaten a diet containing high levels of glycosides, and were therefore likely less able to defend themselves against parasitoids or pathogens.

Sequestration of plant toxins has its associated costs and may, for example, reduce growth rates or have effects on longevity or fecundity. Rothschild et al. (1986) found that various unpalatable (model) swallowtail butterflies (Papilionidae) that sequester nitrophenanthrenes (aristolochic acid) from their *Aristolochia* food plants, also have far higher concentrations of carotenoids than their mimic swallowtails – more than 100 times more in some cases – and carotenoid concentration does not simply mirror that of the food plants. It is suggested that these help to protect the models from the damage that would ensue from free radical oxygen formation caused by oxidation of compounds such as nitrophenanthrenes and salicin in the presence of sun light. They also found very high carotenoid concentrations in the North American viceroy butterfly, *Limenitis archippus* (Nymphalidae), which sequesters salicin and related glucosides from its larval food plant, *Salix caroliniana* (Prudic et al. 2007b). Nevertheless, the picture is complicated because some palatable Lepidoptera also sequester carotenoids (Feltwell & Rothschild 1974).

APOSEMATISM WITH NON-CHEMICAL DEFENCE

The great majority of well-studied and popularly known cases of warning colouration (or other aposematic signals) involve prey that are dangerous in some way – they are perhaps poisonous, can sting, are aggressive, etc. All of these indeed make the prey unprofitable to attack (Mappes et al. 2005). But it is also possible that other forms of unprofitability might lead to the evolution of aposematism. These include such things as rarity, difficulty of capture, resistance to venom and high digestion costs.

Escape speed and low profitability

Batesian mimicry can evolve whenever a predator can recognise one prey type and associate it with a low or negative pay-off. In many popular examples, the negative pay-off

from attacking the model is some sort of physical harm or nuisance, such as when birds or mammals are caused to vomit after eating particular types of unpalatable prey. However, several workers have noted that a predator may do better to avoid attempting to catch certain prey types if the process of chasing or attempting to capture them involves a net energy loss or just a small gain. For example, parasitic carabid beetles of the genera *Lebia* and *Lebistes* appear to mimic flea beetle hosts (*Disonycha* and *Diamphidia* spp.: Chrysomelidae) even though the latter are not known to be unpalatable. Since no lebistines are known to mimic their non-flea beetle chrysomelid hosts, Lindroth (1971) hypothesised that in these mimetic cases, the feature that was important was the ability of flea beetles to escape, an attribute that any entomologist can attest to. Furth (1981,1983) further noted the great similarity between the predatory pentatomid bug *Zicrona coerulea* and the metallic blue, green or bronze-coloured flea beetle, *Altica bicarinata* (Chrysomelidae), the larvae of which they eat. Both co-occur openly on the beetle's food plant, and he proposed that the similarity was not accidental, but rather because of the difficulty that a predator would have catching the adult flea beetle.

D.O. Gibson (1974) followed up Lindroth's (1971) notion by training starfinches (*Neochmia* (=*Bathilda*) *ruficauda*) on aposematic and cryptic seeds which could be removed from view rapidly by the experimenter. In a control situation in which seeds were not removed from view, the birds fed readily on all colour types, but in the experimental situation, the birds learned to avoid the seed that was routinely removed before it could be eaten, showing an increased latency to attack this seed type. In a similar experiment using European robins, *Erithacus rubecula*, and artificially coloured mealworms, *Tenebrio molitor*, as prey, Gibson (1980) further found that 'escape' ability conferred protection and that it was significantly enhanced when the escaping form was aposematic.

Hespenheide (1973) noticed that several Central and South American beetles of several different families (Buprestidae, e.g. various *Agrilus* species; Mordellidae, Cleridae, Curculionidae, Anthribidae) display shapes and colour patterns that closely resemble those of fast-flying larger Diptera, especially various Tachinidae and Stratiomyidae. Larger Diptera were under-represented in the diets of insectivorous birds in areas where the beetles and flies occur (Hespenheide 1971), presumably because of their quick escape reactions, fast flight and manoeuvrability. Thus it appears that the beetles could be mimicking flies because insectivorous birds learn that they seldom succeed in capturing large flies and therefore avoid wasting time and energy in attacking them. The larger flies thus form a category of low-profitability prey rather than unpalatable ones. This mimicry complex is now known to contain a large number of species

(Hespenheide 1995, 2012). Linsley (1959) had earlier reported the apparent mimicry of mosquitos by some cerambycid beetles, and this no doubt represents the same sort of situation – low reward and hard to catch.

The same selection pressure seems likely to explain the striking resemblance of the South-East Asian praying mantis *Metallyticus splendidus* (Metallyticidae) to fast-moving tiger beetles. V. Thompson (1973) suggested that the ability of adults of the common meadow spittlebug, *Philaenus spumarius* (Cercopidae), to jump away very quickly might be related to a supposed aposematic black and white colour morph which occurs at a higher frequency than expected by chance, supposedly due to positive selection as a result of the bug's extremely efficient close-range escape mechanism. It should be noted, however, that there is some debate whether this colour morph might also be a bird dropping mimic (P. Hebert 1974, V. Thompson 1974). It was also suggested that its escape ability might have led to its extreme colour polymorphism, with the plant-hopper benefitting from aspect diversity and/or reflexive selection. Indeed, females are polyandrous and thus their offspring will display greater colour variation than if they mated only once (Yurtsever 2000).

It is well known to butterfly collectors that some taxa are extremely hard to catch. In Africa, members of the nymphalid genus *Charaxes* are especially fast-flying, and good at avoiding capture (Swynnerton 1926). In general it is thought that unpalatable species of butterfly tend to fly quite slowly and in more or less straight lines, whereas palatable ones fly more quickly and erratically, although there is little hard data. Pinheiro & Freitas (2014) reviewed various suggested cases of escape mimicry among the Neotropical butterfly fauna and came out in support of the idea that at least two homeochromatic complexes involving various palatable but hard-to-catch butterflies probably depend on escape mimicry. These were a group with "bright blue bands" comprising *Archaeoprepona*, *Prepona* and *Doxocopa* species and a "creamy bands" ring in species of *Colobura* and *Hypna*. Pinheiro et al. (2016) go on to propose that the reason that so many butterflies, both palatable and unpalatable, are brightly coloured is that this is signalling a general difficulty of capture, which could confound studies of Batesian and Müllerian mimicry.

Some octopuses can change their appearance and behaviour with great rapidity to achieve various cryptic or mimetic effects (Hanlon et al. 1999). At least three species of octopus – the mimic octopus, *Thaumoctopus mimicus*, from the Indo-Pacific, *Macrotritopus defilippi* in the Caribbean/Atlantic and an apparently undescribed species – can mimic flatfish such as flounders by adapting their shape and arrangement of tentacles as well as colouration (Norman et al. 2001, Norman & Hochberg 2005, Hanlon et al. 2010).

The octopuses are cryptic when resting, as are their flounder models, so mimicry is only in effect when they are swimming, and is achieved by the octopuses undulating their bodies and aggregated tentacles in a very similar way to the normal swimming mode of the fish. The fish are, of course, capable of swimming very fast so that they seldom fall prey. *Thaumoctopus mimicus* is particularly versatile: by changing its motion, tentacle arrangement and colour it can also mimic a sea snake or a poisonous lionfish, *Pterois* spp. (Norman & Hochberg 2005).

R.R. Baker & Parker (1979) also invoked the 'unprofitable prey hypothesis' to explain the brightly coloured plumage of some bird species. Some data in support of this was obtained from examining the ringing returns of British birds by R.R. Baker & Hounsome (1983), who found a negative relationship between conspicuous colouration and risk of predation. The conclusion was criticised on both statistical and empirical grounds by Lyon & Montgomerie (1985), who concluded that, in fact, the more standard sexual selection hypothesis was favoured. In a reply, however, R.R. Baker (1985) was successfully able to dismiss most of Lyon & Montgomerie's (1985) counter-examples and drew attention to several studies showing that the more cryptic females of some sexually dimorphic birds did suffer higher levels of predation than their conspecific males. Ohsaki (1995) found that females of some butterfly species also suffered higher predation than males. Of course, other factors might play a role in this.

Morpho (Nymphalidae) butterflies in South America are famous for their large size and the iridescent blue wings of males, which made them popular with collectors, and their wings were often used for decorating ornaments and jewellery. Species vary considerably in the extent of the structural blue colouration and in the degree of male territoriality, with territorial species being especially bright. A.M. Young (1971) postulated that the colour is advantageous because it is easily memorised by birds which have a very low success rate at capturing these species and will therefore more readily learn not to waste time in largely futile attempts to catch them. He further suggests that sometimes the morphos "actually employ pursuit-stimuli to invite birds to attack and be subsequently unsuccessful" but I can see no reason why this would be beneficial as it almost certainly would involve at least a small increase in risk of injury.

A very different example is provided by the American avocet, *Recurvirostra americana*, in which juveniles start to develop an adult-like black and white plumage at the remarkably early age of two weeks, considerably earlier than starting to be able to fly, which occurs around weeks four and five. Several other shorebirds also show some indication of this adult automimicry. The juvenile birds are, in fact, markedly smaller than adults and are far less able to defend themselves from attack. Sordahl (1988) proposes that this is a protective mimicry to persuade avian predators that the very young birds are older and able to fly, and that they would be as good at escaping as fully adult birds.

PARASITOIDS AND APOSEMATIC INSECTS

Aposematic insect larvae may be presumed to gain protection from predation by birds through their advertising colours. However, birds are not the only source of mortality for caterpillars and other insect larvae; parasitic insects and pathogens may frequently be responsible for killing even greater numbers. Without doubt, many aposematic insects that are distasteful to insectivorous birds are also protected against insect parasitoids through the same chemical agents. The brightly coloured black and yellow larvae of the European cinnabar moth, *Tyria jacobaeae* (Erebidae: Arctiinae), which feed on and sequester toxins from ragwort, *Senecio jacobaea* (Asteraceae), have just a few, generally uncommon hymenopteran parasitoids, but also the specialist microgastrine braconid *Cotesia popularis* that can exert control on its populations (Van der Meijden et al. 1998). Although *T. jacobaeae* sequesters dihydropyrrolizine alkaloids from its host plants, it also metabolises them into other toxic compounds (Edgar et al. 1980). In the tropics, it seems that a very large proportion of lepidopteran larvae that feed exposed on their host plants are protected from parasitism by virtue of sequestered secondary plant chemicals, and as a consequence are only attacked by one or a few, often rather specialised wasps (Quicke 2015). This difference between the tropics and more temperate faunas has been suggested to explain the relatively lesser increase in species richness of parasitoid ichneumonid wasps at lower latitudes and is known as the 'nasty host hypothesis'. Whether the anomalous diversity pattern of the Ichneumonidae is as true as was once thought is coming under question, but nevertheless there may be some truth in this trend.

While aposematic and distasteful caterpillars may be attacked by relatively few specialised parasitoids, these may nevertheless be a significant source of mortality and thus the hosts might be expected to have evolved adaptations that increase the chance of their survival. Furthermore, in the case of gregarious species or at least ones that tend to live in family groups, even if a larva has been parasitised and is thus doomed as an individual, genes that cause a reduction in the fitness of its particular parasitoid may be favoured by kin or family selection (Godfray 1994).

Stamp (1981) examined this problem in the case of the North American Baltimore checkerspot butterfly, *Euphydryas phaeton*, which has distasteful aposematic larvae that live gregariously throughout their development, and its primary parasitoid the microgastrine braconid wasp *Cotesia* (as *Apanteles*) *euphydridis*. Stamp (1981) argued that parasitised hosts would be expected to behave differently from unparasitised ones in such a way that would increase the likelihood of them becoming prey to hyperparasitoids, which can lead to more than 50% primary parasitoid mortality; they might also be expected to display behaviours that might increase their probability of falling prey to naïve birds. The study animal used was that suggested by Smith Trail (1980) as well suited for testing the idea of host suicide, because the caterpillars live in family groups throughout their entire development and are attacked by two successive generations of *C. euphydridis*. The first parasitoid generation attack early instar larvae in the autumn and, after a diapause, complete development and emerge during the spring, whereupon they parasitise the remaining unparasitised late instar host larvae which are likely to be siblings of the ones from which they have emerged. However, contrary to the expectations of Smith Trail (1980), Stamp's (1981) results suggested that rather than the hosts having control of this particular host–parasitoid interaction it is the parasitoids that are favoured by the behaviour changes they induce in their hosts.

Bowers (1993) considered the relative importance of various biotic and abiotic factors in the evolution of colouration in larval, pupal and adult Lepidoptera and concluded that whereas adult and pupal appearance may both be strongly influenced by vertebrate predators, larval appearance is likely to be relatively more strongly influenced by invertebrate predators and parasitoids.

DIVERSITY OF APOSEMATIC FORMS

One of the biggest questions that has still not been answered satisfactorily in a global context is the sheer diversity among aposematic patterns. It is true that there are several general themes that are each exhibited by thousands of species, for example the black and yellow stripes of wasp mimics, the startling (at least to humans and probably other animals) combination of red and black seen in so many warningly coloured animals (see Fig 4.1). Interestingly, most insects have little or no visual sensitivity at the red end of the colour spectrum, so to most other insects such species may appear fairly uniformly black, though of course, some reds are more orangey and therefore may be distinguishable, and we have rather little data still on the ultraviolet (UV) reflectance of many such species. However, many insects, passerine birds, many snakes and some other animals can see well into the UV part of the spectrum (see Fig. 2.11). Lyytinen et al. (2001) tested whether UV reflectance can, in itself, serve as a warning colour using wild-caught great tits,

Parus major, as predators and slices of almond glued under pieces of paper with various types of colour and UV reflectance and modified with chalk to ensure that the reflectance spectra only differed in their UV content (<400 nm). However, they found no evidence that the birds avoided UV-reflecting prey and the birds found it hard to learn to associate unpalatability with UV reflectance. Therefore, at least great tits seem unable to use UV signal alone to learn prey avoidance. As Lyytinen et al.'s (2001) experiments used green prey, rather than typical aposematic red ones, their result does not mean that a UV signal might not play a role in learning, only that it does not if it is presented with green.

While most considerations of aposematism have focused on red and black, red and yellow, black and white, etc., there is a growing understanding that iridescence in various colours (blues, greens, oranges) may also act as a recognisable warning signal (Pegram et al. 2013).

Marples (1993) was also able to show that wild birds could learn to distinguish unpalatable pastry baits purely on the basis of size, though some birds (and species) were better at this than others, and birds improved with experience. However, size does not appear to be as good a signal as colour or smell since the birds did not generalise their experience from one day to the next, instead resampling all sizes on a daily basis. Chaffinches, *Fringilla coelebs*, in particular, were not able to distinguish unpalatable prey by size if they differed in linear dimension by a factor of 1.5, though if they were trained with a distinguishable difference with a size factor of 2, they could later distinguish smaller size differences. Marples (1993) notes that many Batesian mimics are somewhat smaller than their models, but in the supposed case of mimicry between the common Palaearctic two-spot ladybird, *Adalia bipunctata*, and the larger seven-spot ladybird, *Coccinella septempunctata* (see later), the size difference is closer to 1.5 than to 2.

Most people will recognise a ladybird, or maybe a ladybird mimic, when they see one, but many species of these beetles display an enormous amount of intraspecific colour pattern variation (e.g. see Fig. 4.1a), and different species can have quite different patterns. Two-spot ladybirds, although having many colour forms, are most commonly either orangered with a single black spot on each elytron (= wing case), or black with a single red spot, and it has been suggested that the predominantly red form is a mimic of the seven-spot ladybird, *Coccinella septempunctata*, which is almost never black,[18] and the black form is a mimic of the four-spot ladybird, *Exochomus* (Brakefield 1985). It seems in the case of the two-spot ladybird that it is only mildly unpalatable compared to the other species (Marples et al. 1989, Marples 1990), and thus might be regarded as a quasi-Batesian mimic (Speed 1993b). However, it seems unlikely that such an explanation could apply to the highly colour-polymorphic species, such as *Harmonia axyridis* (or at least a close relative of that) illustrated in Fig. 4.1a. Dolenská et al. (2009) considered what the ladybird signal actually is and, following a series of experiments with variously modified ladybirds and models, concluded that spottedness was important as well as general appearance. It should be noted, however, that many tropical ladybirds, as well as a few more temperate ones, are striped rather than spotted. One particularly impressive case of ladybird mimicry involves, unlikely as it may seem, a spider, the araneid *Paraplectana duodecimmaculata*.

In the tropics in particular, beetles of the family Lycidae, often called net-winged beetles, are a common sight, and seem to be the principal models in many mimicry rings (Wallace 1867, G.A.K. Marshall & Poulton 1902, Gahan 1913, Darlington 1938, Linsley 1959, Linsley et al. 1961, Selander et al. 1963, Emmel 1965, Wickler 1968, Bocak & Yagi 2010) as they carry high levels of toxins. These have rather soft elytra and pronota that are often markedly expanded lateral to the body, and they are nearly always predominantly red, orange or yellow, with contrasting black such as the elytral tips, bodies (visible when they fly), legs or pronota (see figure 19 in Wicker 1968). Homeochromatic taxa include many other beetles, butterflies and moths, vespid and pompilid wasps, true bugs and include both Batesian and Müllerian mimics and all the spectrum in between.

Hegna et al. (2011) used painted clay models of the allred (Fig. 4.31a) and black-spotted red forms (Fig. 4.31b) of the poison dart frog *Oophaga* (as *Dendrobates*) *pumilio* to test whether the contrast afforded by the black spots enhanced protection, but found no difference between the types. Darst et al. (2006) tried to understand why Central and South American *Dendrobates* species displayed such a wide range of aposematic patterns, both inter- and intraspecifically (Hegna et al. 2013) (Fig. 4.31), despite the fact that they all rely on essentially the same chemical compounds, notably the neurotoxin Batrachotoxin, for protection. Valkonen et al. (2012) combined data from three large-scale field studies on the probability of attack by avian predators on conspicuous and more cryptic viper models and found that buzzards differed from the other predator species (red kite, black kite, short-toed eagle and booted eagle) in being more willing to attack the conspicuously coloured models. Based on an investigation of the fire belly Japanese newt, *Cynops pyrrhogaster*, Michida (2011) suggests that a mixture of selection pressures locally drive the diversity of signals. The extent and hue of the newts' ventral cream to red areas are determined by genetics and diet, respectively; the brighter the red the more carotenoid needs to be in the newts' diet

---

18. One or two reverse pattern individuals have been found out of hundreds of millions.

since they cannot synthesise their own. The newts also have two types of defensive behaviour: a freezing posture that enhances the visibility of their aposematic ventral surface, and running away. Newts on the mainland suffer higher predation by mammals than those on islands, and as mammalian predators are largely nocturnal, they are far less likely to notice the newts' aposematic ventral surface, therefore escape behaviour is more appropriate there. On islands, on the other hand, the main predators are birds and the immobile posture should thus be more effective, and also the degree of redness. In agreement with predictions, Michida (2011) found that colour and behaviour both differed significantly between island and mainland populations, indicating different relative selection pressures.

In terms of the noxious effect itself, there is considerable variation in how rapidly it may be appreciated by the recipient. Sometimes, discomfort or pain may be strong and virtually instantaneous, as with a wasp sting, while in other instances the effect may not be apparent for minutes or even hours, as in the case of some poisonous plants. Immediate effects are clearly advantageous if the desired effect is to avoid immediate injury by a potential predator, but what of slow-acting venoms and poisons?

Nicolaus (1987) provided data obtained from injecting surrogate crane eggs (stained, double-yolked turkey eggs) with a powerful emetic (UC 27867[19]). The eggs of cranes are subject to predation by a range of animals from raccoons and coyotes to magpies and ravens. The experiment showed that, as expected, the presence of noxious eggs in nests protected eggs in the immediate vicinity, but the effect was even more pronounced since the free-ranging predators that had consumed the surrogate eggs also avoided nests over a far wider area than the site where the surrogate eggs had been placed.

Egg load assessment

A number of butterflies that feed on relatively small, often young food plants, inspect those plants first to determine whether another individual has already laid eggs upon them. If they have, they will avoid them since the first individual to lay will necessarily have an advantage and there may not be enough food to go around (Wiklund & Åhrberg 1978, Rausher 1979, Shapiro 1980, Nomakuchi et al. 2001). Such species are often associated with annual plants, such as various pierids that feed upon Brassicaceae in the temperate region (Rothschild & Schoonhoven 1960). Being diurnal and not having a large surface area on their antennae, such assessment is largely or entirely visual. Egg load assessment also appears to be the driving force behind the so-called pierid

red egg syndrome (Shapiro 1981a). In this case, some white butterflies (Pieridae: Pierinae, Euchloinae and Coliadinae) lay conspicuous orange or red eggs rather than the more normal white-green range of most Lepidoptera. Large and small white butterflies, *Pieris brassicae* and *P. rapae*, respectively, which lay yellow eggs, appear to assess whether a host plant has already been laid upon and make oviposition decisions accordingly (Rothschild & Schoonhoven 1977). However, these species lay eggs on leaves, whereas Shapiro (1981a) found that orange and red eggs were predominantly associated with species that oviposited on inflorescences, where their eggs would be more easily visible by a female butterfly searching for a host plant. There is no reason to suspect that where two or more such pierid species use the same host plant species, they would not also see and recognise allospecific eggs, which would represent potential competition, and Shapiro (1981a) provides evidence that this occurs in at least one species-pair, *Euchloe ausonides* and *P. protodice*. (See also Chapter 9, section *Insect egg mimics* regarding plants with butterfly egg-mimicking structures.)

PROOF OF APOSEMATISM

Many organisms are regarded as warningly coloured because to a human observer they seem to be conspicuous in some way. Many of these no doubt are truly aposematic, but rather more has to be done to prove that an organism really is displaying warning colouration. In fact, as well as being conspicuous to natural predators in its natural habitat, at least three other criteria need to be satisfied before one can be certain that an organism is truly warningly coloured. Sundberg (1979) and M. Edmunds (1987) proposed four necessary criteria:

1. they must be conspicuous (e.g. brightly coloured) or stand out in some other manner (e.g. a sound) such that they contrast with the background to predators;
2. they must have sufficiently good primary, and usually also secondary, defence mechanisms (e.g. toxicity, ability to defend itself, bad taste, rapid escape and hence associated low profitability to potential predators) that a predator will learn not to eat them;
3. some predators must learn to avoid attacking them due to their colour pattern;[20]
4. within a species, aposematic individuals are better protected than cryptic ones.

---

19. Trimethacarb; (2,3,5-trimethylphenyl) *N*-methylcarbamate.

20. Edmunds (1987) rules out that the predator can use the organism's shape or smell, referring specifically to 'warning colouration' but I see no reason why these necessarily should not constitute non-visual aposematic signals.

Clearly the last of these criteria can hardly ever be met unless there is relevant natural variation present in the species or an experimenter goes to great lengths to make the comparison through some manipulation. Thus, very few studies have specifically asked all these questions about the organisms that they take to be warningly coloured.

Bioluminescence as a warning signal

Grober (1988) suggested that the bioluminescent flashes of some brittle-stars is a warning signal that deters predation by crabs by comparing handling times and damage caused by three species of crab on the bioluminescent *Ophiopsila riisei*, two non-bioluminescent ophiurids and also an anaesthetised and therefore not luminescing *O. riisei*. To test the function of light production, some trials were carried out with blinded crabs. Crabs learned that *O. riisei* was unpalatable and that bioluminescence deters predation in the cases of those crabs that could see it. However, Guilford & Cuthill (1989) criticised Grober's (1988) experiments and interpretation because although Grober (1988) demonstrated that bioluminescence reduced damage to the brittle-stars, they point out that this is true of any anti-predator defence, and that what Grober (1988) had not shown was that the bioluminescence was an effective anti-predation mechanism. The rate at which sighted and blind crabs learned to reject the brittle-star were virtually identical, and therefore the bioluminescence was not leading to an enhanced learning rate. Indeed, Guilford & Cuthill argue that flashes of light, which are only produced once the brittle-star is grabbed, might actually be a 'noxious' stimulus itself. Grober's (1989) reply presented further evidence that in the combinations of a sighted crab and a bioluminescent *O. riisei*, the reduction in damage caused, over a series of five trials, was always more rapid than with other crab/brittle-star combinations, and also that there was some indication that the crabs' avoidance of the bioluminescent species was innate. Since the brittle-stars only flash in response to being attacked, I think it is not truly aposematism, unless they might be living in aggregations with at least some individuals flashing most of the time, but more like an *aide mémoire* device, reminding the crab that the last time they attacked something that flashed they found it was unpalatable (see Chapter 7, section *Aide mémoire mimicry*).

The significance of bioluminescence by larval and pupal fireflies (Coleoptera: Lampyridae) has long been debated, but recent evidence suggests it is aposematic because the beetles and their glow-worm larvae (e.g. *Lampyris noctiluca*) may produce secretions unpalatable to both invertebrate and vertebrate predators (Tyler 2001a, 2001b, De

Cock & Matthysen 2003, Tyler et al. 2008). Use of bioluminescence as a warning signal has also been suggested for various jellyfish (Cnidaria) such as *Euphysa* and *Hippopodius* species (Mackie 1995) and millipedes (Marek et al. 2011). Another probable instance of bioluminescent aposematism and Batesian mimicry involves the highly unpalatable large New World click beetles of the genus *Pyrophorus* (Elateridae), which at night have two large, putatively bioluminescent areas on the pronotum (Fig. 4.33a). Any vertebrate predator would no doubt find this an unforgettable signal, having discovered their noxiousness through experience. In South America there is a genus of cockroaches, *Lucihormetica*, that are almost certainly Batesian or possibly Müllerian mimics of the click beetles (Fig. 4.33b), but they are so uncommon that no one has seen a live one bioluminescing. Nevertheless, UV fluorescence images strongly suggest that they have bioluminescent patches (Zompro & Fritzche 1999, Vršanský et al. 2012) and although this is contested by Merritt (2013), further supporting physical evidence has been presented (Vršanský & Chorvát 2013).

Warning sounds

People normally think of aposematism, Batesian and Müllerian mimicry in visual terms, and most observations are made during daylight hours. Butterflies are nearly all diurnal and all examples of moths with aposematism belong to day-flying groups (Merilaita & Tullberg 2005). Nevertheless, there are a number of well-documented cases involving sound. Some of these involve diurnal taxa and some nocturnal, and probably all function in the dark as well as during the day. One of the first to be specifically considered was that of the call made by young burrowing owls, *Athene cunicularia*. These sound very like the sounds made by rattlesnakes, which are also likely to be occupying similar ground squirrel burrows in North America (Garman 1882). D.J. Martin (1973) confirmed spectrographically that the rasp sound produced by the owls was very similar to the noise made by rattlesnake rattles, both comprising a wide range of frequencies and having similar mean minimum and maximum frequencies where most of the sound energy was concentrated. For the owl's rattlesnake call these were 0.85–2.3, 3.1–5.7 and 4.7–6.6 kHz ($n=4$), respectively, and for a real rattlesnake (*Crotalus viridis*) rattle, 0.5–1.8, 3.0–4.4 and 4.9–6.4 kHz ($n=3$), respectively. Ground squirrels are probably among the dupes but presumably small terrestrial carnivores are the more important targets but they are harder to experiment with. M.P. Rowe et al. (1986) used natural variation in the occurrence of rattlesnake models to examine the effectiveness of the

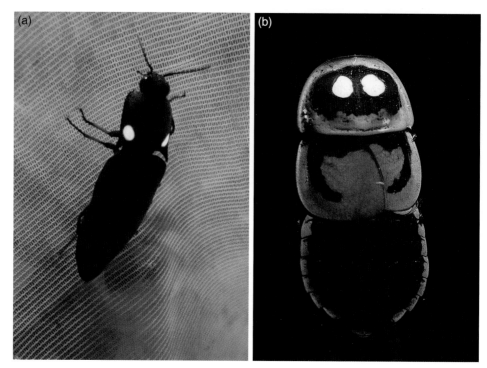

Fig. 4.33 Putative Batesian mimicry system involving bioluminescence in a beetle (model) and cockroach (mimic) [this particular pair of species live far apart but I was unable to obtain a photograph of the actual model which is, however, very similar]. (a) Elaterid beetle of the genus '*Pyrophorus*'[21] from Belize; (b) currently undescribed species of cockroach, *Lucihormetica*[22] sp., from Peru, photograph taken shows UV fluorescence on pronotum which would not normally be visible at night and very bright fluorescence from the presumed bioluminescent pronotal spots. Since the animals are so rare there are only anecdotal observations on actual bioluminescence. (Source: b, Peter Vršanský & D. Chorvát. Reproduced with permission from P. Vršanský.)

owl's mimic calls. Both their study sites had Douglas ground squirrels (*Tamiasciurus douglasii*) and owls, but only one of the sites had rattlesnakes too. Not surprisingly, the ground squirrels from the rattlesnake-infested population were very wary of the owl's vocalisation, but those in the snake-free population were generally less wary and showed considerable variation in their response. It is not known whether these differences are wholly due to evolution or whether some learning might be involved. In any case, it seems clear that the owls are acoustic Batesian mimics, an example of homeophony in Pasteur's (1982) terminology.

---

21. This genus has recently been split into a number of smaller ones.

22. Specimens of this genus are exceedingly rare. Nine of 13 species are known from only a single specimen and an even higher ratio of single specimen species is to be found for undescribed species (one of each of 12 or possibly all 13 other known specimens). So the total number of the *Lucihormetica* species is 26 and in nature probably nearly 100. Only two of them are more or less common (Peter Vršanský, pers. comm.)

Snake-like hissing and posturing by incubating North American chickadees (Paridae) and other hole-nesting birds has also been described as startlingly snake-like (Sibley 1955). Upon being disturbed, a bird performed an apparently stereotyped behaviour: "spread its wings as far as the walls of the [nest] box permitted and swayed slowly from side to side for approximately 10 seconds – then, with explosive suddenness, jumped upward emitting a loud puffing hiss with bill agape". Similar snake-like responses are also known from some other hole-nesting birds, including European tit species, the wryneck (*Jynx torquilla*). Nestlings of the wood warbler, *Phylloscopus sibilatrix*, and the flicker, *Colaptes auratus*, make a similar hissing sound (Sherman 1910, A.H.M. Cox 1930).

Even snakes may mimic snake hisses but, as Cloudsley-Thompson (1981) comments, interpreting snake hissing is not easy: "is the hiss of a grass snake, for example, deimatic behaviour, or is it mimicking the hiss of a poisonous snake in England, an adder?" Aubret & Mangin (2014) investigated this by recording hisses of venomous vipers, mimetic non-venomous viperine colubrids that mimic vipers, as well as non-mimetic colubrid grass snakes. They found that 100% of vipers and 84% of viperine snakes produced inhalation hissing sounds, but only 25% of grass snakes did. In addition, hissing sounds produced by grass snakes differed significantly from those of both asp vipers and viperine snakes, whereas the hisses of the latter two were statistically indistinguishable. Therefore, the evidence seems to show clearly that viperine snake hisses are mimetic, but grass snake hisses are not, or at least they are not very good mimics.

Burying beetles of the genus *Nicrophorus* (see Fig. 4.1e) have a special stridulatory file set at an angle on either side of the fifth abdominal tergite, which they rub against the posterior of the elytra to produce a buzzing sound when handled roughly. They can also produce a "nauseous, stinking irritant froth from the anus" (Lane & Rothschild 1965). The noise sounds like that made by a bumblebee, *Bombus* spp. (or cuckoo bumble bee, *Bombus* ('*Psithyrus*') spp.) and the resemblance is enhanced by their behaviour. Lane & Rothschild (1965) state that "If *N. investigator* ... is poked or seized with a pair of forceps in the head region, and then quickly released, it will often turn on its back and begin to stridulate. In this position the first pair of legs is held upright at an angle to the body and the second and third pairs pushed out like oars." These postures are very similar to those of bumblebees when disturbed in a semi-torpid state. Furthermore, the burying beetle's stridulatory abdominal movements closely resemble the bumblebees' sting-thrusting. Thus they seem likely to be auditory Müllerian or possibly Batesian mimics of the bees (R.M. Fisher & Tuckerman 1986). Similarly, the aculeate mimicking hoverfly, *Spilomyia hamifera*, has a wing beat frequency that is remarkably similar to that of its visual model *Dolichovespula arenaria* (Vespidae), suggesting that sound may play a part in enhancing the mimetic resemblance (Gaul 1952).

C. Rowe & Guilford (1999) carried out experiments with domestic chicks and showed that particular sounds can deter them from pecking at yellow or green food crumbs in various pairwise combinations of yellow, green and brown (general non-parametric ANOVA: $H_1 = 10.6$, $p < 0.01$) with no direct effect of the colour itself ($H_2 = 0.68$, N.S.). This is completely analogous to the way that other experiments showed that particular smells, such as pyrazine (see section *Innate responses of predators* earlier) or just novel smells, interact with aposematic or novel prey types.

Warning colouration in mammals

Pocock (1908) appears to have been the first person to suggest that the prominent black and white markings of various carnivores, notably mustelids, probably evolved as warning colouration; in particular many are paler above than below, contrary to normal countershading. Various skunks and badgers have prominent black and white patterns, quite unlike the more typical browns of less well-defended species, while the quills of many porcupines are boldly black and white banded (Fig. 4.34a) (Inbar & Lev-Yadun 2005). Bold patterning in mammals, although it occurs in various groups, is concentrated among the Carnivora and is often associated with species that produce repugnant anal-gland secretions or are well known for being particularly aggressive. There is evidence that the Norwegian lemming, *Lemmus lemmus*, a medium-sized rodent, is also aposematically coloured, being boldly patterned in black, yellow and white, quite unlike most other rodents (Andersson 1976), because it is more dangerous than might be expected. Furthermore, the Norwegian lemming makes loud calls when threatened whereas the cryptically coloured Alaskan brown lemming, *L. trimucronatus*, seldom does (Andersson 2015). The Norwegian lemming's colouration does appear to have dual functions though, and may also be cryptic under appropriate conditions.

Caro (2009, 2011, 2014) surveyed contrasting dark and light patterns in mammals and concluded that aposematism is the best-supported explanation in most cases, though it may also play a role in intraspecific signalling, while other explanations such as pattern blending, disruptive colouration, reducing eye glare or temperature regulation have hardly been investigated. The phylogenetic distribution of putatively aposematic colouration among the terrestrial carnivores was examined in more detail by Stankowich et al. (2011), who found not only that it had multiple evolutionary origins but also that some pattern features were associated with particular other traits. Specifically, they noted that reverse countershading was displayed primarily by nocturnal anal-gland secretion sprayers, horizontal stripes occur in species such as skunks that can spray anal secretions accurately at targets (Figs 4.34b,d), and that facial stripes occur in burrowing species that "typically leave only their heads exposed to attack" such as badgers (Fig. 4.34c).

In relation to porcupines, Pocock (1908) delightfully quotes the naturalist Charles Kingsley's notes on the Brazilian tree-porcupine or coendoo, *Coendou* sp.:

More than once we became aware of a keen and dreadful scent, as of a concentrated essence of unwashed tropic humanity, which proceeded from that strange animal, the porcupine with a prehensile tail, who prowls in the tree-tops all night, and sleeps in them all day, spending his idle hours in making this hideous smell. Probably he or his ancestors have found it pay as a protection; for no Jaguar or Tiger-cat, it is to be presumed, would care to meddle with any thing so exquisitely nasty, especially when it is all over sharp prickles.

Fig. 4.34 Warning colouration in mammals nearly always involves black and white patterning and often reverse countershading (i.e. paler above than below) or facial masks in those species that tend to defend burrow entrances head first. (a) African crested porcupine, *Hystrix cristata* (Hystricidae) showing bold black and white striping of quills; (b) ratel, also called the honey badger, *Mellivora capensis* (Mustelidae), from Africa; (c) European badger, *Meles meles*; (d) juvenile North American striped skunk, *Mephitis mephitis* (Mephitidae), mature individuals have the white markings extending far further back. (Source: a, Eric Kilby 2014. Reproduced under the terms of the Creative Commons Attribution Share-Alike Licence, CC BY-SA 2, via Flickr; b, Christopher T. Cooper 2011, via Wikimedia Commons.; c, BadgerHero 2003. Reproduced under the terms of the Creative Commons Attribution Share-Alike Licence CC BY-SA 3.0, via Wikimedia Commons; d, Ryan Hodnett 2014. Reproduced under the terms of the Creative Commons Attribution Share-Alike Licence CC BY-SA 4.0, via Wikimedia Commons.)

Skunks are one of the most obvious examples of aposematic mammals which defend themselves by scent. They are able to squirt an exceedingly obnoxious liquid at predators containing a mix of thiols.[23] This, while long-lasting, is harmless, though it often induces nausea and, not infrequently, vomiting in both humans and other mammals.

---

23. Sulphur-containing organic compounds with a carbon-bonded sulphhydryl (–C–SH or R–SH) group that usually have very strong odours.

Weapon advertisement

Many animals with visible defensive structures draw attention to them with contrasting colours, thus making them also a primary defence. This is true for many spiny creatures (Fig. 4.35), as well as often for the claws of crabs, and similar dangerous appendages. The aposematism of porcupine spines is purely advertisement (but see also section *Aposematism in plants* later). Furthermore, when some such species are disturbed, they advertise their aposematism by conspicuous behaviours. Lariviere & Messier (1996) reported that striped skunks, *Mephitis mephitis*, when sensing potential danger, usually carried out one of two conspicuous dis-

Fig. 4.35 Examples of aposematic spines in various animals. (a) Chocolate chip star (horned sea star), *Protoreaster nodosus*, Indonesia; (b) caterpillar of a large unidentified lasiocampid from China 'flashing' two patches of black urticating bristles behind its head, which are normally concealed until it is disturbed; (c) larva of *Leucanella ?hosmera* (Saturniidae) from Peru (identified by Bart Coppens and Carol Bere); (d) caterpillar of *Saturnia pavoniella* showing yellow spots associated with clusters of black spines; (e) caterpillar of *Talima beckeri* (Limacodidae) showing contrasting colours of lobes bearing protective spines. (Source: a, © Linda Pitkin Underwater Photography. Reproduced with permission from Linda Pitkin; b, Itchydogimages. Reproduced with permission from John Horstman; c, Chris Jeffs. Reproduced with permission from Chris Jeffs; d, Harald Süpfle 2012. Reproduced under the terms of the Creative Commons Attribution Share-Alike Licence CC BY-SA 3.0, via Wikimedia Commons; e, Professor Dan Janzen. Reproduced with permission from Professor Janzen.)

Fig. 4.36 Two caterpillars from forests in Vietnam showing clustering of long secondary setae that converge apically to resemble plant thorns. (a) Probably a species of Eupterotidae (identified by Franziska Bauer); (b) caterpillar of the Koh-i-Noor butterfly, *Amathuxidia amythaon* (Nymphalidae) (identified by Keith Wolfe [24]), on a young rattan palm (*Calamus* sp.) with its clusters of yellow secondary setae appearing similar to the yellow thorns of its host plant.

plays that increase the level of advertisement, either a tail-up one (69.1% occurrence) or a stomping one (17.4%). The latter was used more frequently when among taller vegetation where the silent tail-up posture might not be noticed in time to deter a predator's attack. This sort of advertisement may be acting in an *aide mémoire* sense, to remind a predator of a previous encounter (see Chapter 7, section *Aide mémoire mimicry*), or predators may have enough cognitive ability to determine that the animal is potentially dangerous. In contrast, in purely toxic species, conspicuous behavioural display may be involved in convincing predators of the reliability of the signal. Spininess may also be feature that is mimicked (Fig. 4.36).

Speed & Ruxton (2005b) noted that up until then, nearly all theoretical work on the evolution of aposematism had concentrated on species that use chemical defences, even though warning colouration in spiny animals has long been recognised. They used an individual-based selection model to examine both the initial evolution of the protection (spines) and evolution of aposematism. As one might expect, their model showed that evolution of the defence was easy in the absence of warning colouration, but once a spiny defence has evolved, then evolution of aposematism can easily follow as it advertises the threat the spines represent. This then means that the species may experience fewer attacks, which in turn means that they may evolve to put less resource into their spines. Because spines are honest signals of an organism's defensive potential, they may be subject to different selection pressure than, say, the chemical defence of skunks.

Mutualistic aposematism

An excellent example of this, possibly rare, aspect was provided by West (1976) who showed how an edible, black-coloured sponge was protected from predation by

---

24. This may be only the second time that the caterpillar of this spectacular butterfly has been photographed.

the presence within its colony of a very brightly coloured epizoic zoanthid of the genus *Parazoanthus* (Cnidaria) that was avoided by fish browsing on encrusting species. The colour contrast between sponge and zoanthid, black and orange, collectively produces a strong aposematic signal. The zoanthid benefits because its warning colour is most effective against a black background, and the sponge benefits by association, and West (1976) notes that "predation on zoanthid-bearing sponges does not seem to occur".

Aposematism induced by a parasite

When many organisms get infected with parasites or pathogens they change appearance (J. Moore 2002). Humans with a common cold virus get red noses. The evolution of such visual signals is complicated because in larger organisms, obvious signs of illness may reduce chances or mating even if they are still physically capable. Host insects infected by the entomopathogenic nematode *Heterorhabditis bacteriophora* frequently change colour such that exposed membranous areas come to appear pink or red, giving them a generally aposematic appearance which avian predators have been shown to avoid (Fenton et al. 2011). Since *Heterorhabditis* appears to pose no risk to birds, the avoidance presumably reflects a general avoidance of black and pink/red-patterned prey. However, invertebrates might still consume the dead/dying host insect body which would also be fatal to the nematode, and R.S. Jones et al. (2016) demonstrated that nocturnal, non-visual insect predators such as a carabid beetle, also start to avoid feeding on *Heterorhabditis*-infected hosts, suggesting that the nematode also induces the production of olfactory feeding deterrents.

Aposematic commensalism

The evolution of aposematism in undefended prey might be facilitated by close proximity to some other defended aposematic prey (de Wert et al. 2012). Many anemone fish, *Amphiprion* species (Pomacentridae: Amphiprioninae), are patterned with red and white vertical stripes that render them highly conspicuous and thus attractive to aquarists, although when they are swimming among the tentacles of their host sea anemones, their markings may serve more to render them inconspicuous (Y. Leroy 1974). These fish no doubt gain protection by their close association with sea anemones through the latter's ability to sting other types of fish that may come within tentacle range.

While their vertical striping might provide some protection from motion dazzle (see Chapter 2, section *Disruptive and distractive markings*), when they venture away from their host anemones, any piscivore that has attacked them while they are swimming among an anemones tentacles is likely to have got stung, and thus might be expected to learn to associate the rather distinctive *Amphiprion* pattern with a noxious experience, and so perhaps avoid them even when they are in isolation.

POLYMORPHISM AND GEOGRAPHIC VARIATION IN APOSEMATIC SPECIES

As discussed also in Chapter 6, section *Sex-limited mimicries and mimetic load*, a surprisingly large number of aposematic species are polymorphic. Intuitively, this would undoubtedly appear surprising since, as Nokelainen et al. (2012) put it, "As selection should favour traits that positively affect fitness, the genes underlying the trait should reach fixation, thereby preventing the evolution of polymorphisms". Given that aposematism works by predators learning to avoid the pattern, the larger the number of individuals sharing the same pattern the greater the protection ought to be (Müller 1879, V. Thompson 1984, Mallet & Gilbert 1995, Joron & Mallet 1998), however, this does not always appear to be how it works in nature. V. Thompson (1984) suggested that this is because of a balance between selection for aposematic colouration and frequency-dependent selection. Endler (1988), in a more general setting, investigated the likely importance of multiple types of predator each with different visual or cognitive characteristics and proposed a simple genetic model with frequency-dependent selection to show what conditions lead to either fixation or stable or unstable polymorphism – the outcome being highly sensitive to many parameters.

Mallet & Joron (1999) discuss why this should be in terms of Müller's (1879) mathematical model (see Chapter 5, section *Müller's original model*). When the aposematic forms are rare, the selection pressure to minimise variation and hence predator learning will be very strong since the prey-to-predator ratio will be low. However, when the aposematic pattern becomes very common, predator learning will be saturated quickly, and so selection pressure relaxed, and this provides one route to variation, via genetic drift, stochastic events, and ultimately to polymorphism (D. Kapan cited in Mallet & Joron 1999, see also Harper & Pfennig 2007).

Polymorphic forms may occur allopatrically and, more paradoxically, often sympatrically. The aposematic and unpalatable South American butterflies *Heliconius erato* and

Table 4.6 Attributes of various *Heliconius* and other Heliconiini species. (Adapted from Benson 1971.)

	Number of continental races	Occupation of isolated islands	Migratory behaviour	Gregarious roosting	Limited individual home range
H. ethilla	>8	−	−	+	+?
H. melpomene	>13	−	−	+	+
H. erato	>14	−	−	+	+
H. sara	>5	−	−	+	+?
H. dorsis	2	−	−	−	−?
Dryas iulia	1	+	+	−	?
Agraulis vanilla	4	+	+	±	?

H. melpomene[25,26] provide excellent examples of geographic variation. They are both polymorphic for their aposematic patterns, and provide a classic example of geographic variation in colour pattern (Emsley 1964) and it has been shown that regional Müllerian mimicry between these two iconic species is largely or entirely the result of coevolution (A.V.Z. Brower 1996, J.H. Cuthill & Charleston 2012, see Chapter 7, section *Heliconius*), as had been proposed earlier by J.R.G. Turner (1982, 1983) and Sheppard et al. (1985) based on breeding rather than molecular data. Purifying selection is strong in each area, and particularly so at the boundaries between colour morphs. One hypothesis to explain these geographic examples is that during the Quaternary epoch[27] the region now covered by the vast (though dwindling) Amazonian wet forests was far drier and suitable habitats for certain groups, such *Heliconius* butterflies, was highly fragmented (K.S. Brown et al. 1974, K.S. Brown 1982, J.R.G. Turner & Mallet 1996).

However, in the Neotropics there can also be up to nine distinct Müllerian mimicry rings among similarly sized butterflies occurring sympatrically (K.S. Brown 1979). Parallel colour morph variation as well as sympatric polymorphism occurs in *H. numata* and its co-mimic ithomiine *Melinaea* species (A.V.Z. Brower 1994, 1996, Beccaloni 1997b, Joron et al. 2006) (see Figs 7.2 and 7.4). Indeed, Mallet & Joron (1999) refer to *H. numata* as the pinnacle of Müllerian mimicry polymorphism. To explain these, other factors probably need to be considered such as microhabitat specialisation, the occurrence of palatable Batesian mimics (mimetic load and coevolutionary chases), or less extreme variation in the palatability of different species, and the precise relationship between fitness and abundance.

Benson (1971) surveyed the number of races of various aposematic heliconiine butterflies and found several strong associations. Species with the largest numbers of polymorphic forms tended to be more sedentary, with limited home ranges, not migratory and often roosted gregariously (Table 4.6), and they also seemed seldom to have colonised isolated islands. Therefore the less likely the individual butterflies were to stray into other regions the lower the gene flow and the greater the probability of local races evolving.

In Europe, the warningly coloured wood tiger moth, *Parasemia plantaginis* (Erebidae, Arctiinae), has white and yellow background forms, the proportions of each of which vary locally from nearly zero to nearly 100%. Predation on accurate artificial models of these moths also showed marked site-dependent variation, with attacks partially correlated with the abundance of different groups of insectivorous birds in the locality, with the yellow morph generally experiencing less predation when tits (Paridae) were abundant but relatively more when Prunellidae were common (Nokelainen et al. 2013a).

Many of the aposematic and highly toxic poison dart frogs (Dendrobatidae; see Fig. 4.31) of South America are colour polymorphic, and indeed different colour forms

---

25. Telling the two apart is done by reference to a feature of the underside of the hind wing: *H. erato* has four red spots at the base whereas *H. melpomene* has only three (Henderson 2009). When alive they also smell somewhat different. Other characters may be useful locally.

26. There have also been nomenclatural problems. The original Linnaean type of *H. erato* is actually a specimen of *H. doris* but I.C.Z.N. *Opinion 1386* designated a neotype to allow the name to be conserved for *H. erato* in the sense that everyone uses it. There is also a problem with the type material of *H. melpomene* with some, and possibly all, specimens actually being *H. erato*!

27. Approximately the past 2.6 million years, but the separation of races was far more recent.

were often treated as separate species until more information became available. For example, *Oophaga pumilio* has at least 15 different colour morphs (I.J. Wang 2008, Toledo & Haddad 2009), and other well-known examples include various swallowtail and danaine butterflies, ladybird beetles (Coccinellidae) (see Fig. 4.1a) and cantharid beetles and various coral snakes. Different poison dart frogs are involved in Müllerian mimicry complexes along with other toxic frogs and such species therefore provide robust tests of the existence of mimicry if they resemble different co-mimics in different parts of their range (Pough 1988b). Individuals of *Dendrobates imitator*, for example, variously resemble three different *Dendrobates* species in different places (Symula et al. 2001). As with *Heliconius* butterflies, different species are not only involved in Müllerian mimicry complexes, they are also models for various Batesian mimics, some of which track the colour morphs geographically (Toledo & Haddad 2009). Similarly, the non-veomous colubrid snake *Erythrolamprus guentheri* mimics two different species of *Micrurus* coral snakes in different places (Greene & McDiarmid 1981).

APOSEMATISM IN PLANTS

Most plants are green because, apart from a few truly parasitic species, they necessarily contain considerable amounts of the green photosynthesis pigment chlorophyll. However, flowers and fruit are largely exempt from this requirement, especially when nearing maturity (see Chapter 14, section *Fruit and seed dispersal by birds*). Since almost all plants are green or greenish, greenery would normally be considered at most cryptic, matching the background created by other plants, but Lev-Yadun & Ne'eman (2004) propose that in some circumstances, green foliage might also be aposematic. During the hot and dry part of the year in desert, semi-desert and savannah regions, most edible plant leaves have been eaten or shed by seasonally deciduous shrubs, or remaining leaves turned yellow and dry, but a few, deep-rooted plants may remain green throughout the year. Significantly these 'summer green' species are invariably either highly poisonous or very well-defended by thorns. Thus, green may be aposematic when the background is red or yellow, just as red or yellow may be aposematic against a background of wet season greenery.

One of the problems for plants that rely on seed dispersers is that their seeds/fruit should not be eaten by species which will (a) not disperse them appropriately nor (b) attempt to disperse them before the seeds have a good probability of being able to germinate. It is well known that many seeds are poisonous to humans and, while not strictly dangerous, the hot burning flavour of capsaicin in chillies is not relished by mammals other than perverse humans. Capsaicin is an example of a selective antifeedant that is unpleasant to mammals that are not the seed dispersers, but is not repellant to birds that are the seed dispersers of Solanaceae.

The signals that plants give out to indicate unpalatability before a herbivore has even tried to take a bite include colouration and odours and visually enhanced weaponry. The sorts of defences that they are warning against may be toxins, irritating hairs, spines, etc. and will have potentially different meanings to different potential herbivores. Large spines may be effective defences against some mammals but are unlikely to influence Lepidoptera caterpillars. Some Lepidoptera caterpillars, however, seem to be mimicking spines by clustering their irritating setae into more clearly visible structures (see Fig. 4.36b).

Just as in animals, in plants thorns, spines, prickles[28] as well as spinose leaf margins are seldom green, instead they are typically contrastingly coloured (see Lev-Yadun 2001, 2003a, 2003b, 2009b, Lev-Yadun & Ne'eman 2004, Rubino & McCarthy 2004, Ruxton et al. 2004a). Prickles are often either black (Fig. 4.37a) or red (Fig. 4.37d), and the leaf spines of many *Agave* are black and red (Fig. 4.37e,f). In *Acacia* and relatives[29] (Fabaceae) and cacti, as well as various other plants, spines and prickles may be conspicuously white or pale straw coloured (Fig. 4.37b,c) (Midgley 2004). In most cases the whiteness is because of their nature and lack of photosynthesising cells, but with thick thorns and large prickles, other pigments may act as honest signals of their presence. It should be noted that the visual spectra and capabilities of mammalian herbivores are still rather poorly understood. Although many have long been assumed to be colour blind (monochromats) there is growing evidence that at least ungulates are dichromats (Jacobs 1993, Jacobs et al. 1998).

---

28. Technically, thorns are modified branches that arise from the axils of leaves, spines are modified leaves, and prickles are outgrowths from the cortex or epidermis but which lack vascular bundles. Thus those on a rose stem are technically prickles and it is because they lack vascular bundles that they are more easily detached.

29. The commonly known, previously cosmopolitan genus *Acacia* has been shown to be polyphyletic and has been split into five genera (see Kyalangalilwa et al. 2013). To add confusion, *Acacia* itself was retypified so that the name could still be applied to the large Australian contingent.

Fig. 4.37 Contrasting colours and emphasising effects in thorny and spiny plants. (a) Black prickles on an unidentified Fabaceae, possibly *Erythrina* sp., Thailand; (b) rattan palm, *Calamus* sp. (Arecaceae), Vietnam, showing contrasting white spines; (c) bullhorn wattle, *Vachellia* (preciously *Acacia*) *cornigera* (Fabaceae), from North America, showing white, hollowed-out thorns that are domatia for symbiotic ants; (d) prickles of *Rosa* sp. (Rosaceae), showing contrasting red thorns; (e) *Agave victoriareginae*, showing black apical leaf spines, narrowly red at base, accentuated by white zone; (f,g) *Agave* spp., in addition to contrastingly coloured marginal and apical leaf spines, showing the impressions made by spines on adjacent leaves during leaf development which, it is proposed, create a sort of automimicry. (Source: c, Stan Shebs 2007. Reproduced under the terms of the Creative Commons Attribution Share-Alike Licence CC BY-SA 2.5, via Wikimedia Commons; d,g, Professor Simcha Lev-Yadun. Reproduced with permission from Professor Lev-Yadun.)

One way of investigating the likely protective aspect of plant 'spininess' is by looking at the long-term effects of grazing by macroherbivores on ecosystems. Ronel et al. (2009) describe the flora of parts of Israel that have been continuously grazed for 1000+ years as:

> [a]n unbroken blanket of spiny shrubs such as *Sarcopoterium spinosum* (L.) Spach and *Calicotome villosa* (Poiret) Link and many types of thistles of the Asteraceae (e.g., *Carthamus tenuis* (Boiss et Blanche) Bornm., *Centaurea iberica* Trevis et Sprengel, *Echinops adenocaulos* Boiss., *Notobasis syriaca* (L.) Cass., *Scolymus maculatus* L., *Silybum marianum* (L.) Gaertner) cover large tracts of the land.

It would be hard to imagine that this was not the result of selection against non-spiny plants. Similarly, in the savannahs of Africa, a large proportion of the herbs, shrubs and trees are spiny.

Central to whether the contrasting colour of spines can be considered aposematic is whether large herbivores do actually learn to avoid thorny plants by sight. S.M. Cooper & Owen-Smith (1986) studied the effect of various sorts of thorn on the mode and rate of feeding by goats, impalas, *Aepyrocerus melampus*, and kudus, *Tragelephus* spp., and found that thorns reduced feeding rate and often reduced the size of the bite compared to when the animals were feeding on palatable non-thorny plants. However, they did not, in this case, find any particular preference for feeding on non-spiny plants. That 'thorniness' must be a fairly successful anti-herbivory strategy is largely inferred not only because it is very widespread among plants, particularly those growing in more arid environments later in the year, but also because there is not a great deal of experimental evidence. T.P. Young et al. (2003) showed that herbivory may also lead to increased spine length in the African whistling acacia (*Vachellia drepanolobium*), and it is supported further by observations that the parts of trees and shrubs at a height where grazing is higher are often the most spinose, and there is some evidence that under natural conditions, plants with thorns removed suffer greater herbivory than ones without thorns removed (Milewski et al. 1991). Furthermore, Gómez & Zamora (2002) showed that in another plant, the perennial shrub, *Hormathophylla* (*Alyssum*) *spinosa* (Brassicaceae), the thorns are energetically expensive to produce and that, through some intraplant feedback mechanism, exclusion of ungulate herbivores led to a reduction in thorniness over a 3-year period.

The nastiness of being pricked by thorns is something humans are acutely aware of, but clearly many grazers have evolved adaptations to cope with chewing thorny plants, indeed some animals, such as the black rhinoceros,

more or less specialise on doing so. Most readers will also know that when a person gets pricked by a spine or prickle, the wound has a tendency to get infected. This is not just a chance occurrence; it is now apparent that many plant spines, including ones with internal calcium oxalate needles (called raphids), may have evolved to maximise the penetration of pathogenic bacteria into the wounds caused when a herbivore's skin is penetrated either by external spines or raphids (Halpern et al. 2007a, 2007b, Lev-Yadun & Halpern 2008).

The markings on some thorny succulent species may be an example of weapon automimicry among plants (Lev-Yadun 2003b). Several members of the genus *Agave* have very robust thorns along the edges of their succulent leaves (Fig. 4.37f,g) and during growth those on as yet unfurled leaves leave clear impressions on the succulent outer surfaces of less advanced leaves, which when they unfurl appear to have additional rows of spines, potentially enhancing the warning signals of the rows of true spines. Lev-Yadun (2009a, 2009b) thus considered possible Müllerian mimicry among aposematic spiny plants.

Many leguminous plants (Fabaceae) have chemically very well-defended seeds, however that is obviously not the case with the edible pea, *Pisum sativum*. Aviezer & Lev-Yadun (2014) suggest that some of the colour forms of the pods of *sativum*'s wild relatives, *P. humile* and *P. elatius*, are probably mimetic of aposematic caterpillars. In many individuals, the locations of the peas within the mature pods is indicated by a conspicuous red spot against the green background, and the dorsal edge of the pod often bears a conspicuous dark stripe.

Synergistic selection of unpalatability in plants

Following from Guilford & Cuthill's (1991) discussion of the roles of individual, kin and synergistic selection in evolution of warning colouration and unpalatability in animals (see section *Evidence for individual selection* earlier), Tuomi & Augner (1993) questioned whether the same situation would apply to plants. In plants, a predator (herbivore) can sample part of a plant without killing it, therefore herbivores are more akin to parasites than to predators. It is obvious that for a defence mechanism such as unpalatability or spininess to evolve, the cost of the defence mechanism, C, in defended plants must be less than the product of the risk of herbivory, m, and the difference in the cost of herbivory in undefended and defended plants $(H_{ND} - H_D)$, i.e.

$$C < (H_{ND} - H_D)m$$

Table 4.7 The pay-off matrix for a game between defended and undefended plants, with rows being the player and the columns being the opponent. (Tuomi & Augner 1993. Reproduced with permission from John Wiley & Sons.)

	D	ND
D	$-H_D\, m\, \delta_D - C$	$-H_D\, m - C$
ND	$-H_{ND}\, m\, \delta_{ND} - C$	$-H_{ND}\, m$

Tuomi & Augner then constructed a pay-off matrix (Table 4.7) which specified the costs and benefits in a game situation between defended and non-defended plants. In the matrix δ_D and δ_{ND} are the effects that a defended opponent plant has on the herbivory load (Hm) on defended and non-defended players, respectively. The result is that defence should evolve if

$$C < \left(H_{ND} - H_D\right)m + \left[H_D\left(-\delta_D\right) - H_{ND}\left(-\delta_{ND}\right)\right]mp$$

where p is the probability that any plant interacts with a defended plant and thus $(1 - p)$ is the probability that any plant will interact with a non-defended one. The second term on the right-hand side is the condition for synergistic selection, which must be greater than zero, i.e. it is the benefit that an individual gets from interacting with the same type of individual. Unpalatability can therefore only evolve when individual benefit exceeds direct costs. Further analyses shows that if no plants affect the herbivory of their neighbours, then either defended or non-defended wins, depending on the various parameters, but if $0 \leq \delta < 1$ it is possible to find parameter space where a mixed evolutionarily stable strategy is possible.

APOSEMATISM IN FUNGI

It is well known that a number of macro-fungi are highly toxic or even lethal to humans, though they are only a minority of those that occur, and some others are very palatable. Given that the fungal fruiting bodies are the means of reproduction, it would not be surprising if they had defensive mechanisms.

Those that are deadly to humans, such as the 'destroying angels' (*Amanita bisporigera* and *A. ocreata* in North America, and *A. virosa* in Europe) and deathcap (*A. phalloides*) have very conspicuous white caps, but in these cases symptoms often do not appear for a day and death follows a few days after consumption, primarily from liver failure. Although humans and our close ancestors have been foraging for edible fungi for some time, maybe even a few million years, given the diversity of mushrooms, their potential longevity (of their mycelia) and that our ancestors were probably not particularly mycophagous, humans are unlikely to have been a strong driving force in fungal self-protection. Hanski (1989) nevertheless thought that mammals, including humans, have probably been the main selecting force for the evolution of fungal toxins since they would tend to consume entire fruiting bodies and possibly before significant spore release, whereas insects and molluscs tend to consume less and do so at a later stage, usually allowing the fungus ample time to release many of its spores. Lev-Yadun & Halpern (2007) suggest that ergot, *Claviceps purpurea*, which produces a conspicuous sclerotium in the seed heads of rye and some related grasses and cereals, is aposematic. Ergot contains various toxic alkaloids that cause ergotism in people and other animals, as well as lysergic acid, the base structure in the psychedelic drug, LSD (lysergic acid diethylamide).

Camazine et al. (1983) found that *Lentinellus ursinus* produces the chemical isovelleral, a pungent sesquiterpenoid dialdehyde, which is a potent antifeedant towards the possum, *Didelphis virginiana*, a small omnivorous marsupial from North America. Sherratt et al. (2005) demonstrated that poisonous mushrooms were not on average particularly visually aposematic. The famed fly agaric, *Agaricus muscari*, with its red or orange cap spotted with white, despite containing toxins and hallucinogens (muscimol) is far less toxic to humans than many a parent would have their offspring believe. The American Mycological Society has found no reported case of any deaths and posted that they thought the lethal dose to humans would be some 15–20 large caps (Benjamin 1995). That is a lot of mushroom, and no doubt anyone trying to commit suicide that way would be vomiting them up long before they could consume that sort of quantity. However, odour is a different matter, and surveys of European and North American 'mushrooms' showed that there was a highly significant relationship between odour and toxicity (Fig. 4.38), which might be especially relevant if most foraging fungivores are crepuscular or nocturnal. Sivinski (1981) regarded the bioluminescence of certain fungi, either of their mycelia or their fruiting bodies, as a type of aposematism, warning potential crepuscular and nocturnal fungivores of their unpalatability, and animals that might be deterred by their display would seem most likely to be mammals.

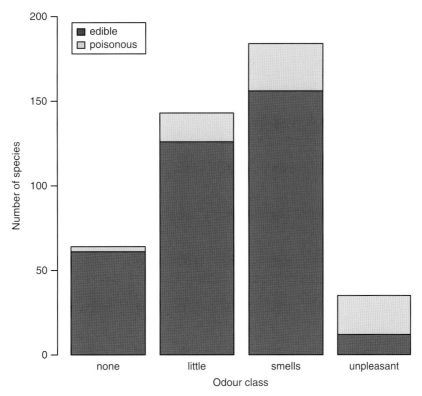

Fig. 4.38 Odoriferous aposematism in fungal fruiting bodies based on combined European and North American data showing that strong odours are given off disproportionately and significantly more by poisonous species ($\chi^2 = 69.8647$, d.f. = 3, $p < 0.0001$; both European and North American data when analysed separately, also showed the same significant relationship) (Source: Adapted from Sherratt et al. 2005.)

WHY ARE SOME UNPALATABLE ORGANISMS APOSEMATIC AND OTHERS NOT?

Despite there being many well-known aposematic poisonous animals, there are also some that are just as unpalatable or dangerous that are highly cryptic, a classic European example being adults of the buff-tip moth, *Phalera bucephala* (Notodontidae) (N. Marsh & Rothschild 1974). Prudic et al. (2007a) have applied phylogenetic methods to investigate what factors may have led some swallowtail butterflies (Papilionidae) to evolve aposematic caterpillars when a number of species that feed on toxic host plants and probably sequester noxious compounds are nevertheless not aposematic. In all, they examined nine dietary characteristics on a phylogeny with 53 species of *Papilio*, plus some separate subspecies and forms using the concentrated changes test (Maddison 1990).

Surprisingly, food plant chemistry features (alkaloids, phenolics, terpenoids, triterpenoids, coumarins) and specialisation on a single host family all showed no significant relationship to larval aposematism. However, whether they fed on herbs[30] and whether their food plants included species with narrow leaves were both highly significantly associated with aposematism (Fig. 4.39). What this finding seems to show is that the signal environment is a strong determinant of whether a noxious prey species will evolve aposematism, at least in the case of caterpillars that are rather sedentary.

In experiments with the gregarious cryptic larvae larch sawfly, *Neodiprion sertifer*, and its relative *Diprion pini* (both Diprionidae), Lindstedt et al. (2011) found that individuals in large aggregations (about 50) were more at risk than

---

30. The correlated feature of having no tree species in their diet was also correlated significantly with aposematism ($p = 0.02$).

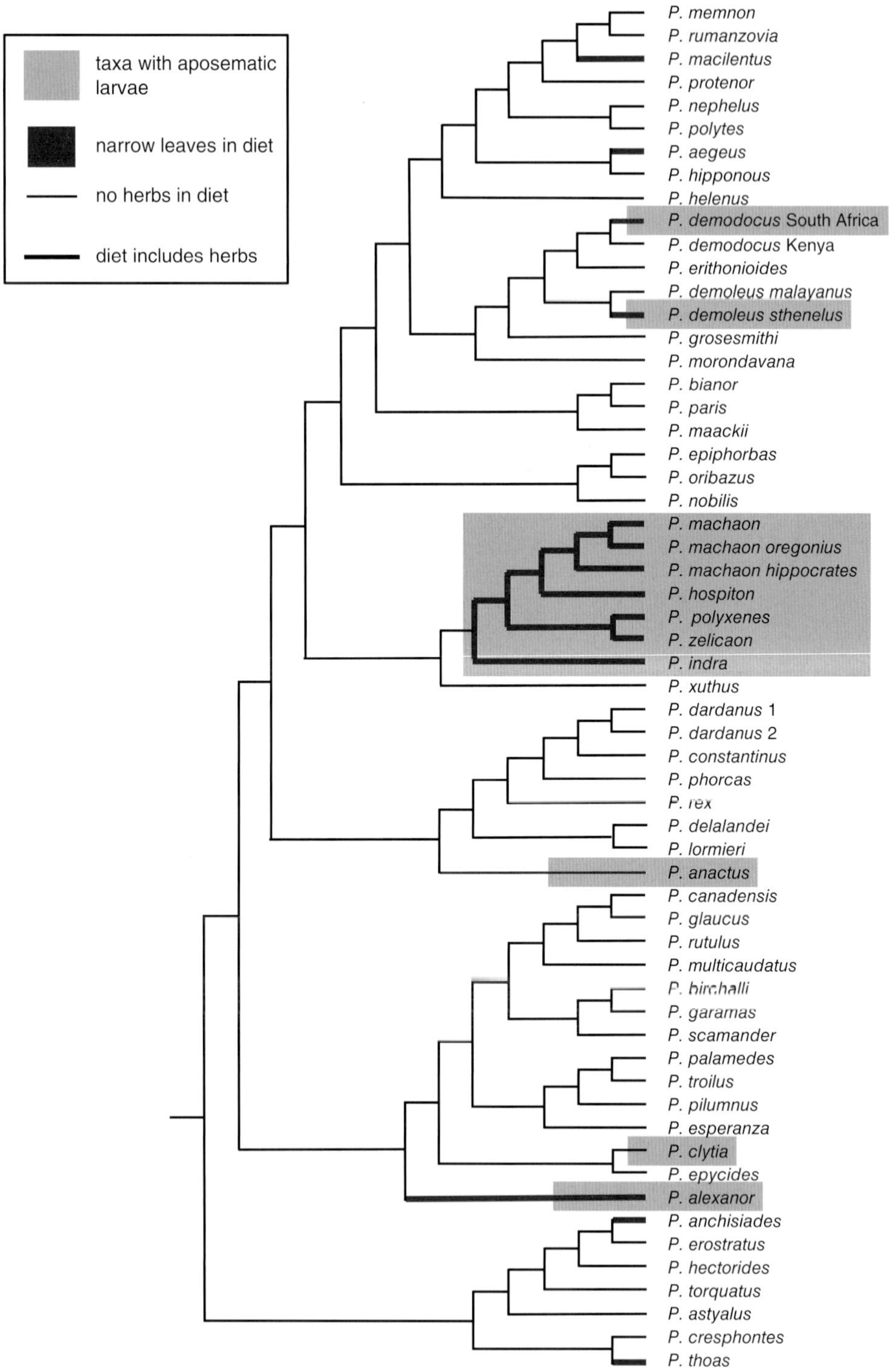

Fig. 4.39 Molecular phylogeny of species of the swallowtail butterfly genus *Papilio*, showing the statistically significant associations between caterpillars being aposematic (grey boxes) and two aspects of their diet: diet includes narrow leaves (red lines) (concentrated changes test, $p = 0.005$) and presence of herbs in diet (thickened lines) (concentrated changes test, $p = 0.03$). (Source: Adapted from Prudic et al. 2007a. Reproduced with permission from National Academy of Sciences, USA.)

those in smaller aggregations (about 15), indicating that large groups posed increased detectability costs. On the other hand, they found no evidence that sequestering pine resins imposed a cost, indeed the opposite, so they concluded that if there were no (or very low) costs to secondary defence but costs of detection, it could be profitable for protected prey not to evolve warning signals.

Endler & Mappes (2004) present a mathematical model of the rather poorly explored phenomenon of well-defended but poorly aposematic prey, in particular in relation to the variability between predators. In their model, weak signalling of unpalatable prey was found to be selected for when predators differed in their probability of attacking unpalatable prey of a given appearance.

Chapter 5

ANTI-PREDATOR MIMICRY. I. MATHEMATICAL MODELS

Visual predators live in a world of deceit.

From J. Diamond & Bond (2013, p. 121)

INTRODUCTION

Müller (1879) framed his theory of why distasteful species would benefit from resembling one another in mathematical terms, which was something new for that time. Then, apart from a few other similarly framed pieces of work, nothing much happened until the 1960s, when papers by Huheey (1964), Holling (1965) and Emlen (1968) really started the ball rolling. Mathematical models of Batesian and Müllerian mimicry have come a long way since these early works. Ruxton et al. (2004a) summarise most of the important modelling papers in two appendices. Here I present a discussion of a selection of types of model to show how new approaches have led to new insights. Increases in computing speed have assisted especially with some areas, initially in population simulations and more recently in spatially explicit simulations.

As with the initial evolution of aposematism, there is a problem of how such mimicry might first evolve because the starting point must be assumed to be that the future mimic is already well adapted to being cryptic or, in the case of Müllerian mimicry, possibly resembling some other model. Therefore any initial small mutation is likely to diverge from that but is unlikely to render it a good mimic. Many people have argued that the process must therefore involve two steps: an initial large phenotypic jump that renders it sufficiently close to its eventual model, followed by further fine-tuning (Nicholson 1927, R.A. Fisher 1930, E.B. Ford 1963, D. Charlesworth & Charlesworth 1975, B. Charlesworth et al. 1982, Turner 1988, B. Charlesworth 1994).

Most theoretical studies have, for simplicity, represented phenotype as a single dimension (see Figs 5.2, 5.7 and 12.14). While this is easy to conceptualise and model for a single feature, in reality variation exists along multiple dimensions and therefore resemblances will involve several genes that each might have large effects on some aspect of the resemblance. This would seem to make it very unlikely that suitable large mutations could occur in all such genes at once. However, based on receiver psychology, Balogh et al. (2010) propose that if predators rely largely on a single feature to classify prey items as palatable or unpalatable this might then provide a way around this. The term they use to describe this is 'feature theory'. Using virtual predators they demonstrate that an initial mutation causing only one feature to move a large amount towards similarity in one dimension facilitates the evolution of mimicry in traits along multiple other dimensions.

Balogh & Leimar (2005) had earlier shown that a gradual evolution of mimicry could be achieved through the simultaneous action of a range of predators that differed in their degree of generalisation and, importantly, this could also work if the mimic species initially differed markedly in

appearance from the model. Franks & Sherratt (2007) extended Balogh & Leimar's (2005) modelling approach and found that gradual evolution of mimicry was possible but became far less likely as the number of important signal components increased. The caveat 'important' is there because of how predators generalise across signals, and greater generalisation reduces the number of separate mimicry components. Overall, they conclude that the two-step model offers the easiest route to the evolution of mimicry, but a gradual evolution may be possible under some sets of conditions.

Tsoularis & Wallace (2005) modelled the probability of a learning predator eating or ignoring a prey, which may be palatable or unpalatable, with its action having either a favourable or an unfavourable outcome. Favourable outcomes are those cases where the predator consumes a palatable prey or ignores an unpalatable one. They considered both a specialist predator faced with noxious models and palatable mimics and a generalist predator faced with the same model and mimic but also with other prey available. Their model used a linear reinforcement algorithm such that consecutive encounters with palatable prey increased the likelihood of eating more individuals, and encounters with unpalatable models reduced the likelihood. The resulting model, they hope, provides a framework for understanding predator decisions, such as when to leave a patch or improve its payoff by adjusting whether it consumes prey of certain types. However, the main problem is that actual data on real predator psychology are largely lacking and so parameterising the models to represent real organisms is really not possible at present.

PROPERTIES OF MODELS, REWARDS, LEARNING RATES AND NUMERICAL RELATIONSHIPS

It is a tenet of many textbooks and even scientific articles that for Batesian mimicry systems to work (i.e. persist in evolutionary time), the number of models should exceed, maybe greatly, the number of mimics. However, this is no longer seen as a prerequisite to interpreting a system as one of Batesian mimicry. If models are highly obnoxious such that numerically important predators will not resample similar-looking types for a very long time after an encounter, the numbers of mimics may exceed the number of models (e.g. J.V.Z. Brower 1960, L.P. Brower & Brower 1962). Of course, relative abundance of models will affect, and limit in the long term, the mean abundance of mimics in any given case, and also the geographic range where mimicry is a viable option (Ries & Mullen 2008). Further, the fidelity (accuracy) of mimicry is likely to be affected by relative abundance of models, with only high-fidelity mimicry being a viable strategy when models are rare, but poorer mimicry being possible if models are common (Lindström et al. 1997, Mallet & Joron 1999, G.R. Harper & Pfennig 2007) (see Chapter 7, sections *The coral snake problem – Emsleyan (or Mertensian) mimicry* and *Inaccurate (Satyric) mimics*).

SIMPLE MODELS AND THEIR LIMITATIONS

While Batesian and Müllerian resemblances have attracted much attention over many years, this has mostly been of a natural history nature, and there have been relatively few detailed mathematical analyses of the relationships between mimic and model population levels. Huheey (1988) points out that an advantage of mathematical models is that they produce predictions that are testable. Furthermore, they can also provide new insights into how evolution works or why particular relationships do or do not occur.

Müller's original model

Müller (1879) proposed a simple mathematical model to explain the similarity between multiple species of aposematic insects, notably Neotropical butterflies, that are all unpalatable. Müller's (1879) use of mathematical notation to render his ideas was clearly a landmark in biology and generally well ahead of its time. He proposed that if a predator kills n individuals of a distasteful species before learning to reject them, and there are two unpalatable prey that are completely different in appearance, the predator will kill/eat n individuals of each. However, if the two species are indistinguishable to the predator and have population sizes a and b, the total number of individuals killed is still n, so the overall proportion killed will be

$$P = \frac{n}{(a+b)}$$

and therefore the number of species a killed (K) will be

$$K_a = a.P = \frac{a.n}{(a+b)}$$

which means that $n - K_a$ fewer of species a will be killed. Similarly, if K_b is the number of b killed in the mimetic situation then $n - K_b$ is the number of b that will not be killed that otherwise would. For both species the degree of

protection is increased the more of the other species are present because the larger $(a+b)$ is relative to a, the fewer of species a will be eaten, and vice versa.

Blakiston & Alexander (1884) put forward a slightly modified version of Müller's (1879) equation to take into account that after the predator has consumed its n prey, the actual number of each mimic remaining alive as part of the mimicry system has been reduced in proportion to their relative abundance. However, when the number eaten by the predator is a lot smaller than the population sizes of the species then it gives near enough identical results to Müller's (Ruxton et al. 2004a). Further developments of Müller's (1879) model that allow for the species to have unequal unpalatabilities (toxin load) by Mallet (2001) are discussed later in this chapter.

Simple models of Batesian and Müllerian mimicry

Holling (1963) outlined a model based on his 1959 disc equation which incorporated unpalatable models and palatable mimics and a total of 22 parameters including memory and forgetfulness, hunger, shape of functional response, occasional predator errors, handling time and so on. At the time of his simulations, computation was exceedingly limited compared with what can be done today. Even so, he showed that mimics might gain considerable benefit from their resemblance even while experiencing some predation. Such parameter-rich models are, however, difficult even today and nearly all work that has yielded fruitful insight takes a far simpler approach, such that only one or a small number of parameters may be varied, enabling easier interpretation of their individual effects and low level interactions.

Huheey (1964) started this approach by developing a simple mathematic model to explore how many of each prey type would be eaten based on the proportions of mimics (p) and models (q) in the population and a parameter n, which is a simple measure of a predator's memory. A predator having once sampled a model will avoid the next $n-1$ individuals that it encounters irrespective of whether they are mimics or models because they are assumed to be indistinguishable. The average number of protected mimics in a predation sequence was therefore found to be $p(n-1)$ and the probability that an individual prey in a random sequence of encounters between a predator and the mimicry complex (P) is given by

$$P = \left(\frac{p}{q}\right) \bigg/ \left[\left(\frac{p}{q}\right) + p(n-1)\right]$$

hence

$$P = \frac{1}{(p+nq)}$$

or

$$P = \frac{1}{p+n-pn}$$

Figure 5.1 shows the effect of increasing the memory parameter n from 2 to 16. As the memory parameter increases, so too does the protection afforded by the mimicry until the proportion of mimics in the population becomes very high. This implies that if the models are sufficiently unpalatable that predators will not sample similar potential prey for a long time, then the number of mimics might greatly exceed the number of models.

Are Batesian and Müllerian mimicry different?

A topic that has attracted considerable attention is the distinction, or lack of distinction, between Batesian mimicry and Müllerian convergence, no doubt resulting in part from the nomenclatural complexity that has been an awkward conundrum for ages. Huheey (1976) developed a model for the advantage of Müllerian resemblance from his earlier work on Batesian mimicry (see previous section) with two models (M_1 and M_2), each with its own level of protection n and m respectively (i.e. the memory parameter), and with population proportions q and p. The proportions of each that are sampled (S) by a predator will be

$$S_1 = q/(qn+pm)$$

and

$$S_2 = p/(qn+pm)$$

From this he deduced that the proportion, P_1, of M_1 eaten is therefore S_1 divided by the frequency of M_1 (q), i.e.

$$P_1 = 1/(qn+pm)$$

By Huheey's convention, M_1 is the more unpalatable model ($n > m$) and it then follows that

$$1/(qn+pm) > 1/n$$

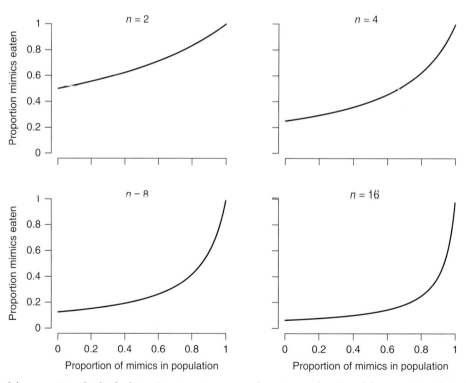

Fig. 5.1 Plots of the proportion of individuals in a Batesian mimicry complex eaten as a function of the proportion of mimics in the population, showing the effect of increasing the unpalatability memory (n) parameter in Huheey's (1964) equation. (Source: Created using equations of Huheey 1964.)

though in his paper the inequality was printed the wrong way around (Benson 1977). Thus, according to Huheey's (1976) model, all Müllerian situations are actually Batesian except for the very special case when palatabilities are exactly the same, i.e. $n = m$. Indeed, from Huheey's (1976) equation, Müllerian mimicry cannot exist at all since there is no benefit of shared predator learning even when model and mimic have equal unpalatabilities. Since this seems to disagree with intuition, it is not surprising that Huheey's (1976) model came under attack (Benson 1977, Sheppard & Turner 1977). The apparent differences of opinion were so large that Huheey (1980a), in his reply, wrote that they had "arrived at conclusions that appear to differ so substantively from mine as to preclude any agreement whatsoever."

Benson (1977) noted that Huheey's (1976) model was dangerously unrealistic in a number of ways, most importantly in lacking a time component, and in not taking into account that the total population size of $M_1 + M_2$ will be greater than for M_1 or M_2 separately. However, when Benson (1977) introduced these corrections, an equally nonsensical inequality was obtained that suggested that all mimicry is Müllerian no matter how palatable one of the species is (Huheey 1976). J.R.G. Turner (1984) claimed that it was unrealistic to suggest that predators counted the number of unpalatable prey encountered, referring to the memory parameters, n and m, but as Huheey (1988) pointed out, he only formulated the equations that way for simplicity and it had always been implicit that these parameters were surrogates for time and would be linearly related to a time parameter.

R.E. Owen & Owen (1984) extended Huheey's (1976) ideas to cover a prey species with a spectrum of relative palatabilities, and also with predators sampling recurrently from a population of identical models and mimics during a single season. Their various models become rather more cumbersome because of the range of parameters they take into account, but it is worth discussing their 'abundance-regulated anamnesis' (ARA) paradigm. This refers to a predator's constant rate of forgetting and therefore makes the absolute abundances of the prey with their individual palatabilities important, rather than just relative abundance as in Huheey's (1976) equations. While the purely Batesian and Müllerian cases they explored followed expectations, their results for an intermediate situation, with the two prey unequally unpalatable, was interesting since it showed that when abundance of the more palatable prey was low it was effectively acting as a Batesian mimic, and so increases in its abundance led to increased predation on the most unpalatable species, but above a critical density the situation becomes Müllerian. Such situations involving two differently unpalatable species are now often referred to as quasi-Batesian. Speed (1993a) showed, using simulations,

that quasi-Batesian mimics might be expected to evolve towards mimetic polymorphism (i.e. different individuals mimicking different models) because predation on them, at least in quite simple models, may be positively density-dependent. As an example, Speed (1993a) cites the common European two-spot ladybird, *Adalia bipunctata*, the African queen butterfly, *Danaus chrysippus*, and the South American butterfly, *Heliconius doris*, as possible examples of this, since each has multiple sympatric mimetic forms (see Chapter 4, section *Diversity of aposematic forms*).

Endler & Mappes (2004) present a mathematical model of the rather poorly explored phenomenon of well-defended but poorly aposematic prey, and in particular, in relation to the variability between predators (both inter- and intraspecific) in terms of their ability to distinguish between unpalatable prey with different levels of conspicuousness. Each of their models, if the proportion of predators that notice and learn to avoid noxious prey was less than 0.5, weak signalling of unpalatable prey was found to be selected for.

Ferreira & Marcon (2014) developed a more sophisticated discrete time model of Müllerian mimicry, formulating the simplest "sufficient" model capable of justifying Müller's theory with two classes of predator – naïve and expert – and exponentially decaying memory loss. Given the 'strength in numbers' notion, i.e. that predation pressure will decline exponentially with increasing numbers of co-mimics, their model is essentially the same as those of J.R.G. Turner (1981, 1984), i.e. if predation pressure is more extreme beyond the total range of variation of the species than it is in some potential zone of phenotypic overlap, then selection will lead to convergence of phenotypes. Turner (1988) represented this graphically (Fig. 5.2). At the top of the figure predators generalise the phenotype of the existing phenotypes they have experienced (grey area) and so would also avoid phenotypes beyond this range if they encountered them, with some non-zero probability. The second part of the figure shows how if predators have learned two, not totally dissimilar, avoidance patterns, though with no current overlap, but with overlap of the generalisation zones, selection will favour convergence. However, if the existing phenotypes are too disparate, as illustrated in the third part of the figure, they will never converge because predators will not confuse them. Finally, at the bottom part of the figure, unequally but different aposematic phenotypes might be able to converge if they are fairly unequally protected. In this case, if a mutation that shifts individuals from the least protected avoidance zone (on right) into the generalisation zone that predators have for the better defended species (specifically within the zone between a and b) it will confer a higher level of protection because between a and b protection would be better than any possible phenotype within the right-hand species or complex (see Leimar et al. 2012).

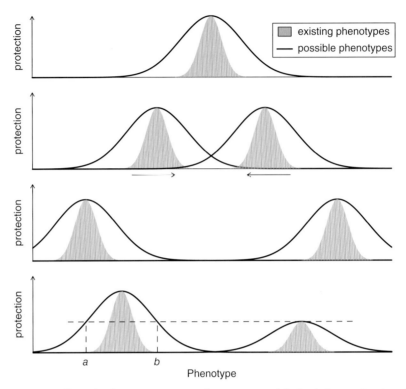

Fig. 5.2 Graphical representation of how the phenotypic variation of two species, and the level of protection they afford, may lead to Müllerian convergence if protection is greater in an area of potential phenotypic overlap, or maintain them separately if they are too distinct, or lead to switching by the predator. (Source: Adapted from Turner 1988 with permission from University of Chicago Press.)

An information theory model

Emlen (1968) created a different type of model based on information theory, in which two species (S_i, S_j) have different palatabilities such that S_i is far less palatable than S_j, and the predator is familiar with these differences and so the frequencies with which it eats them (P_i, P_j) is related to their palatabilities, i.e. $P_i << P_j$. If the encounter rates with the species are μ_i and μ_j, then from information theory, the mean minimum information that a predator needs to describe an encounter is

$$H = -\left(\mu_i \log_2 \mu_i + \mu_j \log_2 \mu_j\right)$$

which depends then on the way the predator 'questions' the prey (i.e. assessing its colour, size, shape, etc.) and because H is a minimum, the 'questions' must, by definition, be optimally structured. If the number of such optimally structured 'questions' is called N, then if $N > H$ the predator will be able to distinguish the two perfectly. But if mimicry is involved, the predator may wrongly interpret the 'answer' to one or more of its 'questions'. For example, if the predator associates the colour red with unpalatable prey S_i but there is a red mimic, it will make the wrong interpretation. Letting the mean probability of receiving a wrong answer to a question be m ($0 < m < 0.5$) then the amount of information needed to correct the misinformation, H_m, is given by:

$$H_m = -\left[m\log_2 m + (1-m)\log_2 (1-m)\right]$$

This changes the criterion that has to be met for the predator correctly to identify the type of prey; instead of N being greater than H, it now has to be greater than $H/(1-H_m)$. Thus, the bigger the mean probability of getting answers (m) the bigger H_m will be and therefore the larger N will be. If the mimicry is very good, most prey signals may have a high probability of being misleading (m approaches 0.5) then the number of optimal questions the predator needs to ask to make a correct decision rapidly becomes very large. However, when Emlen's (1968) equations are used to examine how much protection mimicry will provide given different proportions of models and mimics, the results are quite unexpected, with protection declining at low mimic frequency – very much the opposite of expectation and Huheey's (1964, 1976) type models.

Monte-Carlo simulations

J.R.G. Turner et al. (1984) produced a simple Monte-Carlo model of the behaviour of a 'sit-and-wait' predator such as a flycatcher or jacamar, which sits on a branch and waits for insect prey to fly past. The models and mimics were assumed to be butterflies and the birds were assumed to have alternate prey available so their level of hunger was kept constant. Four types of prey (i.e. butterfly) were included in the model: Solo, Nasty, Model and Mimic. Solo and Nasty were differently patterned, while Model and Mimic were indistinguishable and different from the first two. Nasty and Model were equally unpalatable as were Solo and Mimic. In Fig. 5.3, I show two Monte-Carlo simulations with three types of prey: very unpalatable, mildly unpalatable and alternative palatable. One simulation is with avoidance memory declining linearly with time (Fig. 5.3b), and the second with the probability of the predator avoiding members of the mimicry complex declining non-linearly with time (Fig. 5.3c). In J.R.G. Turner et al.'s (1984) simulations, probability of attack on a given phenotype was altered by multiplying by some modifying constants (a, b), depending on whether it was unpalatable (new probability = probability * a) or palatable (new probability = 1 − old probability * b) with a and b < 1.

Speed's (1993a) Monte-Carlo simulations were time-based and incorporated rules to describe how a predator learns, remembers and is motivated to attack, which was a

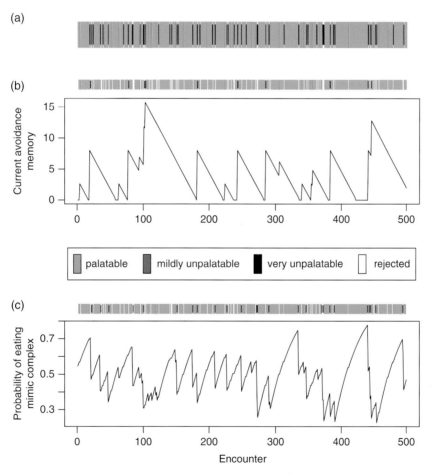

Fig. 5.3 Monte-Carlo simulations with two quasi-Batesian mimetic species and alternative palatable prey, with changes in avoidance memory or attack probability following equation 1 in Speed (1993a). (a) Random sequence of prey types that the predator encounters; (b) simulation with linear decline in avoidance of preying on members of mimicry complex after attacking an unpalatable prey, and with a 10% chance of the predator making a mistake; (c) simulation with probability of attacking members of mimicry complex being moderated after attacking an unpalatable prey. Bars above graphs indicate what prey were actually attacked in the simulations.

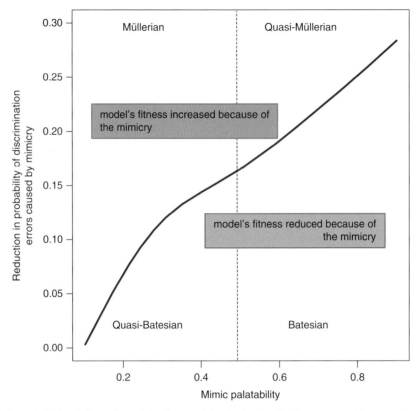

Fig. 5.4 Results of Monte-Carlo simulations of prey taken by a predator that makes decision errors and is exposed to mimics ranging in palatability from 0.1 (nasty) to 0.9 (nice). The blue line shows the values where benefits from a reduction in the predator's discrimination errors equals the costs of an increase in the model's average asymptotic attack probability. (Source: Adapted from MacDougall & Dawkins 1998. Reproduced with permission from Elsevier.)

function of the availability of alternate prey. His model allowed that Müllerian, i.e. true mutualism, mimicry would exist as long as the two species were fairly similar in their unpalatability. Speed (1993b) argued that if one member of the Müllerian complex was consistently only mildly unpalatable/toxic and that by avoiding them a predator lost a considerable potential dietary gain, then selection should lead to predators evolving better discriminatory abilities.

In an informative simulation study, MacDougall & Dawkins (1998) incorporated a range of co-mimics of increasing degrees of unpalatability and a 'sit-and-wait' type predator that makes a fixed proportion of recognition errors. The results showed that pure Batesian, pure Müllerian as well as quasi-Batesian and quasi-Müllerian relationships were all possible outcomes depending on the level of unpalatability of the mimic and the similarity of mimic to model (Fig. 5.4). Quasi-Batesian relationships were described by Speed (1993a) and represent situations in which even highly unpalatable mimics may reduce the fitness of the more unpalatable one,

making it a sort of parasitic Müllerian mimicry (Fig. 5.4 bottom left); from their simulations, MacDougall & Dawkins (1998) found that the presence of even very palatable mimics could also increase predator fitness if it makes sufficiently few discrimination errors (Fig. 5.4 upper right).

Using a virtual predator system with unequally defended prey, Speed (1999a) found that the way memory is modelled is crucial to the outcome. Fixed rates of forgetting always led to monotonic changes in protection (either increased or decreased) with increasing density of the least-defended type. But if memory was inversely related to noxiousness, then there was an unexpected increase in overall predation on the protected prey if only a small percentage of prey was mildly noxious, something that Speed (1999a) terms the Owen–Owen effect after the non-monotonic results obtained by R.E. Owen & Owen (1984). Figure 5.5 shows how this occurs over a range of frequencies of the mimicry complex relative to unprotected prey that are consumed when encountered. Note that there is a

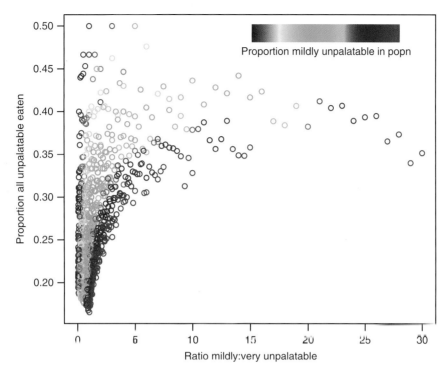

Fig. 5.5 Results of Monte-Carlo simulations showing protection afforded by quasi-Batesian mimetic system when one prey type is five times more unpalatable than the mildly unpalatable one, and both range from between 1 and 20% of the total prey population. The rise to the left of the dip in the swarm of points at the bottom left indicates the initial deleterious effect of having a few mildly unpalatable prey on overall predation of the complex (the Owen–Owen effect).

dip towards the bottom left of the plot, showing that the addition of just a few mildly unpalatable prey to a population of well-defended ones can cause a greater proportion of unpalatable ones to be consumed.

In two related papers, Turner & Speed (1996) and Speed & Turner (1999) compared 30 and 29 sets of predator learning and forgetting rules, respectively, in simulations derived from various people's works. The first paper showed that whether the proportions of members of a Batesian mimicry complex attacked, relative to the proportion of mimics, was linear or curvilinear, depended largely on whether predator learning and forgetting rules were all-or-none or whether they were incremental. Of the systems investigated in the second paper, only two gave rise to 'classical' Batesian–Müllerian systems defined purely by edibility and with increased numbers of mimics causing increased predation on the complex; in both of these, the 'predators' had extreme asymptotic learning and fixed-rate (time-dependent) forgetting. The other systems, with different predator psychology rules, resulted in quasi-Batesian and quasi-Müllerian mimicries. Whether forgetting was time-dependent or encounter-dependent (as with Huheey's (1964, 1976)

memory parameter n, see section *Simple models of Batesian and Müllerian mimicry* earlier) had an interesting effect since in the latter case true Müllerian mimicry became impossible. Neither study allows unambiguous assignment of current experimental data in themselves to support either step-function or gradual models of real predator learning and/or forgetting.

In addition, Speed & Turner (1999) also investigated a suggestion that simply encountering a member of a Batesian–Müllerian complex after having had a bad experience and before that experience was forgotten might act as a memory jogger and so extend the period of avoidance (see Guilford 1990a). This is related to a suggestion from Turner & Speed (1996) that predators might use two types of memory: a short-term one and a long-term one. Depending on the temporal parameters of such memories, a predator might quickly learn to avoid an unpalatable prey type when several are found in succession in, say, one day, but without constant reinforcement, as might occur in a patchy environment, could fail to make a long-term memory out of the experiences without memory jogging within an appropriate time time window.

MORE REFINED MODELS – TIME, LEARNING, FORGETTING AND SAMPLING

How a predator learns may have important consequences for both the predator and the mimic. Estabrook & Jespersen (1974) modelled a predator with a Batesian prey system and assumed that the predator only required a single experience with an unpalatable prey to learn to avoid members of the mimicry complex. After encountering a model, the predator avoided sampling individuals from the mimicry complex for a set number of encounters, N, called a 'waiting time'. The unpalatability of the model, b, represents the profit from eating a mimic divided by the loss it incurs through eating a model. Estabrook & Jespersen's (1974) model also assumes the availability of alternative food, so there is no pressure to eat members of the mimicry complex other than the potential loss due to not eating a palatable mimic. Thus the predator may have six types of encounter: it may eat a model, eat a mimic, and encounter either of these during an initial protected period after having consumed a model, and encounter either of these during a second protected period. Each of these six has a probability of being followed by one of the other types of encounter – most pairs have probability zero but some have probabilities based on the relative proportions of models and mimics. Analysis predicts that optimal predators should have an ignoring period for unpalatable prey and also predicts situations in which mimicry will be advantageous.

Bobisud & Potratz (1976) created a model similar to that of Estabrook & Jespersen (1974) in which the predator required two or more experiences to learn. In their model the number of encounters was set to two, i.e. a predator would show its maximum avoidance (N) after two successive encounters with a model. The optimal predator strategies for all combinations of the conditional probabilities of encountering a model (p) or mimic (q) are shown in Fig. 5.6. When the probability of encountering a model is high, it is always better to avoid all members of the mimicry complex. When that probability is low and the probability of encountering a mimic is high, then the predator should always take the risk and attack. When both probabilities are low, there is some finite value of the avoidance that will be optimal. The two models yielded similar results, but with single-trial learning the predator always fared better. By rendering the probabilities as conditional on what was experienced in the recent past, they can accommodate non-random (i.e. clumped) distribution or encounters with particular prey types.

More recently, R.E. Owen & Owen (1984) developed a more complicated model for mimicry, based on recurrent sampling by predators and short-term avoidance of both model and mimic after encountering a model, and allowing for differential unpalatability and absolute and relative

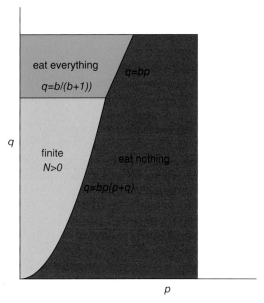

Fig. 5.6 Optimal strategies for a predator encountering a Batesian mimicry system with alternative prey, depending on the probabilities of encountering an unpalatable model (p) and mimic (q), in this case with the number of successive attacks on an unpalatable host required to give maximal avoidance, b, set to 2. (Source: Adapted from Bobisub & Potratz 1976. Reproduced with permission from University of Chicago Press.)

abundances. While the involvement of absolute abundance is desirable, as implemented in their "abundance-regulated anamnesis" (ARA) model, the only very short-term memory of obnoxious prey is highly unrealistic (Ruxton et al. 2004a).

The memory of having an unpleasant, potentially dangerous, experience after eating or attempting to eat an unpalatable model in Huheey's (1964) formulae was very simplistic – a fixed window of numbers of encounters before a predator is willing to take the risk again. This is not very different under most circumstances from having a fixed time window if there were rather more alternative prey available, though the precise outcomes of model systems will vary. Of course, learning and forgetting in real organisms is far more complicated. The first sorts of modelling innovations involved additivity of successive obnoxious experiences (up to a limit) as well as some sort of linear or exponential decline in the probability of a predator's avoidance of a given prey type. Even these are simpler than in real life. Balogh et al. (2008) have perhaps gone the furthest by applying a variant of the Rescorla–Wagner learning model (Rescorla & Wagner 1972), which allows for variation in the strength of a learning stimulus (e.g. degree of unpalatability) within a species. The earlier models assume that all individuals of a given prey cause the same effect on the predator's learning given its state at the time, whereas variation, and hence unpredictability, could, and in their models does, have an effect of its own – because variation might lead to surprise and predators might actually be more wary of prey whose potential unpleasantness is unpredictable. In Balogh et al.'s (2008) simulations, which are too specialised to go into detail about here, this even led to the discovery of super-Müllerian effects with the protection afforded increasing disproportionately with the number of unpalatable individuals as prey unpalatability became more unpredictable.

Importance of alternative prey

Holling (1965) recognised the tremendous importance of alternative prey on a model–mimic system, although thorough and explicit investigations of what effects they have on the mimicry system did not commence until the 1980s, though Emlen's (1968) models did show that in their absence mimetic advantage is also lost at high model frequencies. Luedeman et al. (1981) presented a Markov chain model that explicitly takes into account the availability of alternative palatable prey. The model requires the probabilities of encountering each given prey type after having encountered each type in the preceding encounter to be defined explicitly; these are called the conditional probabilities.

$$T = \begin{array}{c} \\ A \\ M \\ X \end{array} \begin{array}{c} A \quad\quad M \quad\quad X \\ \begin{bmatrix} 1-p-y & y & p \\ q & 1-q-w & w \\ z & r & 1-r-z \end{bmatrix} \end{array}$$

where M is the model, X is the mimic and A is the alternative prey, y is the probability of encountering a model after having encountered an alternative prey, p the probability of encountering two alternate prey in a row, q the probability of encountering an alternate prey following a model, etc. The benefit to a predator of eating a mimic is defined as one unit, the benefit of encountering (and eating) an alternative prey was defined as a units, while b was defined as the loss resulting from eating a model. Finally, a parameter N was defined as the number of models and mimics that a predator would avoid (not eat) after having consumed a model. Using this model, Luedeman et al. (1981) explored the conditions that would favour either of two simple strategies, an 'ignore models and mimics' strategy and an 'eat everything or one' strategy, the latter being equivalent to $N=0$, and found that eating everything would be favoured if the following criterion was satisfied:

$$b < (pq + pw + yw) / (rp + ry + yz)$$

Signal detection theory

Swets (1964) formulated a framework for understanding how observers recognise and discriminate things they see, based on the true patterns of the signal, background noise, and 'noise' within the viewer's nervous system. Several workers have employed it to help understand the evolution of Batesian mimicry and other adaptive resemblances (e.g. Oaten et al. 1975, Getty 1985, J.J.D. Greenwood 1986, Sherratt 2002a, Lynn et al. 2005, Pie 2005, Speed & Ruxton 2010). Pie (2005) illustrates graphically the zones of acceptable signal in relation to background noise on the one hand and potentially harmful or obnoxious prey on the other (Fig. 5.7). Signal detection theory is usually set in a Bayesian framework such that the observer has some initial expectation (the *prior probability*) that a given signal represents a particular item, say a prey, and after sampling it, the probability of the signal representing the prey is updated (the *posterior probability*). Mathematically, the decision of whether or not to attack a prey can be formulated as follows:

$$bp\, f_{\mathrm{mimic}}(s) > c(1-p)\, f_{\mathrm{model}}(s)$$

where b and c are the benefits and costs of attacking a palatable mimic or unpalatable model, respectively, p is the proportion of mimics (hence $1-p$ is the proportion of models), and f_{mimic} and f_{model} are the probability densities of mimic and model being associated with a given signal (s). Thus if, say, a prey looks very much like a mimic ($f_{\mathrm{mimic}}(s)$ is large), or there are many mimics to models (p is large) or the benefit of attacking a mimic greatly exceeds the cost of attacking a model ($b >> c$) then the predator ought to attack. But particularly good mimics whose signals might easily be confused with a model ($f_{\mathrm{mimic}}(s)$ low compared to $f_{\mathrm{model}}(s)$) then they should not attack.

To test whether, in a Batesian mimicry context, the above equation fitted how human subjects foraged, McGuire et al. (2006) used computer-generated model and mimic images, each of which were sampled from a normal distribution with some overlap and associated costs and benefits. Depending on what the subjects chose to attack they received a score based on the costs and benefits of what they attacked, and their object was to maximise their scores by making the best decisions, i.e. attacking mimics but not models. McGuire et al. found that not only did the humans adopt strategies that were close to those predicted by signal detection theory, but also that they adopted (in a given situation) threshold levels of similarity between model and mimic to make attack decisions. McGuire et al. also allowed mimics and models to evolve in appearance in response to the predation and, not surprisingly, mimics

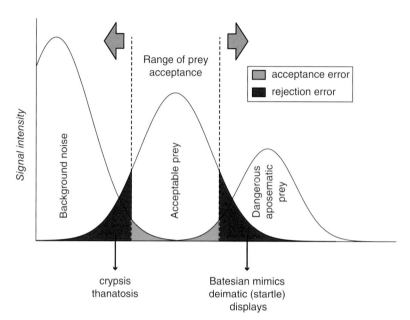

Fig. 5.7 Model of how potential prey appear to predators in a spectrum from being undetectable (crypsis), through acceptable to being aposematic. (Source: Adapted from Pie 2005 with permission from Elsevier.)

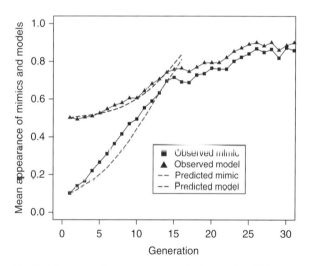

Fig. 5.8 Responses of computer-generated mimic and model images (varying in a single signal dimension: *y*-axis) to 'predation' selection by human subjects and the predictions from the signal detection theory model; predictions were no longer appropriate after mimics had evolved to resemble the models closely. (Source: Adapted from McGuire et al. 2006. Reproduced with permission from Elsevier.)

evolved quickly to look more like models, and models evolved more slowly 'to escape' (Fig. 5.8).

Most models predict that as the abundance of an aposematic protected species increases, predator learning and avoidance will increase which means that the proportion eaten will either remain level or decline. This is akin to saying that selection on the aposematic species is likely to be antiapostatic because when they are rare they are more likely to be encountered by a naïve predator or one that is well along the pathway of forgetting an earlier noxious experience with the prey type and might therefore be inclined to sample again. This antiapostatic effect does not necessarily depend on predators showing associative learning of pattern and unpalatability, however, as it can also be demonstrated in models based on signal detection theory (Lynn 2005). Indeed, the latter may explain why some empirical studies have found higher rates of attack on aposematic prey when the latter are common. Signal detection theory takes account of the uncertainty that predators are faced with during the process of discrimination learning, since there may be a time when they recognise the prey both as different (e.g. colour) but also the same as ones it normally eats (e.g. being an insect).

Genetic and evolutionary models

Genetic modelling involves determining the conditions such as selection coefficients, linkage, dominance, modifier genes, etc. that will lead to the evolution of a trait or final gene frequencies. As quoted earlier by L.P. Brower et al. (1971), the founder of modern population genetics,

E.B. Ford (1964, p. 245), in the context of the evolution of mimicry wrote:

> [S]mall changes in the original pattern ... are almost certain to be harmful. Only a considerable step, producing something near enough in appearance to a protected form to give advantage, is likely to become established. This could then be perfected by selection acting on the gene-complex.

It should be self-evident that for Batesian mimicry to exist at all, the mimic must be able to approach the model phenotypically faster than evolution will allow the model to diverge away from it (R.A. Fisher 1930, Nur 1970, J.R.G. Turner 1977, 1987). This implies that the selective advantage to the mimic must exceed the selective disadvantage to the model.

Matessi & Cori (1972) provided a detailed discrete time genetic model to investigate the effect of selection on a gene for Batesian mimicry. Such a model necessarily involves density-dependence, since the value of mimicry depends on the proportions of models to mimics and hence the proportion of the mimic species' population possessing the gene. M and M' were defined as the phenotypic characteristics of the model and mimic, respectively, that are relevant to the mimicry and W, a non-mimetic morph. They considered situations in which both sexes could be mimics of an unpalatable model or only the female sex, but for simplicity had to assume that the ratio of model to mimic populations in a given place was a constant. Thus, the mimicry allele is expressed in both sexes, p_n is defined as the frequency of the mimetic allele at the beginning of the nth generation,

$$p_{n+1} = \frac{P_n F}{\left[F - (F - f)(1 - P_n)^2\right]}$$

whereas, if mimicry is female limited,

$$p.\text{male}_{n+1} = \left(p.\text{male}_n + p.\text{female}_n\right)^2$$

$$p.\text{female}_{n+1} = \frac{F\left(p.\text{male}_n + p.\text{female}_n\right)/2}{\left\{F - \left[(F - f)(1 - P.\text{male}_n)(1 - p.\text{female}_n)\right]\right\}}$$

where F and f are the survival probabilities due to predation of mimetic (M') and non-mimetic (W) forms, respectively. Running simulations of both models gave stable oscillations of M' and W allele frequencies, sometimes producing limit cycles, sometimes bi- or tri-stable cases.

Charlesworth & Charlesworth (1975) provided a different genetic model of Batesian mimicry which was based on a single locus controlling mimicry with the initial population

of the 'mimic' (butterfly) species before the mimicry mutation occurs subject to density-dependent survival, thus:

$$\text{Survival rate} = \frac{1}{1 + \text{constant} * \text{larval density}}$$

They then propose that the populations are subject to predation such that (in their words):

> (a) After eating a number of butterflies, the presence of distasteful models among the butterflies eaten causes the predator to avoid this pattern in future. The predator is assumed to reject the models with higher probability the more of them it has eaten in the past.
> (b) The predator is assumed to forget its past experience to some extent. It therefore acts on the basis of the recent rather than the distant past.

and the predator does not otherwise show any tendency to apostatic or antiapostatic behaviour. If, then, a mutation occurs in the cryptic species that renders it somewhat more similar to the model, assuming that initially the mimics will be at very low frequency, the criterion that must be satisfied for the mutant allele to increase in frequency is:

$$P_{\text{mimic being eaten}} > \frac{P_{\text{mimic being detected}} - P_{\text{cryptic being detected}}}{P_{\text{mimic being detected}}\left(1 - P_{\text{model being eaten if detected}}\right)}$$

and because the proportion of mimics in the population is, at the beginning, very low, the probability of a model being eaten if detected in the equation is effectively what it would be if no mimics were present. The results of solving this for a variety of values of the probability of the cryptic form of the mimic species being detected, and with the probability of a model being eaten set to 0.05 (i.e. the model is quite well protected) are shown in Fig. 5.9. The curves show that if the cryptic forms are hard to detect, as they "probably would be in nature", a mutant allele will only be successful if it renders the individual very similar to the model, i.e. their results suggest that only a 'major gene mutation' rendering a sufficient degree of resemblance to the Batesian model will be successful as, indeed, had been postulated by various earlier workers. The lower slopes of the curves with the cryptic form not being particularly cryptic (Fig. 5.9, higher values in boxes) suggest that mimicry will evolve more easily in such species. Their results also showed that if the model was sufficiently unpalatable, the mimic populations could be higher than those of the models, something that early on was thought impossible.

O'Donald & Barrett (1973) compared two models for the evolution of mimicry, involving a mimic locus and a modifier locus determining whether the Batesian mimetic allele will evolve to dominance. The models differed in whether

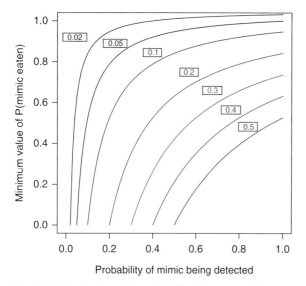

Fig. 5.9 Critical values of the probability of a rare mimic form being detected (x-axis) and eaten if detected (y-axis) and for the mutant allele to be at an advantage for seven values of the degree of crypticity of its base population, i.e. probability of being detected by a predator if encountered (values in boxes). (Source: Created using equations of Charlesworth & Charlesworth 1975. Reproduced with permission from Elsevier.)

the modifier locus displayed full dominance or whether its allelic effects were additive. With the modifier gene showing full dominance, the model showed that a mimic allele will inevitably evolve to dominance, but if the modifier's effects were additive, dominance would only evolve if the modifier allelic frequency was already at sufficiently high frequency.

Charlesworth & Charlesworth (1975) explored how complete or incomplete dominance of the mutant mimetic allele would affect things. With complete dominance they found that either the mimic allele would go to fixation in the population, or if a mimic/cryptic polymorphism persisted, then predation rates on mimic and cryptic forms would be equal at equilibrium. With incomplete dominance, polymorphism will remain but can be due to heterozygous advantage whereby the increased similarity of the mimic homozygote to the model does not outweigh the penalty of increased conspicuousness, *or* disruptive selection (heterozygous disadvantage) with the conspicuousness of the homozygous mimic not outweighing its advantage from similarity to the model.

Charlesworth & Charlesworth (1976a) expanded their earlier models to examine the roles of various sorts of modifier genes which may

1. simply enhance similarity to the model caused by the mimicry allele,

2. add features that may improve or reduce the mimetic resemblance caused by the mimicry allele and may have deleterious effects on individuals of the non-mimetic form,

3. bring about resemblance to a second type of unpalatable model.

Furthermore, they explored the effects of recombination between the mimicry and modifier loci. With the simple similarity-improving second locus they found that there would be very little, if any, selection pressure for them to become more closely linked. Such systems show little or only transient linkage disequilibrium.

With both the second and third types of modifier allele, especially if they were already located fairly close on the same chromosome, linkage disequilibrium would be expected and also selection will favour increased linkage between the gene loci either by them becoming physically closer or rates of recombination being reduced. However, if the modifier and mimicry genes are on different chromosomes or are only very loosely linked, they will not be selected to become more closely linked, and under some circumstances either the modifying allele may be eliminated or the mimicry allele will become fixed, depending upon relative numbers of models.

In butterflies, sex-limited mimicry invariably involves the female and as they are the heterogametic sex (XY or XO) and this means that they have greater likelihood of displaying phenotypic variation for features encoded by alleles on sex chromosomes, and, as is well known among butterfly collectors, females show greater intraspecific variation than males. Obviously also, there would be no opportunity for recombination to occur between genes located on a female Y chromosome, but sex-linked genes are hardly known in the Lepidoptera. However, it appears that in the Lepidoptera in general, females do not show any recombination at oogenesis, no chromosomal chiasmata being visible (Suomalainen et al. 1973, J.R.G. Turner & Sheppard 1975).[1] Charlesworth & Charlesworth (1976a) suggest that marked variation in the degree of mimetic fidelity shown by females of the race *polytrophus* of *Papilio dardanus* could be explained by a modifier locus that confers some disadvantage if the mimetic allele is absent, where some females display hints of male-like, yellow colouration (see Fig. 7.6a). This latter case might also reflect a finding of Charlesworth & Charlesworth (1976b) that when only one model is involved, evolution might lead to the mimicry allele evolving either to be dominant or recessive depending on model parameters.

---

1. The chromosomes of butterflies are, however, typically small and hard to study cytologically.

J.R.G. Turner (1980) developed genetic models of Batesian mimicry further by considering three different genetic situations, where the allele conferring mimicry is sex-linked (Y-linked), recessive or dominant. These cases correspond respectively to the situations known in *Papilio glaucus,* Ethiopian *P. dardanus* and most of the other examples we currently understand.

The mimicry supergene complex in *Heliconius* butterflies is associated with a polymorphic chromosomal inversion and gene rearrangements (Joron et al. 2011). Le Poul et al. (2014) showed that with sympatric morphs there was complete dominance, but when they were allopatric the relative dominance of genes responsible for different pattern elements was quite variable. This suggests that the strict dominance had been strongly selected for only when natural selection was favouring local mimetic polymorphism and therefore against intermediate forms.

Coevolutionary chases

Coevolutionary modelling by Franks et al. (2009) produces the same result, that warning signals in unprofitable prey can evolve away from Batesian mimics, and puts a slightly different spin on it in that this is to be expected, especially when there are more ways to look different from the background rather than matching it, which will usually be the case. Thus, the possibility that defended prey might evolve to be different from their chasing Batesian mimics by becoming slightly more cryptic does not appear to be a likely strategy. They also show that a peak shift in the nature of the warning signal is to be expected during the evolution of aposematism. A further aspect of Franks et al.'s (2009) models is that warning signals can evolve through slow gradual change rather than requiring a two-step process (see Chapter 4, section *Mathematical models and ideas of warning colouration evolution*).

Gavrilets & Hastings (1998) created a coevolutionary model based on two haploid species with a single locus with two alleles and which incorporated the ability of each species to evolve. Applying their results to the Batesian–Müllerian mimicry spectrum they discovered a great range of possible evolutionary dynamics. Critical to their results was how much the fitness of individuals in one species depended on the genetic composition of members of its own species (within-species interactions) and how much on gene frequencies of the other species (between-species interactions). Their final model has four important parameters (a, b, c, d) for each species, two to indicate the direction and magnitude of the within-species interaction (a, c), and two the between-species interaction (b, d). Some of their results were very similar to the population dynamics of predator–prey systems, i.e. increases in one species' abundance leads to an increase in the abundance of a mimic, which then increases overall predation on both, and so cyclical dynamics ensues. For classical Müllerian mimicry (i.e. $a, b, c, d > 0$), their model showed that both species should evolve to be monomorphic and resembling each other. For classical Batesian mimicry (i.e. $a, c < 0, b, d > 0$) a rich range of dynamics was recovered (both species monomorphic, both polymorphic or the mimic polymorphic but the model monomorphic). Mutations could also lead to change in the shape of the population dynamic cycles with the system alternating between one species greatly outnumbering the other and then quite rapid switching to the opposite situation, and thus a constant evolutionary chase. This was also the case when one species was less palatable than the other, such that the most unpalatable species suffers when its weaker counterpart increases in frequency ($a, b, d > 0, c < 0$).

Models involving population dynamics

Hadeler et al. (1982) introduced population dynamics into the study of Batesian and Müllerian mimicry using a model that incorporated intrinsic population growth rates and carrying capacities of the mimetic pair of species, their unpalatabilities, recruitment and training of the predators and the degree of similarity of the two mimetic species. Their model incorporated population parameters based on real mimetic butterfly systems and keeps the overall level of predation constant. The relative increase in equilibrium density of the mimic species was used to represent the protection due to mimicry and the model generated the full spectrum of mimicry types depending on unpalatability, from pure Batesian to pure Müllerian with intermediate quasi-Batesian results, in which a moderately unpalatable mimic can be parasitic on the more unpalatable model.

Yamauchi (1993) also examined Batesian mimicry from the point of view of population dynamics. The population sizes of model and mimic species were similarly limited by intrinsic growth rate and carrying capacity and, in addition, by a predation level which was determined by the relative proportions of models to mimics using a modification of Huheey's (1964) equation – instead of using a variable memory parameter the term was raised to a power, s, so as to modify the predator's sensitivity to the proportion of models and mimics. Three separate models were investigated. In one, the population of the model species displayed density-dependent mortality, in the second, that of the mimic displayed density-dependent mortality, while in the third, both did. Yamauchi (1993) was only able to solve the first two analytically. In the case with the model species' population displaying density-dependence, both species could coexist or both go extinct, whereas without density-dependence for the model, only one species could persist or both would go extinct. However, to examine the more realistic case of both species showing density-dependence, simulations had to be employed based on the following equations.

$$M_{t+1} = M_t \exp\left\{ r_1 \left[1 - \frac{M_t}{K_1} - P\alpha \left(\frac{N_t}{N_t + M_t} \right)^s \right] \right\}$$

$$N_{t+1} = N_t \exp\left\{ r_2 \left[1 - \frac{N_t}{K_2} - P\beta \left(\frac{N_t}{N_t + M_t} \right)^s \right] \right\}$$

where P is an overall rate of predation on the complex (set to 1), M and N are mimic and model population sizes at times t and $t + 1$, r_1 and r_2 are their respective intrinsic rates of increase, K_1 and K_2 their carrying capacities, and α and β are predation coefficients that represent a combined probability of being attacked and being killed. The results, unsurprisingly, can be quite similar to the outputs of discrete time predator–prey or host–parasitoid population model simulations since the mimic is having a negative effect on the host's survival. Depending on parameters, either species can be driven to extinction (e.g. Fig. 5.10a), stable equilibria can be reached, sometimes smoothly (Fig. 5.10b,c) and sometimes as a result of damped oscillations (Fig. 5.10d), limit cycles (Fig. 5.10e) or complicated and chaotic persistent dynamics (Fig. 5.10f).

Getty (1985) took a different approach by examining the effects of incorporating model/mimic systems in an optimum diet model, which assumes the presence of alternative prey. The results suggest that predators could optimise their profitability of the prey they attack by displaying frequency-dependent partial preferences, and these can enhance the common sigmoid functional response which in turn can stabilise model–mimic complexes. An important feature of the results is that the predator's search rate should affect its attack decisions. When search rate and encounter rate are low, the predator should nearly always attack members of the mimicry complex, but when search rate is high the predator should avoid attacking the models and mimics, as well as avoiding all low-profitability prey, because high-profitability prey will be encountered often. Making the model highly unpalatable, i.e. of low value to the predator, leads to both model and mimic populations reaching equilibrium densities close to their carrying capacities, and this also happens when predator density is low or both model and mimic have high intrinsic rates of increase, as would one would intuitively expect.

The model assumes that the fitness of a Batesian mimic is maximal when it is rare and declines as it increases in frequency. If a butterfly, for example, is polymorphic with a mimetic and a non-mimetic morph, then the fitness of the mimic can be set as unity when rare and to decline by a small amount (S) with the addition of every additional mimic individual to the population, and the fitness of the non-mimic morph is assumed to be unaffected by its abundance can be given as $1 - t$, where t is the reduction in fitness due to the mimic's relative abundance. The equations derived are simplified by assuming some linearity of the effect of abundance on the mimic's fitness increment but the effect of this simplification is trivial. J.R.G. Turner (1978)[2] showed that given this simple situation, model and mimic morphs will reach a stable equilibrium when

$$1 - NMS = 1 - t$$

where M is the frequency of the mimic phenotype and the total population size is N. From this the equilibrium frequency (M') of the mimic is

$$M' = t / SN$$

Using this, J.R.G. Turner (1980) investigated the stability of the equilibrium under various conditions of genetic dominance of the mimetic allele: Y-linked (i.e. haploid), diploid with the mimic allele either dominant or recessive. The results showed that in all cases it was possible to get oscillations in mimetic allele frequency but only if the selection coefficients were very large and, also, in the diploid cases, if mimicry could be displayed by both sexes. In the real world this was shown to be more likely to occur if there were large fluctuations in population size because the selection pressure is also density-dependent.

Mallet (2001) proposed a model to explain the evolution of Müllerian mimicry starting with two unpalatable species that are initially very different in appearance, though it is simplified in that it assumes that a single mutation in one can render it looking effectively like the other. Suppose the two species have abundances A and B, and they are both non-lethally toxic, and have per capita toxin loads of d_A and d_B, respectively, to a predator. A hungry predator is likely to eat a number of each determined by the toxicity of each prey and the dose of toxin it can risk consuming, D. Thus the predator, in a given time frame, would be expected to eat D/d_A of species A and D/d_B of species B, before stopping. Now if a mutation arises in one of them, say species A, making it now resemble species B, and this has a higher probability of surviving than non-mutant forms, then evolution will lead to the mutation spreading, ultimately resulting in Müllerian mimicry between A and B. The relative fitness of the mutant will depend on the numbers of individuals at any one time displaying the mutation, a number we will call m.

---

2. But see J.R.G. Turner (1979) for a correction to the equations.

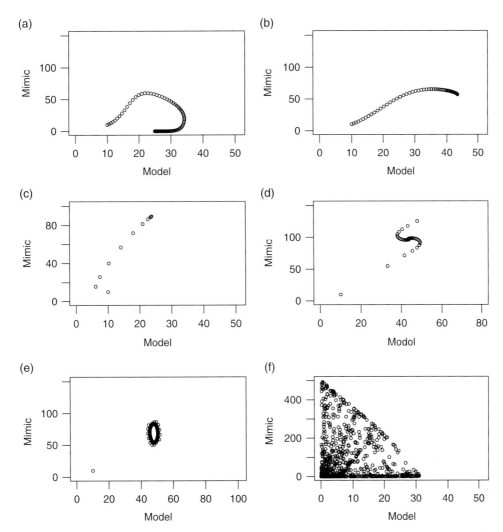

Fig. 5.10 Phase plots of number of mimics versus number of models from Batesian mimicry simulations starting at 10 of each, with respective carrying capacities of 100 and 50, in which predation on each is increased as a result of increased proportions of mimics in the population. The parameters varied are the intrinsic rates of increase of mimic and model (r_1, r_2) and the predation coefficients of the model and mimic (α, β) which represent their probabilities of being attacked and killed. (a) $r_1 = 0.8$, $r_2 = 0.1$, $\alpha = 1.5$, $\beta = 0.5$; (b) $r_1 = 0.3$, $r_2 = 0.1$, $\alpha = 1$, $\beta = 0.3$; (c) $r_1 = 0.7$, $r_2 = 1.1$, $\alpha = 0.5$, $\beta = 2.5$; (d) $r_1 = 2$, $r_2 = 2$, $\alpha = 0.1$, $\beta = 0.4$; (e) $r_1 = 4$, $r_2 = 2$, $\alpha = 0.75$, $\beta = 0.1$; (f) $r_1 = 4$, $r_2 = 1$, $\alpha = 0.5$, $\beta = 2$.

The per capita survivorships – PCS_A for the normal type of species A and PCS_M for the mutants – are then as follows:

$$\text{PCS}_A = \frac{A - m - (D/d_A)}{A - m}$$

which is the proportion of typical-looking individuals of species A ($A - m$) surviving after D/d_A have been predated, and

$$\text{PCS}_M = \left\{ m - \left[\frac{m(D/d_A)(D/d_B)}{m(D/d_B) + B(D/d_A)} \right] \right\} / m$$

which is the proportion of the mutants (m) surviving. The condition for PCS_M to be greater than PCS_A then comes out as:

$$Ad_A < Bd_B$$

In other words, if abundance times toxicity is greater for species B than for species A, a mutation making A look like B will spread. The converse is obviously also true, as abundance times toxicity must always be greater for one or other of the species, and this leads to the conclusion that, appropriate mutations permitting, all toxic species will come to look like the species with the highest product of abundance and toxicity.

Mallet (2001) also importantly showed that the relative fitness gain due to mimicry is proportional to squares of their relative palatability and to the squares of their relative abundances. This means that the rarer, less palatable mimics will be at a particular advantage.

Dill (1975) used the well-known Holling's (1959) disc equation as the basis for simulations of predation on model, mimic and alternative prey. In Holling's (1959) original experimental investigation, a blind-folded human 'predator' searched an arena for food items or equivalents, by prodding the arena to find and pick up sandpaper discs. Dill (1975) modified this, with some discs representing mimics, some alternative prey and some models. To quote,

> Mimics and alternate prey could be eaten without penalty. However, any attempt to eat a model resulted in the subject being given 1.0 ml of vinegar via a tygon tube held in the mouth. This stimulus, administered by an experimenter with a syringe, was intended to represent the noxious taste of a model, and caused the "predator" to put the prey down.

An analytical approach by Kokko et al. (2003) showed just how important alternative prey are to the optimal decisions made by a predator when it can choose to attack members of a quasi-Batesian mimicry system (i.e. with the mimics being less toxic than the model) or cryptic edible alternatives, which are harder to locate. When potential prey are rare, the predator's optimal strategy is to test the members of the mimicry complex, because the threat posed by starvation is greater than that of poisoning as long as the model is not totally lethal. However, if there are abundant alternative prey it should ignore the mimicry complex species even if the mimic is completely palatable. Thus their model shows that even when the mimic in a quasi-Batesian system is considerably less well defended (toxic) than the model, the system can still behave in a perfectly Müllerian way with the more palatable mimic protecting the model. Conversely, if alternative prey are rare and starvation of the predator becomes a significant risk, even when the 'mimic' is as toxic as the model, it can still create a quasi-Batesian system such that increases in the numbers of either is detrimental to the other. An important parameter in Kokko et al.'s (2003) model is whether the predator can increase efficiency of capturing the cryptic alternative prey, for example, by forming a search image. If the predator can find

enough alternative food, the mimic can be considerably less toxic than its model (quasi-Batesian mimicry) without leading to greater mortality of either, whereas if the predator is forced by hunger to prey on members of the mimicry complex, selection will favour the mimic becoming as toxic as the model, thus creating a classical Müllerian system.

Neural networks and evolution of Batesian mimicry

Holmgren & Enquist (1999) used an artificial neural network approach to consider the problem of how many dimensions of variation there are in model and mimic species (using nine in their runs). Their system comprised three coevolving simulated populations: the signal receiver, the model and the mimic. Individuals within the model and mimic populations varied in nine 'dimensions' and random variation was allowed to modify these during the simulations. One of the findings from this 'simple' system was that the mimic always approached the phenotype of the model (i.e. the set of values in the nine-dimensional space) because the model, although always moving away, did so more slowly than the mimic evolved to approach it, the result being that once a good mimetic match had been reached, the actual, virtual mimetic pattern constantly shifted.

AUTOMIMICRY IN BATESIAN/MÜLLERIAN MIMICRY

Whereas all the classic examples and mathematical models deal, largely for simplicity, with an allospecific pair – one an unpalatable model, the other an edible mimic – automimicry involves variation in palatability within an aposematic species (Browerian mimicry). Inevitably, not all members of a model species are ever going to be equally unpalatable, but sometimes the variation is considerable, even ranging from fully palatable to highly toxic. In these cases, the most palatable individuals are acting as Batesian mimics of the least palatable ones. The best-known and studied example is that of the North American monarch butterfly, *Danaus plexippus*. Its food plants, a range of milkweed species (Asclepiadoideae), vary greatly in their cardiac glycoside (CG; cardenolide) content, both inter- and intraspecifically (S.B. Malcolm & Brower 1989), but similar food plant-related variability is also known in other *Danaus*, for example, the African queen *D. chrysippus* (L.P. Brower et al. 1975, 1978) and the queen butterfly, *D. gilippus* (Ritland 1994). Further species of *Danaus* can also regulate the amount of cardenolides sequestered, often showing highly efficient sequestration on plants with only low concentrations, but saturating their uptake when food plants are rich in the compounds. There is also a relationship between butterfly glycoside content and their annual migration to Mexico, with individuals migrating south having generally lower levels of sequestered glycosides than the spring generation due to the nature of the food plants; milkweed plants in the warmer, southern part of the USA have, on average, higher CG content than those in the northern part of the monarch's range, and these are fed upon by the later brood which migrate southwards (Seiber et al. 1980, S.B. Malcolm et al. 1989, Nishida 2002). Within the monarch, there is a negative correlation between adult body size and the concentration of CGs in their tissues, which suggests that there could be a physiological cost to storing the toxins (Cohen 1985). Individuals of the African queen butterfly, *D. chrysippus*, and another mimetic danaid, *Euploea core*, often discard the glycosides before moulting to the adult stage (S.B. Malcolm & Rothschild 1983), but *D. chrysippus* larvae grow faster and attain a significantly greater size when fed on milkweeds with CGs than when fed on plants lacking CGs (D.A.S. Smith 1978). As with *D. chrysippus*, *Euploea core* also does not sequester toxins from its Asclepiadoideae food plants efficiently, but the adults synthesise toxins *de novo* (Rothschild et al. 1978).

L.P. Brower et al. (1970) and, in a more refined version, Pough et al. (1973), explored automimicry with mathematical models. In the first paper the prey was assumed to be so unpalatable that predators avoided them after a single encounter for a certain fraction of foraging time, *n*, which is linearly related to Huheey's (1964) memory parameter *n*. Their key equations were as follows: if *m* is the ratio of prey per predator, then n/m will be a measure they term 'predation potential' and if all prey are equally palatable then the proportion of prey surviving over a time period will be $1-n/m$, which they call J_0. However, if the memory constant is larger than the number of prey per predator, J_0 will be negative, which makes no biological sense, and therefore they set $J_0 = 0$ for such cases.

If k'[3] of the prey encountered are unpalatable, and $1-k'$ are palatable, and all prey deemed palatable are eaten on discovery, then the proportion of prey surviving being eaten, *J*, will be given by

$$J = \left[1 - \left(1 - k'\right)^n \right] / mk'$$

The relative survival advantage in the case of automimicry would therefore be the difference between this value, *J*, and

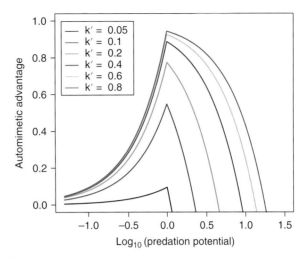

Fig. 5.11 Showing the relationship between the relative advantage of automimicry and predation potential (which is defined as the ratio of the avoidance memory parameter, *n*, and the mean number of prey per predator) for six different proportions of automimics, *k*, in the system. (Source: Created using equations of L.P. Brower et al. 1970. Reproduced with permission from Professor L.P. Brower.)

survival when all are palatable, i.e. J_0 (see above). A typical series of plots is shown in Fig. 5.11 – the sharp change at $n/m = 1$ is because negative values of n/m are nonsensical. The plots show that the advantage of automimicry is maximal when $n/m = 1$ (i.e. log $n/m = 0$), and declines progressively more slowly as predation potential decreases. The level of protection afforded by automimicry increases with both the proportion that are unpalatable (k') and the degree of unpalatability (*n*), but once k' reaches approximately 0.25 and/or *n* reaches about 20 there is very little further advantage to be gained by further increases because almost all the potential prey are protected virtually as if they were all unpalatable.

In the second paper (Pough et al. 1973), prey were assumed to be less unpalatable, thus requiring a predator to sample more than one before rejecting them, either two with any number of palatables in between or two unpalatables consecutively. All results were broadly similar and in line with most other simple models, but the requirement of Pough et al.'s (1973) third case, i.e. a predator having to sample two unpalatable prey consecutively before learning to avoid them, gives insight about how clumped the distribution of unpalatable prey in the landscape can be, thus they concluded that species that were "uniformly and highly unpalatable can afford to be more dispersed than automimetic species".

---

3. The authors take *k* to be the real proportion of unpalatable prey but accept that these might behave differently from palatable ones in a way that affects their encounter rate with predators, hence k' is the proportion of unpalatable prey that a predator actually encounters.

Svennungsen & Holen (2007) modelled whether auto-mimicry can be evolutionarily stable not just when there are two fixed levels of unpalatability, but a spectrum, as indeed is the case with most cases of automimicry, and found that it was indeed a possible stable strategy under many conditions and therefore not simply a transient stage. Another aspect of automimicry is that predators might still attack aposematic prey but do so cautiously, so that before actually consuming them they can gauge how noxious they are going to be. This would explain, for example, the high levels of cardiac glycosides in the wings of monarch butterflies (L.P. Brower & Glazier 1975; see above) and would therefore enable predators to discriminate between edible and inedible prey. Ruxton & Speed (2006) show that even under these conditions 'non-trivial' levels of automimicry might still be possible for a variety of reasons. Most interesting among these is the idea that by carefully testing each potentially unpalatable prey the predator may suffer opportunity costs, i.e. the time spent might be more profitably used by concentrating on prey that are unlikely to be unpalatable.

In experiments using domestic chicks, Gamberale-Stille & Guilford (2004) found that the birds could correctly differentiate between automimics and automodels a large percentage of the time, calling into question how automimics can be retained in a population. However, Skelhorn & Rowe (2007b) pointed out that the treatments of automimic and automodel in the Gamberale-Stille & Guilford (2004) experiments were not the same, so the chicks might have been able to distinguish the food types visually or texturally. Using a more carefully controlled experimental design, Skelhorn & Rowe (2007b) investigated how domestic chicks selectively rejected aposematic prey (dyed red food crumbs) when there were various proportions of automimics present. The tests varied the automimic frequency between 0 and 75%, which was thought to cover the natural range in taxa such as *Danaus chrysippus* (cf. L.P. Brower et al. 1975). The unpalatable models were found to be tasted and rejected at the same rate regardless of automimic frequency, but the automimics were accepted more as their frequency increased above 25%. This means that at low frequency the automimics should have an advantage over the automodels because they avoid the costs associated with sequestering/producing toxins and therefore automimics ought to be abundant in nature. Skelhorn & Rowe (2007b) suggest that it might actually be a lot more common than is reported because researchers have not generally been searching for it. However, if it is rather rare then it might be because insects do not have the opportunity to be automimics – perhaps their food plants are uniformly toxic – or if they produce their own toxins, perhaps predators might be able to distinguish producers from non-producers and so actively select against automimics. In this latter respect it should be noted that protected prey species that do defensive things such as reflex-bleeding, foaming or regurgitating foul substances can hardly be likely to survive an attack if they do not show those types of secondary defence.

PREDATOR'S DILEMMA WITH POTENTIALLY HARMFUL PREY

When a predator is faced with the choice of not eating a potentially harmful prey because it cannot reliably distinguish the harmful one from its model, but faces hunger with its associated negative fitness consequences, it then has the dilemma of what strategy to use. If it adopts a generalist strategy (i.e. not discriminating against complexes including harmful prey types), it risks feeding on more of the harmful prey and suffering. But if it is a specialist that avoids such complexes, it risks getting less nutrition and therefore might be outcompeted by individuals that adopt the generalist strategy. Heller (1980) explored this mathematically assuming that there are four different niches varying in the combinations of prey that they contain, and the prey fall into two categories, one of which is the equivalent of a Batesian mimicry system and the other a category that is distinguishable from the mimicry system. The model's results show that a predator's evolutionarily optimal strategy can be to produce broods comprising a mix of offspring with generalist and specialist strategies. Three species of reptile are suggested as displaying such a balanced polymorphism of risk-taking: the Jamaican lizard, *Anolis linearopus* (von Brockhusen & Curio 1975) and the garter snakes, *Thamnophis sirtalis* (Burghardt 1975) and *T. elegans* (S.J. Arnold 1977), but it is not hard to imagine that such feeding polymorphism might be far more widespread and go largely unnoticed.

Chapter 6

ANTI-PREDATOR MIMICRY. II. EXPERIMENTAL TESTS

We seek him here, we seek him there,
Those Frenchies seek him everywhere.
Is he in heaven? – Is he in hell?
That damned, elusive Pimpernel.

The Scarlet Pimpernel, play and novel by Baroness Emma Orczy

INTRODUCTION

In this chapter I will describe a range of experiments that have been used to demonstrate that the principles of Batesian and Müllerian mimicry actually work, both in the laboratory and in the wild, followed by experiments that examine other important aspects related to the evolution of mimicry, such as how much do predators generalise, how quickly they learn or forget avoidance, and how predation can lead to evolutionary chases.

EXPERIMENTAL TESTS OF MIMETIC ADVANTAGE

Although mimetic advantage to Batesian mimics seems obvious, experimental tests only really started in the 1960s, predominantly with the work of J.V.Z. and L.P Brower. To show that a species is a Batesian mimic it is necessary to show that it is itself edible, that its assumed model is not edible and that predators cannot, or do not, distinguish between the two. Several early studies demonstrating Batesian mimicry are worth noting. For example, J.V.Z. Brower (1958a, 1958b, 1958c) investigated various famous North American butterfly examples and L.P. Brower et al. (1960) demonstrated effectively that naïve toads, *Bufo terrestris*, would readily attack and learn to reject bumblebees, *Bombus americanorum*, and subsequently refuse to attack the robberfly mimics, *Mallophora bomboides*, but would accept bumblebees from which the stinging apparatus had been removed and then also attack the robberfly, demonstrating that the bee's sting was their only protective system and that they were otherwise palatable, as well as confirming that the robberfly's resemblance to the bumblebee afforded it protection from experienced predators. In this instance, the mimicry is perhaps more complicated since the robberfly, although not a complete specialist, does predate quite heavily on its model bumblebees. In a similar test, Huheey (1980c) presented toads and a tree frog, *Hyla cinerea*, with honey bee, *Apis mellifera*, models and de-stinged mimic bees in different relative abundances and for both species of predator the frequency of models + mimics eaten was a function of the proportion of models and fitted his simple equations for Batesian mimicry (see Chapter 5) remarkably well (Fig. 6.1).

Bowers (1983) wanted to determine whether the North American Harris's checkerspot butterfly, *Chlosyne harrisii*, was a Batesian mimic of the Baltimore checkerspot, *Euphydryas phaeton*, which is highly unpalatable and causes vomiting in birds (Bowers 1980). On the first of a three-day test period, blue jays, *Cyanocitta cristata*, were presented

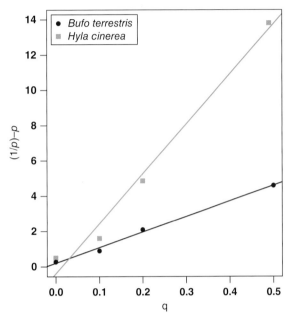

Fig. 6.1 Results of simple experiments to see how the proportion of models in a population of mimic and model prey affected predation on the mimicry complex performed using two toad species as predators and honey bees with and without their stings as models and mimics. The axes are derived from Huheey's (1964) equation (see Chapter 5) rearranged so as to give a linear relationship. (Source: Adapted from Huheey 1980c with permission from the Society for the Study of Amphibians and Reptiles.)

with the edible common wood nymph, *Cercyonis pegala*,[1] which looks quite unlike the supposed mimics and these were almost all eaten (Fig. 6.2). On the following two days they were given *C. harrisii* followed by *E. phaeton*, or vice versa. None of the *E. phaeton* were eaten and when presented first, several of the blue jays rejected *C. harrisii*, at least at first (Fig. 6.2 left-hand series). However, blue jays that had not experienced *E. phaeton* mostly readily accepted *C. harrisii*, though one bird rejected them completely, and subsequently most birds rejected the unpalatable *E. phaeton* (Fig. 6.2 right-hand series). Needless to say, there are not many such precise studies of mimicry systems because of the effort involved.

---

1. Bowers & Wiernasz (1979) found that surprisingly high proportions of this species caught in the wild showed signs of bird damage, but there was no one particular area of the wings involved, despite the highly conspicuous forewing upper and underside eyespots.

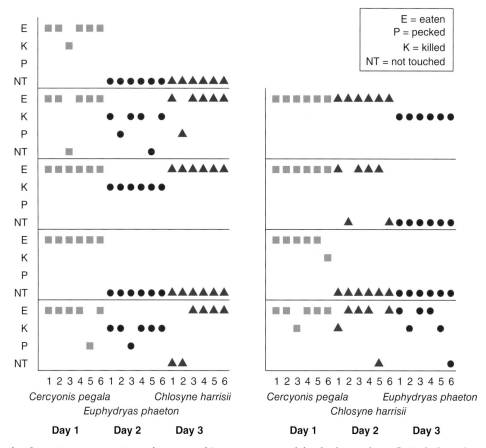

Fig. 6.2 Results of experiments to investigate the suggested Batesian mimicry of the checkerspot butterfly *Euphydryas phaeton* by *Chlosyne harrisii*, using blue jays as predators. Following training on edible *Cercyonis pegala*, birds in the left-hand set of graphs were presented first with unpalatable *E. phaeton* then *C. harrisii*, and the birds in the right-hand graphs were presented with the test species in the opposite order. (Source: Adapted from Bowers 1983. Reproduced with permission from John Wiley & Sons.)

Slobodchikoff (1987) carried out an experiment in a desert region around the Arizona–New Mexico border using the noxious tenebrionid beetle, *Eleodes longicollis*, as model, and the edible longhorn beetle,[2] *Moneilema appressum*, as mimic and exposed them to nocturnal predators such as coyotes and various rodents, though the exact range of predators was not known. His results were very clear – the beetles were sampled by the predators over the first four days (peaking on day two) and thereafter they were virtually completely ignored; the few that were taken might well have been due to encounters with new naïve predators. This study revealed the effectiveness of a highly noxious model in

deterring subsequent predation on the mimicry complex. It also shows that such protection works across a whole complement of predators which, while probably having different responses to the *Eleodes* protective chemicals, were all sufficiently deterred by it. Hetz & Slobodchikoff (1988) carried out a similar experiment using unpalatable *Eleodes obscura*, a palatable mimic, *Stenomorpha marginata*, and house crickets, *Acheta domesticus*, as alternative prey, and found that the latter were consumed at a higher rate than would be expected by chance and the model at a lower rate, whereas the mimic was consumed in proportion to their abundance, indicating that at least some of the guild of mammalian predators could distinguish between models and mimics.

Another early experiment by L.P. Brower et al. (1964) involved the release of artificially painted day-flying moths,

---

2. This is a well-known mimicry example and involves behavioural components as well as overall similarity of size, shape and colour (Raske 1967).

some coloured so as to resemble a local model, the others painted as controls, to see what advantage might be gained by mimetic resemblance in a field situation. They used the day-flying North American moth *Callosamia* (as *Hyalophora*) *promethea* (Saturniidae), which they painted either to resemble the unpalatable and conspicuous butterflies *Heliconius erato* or *Parides anchises* (Papilionidae). To their surprise, they recaptured virtually the same proportion of control and falsely mimetic moths, thus showing no Batesian mimetic advantage. They then painted some moths to be extremely conspicuous but not resembling the Trinidadian butterfly, and in this case they recaptured a lower proportion of the more conspicuous experimentals than painted controls, which they reported as being significant at the $p < 0.05$ level.[3] In work carried out over the following few years on the same system, L.M. Cook et al. (1969) concluded that local birds could adapt rapidly to conditions and when alternative prey became scarce, their ability to discriminate the artificially contrived mimic improved. L.P. Brower et al. (1967a) described a similar experiment, painting the moth either to resemble *P. anchises* or painted dark as controls, and released them over a period of 24 days at various localities in Trinidad, and assessed predation by recapture.[4] The results, however, were not quite as expected in that, by the end of the period, significantly more non-mimetic controls had survived to be recaptured, though in the early part of the study they fared marginally better. The authors suggested that this might have been due to the local predators coming to learn that the pseudomimics were indeed palatable. In a follow-up paper based on further experimental releases over the following 2 years, L.M. Cook et al. (1969) did find some overall evidence of selective advantage to the pseudomimics but again noted that this became lost as local birds learned of their palatability, and they also found that they suffered more when alternative prey were less abundant. Waldbauer & Sternburg (1975) suggested a different interpretation of L.P. Brower et al.'s (1964, 1967a) results because the black-painted *Callosamia* moths might actually have been an effective mimic of highly unpalatable *Battus* swallowtail butterflies, especially *B. polydamas*, and so in their words, "most of the experiments actually compared mimics of one model with mimics of another model". All this just goes to show that there are many potential, and often unexpected, difficulties in carrying out any experiment in the field.

Quite probably a similar problem was encountered by Silberglied et al. (1980), who released and recaptured individual specimens of the butterfly *Anartia fatima*, some of which had had their putatively distractive dorsal white and yellow markings obliterated. No difference in recapture rate was found between experimentals and controls. However Waldbauer & Sternburg (1983) commented that the first authors had, in painting out the dorsal white marks, probably inadvertently created individuals that closely resembled *Parides sesostris*, *P. erithalion* and *P. arcas*, all of which are probably distasteful.

Sternburg et al. (1977) performed a sort of reverse version of L.P. Brower et al.'s (1964, 1967a) experiment with *Callosamia promethia* in Illinois and did demonstrate a selective advantage of Batesian mimicry. The releases involved both black-painted moths and yellow stripe-painted ones so both treatments involved manipulating the moths, the former resembling the toxic *Blattus philenor* and the latter the yellow morph of edible *Papilio glaucus*, both of which occurred abundantly locally. In this experiment, significantly more of the artificial black Batesian mimic moths were recaptured ($\chi^2 = 14.7$, d.f. = 6, $p < 0.025$) and the recaptured moths differed markedly in the amount of wing damage due to bird attacks they had experienced while in flight[5] (Fig. 6.3). Jeffords et al. (1979) extended the work and included male promethean moths painted quite accurately to resemble monarch, *Danaus plexippus*, patterns, with black and orange (monarch)-coloured individuals having higher survivorship and again with survivors suffering less dramatic wing damage. Jeffords et al. (1980) varied the male moth release time until after females had stopped calling pheromonally and confirmed that most predation on the butterflies occurred during the late afternoon by birds while the moths were in flight, and again found that predation on yellow forms was higher than on black-painted ones.

How similar do mimics need to be?

Soon after J.V.Z. Brower's (1958a, 1958b, 1958c) papers, Sexton (1960) conducted a neat experiment using carefully modified adults of the edible mealworm (*Tenebrio molitor*), gluing onto them the prothorax[6] and/or the elytra from freshly killed, similar sized, unpalatable fireflies (*Photinus pyralis*), and presenting them to caged anole lizards (*Anolis carolinensis*). The anoles had been slightly starved before

---

3. The Pearson χ^2 uncorrected for continuity is 3.98, d.f. = 1, $p = 0.046$; however, a more accurate estimated probability would use Yates' correction for continuity in which case $\chi^2 = 3.08$ and p is then not significant.

4. Males only were recaptured using traps baited by virgin *Callosamia* females that release attractive pheromones.

5. The few individuals recaptured with symmetrical wing damage, indicating that they were attacked while resting, were excluded from the analysis and Fig. 6.3.

6. Sexton is unclear about how this was done, but it seems likely that only the pronotum, cleaned of soft tissue, was used.

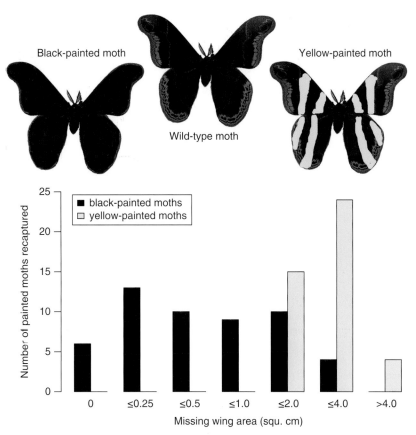

Fig. 6.3 Day-flying male *Callosamia promethia* moth (top centre), artificial Batesian mimic (black-painted to resemble unpalatable *Battus philenor*) and control (yellow stripe-painted to resemble the yellow morph of edible *Papilio glaucus*), and graph of wing damage on recaptured individuals showing that the controls probably experienced an increased number of, and more severe/persistent, bird attacks. (Source: Adapted from Sternburg et al. 1977. Reproduced with permission from AAAS. *C. promethia* image, Megan McCarty, public domain.)

testing, just enough to ensure that they would approach any moving potential prey. On the whole, the anoles did not attack the fireflies and out of 67 presentations, only three fireflies were attacked (4%) and of those, only one was eaten, the other two being rejected. In the experimental series, the *Tenebrio* and either an unmodified *Photinus* or a modified *Tenebrio* were offered in pairs to a caged lizard. The results of Sexton's pairwise tests are shown in Fig. 6.4. When tested against unmodified *Photinus*, only the artificial mimics with both prothorax and elytra escaped some predation, and when tested against unmodified *Tenebrio*, artificial mimics with elytra only and with both elytra and prothorax both experienced some protection. The prothorax pattern alone therefore had no effect.

A similar result was found in a quite different situation. The toxic pufferfish *Canthigaster valentini* (Fig. 6.5a) is putatively mimicked by the edible reef fish *Paraluteres prionurus* (Fig. 6.5b). Using various models with progressively lower similarity to the pufferfish (Fig. 6.5c), Caley & Schluter (2003) found that there was a "fairly broad region of protection", with various piscivorous species avoiding the models, but non-piscivorous reef fish were not deterred (Fig. 6.6). Even moderate resemblance to the pufferfish's pattern afforded some protection, suggesting how the initial selection for the mimetic resemblance could have occurred. Although not experimental, Frisch's (2006) survey suggests that juvenile coral-trouts, *Plectropomus*, are also mimics of *Canthigaster*.

To test whether Müllerian mimicry works in the wild, Pinheiro (2003) released a range of butterflies, including members of Müllerian mimicry rings and others known to be generally non-mimetic and palatable wild insectivorous

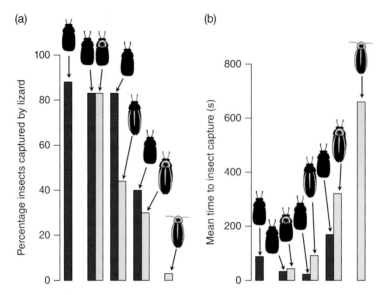

Fig. 6.4 Predation by lizards (*Anolis carolinensis*) on unpalatable fireflies (*Photinus pyralis*) and palatable mealworm adults (*Tenebrio molitor*) and on three mealworm variants, one with *Photinus* pronotum glued on to it, one with *Photinus* elytra glued on, and one with both *Photinus* pronotum and elytra glued on (as illustrated). Pairwise tests results and values for *Photinus* and unmodified *Tenebrio* alone. (a) Percentage of each prey type attacked in pairwise trials; (b) mean time until the lizard attacked each type. (Source: Adapted from Sexton 1960.)

birds, kingbirds, *Tyrannus melancholicus*, and cliff-flycatchers, *Hirundinea ferruginea*, perched on wires in three Amazonian habitats. He found that the birds rejected members of the mimicry rings that occurred near each given release site, but readily attacked butterflies belonging to Müllerian rings elsewhere, apparently demonstrating that the birds had learned from their experience with the local butterfly fauna but were not put off attacking butterflies with patterns that were new to them.

Unambiguous experimental proof of the protective advantage afforded by Müllerian resemblance was presented by Rowland et al. (2010) using great tits, *Parus major*, in a 'novel world' situation. They used baited shapes and importantly demonstrated first that the birds found them equally visible, could be equally easily learned and could be discriminated between. In the critical tests birds were presented with models (targets with small pieces of almond rendered unpalatable by soaking in a chloroquine solution) alone, or with identical patterned mimics with unpalatable almond pieces, or dissimilar targets ('distinct mimics') with unpalatable almonds. After a first training trial to avoid the target with unpalatable almond, the risk of predation in the paired situations was not significantly reduced when accompanied by imperfect mimics (model alone versus model plus distinctive mimic: Tukey's post hoc test: $p = 0.994$) but was significantly reduced when the models were accompanied by perfect unpalatable mimics (Tukey's post hoc test: $p = 0.005$).

There is considerable field evidence that the presence of potential co-models in an area permits species to converge on the same phenotype. A nice example is provided by the burnet moth, *Zygaena ephialtes*, which in northern and central Europe displays contrasting bright red markings and metallic green-black, whereas in southern Europe it is has a white and yellow-marked morph that closely resemble unpalatable *Amata* (Lepidoptera: Erebidae: Arctiinae) moths that only occur in the south (Bullini & Sbordoni 1971, Sbordoni & Bullini 1971, Sbordoni et al. 1979). However, Kapan (2001) was the first person to demonstrate the advantage of Müllerian resemblance in a field situation experimentally, making use of the mimetic polymorphism found in the butterfly *Heliconius cydno* in western Ecuador. At some sites a yellow morph occurs that mimics *H. eleuchia* alongside a white morph that mimics *H. sapho*. Pairs of the two different morphs were collected at a site where *H. cydno* was polymorphic and they were then marked and transported to separate sites where either one or the other of the co-mimic species were relatively much more abundant. As expected, relative survival of the transported *H. cydno* morphs at each release site was positively related to the relative abundance of the co-mimics.

Beatty et al. (2004) employed a computer game system with human hunters and a range of computer-generated prey to investigate whether the hunting process could lead to Müllerian convergence of members of the unprofitable prey category and, indeed, found intense selection pressure

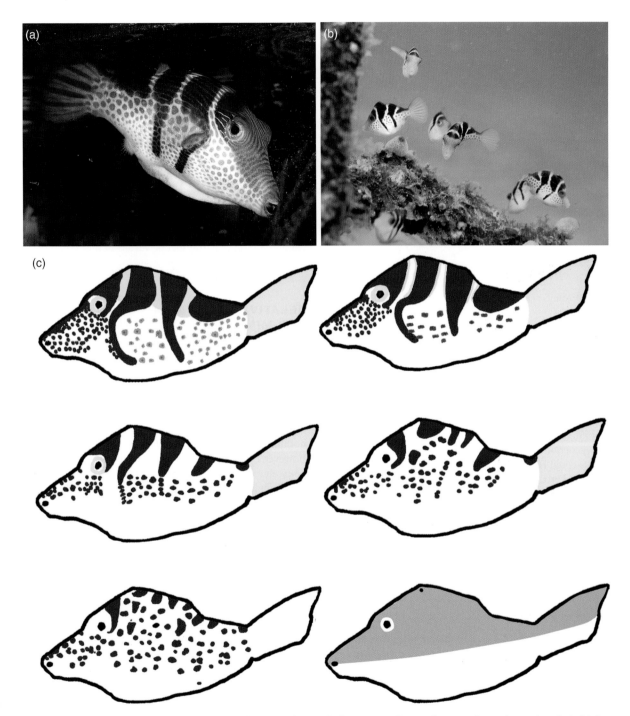

Fig. 6.5 Mimetic pair of coral reef fish species. (a) The toxic Valentinni's sharpnose puffer, *Canthigaster valentini* (Tetraodontidae); (b) the non-toxic, black-saddle filefish, *Paraluteres prionurus* (Monacanthidae), otherwise known as the false puffer; (c) drawings of the different models of the pufferfish of various degrees of fidelity, used by Caley & Schluter (2003) to assess the potential benefit to it and its putative mimic, the black-saddle filefish. (Source: a, Haplochromis 2009. Reproduced under the terms of the Creative Commons Attribution Share-Alike Licence CC BY-SA 3.0, via Wikimedia Commons; b, Jens Petersen 2006. Reproduced under the terms of the Creative Commons Attribution Share-Alike Licence CC BY 2.5, via Wikimedia Commons; c, adapted from Caley & Schluter 2003. Reproduced with permission from The Royal Society.)

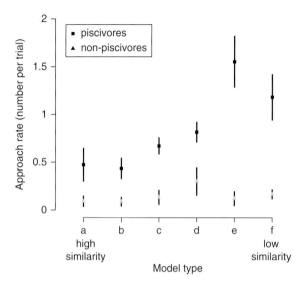

Fig. 6.6 Numbers of piscivorous and non-piscivorous fish approaching *C. valentini* models (see Fig. 6.5): mean numbers per trial (± S.E.). (Source: Adapted from Caley & Schluter 2003. Reproduced with permission from The Royal Society.)

in that direction. That in itself is no big surprise, but the mechanism behind it is interesting. The players were not evaluating the relative profitabilities of the different prey types independently but instead they learned better to avoid unprofitable phenotypes that were similar to one another. This generalisation process meant that rare imperfect mimics could be favoured, such that their genotypes could spread and progressively evolve better resemblance, and also indicates that the evolution of Müllerian mimicry should be facilitated in communities with many types of prey.

Is a two-step process necessary?

Most researchers and theoreticians have taken the view that the evolution of Batesian mimics from a cryptic ancestral state must involve a two-step process (see Chapter 5, section *Introduction*) because an initial slightly more conspicuous but not very mimetic form would gain almost no protection from mimicry but suffer a huge cost from no longer being very cryptic, i.e. there is an adaptive fitness valley that cannot be crossed gradually (Sheppard 1959, C.A. Clarke & Sheppard 1960a, 1960d, Turner et al. 1984, Turner 1987, 1988).

Kikuchi & Pfennig (2010) propose that a gradual evolution towards Batesian mimicry might be possible if models are very abundant. Using accurate replicas of coral snakes, *Micrurus fulvius* (see Fig. 7.16a) and mimic kingsnake,

Lampropeltis species (see Fig. 7.16d,f; *L. alterna* looks quite similar to the non-mimetic *L. triangulum* of North Carolina), at different localities with varying abundance of real coral snakes (Florida with high abundance and North Carolina with low abundance) they found that intermediate replica 'phenotypes' were significantly more heavily attacked in North Carolina, i.e. there is an adaptive fitness valley there, but in Florida they suffered an intermediate level of attack (Fig. 6.7), indicating that there may be no adaptive valley.

Critical in this is whether the degree of generalisation by predators depends on the level of noxiousness of an experience with a model. Some studies (e.g. L.P. Brower et al. 1971) have demonstrated that avian predators that have had experience of a highly unpalatable prey may be reluctant to attack species/individuals that are only vaguely similar, i.e. the sort of variation that could result from a single gene mutation.

RELATIVE ABUNDANCES OF MODELS AND MIMICS IN NATURE

It is still often stated that for Batesian mimicry to be successful there ought to be a higher abundance of models than mimics. These early notions have largely been dismissed because of the realisation that encounters by predators with really noxious models may cause them to avoid things that look like the model for a long time. There is certainly nothing unique about a 50:50 ratio (Huheey 1964). However, the mimic will obviously gain more protection, the more abundant its model is. L.P. Brower & Brower (1962), for example, found that the abundance of the female form of the mimetic *Papilio glaucus* closely corresponded to that of its putative model *Battus philenor* over much of the range (see also Platt & Brower 1968). This also means that if the model is absent from an area, there will be no protection afforded by the mimicry and therefore either the mimic may go extinct in such places or, as described earlier, evolve locally to be more cryptic.

SEX-LIMITED MIMICRIES AND MIMETIC LOAD

Many examples of Batesian mimicry are known in which only one sex is mimetic, the other sex usually being fairly similar to related non-mimetic taxa. This bias is usually referred to as sex-limited mimicry. In many species of mimetic butterfly, both sexes look similarly like the same model, but in quite a few cases, the males and females differ quite markedly, with either one or both of them resembling

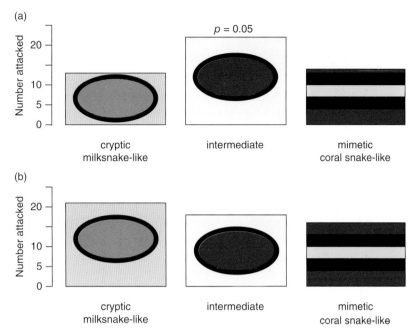

Fig. 6.7 Predation on model snakes that are either non-venomous coral snake mimics and rather cryptic in their habitat, aposematic venomous coral snake mimics or intermediate, at two localities, one (a) in North Carolina where coral snake models are rare, the other (b) in Florida where coral snakes are common, showing that poor mimics do not suffer significantly higher mortality if the models are common, which they do when coral snakes are rare. (Source: Adapted from Kikuchi & Pfennig 2010.)

a particular model (Fig. 6.8) (D.F. Owen 1971b, D.F. Owen & Chanter 1972, Long et al. 2014). In most cases of sex-limited mimicry, i.e. where the sexes have different phenotypes, the male displays an 'archetypic' morph that is non-mimetic and females have phenotypes that mimic one or more unpalatable models (Vane-Wright 1971). That this universal trend across many unrelated butterfly taxa strongly indicates that it is the mimetic forms that are derived (C.A. Clarke & Sheppard 1960d).

Sometimes, notably in the genus *Papilio*, different female forms of the same species mimic different models (see Fig. 6.9) while the males nearly always display the more or less typical shape and colouration of the family. An exception is provided by the African false wanderer butterfly, *Pseudacraea eurytus*, in which both males and females are polymorphic, though some forms are only found in females (G.D.H. Carpenter 1936, 1949, D.F. Owen & Chanter 1972) which Carpenter (1936) suggests could be considered a rival to *Papilio dardanus* (see Chapter 7, section *Papilio dardanus*) as the "most interesting butterfly in the world".

While sex-limited mimicry is best known among the butterflies, examples are also to be found among beetles (see later), pompilid wasps (H.E. Evans 1968), true bugs (Gnezdilov & Viraktamath 2011; see Chapter 7, section

Ant mimicry as defence against predation), spiders (Reiskind & Levi 1967), Orthoptera, Odonata (see Chapter 11, section *Androchromatism and male mimicry by females*), the hoverfly, *Eristalis tenax*, and even a few day-flying moths (Aiello & Brown 1988). Unfortunately, there has been almost nothing done on the genetics or biology of any of these (e.g. McIver 1987). Stamps & Gon (1983) survey the occurrence in sex bias in different arthropods and even note some examples among reptiles and amphibians.

Vane-Wright (1971, 1975, 1979, 2009) classifies butterfly mimicry as 'unimodal', 'dual' and two forms of 'sex-limited' mimicry depending on whether the females are mono- or polymorphic (e.g. Figs 6.9 and 7.6). When the polymorphism differs between the sexes it is, in Vane-Wright's terminology, 'bimodal' and may be complete, i.e. there are no shared morphs between the sexes, or partial. Both bimodal morphs may be mimics, as in *Papilio paradoxa*, the males and females mimicking the male and female colour morphs of the nymphalid *Euploea mulciber* (see Fig. 6.8), or only one sex may be mimetic, as in the Neotropical pierid *Perrhybris pyrrha*, in which males are always typically white butterfly-like and females are of the tiger pattern forms of heliconiine and ithomiines. More often, bimodal polymorphism is partial, with most females being mimetic but a few

Fig. 6.8 Sex-limited mimicry between the Indian swallowtail butterfly, *Papilio paradoxa* (Papilionidae), and the striped blue crow butterfly, *Euploea mulciber* (Nymphalidae). Larvae of the latter feed on various *Nerium*, *Aristolochia*, *Toxocarpus* and *Ficus* species. (Source: Dr Krushnamegh Kunte. Reproduced with permission from Dr Krushnamegh Kunte.)

Fig. 6.9 Female-limited mimetic polymorphism in the swallowtail butterfly, *Papilio polytes*, showing the male-like form '*cyrus*', and mimetic forms '*romulus*', '*polytes*' and '*theseus*' and their respective models. (Source: Dr Krushnamegh Kunte. Reproduced with permission from Dr Krushnamegh Kunte.)

resembling non-mimetic males. This latter phenomenon has attracted a lot of attention and much effort has gone into trying to understand its genetic basis.

L.P. Brower (1963) discusses the history of ideas about why females should be the mimetic sex, starting with proposals first put forward by Wallace (1865), for example that it would be more beneficial to females that may be burdened by being heavily laden with eggs. While that might be true, Darwin (1871) could not see any reason why such mimicry might not also be advantageous to the males, and this view was supported by Belt (1874), who first proposed that if female butterflies chose mates by their appearance rather than the other way around, then males departing from the ancestral, or in his language "primordial", colour pattern would be at a disadvantage. Evidence about female mate choice, however, was lacking at that time.

The above were not the only possible explanations. Others included Bates' (1862) observation that males and females often spent most of their time in different locations (e.g. Joron 2005, Kunte 2009a), and Stehr's (1959) gene dosage hypothesis[7] because of a lack of dosage compensation in female butterflies (J.R.G. Turner & Johnson 1980). As an interesting and possibly pertinent aside, Brower (1963) notes that most butterfly collections have more males than females, despite the sex ratio of adults at emergence being approximately equal, possibly indicating, at least in part, that males are generally easier to capture. Nevertheless, examination of wings of the North American checkerspot butterfly, *Euphydryas chalcedona* (Nymphalidae), discarded by birds showed that a higher proportion of females had been attacked (Bowers et al. 1985), and Ohsaki (1995) found that for many tropical butterflies, though not all, females suffered higher levels of predation in the field.

Hespenheide (1975) expressed the view that the mate recognition hypothesis was weak because male choice, male courtship effort and sex pheromones would seem likely to be far more important in determining mating success than female choice based on colour pattern alone. Instead, he proposed that the reason for the female bias in sex-limited mimicry probably lies with behavioural differences between the sexes, such as how much they fly or where they spend most of their time. For example, in the case of the reversed sex-limited mimicry in the buprestid beetle *Chrysobothris humilis* described by Hespenheide (1975), the males (which are the mimics) spend much of their time in the same exposed microhabitat as their supposed models, the clythrine chrysomelid beetle *Saxinus deserticola*, whereas the females spend most of their time more concealed. This situation creates a selective differential for mimicry between the sexes.

From this it can be surmised in general that examples of sex-limited Batesian mimicry would be most likely to occur in situations where the two sexes of the mimic species display marked differences in microhabitat occupation, and that cases of reversed sex-limited mimicry in particular should be most common when the male is the sex that is most exposed to predation.

Magnus (1963) describes how, in all butterflies, males are the ones that initiate interactions with the opposite sex, either searching with zig-zag flight through appropriate habitat, or sitting, waiting and watching for potential mates to fly past. To quote, "the male approaches all optical stimuli which appear to him to be sexually positive". He also notes that "the optical stimulus pattern produced by the female ... is not necessarily the optimal one". Therefore male butterflies seem to be predisposed to investigating potential mating opportunities with a range of potential female forms. Experimental evidence for the importance of female choice in butterfly courtship and mating has since been obtained for several species. For example, L.P. Brower et al. (1965) released variously painted and control Florida queen butterflies, *Danaus gilippus berenice*, in an area with plenty of males and recorded their mating success and, as expected, painted individuals, even highly modified ones, showed no lower mating success than painted controls, although marginally lower than unmodified controls.

There is also evidence that male appearance is important. R.A. Krebs & West (1988) studied courtship in the North American tiger swallowtail, *Papilio glaucus*. The reared males used were either unpainted, yellow-painted controls or experimentally painted black to represent a new mimetic form that resembles the female. Unpainted and yellow-painted control males had very similar mating success but black-painted males received significantly fewer matings (Fig. 6.10). Furthermore, because males can mate multiple times but females only need to mate once, sexual selection will favour conservatism in male patterns and thus the evolution of sex-limited mimicry only in females. Nevertheless, male choice also plays a role, and in the *P. glaucus* system, males show preferences for different female colour forms but the preference varies from region to region (Scriber et al. 1996a). Genetic modelling, however, suggests that neither assortative mating nor differential selection favouring the yellow female form are absolutely necessary for maintaining the polymorphism (J.A. Barrett 1976). In an aposematic, polymorphic European arctiine moth, the wood tiger, *Parasemia plantaginis*, there appears to be a trade-off between protection and mating success in the two male colour morphs (Nokelainen et al. 2012). Yellow-and-black males appear to have higher protection from predation, but lower mating success compared to white-and-black morphs.

---

7. Male butterflies have two X chromosomes, females only one.

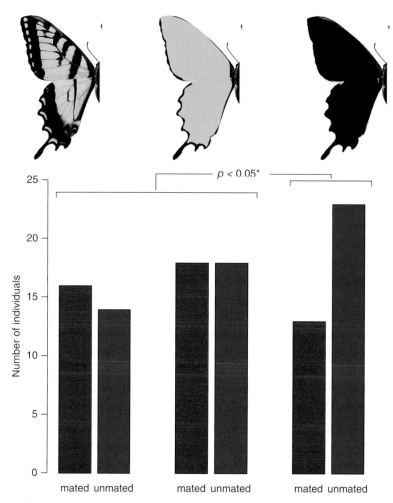

Fig. 6.10 Demonstration of female choice in the tiger swallowtail, *Papilio glaucus*, using untreated, yellow-painted control, and experimental female-like black-painted males. Unaltered and control painted males did not have significantly different mating successes but black-painted ones were significantly less successful than yellow-painted ones ($\chi^2 = 10.8627$, d.f. = 3, $p = 0.01249$). (Source: Adapted from Krebs & West 1988 with permission from John Wiley & Sons; images of *Papilio glaucus*, Megan McCarty 2011, Reproduced under the terms of the Creative Commons Attribution Share-Alike Licence CC BY-SA 3, via Wikimedia Commons.)

J.R.G. Turner (1978) suggested that if a Batesian mimetic mutation is initially expressed in both sexes of a butterfly (i.e. unimodal incipient mimicry, see earlier), it will rise to an equilibrium frequency and then modifier genes suppressing expression of mimicry only in males and ones enhancing the mimicry only in females will be favoured. It should be noted, however, that the strong visual similarity of many Batesian and Müllerian mimetic butterflies may cause confusion; males often engage in courtship behaviour or territorial disputes with members of other species (Vane-Wright & Boppré 1993). Indeed, interspecific confusion among mimics is even known in aposematic poison dart frogs (Mallet 2014). From an evolutionary point of view, although these attempts are pretty futile, they illustrate the importance of colour pattern in maintaining reproductive isolation between species.

Sex-limited mimicry is exhibited by only a few temperate but numerous tropical butterflies. In Africa, *Hypolimnas misippus* (Nymphalidae) has four female forms that mimic a colour form of the unpalatable *Danaus chrysippus*, while the male is very different from any of the females (D.A.S. Smith 1976). The most famous examples, however, involve various species of swallowtails, e.g. *Papilio aegeus, P. dardanus* (Vane-Wright et al. 1999), *P. glaucus* (Scriber et al. 1996b),

P. polytes (Kunte 2009a, 2009b), and *P. polyxenes* (Hazel 1990, Herrel & Hazel 1995). Indeed, for a long while it was thought that the different colour forms of *P. dardanus* represented different species, until Trimen (1869) discovered that the variation was a polymorphism.

In North America, the eastern tiger swallowtail, *Papilio glaucus*, has males that are typically swallowtail patterned with black stripes, and females that may be male-like in appearance or which are entirely black and mimic the unpalatable *Battus philenor*. The question arises as to how this female polymorphism is maintained, especially in areas where the model is abundant, and indeed also in nearly all the other cases of sex-limited mimicry. J.M. Burns (1966) found a possible solution in that males preferentially mated with the non-mimetic female forms and that, in the wild, pale non-mimetic females had received significantly larger numbers of spermatophores than mimetic ones at the same locality. Whether similar mating preference occurs in other cases of sex-limited mimicry is largely unknown, but it is also present in the Asian and East Palaearctic butterfly *Papilio polytes* (see Fig. 6.9). Furthermore, in this species, not only the mating success of the male-like (f. *cyrus*) but also the larval growth rate are higher than in the mimetic (f. *polytes*) form. Sekimura et al. (2014) created a mathematical model of the population dynamics of a predator, the model (*Pachliopta aristolochiae*) and the mimetic and non-mimetic forms of *Papilio polytes* in the system using three difference equations, and found, satisfyingly, that observed frequencies from the model closely matched those from the field.

Kunte (2008), using an independent phylogeny of the genus *Papilio*, showed that sexual dimorphism was significantly correlated with female-limited Batesian mimicry, where females are mimetic and males are not but retain the plesiomorphic state, and he was able to reconstruct ancestral states. Kunte (2009a) discusses the possible evolutionary pathways between non-mimetic (monomorphic), non-mimetic (sexually dimorphic), monomorphic (mimetic) and female-limited polymorphic mimicry. Perhaps initially surprisingly, he found that sexually monomorphic mimicry and female-limited mimicry have evolved repeatedly but predominantly independently in different clades, but this lack of evidence for step-wise transitions can be explained by the scarcity of transitional forms and no doubt extinctions and the inevitable lack of complete historical accuracy of phylogeny estimates. One of the interesting findings was that there was no evidence of mimetic forms having evolved back to non-mimetic forms. C.A. Clarke et al. (1985) did show, however, that the male-like form of female of *P. phorcas* was atypical in being dominant to the mimetic female form, something which is usually associated with a derived state. Given that there are instances among other

insects where females mimic males (see Chapter 11, section *Androchromatism and male mimicry by females*), the possibility that some female butterflies might have evolved to do the same transvestitism as proposed by Vane-Wright (1984) cannot, as yet, be completely excluded.

Although in the vast majority of cases of sex-limited mimicries, the female is the mimic (Vane-Wright 1971), very rarely the opposite situation occurs, in which the male is the mimic and the female 'typical', termed known as 'reverse sex-limited mimicry'. One of the few known examples is that of the beetle *Chrysobothris humilis* from California, mentioned above (Hespenheide 1975). In this species the female is uniformly metallic green but the male is black with red shoulders and and mimics the aposematic and unpalatable *Saxinus deserticola*. Indeed, the two sexes of *C. humilis* were originally described as separate species because of their colour differences. Vane-Wright (1971) cites only one other case of reversed sex-limited mimicry, the Pantropical spider *Coleosoma floridanum* in which males are ant-like, but males spiders are often smaller than females.

The underlying genetics of sex-limited mimicry has not been easy to study, though modern genomics approaches are starting to make major advances in our understanding. Since butterflies have heterogametic females it has been suggested that female-limited characters should be predominantly controlled from the autosomes, but that male-limited characters should often be X-linked. Kunte et al. (2014) were able to demonstrate that *doublesex* was the supergene responsible for the female mimetic polymorphism in *Papilio polytes*, and even more recently, Nishikawa et al. (2015), who generated whole-genome sequences for *P. polytes* and the related *P. xuthus*, have identified a single ~130 kb autosomal inversion separating mimetic (*H*-type) and non-mimetic (*h*-type) chromosomes, and indeed the region includes the *doublesex* locus. Using a gene knockdown technique[8] to suppress expression of the *H*-type (mimetic) *doublesex* allele, Nishikawa et al. showed that this inverted region was responsible for generating the mimetic patterns, and the non-inverted *h*-type does not alter the pattern.

Mimetic load

The more common Batesian mimics are relative to their model(s), the more often they will tend to be attacked by predators who have not encountered sufficient numbers of the model to form an avoidance reaction. Polymorphism of

---

8. By introducing small complementary RNA (siRNA) molecules into cells by electroporation, it is possible to downregulate gene expression and thereby study their actions.

the mimic species is one way in which this can be overcome if there are appropriate alternative models (J.R.G. Turner 1975, J.J.G. Greenwood et al. 1981). As R.A. Fisher (1930) pointed out, polymorphisms will be stable if an increase in the abundance of one of the morphs reduces its relative advantage over another such that at some relative abundance they have equal fitness.

APOSTATIC SELECTION AND BATESIAN MIMICRY

In contrast to Batesian mimetic situations in which polymorphism of the mimic is common, it is often noted that Müllerian mimics are generally locally monomorphic, though with some well-known exceptions. Monomorphism is usually thought to result from antiapostatic selection, that is the rare form will be subject to relatively higher levels of predation because a predator will have to sample a disproportionately larger number of a rarer prey type before learning to avoid it because the encounters with the rarer type will be more widely spaced in time. Benson (1972) appeared to provide some evidence for this in an experiment in which some individuals of an aposematic and unpalatable Neotropical butterfly, *Heliconius erato*, were painted so as to produce a different aposematic pattern. Using mark–release–recapture methodology he found that individuals with the natural red colour obscured had lower survival rates than unmodified butterflies. However, as J.J.G. Greenwood et al. (1981) pointed out, Benson's (1972) modified *H. erato* might simply have had a less effectively aposematic pattern than the natural ones, and so his results could have reflected this rather than the predator's encounter rate and pattern learning per se.

Experiments with wild passerine birds have yielded mixed results. J.A. Allen & Clarke (1968) and Bantock & Harvey (1974) found evidence for apostatic selection, but both Horsley et al. (1979) and J.A. Allen & Anderson (1984) observed antiapostatic behaviour when food was presented at high density and Lindström et al. (2001a) found strong antiapostatic selection in an experiment with captive blue tits, *Cyanistes caeruleus*, against rare aposematic prey. Such antiapostatic selection will tend to lead to the evolution of monomorphism and situations where prey density will always be high, include schooling fish and flocking birds (see Chapter 8, section *Schooling, flocking and predator confusion*). Greenwood et al. (1989) found that wild passerines displayed antiapostatic selection on mildly distasteful coloured pastry prey, but showed no frequency-dependent effect when they were palatable. However, it is hard to see the selective advantage of antiapostatic predation to the predator simply because rarer food would be expected to be harder to find (J.J.G. Greenwood et al. 1984).

J.J.G. Greenwood et al. (1981) carried out two sets of reciprocal experiments using coloured food pellets, one set with wild passerine birds, the other with domestic chicks. Contrary to expectation, the first experiment provided no evidence for frequency-dependent predation, and in the second, more surprisingly, the rarer form experienced disproportionately less predation.

J.J.G. Greenwood et al. (1984, 1985) showed that laboratory mice preferred the rarer of two types of food when presented in 9:1 and 1:9 ratios. In their experiments, which were carried out in total darkness, they used rat cake that had been flavoured with either peppermint or vanilla, and additionally, as a control, they dyed the food either green or brown with food colouring since the mice could not see the colours. However, irrespective of the experimental conditions, the mice showed a significant preference for green at least in the first few days of trials, indicating that although the food colouring was tasteless to humans, it was not to the mice. Further, the antiapostatic selection was enhanced when the rare food items differed both in colour and taste.

Positive density-dependent predation by fawn-breasted tanagers, *Pipraeidea melanonota*, on roosting aposematic ithomiine butterflies in Brazil was noted by K.S. Brown & Vasconcellos-Neto (1976). Bantock & Harvey (1974) suggested that naïve song thrushes (*Turdus philomelos*) preying on banded *Cepaea nemoralis* snails may regularly form search images and thus selectively predate one of its colour forms.

Müllerian mimicry and unequal defence

Despite its long history as a concept there has been surprisingly little experimental work investigating how predators respond to aposematic distasteful mimics that, as will almost always be the case, differ in some aspect of their unpalatability. Various models pertaining to this have been discussed in Chapter 5.

Speed et al. (2000) used wild birds feeding on coloured pastry 'prey' presented on variously coloured card triangles to test the protective nature of unequally defended prey. Prey presented were either palatable, moderately defended or highly defended by addition of different amounts of a combination of quinine hydrochloride and mustard.[9] Use of two unpalatable compounds was intended to increase the potential range of effect across the multiple wild-bird species that might show different degrees of aversion to each. In two experiments, each conducted over a 40-day period,

9. Treatments were: moderately defended 'prey' (0.5 g quinine hydrochloride and 1 g of mustard per 500 g of pastry) or highly defended 'prey' (1.25 g quinine hydrochloride and 2.5 g of mustard per 500 g of pastry).

Speed et al. found that moderately defended prey diluted the protective effect of the highly defended model (or co-mimic). Speed (1993b, 1999a) termed this situation "quasi-Batesian mimicry" since some, but not all, individuals are Batesian mimics of the most unpalatable ones, indicating that the relationship between model and mimic lies somewhere between pure Batesian mimicry and strict Müllerian resemblance. Depending on density, the mimic could also act in a Müllerian way to equally moderately defended prey. Using great tits, *Parus major*, as predators, Lindström et al. (2006) found that the effect of quasi-Batesian mimics did not always increase the overall number of more defended models eaten, and instead the effect depended on the visual appearance of the prey.

It has been suggested that naïve predators that may later develop the ability to differentiate between more and less palatable prey that are not perfect mimics might be important for the initial evolution of Müllerian mimicry, and that they might respond to the mean palatability of the prey. Ihalainen et al. (2007) varied the level of toxicity of artificial prey to great tits, *Parus major*, and found that mortality experienced was the same as when all prey had the same mean level of toxicity, and Balogh et al. (2008) note that this is consistent with predators having a higher rate of learning when there is variation in unpalatability.

Eurasian starlings, *Sturnus vulgaris*, have been shown to regulate their acceptance of distasteful and poisonous prey according to their own internal toxin load. Skelhorn & Rowe (2007a) demonstrated this by training captive starlings with mealworms, *Tenebrio melitor*, injected through the mouthparts with quinine sulphate to make them toxic and coated with Bitrex, a non-toxic bitter compound that birds can taste, to make them distasteful (since evidence suggests that despite being toxic at high dose, birds cannot taste quinine itself). The birds would consume a proportion of the defended prey under normal circumstances, but this proportion was reduced when they were administered separately with a low dose of quinine. Thus the birds seem to be balancing the potentially harmful effect of the quinine with the benefit of eating a protein-rich prey insect. However, they also showed in choice experiments that the birds learned to recognise the defended mealworms through association with colour cues presented in the learning session, but did not learn to avoid them through the Bitrex taste association. Starlings also moderate their intake of distasteful mealworms according to the nutritional value they have. Barnett et al. (2007), again with starlings, examined the effect of their nutritional status by putting them on restricted diets for two periods (Fig. 6.11), adjusting each individual bird's food ration based on their body mass such that they stabilised at the desired reduced mass (Fig. 6.12a). They found that the birds not only increased their attack rates on chemically defended (quinine-injected and coloured mealworms) prey, as might well be expected if they were desperate, but did so in a way that showed they were balancing quinine and energy intakes (Fig. 6.12b).

Halpin et al. (2014) presented European starlings, *Sturnus vulgaris*, with distasteful (quinine-injected) mealworms over a series of 15 sessions. For the first five, the mealworms had had their nutritional value elevated by injection with a protein-rich food supplement, for the next five sessions no extra protein was added, and for the final five protein-enriched prey was used again. When the more nutritious mealworms were replaced with less nutritious ones, the starlings started eating significantly fewer, but they returned to a higher level of predation when the prey were again protein enriched (Fig. 6.13). This extended Halpin et al.'s (2012) earlier work, which had shown that because predators should limit their intake of a toxic substance over a given period of time (or number of feeding sessions), so-called 'saturation theory', dissimilar prey that are unequally defended may still have a mutualistic relationship.

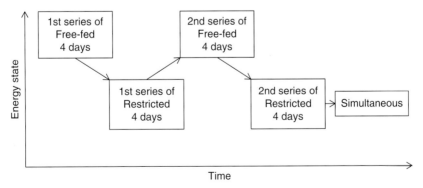

Fig. 6.11 Experimental protocol used by Barnett et al. (2007) to examine the effect of condition of a starling predator on decision making when presented with a mix of defended or undefended prey signaled by coloured discs. After each 'free-fed' series the birds were put on a restricted diet to reduce their body mass and fat reserves before the 'restricted' tests. (Source: Adapted from Barnett et al. 2007. Reproduced with permission from Oxford University Press.)

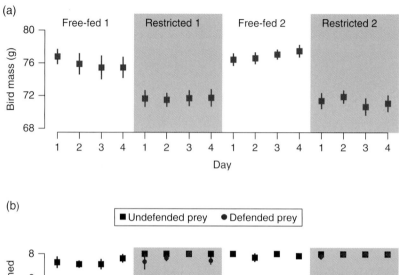

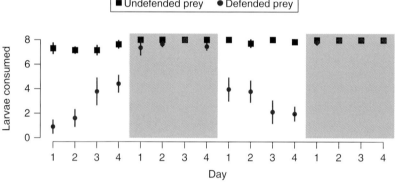

Fig. 6.12 Weights and consumption of undefended and defended prey (quinine-injected mealworms) by starlings through two periods of diet restriction to bring the birds' body masses and fat reserves to reduced levels. (Source: Adapted from Barnett et al. 2007. Reproduced with permission from Oxford University Press.)

Some workers have argued that quasi-Batesian mimicry is likely to be rare in nature because there will be a cost to being able to have moderate defence, so why not be a Batesian mimic (e.g. Joron & Mallet 1998). Getting empirical data for wild predators on natural prey is not particularly easy, nevertheless, Sargent (1995) provided evidence that aposematic insects show a range of degrees of unpalatability to wild birds, and it is clear that many aposematic species show considerable intraspecific variation in unpalatability (see Chapter 5, section *Automimicry in Batesian/Müllerian mimicry*). Intermediately palatable prey are not undefended, as given a choice between them and totally edible prey, Pavlovian predators will tend to preferentially attack the edible ones. A Batesian mimic that is relatively abundant compared to its model, or one that is predated on by an animal that is not repelled by the model's defence, might gain benefit by evolving some of its own defence. Indeed, if it acquired a different defensive capability (as, for example, the arctiine moth/aculeate wasp system described by R.B. Simmons & Weller 2002; see Chapter 4, section *Is the nature of the protective compound important?*) they

might increase the range of protection to both species. The protection afforded by quasi-Batesian mimicry is likely also to depend a lot on the relative availability of alternative prey as well as on the relative abundance of highly protective models. If both these other categories are rare then predators are likely to trade off the relatively milder toxicity of the quasi-Batesian mimics against the nutrition that they provide. Also, if the defence of a quasi-Batesian mimic is not immediately apparent (i.e. there is no association with some widely recognised *aide mémoire* feature such as pyrazine, and no immediate pain or sickness) then naïve predators may exert some considerable toll on them before they come to associate the aposematic pattern with unpalatability.

IMPERFECT (SATYRIC) MIMICRY

The satyrs of mythology were ambiguous creatures and hence Howse & Allen (1994) proposed the term for situations of imperfect Batesian mimicry in which a predator is at least temporarily confused as to whether a prey is itself

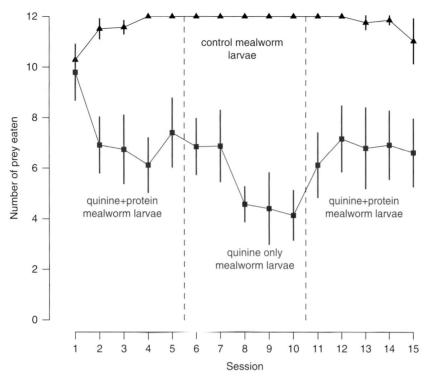

Fig. 6.13 Experimental demonstration that a predator will adjust its consumption of distasteful prey according to the prey's nutritional value, being more tolerant of unpalatability if the prey is richer in protein. (Source: Adapted from Halpin et al. 2014. Reproduced under the terms of the Creative Commons Attribution Licence, CC BY 3.0, via Proceedings of the Royal Society of London, Series B: Biological Sciences.)

unpalatable or a poor mimic of one, which would therefore allow the prey additional moments to escape. The example that was suggested was rather poor resemblances (at least to humans) of some hoverflies (Syrphidae) to aculeate hymenopteran models, though indeed it can be argued that most mimicry is rather poor (Getty 1985) and that it is primarily the most accurate examples that draw the most attention. The existence of so much apparently imperfect mimicry requires explanation. Kikuchi & Pfennig (2013) summarise various hypotheses that have been put forward and show that they fall into four groups:

1. an artefact of human perception due perhaps to the critical features being quite satisfactorily mimetic to the dupes that count (see also exploitation of sensory biases [EPB] in Chapter 9, section *Crypsis and protection in plants*),
2. relaxed selection, where imperfect mimicry is as adaptive as perfect mimicry because important predators do not make finer discrimination,
3. imperfect mimicry is more adaptive than perfect mimicry because of other trade-offs,

4. a result of genetic, developmental or time-lag constraints that prevent selection for more accurate perfection (at least until appropriate mutations arise).

Most workers currently appear to favour the idea of relaxed selection, and this was supported by a study of intraspecific phenotypic variation in some vespid-mimicking hoverflies by Holloway et al. (2002), who found that some of the best mimics were also the most variable The second hypothesis can excluded for the hoverfly example because they appear virtually equally poor mimics in terms of an avian visual model as they do to human perception (Taylor et al. 2016).

The degree of mimetic similarity needed to obtain sufficient protection from predation to outweigh the increased costs of conspicuousness is, not surprisingly, affected by the degree of noxiousness of the model (e.g. C.J. Duncan & Sheppard 1965): the more dangerous/unpleasant an encounter with a model, the less willing the predator is to take chances with even vaguely similar mimics.

Both laboratory (Mühlmann 1934, R.S. Schmidt 1960, Sexton 1960, Mappes & Alatalo 1997) and field (Mostler 1935, G.M. Morrell & Turner 1970, H.A. Ford 1971)

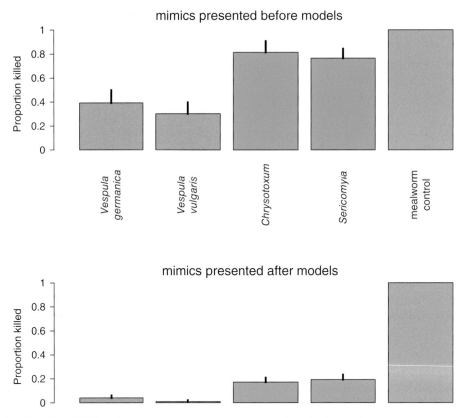

Fig. 6.14 Results of Mostler's (1935) early large-scale experiment on the protection afforded by relatively poor wasp mimic hoverflies (*Sericomyia* and *Chrysotoxum*) from their resemblance to wasps (*Vespula* spp.) carried out by allowing wild birds to fly into a room through a large window where live potential prey were released. When birds experienced the wasps first, they avoided both models and mimics far more than when they had first experienced the mimics. (Source: Adapted from F. Gilbert 2005. Reproduced with permission from CAB International.)

experiments have demonstrated that poor mimics can still gain an advantage from predation compared to non-mimetic individuals. In the first large experiment to demonstrate that imperfectly mimetic hoverflies gained benefit from their resemblance to *Vespula* wasps, Mostler (1935) left a large room window open to allow wild birds to enter, furnished the room with natural type perches and released various live hoverflies or wasps, either first wasps then hoverflies or vice versa, and recorded what was eaten. F. Gilbert (2005) presented Mostler's (1935) data graphically and I have replotted the data in Fig. 6.14. It is clear from Mostler's (1935) results that if the birds had had previous experience with a wasp, they killed fewer of both the wasps and the relatively poor mimetic hoverflies.

Dittrich et al. (1993) put forward the idea that the imperfection resulted from the limitations of a predator's learning ability. They trained retired racing pigeons, *Columba livia*, to peck at a key to get a food reward in response to projected

images either of *Vespula* wasps or of non-wasp-like flies.[10] Some birds were trained to respond positively to wasp images with rewards and others to respond positively to fly images to get rewards. Then for the testing of discrimination the birds were presented with images randomly from a batch of 260 slides of wasps, non-mimetic flies, mimetic hoverflies and control shapes. The pigeons showed a strongly stepped (logistic) response which was broadly similar, irrespective of which group they had been trained on (Fig. 6.15), with most responses being to the high-fidelity

10. For the wasps, they presented images randomly of 40 different *V. vulgaris* and *V. rufa* wasps to achieve a generalised response; for the flies, birds were presented with random images of 40 different flies belonging to several families including Tachinidae, Sarcophagidae and Tabanidae, as well as some non-wasp-like Syrphidae (e.g. *Eristalis tenax*, *Xylota sylvarum* and *Chrysogaster* sp.).

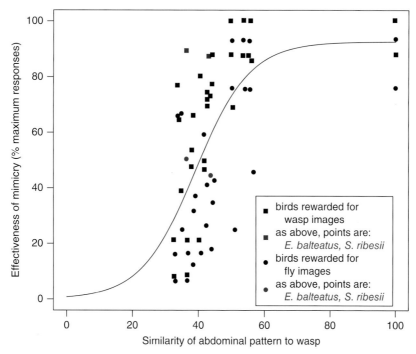

Fig. 6.15 Responses of pigeons trained to peck at a target for a food reward in response to either a wasp image (squares) or fly image (circles) to images of various hoverfly species displaying different degrees of similarity to the wasps as measured by a complex algorithm. Images of wasps and flies were all in natural settings, showing that the birds, even though they are not insectivores, are capable of generalising wasp or fly-like features from the images they were trained upon, and subsequently mistake for wasps those mimetic hoverflies with the greatest similarity. Points representing two hoverflies that, to the human eye, appear as rather poor mimics, are differentiated, because, at least when pigeons were trained for pecking when wasp images were presented, these two hoverflies were treated as if they were very good wasp mimics. The curve was fitted with a scaled two-parameter logistic model and approximates closely that presented by Dittrich et al. (1993), which was 'forced' by the addition of extra points at the lower left-hand corner. (Source: Adapted from Dittrich et al. 1993. Reproduced with permission from The Royal Society.)

mimics, though two species that do not appear particularly wasp-like to humans (*Episyrphus balteatus* and *Syrphus ribesii*) were responded to by the birds as being quite wasp-like. I.C. Cuthill & Bennett (1993) suggested that these inconsistencies might have been partly due to the differences in visual sensitivity of birds and humans, as, in the field, insectivorous birds especially will no doubt also be using information in the UV part of the spectrum, but this has been shown unlikely to be the case (Taylor et al. 2016). Bain et al. (2007) returned to Dittrich et al.'s (1993) images and data and used an empirical 'reverse engineered predator' model on a feed-forward neural network to extract the features the pigeons had probably been using in their original discrimination. The final iterative results closely matched the pigeons' behaviour and revealed a somewhat unexpected set of morphometric characters the birds might have been using. Of note was that antennal length, a feature humans often use to distinguish wasps from flies and

which various tropical hoverflies have evolved to mimic (see Fig. 7.8b,d), was only of minor importance for the pigeons' discrimination. Green et al. (1999) have started to develop an apparatus to permit a slightly more natural way of conditioning pigeons, which uses pinned insect specimens under natural lighting conditions, though that still leaves the issue of what backgrounds to present them on.

H.A. Ford (1971) used coloured pastry 'caterpillars' in a garden bird situation in a simple but effective series of experiments. Over a period, the wild garden bird population were trained to avoid eating red, quinine-containing prey. Then the red pastries were presented simultaneously along with edible purple ones. After a period of learning that the novel purple 'caterpillars' were palatable, these were replaced with bluish-purple and reddish-purple ones. The results, shown in Table 6.1, are interesting, with the birds attacking significantly fewer of the baits that were intermediate between purple and red (i.e. poor mimics of red) than

Table 6.1 Numbers of pastry baits eaten and pecked (brackets) after training wild garden birds on unpalatable red and palatable intermediate purple prey, and associated statistics. (Source: H.A. Ford 1971. Reproduced with permission from Macmillan.)

	Blackbird (*Turdus merula*)	Song thrush (*Turdus philomelos*)	Other birds	Total
Red	0 (1)	0 (0)	2 (8)	2 (9)
Reddish-purple	10 (6)	8 (3)	145 (35)	163 (44)
Bluish-purple	54 (2)	46 (1)	408 (41)	508 (44)
χ^2 (d.f. = 1)	30.2***	26.7***	125.2***	178***

*** *p*-values < 0.001.

the bluish ones, which were novel, and this was true for two focal species (blackbird, *Turdus merula*, and song thrush, *T. philomelos*) as well as the mix of other species present. This suggests that birds could associate reddish-purple with red and generalise it to the unpalatability of the latter.

In contrast to H.A. Ford's (1971) results, Sillén-Tullberg et al. (1982) found that great tits, *Parus major*, were not good at generalising striking features, in particular, the aposematic, red-and-black adults and nymphs of the bug *Lygaeus equestris*. Their results therefore suggest that birds such as great tits would continue selecting for very close resemblance both between Batesian mimics and their models or between co-Müllerian mimics.

L.P. Brower et al. (1971) noted that some Neotropical birds that eat butterflies show discriminatory abilities that ought to make them able to distinguish unpalatable *Heliconius erato* from imperfect mimics such as *Anartia amalthea* and *Biblis hyperia*, but "appeared reluctant to do so". Rowland et al. (2007b) used wild great tits, *Parus major*, and artificial prey in a 'novel world' context to explore the effect of imperfect mimicry when the prey mimics were weakly defended rather than completely palatable, and found a mutualistic relationship between prey types.

Sourakov (2013b) illustrates some putative vague resemblances between some tiger moths (Erebidae: Arctiinae) and some spiders. There is a general similarity in spotted-ness and the colours, but the resemblance is definitely limited. He argues that this is a form of *aide mémoire* or satiric mimicry, whereby the general suggestion of a resemblance causes predators to recall a previous past experience.

The main hypotheses that have been put forward to explain the clearly long-term persistence of imperfect mimicry, such as that of many hoverfly species and their wasp models, are:

1. the imperfection is only due to human cognitive features (perhaps disproportionately ranking just one or two features),

2. the mimics are an average of a group of potential models and therefore do not resemble any one model very closely (multimodal model), and

3. selection for a predator to be able to discriminate between model and mimic will not be strong enough if the mimic is rare and so seldom encountered.

These are discussed here.

M. Edmunds (2000) put forward what he called the "multimodal model of mimetic fidelity". In that model there is a trade-off between accuracy of resemblance to a single, very unpalatable model but then being restricted to only where the model lives, or being a rather poorer mimic of several species, which means that the mimic's potential range might be expanded, but at the cost of a higher predation rate.

Pilecki & O'Donald (1971) suggested that imperfect mimicry might also be limited by an upper bound on mimic frequency, with predators unlikely to learn to distinguish between imperfect mimics and the models if the encounter rate with them was very low. Using wild-caught great tits, *Parus major*, Lindström et al. (2004) found that the survival of imperfect mimics in a Batesian system was strongly positively related to the abundance of alternative prey, and that when the latter were scarce the imperfect mimics were rapidly selected out of the population.

Penney et al. (2012), using a morphometric approach, were able to show that none of the European imperfect mimic hoverflies investigated ever clustered somewhere between wasps in morphospace and thereby effectively disproved the average mimicry hypothesis. Significant positive correlations were found between human estimates of mimetic fidelity and morphometric measures, and between humans' and birds' rankings, thereby largely ruling out the possibility that human perception was largely at fault. However, they did find a strong positive correlation between mimetic fidelity and hoverfly body size. Penney et al. (2014) examined the relationship between behavioural mimicry and visual similarity of hoverflies to aculeate wasps and found a positive trend, albeit not statistically significant (logistic regression, $0.1 > p > 0.05$), although it seems likely that there ought to be a relationship between them. Chittka & Osorio (2007) suggest that cognitive aspects of the predator might facilitate the perpetuation of such imperfect mimicry systems; in particular, if the predator spends too long making the decision to attack, then the imperfect mimics might escape.

Sherratt (2002a) considered using a modelling approach to investigate the frequent 'problem' observed in nature that many apparent cases of mimicry are distinctly imperfect to the human eye, which poses the question as to why the selecting agent has not led to the loss of more distinguishable imperfect mimics. As predicted by R.A. Fisher in 1930, one would expect evolution always to lead to very close, near perfect, similarity between mimic and model, and it seems implausible that we are observing so many cases that are simply on that trajectory but are still a long way from perfection – natural selection is a very powerful force. From Sherratt's (2002a) models it appears, as would be expected, that selection for resemblance is asymptotic, but also that as the mimic approaches the model the strength of selection for improvement diminishes, and that beyond a certain degree of resemblance it drops almost to zero. The point at which selection diminishes depends on "if the model species is costly to attack, if the mimic species is not particularly profitable (e.g. hard to catch), or if the mimic is relatively rare." A further and very interesting result was that if there were two similar but distinct models in an area, the mimic should evolve towards just one of them, whereas if there were several models, then selection may favour a "Jack of all trades" result with the mimic being a compromise between the patterns of the available models. Such a result fits well with the observation of imperfect mimicry in that if, for example, there were several

different ladybird beetles acting as models in an area, each with a somewhat different spot pattern, a mimic might gain an advantage simply by being ladybird-like, i.e. red with some black spots. M. Edmunds (2006) tried to test his own multimodal hypothesis and Sherratt's (2002a) by examining the fidelity of ant-mimicking *Myrmarachne* spiders, but the results were inconclusive.

Speed & Ruxton (2010) used a deterministic evolutionary model which allowed competition between a large number of Batesian mimetic forms ranging from cryptic to even more conspicuous than the model, with increased conspicuousness leading to increased detectability and death. Depending upon parameters, their model predicted several possible evolutionarily stable outcomes, including wide variation in mimetic forms (polymorphism) and dimorphism with one form maximally cryptic. Similar (simulation) results are found in metapopulation models as shown in Chapter 4 (section *Spatial models and metapopulations*), but as noted by Speed & Ruxton (2010) there are very few such examples known in nature, the best known being in some *Papilio* butterflies displaying sex-limited mimicry (e.g. Ohsaki 2005). A possible explanation for the dearth of real-life examples is that mimicry of aposematic models is only likely to occur in species whose crypsis is in some way compromised by their having to be relatively noticeable anyway, such as through their need to move around as a consequence of their foraging behaviour.

Pfennig & Kikuchi (2012) proposed another mechanism that might mean that systems remain at imperfect mimicry based on competition. In their model there is a balance point between low- and high-fidelity mimicry. Below this point, decrease in resemblance to the model leads to increased predation, however, under some circumstances they propose that above this point, competition between the mimic and model or co-mimics increases, thus reducing their fitness.

Chapter 7

ANTI-PREDATOR MIMICRY. III. BATESIAN AND MÜLLERIAN EXAMPLES

Even if a snake is not poisonous, it should pretend to be venomous.

Chanakya, Indian politician (350–275 BC)

INTRODUCTION

It is perhaps no coincidence that the early ideas about the adaptive value of mimicry came from naturalists working in the tropics rather than in higher latitudes (Bates 1862, Wallace 1865, Müller 1879). There are several reasons for this. First, there are many more species in those latitudes and therefore more opportunities for mimetic systems to evolve there. The lack of winters and often the ability of insects to be active adults all year round means that there are more generations in a given period of time for evolution to act upon. The higher temperatures and sunlight availability seems to be correlated with plants generally showing increased levels of chemical protection, and therefore probably greater opportunity for insects to sequester protective compounds. There are also some arguments that predation pressure may, on average, be greater at lower latitudes (Rathcke & Price 1976). The diversity of host plants in a given area may also mean that there is more spatial overlap between species.

A great deal of mimicry research is also centred around butterflies because of their historical association with mimicry theory, size, diurnal behaviour, conspicuousness and aesthetic appeal. It has often been considered that tropical butterflies are more brightly coloured than temperate ones, attributed by some to greater predation pressure or stronger selection for attracting mates in the tropics. However, up until recently, this had not been well-tested. Adams et al. (2014) compared dorsal wing surfaces of male butterflies from three different New World localities (Maine and Florida, USA, and Ecuador). While their results are rather complicated and their analyses not controlled for phylogeny, they did find that their Ecuadorian tropical sample differed from the two US local faunas in several measures of colouration – Ecuadorian butterflies overall displayed a greater level of variation in both intensity and hue within a specimen. So perhaps, as previously assumed, tropical butterflies are more colourful.

TYPES OF MODEL

In animal mimicry, the most widespread and best-studied Batesian models are other animal species that are toxic, venomous or potentially dangerous. Among insects and other terrestrial arthropods, models include many toxic and stinging species (the aculeate Hymenoptera: bees, wasps and ants). In the sea, despite there being many brightly coloured, putatively aposematic creatures (see Fig. 4.32), cases of Batesian mimicry seem surprisingly uncommon, the best known being among some fish. Many snakes are venomous, making them some of the most likely vertebrates to be mimicked, and their large size makes it almost inevitable that many of their (visual) mimics will also be vertebrates.

Mimicry of slow flight in butterflies

Since distasteful aposematic butterflies (and indeed bees and some wasps) sometimes fly slowly and sluggishly, this behaviour might also be mimicked by palatable species as part of their way of signalling to predators that they are likely to be unpalatable (e.g. Beccaloni 1997b). In any case, the slower wing flap rate gives predators a better view of their wing patterns. Srygley (1994) analysed the morphology and body mass of a large number of Neotropical butterflies and found that unpalatable species lie at one corner of an adaptive landscape of wing shape features, and palatable species at another, where the characters are suited to rapid erratic escape flight. In order to resemble unpalatable models more closely, the flight of mimics ought to be expected to converge on that of their models, called 'locomotor mimicry' by Srygley (1994), but the constraint of needing to evade predators would act as a filter for whatever mimetic flight features evolved first. A.V. Z. Brower (1995) dismisses the possibility that palatable but unprofitable prey could become Batesian models simply through the way they fly, at least in the case of Neotropical butterflies, because Srygley's (1994) analysis failed to demonstrate that the fast escapers really would deter predators because of low profitability and because the analysis was not phylogenetically controlled. Srygley & Chai (1990a) focus on the way evolution of slow, mimetic flight by palatable butterfly species conflicts with their need to escape quickly if pursued by a bird because their mimetic wing shapes will be less well adapted to rapid escape. Srygley (2004) showed that flight in those palatable Batesian mimic butterflies whose normal flight mimics the slow flight of their models is more costly than in non-mimetic species. This is because although wing beat frequency is less, their mimetic wing shape increases aerodynamic costs. Indeed, flight by the mimics is also more costly than that of the models, and because of this Srygley (2004) goes on to suggest that slow flight and poor aerodynamics of the models is actually a handicap signal in the sense of Zahavi (1993), in that in the models it is costly to produce and an honest reflection of their unpalatability (see Chapter 4, section *Handicap and signal honesty*).

Kitamura & Imafuku (2010) explored locomotor mimicry in the polymorphic Batesian mimic butterfly *Papilio polytes*, one of its unpalatable models *Pachliopta aristolochiae* (see Fig. 6.9) and the palatable non-mimetic *Papilio xuthus*, and found that both the minimum positional angle (ϕ) and, to a slightly lesser extent, wing beat frequency, were more similar between the model and the mimetic *Papilio polytes* f. *polytes* than between the model and either the non-mimetic *P. polytes* f. *cyrus* or *P. xuthus*, which appears to be good supportive evidence for locomotor mimicry. Locomotor mimicry also appears to be involved in the Batesian mimicry between Polythoridae damselflies (the mimics) and their ithomiine butterfly models studied by Outomuro et al. (2016).

Among butterflies, stouter bodies are associated with relatively shorter wings and smaller wing areas. These stout-bodied species can fly faster, are more agile and are better at escaping attack. Chai & Srygley (1990) correlated the attack success of caged rufous-tailed jacamars, *Galbula ruficauda melanogenia*, against body proportions of a large number of Panamanian butterflies over a 2-year period and found a highly significant, though not phylogenetically controlled, negative relationship between capture success and body stoutness and a positive correlation between palatability and stoutness. Potential flight escape ability was compared between 124 species of Neotropical butterfly by Marden & Chai (1991), using estimates of potential acceleration capability, which is related to the strength, and hence size, of the flight muscles relative to the mass of the whole insect. Unpalatable species and their Batesian mimics had relatively smaller flight muscles and hence lower escape potential, whereas nearly all palatable as well as non-mimetic species had far larger flight muscles and indeed sufficient to enable them to out-accelerate avian insectivores (trogons, motmots and flycatchers) (Fig. 7.1). Importantly, the authors did not just compare the values depicted in the figure directly but took into account phylogenetic relatedness between the species. They also discovered other likely trade-offs by dissecting and weighing organs and measuring lipid content. Both gut mass and ovary mass were significantly negatively related to the flight muscle mass:body mass ratio in both palatable and unpalatable species. Taking body mass into account as a covariate, and palatability as a factor, in an ANCOVA, they showed that unpalatable species invested

significantly less than palatable ones in flight muscle relative to their gut and ovaries ($p < 0.0001$ and $p < 0.008$, respectively). However, there could be several interpretations of the reason for this, because the relationship between flight muscle mass and body mass could be a spurious result of selection acting on abdominal features (gut and ovary size). The authors were largely able to eliminate this by examining a range of other correlations.

It should be noted that slow movement and general lack of escape responsiveness is not limited to unpalatable butterflies but is a general feature of many, if not most, well-defended aposematic species.

THE BATESIAN/MÜLLERIAN SPECTRUM

Following Bates' (1862) exposition of mimicry of unpalatable models by palatable mimics, Fritz Müller (1878) put forward the idea that multiple unpalatable species can benefit by evolving mutual aposematic resemblance. Müllerian resemblance not only benefits the prey because they share the cost of predator learning, but also may benefit the predator, which has to spend less time learning (Speed 1993b).

Pure Batesian and Müllerian mimicry represent the extremes of a spectrum of relative palatabilities (Turner 1984). They also imply different sets of relationships between, and attributes of, the different signal emitters (Table 7.1) (see also table 1 in Huheey 1988). Consideration of these near-pure forms of resemblance is generally relatively easy, and first I will treat these as separate. However, it has long been realised that intermediates with a range of palatabilities or with different relative palatabilities to different predators, is likely to be a more accurate scenario in many situations. Indeed, there are some well-known insectivores that have evolved behaviours that allow them to eat well-defended prey. For example, adult cuckoos readily

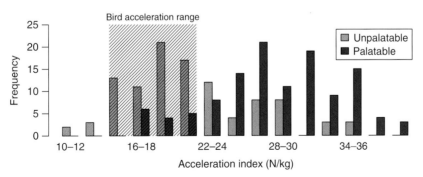

Fig. 7.1 Potential flight acceleration relative to body mass in flight Neotropical butterflies and their avian predators showing that for both unpalatable and palatable species, non-mimetic species had significantly greater acceleration indices than mimetic ones. (Source: Adapted from Marden & Chai 1991. Reproduced with permission from University of Chicago Press.)

Table 7.1 Main differences generally postulated between Batesian and Müllerian mimicry. (Turner, 1987. Reproduced with permission from John Wiley & Sons.)

Batesian mimic	Müllerian mimic
Palatable	Unpalatable
Model suffers	Model benefits
Often polymorphic	Monomorphic
Mimic converges; model escapes	Species mutually converge
Evolves by major mutations	Evolves gradually
Supergenes	Unlinked genes

consume many toxic and hairy caterpillars or scorpions, and bee-eaters and various other birds have evolved specialised de-stinging behaviours, which they employ before swallowing their otherwise edible hymenopteran prey. In the case where a mimetic prey is unpalatable but markedly less so than the model and can have a negative effect on it, it is termed quasi-Batesian (Speed 1993a, 1999b, MacDougall & Dawkins 1998). There has been considerable debate and modelling about whether quasi-Batesian systems can be stable and whether the quasi-Batesian mimics enhance or reduce protection from predation. Chapter 5, section *Monte-Carlo simulations* deals with this in more detail.

There is doubtless a general presumption that Batesian mimicry comes about by some palatable species being selected for to look more like an aposematic unpalatable one, and indeed this must happen frequently, despite the aforementioned difficulties imposed by increased conspicuousness. However, Holloway (1976, 1984) points out that it could easily come about the other way around, with a Müllerian mimic losing its defences. This is really what is happening in the case of automimicry, and should circumstances lead to an aposematic butterfly feeding on a non-toxic plant speciating away from its previous unpalatable conspecifics, a new Batesian mimic will evolve.

FAMOUS BUTTERFLIES: ECOLOGY, GENETICS AND SUPERGENES

The lack of intermediate patterns and blending in some species that are polymorphic for mimicry patterns led to a very large number of genetic studies to try to understand the mechanism by which the patterns were controlled. Initially these were breeding experiments and more recently the use of modern genomic techniques have focused on the critical gene regions.

Heliconius

The first genetic studies on intraspecific colour polymorphism in classic mimicry systems was that of Beebe (1955), who presented illustrations of the results of four crosses between different colour forms of *Heliconius erato* and *H. melpomene*. It is clear that in this species heterozygotes do occur both in captive breeding and in the wild, displaying various combinations of characters, though predation on intermediates in the wild is high. *Heliconius* butterflies, comprising approximately 40 species in total, come from Central and South America, and have been particularly well studied because of their regional colour races that are mimicked in parallel by other species of *Heliconius* and by other butterflies, both as Batesian and Müllerian mimics (Emsley 1964). For example, both *Heliconius melpomene* and *H. erato* are widely distributed all over the Neotropics and in any given place you will find that the two species have evolved identical colour patterns, but the patterns they display differ from place to place (Fig. 7.2). *Heliconius* itself started radiating rapidly approximately 10 million years ago (Kozak et al. 2015), but the phylogenetic picture is complicated because, throughout their history, species have frequently hybridised with subsequent gene introgression, and this has played an important role in their mimicry (L.E. Gilbert 2003, Nadeau et al. 2013, J. Smith & Kronforst 2013) (see later). The close parallel evolution between *erato* and *melpomene* colour morphs was long suspected and supported by some earlier studies, but the discovery of discrepancies (at least partly due to introgression) cast some doubt on the hypothesis. The initial hypothesis has recently been corroborated in a molecular phylogenetic analysis by Cuthill & Charleston (2012) (Fig. 7.1).

Unlike *H. erato* and *melpomene*, some species display local polymorphism, for example, *H. numata* can have up to seven different colour morphs occurring at the same site. K.S. Brown & Benson (1974) thought that this local polymorphism was probably driven in part by spatial and temporal variation in their far more abundant Müllerian co-mimic ithomiines. Several studies of *Heliconius* and Ithomiini butterflies have indicated that mimicry complexes may be differentially segregated between either habitat types or flight heights (Mallet & Gilbert 1995, Beccaloni 1997a, DeVries et al. 1997, 1999, Hill 2010). In a phylogenetically controlled study, K.R. Willmott & Mallet (2004) also showed that they shared the same larval food plants significantly more often than would be expected by chance, though whether this is cause or effect is not clear.

Heliconius butterflies make an excellent system for the study of ecology, natural selection, speciation and mimicry (J.R.G. Turner 1971a, 1971b, 1981, K.S. Brown & Benson 1974, K.S. Brown et al. 1974, Sheppard et al. 1985,

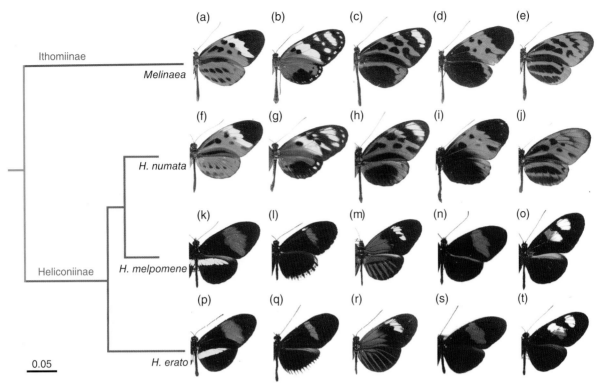

Fig. 7.2 Selected mimetic forms of *Mellinaea* (Ithomiinae) and three species of *Heliconius* (*H. numata*, *H. melpomene* and *H. erato*) (Heliconiinae) from South America. (a–e) *M. menophilus*, *M. ludovica ludovica*, *M. marsaeus rileyi*, *M. marsaeus mothone* and *M. marsaeus phasiana*; (f–j) *H. n. f. tarapotensis*, *H. n. f. silvana*, *H. n. f. aurora*, *H. n. f. bicoloratus* and *H. n. f. arcuella*; (k–o) *H. m. rosina*, *H. m. cythera*, *H. m. aglaope*, *H. m. melpomene* and *H. m. plesseni*; (p–t) *H. e. cf. petiveranus*, *H. e. cyrbia*, *H. e. emma*, *H. e. hydara* and *H. e. notabilis*. (Source: Joron et al. 2006. Reproduced under the terms of the Creative Commons Attribution Licence CC BY, via PLoS ONE.)

Mallet & Singer 1987). The colour patterns of *Heliconius erato* and *H. melpomene* in particular have been subject to intensive genetic research for many years (J.R.G. Turner & Crane 1962, Mallet 1989). Both species are widely distributed over the forested Neotropics, where they are Müllerian mimics of one another, yet in different regions they have quite different colour patterns that meet in narrow boundary zones where hybridisation occurs (Mallet et al. 1990, 1996).

Hybrid zones

There has been a great deal of discussion about how the polymorphisms initially came about and how they are maintained, the main hypotheses being summarised by Mallet & Gilbert (1995) and Mallet & Joron (1999). Particularly interesting from genetic and evolutionary points of view is what happens in those narrow hybrid zones between main forms and subspecies where populations with different patterns overlap (Mallet 1989, Mallet & Barton 1989). These zones can persist for a long time and

are often only a few kilometres wide, partly because the butterflies are not generally that mobile, though neutral genetic markers move freely across the boundaries (Hines et al. 2011). The hybrid zones are often very clearly defined and occur between most of the races where their distributions overlap. In them there can be many distinctly recognisable forms as well as butterflies with various intermediate patterns, all of which can be explained by the segregation of alleles at four important loci (J.R.G. Turner 1971b).

Nadeau et al. (2012) using targeted next-generation sequence capture showed that the gene regions controlling colour pattern (*HmYb* and *HmB/D*; Fig. 7.4) are very strongly differentiated between species across hybrid zones where selection for maintaining the differences is strongest. In contrast, other gene regions not linked to the colour pattern showed no such increase in difference in these hybrid zones. This supports Mallet & Barton's (1989) earlier results of mark–release–recapture experiments of colour morphs of *H. erato* across a hybrid zone in Peru which showed that selection against released 'foreign' morphs was approximately 50%. In other words,

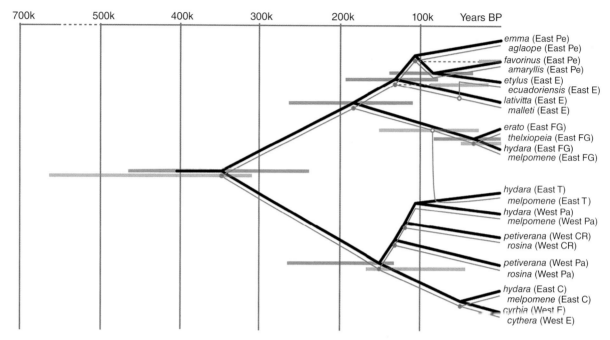

700k   500k   400k   300k   200k   100k   Years BP

emma (East Pe)
aglaope (East Pe)
favorinus (East Pe)
amaryllis (East Pe)
etylus (East E)
ecuadoriensis (East E)
lativitta (East E)
malleti (East E)

erato (East FG)
thelxiopeia (East FG)
hydara (East FG)
melpomene (East FG)

hydara (East T)
melpomene (East T)
hydara (West Pa)
melpomene (West Pa)

petiverana (West CR)
rosina (West CR)

petiverana (West Pa)
rosina (West Pa)

hydara (East C)
melpomene (East C)
cyrbia (West E)
cythera (West E)

Fig. 7.3 Phylogenetic trees showing high level of co-phylogeny between races of *Heliconius erato* and *H. melpomene* butterflies. Note the apparent colonisation of the western Neotropics somewhere around 180,000 years BP. Horizontal bars indicate 95% Bayesian confidence intervals for divergence times. Countries are indicated as follows: C, Colombia; CR, Costa Rica; E, Ecuador; FG, French Guyana; Pa, Panama; Pe, Peru; T, Trinidad. (Source: Cuthill & Charleston 2012. Reproduced under the terms of the Creative Commons Attribution Licence CC BY, via PLoS ONE.)

mimetic individuals in these places survive approximately twice as long as non-mimetic ones. Since only three major gene loci are involved in differentiating the patterns of these forms, this corresponds to a selection coefficient of 0.17 per locus, which is very high indeed. Langham (2004) similarly found that the behaviours of rufous-tailed jacamars, *Galbula rufi-cauda*, a relatively sedentary specialist predator of butterflies, would contribute to stability and maintenance of the hybrid zone boundaries because they avoided those *Heliconius* with the local colour morph patterns, but readily attacked novel ones as might occur if a hybrid butterfly or an individual of an adjacent morph transgressed into its territory. This tendency to attack novel colour morphs is apparently positively age-dependent (Langham 2006) rather than negatively, which is what one would expect if sampling was predominantly being carried out by naïve juvenile birds.

Hybrid zones between areas with different colour morphs of aposematic species are not restricted to butterflies, and Chouteau & Angers (2012) describe analogous situations in the poison dart frog, *Ranitomeya imitator* (Dendrobatidae), which has radiated to mimic three other dendrobatid species in different parts of its range (Stuckert et al. 2014). In

this species, the hybrid zones are characterised by high levels of phenotypic variation and apparently reduced selection pressure, which might provide an increased opportunity for genetic drift. The authors propose that new phenotypes that arise in this way might become locally fixed, something that never seems to occur in *Heliconius*.

Interestingly, although the mimicry often appears near perfect to a human observer, spectroscopic analysis of wings has shown that there are often differences between models and mimics. In the case of *H. numata* and *Melinaea* species, these differences are most notable in terms of black and yellow contrast, and visual modelling suggests that they may, as with humans, be almost undetectable by birds but would probably be distinguishable to the butterflies themselves and therefore may aid them in distinguishing potential mates from mimics (Llaurens et al. 2014). T. J. Thurman & Seymoure (2015) similarly found that the pair *Mimoides pausanias* and *Heliconius sara* displayed marginally better colour matching for avian predators than for humans. Both *H. erato* and *H. melpomene* have been shown to spend a considerable amount of time courting females of the other, co-mimicking species, so they are presumably confused by the

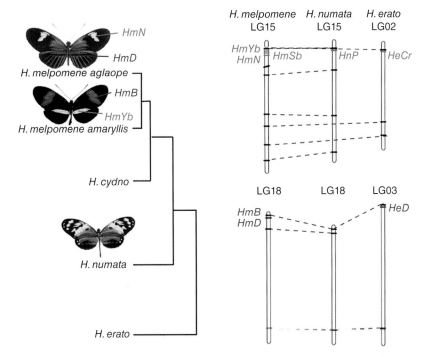

Fig. 7.4 Chromosomal locations of colour pattern genes in the butterflies *Heliconius erato*, *H. melpomene* and *H. numata*. (Source: Dr Nicola Nadeau. Reproduced with permission from Dr Nadeau.)

mimic too (Estrada & Jiggins 2008).[1] Conversely, assortative mating in the recently diverged species pairs *H. melpomene* and *H. cydno*, and *H. erato* and *H. sapho*, where they occur sympatrically, means that hybridisation in the wild is rare, and almost certainly facilitated the speciation of the latter two, which are both largely black and white and very different in appearance from other *melpomene* and *erato* morphs (Jiggins et al. 2001).

Sherratt (2006) created a spatially explicit simulation model for the evolution of Müllerian mimicry which showed that a mosaic of phenotypes separated by narrow hybrid zones could be formed if the grid was seeded randomly with each species. These spatial groups arose primarily through the occasional establishment of a new mutant morph in unoccupied cells, which then gained a foothold through frequency-dependent selection. The mosaics in the model were generally highly stable and although hybrid zones could move, they did so slowly or only under particular conditions.

Wing pattern genetics

Traditional genetic breeding experiments yielded a lot of basic information about the underlying genetics of the *Heliconius* and up to 20 separate gene regions have been identified within a given species that influence colour patterns (Sheppard et al. 1985, Mallet 1989, Jiggins & McMillan 1997, Naisbit et al. 2003). However, the major features, such as the presence of a red bar on the fore wing or a yellow bar on the hind wing, are controlled by just a few major loci, each of which causes a large shift in some phenotypic feature.

Genomic studies have been carried out in particular on *Heliconius erato*, *H. melpomene* and *H. numata* (Joron et al. 2006, 2011, Baxter et al. 2008, 2010, Hines et al. 2011, Supple et al. 2013). Kapan et al. (2006) created a dense linkage map for mimicry genes in *H. erato* using amplified fragment length polymorphisms (AFLP), microsatellites and single-copy nuclear loci, and utilising crosses between *H. e. etylus* and *H. himera*. They located the positions of two genes, one called *D*, which is responsible for red and orange elements of the colour pattern, and one called *sd*, which controls fore wing melanic markings. Joron et al. (2006) compared linkage maps controlling pattern elements in

---

1. Heterospecific courting as a result of mimicry has also been observed in some poison arrow frogs (Mallet 2014).

H. melpomene, *H. erato* and *H. numata* and found that the gene locus called *Yb*, which controls a yellow band in the first, maps to the same chromosomal location as the *Cr* locus in the second, which also controls a yellow pattern element (Fig. 7.4), thus indicating conservation of the locus across species. Baxter et al. (2008) found that subtle differences in the gene regions responsible for the red wing phenotypes of *H. erato* and *H. melpomene*, despite being located on homologous chromosomes and controlled by the same switch, are actually convergent and not homologous.

The Heliconius Genome Consortium[2] (2012) found that past hybridisation events and back-crossing had led to the introgression of gene regions involved in mimicry control between different species within *Heliconius*. These introgressed regions in *H. timareta* that are thought to have originated in *H. melpomene amaryllis* display less sequence variation than other comparable genomic regions in *H. timareta*, indicating that they entered its genome relatively recently, thus excluding the possibility that the similarity between the gene region in the two species had been the result of retention of an ancestral polymorphism predating the species split (J. Smith & Kronforst 2013).

Pardo-Diaz & Jiggins (2014) have used quantitative, real-time PCR (qTT-PCR) to investigate gene expression in *H. erato* and *H. melpomene*, and found two closely linked candidates that are expressed at the time of red wing pigment deposition in the early pupal stage. Expression of one of them, *optix*, correlates perfectly with the red pattern elements in both species, whereas *kinesin* correlates with the development of the red fore wing band.

Modelling polymorphism

Joron & Iwasa (2005) used analytical modelling to assess what parameters of a Müllerian mimic with multiple potential aposematic models and patchy environment would permit the stable coexistence of two different mimic forms. Their model assumes spatial heterogeneity with two different unpalatable species whose proportions vary spatially, and a polymorphic quasi-Müllerian mimic that can migrate between patches. The model outcome depends on the relative unpalatability of the mimic (λ) and the amount of migration between patches (μ). When there is no migration between patches ($\mu = 0$) coexistence of the two mimic forms is always possible, but as unpalatability increases there is a region where it is not attainable from a monomorphic starting point (Fig. 7.5, top). As migration between patches becomes possible ($\mu > 0$) (Fig. 7.5, middle) and increases, there becomes an upper range of unpalatabilities where

coexistence is no longer possible and a range where both coexistence or monomorphy can occur, depending on the initial conditions. But as the level of inter-patch migration increases towards unrestricted movement, the zone where both coexistence and monomorphy can both be possible outcomes becomes ever smaller, such that coexistence can only occur if the model is totally palatable (Fig. 7.5, bottom).

Danaus and *Hypolimnas*

The African nymphalid butterfly *Hypolimnas misippus* and its model, the African queen butterfly, *Danaus chrysippus*, both display considerable pattern polymorphism and multiple forms often coexist at a given site (M. Edmunds 1969). Unlike *P. dardanus*, however, in addition to having forms that are excellent mimics (D.A.S. Smith 1973a), *Hypolimnas misippus* also produces approximately 24% of individuals that are intermediate between model forms and are quite poor mimics to the human eye (D.A.S. Smith 1976). The genetics of the wing colour patterns in these species have been quite well studied (*H. misippus*: D.A.S. Smith 1981, D.A.S. Smith & Gordon 1987, Gordon & Smith 1989, 1998, Gordon et al. 2010; *D. chrysippus*: D.A.S. Smith 1975a, 1975b, 1981). Gordon & Smith (1989, 1998) report some interesting differences in the genetics of *H. misippus* and *D. chrysippus*'s matching (mimetic) patterns. In *H. misippus* the same wing pattern can be produced by several different genotypes, whereas in *D. chrysippus* each pattern is produced only by a single genotype. Also, interestingly, genes controlling fore and hind wing pattern in *H. misippus* are closely linked and probably represent a supergene, whereas in *D. chrysippus*, fore and hind wing patterns are uncoupled – the genes controlling them are not linked, and are presumably on separate chromosomes. The polymorphic patterns of *H. misippus* are female limited whereas those of *D. chrysippus* are shared by both sexes. Along with a number of other differences, these features suggest that mimetic resemblance between *H. misippus* and *D. chrysippus* has arisen independently on a number of occasions, when selective forces have been very strong. However, these periods appear to be intermittent and in many places now they are considerably relaxed. C.A. Clarke & Sheppard (1975, 1977) studied the genetics of the colour forms of the widespread Asian *H. bolina* and similarly found an absence of linkage between major loci and also emphasised the importance of modifier genes.

In a review article, Joron & Mallet (1998), citing D.A.S. Smith et al. (1993), suggested that considerable polymorphism of colour patterns in *D. chrysippus* was due to Batesian overload by *H. misippus*. However, Gordon & Smith (1999) responded that Joron & Mallet (1998) had misinterpreted

---

2. Comprising 80 authors (if I counted them all correctly).

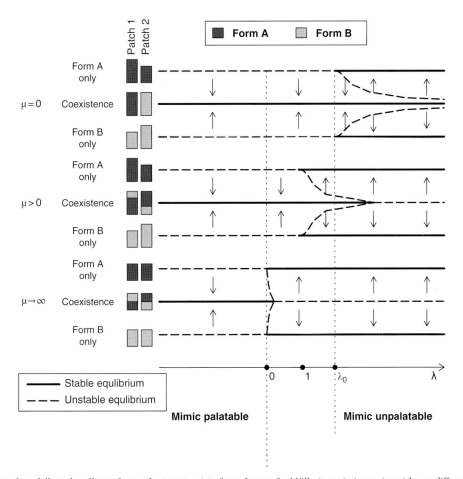

Fig. 7.5 Results of modelling the effects of rate of migration (μ) of two forms of a Müllerian mimic species with two different model species, between two patches within a heterogeneous environment, and degree of unpalatability (λ) on the possibility of stable coexistence of the two forms. The rectangles towards the left-hand side of the figure show the proportions of each form of mimic that will occur in each of the patches either at monomorphy or in coexistence. (Source: Adapted from Joron & Iwasa 2005. Reproduced with permission from Elsevier.)

what they had written and oversimplified their views. Rather than explaining the diversity of aposematic forms in *D. chrysippus* as being due to the presence of too many Batesian mimics, they proposed that they were the result of hybridisation events between previously geographically separate (allopatric) forms, and that colour patterns, even in such species, may not be determined purely by mimicry and predation and that other factors might sometimes be equally, if not more, important. Indeed, C.A. Clarke et al. (1989) note that both *H. bolina* and *H. misippus* often exist in the absence of models, may even be unpalatable themselves depending on larval food plant, and that in most places functional Batesian mimicry is currently absent or unimportant.

Papilio dardanus

Poulton (1924) described the mocker swallowtail, *Papilio dardanus*, with its female-limited mimicry and numerous female forms as "the most interesting butterfly in the world". C.A. Clarke & Sheppard (1960a, 1960b, 1960c) showed that only two loci were responsible for the various mimetic forms: one controlling the presence and absence of hind wing tails (present in males and male-like females) and the other controlling all the various colour forms (Fig. 7.6a). At that time they did not, of course, consider that a single gene controlled all the different colour pattern features but rather that the alleles acted as switches and that the switch mechanism itself is probably also an aggregation of

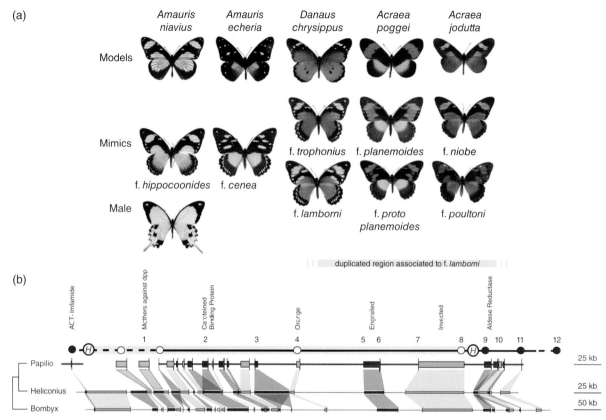

Fig. 7.6 The *Papilio dardanus* mimicry system and partially mapped gene region associated with mimetic forms showing the location of the *H*-locus that controls the major mimetic female forms. (a) Phenotypes of *P. dardanus* and their presumed models; (b) linkage maps of the components of the *H*-locus mimicry supergene in *P. dardanus*, *Heliconius melpomene* and the non-mimic silk moth, *Bombyx mori*. Numbers 1–12 are locations of particular single nucleotide polymorphisms (SNPs, pronounced snips) used in the study; thick red and green lines show predicted protein-coding genes with thin vertical lines indicating their 3′ ends, and alternate colours demonstrating the high degree of conservation in gene order across such widely divergent species. (source: Timmermans et al. 2014. Reproduced under the terms of the Creative Commons Attribution Licence, CC BY 3.0, via Proceedings of the Royal Society of London, Series B: Biological Sciences.)

separate but closely linked genes called a supergene (Clarke & Sheppard 1960d). Since then, a great deal of effort has gone into trying to understand the control mechanism in detail, and much progress has been made, although the tight linkage between supergene loci has not made it easy.

The colour variation is controlled by a small gene region called the *H*-locus, with at least 14 different 'alleles' controlling 14 different female wing pattern phenotypes (Fig. 7.6) (Nijhout 2003). These different mimic alleles display a range of relative dominances. For example, in South Africa the forms *trophonius* and *leighi* are dominant to *hippocoonides* and *cenea*, though the dominance pattern is not complete as any hybrid between *leighi* and *trophonus* forms are intermediate in pattern. When C.A. Clarke & Sheppard (1959a, 1959b, 1960a, 1960b, 1960c, 1962a, 1963)

wrote their many papers they were only able to determine the relative dominances of the different alleles and to show that they were not sex-linked, i.e. the locus controlling them was on an autosome rather than on a sex chromosome. Even today, after many more thousands of hybridisation experiments with *Papilio dardanus*, no recombination within the supergene (*H*-locus, with various dominant *H* and recessive *h* alleles) has been found. The chromosomal position of the *H*-locus has, however, gradually been narrowed down. Clark et al. (2008) found it was close to a transcription factor called *invected*, and M.J. Thompson & Timmermans (2014) and Timmermans et al. (2014) have narrowed it down further to a region comprising some 24 genes that are centred around a transcription factor locus called *engrailed*. All the phenotypic variation within this species appears to have arisen within, rather than having been the result of, introgressive hybridisation or carryover of polymorphisms from deep ancestors (M.J. Thompson et al. 2014). The chromosomal region within the *H*-locus is also interestingly co-linear with (i.e. has the same gene order as) the homologous regions of *Heliconius* butterflies and the far more distantly related silkworm, *Bombyx mori* (Bombycidae) (Fig. 7.6).

Papilio glaucus

The American eastern tiger swallowtail, *Papilio glaucus*, displays sex-limited mimicry, with a black form resembling the more toxic *Battus philenor*. Its close relative and highly similar *P. canadensis* is, in contrast, monomorphic without a black form. Clarke & Sheppard (1962b) carried out breeding experiments which strongly suggested that the black colour gene was Y-linked (i.e. female-linked in butterflies), and additionally supported this by reference to the discovery made a long time previously of a female which was black on one side and typical yellow on the other – something that would be expected during the first cleavage of the egg if the Y chromosome was omitted from one of the daughter cells.[3] The two species will occasionally hybridise, allowing Scriber et al. (1996b) to determine that the variation in *P. glaucus* is primarily controlled by two sex-linked genes, one on the Y chromosome (i.e. the female-determining sex chromosome in butterflies) which is polymorphic and controls the typical and black colour morphs, and a black morph suppressor gene on the X chromosome, which is at high frequency in *P. canadensis* populations but low frequency in *P. glaucus*.

Papilio memnon

The Asian great Mormon swallowtail, *Papilio memnon*, is similar in many respects to *P. dardanus* in that it has a monomorphic male and numerous mimetic female forms (C.A. Clarke et al. 1968, C.A. Clarke & Sheppard 1971, 1973, J.R.G. Turner 1984), as does *P. polytes* to a lesser extent (C.A. Clarke & Sheppard 1972a). However, Clarke & Sheppard were able to show that although closely linked, the loci involved in the polymorphism were, unlike in *P. dardanus* and *P. polytes*, separable. Breeding experiments showed

that there are at least six separate genes controlling wing pattern, colour and morphology, with crossing-over occasionally occurring between them (Turner 1984). This means that when crosses were made between races, there were no instances when mimetic resemblance to a model was improved, and in three-quarters of cases the mimetic resemblance was reduced. Evolution will therefore tend to tighten the linkage between these supergene loci.

Supergenes and their origins

There has also been debate about how supergenes came about. The two hypotheses are gene aggregation and gene function co-opting. In the first, it is assumed that different loci controlling, for example, aspects of fore wing pattern, hind wing pattern, body colour, and presence or absence of tails were initially dispersed around the genome, and that chromosomal rearrangements led eventually to them becoming tightly linked (C.A. Clarke & Sheppard 1971) because recombination between them, which would certainly lead to maladaptive intermediate phenotypes, would be exceedingly rare. However, this hypothesis, which was initially popular, has an associated genetic difficulty, that is, for selection to favour closer linkage, there must initially be linkage disequilibrium, which means that natural populations have an excess of some allelic combinations and a deficit of others. J.R.G. Turner (1984, p. 155) proposed the second mechanism, whereby gene loci that are already closely linked to a locus controlling a major aspect of the mimetic resemblance can remain polymorphic and so the species can evolve to be a polymorphic mimic of two different models. Certainly all cases are not going to be the same, but since the days of C.A. Clarke, Sheppard and J.R.G. Turner we have gained a far greater understanding of the way genomes work. We now understand that the genes controlling the different aspects of colour, pattern and shape do not have to be physically close (linked) on a chromosome, since dispersed genes can be co-regulated by transcription factors. These gene products, produced by a single locus, diffuse and interact with the promotor regions of other genes.

More or less in parallel with the studies on *Papilio dardanus* and other members of the genus, workers also began to study the genetics of *Heliconius* butterflies, in particular *H. melpomene*, *H. erato* and *H. numata*. Work by J.R.G. Turner and co-workers, and more recently by Joron et al. (2011), found that chromosomal rearrangements had been important in generating and maintaining the colour polymorphism at some of the mini-supergene loci, though in *Heliconius* there are a number of separate clusters of tightly linked colour and pattern genes (see later).

---

3. Insects, together with other Ecdysozoa, have deterministic development such that each cell lineage in early embryos always produce the same structures, organs, etc. – the first division separates lineages that will form the left and right sides of the body.

MIMICRY BETWEEN CATERPILLARS

Bowers (1993) has noted that the number of instances of mimicry between caterpillars is probably far fewer than those recognised for adult Lepidoptera, even after making allowance for the greater number of studies on the latter. Bowers' (1993) list of examples which, incidentally, missed Rothschild's (1981) mention of aganaine (formerly Hypsidae) moth larvae, was expanded somewhat by K.R. Willmott et al. (2011) but in total there seemed to be fewer than 20 examples. Berenbaum (1995) proposed four possible factors that could explain this:

1. adults may have to be brightly patterned for other reasons, for example mate recognition, territoriality, etc. whereas intraspecific communication by larvae is typically small or non-existent,
2. larvae are subject to different sorts of predation and parasitism than are adults,
3. caterpillars are less mobile than adults, and
4. larvae, generally being soft-bodied, may be more prone to injury in attacks by predators.

A.V.Z. Brower & Sime (1998), while accepting that all of the above might be true, further suggested that Richard Dawkins' (1999) concept of extended phenotype might apply since caterpillars are generally associated with their host plant, and in most cases, acquire their toxicity or distastefulness from that plant. In this context, D.O. Gibson & Mani (1984) showed that predatory birds were able to distinguish externally identical larvae of the African queen butterfly, *Danaus chrysippus*, of different palatabilities based on the food plant species upon which they were feeding. Furthermore, certain species of toxic plant have associated with them multiple species of herbivorous insects, many, if not all, of which may sequester chemicals from them and so might all be unpalatable (see also Rothschild 1964b). The protection afforded by a predator's experience with a particularly unpleasant prey may also extend to the actual place where the encounter took place. Hansen et al. (2010) found that great tits, *Parus major*, were able to learn two different context-dependent sets of food preferences and aversions. Uesugi (1996) found that brown-eared bulbuls, *Hypsipetes amaurotis*, having experienced the unpalatable papilionid butterfly *Pachliopta aristolochiae* at a bird feeder, subsequently reduced its frequency of taking its normal types of palatable food from that particular feeder but carried on feeding at other feeders where they had not encountered the *Pachliopta*.

One possible example of Müllerian mimicry involving caterpillars is that between the larvae of two geometrids, *Meris alticola* and *Neoterpes graefiaria*, both of which feed on glycoside-containing *Penstemon* species (Poole 1970, Stermitz et al. 1988). Berenbaum (1995) has suggested the possibility that the banded caterpillars of the North American monarch butterfly, *Danaus plexippus* (Fig. 7.7a), and those of *Papilio polyxenes* (a member of the *P. machaon* group; see Fig. 4.39) (Fig. 7.7b) might well be Müllerian mimics. A.V.Z. Brower & Sime (1998) suggested that just because a caterpillar (or any animal for that matter) appeared brightly and distinctively coloured to the human

Fig. 7.7 Larvae of two conspicuous North American butterflies that have been suggested to be mimics (but see text). (a) The monarch, *Danaus plexippus* (Nymphalidae), feeding exposed on *Asclepias* sp. (Apocynaceae); (b) black swallowtail, *Papilio polyxenes* (Papilionidae), feeding on fennel, *Foeniculum vulgare* (Apiaceae). (Source: a, Ettore Balocchi, Hectonichus 2009. Reproduced under the terms of the Creative Commons Attribution Share-Alike Licence CC BY-SA 3.0, via Wikimedia Commons; b, Beatriz Moisset 2005. Reproduced under the terms of the Creative Commons Attribution Share-Alike Licence CC BY-SA 3.0, via Wikimedia Commons.)

observer, it did not necessarily mean that it was aposematic. They noted that monarch caterpillars feed openly and exposed in full view of potential predators, whereas the larva of the black swallowtail appeared to behave in a cryptic way, despite both being aposematically coloured. Field data suggested to A.V.Z. Brower & Sime (1998) that *P. polyxenes* caterpillars, despite feeding on fennel, *Foeniculum vulgare* (Apiaceae), are not particularly unpalatable to birds, and that for their black and yellow stripes to be considered as aposematic it would have to be shown that predators learned to avoid them. A.V.Z. Brower & Sime (1998) also argue that to demonstrate that the swallowtail and monarch caterpillars are Müllerian mimics, it would have to be shown that birds, having learned to avoid one of them, most likely the monarch, would then avoid the other. Instead, they consider that the black swallowtail caterpillars are exhibiting the plesiomorphic state within the *P. machaon* group and that their colour and banding are purely cryptic.

Holen & Johnstone (2006) present a mathematical optimality model to explore how predators should behave when models and mimics are each more preferentially associated with some particular but overlapping contexts, such as food plant, time of day of activity, etc. They found a whole range of outcomes, with all combinations of mimics and models benefitting or being harmed. Furthermore, if mimicry had a high inherent cost (e.g. Holen & Johnstone 2004), context-dependence could have profound influences on the evolution of mimicry.

SOME SPECIFIC TYPES OF MODEL AMONG INSECTS

Most of the earlier examples discussed here have involved butterflies or snakes, which is where understanding of protective mimicry began, mainly because they are generally large, conspicuous and colourful. But among insects there are three other major groups that are protected by their ability to sting or by their aggressiveness, or both, that form the models for numerous cases of mimicry and afford opportunity for studying other evolutionary aspects. These three groups are the bees, vespoid wasps[4] and ants, which belong to the major monophyletic assemblage of Hymenoptera called the Aculeata. The aculeates are characterised by a change in use of the ancestral ovipositor, an organ system used by the great majority of hymenopterans for laying eggs, to a device used purely for stinging.

The transition is not that surprising because the ovipositor system already had associated venom glands (also called acid glands, particularly in older literature) that produced a range of physiologically active compounds for modifying the physiology of the (predominantly insect) hosts of the parasitic Hymenoptera from which the aculeates evolved (Quicke 1997, 2015). Thus when other life history changes meant that prey for the aculeate's larvae no longer needed to be reached by a long ovipositor and the females' eggs could just be 'popped' out at the ovipositor's base straight onto the food, the ovipositor system was ready to become modified by evolution more for defensive roles. Some ants (the Formicinae) have lost the stinger but not the associated venom gland and this is now used to produce formic acid,[5] which is squirted in defence. This can also be quite unpleasant. Because the 'stinger' of aculeate hymenopterans is a modified ovipositor, only females can sting, but of course in the social bees, wasps and ants, the vast majority of individuals (workers) are female and males are generally only produced at particular times of the year and are generally short-lived.

Wasp (and bee) mimicry

Many aculeate wasps are capable of delivering a potent sting, and some do so very readily when accidentally touched or their nests are threatened.[6] Not many people outside of the Arctic will have gone all their lives without at least one painful encounter with them. Many wasps and their mimics are also conspicuously coloured – there are the iconic black and yellow striped vespids (Fig. 7.8a–c) but also various other Müllerian complexes, including those centred around pompilids such as *Pepsis* species, which typically have black bodies and boldly contrasting orange wings (see figure 4 in Owada & Thinh 2002). In temperate species, such as bumblebees, aposematic patterns are may be balanced by thermal considerations, especially in queens that emerge from hibernation early in the spring and have to raise their body temperature above ambient conditions for efficient flight (Plowright & Owen 1980), but Perrard et al. (2014) found no temperature or climatic correlation for colour pattern in the polymorphic hornet *Vespa velutina*.

Wasp and bee mimics include various moths (Fig. 7.8f), beetles (Fig. 7.8e) and, of course, flies (Figs 7.8d and 7.24)

---

4. The vespoid wasps that are the models comprise three major families: the social Vespidae, familiar as yellow-jackets and hornets, the solitary velvet ants (Mutillidae) and spider-hunting Pompilidae.

5. Formic acid (methanoic acid) is so called because it was first isolated from bodies of formicine ants.

6. Members of the large family of solitary Sphecidae are generally far less likely to sting when handled than similar-sized vespid wasps.

Fig. 7.8 Photographs of various Batesian mimics of vespid and pompilid wasps. (a) *Chrysotoxum* sp.? (Diptera: Syrphidae); (b) potterwasp mimicking hoverfly, *Ceriana* sp, from South-East Asia, showing elongated antennal sockets and rather large antennae (for a fly) as well as very pronounced wasp waist; (c) non-stinging British tenthredinid sawfly (Hymenoptera); (d) unidentified Neotropical syrphid fly mimicking the social wasp, *Polybia dimidiata;* (e) *Scalenus sericeus* or *S. auricornis* (Coleoptera: Cerambycinae: Callichromatini) mimicking a pompilid wasp; (f) *Myrmecopsis* sp. (Erebidae: Arctiinae: Euchromiini), mimic of *Agalaia panamensis* (Vespidae), from Costa Rica. (Source: a, Conrad.P.D.T. Gillett. Reproduced with permission from Conrad P.D.T. Gillett.; b, Steve Chuang. Reproduced with permission from Steve Chuang.; c, Andrew Green. Reproduced with permission from Andrew Green; d, © Professor Phil deVries. Reproduced with permission from Professor deVries; e, Itchydogimages. Reproduced with permission from John Horstman; f, Kenji Nishida. Reproduced with permission from Kenji Nishida.)

(Vane-Wright 2010), perhaps the most conspicuous of which are the large mydids, the largest flies on earth,[7] with wingspans up to 9 or 10 cm. Various hoverflies are perhaps the best known in the temperate region (Maier 1978, Howarth et al. 2000) and have been the subject of a considerable number of experiments. In the north temperate region, many species of hoverfly appear to be aculeate mimics with vespid wasps, honey bees and bumblebees as models. Far less is known about tropical and southern mimicry complexes and little detailed observation or experimentation has been made (Polidori et al. 2014). Hoverflies themselves are not particularly tropicocentric. Millot (1946) even suggests that the caterpillar of a Madagascan moth,[8] with its highly spatulate setae, mimics the small nest of a wasp such as the vespid *Belonogaster*. Having once walked accidentally into such a nest, I can attest that I certainly would not wish to do so again, but I fear the suggested mimicry is a little fanciful.

The likely deterrent effect of being stung by an aculeate hymenopteran is probably fairly well estimated by their effects on humans because a lot of their pain-causing venom components have very direct physiological actions on the receiver's peripheral nervous system. In general, bee venom contains histamine, serotonin, dopamine, noradrenaline and several potent small peptides. Vespid wasp venoms contain fewer peptides but do contain acetylcholine and various kinins, whereas ant venoms contain various components including, notably, formic acid, piperidine alkaloids and various enzymes. Perhaps it is no coincidence that the person who devised the pain scale, Justin Schmidt, works on Hymenoptera, mainly on honey bees. The sting pain scale initially published by J.O. Schmidt et al. (1983) and refined by J.O. Schmidt (1990) ranges from 1, meaning barely noticeable and maybe just irritating, to 4+. Honey bee, *Apis mellifera*, stings score a mere 2, effectively the same as the larger, but generally tamer, European hornet (*Vespa crabro*). At the top end of the scale come the Central American bullet ant, *Paraponera clavata*, whose sting is described as "Pure, intense, brilliant pain. Like fire-walking over flaming charcoal with a 3-inch rusty nail grinding into your heel". One particularly painful North African mutillid wasp (also known as velvet ants) has earned the name 'camel killer', and a North American *Dasymutilla* sp. is commonly called the 'cow killer'! It is hardly surprising that such effects make these animals the models of many mimicry systems and form mimicry rings of their own (Manley 2000, J.S. Wilson et al. 2012, 2015). Indeed mutillids are mimicked by quite a number of beetles, notably clerids, cerambycids and curculionids, as shown by several earlier more or less anecdotal reports (de Gunst 1950, Brach 1978, G. Edwards 1984, Mawdsley 1994, del Río & Lanteri 2013), and also by spiders, notably Clubionidae, Salticidae and Gnaphosidae (Nentwig 1985a) and even by the larva of an antlion (Myrmeleontidae) (Brach 1978). Furthermore, some arborial mutillids are Müllerian mimics of true ants (Yanega 1994) and not an aggressive mimic as G.C. Wheeler (1983) had proposed. Research on the mimetic relationships involving this family has only recently started to gain momentum (Lanteri & del Río 2005, Rodriguez et al. 2014) but is hampered by them not being easily bought into culture since they are obligate ectoparasitoids within the nests of various aculeate hymenopterans, notably Sphecidae.

R.B. Simmons & Weller (2002) used a phylogenetic analysis of wasp-mimicking tiger moths (Erebidae: Arctiinae) of the *Sphecosoma* group, to test whether the characters involved in the three main mimetic complexes (yellow *Polybia*, black *Polybia* and *Parachartergus* mimics) evolved individually and independently or in a concerted fashion, as would be the case if they were the result of a supergene or tightly linked. Indeed, they found highly significant conservation of mimicry patterns with very few transitions between them (Fig. 7.9). They then created two further, but constrained, phylogenetic tree reconstructions for the social wasp-mimicking *Pleurosoma*, *Sphecosoma* and *Myrmecopsis* species, one to fit a topology consistent with Müllerian mimicry with each mimicry type evolving only once, with the yellow *Polybia* type evolving first, and the other consistent with a quasi-Batesian system which maximised changes between mimicry types. Comparisons of the lengths of these trees[9] with lengths of a large sample of random trees shows that while the Müllerian one was not statistically different from random (408 versus 415 steps), the quasi-Batesian one was (527 steps, $p < 0.05$). This means that the data can significantly reject the quasi-Batesian hypothesis but are not inconsistent with the idea that the moths are in Müllerian mimicry complexes with the wasps, despite having very different means of defence.

While it is generally assumed that the main dupes in the hoverfly–aculeate mimicry systems are visually orientated vertebrate predators, there is some evidence that the black and yellow patterns they typically display may also be involved in associative avoidance learning by some invertebrate predators such as crab spiders (Thomisidae) (Morris & Reader 2016).

---

7. Members of the Pantophthalmidae are of similar size.

8. Millot suggests that it is "Sans doute appartient-elle à la famille des Noctuidae" but it looks quite like some geometrid larvae I have seen in Africa.

9. The minimum number of evolutionary changes between character states in the phylogeny reconstruction data set that a given tree topology requires.

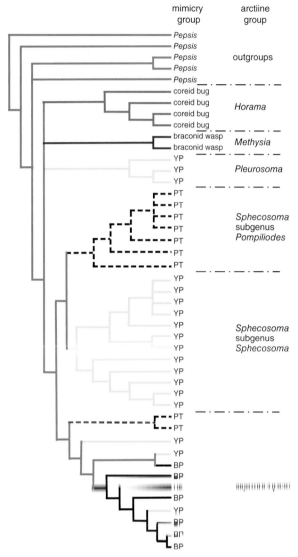

mimicry group / arctiine group

Pepsis
Pepsis
Pepsis — outgroups
Pepsis
Pepsis
coreid bug
coreid bug — *Horama*
coreid bug
coreid bug
braconid wasp — *Methysia*
braconid wasp
YP
YP — *Pleurosoma*
YP
PT
PT
PT
PT — *Sphecosoma* subgenus *Pompiliodes*
PT
PT
PT
YP
YP
YP
YP
YP
YP — *Sphecosoma* subgenus *Sphecosoma*
YP
YP
YP
YP
YP
PT
PT
YP
YP
BP
BP
BP
BP
YP
BP
BP
BP

Fig. 7.9 Reconstruction of mimicry groups on estimated phylogeny of tiger moths of the *Sphecosoma* genus group, showing considerable phylogenetic conservation of mimicry patterns with only a few transitions between mimicry complexes. The main ingroup models are abbreviated as follows: BP, black *Polybia* wasps; PT, black and white *Parachartergus* wasps; YP, yellow *Polybia* wasps. A *Myrmecopsis* species is illustrated in Fig. 7.8 f. (Source: Adapted from Simmons & Weller 2002 with permission from The Royal Society.)

How to look like a wasp

Wasps are mimicked by a wide range of other insects and, in addition to many flies, non-stinging Hymenoptera and moths mentioned above, the list includes beetles such as the cerambycids (*Clytus*, *Strangalia* and *Megacyllene robiniae*), bugs (Hemiptera, e.g. *Zelurus* spp.), Neuroptera (notably various Mantispidae) and Orthoptera (e.g. *Aganacris pseudosphex*) (see section *Batesian–Poultonian (Predator) mimicry* later). Each mimic group has its own general body form which imposes limitations on how effective mimicry might be achieved. In addition to colour pattern, many wasp mimics have evolved morphological features not typical of their own groups as a whole, for example narrow waists and long antennae in flies (Waldbauer 1970; compare Fig. 7.8a with 7.8b,d). Although the mimetic sawfly illustrated (Fig. 7.8c) does not show a wasp waist, some others, such as *Athlophorus* species, do.

As Townes (1972) pointed out, many flies mimic stinging Hymenoptera but one conspicuous hymenopteran feature, their prominent antennae, is not evolutionarily easy for a fly to emulate. A number of species of syrphid, rhagionid and micropezid flies wave their fore legs in a manner that emulates a wasp's antennae, and some syrphids, stratiomyids and conopids in particular have evolved highly modified antennae that are far larger than usual for higher Diptera (Fig. 7.8b,d). Townes (1972) proposed that the elongate eye-stalks of the stalk-eyed fly family Diopsidae may also be wasp antennal mimics that, at a quick glance, add to the generally wasp-like habitus and behaviour these flies display. He also suggested that the thoracic spines might give a predator that grabs one the fleeting impression that it is about to be stung, a sort of aide mémoire mimicry, and therefore might let the fly escape.

Time of appearance of aculeate mimics

In annual or other cyclical climatic systems, the relative time of first appearance of models and their mimics would seem likely to be important in the success of the mimic, especially if the majority of potential predators are naïve or effectively so (i.e. they have forgotten anything learned from previous years). Huheey (1980b) stated:

> It is obviously to the benefit of a Batesian mimic to retard its emergence until after the appearance of its model, thus allowing the predator to be "educated" before being exposed to the mimics, other factors remaining constant.

A number of studies have provided evidence that this is the case in some systems, for example, in various moths (Frazer & Rothschild 1960, Rothschild 1971, Sbordoni et al. 1979). This expectation was also supported by Bobisud (1978), who used simple difference models to examine what temporal relationship between model and mimic populations would give maximum benefit to the mimics.

If a palatable and conspicuous Batesian mimic is abundant, yet its obnoxious model is very rare, naïve predators are likely to exert a heavy toll on them, so the question arises as to whether the phenology of mimics and models are the same or whether the mimics might become most abundant only after the year's cohort of new predators, such as young birds, have learned their lessons from attacking models. D.L. Evans & Waldbauer (1982) and D.L. Evans (1984) compared responses of wild-caught, presumably experienced, red-winged blackbirds, *Agelaius phoeniceus*, and common grackles, *Quiscalus quiscula*, to bumblebees, *Bombus pennsylvanicus*, and the mimetic hoverfly *Mallota bautias*, and found that naïve fledglings attacked both model and mimic, whereas wild adults avoided both. This suggested that it would be advantageous for the mimic to appear after insectivorous birds had fledged and gained experience.

The phenological relationships between dipteran mimics of aculeate Hymenoptera have been studied in a number of papers (Waldbauer & Sheldon 1971, Waldbauer et al. 1977, Waldbauer & LaBerge 1985, Waldbauer 1988b) and there is a considerable degree of variation in the degree of temporal synchrony between models and mimic, some actually showing almost no overlap at all (Fig. 7.10). However, at first glance it appears that the mimics are appearing before their models rather than after them, and Waldbauer (1988b) points out that some vertebrate predators are long lived and may remember patterns from previous years. Thus the mimics, as expected, are really peaking after their models, albeit in the subsequent year. As Huheey (1980b) points out, in some of the hoverfly/wasp mimicry systems the almost complete lack of temporal overlap means that the naïve predator is likely to encounter multiple unpalatable models in sequence, which could result in very long-term rejection of members of the mimicry complex; even if they occasionally sample a palatable individual it might be unlikely to change their overall avoidance behaviour.

Another issue with the general resemblance of wasp mimics such as hoverflies and their models is that the adult hoverflies of particular species may only be present for a relatively short time window each year, whereas the aculeate models are almost always present throughout most of the growing season, their numbers increasing almost exponentially as time goes by. In the highly seasonal temperate zone, there is also a high degree of synchrony in the timing of bird nesting, so there is a relatively short window when there will be many recently fledged, but inexperienced birds starting to forage for themselves. The data shown in Fig. 7.10 are fairly typical of other sites in North America (Waldbauer & Sheldon 1971, Waldbauer et al. 1977, Waldbauer 1988b) and show that the mimics are most abundant before many birds have fledged, whereas the models are more abundant during and after the birds have fledged so they are far more likely to experience models

before they encounter any mimics. With the possible exception of two hoverfly species, Howarth & Edmunds (2000) were unable to replicate Waldbauer's patterns in north-west England. Waldbauer points out, however, that mimicry and predation are not likely to be the sole forces determining the phenology of the mimics, and other things, such as food and oviposition site availability, are probably important as well. Howarth et al. (2004) looked at the within-day activity patterns of various hoverfly species and their putative hymenopteran models, and found generally positive correlations, which was especially so for abundant mimics compared to rarer ones. Interpreting their results is not straightforward, but they suggest that this difference could be because when mimics are rare, predators are likely to have had prior experience with the noxious models and are therefore likely to show general wariness, but when mimics are common, predators would be more likely to be less wary and so better temporal synchrony would be favoured. This is supported by dissections and examination of the stomach contents of buff-backed egrets, *Ardea ibis*, which sometimes consume the wasp-mimicking hoverfly *Syrphus corollae*. Of 10 stomachs, eight contained only one or two of the hoverflies, one had 27 and the other contained 90. This highly non-random distribution is interpreted as showing that two of the birds had sampled the hoverflies sufficiently to realise that they were quite palatable and had started feasting on them, whereas the other birds were wary and had perhaps had recent encounters with stinging vespids (Kirkpatrick 1957 cited in Rettenmeyer 1970). However, as Carrick (1936) pointed out, stomach content analyses are only truly meaningful with a parallel knowledge of what prey were actually available.

Although it might be suspected that the similarity between vespid[10] wasp species (since all females can sting but males cannot) is always Müllerian between females of different species and Batesian between males and females, this may not always be the case. *Mischocyttarus mastigophorus* (Vespidae: Polistinae) from Costa Rica is interesting in this respect since, despite being eusocial and capable of stinging, it appears to be a mimic of other polistine vespids of the genus *Agelaia*, since both males and females of *M. mastigophorus* are polymorphic, each morph resembling one of two different sympatric *Agelaia* species (Joyce 1999). The *Agelaia* 'models' are abundant and highly aggressive, which would be reason enough for the mimicry, but maintenance of the polymorphism is perhaps best explained by differences in

---

10. Potentially members of the traditional Sphecidae can also sting, but most are quite reluctant to do so, and thus may be more Batesian mimics than Müllerian ones. The Sphecidae, in its old sense at least, appears to be paraphyletic with respect to the bees (Apidae) and no species are eusocial.

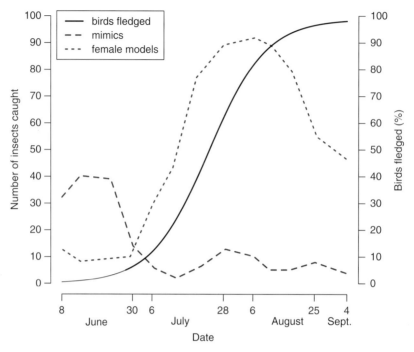

Fig. 7.10 Phenological relationships between high-fidelity hoverfly mimics of aculeate bees and wasps, the female aculeates (i.e. the ones capable of delivering a sting), and the proportion of insectivorous birds that have fledged that season. Data are from lower Michigan state, and the fledgling data have been shifted 10 days to the right to allow for the days after fledging when the young birds are still being fed by their parents rather than searching for food by themselves. (Source: Adapted from Waldbauer & LaBerge 1985. Reproduced with permission from John Wiley & Sons.)

the elevational distributions of the model species, which are not perfectly sympatric, but the supposed mimic overlaps with both

Pseudostings and pseudostinging behaviour

Females of many Hymenoptera, notably members of the Aculeata but also in some parasitic species, i.e. various larger Ichneumonoidea, can deliver rather a painful sting. In a few cases the effects can be quite devastating, reaching level 4+ on Schmidt's (0–4) pain scale (J.O. Schmidt et al. 1990). Male hymenopterans, on the other hand, are unable to sting since they lack both ovipositors or stingers and have no venom glands. However, in the vast majority of these species, the males will, when handled or grabbed by a predator, probe around with the apex of the abdomen in a behaviour that is very reminiscent of a female hymenopteran stinging movement. The effect is enhanced in some cases by the male wasps having their external genitalia modified so as to be quite stinger-like, e.g. some Tiphiidae (Fig. 7.11).

Several species of the New World sesiid moth genus *Alcathoe* and erebid genus *Trichura* are aculeate hymenopteran

Fig. 7.11 Pseudosting on a male five-banded tiphiid (*Myzinum quinquecinctum*). (Source: Loren J. Padelford. Reproduced with permission from Loren J. Padelford.)

mimics, some bearing a strong resemblance to pompilid spider-hunting wasps. However, males of *T. cerberus* and a few other species have a long conspicuous tail from the tip of the abdomen. This has been likened to mimicking the trailing hind legs of a pompilid in flight, but the colour pattern of this, and certainly of some other species, could also be mimetic of various ichneumonid or braconid wasps, females of which have long ovipositors, so I suggest that it is probably a pseudo-ovipositor. Van Achterberg (1989) described a new genus of braconid wasp, *Pheloura*, from South America that has a remarkably long pseudo-ovipositor, comprising three long strands, some 14 times the length of the body (the true ovipositor is quite short), which emerge from the anal area. These can really only be mimetic of the truly long ovipositors of some other ichneumonoids, though the reason remains unknown as there are no observations of living specimens of this apparently very rare species.

Wasmannian (or ant) mimicry

Ant or Wasmannian mimicry should be considered as a spectrum of deceptions ranging from animals that mimic ants to deceive the ants (aggressive mimicries) to ones that mimic ants to deceive potential predators (protective mimicries), and in many cases the mimicry plays both these roles, though through largely different senses since ants have poor vision and perceive the environment largely through chemical and tactile senses. Because these different functions have not always been distinguished and because of the complexity of many of the relationships, I am deviating here from the normal functional plan of this book and dealing with all ant mimicry in one section.

Ant mimicry as defence against predation

Ants are virtually ubiquitous; many can sting, bite and/or squirt out formic acid, and their numbers in tropical wet forests can be astounding, sometimes constituting the largest component of animal biomass in a given area. Furthermore, if a predator or other animal disturbs an ant, that ant will often release an alarm pheromone which can quickly attract large numbers of its sisters who will join the attack on the intruder. This applies to both vertebrate and invertebrate intruders, and while an ant attack against a vertebrate is disagreeable, attack against a small invertebrate may be lethal. Not surprisingly, therefore, many insects and spiders have evolved to mimic ants. Furthermore, a number of insectivores have an innate aversion to ants because of their defensive capabilities (X.J. Nelson & Jackson 2006a, X.J. Nelson et al. 2006). Lev-Yadun & Inbar (2002) also suggest that some plants might have ornamentation that mimics ants (see Chapter 9, section *Ant mimicry in plants*). Anyone who has had encounters with nests of some aggressive tropical ants, such as *Oecophylla smaragdina*, will know how alarming and, to some extent, damaging an encounter with a plant covered in ants might be.

Ants are mimicked morphologically (myrmecomorphy), behaviourally and chemically by a wide range of insects and particularly, perhaps, by various spiders. Ant mimics are known in at least 31 arthropod families belonging to 10 orders and involve thousands of species (Pocock 1910, Lenko 1964, Papavero 1964, Cobben 1986, McIver 1987, McIver & Stonedahl 1993), including other species of ant (Ito et al. 2004, Gallego-Ropero & Feitosa 2014) and even adult moths (e.g. Stathmopodidae).

Myrmecomorphs are often animals that live outside of ant nests and their body shape resemblance towards ants that forage during the day is nearly always Batesian. Because ants are rather small, the insects that mimic them physically are either small or juveniles of larger species. The early instars of several tropical praying mantids are very ant-like (Fig. 7.12) and unlike their often cryptic more fully grown individuals, sit or hunt openly on the tops of leaves. They are quite agile and no doubt adept at escaping predators but I wonder whether they might also possess some chemical deterrence. McIver (1987) has conducted a detailed survey of a North American ant-mimetic system based around the insect community of lupins (*Lupinus*; Fabaceae). In this system, the mirid plant bug *Coquillettia insignis* mimics several different species of ant during its different nymphal and adult instars. Perhaps one of the most interesting aspects of McIver's (1987) findings is that spiders constitute one of the major groups of predator that are deceived by the ant mimics.

One of the strangest instances of myrmecomorphy must be that by the male (and not the female) of the recently discovered hemipteran *Formiscurra indicus* (Caliscelidae). This plant hopper has a large lobe protruding from the front of its head which mimics the head of an ant (Fig. 7.12c) (Gnezdilov & Viraktamath 2011), though why it should only be the males that are mimetic is unclear. In some taxa the ant's narrow waist(s) are mimicked by similar morphological changes (Fig. 7.12a,c,d), but in others it is at least a partial illusion created by pale marks (Fig. 7.12b,d). Some parasitic wasps with wingless females, such as members of several bethylid and ichneumonoid genera, such as *Gelis*, are often described as ant mimics, but for these, their basic body shape is ant-like as they naturally have a wasp waist, so at least some of their supposed myrmecomorphy is due to phylogenetic relatedness (they are all apocritan Hymenoptera), and many of their close relatives are predominantly ant-coloured black or brown. So while they may gain some additional mimetic protection from their ant-like form, it would be difficult to dissect this from their natural resemblance. It should also be noted that despite there being very convincing ant mimicries, there have only been a few experimental demonstrations that ant mimicry is actually protective (e.g. Pie & Del-Claro 2002). It is

Fig. 7.12 Examples of ant-mimicking (myrmecomorphic) insects and a plant. (a) *Macroxiphus* sp. (Orthoptera: Tettigoniidae); (b) nymph of *Himacerus mirmicoides* (Hemiptera: Nabidae); (c) male of *Formiscurra indicus* (Hemiptera: Caliscelidae) from India, showing protuberance from head that mimics an ant's head; (d) juvenile of an unidentified species of praying mantid, Thailand; (e) unidentified tropical spider from Vietnam that runs over foliage in an ant-like fashion and bears an approximate resemblance to a large black ant; (f) dark markings forming approximately lines of the stem of cocklebur, *Xanthium strumarium* (Asteraceae) that resemble columns of foraging ants. (Source: a, Muhammad Mahdi Karim (www.micro2macro.net) 2009. Reproduced under the terms of the GNU Free Distribution Licence GFDL 1.2, via Wikimedia Commons; b, Alfred, 2007. Reproduced under the terms of the Creative Commons Attribution Share-Alike Licence CC BY-SA 2.5; c, L. Shyamal 2013. Reproduced under the terms of the Creative Commons Attribution Share-Alike Licence CC BY-SA 3.0, via Wikimedia Commons; f, Professor Simcha Lev-Yadun. Reproduced with permission from Professor Lev-Yadun.)

particularly difficult to know whether mimicry is involved when the 'mimics' are taxonomically close to ants. Thus rather a lot of small, apterous parasitic wasps are described as being ant-like, inferring that they may be gaining protection from their resemblance. Of course their pattern of loco-motor activity may give a strong indication, but such wasps are innately ant-like in any case with their narrow apocri-tan wasp waists. Malcicka et al. (2015) provide fairly con-vincing evidence that in the case of the wingless cryptine ichneumonid *Gelis agilis*, protection from a spider predator is afforded by a combination of appearance, motion and chemical mimicry of an ant alarm pheromone.

Ant behaviour is largely directed by chemical senses and so to be able to co-habit with, or perhaps just approach their ant models easily, many associated insects and spiders need to have cuticular hydrocarbon profiles that the ants will not recognise as foreign, thus eliciting attack. Mostly though, these come into play only when the mimic and ant come into direct physical contact. Species that are additionally gaining protection from a different signal receiver also often evolve an ant-like body form. Table 7.2 lists a range of examples of ant mimicry and indicates, when known, both function and modalities involved.

Poulton (1891) described a rather nice example from South America involving mimicry of leafcutter ants of the genus *Atta* (=*Oecodoma*) *cephalotis*, which are conspicuous because they form trails of individuals carrying roughly 'D'-shaped sections of leaves to their nests. Among them were also individuals of a laterally compressed, green membracid plant hopper, probably a *Stegaspis* species, whose form mimicked not the ant but the combination of ant and the piece of leaf it is carrying. Poulton (1891) suggests that the mimic probably frequents trees where the ants are abundant.

S. Powell et al. (2014) describe how the mirror turtle ant, *Cephalotes specularis*, uses visual posturing mimicry with its raised gaster, probably in conjunction with chemi-cal mimicry or crypsis, to invade the territories of the hyper-aggressive *Crematogaster ampla*, which enables it to forage therein. In addition, *C. specularis* parasitises the host ant through its highly efficient ability to follow its trails, thus 'eavesdropping' on the host's chemical com-munication network. S. Powell et al. (2014) refer to this as a case of Batesian–Wallacian mimicry, but as the mimic does not eat the model ant, this seems an inappropriate label for this relationship.

Table 7.2 Some examples of mimicry of ants, also called Wasmannian mimicry.

Mimic	Ant model	Type of interaction	Hydrocarbon similarity	References
Cosmophasis bitaeniata	*Oecophylla* spp.	Predator on ant larvae	Obtained from ants consumed	Allan et al. 2002, Elgar & Allan 2004, 2006
Paralipsis eikoae	*Lasius niger* and *L. sakagamii*	Parasitoid of aphids 'farmed by ants'. Also obtains food from ants by trophylaxis	Obtained by direct contact; tries to ride on ant. Innate cuticular hydrocarbons are fairly neutral *n*-alkanes	Takada & Hashimoto 1985, Akino & Yamaoka 1998
Lysiphlebus cardui	*Lasius niger*	Parasitoid of aphids 'farmed by ants'	apparently evolved similarity	Liepert & Dettner 1996
Aphids farmed by ants	Many species	Protection from parasitoids and predators in return for providing honeydew	Rear view with siphuncles may resemble ant head morphologically	Kloft 1959, Way 1963
Myrmarachne	Many species including *Oecophylla smaragdina*	Probably Batesian, may be aggressive		R.R. Jackson 1986, X.J. Nelson et al. 2005, 2006
Amyciaea	*Oecophylla smaragdina*	Predator of adult ants	Visual, probably Batesian	Leong & D'Rosario 2012
Pranburia mahannopi	*Diacamma* spp.	Probably Batesian		Deeleman-Reinhold 1993
Pertyia sericea (Cerambycidae)	*Camponotus sericeiventris*	Probably Batesian	Unknown	Lenko 1964

Ant mimicry by spiders

Ant-mimetic spiders are mainly tropical, with only a few examples in the temperate region (Cutler 1991), which has no doubt limited experimental work. Nevertheless, ant mimicry has evolved in at least 10 families of spiders, such as the Aphantochilidae, Araneidae, Clubionidae, Corinnidae, Eresidae, Gnaphosidae, Salticidae, Thomisidae, Theridiidae and Zodariidae, and in a number of these on multiple occasions (Reiskind & Levi 1967, Reiskind 1977, McIver 1987, Elgar 1993, McIver & Stonedahl 1993; see Cushing 2012 for a review). Ant mimicry by spiders may be Batesian (mainly in terms of their visual appearance and behaviour) or aggressive (mainly in terms of odours but also behaviour), and sometimes both. Since the numerous workers of most ant species are fairly restricted in their size range, insects and spiders that mimic ants may show transformational mimicry, shifting resemblance to progressively larger model species as they grow or, in some cases, modify to resemble other types of models (Mathew 1934, 1935, J.F. Jackson & Drummond 1974).

None of the 200 or so species of the common ant-mimicking salticid spider genus *Myrmarachne* (Fig. 7.13b–d) appear to prey on ants, at least not often, and are therefore just Batesian mimics. Not only do *Myrmarachne* look very much

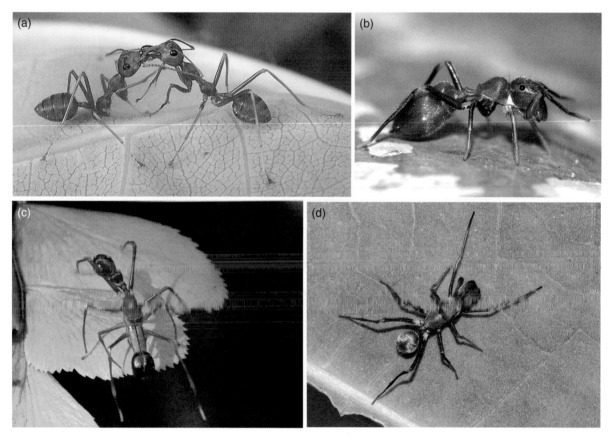

Fig. 7.13 Ant and ant-mimicking jumping spiders (Salticidae). (a) Green tree ants, also called weaver ants and Karengga ant, *Oecophylla smaragdina*, which are common and particularly aggressive throughout much of Asia; (b) female *Myrmarachne* sp. showing normal-sized chelicerae, cephalothorax with strong constriction, petiolate abdomen and anterior legs coloured and held to resemble the model ant's antennae; (c) male *Myrmarachne plataleoides* from South-East Asia, a mimic of *O. smaragdina*; (d) male of an unidentified *Myrmarachne* sp. mimicking dark ant. (Source: a, Sean Hoyland 2005, via Wikimedia Commons; b, Charles Lam 2009. Reproduced under the terms of the Creative Commons Attribution Share-Alike Licence CC BY-SA 2.0, via Flickr; c, Jeevan Jose, Kerala, India 2010. Reproduced under the terms of the Creative Commons Attribution Share-Alike Licence CC BY-SA 4.0, via Wikimedia Commons; d, Sean Hoyland 2007, via Wikimedia Commons.)

like ants, but X.J. Nelson & Card (2016) showed that these ant-like salticids also moved in a very similar way to ants, and quite differently from non-ant-mimicking salticids, i.e. they display locomotory mimicry too.

Juveniles and females of *Myrmarachne* are particularly ant-like (Fig. 7.13b). While male *Myrmarachne* have enlarged chelicerae, which may detract from the mimetic resemblance (Fig. 7.13c), X.J. Nelson & Jackson (2006b) suggest that these may in fact be mimicking an ant carrying some other object, something they call 'compound mimicry'. X.J. Nelson & Jackson (2006b) experimentally tested whether spiders that feed on spiders and those that feed on ants showed different responses to male and female *Myrmarachne*. The spider-hunting specialist salticid, *Portia fimbriata*, was found to avoid both unencumbered and encumbered ants and both female and male *Myrmarachne*. In contrast, the specialist ant-predator spider *Chalcotropis gulosus*, which preferentially attacks encumbered ants, attacked male *Myrmarachne* significantly more often than females. Cutler (1991) also demonstrated that ant mimicry by the North American salticid *Synageles occidentalis* provides significant protection from predation by other spiders.

Assessing predation by birds and lizards in the field is difficult and there is only limited evidence on it, but ant-mimicking spiders are also the prey of various specialist spider-hunting sphecid wasps. M. Edmunds (1993) considered the question as to whether ant mimicry by salticid spiders offered protection from the specialised spider-hunting wasp *Pison xanthopus* in Ghana. *P. xanthopus* is a sphecid wasp that constructs mud nests with one to six cells, arranged in a row. Into each cell the wasp places from five to ten paralysed salticid spiders. Samples of *P. xanthopus* nests were collected and their spider contents compared with the salticid spider fauna of the nearby vegetation where the wasp naturally searches for its prey. Edmunds' (1993) results were highly significant, with wasp cells containing far fewer good ant mimics than were present in the samples of apparently available spiders (Fig. 7.14). Durkee et al. (2011) then found that ant mimicry by the salticid spider *Peckhamia picata*, which resembles the ant *Camponotus nearcticus*, not only protects it from avian

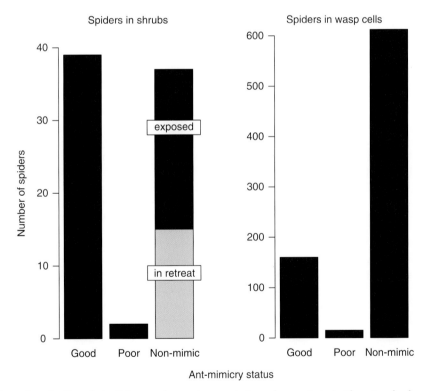

Fig. 7.14 Proportions of good spider mimics (*Myrmarachne* spp.), poor mimics and non-mimics in Ghanaian shrubs and in the solitary sphecid predatory wasp, *Pison xanthopus*. Wasp nest cell showing that ant mimicry protects against predation by the visually hunting wasp, as indicated by an excess of non-mimetic spiders in the wasp's brood cells compared to the numbers on the vegetation that the wasps were searching for prey on. (Source: Data from Edmunds 1993.)

and hymenopteran predation but also from predation by other (larger) spiders. Uma et al. (2013) went on to show that it also experiences reduced predation by the mud-dauber wasp, *Sceliphron caementarium* (Sphecidae), which relies more heavily on chemical cues to identify prey. In this case, it seems to be not through mimicry of ant cuticular hydrocarbons, but a lower concentration of cuticular hydro-carbons than those of other spider prey of the *Sceliphron*, and therefore a chemical crypsis. Another invertebrate predator, the ant-mimicking praying mantis, *Euantissa pulchra*, has been found to differentiate between two closely related, ant-mimicking *Myrmarachne* species. It approached the species mimicking the more aggressive ant model, *Oecophylla smar-agdina*, significantly less than the one mimicking the less aggressive *Campanotus sericeus* (Ramesh et al. 2016).

The East African spider *Myrmarachne melanotarsa* is gre-garious and associated with the aggressive *Crematogaster* ants, and not only does its myrmecomorphy confer protection against predation by large tree-ant-averse salticids that other-wise are arachnophages, but it gains greater protection when the spiders form a group resembling a group of ants (X.J. Nelson & Jackson 2009a). The authors further concluded that the predatory spiders they tested were showing not just an innate aversion to ants and their mimics, but an enhanced innate aversion to groups of ants (or those of ant mimics) (see also Chapter 10, section *Mimicking danger as a flushing device*) about the use of aggressive mimicry in this species).

Spiders that feed on ants

Specific predation on the model ants, i.e. potential aggres-sive mimicry, is only demonstrated in a few cases (Wing 1983, Cushing 1997, Pekár & Král 2002). P.O. Oliveira and Sazima (1984) found that the Brazilian spider *Aphantochilus rogersi*, which, like members of the other three genera in its family, are mimics of black formicine ants of the tribe Cephalotini (various *Zacryptocerus* species), not only preys specifically on these ants but also rejects other potential prey offered to them, though they would also feed on the non-model yellow ant *Zacryptocerus clypeatus*. In this case it appears that the mimicry has both aggressive and Batesian components. Pekár & Král (2002) describe how the European zodariid spider *Zodarion rubidium* is an aggressive mimic of ants such as *Myrmica sabuleti*, as having captured and killed an ant, the spider successfully (75% of encoun-ters) deceives any other worker ants that approach it by imitating their antennal 'greeting' behaviour, stroking the ant's antennae with its own fore legs. If this fails to work, the spider either tries to flee with its prey or drops the prey to distract the ant worker. The ant-like jumping spider *Tutelina similis* (Araneae: Salticidae) and a few other mimetic species also prey on ants (Wing 1983).

The Australasian salticid *Cosmophasis bitaeniata* is closely associated with aggressive *Oecophylla* green tree (or weaver) ants (Fig. 7.13a), and feeds on them, but does not bear any resemblance to them, and thus seems to rely largely on avoiding detection by the ants by using acquired chemical mimicry (Allan & Elgar 2001, Allan et al. 2002, Elgar & Allan 2004). The cuticular hydrocarbon profiles of *C. bitae-niata* have been shown to be ant colony-specific but they do not match those of the major workers, and Elgar & Allan (2006) suggest their role is primarily to facilitate stealing food from the host's minor workers rather than to reduce aggression from the major workers.

Pekár & Jiroš (2011) also investigated whether chemical mimicry plays a part in protecting ant-mimetic spiders from their models. The five species of spider they investigated generally had cuticular hydrocarbon profiles that were quite different from those of their respective ant models, except for two compounds – octacosane and nonacosane. By experimentally exposing the spider species as well as various non-mimetic species to ant predation, they found that ants readily attack non-mimics and three of the mimic species, but that two of the mimics, *Myrmarachne formicaria* and *Zodarion alacre*, were attacked less, and they attributed this to their possession of long-chain hydrocarbons. The other species must presumably rely largely on behavioural avoidance of their dangerous models.

Pekár & Král (2002) posed the interesting question of how combined ant predation and ant mimicry in spiders evolved. There are two possibilities: the spiders might initially have evolved to be Batesian mimics of the ants, and perhaps, because this would be favoured by those living in places with numerous potential model ants, they then started eating more of the model ants, with natural selection then leading to the evolution of features improving their aggres-sive mimicry. Alternatively, the spiders may have started as specialist ant predators but without any mimetic resem-blance to the ants, but then because of their physical prox-imity to these aggressive prey, natural selection favoured increased morphological similarity and associated Batesian protection. Another evolutionary question was addressed by Ceccarelli & Crozier (2007) for *Myrmarachne*: Have these ant-mimicking spiders co-speciated with their models? Using molecular phylogenies they found that not to be the case, but rather that there has been a very high level of switching to mimic quite distantly related ant taxa.

How to look like an ant or an ant carrying something?

There is a particular 'difficulty' in looking like an ant in the case of spiders, because they have only two tagmata (i.e. major body regions) – a cephalothorax and an abdomen (or postsoma) – whereas their ant models conspicuously have three: head, mesosoma[11] and metasoma. They also sometimes have modified scale-like second and third abdominal segments called petiolar segments, which afford the posterior of the abdomen even greater flexibility for stinging. Physical resemblance has most commonly been achieved through evolution of an elongated cephalothorax with a distinct constriction, very enlarged chelicerae (in males), and a distinct petiole between cephalothorax and abdomen (Fig. 7.13a,b), as well as running in a very ant-like fashion. The recently described South-East Asian corinnid spider *Pranburia mahannopi* achieves the effect in a quite different way. It is somewhat elongate but the cephalothorax is not divided, instead the femora of the fore legs have a dense brush of long, dark setae such that when they are held close together, as they normally are, they create the impression of an ant's head (Deeleman-Reinhold 1993).

R.R. Jackson (1986) provides a detailed account of the ant-like, cursorial, salticid spider *Myrmarachne*[12] in Australia. These spiders walk on only their rear six legs and hold up and wave their anterior pair, mimicking an ant's antennae, and unlike most related spiders, their bodies are largely smooth and ant-like rather than setose.

Myrmecomorphy by caterpillars

One of the most surprising cases of ant mimicry involves caterpillars of the South-East Asian noctuid moth genus *Homodes* (Kalshoven 1961, Leong & D'Rozario 2012), whose anterior end has a lot of long bent tentacle-like spatulate setae. *Homodes* caterpillars mimic *Oecophylla* tree ants which are notoriously aggressive, and indeed have been given a special name by various indigenous peoples (e.g. the ngrangrang ant in Borneo and karinga in Malaysia). Shelford (1902), also quoted in Kalshoven (1961), noted that *Homodes* caterpillars only seem to be found in trees with the *Oecophylla* present and not in trees with other ant species, though whether this is cause or effect is not known. Shelford (1902) stated:

> It is very difficult to convey by a description or even a drawing, the very startling resemblance of this caterpillar to an ant, yet the resemblance will not really bear a close examination.

Shelford's (1902) drawing really does not do the job very well either, but photographs really do (Fig. 7.15a,b) (also see those in Leong & D'Rozario, 2012).

Larvae of the European lobster moth (*Stauropus fagi*: Notodontidae) and related East and South-East Asian species are particularly strange (Fig. 7.15c–e). Professor Poulton (1890) wrote of the lobster moth caterpillar:

> The larva of *Stauropus fagi* therefore bristles with defensive structures and methods. At rest, it is concealed by a combination of the most beautiful Protective Resemblances to some of the commonest objects which are characteristic of its food-plant. Attacked, it defends itself by a terrifying posture, made up of many distinct and highly elaborate features, all contributing to this one end. Further attacked, it reveals marks which suggest that it can be of no interest to an insect enemy, for another parasite is already in possession.

They have extremely elongate true legs for caterpillars and can run quite quickly, and together with their strange body shape and posture, can resemble ants, though as summarised by McAtee (1912) different observers have come to various conclusions as to what this unusual insect may be mimicking, including leaves, twigs, earwigs and staphylinid beetle. I personally have thought them to be rather similar to rearing spiders, but when seen on vegetation with many ants, especially in the tropics, there is no doubting that to this human observer they do have a certain similarity to ants. What is interesting is that the legs of larger larvae can be held tucked into the body (Fig. 7.15e) or splayed out and the knobbiness is not dissimilar to an ant's elbowed antennae (Fig. 7.15d). As with *Homodes*, some species also have elongate posterior protuberances. It is quite possible, of course, that these uncaterpillar-like caterpillars resemble different things to different predators and maybe their overall strangeness simply causes potential predators to be wary of them.

Ant chemical mimicry by parasitoid wasps

The rare European butterfly *Phengaris* (=*Maculinea*) *ribeli* lives as a caterpillar within the nests of *Myrmica* ants, where its fourth instar larvae, which have had no previous contact with ants, are taken by the host *Myrmica* who appear to mistake the caterpillar for their own larva. Once there, the larva feeds on the ants' brood (Elmes et al. 1991) and in the nest the butterfly caterpillar escapes subsequent recognition by the ants through chemical mimicry and camouflage (Akino et al. 1999). In this case some key compounds are synthesised while others are acquired. Other species of

---

11. The middle body region of aculeate Hymenoptera called the mesosoma comprises the thorax and first abdominal segment, and the narrow waist is located between the first and second abdominal segments, an adaptation that allows them to have large longitudinal flight muscles. The third body region is therefore abdominal segments 2 onwards and so to distinguish it from the true whole abdomen it is referred to as the metasoma or gaster.

12. Myrmarachne is a large genus with some 200 described species that are distributed mainly in the tropical forests of Africa, Asia, Australia and the New World.

Fig. 7.15 Caterpillars with ant-like appearance. (a,b) anterior views of *Homodes* larva (Erebidae) from Singapore; (c–e) caterpillars of lobster moth, *Stauropus* spp. (Notodontidae) which are variously described as resembling ants, dead leaves or leaf scales and, I think, sometimes, spiders, but certainly not typical caterpillars. (Source: a,b, Craig Williams. Reproduced with permission from Craig Williams; c, Itchydogimages. Reproduced with permission from John Horstman.)

lycaenid that are associated with ants in various ways also produce mimetic chemicals that modulate or change the behaviours of their associated ants (S.F. Henning 1983, 1997, Pierce et al. 2002). There are also specialist, and consequently similarly endangered, ichneumonid wasp parasitoids of the *P. ribeli* larvae. Wasp females first locate nests of the ant, presumably by a combination of olfaction and vision, but only enter nests that contain a *Phengaris* caterpillar, which they then seek out to oviposit within. The *Phengaris* is killed as a pupa by the ichneumonid, which subsequently emerges as an adult within the ant colony.

Several other parasitic wasps in the family Eucharitidae are a brood parasite of various ants and presumably all have some chemical camouflage system, though only in a couple of cases is there any chemical evidence. An unidentified *Orasema* species, parasitic within the nest of *Solenopsis invicta* fire ants, appears to acquire its camouflaging chemicals from the host ants passively (Vander Meer et al. 1989). R.W. Howard et al. (2001) examined the cuticular hydrocarbons of adult *Kapala sulcifacies*, which attacks the ponerine ant *Ectatomma ruidum*, and found they shared 40 cuticular hydrocarbons in common with their hosts. The wasps were generally left unattacked for a while post-emergence in the colony, but were eventually evicted by the ants. It seems likely that *Kapala* adults synthesise their own mimetic cuticular hydrocarbons.

PROTECTIVE MIMICRIES AMONG VERTEBRATES

Fish

Batesian mimicry among fish

Probably the best-known, and earliest reported, supposed instance of Batesian mimicry among fish is that between the normally cryptic common sole, *Solea vulgaris*, and venomous weaver fish of the genus *Trachinus*, including *T. draco* and *T. vipera*. The weaver fish has a black dorsal fin that is erected when the fish is threatened, and when the sole is threatened it raises its pectoral fin in a similar way (Masterman 1908). Evidence that this is true mimicry comes from the fact that other species of sole lack the black-marked pectoral fin and do not raise those fins when disturbed. Furthermore, the common sole has a near-identical distribution to the two models, with young sole living in shallow waters where *T. vipera* occurs and mature fish in deeper waters where *T. draco* lives. Since this first observation, several other instances have been discovered, particularly involving *Scorpaenodes* (Scorpaenidae) as models (Whitley 1935, Randall & Randall 1960, Seigel & Adamson 1983).

Banded snake eels, *Myrichthys colubrinus*, are excellent Batesian mimics of banded sea snakes, *Laticauda colubrina*, and Dudgeon & White (2012) suggest that juvenile zebra sharks, *Stegostoma fasciatum*, probably are too, and if so this is perhaps the first case of Batesian mimicry in an elasmobranch. Although the authors note that it still needs empirical testing, the juvenile sharks often swim close to the surface where the snakes occur and they possess "a very long, single-lobed caudal fin that remarkably resembles the broad, paddle-like tail of sea snakes".

Müllerian mimicry among fish

There are surprisingly few examples of Müllerian mimicry among fish, given that there are perhaps some 30,000 species of them. Undoubtedly this is because the vast majority of fish are palatable and lack venoms and those that do have venom are typically cryptic, such as stonefish (Synanceiidae) and weaver fish – though of course there are exceptions. Maybe there is something about vertebrate physiology or microstructure that makes it difficult to store large amounts of toxins, but perhaps more likely is the fact that as you ascend the food chain the numbers of different major groups of predators diminishes, and the main predators of fish are other fish.

The freshwater *Corydoras* and related catfish in South America are a species-rich group, often with up to three similarly coloured species inhabiting the same stretch of water, and often associating in mixed shoals. Some species are cryptically coloured, some have disruptive patterns and some appear to be aposematic. However, the colour patterns they show vary from region to region, and Alexandrou et al. (2011) postulate that at each site the species form Müllerian rings. Evidence that some sort of mutualistic mimicry is involved comes from the fact that the similarly patterned species in a given locality are often not closely related and that sympatric species in local communities usually differ in their snout morphology and size, indicating that they are probably partitioning the food resource and therefore not competing in that way.

Batesian and Müllerian mimicry among terrestrial vertebrates

Examples of Batesian and Müllerian mimicry among terrestrial vertebrates are generally rather rare and taxonomically sparsely distributed. The best-known examples involve various frogs (e.g. Summers & Clough 2001, Symula et al. 2001, Hagman & Forsman 2003) (see Chapter 4, section *Polymorphism and geographic variation in aposematic species*), salamander (e.g. Brandon et al. 1979a, 1979b) and snake-based systems. First, I will briefly mention some salamander and mammal cases and then devote rather more space to coral snakes and their mimics, which have been a major subject of debate and experimentation.

Dunn (1927) had noticed the marked similarity between various distantly related North American salamanders, and subsequent discoveries of races resembling other species with different patterns made this a particularly interesting example of potential mimicry rings in a terrestrial vertebrate, and therefore became subject to a series of investigations (Huheey 1960, Huheey & Brandon 1961, Orr 1962, 1967, Cody 1969). Whereas earlier workers had found no apparent aversion to the salamanders by reptile and mammal predators, Brodie & Howard (1973) investigated the responses of wild-caught potential avian predators. They found that *Plethodon jordani*, which has red cheeks and produces noxious skin secretions, deterred birds from eating them and that, once conditioned, the birds also avoided red-cheeked individuals of the otherwise palatable *Desmognathus ochrophaeus*, thus apparently finally showing that this is most likely a case of Batesian mimicry.

When it comes to mammals there are decidedly few convincing cases. Gingerich (1975) suggested that the aardwolf (*Proteles cristata*) is a Batesian mimic of the striped hyena, *Hyaena hyaena*, and R.L. Eaton (1976) proposed that cheetah cubs, *Acinonyx jubatus*, which differ markedly in their colour pattern from adults, may be mimics of the ratel (also called the honey badger), *Mellivora capensis*, both having a conspicuous white dorsum (see Fig. 4.34b). In the first of these examples, it should be noted that both are species of Hyaenidae, so the resemblance could be plesiomorphic, although H.W. Greene (1977) considered this far from proven. Furthermore, Gingerich's (1975) example requires that predators are unable to detect, from a distance, the marked disparity in size between the species. However, I find the second example rather convincing since ratels have classic mammal warning colouration and are aggressive (see Chapter 4, section *Warning colouration in mammals*). Eaton (1976) also puts forward the possibility that ratels might also be dupes (see section *Batesian–Poultonian (predator) mimicry* later) if they sometimes recognise visually, and avoid encounters with, potentially highly aggressive conspecifics.

The coral snake problem – Emsleyan (or Mertensian) mimicry

One of the most famous mimicries, probably because getting it wrong can be fatal to humans, involves various species of usually brightly aposematically coloured coral snakes (Fig. 7.16a–c), which Wallace (1867, 1870) considered to be the prime example of mimicry and warning colouration among vertebrates (H.W. Greene & McDiarmid 2005). Coral snakes belong to the front-fanged, highly venomous and often deadly, family Elapidae (which also includes cobras

and kraits). The most notable coral snake genus is *Micrurus*, which has over 65 species recognised in the New World, mostly tropical parts but with several species in Mexico. The ranges of *Micrurus fulvius* and *Micruroides euryxanthus* extend into the southern parts of the USA where they are mimicked by the harmless scarlet snake, *Cemophora coccinea* (Fig. 7.16e), *Chionactis palarostris*, *Rhinocheilus lecontei*, the often much larger, scarlet kingsnake, *Lampropeltis elapsoides*, and eastern milksnake, *L. triangulum* (Fig. 7.16d), all Colubridae. There are numerous other good coral snake mimics in the Neotropics. An important issue in interpreting the coral snake mimicry situation is that the geographic range of the kingsnake greatly exceeds that of the putative coral snake models, which begs the question of why are they always still so conspicuously coloured, even if not precise coral snake mimics?

The major question regarding the evolution of the coral snake warning pattern, and the one that has made this such an interesting and debated topic, is what advantage is there to an animal in advertising itself if it was so deadly that its predator dies before it ever encounters another one that it might then avoid? This became known as 'the coral snake mimic problem' (Dunn 1954, Brattstrom 1955), and not long thereafter it was postulated that in fact the coral snakes are not the real models at all but are themselves mimics of more mildly venomous species, and probably ones whose bite is more instantly painful, thus affording them a chance of escape before being killed. Wickler (1968, pp. 111–21) discussed this at length and named the phenomenon after Robert Mertens, who published on coral snake mimicry (Mertens 1956). However, as noted by Pasteur (1982), Mertens' paper did not explicitly state the hypothesis which was first properly expressed by Emsley (1966), and indeed a similar idea had been put forward earlier by Hecht & Marien (1956). Although it is most widely referred to as Mertensian mimicry because of the influence of Wickler's (1968) book, many people now refer to it as Emsleyan mimicry.

It has been estimated that nearly 18% of New World snakes, more than 115 species, are involved in the coral snake mimicry system and there is good geographical concordance between the patterns of the elapids and those of the non- or less venomous taxa (Savage & Slowinski 1992). Coral snakes in general (they occur from the southern states of the USA to Argentina) and the mimics in this case, are boldly patterned with red, black and yellow stripes. This led to various adages such as "Red then yellow: kill a fellow; red then black: venom lack", or "...OK chap" or "...friend of Jack" (the version with the last of these endings appearing in the *Boy Scout Handbook*; Huheey 1988). However, as you can see from Fig. 7.16b,c, the rule only applies to the species with triplet bands, and in Central and South America there are exceptions (Pough 1988b). For example, the deadly

Fig. 7.16 Coral snakes and their snake mimics. (a) Highly venomous eastern coral snake, *Micrurus fulvius* (Elapidae), from the eastern USA and northeastern Mexico (note that the red and yellow bands are adjacent); (b) *Micrurus pyrrhocryptus* (Elapidae); (c) variable coral snake, *Micrurus diastema*, from southeastern Mexico and Central America; (d) non-venomous red milksnake, *Lampropeltis triangulum syspila* (Colubridae); (e) non-venomous scarlet snake, *Cemophora coccinea*, from southeastern USA; (f) grey-banded kingsnake, *L. alterna*. (Source: a, John 2008. Reproduced under the terms of the Creative Commons Attribution Licence, CC BY 2, via Flickr; b, CHUCAO 2013. Reproduced under the terms of the Creative Commons Attribution Share-Alike Licence CC BY-SA 3.0, via Wikimedia Commons; c, Ruth Percino Daniel 2012. Reproduced under the terms of the Creative Commons Attribution Share-Alike Licence, CC BY-SA 3.0; d, Mike Pingleton. Reproduced with permission from Mike Pingleton; e, Glenn Bartolotti 2013. Reproduced under the terms of the Creative Commons Attribution Share-Alike Licence CC BY-SA 3.0, via Wikimedia Commons; f. © Patrick JEAN, Muséum d'Histoire Naturelle de Nantes.)

black and red striped *Micrurus frontalis* (Elapidae) has a red–black–yellow–black–yellow–black–red[13] repeat and is mimicked almost perfectly by the harmless colubrid *Simophis rhinostoma* (see figure 24 in Wickler 1968). Coral snake bites usually affect the extremities, but their local effects in humans are typically minimal, i.e. not much instant pain, the real killer is in the systemic action of their venom neurotoxins, which lead to skeletal and cardiac muscle inhibition through blocking neuromuscular transmission, followed by convulsions and then headaches, difficulty swallowing and change in skin colour; ptosis is a common first sign. Muscle weakness progresses to respiratory failure and death. Because there is usually no pain associated with the bite, victims often think they will be okay and do not therefore seek the immediate medical attention which they urgently need.

S.M. Smith (1975, 1977) showed that two potential coral snake predators – the turquoise-browed motmot, *Eumomota superciliosa* (Momotidae), and great kiskadee, *Pitangus sulphuratus* (Tyrannidae) – both had innate aversion to coral snake banding patterns. Despite their clear deadliness to mammals, it is odd that Beckers & Leenders (1996) did not find any avoidance of live snakes by mammalian predators.

Much evidence has been accumulated to show fairly conclusively that the coral snakes themselves are the principal models (e.g. S.M. Smith 1975, H.W. Greene & McDiarmid 1981, Pough 1988b, Campbell & Lamar 1989, Brodie 1993, Brodie & Janzen 1995, Brodie et al. 1995). A recent study by Davis Rabosky et al. (2016), using phonetic, distributional and phylogenetic data, has shown that shifts in colouration of non-venomous mimics is highly correlated with the presence of coral snakes in space and time. Nevertheless, that is not to say that less venomous snakes might not be Müllerian (i.e. quasi-Batesian) mimics of them, contributing to general predator avoidance. Some of the strongest evidence must be the innate aversion some birds exhibit to coral snake ringed patterns (see Chapter 4, section *Innate responses of predators*) and also the fact that aversion by predators is correlated with the presence of coral snakes in an area.

Using three types of plasticine snake replica baits in different parts of the USA, Pfennig et al. (2001) found that the proportion of attacks on the coral snake replicas was highest where coral snakes themselves were absent (i.e. at high latitudes and altitudes). The accuracy of mimetic resemblance of kingsnakes,[14] *Lampropeltis elapsoides* and *L. triangulum*, to coral snakes in an area where the models used to be present but apparently went extinct approximately

50 years ago, the North Carolina sandhills, shows a significant reduction over the past 40 or so years ($F_{1,26} = 6.997$, $p = 0.014$), whereas no equivalent change was observed in Florida populations where model coral snakes, *Micrurus fulvius*, still occur (Akcali & Pfennig 2014). Fifty years corresponds to a relatively low number of generations of kingsnakes, which have a maximum longevity of 38 years, therefore it seems that selection for them to become less conspicuous in the absence of models is strong. The effects of relative abundance of models and mimics on the accuracy of mimicry in this system are also discussed in Chapter 6, section *Relative abundances of models and mimics in nature*. Hinman et al. (1997) carried out similar plasticine replica experiments to those of Brodie (1993) (see later) and at the same site – the La Selva Biological Station in Costa Rica. They found that the width of the combined black plus yellow bands relative to the red background colour of artificial models was an important factor in whether they were attacked. Significantly, the poor mimic replicas differed from any Neotropical coral snakes in that they lacked any repeated bands, so they just comprised red–black–yellow sequences, but nevertheless the type that had the same red versus black + yellow region spacing received almost the same level of protection as the completely mimetic models. Brown models with bands (similar to the snake shown in Fig. 7.16f) were attacked as frequently as plain brown ones.

Other snakes are not necessarily the only possible models for the original coral snake banding pattern mimicry complexes, and it has been proposed that it could have been large, black and red ringed millipedes that are highly unpalatable, as these are not only common but some species are mimicked by lizards (Vitt 1992) and form Müllerian mimicry rings with other millipedes (e.g. D.R. Whitehead & Shelley 1992, Marek & Bond 2009). However, the width of the rings in these is far narrower than in the coral snakes. Brodie & Moore (1995) tested this idea using models, but the millipede-like ones suffered a far greater number of attacks not only than wider red–black ringed coral snake replicas but also than simple brown snake replicas ($G = 11.26$, $p < 0.01$). This, of course. does not necessarily mean that snake predators were not duped as there could be specialist millipede predators involved.

A lot of coral snake mimics get the pattern wrong and, given that predators such as birds have selected for extremely precise mimicry in many butterfly mimicry systems, it is pertinent to ask whether the inaccuracy with snakes has any biological significance. Pough (1988a, 1988b) suggested that with the coral snake system, the inaccuracy is permitted because the potential consequence of making a mistake and attacking a model is so severe. Gans (1961, 1964) suggests that group learning by socially hunting predators could offer a way out of the conundrum. Many

---

13. Or red–black–white–black–white–black–red repeat.
14. Based largely on museum specimens.

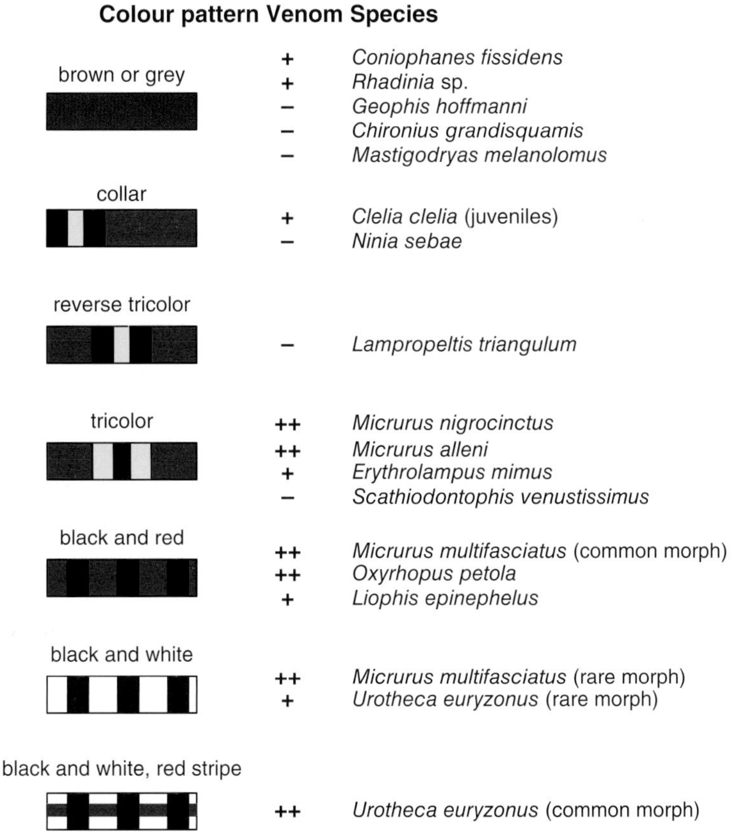

Colour pattern Venom Species

brown or grey	+	*Coniophanes fissidens*
	+	*Rhadinia* sp.
	−	*Geophis hoffmanni*
	−	*Chironius grandisquamis*
	−	*Mastigodryas melanolomus*

collar	+	*Clelia clelia* (juveniles)
	−	*Ninia sebae*

reverse tricolor	−	*Lampropeltis triangulum*

tricolor	++	*Micrurus nigrocinctus*
	++	*Micrurus alleni*
	+	*Erythrolampus mimus*
	−	*Scathiodontophis venustissimus*

black and red	++	*Micrurus multifasciatus* (common morph)
	++	*Oxyrhopus petola*
	+	*Liophis epinephelus*

black and white	++	*Micrurus multifasciatus* (rare morph)
	+	*Urotheca euryzonus* (rare morph)

black and white, red stripe	++	*Urotheca euryzonus* (common morph)

Fig. 7.17 Schematic illustrations of the colour patterns of the various ground-dwelling snakes at the La Selva Biological Station in Costa Rica and the relative toxicities of their bites. (Source: Adapted from Brodie 1993. Reproduced with permission from John Wiley & Sons.)

avian snake predators are specialists and may have evolved methods for handling dangerous prey that minimise the risk of getting bitten or they may have evolved some degree of immunity to the venoms of coral snakes etc. But some predators, for example baboons, might be expected to learn from observing the effects on another individual of, for example, eating something poisonous or getting bitten by a deadly snake, something that Gans (1961, 1964) referred to as 'empathic learning', but is more commonly known as 'observational learning'. Jouventin et al. (1977) demonstrated this in principle with their experiment with mandrill baboons, *Mandrillus sphinx*, and banana slices, some of which had been rendered unpalatable by quinine treatment (actually just impregnating them with ground quinine-based antimalarial tablets, but it did the trick).

Savage & Slowinski (1992) created a system for codifying the various banding, striping and blotch patterns of the 56 species of coral snake and their mimics and recognised four major dorsal patterns, and, for the tricolor patterns, five

subdivisions (monads, dyads, triads, tetrads and pentads), depending upon the number of black bands or rings separating the red ones. In any one locality in Central and South America there are likely to be species of both models and mimics with various patterns. Brodie (1993) carried out experiments at the La Selva Biological Field Station in Costa Rica using snake replicas constructed of segments of coloured, non-toxic plasticine, to represent the snake patterns present there (Fig. 7.17). The patterns that suffered the fewest avian attacks, as indicated by beak marks left on the plasticine replicas, were the four that are associated with highly venomous species (Fig. 7.18).

Janzen (1980) proposed not only that other snakes may mimic coral snakes, but also that some other reptiles and even large and gaudy moth caterpillars might too. His two particular examples from Santa Rosa dry forest in Costa Rica were the turtle *Rhinoclemmys pulcherimma* and the large caterpillars of the hawkmoth *Pseudosphinx tetrio* (Sphingidae) (see Fig. 4.26f). The turtle, especially in young

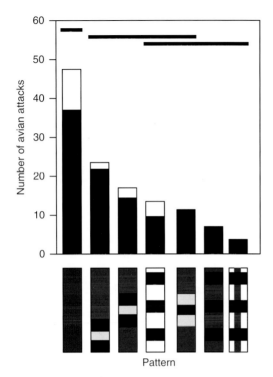

Fig. 7.18 Results of experiment using coloured plasticine snake dummies laid out in the field to monitor bird attacks, showing that the four least-attacked patterns corresponded to the most venomous snake models. Black bars indicate numbers of attacks at the middle part of the dummy, white zones indicate attacks near one of the dummy's ends; horizontal lines indicate sets of ranked categories whose attack rates differ from the rest. (Source: Adapted from Brodie 1993. Reproduced with permission from John Wiley & Sons.)

individuals has the ventral side of the scutes alternately striped with "intense red, yellow and black" (Fig. 7.19b,c) and the dorsal carapace is concentrically marked with a black–red–black–yellow–black pattern, though far less vividly than around the margin (Fig. 7.19a).

In tropical Asia, several species of krait, highly venomous elapid snakes of the genus *Bungarus*, are conspicuously black and white or black and yellow banded[15] (see Fig. 4.3e) and are mimicked by various other taxa (see Fig. 4.3f). For example, in Sri Lanka, the local krait is mimicked by the non-venomous colubrid *Lycodon* carinatus,[16] and in Malaysia, the southern Titiwangsa bent-toed gecko, *Cyrtodactylus australotitiwangsaensis*, has a very similar pat-

tern. It is arguable that this mimicry complex also includes various other animals since that pattern is widespread in the region; it is exhibited by various caterpillars (see Fig. 4.3b,d) and even terrestrial flatworms such as *Bipalium* spp. (see Fig. 4.3a) and even millipedes. In the Philippines, a differently patterned krait (also called a coral snake), *Hemibungarus calligaster* (Elapidae), is apparently the model for caterpillars of geometrid moths of the genus *Bracca* (R.M. Brown 2006).

Goodman & Goodman (1975) noticed that kingsnakes actively advertise their presence to birds, and nesting birds will often attack them. If the snake is spotted by the birds a good distance from their nests, their attacks are rather muted, but as the snake approaches the nest, the attacks become more ferocious, and therefore the birds, by changing their behaviour, are quite probably signalling the whereabouts of their nests which might then fall prey to the snake.

Davis Rabosky et al.'s (2016) study of the New World coral snake mimicry complex interestingly revealed unexpectedly high rates of evolutionary shifts between cryptic and mimetic colouration in the mimics that may indicate that the evolutionary dynamics in this system differ from that apparent among insect Batesian mimics, in which almost no reversals from mimetic to cryptic states are known.

Other snakes, zig-zag markings and head shape

Coral snakes are not the only snakes that could fall into the class of Emsleyan/Mertensian models. Werner (1985) discussed similar resemblance between the colubrid snakes *Spalerosophis* and *Pythonodipsas* and vipers, though suggestions that non-venomous snakes might be mimicking viperids are not recent and was first put forward as an idea by Sternfeld (1910) and Ditmars (1937). Werner (1983) suggests that bites by some Eurasian and North African vipers, such as the saw-scaled viper, *Echis carinatus* (Viperidae) (Fig. 7.20a), are so highly venomous that they would undoubtedly kill most predators with a bite. Thus, not only is *E. carinatus* a model for non-venomous mimics such as the common egg-eater, *Dasypeltis scabra* (Colubridae), but also several less venomous sympatric species are potentially Emsleyan mimics. Werner suggests that *Telescopus dhara* and *E. carinatus*, and *T. fallax* and *Vipera palaestinae* might be two such mimic–model pairs, in each case the *Telescopus* species being only mildly venomous.

Snakes, such as many vipers, are undoubtedly cryptic but their markings can also be quite distinctive. Valkonen et al. (2011a) found no evidence that European viper markings were disruptive and, significantly, their zig-zag pattern does not meet the edge of the snake when viewed from above so their body outline is not broken by them. In the case of the

---

15. Some other krait species are quite differently marked.

16. Often referred to in the literature under the generic name *Cercaspis*.

Fig. 7.19 Three views of the tortoise *Rhinoclemmys pulcherrima* from Area de Conservación Guanacaste, Costa Rica, showing the red, yellow and black banding laterally and around the margin ventrally, strongly reminiscent of coral snake warning pattern. (Source: Professor Dan Janzen. Reproduced with permission from Professor Janzen.)

mimicry complex involving desert-dwelling species, there is often quite close background colour matching, so the mimicry includes behavioural components (see section Aide mémoire *mimicry*). They seem likely to have been

constrained not to evolve classic aposematic bright warning colours and patterns, perhaps because they need to remain cryptic to avoid specialist predators and/or to catch their prey. Wüster et al. (2004) tested whether, despite not being

Fig. 7.20 Eurasian snakes involved in mimetic relationships, both showing their heads pulled back ready, or as if ready, to strike. (a) Highly venomous, front-fanged, *Echis carinatus* (Viperidae); (b) catsnake, *Telescopus fallax* (Colubridae), which, although moderately venomous, is rear-fanged and therefore unable to envenomate aggressively with ease. (Source: Omid Mozaffari. Reproduced with permission from Omid Mozaffari.)

classically aposematic, they might be aposematic to relevant predators by placing plasticine snake models out in the field, some plain-coloured (grey or terracotta) and others painted with a dark zig-zag pattern such as in the adder (*Vipera berus*). Their results (Fig. 7.21) clearly show that wild birds, such as buzzards, *Buteo buteo*, that feed upon snakes avoid the zig-zag patterned replicas, so this marking appears to act as an aposematic signal. The repetitive nature of the pattern and general failure to disrupt the snake's outline are also consistent with aposematic signals (Stevens 2007).

Several other possible mimicry examples involving snakes, apart from those suggested as cases of Emsleyan (=Mertensian) mimicry are likely to be purely Batesian, for example, the resemblance between gopher snakes and rattlesnakes in North America (Kardong 1980, Sweet 1985). Purely Batesian mimicry of the zig-zag pattern of the European adder, *Vipera berus* (Viperidae) (Fig. 7.22b), by the completely non-venomous and endangered smooth snake, *Coronella austriaca* (Colubridae) (Fig. 7.22a), has no doubt

led to high mortality of the latter at the hands of humans who have failed to recognise it for what it is (Valkonen & Mappes 2014). R.H. Smith (1974) also suggests that the colour pattern of young, as well as one adult female morph, of the harmless slow worm, *Anguis fragilis*, a Eurasian legless lizard are also Batesian adder mimics.

Valkonen et al. (2011b) used model snakes and free-ranging raptor predators in Spain to test whether triangular heads resembling those of vipers offer protection against predation. No significant difference in attacks was found between wide- and narrow-headed models bearing typical viperid zig-zag markings, but with plain-coloured models, the wide-headed ones experienced significantly fewer attacks ($\chi^2 = 5.04$, d.f. = 1, $p = 0.025$), showing that head shape can be important (Fig. 7.23). It would be interesting to repeat the experiment in locations where there is a greater diversity of viper species and their mimics.

Mimicry of invertebrates by terrestrial vertebrates

A South African lizard called *Eremias lugubris* is of interest because, while the adult is cryptically coloured, their juveniles have a very different appearance and gait. The anterior of their body is black with prominent white spots and they are very active foragers that move with their backs arched, mimicking the well-defended "oogpister" or tyrant ground beetles, *Anthia* spp. (Carabidae) (Huheey & Pianka 1977). The geckos *Chondrodactylus angulifer* and *Coleonyx variegatus* raise their tails over their backs in a manner resembling the way that scorpions do (Fitzsimons & Brain 1958, W.S. Parker & Pianka 1974, Pough 1988b), but some experimental evidence suggests that the main reason is that the tail acts as an attack deflector (Congdon et al. 1974). Scorpions generally do not make good models because they are predominantly nocturnal and sombre-coloured, but some South American species, from the Brazilian Cerrado, are rather aposematic and circumstantial evidence suggests that they may be the models for another gecko species (Brandão & Motta 2005).

Another arthropod-mimicking vertebrate system that has recently been discovered and made the news headlines involves the amazing nestlings of the South American bird *Laniocera hypopyrra* (Tityridae), also known as the cinnamon mourner (Londoño et al. 2015). The nestlings, which remain so for a relatively long duration, are covered with bright orange modified feathers with occasional black-tipped ones, and closely visually resemble large aposematic hairy caterpillars. The resemblance is further effected behaviourally because when disturbed, the chicks slowly sway their heads from side to side in a caterpillar-like fashion. Furthermore, unlike most birds, the nestlings do

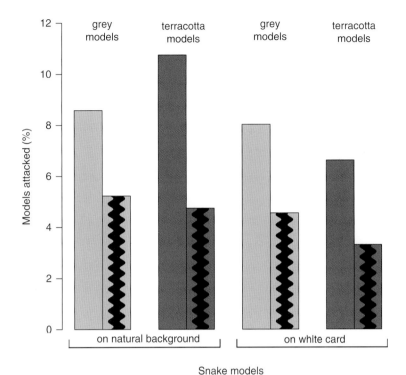

Fig. 7.21 Bird attacks on snake-shaped models made out of plasticine with and without dorsal 'zig-zag' pattern painted on them, placed out at 12 sites across the UK and one site in Finland where adders (*Vipera berus*) occur naturally, showing that despite not being brightly coloured, the zig-zag marking acts as an aposematic signal to wild predator birds. (Source: Adapted from Wüster et al. 2004. Reproduced with permission from The Royal Society.)

Fig. 7.22 Batesian mimicry between (a) European smooth snake, *Coronella austriaca*, and (b) the European adder (viper), *Vipera berus*. (Source: a, Frank Vassen 2010. Reproduced under the terms of the Creative Commons Attribution Licence CC BY 2.0, via Flickr; b, Benny Trapp. Reproduced under the terms of the Creative Commons Attribution Licence CC BY 3.0, via Wikimedia Commons.)

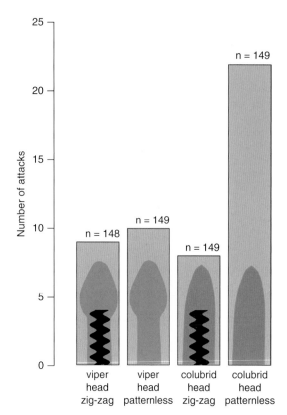

Fig. 7.23 Model snakes with triangular viper-like heads set out in the wild received significantly fewer attacks by raptors than models with simple heads, which are more typical of less venomous species. Triangular heads afforded the same amount of protection as did viper-like zig-zag markings and there was an additive effect of having both types of indicator of potential danger. (Drawn. Adapted from Valkonen et al. 2011. Reproduced under the terms of the Creative Commons Attribution Licence CC BY, via PLoS ONE.)

not beg immediately when a parent bird returns to the nest but remain caterpillar-like until they are sure that it really is a parent and not a predator.

Although data are, at present, rather sparse, McCallum et al. (2008) describe a probable case of mimicry between a North American newt called *Notophthalmus viridescens* and the rather large and strikingly similarly coloured Ozark Highlands leech, *Macrobdella diplotertia*, though it is not yet certain what their natural predators would be and which species might be the model.

These examples appear to be the only known cases of terrestrial vertebrates mimicking invertebrates. Although perhaps not the sole reason for this, size discrepancy must be a major factor (Wallace 1870); there is considerable size overlap, however. Pough (1988b) asks whether there are

different rules for vertebrates than invertebrates when it comes to mimicry. Most zoologists will be far more aware of the plethora of examples involving arthropods, and especially insects, but it is also important to note the huge disparity in the number of known species in the groups, and certainly the vast majority of arthropods are cryptic rather than aposematic. Another important factor that Pough (1988b) notes is that humans are highly visual animals, whereas many terrestrial vertebrates, especially mammals, rely more heavily on sound and smell to interpret their environment. Humans may, therefore, simply be missing other examples.

INACCURATE (SATYRIC) MIMICS

While it is certain that most cases of both Batesian mimic and Müllerian resemblance originate as only vaguely similar forms (see also Chapter 4, section *Initial evolution of aposematism*), there are a number of cases where this inaccurate mimicry is probably long-standing and not for want of appropriate genetic variation upon which selection could act if it were strong enough. When models are particularly unpalatable, predators will often avoid prey that are only vaguely similar for a long time (J.V.Z Brower 1958c, 1963, L.P. Brower et al. 1971, Platt et al. 1971, Shideler 1973). Therefore, poor mimics, in the absence of high levels of predation by good discriminators, may gain significant advantage.

Ihalainen et al. (2012) found that selection against inaccurate mimics was affected by the complexity of the prey community in which a predator learned. Great tits, *Parus major*, that had been trained to forage in a community with only a few prey types successfully discriminated against inaccurate mimics in a subsequent generalisation test, but birds that had prior experience feeding in a complex system did not. The rationale of their results was that complex prey communities lead to predators generalising more and therefore, in the case of incipient Batesian mimicry, might avoid prey that only vaguely resemble a model.

D.C. Marsh et al. (1977) describe the case of the nymphalid butterfly *Hypolimnas bolina*, which is widely distributed east to west between Easter Island and Madagascar and north to south between Japan and Australasia. While in Madagascar and other western parts of its range the female is monomorphic and a good mimic of various unpalatable danaine *Euploea* species, further eastwards, the females are often polymorphic, but without any strong resemblance to any distasteful models. The answer in this case may lie in what its larvae eat in different places, with some populations that feed, for example, on *Ipomoea batatas* (Convolvulaceae) sequestering cardiac-active compounds,

Fig. 7.24 Anciently recognised mimicry of (a) the honey bee, *Apis mellifera* (Hymenoptera: Apidae), by (b) the dronefly, *Eristalis tenax* (Diptera: Syrphidae). (Source: a, John Severns, Severnjc, via Wikimedia Commons.; b, Airwolfhound 2011. Reproduced under the terms of the Creative Commons Attribution Share-Alike Licence CC BY-SA 2.5, via Flickr.)

and others not. Thus in some localities it may be as unpalatable to predators as its normal Batesian model. However, it is unclear why it is not part of any Müllerian complex in those places where it is unpalatable – maybe insufficient evolutionary time or opportunity.

MIMICRY OF MODEL BEHAVIOUR

For Batesian mimicry to be most effective, all possible ways that a predator can learn to differentiate model and mimic at a distance need to be minimised, and the way animals move is something that predators can also potentially learn. Droneflies, *Eristalis* spp. (Syrphidae), look a lot like honey bees, *Apis mellifera* (Fig. 7.24) (Heal 1979, 1982, Vane-Wright 2010). Adult droneflies spend much of their time foraging on flowers and are often mistaken for honey bees. Indeed, in ancient Egypt, presumably because the dronefly larvae (rat-tailed maggots) live in highly polluted water, e.g. around carcasses, it was believed that honey bees spontaneously generated from dead animals, particularly oxen.[17] That *Eristalis* are effective Batesian mimics of honey bees was shown by J.V.Z. Brower & Brower (1962) using common toads, *Bufo terrestris*, as predators in one of the early scientific tests of protective mimicry. In addition to their rather

accurate physical resemblance to honey bees, the amount of time droneflies spend both visiting individual flowers and flying between them is more similar to the behaviour of honey bees than to that of other species of hoverfly or wasp (Golding & Edmunds 2000, Golding et al. 2001). However, in further work on other aculeate-mimicking syrphids, Golding et al. (2005b) found virtually no similarity between the hoverfly's flight behaviours and that of their presumed *Vespula* models.

Behavioural mimicry of flight behaviour is, nevertheless, prevalent in many insects. Brightly coloured tropical parasitic braconid wasps tend to have a slow, slightly bobbing flight as they search over vegetation, presumably using olfaction to locate possible host sites, and this is mimicked very well by various small day-flying moths, hemipterans, beetles and some flies. I have been repeatedly fooled into thinking I had captured a wasp only to discover otherwise (Quicke 1986a, unpublished personal observation). Silberglied & Eisner (1969) wrote about how wasp-mimicking buprestid beetles of the genus *Acmaeodera* have a highly modified morphology and form of flight that makes their resemblance to aculeate wasps very convincing. Nearly all beetles fly with their elytra[18] held out to the side, while the flight power comes from the membranous hind wings behind them. However, in *Acmaeodera*, the black and yellow

---

17. This became widely known as the *Bugonia* hypothesis and is even mentioned in the Bible (Judges xiv: 8) (E.L. Atkins 1948).

18. Wing cases, i.e. modified tough fore wings in Coleoptera and Dermaptera.

elytra are fused and in flight they are raised up, allowing the hind wings to spread out; they appear only to have membranous wings and their elytra mimic a wasp's metasoma.[19]

A nice example of the importance of behaviour is provided by McCosker (1977), who describes the defensive posture adopted by the coral reef-dwelling comet fish, *Calloplesiops altivelis* (Plesiopidae), which is popular with aquarists. When threatened, the fish darts to a nearby piece of coral and assumes a special intimidation posture, in which it reverses direction and expands its normally relaxed caudal, anal and dorsal fins, making its eyespot, which is located near the posterior base of the dorsal fin, fully visible. In this posture, its rear end and eyespot now closely resemble the front end of the potentially dangerous turkey moray eel, *Gymnothorax meleagris* (Muraenidae). McCosker (1977) discounted the possibility that the comet fish might be a Müllerian mimic by tasting its skin to eliminate the possibility that it contained toxins or bitter components, and showed that keeping the fish in close proximity to other species caused the latter no ill-effects.

Aide mémoire mimicry

Rothschild (1984) proposed the term *aide mémoire* mimicry to describe certain behavioural aspects of mimicry that serve to help remind a predator of a past unpleasant encounter with a particular sort of prey. As examples of such behaviours, she cited the buzzing sounds and upside-down posture with waving legs of sexton beetles, *Nicrophorus* spp. (Silphidae), which are believed to mimic bumblebees; similar buzzing by the giant fly *Mydas heros* (Mydidae), a mimic of large *Pepsis* (Pompilidae) wasps (Hanson 1976, W.E. Cooper 1981, J.W. Nelson 1986); and the abdominal probing of male wasps (that themselves do not possess stings) which might enhance their mimicry of female wasps which can, of course, deliver a sting (Quicke 1986b). Stridulatory buzzing sounds have been shown to help protect venomous velvet ants (Mutillidae) from wild-caught *Peromyscus floridanus* mice and possibly do so either by the *aide mémoire* principle or through an auditory startle effect (W.R. Masters 1979; see also G.A.K. Marshall 1902a, 1902b, Dunning 1968, Claridge 1974). The alarm buzzing of bumblebees is significantly different from their flight sounds, as is that of the bee-mimic hoverfly *Cheilosia*

illustrata, but whereas wild birds are deterred from attacking artificial prey that are associated with the bumblebee sound, the hoverfly alarm buzz had no deterrent effect, despite sounding very similar to human hearing (Moore & Hassall 2016). The aspect of the sounds responsible is not yet known, but the authors did provide evidence that the bird avoidance response is learned rather than innate.

The literature is replete with other examples that fall into Rothschild's (1984) *aide mémoire* mimicry concept. For example, the non-venomous Asiatic green keelback snake, *Macropisthodon plumbicolor*, and the mildly venomous false cobra, *Rhagerhis moilensis*[20] (both Colubridae), both flatten their necks, reminiscent of a cobra, when threatened, and the first of these also emits a noxious smell from nuchal glands in its neck (Mertens 1960; see also Mori et al. 2012). The smell may thus serve to emphasise it as an undesirable prey. In this species the flattening is achieved by having mobile neck vertebrae. The head-flattening mimicry of cobras is extreme, but other snakes may do a less extreme version called head triangulation (see earlier), which makes them appear more like poisonous, front-fanged vipers (Viperidae); this is true mimicry (Valkonen et al. 2011b).

Werner (1983) described viperid mimicry in three Israeli rear-fanged colubrid snakes (*Telescopus fallax* and *dhara*) which, when threatened, coil, and pull their head and neck back in a posture that makes them appear to be about to strike (see Fig. 7.20b) though their fang location makes an envenomating aggressive strike difficult. They can also triangulate their heads, thus enhancing their resemblance to dangerous vipers.

It is also probably pertinent here that many well-defended but otherwise largely cryptic species, especially amphibians and reptiles, normally have their bright, aposematic patterns hidden but display them when threatened, for example, the throats or whole undersides of many frogs and newts are brightly coloured but these are not normally visible (e.g. C.R. Williams et al. 2000).

A strange case of Batesian mimicry was described by A. González et al. (2004) which might fit into this category. The harvestman, *Parampheres ronae* (Arachnida: Opiliones), has a pair of dorsal orange spots on its carapace positioned where several other harvestman species have defensive glands that secrete quinone-rich orange-coloured secretions. *P. ronae* also has defensive glands but they release a translucent, colourless, quinone-free secretion, which would not be a good visual advertisement of its defensive potential, and suggests therefore that the orange spots are mimicking the released orange defensive secretions of other species.

---

19. In Hymenoptera with a wasp waist, the narrowing is not between thorax and abdomen but between the 1st and 2nd abdominal segments, therefore the part behind the waist is referred to as the metasoma and not the abdomen. Some authors use the term gaster instead.

---

20. In most literature referred to under the combination *Malpolon moilensis*.

BATESIAN–POULTONIAN (PREDATOR) MIMICRY

Pasteur (1982) proposed this name for S1R/S2 systems in which the model is agreeable to the dupes, and cited the example, as previously reported by both Müller (1878) and Wallace (1891), of South American bush crickets of the genus *Aganacris* (as *Scaphura*) which mimics a sphecid wasp that provisions its nests with grasshoppers to feed its young. However, it seems more likely to me that both might be Batesian mimics of stinging spider-hunting wasps (Pompilidae), since sphecids are generally not that aggressive; *Aganacris* are also capable of delivering a strong bite.

However, a number of workers had discovered several apparently mimetic systems between tephritid flies and jumping spiders in which the patterns on the fly wings resemble the legs of the spiders when viewed from the front (Eisner 1985, Mather & Roitberg 1987, E. Greene et al. 1987, Whitman et al. 1988). Zolnerowich (1992) described

another example involving the nymph of an *Amycle* plant bug (Fulgoridae) from Mexico, whose modified posterior closely resembles the face of a jumping spider (Salticidae). In the example described by E. Greene et al. (1987), the tephritid fly additionally mimics the spider's territorial display. More recently, attention has been drawn to a number of other insect species belonging to other orders, notably Lepidoptera (Solis et al. 2005, Rota & Wagner 2006) but also Hemiptera (Floren & Otto 2001). Most of the moth examples involve metalmark moths, *Brentia* spp. (Choreutidae) (Fig. 7.25a), but a very recent example is *Siamusotima aranea* (Fig. 7.25c,d), which belongs to the Crambidae. The latter is also known as the *Lygodium* spider moth because it feeds on invasive climbing ferns (*Lygodium* spp.) in various parts of the world. Boppré et al. (2016) further suggest that much or all aculeate wasp mimicry by hoverflies (Syrphidae) might fall into this category rather than it being a combination of Batesian and Müllerian mimicry.

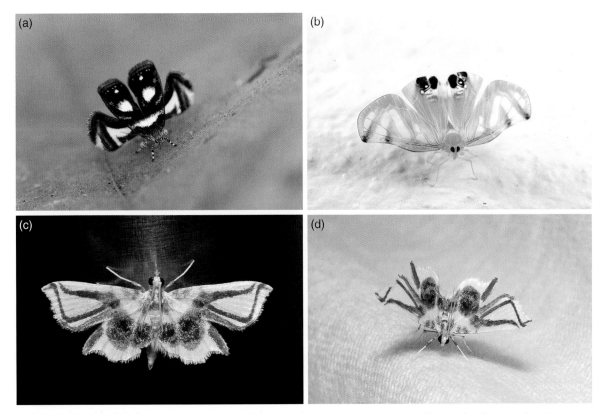

Fig. 7.25 Batesian–Poultonian spider mimicry by small insects directed towards spiders, especially salticid spiders, to reduce the probability that they will attack. (a) Metalmark moth, possibly *Brentia* sp. (Choreutidae); (b) unidentified hemipteran from China; (c,d) *Siamusotima aranea* (Crambidae) as seen from above and in head-on display mode, respectively. (Source: a,b, The Budak, M. Ng. Reproduced with permission from M. Ng; c,d, Itchydogimages. Reproduced with permission from John Horstman.)

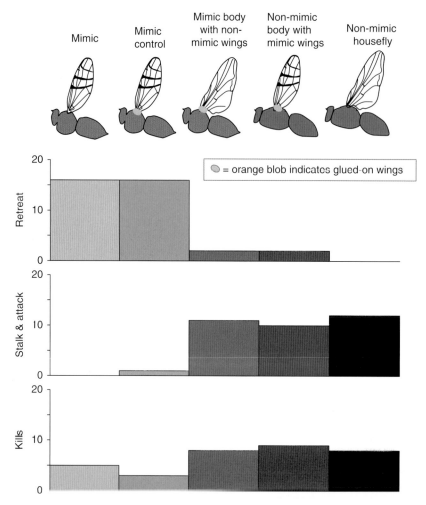

Fig. 7.26 Demonstration of the protective role of the combination of leg-like markings on the wings of the tephritid fly *Zonosemata vittigera* and the fly's behaviour, as defence from the salticid spider *Phidippus apacheanus*. The treatments were, from left to right, as indicated in the drawings, intact *Zonosemata*, *Zonosemata* with wings amputated and glued back on, *Zonosemata* with housefly wings glued on, housefly with *Zonosemata* wings glued on, and intact housefly. In all three graphs, the two left-hand treatments differed significantly ($p < 0.01$) from the three on the right. (Source: Adapted from Greene et al. 1987. Reproduced with permission from AAAS.)

Several of the above studies proposed that the function of this spider mimicry was to deceive spiders, particularly members of the actively hunting jumping spider family Salticidae, which would generally treat other salticids as unsuitable (dangerous) prey – they might get eaten themselves. Two experimental studies have been carried out, both of which confirm the protective nature of the mimicry. The first, by E. Greene et al. (1987), dealt with the tephritid fly's wing pattern, and was demonstrated experimentally by removing and gluing on wings of the tephritid (*Zonosemata vittigera*) and houseflies (*Musca domestica*) in various combinations (Fig. 7.26). Tephritids with tephritid wings, whether intact or glued back on as a control, caused salticid spiders to retreat, attack less and kill fewer flies than they did tephritids with housefly wings glued on or houseflies with tephritid or housefly wings. The tephritid wing pattern alone was not sufficient to afford protection, since houseflies with tephritid wings glued on fared the same as intact houseflies, but behaviour alone was not enough either, since tephritids with housefly wings glued on suffered as badly as houseflies. In the second study, Rota & Wagner (2006) carried out controlled experiments with *Brentia* sp. (Choreutidae), using the spider model *Phiale formosa* (Salticidae), and showed that the spider mistook the mimic for another spider and instead of attacking it, performed their territorial display. These appear to be antergic protective mimicries (see Chapter 1, section *Vane-Wright's system* and Table 1.4) because the spiders avoid attacking them, probably interpreting the moths as other spiders, and therefore miss out on a meal.

This sort of mimicry appears to be nothing new: Shcherbakov (2007) describes some new fossil fulgoroid plant bugs from the Early Cretaceous that are so well preserved their quite distinct wing markings can be seen. They are exceeding similar to those of the *Amycle* bug described by Zolnerovich (1992).

MIMICRY WITHIN PREDATOR–PREY AND HOST–PARASITE SYSTEMS

Some authors have commented on the abundance of examples of mimicry between predators and their prey. However, a search of the literature does not leave one with the impression that there are proportionately more examples in this category than would be expected by chance, indeed quite the opposite. Quicke et al. (1992) proposed a number of possible reasons why mimicry between organisms with connected life cycles might be expected to be common:

- model and mimic will necessarily occupy the same territory;
- in many cases the species will be adult at around the same time of year;
- in many cases with solitary parasitoids, both species are likely to be of similar adult size.

While there may or may not be any particular excess of mimicry between predators and their prey etc., Quicke et al. (1992) did show that among the probably Batesian examples that do exist, there is a considerable asymmetry in terms of whether it is the predator or the prey that is the likely model, with a marked excess of the parasitoid or predator being the mimic, as with various large robberflies and their aculeate prey (Poulton 1906, J.V.Z. Brower 1960, Linsley 1960, Tsacas et al. 1970). In the case of the *Hyperechia* fly and *Xylocopa* model described by Tsacas et al. (1970), not only is the adult fly predatory on adult models, but the fly's larvae are predators within the *Xylocopa* nest. In many other cases, the relationship appears likely to be purely Batesian or Müllerian (see also A.C. Harris 1978), although the similarity of some conopid flies to their probable aculeate hosts might enable them to approach places where adult hosts are present more easily (see Vane-Wright 2010).

One possible explanation for this bias towards parasitoid or predator being the mimic could be that the populations of a parasitoid and its host (or predator and specific prey) are connected through standard population dynamics models, and may show stability, limit cycles or chaotic behaviour, and that this connection between the species, relative population sizes might or might not favour the evolution of mimicry between them, assuming one could be an appropriate model. To investigate this further, Nicholson–Bailey parasitoid host discrete generation population dynamics models (Nicholson & Bailey 1935) were modified to include a Batesian mimicry feedback with adult mortality defined such that it increased as the ratio of edible mimics to mimics + models increased, according to Huheey's (1964) equations. As the strength of the mimetic protection increased (i.e. the more unpalatable the model) this feedback was found to have an increasingly large effect on the population dynamics of the system, but differed according to whether the host or its parasitoid was the mimic. Furthermore, longterm persistence of the modelled host–parasitoid systems differed, with a smaller region of parameter space permitting persistence when the host is a Batesian mimic of its parasitoid – this is likely due to the palatable host having a higher general abundance, leading to higher levels of predation both on it and on its less-abundant parasitoid model. This in turn leads to the parasitoid going extinct in many cases, especially when mimetic protection strength is strong. Across a large range of parameter values it was found that the host as mimic it had a larger mean population than if the same amount of overall predation was exerted as a mean value (Fig. 7.27a), and the parasitoid had far lower mean population size (Fig. 7.27b) and, indeed, frequently became extinct. In the opposite mimicry configuration (host as model) the effect had a far lower magnitude (Fig. 7.27b,c). This asymmetry results from the fact that the host's population is constrained by its carrying capacity term, whereas the parasitoid's population is only constrained by that of the host. The results are also consistent with the bias in observed cases, i.e. that parasitoid or predator is almost always the mimic.

BLUFF AND APPEARING LARGER THAN YOU ARE

Bluff is a mimicry that deceives in terms of the level of defence an individual has. It may be an S1S2/R or a conjunct system in Wickler's (1965) and Vane-Wright's (1976) notation. Small animals may adopt postures or changes in body size or silhouette so as to appear larger and more dangerous than they are. Pufferfish and swellfish (Tetraodontidae) and porcupinefish (Diodontidae) are renowned for their ability to swell when threatened; as well

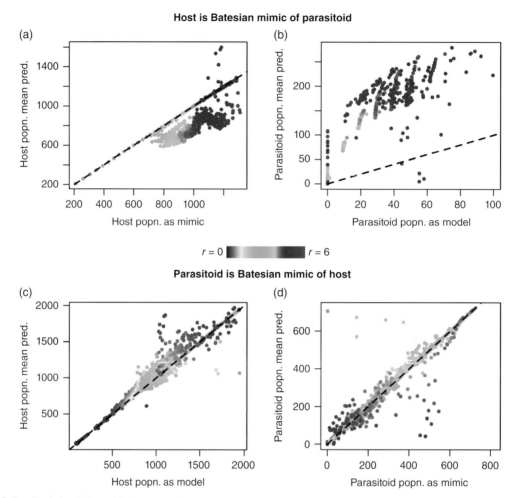

Fig. 7.27 Results of simulations of host–parasitoid population dynamics in which the two species are also Batesian mimics of one another. Each point is the mean population size of a 5000 generation simulation, each run with a different combination of host reproductive rate (r, shown) and parasitoid attack rate. The x-axis shows values for when they were mimics of one another (in which case protection from mimicry varied from generation to generation in accordance with relative model mimic frequencies) and the y-axis, corresponding simulations in which the mean predation level was applied to each generation. (a,b) With parasitoid as the model; (c,d) with host as the model.

as making them appear larger, which might be a deterrence, this can have direct survival value by making them hard to handle or swallow. If a predator finally has to reject them, the predator may learn to avoid them as unprofitable. Similarly puff adders (*Bitis* spp.) can expand their bodies considerably by filling their lungs as well as an avascular air sac. Together with their loud hisses on exhalation, this constitutes an intimidatory behaviour. Their bites are very quick and dangerous and I wonder whether Gans' (1964) suggestion of observational (empathic) learning might have played a role (see section *The coral snake problem – Emsleyan (or Mertensian) mimicry*). Other well-known

cases of intimidatory display include the neck-ruff of the Australian and New Guinean frilled or frill-necked lizard, *Chlamydosaurus kingii*, which is expanded in a direct confrontation with a predator, and hood-spreading in cobras.

In the above cases the response is largely or entirely aimed at potential predators, whereas in various mammals, raising of the hackles by cats, dogs and wolves is mostly involved in intraspecific social interactions (see Chapter 13) but it could also be protective here as well through deterring a potential aggressor.

The founder of the Scout Movement (www.scouts.org. uk), Lord Baden-Powell, was a master of bluff. Famously at the battle of Mafeking during the Second Anglo-Boer war (1899–1900) in South Africa, where the British forces were severely out-gunned, he was put in charge and started deploying barbed wire as protection. However, when his supply of wire ran out he realised that the wire itself was not visible from a distance, but that the enemy interpreted it as being there because they could see soldiers crawling under it. He therefore instructed his men to continue erecting the wooden posts and pretend to be stringing wire between them and to crawl underneath the imaginary wire, thus deceiving the enemy that the British troops were far better defended than they really were (Latimer 2001).

COLLECTIVE MIMICRY INCLUDING AN AGGRESSIVE MIMICRY

Several cases of collective mimicry have been suggested, with various levels of credibility for insects and fish. One of the best known is that of flatid bugs (China 1929, Wickler 1968), which, when resting on plant stems, make it appear to be a brightly coloured flower spike. Wickler's (1968, his Fig. 9 therein) particular example involved *Ityraea gregoryi*, whose aggregations resemble a flower spike or inflorescence of a plant of the family Vitiaceae, though it could be argued that this is not collective mimicry at all since each single flatid could mimic a flower. These assemblages are difficult to observe because the bugs jump for safety at the slightest hint of danger. He also cites a rather beautiful example that I will present *verbatim*.

> This was pointed out to me by Professor Koenig of Vienna. He once collected sand from the Mediterranean coast in the vicinity of Portofino in Northern Italy for the aquarium at his biological station. Later, he discovered a beautiful marine anemone sitting in the sand in one aquarium and had no idea where it came from. The next day there was a further surprise – there were two anemones each half the size of the first. On the following day there was again one large anemone. ... It then emerged that it was not an anemone at all, but consisted of numerous worms, each resembling the common Tubifex, which crept through the sand in a chain like that formed by some caterpillars.[21]

Most importantly, Professor Koenig noted that small fish which would normally eat solitary worms avoided these sea anemone-like assemblies. Richard Dawkins (1999) points out that in such cases you should not refer to selection favouring groups of worms but rather favouring-individuals that join groups of ring-formers. Another example,

described by Fricke (1970), involved a group of *Siphamia* fish (Apogonidae) that closely resemble a sea urchin; many *Siphamia* spp. also live commensally with sea urchins.

Similar collective mimicry of sea anemones was also proposed by Knipper (1953, 1955) for juveniles of the marine catfish *Plotosus anguillaris*, but this opinion was not shared by D. Magnus (1967). It seems likely that aggregations of this fish resemble a larger organism of some sort, even if it is less defined. Recently, Hafernik & Saul-Gershenz (2000) have demonstrated that the first instar larvae (triungulins) of the blister beetle, *Meloe franciscanus*, cooperate to form an assemblage near the end of a twig which roughly resembles a female bee and makes an adequate visual target for sexual attention of a male bee if it comes sufficiently close. It has subsequently been shown that sex pheromone mimicry is also involved, as was originally suspected (Saul-Gershenz & Millar 2006). Thus, males of the anthophorine bee *Habropoda pallida* attempted to copulate with these blobs of beetle larvae, whereupon most of the larvae boarded the bee. This venereal transmission is just the intermediate stage because when the male finds a real mate, the *Meloe* larvae transfer to her, and subsequently to her nest, where they parasitise her larvae. Similar sexual deception with very similar larval aggregations and use of bee sex pheromones was found by Vereecken & Mahé (2007) in another blister beetle, *Stenoria analis*, which is a parasitoid of the European bee *Colletes hederae* (Fig. 7.28).

Pasteur (1982) reproduces a drawing of berry butterfly caterpillars, though it is actually a group of *Asota* (as *Hypsa*) *monycha* (Noctuidae) moth caterpillars from Singapore, which are well-known to form aligned clusters at the tops of stems that resemble an opened fruit. The caterpillars are crimson with white bands and in their aggregations the white bands are aligned. It is supposed that this deception protects them from predation by insectivorous birds but it would seem probable that it might attract seed-dispersers or frugivores which might also pose a threat. As the caterpillars feed upon poisonous plants and probably sequester toxins from them, I suspect the collective display is mostly likely aposematic, and the resemblance to a fruit noted by one observer might be a bit fanciful. Since there is virtually no information on the larval biologies of these moths, other than some rearing records, it is hard to determine what selective forces might be behind this alignment behaviour.

JAMMING

A particularly difficult case of adaptive resemblance to classify is that of the clicks emitted by some tiger moths (Erebidae: Arctiinae) in response to the echolocating ultrasonic pulses emitted by predatory bats as well as in other

---

21. Alluding to the behaviour of caterpillars of the processionary moth, *Thaumetopoea processionea* (Lepidoptera: Thaumatopoeidae).

Fig. 7.28 An aggressive mimicry example of collective mimicry by aggregation of early (triungulin stage) larvae of the blister beetle, *Stenoria analis* (Meloidae), forming a blob approximately the size and colour of the host bee, *Colletes hederae* (Apidae). (a) Blob of triungulins; (b) two male *C. hederae* bees attracted to a similar blob of larvae by their mimetic bee sex pheromone, one bee trying to copulate with the blob. As a result, the triungulins will climb onto the bee, who will transport them to a real female and thus they will gain entry to a new *Colletes* nest. (Source: N.J. Vereecken. Reproduced with permission from N.J. Vereecken.)

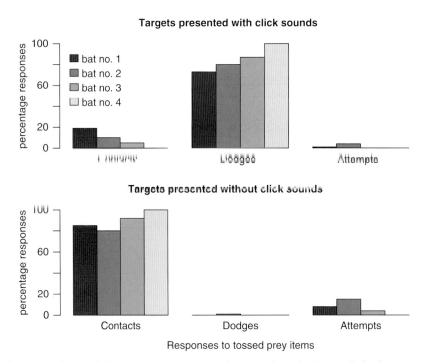

Fig. 7.29 Effect of playing arctiid moth clicks on prey capture attempts by captive bats, showing marked reduction in successful contacts with the prey (mealworms tossed into the bats' flight paths) which was thought to indicate a jamming role for the insect noises, but may be because the bats interpreted them as aposematic sounds. Sample sizes for bats 1 to 4 in upper figure 59, 48, 178 and 92, and in lower figure, 71, 68, 188 and 94, respectively. (Source: Adapted from Dunning & Roeder 1965.)

stressful situations (Blest et al. 1963). Many arctiine ere-bids also produce audible and, in some cases, ultrasonic sound when handled roughly (Blest 1964), and it seems most likely that this originated as part of an aposematic dis-play, or possibly of an *aide mémoire* type (see earlier) and then got selected for a role in bat predation prevention. Possible functions of the clicks have been debated since their discovery, and suggestions as to their roles have included protean display, startle responses and aposematic signals (Fullard 1977). However, the demonstration by Fullard et al. (1979) that the clicks bear a close resem-blance to bat echolocation clicks in their frequency–time structure and their power spectra lends strong support to the notion that these pulses serve to jam the bat's echoloca-tion system, and Fullard et al. (1994) showed that the moth's response is timed to coincide with the bat's terminal attack calls. The low power of the moth clicks relative to those of bats may be relatively unimportant because the moths tend to emit their sounds when an attacking bat has approached to within about 1–1.5 m. Furthermore, while different families of bats such as the Emballonuridae, Molossidae and Vespertilionidae may employ quite differ-ent orientation sounds during general flight, the pulses produced during the closing moments of an attack are remarkably similar and therefore arctiine pulses are likely to be effective against a wide range of potential predators. In response to the moth clicks the bats veer away rapidly from their pursuit course. This lends further support to the idea that the moth pulses directly interfere with the bats' spatial perception rather than simply signifying that the tiger moth is distasteful, in which case the bats might be expected to simply cease the chase rather than undertaking any violent change in direction.

Dunning & Roeder (1965) trained five captive bats (*Myotis lucifugus*) to catch live mealworms tossed into the air, and investigated the effects of arctiine click trains on their suc-cess. They found an extremely strong effect although only with four of the bats (Fig. 7.29). Arctiine sounds greatly increased the number of dodges and reduced the number of contacts with the tossed mealworms. However, when the bats were uninterested in the mealworms or did not detect them, they carried on as normal irrespective of moth clicks. They also showed that arctiid clicks had a similar, though far more profound, effect on prey capture behaviour than other, conspecific bat sounds presented simultaneously with prey capture attempts.

In marked contrast to Dunning & Roeder's (1965) results, Surlykke & Miller (1985) provide evidence that the arctiine clicks are nothing to do with jamming but are aposematic warning of the noxious nature of the moths themselves. Using trained pipistrelle bats, *Pipistrellus pipistrellus*, with the garden, *Arctia caja*, and ruby, *Phragmatobia fuliginosa*,

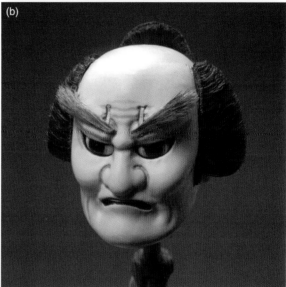

Fig. 7.30 Resemblance between (a) carapace sculpture and markings in the Japanese samurai crab, *Heikeopsis japonica*, and (b) the samurai face on a Bunraku mask. Trepidation by crab fishermen and release of more samurai-like crabs is thought to have inadvertently selected for more human-like carapaces. (Source: a, Shimonoseki Marine Science Museum. Reproduced with permission from T. Ishibashi; b, Reproduced with permission from National Bunraku Theatre, Japan.)

tiger moths respectively, they found that clicks had no effect on the bats' discriminatory ability to find palatable targets, nor did they affect the bats' behaviour in any obvious way. However, the bats readily learned to avoid the clicks after associating them with the unpleasant cervical secretion of *A. caja*, which contains choline esters.

It certainly seems highly likely that the arctiid clicks could be aposematic signals, however their similarity to bat clicks makes jamming a possibility also. Perhaps it depends on the species of bat, the species of arctiine or both.

MAN AS MODEL – THE CASE OF THE SAMURAI CRAB

One of the most bizarre cases of protective resemblance involves a species of crab that occurs in the waters around Japan. *Heikeopsis*[22] *japonica*, commonly called the samurai crab, have a bold pattern on their carapaces that shows an uncanny resemblance to the masks of samurai warriors. Figure 7.30a shows an example of the crab in question and Fig. 7.30b a Bunraku puppet theatre mask of the character Ki-Ichi. Huxley (1957) proposed that this likeness is not

merely a coincidence but rather it is the result of selection by generations of Japanese fishermen who, upon encountering crabs whose carapaces more or less resembled human faces, would throw these back into the water so as to avoid reprisals by evil spirits. He further suggested a test of this explanation would be to compare the carapace markings of crabs between sites where there are many human crab-fishers and those where there are few or none. The origin of the fear is based on historical events; in 1185 the Heike clan suffered a terrible massacre by the Miyamoto clan at the inlet of Dan-no-ura (J.W. Martin 1993), and a vast army of Samurai were killed or drowned. Obviously the more independent comparisons the better, but no one seems to have done this. With modern computer face-recognition type algorithms it ought to be relatively easy to get an objective score of facial similarity.

----

22. Called *Heikea* in some literature, but this was found to be a junior homonym of a gastropod genus.

Chapter 8

ANTI-PREDATOR MIMICRY. ATTACK DEFLECTION, SCHOOLING, ETC.

When one chances upon the nest and goes to catch them, the partridge rolls along in front of the hunter as though easy to catch, and draws him on towards herself as though he is about to catch her, until all the nestlings have scattered; after that she herself flies up and recalls them.

Aristotle (translatcd by Balme), concerning the partridge, probably written between 345 BC (arrival Lesbos) and 322 BC (his death) and is based primarily on second-hand reports

INTRODUCTION

In this chapter I consider a rather diverse set of defensive features and behaviours that are, in themselves, harmless to potential predators but which generally increase the chance of them missing their prey, doing it lesser harm if they do catch it, and sometimes giving it more time to escape. These include both intimidating ('deimatic' *sensu* M. Edmunds 1974a) displays, such as flash colouration and false eyes, devices that make a potential prey appear to be the wrong way around which might lead to easier escape or less damage being caused if the predator makes contact or, in the case of a bird's broken wing display, lead a predator away from a vulnerable nest. The cover illustration of this book shows shows an example of flash colouration. In it one can see three great orange tip butterflies, *Hebomoia glaucippe* (Pieridae), in flight, but if you look carefully there are at least 15 others drinking from the damp soil, but only showing their camouflaged under wing surfaces.

I also include within this chapter 'arithmetic' mimicries, such as schooling and flocking, where the high degree of similarity between individuals thwarts a predator's attempts to follow a given individual and make a successful strike. The mud-puddling butterflies on the cover also illustrate this (see section *Schooling, flocking and predator confusion* later).

ATTACK DEFLECTION DEVICES

Eyespots

Many insects, but particularly adult and larval butterflies and moths, as well as a number of fish, display remarkable eye-like markings, and many animals also have an aversion to eye-like markings which is clearly innate as the responses are usually particularly apparent with naïve individuals. Eyespots may function as intimidation devices or attack deflectors (see also Fig. 4.25), or may guide predators to attack from a different direction, from which a potential prey can see them more easily. If they deflect attack to a posterior part of the body, or to somewhere on an insect's wing away from the body, then they may allow the prey an easier escape route without having to spend valuable time turning away. There is still a great deal to be discovered and understood about eyespots, and it is probable that they play multiple roles and, even in deflection cases, may not all act in the same way (Kodandaramaiah 2011).

Many recent pieces of research have shown that it is not only a similarity to eyes but also general features of conspicuousness that are important (e.g. Stevens 2005, Stevens

et al. 2007, 2008a, 2008b, 2009a, 2009c; see later). However, there can be little doubt that in many instances eyespots are mimics of real eyes (perhaps especially so in caterpillars; see Figs 2.27c,d and 8.12), those on the undersides of the Neotropical owl butterfly *Caligo* (Nymphalidae) being a classic example (Stradling 1976). Eyespots, unlike some other circular markings such as egg spots/dummies in cichlid fish (see Chapter 11, section *Egg dummies on fish*), always have a dark central circle, but otherwise display a great range of variation in number, size and detail. Some are rather abstract, some far more elaborate and realistic, and some present apparently super-normal stimuli. However, we must always be open to plausible alternative explanations. Wickler (1968, p. 65) notes, for example, that some markings on fish might be bubble mimics, aiding camouflage against bubble-rich turbulent water, and the Russian naturalist Porschinsky (cited in Davis 1903, p. 135) thought that all eyespots in insects might be mimicking gland openings from which noxious liquids might be expressed (see Chapter 7, section Aide mémoire *mimicry*, and especially A. González et al. 2004). The great majority of eyespots in the Lepidoptera have a small white spot in the centre or an asymmetric crescent, called 'sparkles' which seems to make the marking more eye-like by mimicking the natural total reflection of ambient light off the cornea of a vertebrate eye (Blut et al. 2012; see later). Ones used in startle displays are often associated with larger bright-coloured areas. This variation alone might suggest that they may function in different ways in different animals or circumstances. In some cases they may be intimidatory, as when suddenly exposed when various saturniid giant silk moths are disturbed (Fig. 8.1b) and in others they may act as attack deflectors, or perhaps cause predator confusion. Some workers have even questioned whether all the markings that fall under the category of 'eyespot' are really mimicking eyes at all (Stevens & Ruxton 2014).

Before going on to discuss many particular experiments, I want to point out that there is a bit of a mismatch between one of the most commonly used experimental set-ups and what happens in nature, though that does not necessarily invalidate any of the findings. Many eyespot studies, especially from Martin Stevens' group have employed baited, variously patterned, triangular-shaped moth-like prey on tree trunks and compared the survivorship of different types. However, moths with eyespots that rest on tree trunks, such as the iconic eyed hawkmoth, *Smerinthus ocellatus* (see Howse 2013), or various giant silk moths, have the eye-like markings on their normally concealed hind wings and only display them when disturbed. Also, the vast majority of moths with eyespots are rather large and predominantly tropical, mostly far larger than the average European woodland bird would attack, though the

Fig. 8.1 A variety of flash and eyespot marking on Lepidoptera. (a) Emperor moth, *Saturnia pavonia* (Saturniidae); (b) *Lobobunaea acetes* (Saturniidae); (c) peacock butterfly, *Inachis io* (Nymphalidae); (d) peacock pansy butterfly, *Junonia almana* (Nymphalidae) from India, also showing likely bird damage to posterior margin of hind wing; (e,f) female *Gastrophora henricaria* (Geometridae), views from upper and underside showing flash colouration of hind wing dorsal surface and eyespot on forewing underside. (Source: a, Jean-Pierre Hamon 1991. Reproduced under the terms of the Creative Commons Attribution Share-Alike Licence CC BY-SA 3.0, via Wikimedia Commons; b, bayanga85 2010. Reproduced under the terms of the Creative Commons Attribution Share-Alike Licence CC BY-SA 2.0, via Flickr; c, Andreas Eichler 2014. Reproduced under the terms of the Creative Commons Attribution Share-Alike Licence CC BY-SA 4.0, via Wikimedia Commons; d, Vinayaraj 2013. Reproduced under the terms of the Creative Commons Attribution Share-Alike Licence CC BY-SA 4.0, via Wikimedia Commons.)

experiments have nearly all been carried out in temperate regions with native predators. In the UK, at least, the only tree trunk-resting moth with eyespots is the eyed hawk-moth, and this is far larger than most others, such as typical noctuids or notodontids, and therefore when the eyespots are revealed they are far more likely to resemble a real potential threat. I think these experiments show rather basic features of the predatory birds' cognition and/or basic neuronal wiring, and as such it might well be at least partially representative of those birds in the tropics that are far more likely to encounter prey with eyespots. Perhaps the many experiments of Stevens' group simply illustrate the potential for eyespottedness to modify survival rate in the absence of other factors, such as overall detection rate. A.M. Young (1979) suggests that one reason why large eyespots may be particularly prevalent among certain tropical butterflies, notably Nymphalidae, is that these insects frequently visit rotting fruit and can become quite intoxicated, reducing their general escape ability. Intimidatory or attack-deflecting devices might have particular value under such circumstances.

Experimental tests of importance of eyespot features

Numerous workers have investigated the features of eye-like markings that elicit responses by potential predators or prey, notably shape, relative position and colour, mostly with adult lepidopteran-like models and birds as the experimental subjects. Some studies have used wild birds, others wild-caught captive ones, or hand-reared naïve ones or domestic chicks. The last two of these categories are important because they allow experimenters to determine what responses are innate rather than learned. Of the wild bird studies, the great majority have been conducted in the temperate northern hemisphere, during spring or summer, and the main wild predators have been blue tits, *Cyanistes caeruleus* (Paridae). While such studies are no doubt informative, it would be useful to have more data from the tropics where eyespots are perhaps far more prevalent (Janzen et al. 2010).

Blest (1957b) started the experimental study of eyespot function by presenting images of eye-like shapes to hand-reared birds (chaffinches, yellow buntings and great tits) just as they were about to attack a mealworm prey, thus simulating the sudden flash appearance of eyespot marks on hind wings as occurs when many of the large saturniid moths he studied are suddenly disturbed (see Blest 1957a for details of the often complex defensive displays of "saturnioid" and sphingid moths). Circular patterns were more effective in enabling escape than non-circular ones, and increased numbers of concentric rings increased the effect, making the flat image more three-dimensional in appearance. Additionally, adding shading and eccentricity further

enhanced their impact, but in all cases, repeated presentation of the images led to the birds rapidly learning to ignore them. Blest (1957b) also showed that, at least in the chaffinch and great tit, the differential response to circular rather than non-circular patterns was innate, suggesting evolution to avoidance of eye-like marks that could signal potential danger.

In another interesting series of tests, R.B. Jones (1980) explored what features of eye-like markings were aversive to domestic chicks. The shapes presented varied in roundness, the presence of a distinct 'pupil', association with a 'mask' (i.e. eyebrow-like arches) and whether or not they were presented horizontally or vertically, and whether singly, in pairs or in triplets (Fig. 8.2). R.B. Jones (1980) carried out three sets of comparisons corresponding to the presentations in the three large boxes shown in Fig. 8.2 and recorded many parameters, of which I show latency to the chick's first step on the left of each 'eye' cartoon, and numbers of chicks entering the half of the arena with the stimulus, on the right of each cartoon. In the upper left set of trials there is a clear trend, with greatest latencies and fewest entries when the masks had a pair of circular concentric eye shapes. The next two sets lacked masks. In the upper right box, the only significant deterrence was provided by a pair of circular concentric eye shapes, even though they were arranged vertically rather than horizontally. In the lower set of trials, R.B. Jones (1980) found the greatest deterrence was afforded by a pair of small black spots that are perhaps pupil-like and this was far greater when they were surrounded by a shape, even though that shape was square.

European starlings, *Sturnus vulgaris*, have been shown to have an innate aversion to eye-like shapes (Inglis et al. 1983). Using wild-caught birds, Brilot et al. (2009) explored how their level of anxiety would affect their responses to eyespots or ambiguous eyespots. Anxiety was manipulated by playing one of four sounds to the birds: starling 'threat call' (control manipulation), a sparrowhawk call (i.e. predator), starling alarm call or white noise, the latter three being expected to increase anxiety regarding potential predators. However, their initial hypothesis that increased anxiety would make the birds more eyespot-averse was not supported, and they concluded therefore that starling aversion to eyespots might not be because they are interpreted as potential predators.

Stevens et al. (2007) used artificial baited prey marked with various shapes, numbers and sizes of eyespots, presented to avian predators in the field. They compared various levels of contrast, shapes (circles and triangles) and concentricity rather than being divided into white/black halves, and eyespots with dark centres or light centres. Prey with high-contrast markings survived better than low-contrast marked ones with circular patterns. Single and double eyespot-marked baits survived better

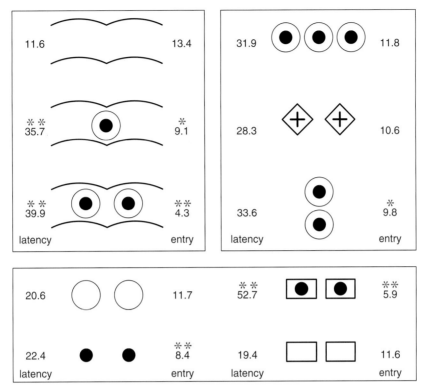

Fig. 8.2 The responses of 7-day-old male domestic fowl chicks to two-dimensional eye-like shapes. Mean values and significance levels for two selected parameters are shown on either side of each eye-like cartoon. Latency (seconds) to first step after presentation is shown on the left and numbers of chicks entering half of presentation arena with the 'eyes' is shown on the right. (Source: Adapted from R.B. Jones 1980 with permission from Elsevier.)

than triple-spotted ones. Stevens et al. (2008a) set out to determine whether survival of artificial prey depended on number (1, 2 or 3) or size of eyespot markings, or whether they were circular or square, and found that the best protection was afforded to baits with large eyespots or larger numbers of eyespots, but they found that changing the spatial arrangement of spots had no significant effect on survival. Stevens et al. (2008b) tested whether eyespots had different effects on camouflaged and conspicuous prey and found, very interestingly, that they enhanced protection of otherwise conspicuous targets but increased predation rates on cryptic ones. Stevens et al. (2009a) used the same methodology to distinguish between eye mimicry and conspicuousness theories for eyespots. In their first experiment they varied shape and position of the markings (Fig. 8.3) and found no significant difference in survival due to shape or arrangement – all markings reduced predation relative to controls (Fig. 8.4). In their second experiment they varied the eyespot colour (Fig. 8.5) and again found no effect of eyespot shape, and no advantage of yellow and black combination compared to other

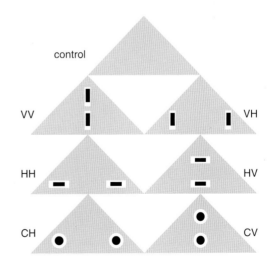

Fig. 8.3 Eye models used for Experiment 1 varying in shape and relative positions on triangular, moth-like mask. (Source: Adapted from Stevens et al. 2009a. Reproduced with permission from Editorial Office Current Zoology.)

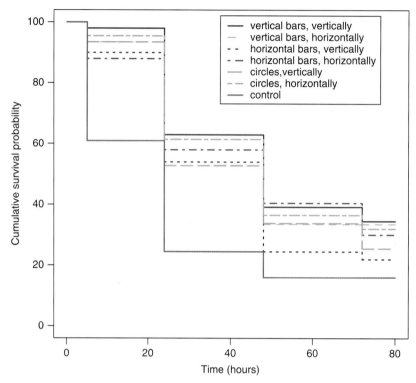

Fig. 8.4 Survival of different baited models with various eye-like representations (see Fig. 8.3). Targets with markings survived significantly better than unmarked controls ($W = 8.154$, d.f. = 1, $p = 0.004$) but there was no significant difference between circles and rectangles, and horizontal or vertical arrangements in any pairwise comparison. (Source: Adapted from Stevens et al. 2009a with permission from Editorial Office Current Zoology.)

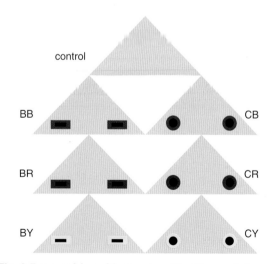

Fig. 8.5 Eye models used for Experiment 2 of Stevens et al. (2009a) varying in shape and 'iris' colour on triangular, moth-like mask. (Source: Adapted from Stevens et al. 2009a. Reproduced with permission from Editorial Office Current Zoology.)

conspicuous colours (Fig. 8.6). Therefore they concluded that the advantage conferred by the markings were not due to their resemblance to eyes but simply their conspicuousness in these circumstances.

In a far more realistic experiment, De Bona et al. (2015) presented manipulated and unmanipulated computer screen images of an owl butterfly, *Caligo martia*, as well as images of a similar owl face (*Glaucidium*) to great tits that were foraging for a mealworm bait. In their set-up, the image of the owl or butterfly was made to open up from the midline where the mealworm bait had been placed over the computer screen, just as if the butterfly was spreading its wings. They found that a reduction in the eye-like features of the eye markings, while maintaining the same level of contrast/conspicuousness, reduced the aversion effect, but that unmanipulated owl butterfly images had the same effect as did images of real owls. I am sure the authors are correct in their conclusion that the results demonstrate that at least in this, and probably many other cases, it really is mimicry of potential predator eyes that has led to the evolution of these eyespots.

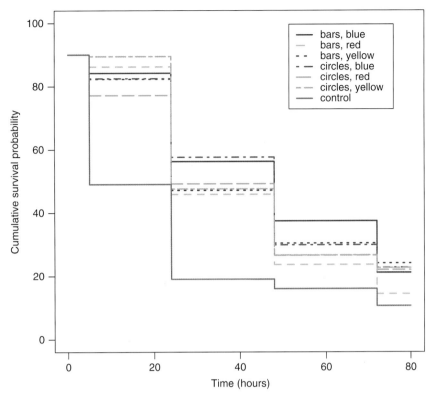

Fig. 8.6 Survival of different baited models with various eye-like representations (see Fig. 8.5). Targets with markings survived significantly better than unmarked controls ($X = 27.495$, d.f. = 1, $p < 0.001$) but there were no significant differences between circles and bars or any of the colour types in pairwise comparisons. (Source: Adapted from Stevens et al. 2009a with permission from Editorial Office Current Zoology.)

Scaife (1976a) investigated the visibility of eyes on dummy birds on the behaviour of domestic chicks. The chicks were placed in an arena with either a dummy bird of prey (in this case a kestrel) which would be a potential predator, and with a dummy of a bird that could not have been evolutionarily familiar to them and would not be associated with any threat, vis-à-vis a kiwi from New Zealand.[1]

As expected, the chicks showed strong avoidance of the kestrel but readily approached the kiwi. The dummies were then manipulated, the kestrel by having its eyes covered, and the kiwi, which normally has rather inconspicuous eyes, had its eyes accentuated, and all possible combinations of variables were tested. The effect was that the chicks showed less aversion towards the dummy hawk but greater avoidance of the enhanced kiwi. Scaife (1976b) followed this up with an investigation of the effect of shape, pairedness and tracking of three-dimensional model 'eyes', yellow circle with central black pupil, or a yellow rectangle with central black bar. The eye or eyes could be moved by the experimenter to track a chick released at the other end of the test arena as if a potential predator was following its movements. Not surprisingly, the combination of eye-like shape, pairedness and tracking caused the greatest avoidance in the chicks, followed by just one single eye tracking the chicks. There were no significant interaction terms, and round shape versus rectangular had the greatest explanatory power (variance explained 40.6%; $p < 0.01$),

---

1. We have presented a small dummy bird with orange eyes and large black pupils, named 'Peck', resembling something like a cross between a snipe and a kiwi, to various birds in aviaries, with surprisingly varied results. A Victoria ground pigeon male courted 'Peck', Amazon grey parrots were, there is no better term for it, terrified, and a lesser sulphur-crested cockatoo attacked it immediately, trying to take out one of its glass eyes. Peck was rescued to investigate further, but this indicates that perhaps results from domestic chicks should not necessarily be taken as being truly representative of all other potential avian predators. Furthermore, it should be noted that chickens, and their wild relatives, the junglefowl, are mainly granivores and only opportunistically vermiphagous or entomophagous.

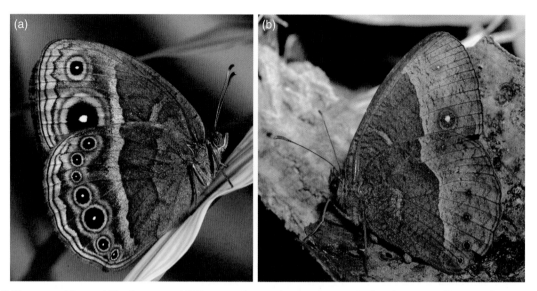

Fig. 8.7 The seasonal forms of *Bicyclus anyana*. (a) Wet season form with very prominent underside eyespots; (b) cryptic dry season form with reduced eyespots. (Source: William Piel, National University of Singapore. Reproduced with permission from António Monteiro, National University of Singapore.)

followed by pairedness (27%; $p < 0.01$), and then, but still significantly, tracking (16.6%; $p < 0.05$).

In recent pairwise choice experiments with domestic chicks to test the effect of eyespots on artificial paper models based on the common buckeye butterfly, *Junonia almata*, Mukherjee & Kodandaramaiah (2015) found that the shape of the eyespot was not overridingly important because fan-shaped markings derived by distorting the concentric real markings were no less effective in deterring attack. However, pairedness provided significantly greater protection to models than ones with an eyespot on a single wing (the single eyespot dimensions being twice those in pairs to control for visual impact) (binomial test, $n = 75$, $p = 0.0052$). Chicks also took significantly longer to attack models with a pair of eyespots.

The marked difference in results between those of Stevens and co-authors (2008a, 2008b) and those of Scaife (1976a, 1976b) and Mukherjee & Kodandaramaiah (2015) does not surprise me since the predators involved (wild birds foraging on tree trunks for normally cryptic moths versus butterfly-like models and domestic chicks) are quite different. As mentioned above, cryptic moths on trees do not display eyespots even if they have them, unless they are part of a flash display (see section *Flash and startle colouration* later). Butterflies that may be basking on the ground which might easily be expected to be encountered by junglefowl (the ancestors of domestic chickens) often have the eyespots on their hind wings exposed, and certainly will show them when disturbed if they were initially resting with their wings folded above

them. Thus it may be that the features of eye-like markings that are important are highly context-dependent.

Eyespots in butterflies

Lyytinen et al. (2003) looked at the role of eyespots in butterflies in relation to whether an attack was coming from the air or from a ground-based predator. Their experimental butterflies were living, palatable *Bicyclus anynana*[2] (Nymphalidae: Satyrinae) and their experimental predators the lizard *Anolis carolinensis*, and the pied flycatcher, *Ficedula hypoleuca*. *B. anynana*, as with many nymphalid butterflies, often feeds on fallen fruit, sometimes in numbers, so is susceptible to predation by both birds and lizards. As with several other tropical nymphalid butterflies (Brakefield & Larsen 1984), *B. anynana* shows seasonal plasticity in its underside eyespots, with the wet season form possessing them and the dry season form not (Fig. 8.7) and both forms were presented to both predators. However, contrary to expectation, neither predator appeared to have its attack deflected by the presence of eyespots in resting butterflies and there was no difference in escape probability once attacked between forms with and

---

2. *B. anynana* has been the subject of a large genetics research programme by Paul Brakefield and collaborators, in particular on understanding the control and development of wing patterning (e.g. Holloway et al. 1993, Brakefield & French 1993, Brakefield 2003) and has become a butterfly version of the 'lab rat'.

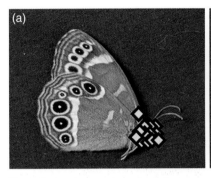 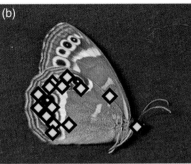 

Fig. 8.8 Distribution of attacks against the underside of the woodland brown butterfly, *Lopinga achine*, by blue tits, *Cyanistes caeruleus*, under three different lighting conditions, showing that marginal eyespots can significantly deflect attacks away from the butterfly's body under low lighting conditions when there is still a UV component present. (a) Under high light intensity with UV present; (b) under low light intensity with UV present; (c) under low light intensity in the absence of UV. (Source: Olofsson et al. 2010 under the terms of the Creative Commons Attribution Licence CC BY, via PLoS ONE.)

without eyespots. Vlieger & Brakefield (2007) also manipulated the eyespots of *B. anynana* and, again using *A. carolinensis* as a predator, found no effect on lizard attacks due to either the presence of eyespots or their position on the wings though, as the authors note, even occasional successful attack deflection could be sufficient to maintain eyespottedness in a species.

Lyytinen et al. (2004) showed that the wet season form was highly cryptic against typical brown dry season background (see Chapter 2, section *Seasonal colour polymorphism*) and was less heavily predated upon by birds than the eyespot-possessing, but otherwise similar, wet season form against the same background, as the eyespot rendered them less cryptic. In the wild, the dry season form preferentially rests on appropriate brown backgrounds, thus maximising their crypsis (Brakefield & Reitsma 1991). However, in the wet season, neither form was cryptic and naïve birds learned to attack the eyespotless dry season form more quickly, and butterflies with more marginal eyespots had a higher probability of surviving attack through escape with just some wing damage, suggesting that the eyespots were effective as an attack deflection device. Thus different seasonal selection pressures are acting to maintain the seasonal eyespot polymorphism.

Kodandaramaiah et al. (2009) experimented with great tits, *Parus major*, and butterfly targets comprising the real wings of the unfamiliar palatable peacock pansy butterfly, *Junonia almana* (Nymphalidae) (see Fig. 8.1d) glued to a card body baited with mealworm. As the butterfly's common English name suggests, both fore and hind wings are decorated with conspicuous eyespots with sparkles. In paired trials, birds were presented with one butterfly with intact eyespots and one with them obliterated. As these 'prey' cannot move, any aspect of startle behaviour was eliminated.

They found a clear advantage to having eyespots: baits with them were attacked significantly fewer times than those without (binomial test: $n = 35$, $p = 0.017$), and the time between first and second attacks was longer when the prey without eyespots was attacked first (Mann–Whitney U-test: $U = 53$, $N = 10$, 22, $p = 0.02$), showing that even after the mealworm reward from the first attack they were deterred by eyespots of the second.

Wing marginal eyespots

In a very effective experiment using blue tits, *Cyanistes caeruleus*, as predators, and the woodland brown butterfly, *Lopinga achine*, which has almost complete rows of underside marginal eyespots on both fore and hind wings, Olofsson et al. (2010) showed that lighting conditions had a major influence on their deflective role. When the lighting was high and natural with UV present the birds correctly identified the butterflies' bodies (Fig. 8.8a) but when the same lighting was dimmed, a very large proportion of attacks were deflected towards the eyespots especially when UV was still present (Fig. 8.8b,c).

Stradling (1976) noted that marginal wing markings on several larger butterfly and moth wings resemble, to a greater or lesser extent, the heads of various non-specific lizards, snakes or amphibians. Such patterns, if recognised by potential avian predators in a similar way as they are perceived by human observers, might serve a frightening or intimidating role that could deter the potential predator.

Eyes with sparkles

The 'sparkle' marks of adult Lepidoptera wing eyespots have not received so much experimental attention, but Blut et al.'s (2012) research provides a lot of useful information.

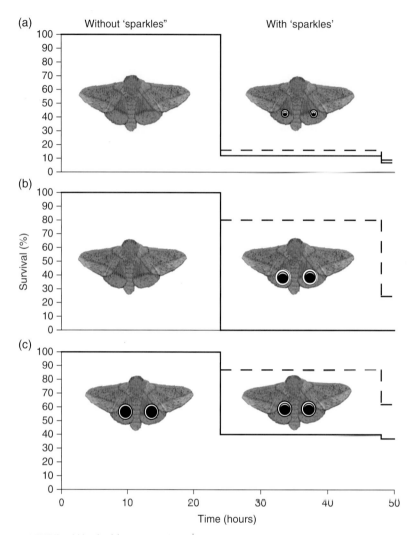

Fig. 8.9 Protective effects against wild birds of large eyespots and eyespots with 'sparkles' on moth models baited with mealworm. (a) No eyespot versus small eyespot (n.s.); (b) large eyespot versus none (p ⪏ 0.0001); (c) large eyespots with and without 'sparkles' (p ⪏ 0.044) (Source: Adapted from Blut et al. 2012 with permission from John Wiley & Sons; the base moth image is not the one used by Blut et al. 2012 but is similar and is modified after image of *Anisota stigma*: adapted from J.D. Roberts 2006 under the terms of the Creative Commons Attribution Share-Alike Licence CC BY-SA 2.5, via Wikimedia Commons.)

First they surveyed the types of 'sparkles' and found that of fore wing eyespots >1 mm diameter, 53% had a central 'pin-point' type 'sparkle', 13% a semicircular 'sparkle', 12% a marginal crescent-shaped 'sparkle', and 22% something of intermediate shape between a semicircular and a crescent-shaped one. These add up to 100%, i.e. 'sparkles' of one sort or another are effectively ubiquitous in lepidopteran eyespots. They also found that not only are the 'sparkles' white to the human observer, but they all reflect UV light, meaning that they would be 'white' also in the eyes of birds. Blut et al. (2012) then set up field trials measuring survival

of pairs of model moths baited with mealworms, with and without small (7 mm) eyespots (Fig. 8.9a), with and without large eyespots (12 mm) with sparkles (Fig. 8.9b), and with large eyespots with and without 'sparkles' (Fig. 8.9c). The small eyespots provided no significant survival benefit, whereas large eyespots with 'sparkles' did, but most interesting is the big difference in survival between moths with large eyespots depending on whether a 'sparkle' was present. In addition, Blut et al. (2012) also found that the orientation of a marginal crescent-shaped 'sparkle' had an effect on survival of moths with a dorsal, natural position suffering significantly less predation than those with a ventral, unnatural position, in their field trials ($\chi^2 = 7.2$, d.f. = 1, $p = 0.007$, $n = 250$).

Eyespots on caterpillars

Considerably more work has been carried out on eyespots on adult insects (almost entirely Lepidoptera) than on eyespots on caterpillars. Most cases of caterpillar eyespots involve larvae that are otherwise rather cryptic (see Fig. 2.27c,d) so the eyespots have presumably not evolved to advertise the presence of the larva. Hossie & Sherratt (2012) presented results of the first field experiment to test whether eyespots on a caterpillar had a significant protective effect. They used green pastry caterpillar models[3] that were loosely based on the larva of *Papilio canadensis*, a swallowtail butterfly which has a pair of eyespots on its third thoracic segment. Four types of models were used, which varied in whether or not they were countershaded (see Chapter 2, section *Countershading*) and whether or not they possessed eyespots, and they were set out on trees in fours, one of each type randomly assigned to each compass quadrant, the whole being replicated many times. Here I show only their results for when the models were placed on the natural host tree of the imitated butterfly as they are more clear-cut despite the smaller sample size (Fig. 8.10). These data showed no simple protective effect of either countershading or eyespots alone (note that the two vertical arrows point in opposite directions, as do the two horizontal ones) but instead a synergistic effect, such that caterpillar models that were both countershaded and had eyespots were attacked significantly less often than ones with only one of those postulated protective attributes. Indeed, eyespots alone (albeit artificial ones) significantly increased risk when presented on non-countershaded pastry caterpillars.

Hossie & Sherratt (2012) also analysed whether the presence of an eyespot affected where birds pecked at the caterpillar model but what they found was slightly complicated. The zone with the eyespot was attacked more than the corresponding zone in eyespotless models but also the far end away from the eyespot was attacked more when an eyespot was present. The most likely interpretation seems to be that some avian predators respond by preferentially attacking the eyespot, whereas others may deliberately avoid it. What the overall consequence of this for survival of real larvae might be is completely unclear; possibly there is no overall benefit from that aspect.

---

3. Pastry was made from a 3:1 white flour and lard mixture and coloured by adding Leaf Green AmeriColor Soft Gel Paste.

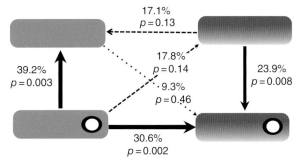

Fig. 8.10 Pairwise Cox proportional hazard regression results of bird attacks on pastry caterpillar model prey (based on *Papilio canadensis* caterpillars) placed out on the host plant *Populus tremuloides*, with and without eyespot marking and with and without countershading. Solid arrows indicate statistical significance at the 0.05 level, arrow width is proportional to the appropriate Wald statistic, and arrow direction leads from most at risk of attack to least at risk. (Source: Adapted from Hossie & Sherratt 2012. Reproduced with permission from Elsevier.)

Importance of eyespot conspicuousness

Stevens et al. (2007) suggest that one of the problems with all the previous studies using artificial moth-like baits is that predators might have been responding to the eyespots as if they were eyes of a real potential threat, i.e. a predator of the predator, or because they are highly conspicuous. Stevens et al. (2007) carried out a series of experiments aimed at disentangling these possibilities using a variety of shapes and also using grey-scale markings to reduce the contrast. Their results show that concentric circles in eyespots have the strongest effect and this is greatest when the contrast is high. The retinal 'wiring' of vertebrate eyes includes 'centre-surround' arrangements of receptive fields that will respond strongly to concentric features, and Stevens et al. (2007) conclude that eyespots are effective predator deterrents but not because they are mimicking vertebrate eyes. However, none of their tested shapes included sparkles, and one might also imagine that retinal centre-surround arrangement of receptive fields may have evolved partly because they were useful for detecting predators' eyes.

Eyespots and fish

Many fish possess eyespots on various parts of their flank, tail or other fins, the function of which has long been subject to debate though only comparatively recently put to experimental test. Of course, there is no necessary reason why all eyespots on fish should have the same function or even why they should not serve multiple functions on the same fish. Indeed, not everything that resembles an eye to a human observer may function as an eye-mimic in the wild.

Members of the coral reef-inhabiting fish genera *Chelmon* (see Fig. 2.34a) and *Chaetodon* (Chaetodontidae), as well as many other fish (see Fig. 2.34b), possess very conspicuous, generally postero-dorsal eyespots, whereas the fish's own eyes are partially disguised by dark stripes running either horizontally or vertically through them (see Chapter 2, section *Disguising your eyes*). Using a molecular phylogeny of the butterfly fish, J.L. Kelley et al. (2013) found that eyespots, eye stripes and other spots were of relatively recent origin and show a great deal of homoplasy. This suggests that the balance of costs and benefits must swing to and fro as predators and environment change over evolutionary time.

Although most people have assumed that pericaudal eyespots in many coral reef fish serve to distract attacks away from more vital anterior parts of the body (Neudecker 1989), this has also been widely debated. Winemiller (1990) provided evidence that the caudal eyespots of the South American cichlid *Astronotus ocellatus* deterred fin predators, and Coss (1979) also presented evidence that they can have an intimidating effect.

To test the attack deflection and intimidation hypotheses, Kjernsmo & Merilaita (2013) used artificial prey with pericaudal eyespots of various size and conspicuousness with naïve three-spined sticklebacks, *Gasterosteus aculeatus*, as predators. The sticklebacks' attacks were significantly drawn towards the eyespots when these were smaller than the sticklebacks' own eyes ($\chi^2 = 11.64$, d.f. = 1, $p < 0.001$), but when the artificial prey had larger eyespots, no significant deterrent effect was found in any of the four trials (Cox regression, Wald = 0–2.91, $n = 41$, $p = 0.09$–0.98). Therefore, with this particular predator there was no evidence that prey eyespots were intimidatory. However, three-spined sticklebacks, while not immune to predation themselves, are well known to be well defended by their dorsal fin spines. Perhaps experiments with more vulnerable predators might yield a different result. In any case, it is likely that there will be marked interspecific differences.

It is also possible that the attack diversion role is not important in some cases. For example, Gagliano (2008) showed that eyespots in juvenile Indo-Pacific damselfish, *Pomatocentrus amboinensis*, have no detectable effect on survival but may be involved in interactions with adults of the same species that lack them. Using mark–release–recapture, after one month she found no evidence that juveniles had thwarted attacks, as indicated by bite marks near the posterior part of the dorsal fin where the eyespot is located, and there was no evidence that juveniles with relatively larger eyespots compared to their own eyes had any greater survivorship. As with all such studies, sample size would probably preclude detection of small selection pressures

that might still be highly significant over evolutionary time. In this species, a few males retain the spot into adulthood and could be involved in deception of other males (see Chapter 11, section *Female mimicry by males*).

The caudal and pericaudal eyespots displayed by some species of cichlid fish have, at various times, been proposed to have a range of functions, most of which are mutually compatible:

1. function in courtship and species-recognition,
2. in juvenile orientation – perhaps reducing the risk of cannibalism by encouraging the offspring to avoid the parental fish's mouth region,
3. to deter fin-predatory piranhas from approaching the cichlid's tail, or
4. to encourage prey to move away from the cichlid's tail and hence closer to the cichlid's head end – the true eye of the cichlid being disguised by an eye stripe.

C.G.M. Paxton et al. (1994) used models of the predatory pike cichlid, *Crenicichla* sp., with or without caudal eyespot markings, to examine their effect on the behaviour of prey fish (guppies, *Poecilia reticulata*). When false eyespots were present, the guppies spent significantly less time near the tail of the model, compared to controls without a caudal eyespot (Fig. 8.11), which would place them in greater danger from a live *Crenicichla* because they would be nearer to its mouth.

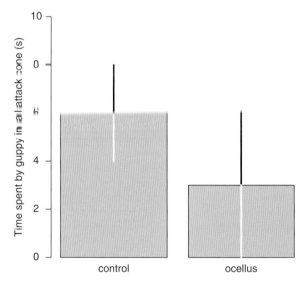

Fig. 8.11 Effect on potential prey guppies of caudal eyespots on models of the predatory fish, *Crenicichla*, showing that the guppies spent less time in the tail attack cone region when an eyespot was present. (Source: Adapted from Paxton et al. 1994. Reproduced with permission from John Wiley & Sons.)

Not just an eyespot but a whole head, winking and other enhancements

Eyespots range from rather abstract entities, which may or may not be representations of real eyes, to very convincing mimics. Sometimes they are just black circles with a pale centre, as in the marginal eyespots of many butterfly wings (see Fig. 8.7a), or they can be far more elaborate patterns that can be remarkably eye-like to a human (see Figs 8.1b,d,f and 8.12d,e). Eyes are normally perceived in the context of a whole head or face, and there are a few quite remarkable examples where evolution has utilised available nearby structures to enhance the realism. The number of eyespots in some cases may have evolutionarily reduced to just one bilateral pair – some swallowtail caterpillars are furnished with numerous eyespots over much of their body, others have only a few located more anteriorly, and then others just a pair near their anterior end, and almost the same range of variation is seen in various hawkmoths (Sphingidae) (Fig. 8.12).

The great similarity of some caterpillars with eyespots to snakes has been recognised for a long time (Bates 1862, R. Morrell 1954, 1969), and is, of course, facilitated by both snakes and caterpillars having elongate cylindrical bodies. Many sphingid caterpillars swell their anterior ends when threatened and thus expand their eyespots, typically assuming the sphinx-like posture which gives the family its scientific name. It is sad that some people, even in the UK, mistake harmless larvae such as those of the large elephant hawkmoth, *Deilephila elpenor*, for snakes and kill them. Without doubt the most impressive eyespot display, in my opinion, is made by the caterpillars of the Neotropical hawkmoths *Hemeroplanes* (=*Leucorampha*) *ornatus* and *H. triptolemus*, which not only expand their anterior end and eyespot when threatened but also assume a remarkably snake-like posture, with most of the body bent away from its foothold and turned to face towards the threat (Fig. 8.12d). The various other markings of its anterior end give the strong illusion of scales and mouth. Clearly, few insectivorous birds or reptiles upon seeing this are likely to pursue the caterpillar much further.

Hossie & Sherratt (2013) investigated the effectiveness of the supposed snake-like defensive posture of papilionid and sphingid-type caterpillars that typically bunch-up and swell their head ends when threatened (see Fig. 2.7c,d). As in their experiment described earlier, they used artificial caterpillars set out on branches, with and without eyespot and with and without a swollen head end, in a 2 × 2 factorial design. They found that both a swollen defensive shape and the presence of eyespots conferred increased protection, but they detected no synergistic effect. They concluded that the protection afforded by a swollen head end was probably due to a combination of reasons, including emphasis of the eyespot, bluff through making the caterpillar seem larger than it is and probably increasing its resemblance to a potentially harmful snake. Hossie & Sherratt (2014) then used landmark analysis to compare the effects of the distortion of various putative snake-mimicking caterpillars' head ends and eyespots during adoption of the swollen head defensive pose. The effect on resemblance was found to depend on the viewing angle, with the caterpillars becoming more snakelike when viewed from above but less snake-like when viewed laterally. Another remarkable Neotropical sphingid caterpillar adaption is displayed by two species of *Eumorpha*, in whose final instars the typical postero-dorsal horn of the hawkmoth caterpillar (see Fig. 2.7) is replaced by a third eyespot, and when the caterpillars are disturbed they contract, giving the appearance of a blinking eye, surely something that could only be mimicking a vertebrate eye (Hossie et al. 2013) (Fig. 8.12f). There is no doubt that many such pseudo-snakes get killed by frightened people.

Reverse mimicry

Insects

Features that make an animal appear to be facing in the opposite direction is generally called reverse mimicry, though the term 'symmetry deception' has also been used (Hailman 1981). Some examples with eyespots in fish have already been discussed earlier. Reverse mimicry may have two non-exclusive functions: it may lead to attacks being targeted to less vulnerable parts of the body, i.e. not the head, and if such an attack occurs the prey is facing away from the predator so may have a better chance of escaping. While eyespots are widespread in the animal world, a number of taxa display far more sophisticated attack deflection devices. False heads, usually situated at the tail end of the animal, are present in several butterflies. They are especially prevalent among the 'blues' (Fig. 8.13; Lycaenidae) but very similar modifications have evolved independently in the Riodinidae (Fig. 8.13f) (Robbins 1985). In these butterflies there are often conspicuous narrow tails that mimic antennae, as well as eye-like markings (pseudo-eyes) on the underside of the wing. The true antennae are usually held concealed along the fore edge of the fore wing, while movements of the hind wings additionally draw a potential predator's attention to the posterior. In many species, dark lines of scales give the impression that the wings are the other way around, and this impression may be further augmented by false leg-like markings. The effect of this distraction may also be enhanced by the butterfly moving its wings alternately on opposite sides. These butterflies are often quick to

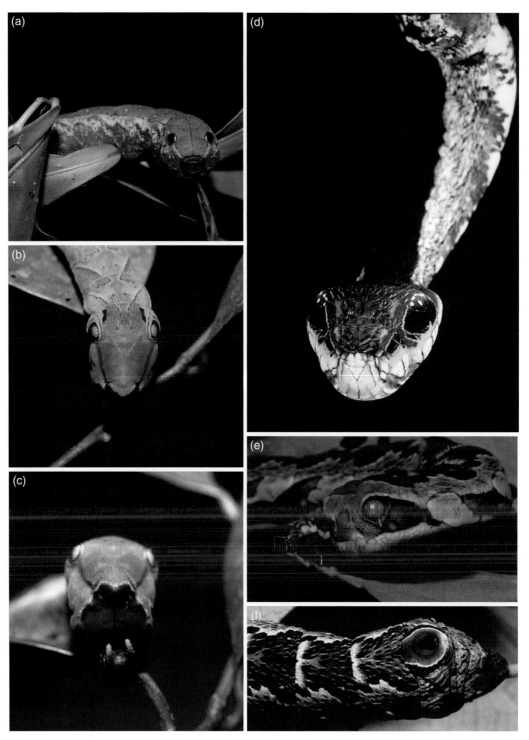

Fig. 8.12 Eyespots and snake-like appearance and posturing in some Lepidoptera caterpillars. (a) Unidentified hawkmoth (Sphingidae) from Costa Rica, possibly *Xylophanes cyrene*; (b,c) caterpillar of an *Archaeoprepona* sp., most likely *A. demophoon*, from Costa Rica (identified from photo by Alberto Zilli); (d) *Hemeroplanes* sp. (Sphingidae) in snake-like posture; (e,f) final instar larva of *Eumorpha labruscae* (Sphingidae), showing (e) anterior end with lateral eyespots and white markings that could represent fangs, and (f) posterior end dorsal view showing blinking eyespot that had replaced the typical sphingid caterpillar horn of earlier instars. (Source: a,b,c,e,f, Professor Dan Janzen. Reproduced with permission from Professor Janzen; d, © Professor Phil deVries. Reproduced with permission from Professor deVries.)

Fig. 8.13 Examples of butterflies displaying reverse mimicry with false heads, including antenna-like thin tails on the hind wings, often with a white tip, and striping diverging from hind wing false head/eye marking. (a) *Thereus pedusa* (Lycaenidae), from Panama; (b) *Arawacus* sp. (Lycaenidae); (c) *Arawacus sino* (Lycaenidae); (d) *Ticherra acte* (Lycaenidae); (e) *Spindasis lohita* (Lycaenidae); (f) *Abisara neophron* (Riodinidae). (Source: a,b, © Professor Phil deVries. Reproduced with permission from Professor deVries; c, Kenji Nishida. Reproduced with permission from Kenji Nishida; d–f, Itchydogimages. Reproduced with permission from John Horstman.)

Fig. 8.14 An unidentified species of the fulgoroid bug genus *Ancyra* (Hemiptera: Eubrachyidae) from Thailand, showing the false antennae that are specialised filaments arising from the apex of the fore wing; it walks backwards when threatened, adding to the illusion of its reverse mimicry. (Source: Les Day. Reproduced with permission from Les Day.)

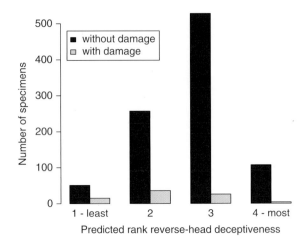

Fig. 8.15 Occurrence of symmetric bird beak mark damage on sampled specimens of Neotropical lycaenid butterflies in relation to their predicted reverse-head deceptiveness, with those of rank 1 being most deceptive. (Source: Data from Robbins 1981.)

take flight when they detect danger, and quite frequently wild individuals are found with bird peck marks on their tails or false heads. In some instances, the mimicry is much more than a simple false head, and the form and markings involved can give the impression that the animal is actually orientated the opposite way around. Similar to the lycaenids, hemipterans of the family Eurybrachyidae have a pair of false antennae posteriorly (Fig. 8.14) that are, in fact, filaments arising from the tip of the fore wing. Some species also have a conspicuous eyespot close to the base of the filament (see figure 14 in Wickler 1968), and when threatened they walk or jump backwards (note also the enlarged fore legs which are more conspicuous than the real jump-providing hind ones).

Robbins (1980, 1981) carried out a series of experiments to test the hypothesis that the false heads of lycaenids act to deflect bird attacks away from the butterfly's true head. Robbins (1980) showed that the false head area of hind wings of lycaenid butterflies are weaker than the leading edge of the fore wing (see Chapter 4, section *Of birds and butterflies*) but also noted that many specimens had also been grabbed from the front where the wings are stronger. In that paper he therefore concludes that the false head area is four times weaker than the fore wing and so would be much more likely to tear in a bird attack, as would be expected if the attack deflection hypothesis was correct (see later). Robbins (1981) used two samples of lycaenid butterflies

belonging to the tribe Eumaeini, the first comprising more than 1000 specimens representing some 125 species from a site in Colombia, the second, some 400 specimens and 75 species from Panama. Species were ranked according to how many 'false head' attributes[4] they displayed (rank 1 for species with all four attributes, etc.). He then scored specimens for evidence of unsuccessful bird attacks directed towards the caudal angle of the hind wing. The distribution of bird attacks among ranks was highly significantly different from random, with 22.7% of those specimens with the most convincing false heads having been attacked at the rear end, whereas only 3.6% of the least convincing false head mimics had been attacked there (and survived) ($\chi^2 = 38.3$, d.f. = 3, $p < 0.00001$) (Fig. 8.15). Robbins' (1981) conclusions have been criticised (Ruxton et al. 2004a) because of the common criticism of beak mark data, i.e. the assumption that increased signs of attack at the tail end means greater survival is not validated because the number attacked and eaten is not known. As Ruxton et al. (2004a) point out, another explanation of the data could be that the false heads attract a higher level of predation.

Wourms & Wasserman (1985b) presented captive blue jays, *Cyanocitta cristata*, with modified dead *Pieris rapae* butterflies to try to determine the roles of the underside

---

4. Presence of two or more contrasting lines converging towards the posterior of the hind wing; anal angle of hind wing less than 65 degrees; anal angle colouration contrasting with ground colouration; the presence of tails.

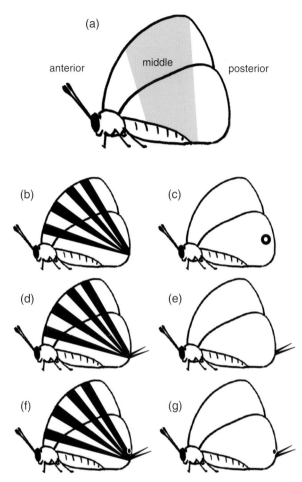

Fig. 8.16 Stylised diagrams illustrating the types of experimental modifications made to edible *Pieris* butterflies to test their effectiveness at distracting bird attack from the real head of the butterfly. (a) Approximate wing zones scored for attack; (b) with converging lines directed posteriorly (see also Fig. 8.13); (c) with caudal eyespot; (d) with tails; (e) with lines and tails; (f) with tails and a black posterior head spot; (g) with converging lines, tails and a black posterior head spot. (Source: Adapted from Wourms & Wasserman 1985b with permission from John Wiley & Sons.)

caudal eyespot, hind wing head spot, hind wing tails and dark lines converging posteriorly (see Fig. 8.13a–c,e). Their experiments examined the distribution of bird strikes on the anterior, middle or posterior of the butterflies (Fig. 8.16a) to see what effect the modifications had in various combinations blue jay attacks on the controls were also significantly biased towards the true head end ($\chi^2 = 6.9098$, d.f. = 2, $p = 0.03159$). They found that the only modification that led to a significant increase in the proportion of strikes at the tail end compared to controls was the presence of a

caudal eyespot (Fig. 8.16c; $\chi^2 = 11.6403$, d.f. = 2, $p = 0.002967$), though they also found that each of the other five marking combinations led to non-significant increases.

Sourakov (2013a) tested the false head attack deflection hypothesis using the hairstreak butterfly *Calycopis cecrops* (Lycaenidae) and other similarly sized butterflies as controls, with a spider predator, the salticid *Phidippus pulcherrimus*. His results were very conclusive that the extravagant false head at the posterior of the hind wings was an effective defence against this predator and also, relevant to escape speed mimicry hypothesis (see Chapter 4, section *Escape speed and low profitability*), and he observed some degree of predator fatigue resulting from the numerous unsuccessful attacks.

Reverse mimicry in flight

Quite a lot of species of large-bodied, predominantly tropical, giant silk moths (Saturniidae) have long hind wing tails that have spatulate ends (notably *Actias*, *Anthistathmophtera*, *Argema*, *Copiopteryx* and *Eudaemonia* species). Those of *Anthistathmophtera* and *Eudaemonia* from Africa reach more than two and five times the fore wing length, respectively. Very similar tails are also found in some African moths belonging to the family Himantoperidae (*Semioptila* species), though these are poorly known and no one has yet studied their behaviour. Barber et al. (2015) have shown that the tails of the North American luna moth (*Actias luna*) are also attack deflection devices, but rather than visual ones they deceive bats acoustically, diverting bat attacks to the tails and away from the moth's body in approximately 50% of attempts. These moths have a relatively slow flapping flight but the tips of their wing tails 'spin' in the air vortices created by flight, and probably generate stronger and more consistent acoustic reflections and I wonder if these may be similar to acoustic reflections from small-bodied moths with faster wing beats. Barber et al. (2015) also show that long hind wing tails in giant silk moths have evolved on several separate occasions rather than defining a monophyletic group, indicating the strong selective advantage of this type of deflective anatomy.

Reverse mimicry in terrestrial vertebrates

A number of snake species and some other reptiles may have the tips of their tails contrastingly coloured while their heads are rather inconspicuous (H.W. Greene 1973). The Malaysian banded coral snake, *Calliophis intestinalis* (Elapidae), which is longitudinally striped black with thin white and red lines, when disturbed sticks its tail in the air and curls it, revealing a very boldly red and black and white and black set of underside stripes, the latter very definitely resembling that of the kraits (*Bungaros* spp. q.v.). While not being a permanent false head, this will almost certainly draw any potential predator's attention to its less vulnerable tail, while simultaneously acting as a warning – a sort of double protection. Wickler (1968) illustrates how the rather drab *Cylindrophis rufus*, when disturbed, raises its tail which is bright red and white underneath to distract a potential predator or prey. Gehlbach (1972) notes that the tail-raising of some elapids is also accompanied by tail-flattening so that it achieves an even more head-like appearance and at the same time, the snakes hide their true heads out of sight. Plates 63 and especially 64 in Mertens (1960) illustrate the same behaviour in some colubrid, cylindrophiid and dipsadid snakes.

R.A. Powell (1982) initially considered that the black tip to the tail of a stoat, *Mustella erminea*, in winter (ermine; see Fig. 2.16b) might be a distracting feature that would deflect attacks by hawks to a less vulnerable or harder-to-hit target than the body. Conversely, the absence of a black tip to the tail of the smaller weasel, *M. nivalis*, might be because it is (a) smaller, and (b) has a shorter tail, so birds of prey might easily catch the animal anyway. Using white dummies of stoat and weasel size and proportions, pulled across the floor of the aviary of captive hawks, R.A. Powell (1982) investigated the effects of the black tip as opposed to either no markings or a black ring on the dummy's body. The results were quite surprising. The naturalistic stoat model (with black-tipped tail) and the weasel (without black) were missed far more often than any other combination, by almost a factor of 10 (see also Ruxton et al. 2004a). While the improved escape of the stoat-like model with black-tipped tail could readily be explained by it throwing the bird of prey off-target, the fact that weasel-like targets escaped, even when they had a black mark, making them presumably more conspicuous, is a bit of a mystery.

Three other terrestrial vertebrates also provide very convincing examples of reverse mimicry. The small Neotropical frog *Physalaemus* (=*Eupemphix*) *nattereri* (Leiuperidae) has a most convincing pair of false eyespots on its rear. These are concealed when it is in its normal sitting posture, but are exposed when, upon being threatened, it raises its rear end (Toledo & Haddad 2009). The serval cat from Africa, *Leptailurus serval* (Felidae), has conspicuous black and white patterned reverse sides to its ears which, from a distance, make it appear to be looking at you when in fact it is facing

Fig. 8.17 Rear-facing false eye markings in terrestrial vertebrates. (a) Prominent eye-like markings on the backs of the ears of a serval, *Leptailurus serval* (Felidae); (b) rear view of a northern pygmy owl, *Glaucidium californicum* (Strigidae), showing eyespots on the back of its head. (Source: a, Lee R. Berger 2007. Reproduced under the terms of the Creative Commons Attribution Share-Alike Licence CC BY-SA 3, via Wikipedia.; b, Seabamirum 2009. Reproduced under the terms of the Creative Commons Attribution Licence, CC BY 2.0.)

away (Fig. 8.17a). Similarly, the North American pygmy owls *Glaucidium californicum* and *G. gnomum* have a pair of eye-like feather patches on the back of the head, and when they are perched it is hard to tell whether they are facing towards or away from you (Fig. 8.17b). Deppe et al. (2003) used wooden pygmy owl replicas with or without painted eyespots and found that eyespots significantly reduced the number of close passes (i.e. within 0.5 m) (t-test: $t = 2.63$, d.f. $= 15$, $p = 0.019$), providing the first empirical evidence for a beneficial function of the markings.

OTHER DEFLECTORS

G.G. Grant & Miller (1995) suggested that the marginal spots of many butterfly and moth wings may represent a form of non-specific mimicry of Lepidoptera larvae that could function as an attack deflection device once the bearer has initially been seen. In my opinion, most of the postulated mimetic resemblances are, for the most part, not at all obvious. Their illustrations show spread specimens which do not accurately represent what a bird might see, and an experimental approach is needed to test whether these markings really do represent mimicry or indeed act as a defensive system. The Neotropical nymphalid butterfly *Pierella astyoche*, and some other species of the genus, has extremely conspicuous white markings close to the rear margin of the hind wing (Fig. 8.18). Hill & Vaca (2004) tested whether these were especially likely to be easily torn by comparing their tear weights (see Chapter 4, section *Of birds and butterflies*) to the same regions of the wing in closely related species that lack the white patches and found that tear weights were significantly lower than in *P. lamina* and *P. lena*, and concluded that *P. astyoche* almost certainly has a better chance of escaping a bird attack on this part of their wing. Many other tropical nymphalids of the subfamily Limenetidinae have similarly conspicuous, white posterior margins to the hind wing, and it is not uncommon to see that these have also been attacked by birds and often the area of wing may be neatly torn off (Fig. 8.19).

Injury feigning in nesting birds

Often referred to as Aristotelian mimicry (Pasteur 1982) because Aristotle first described it based on hunters' anecdotes (almost certainly referring to plovers – see quotation at the beginning of this chapter), a number of ground-nesting birds, upon detecting a potential predator near their nest, assume a behaviour that has a good chance of distracting the predator away from their offspring, specifically mimicking the behaviour of having an injury, such as a broken wing

(Fig. 8.20). This ought, if it were true, to render them an easy and profitable target for attack. Of course, they are not disabled, and thus lure the predator further and further away from their young. This behaviour is well known in various plovers (Charadriidae) (Gochfeld 1984, P.W. Bergstrom 1988, Byrkjedal 1989) and has recently been described in black-winged stilts, *Himantopus himantopus* (Recurvirostidae), which belong to the same suborder as plovers (Wijesinghe & Dayawansa 1998). However, it also occurs among many other groups of birds including ducks (T.W. Johnson 1974), Neotropical warblers (Parulidae) (G.W. Cox 1960, Jablonski et al. 2006), grosbeaks (Cardinalidae) (Moermond 1981), night hawks (Caprimulgidae) (Eisenmann 1962), antbirds (Thamnophilidae) (Greeney et al. 2004) and owls (Strigidae) (Bent 1938).

Armstrong (1954) surveyed the occurrence of broken wing displays and concluded that there were six main factors that predisposed species to evolve this sort of distractive behaviour:

1. their young are in open terrain because prey can see predators from a distance,
2. the nest is accessible to non-avian predators,
3. the nest is inconspicuous or insubstantial,
4. birds nest in relative isolation from conspecifics,
5. predation occurs in daylight by mammals or reptiles,
6. species that live at high latitudes which have long day-lengths in the summer.

That such a high diversity of species perform broken wing displays, in many parts of the world, in many habitats and with many different predators being deceived, may indicate that at least mammalian and avian predators have cognitive ability to recognise when another vertebrate is injured and therefore easy prey (Ristau 1991). It would be hard to imagine that it could be hard-wired and/or that it was selected for in some earlier common ancestor, or a few ancestors, as that would require that injured prey would be such an important part of the predators' diet that not being able to recognise them would put a predator at a considerable disadvantage. Indeed, ignoring the broken wing display of an adult plover, for example, would probably give a predator a good chance of finding chicks or eggs to consume instead.

Tail-shedding (urotomy) in lizards and snakes

An important and partially deceptive form of attack deflection occurs in many lizards which shed their tails when chased or threatened (Zug et al. 2001). It is well known, for example, in the Eurasian legless lizard called the slow worm, *Anguis fragilis*, as reflected in its scientific species name. Shed tails often thrash about vigorously and for a long time, their muscles being particularly well adapted to anaerobic

SATYRIDÆ.

HÆTERA.

W. C. Hewitson, del et lith. 1859.

Printed by Hullmandel & Walton

1. HÆTERA HYCETA.

2. HÆTERA LENA. var. 4. HÆTERA HELVINA.

3. HÆTERA LUNA. 5. HÆTERA ASTYOCHE. var

Fig. 8.18 Illustrations of various Neotropical butterflies including *Pierella* species with conspicuous, presumably attack-deflecting, markings on their hind wings, with *P. astyoche* at the bottom left. (Source: Hewitson 1856.)

Fig. 8.19 Males of the common earl butterfly, *Tanaecia julii* (Nymphalidae), from Thailand. (a) With peck damage in tougher brown part of fore wing, and (b) with apparently sacrificial softer pale blue part neatly lost.

metabolism, and this thrashing obviously serves to attract the attention of the pursuer. From the pursuer's point of view, the tail is a lesser reward than the whole lizard, but it is also a virtually certain reward, whereas there is a good chance that the lizard may escape completely even if the pursuer carries on the chase. Tail-shedding is also exhibited by a few snakes, such as the coral snake mimic *Scaphiodontophis*, but in this species there is no tail regeneration (Savage & Slowinski 1996). Hoogmoed & Avila-Pires (2011) recently described the phenomenon in another colubrid, *Dendrophidion dendrophis*, which is notable for having a relatively long tail for a snake.

E.N. Arnold (1984, 1988) described an experiment carried out by Jonathan Congdon on the western banded gecko, *Coleonyx variegatus*. Using spotted night snakes, *Hypsiglena torquata*, as predators, Congdon showed that while 66% ($n = 30$) of lizards with initially intact tails were captured by snakes, 100% ($n = 12$) of lizards lacking their 'disposable' tails were consumed in the control experiment. Interestingly, in most lizards, although the tail that regenerates looks superficially like the original, it is morphologically quite different, and includes no vertebrae, being instead a cartilaginous structure (Pianka & Vitt 2003). Thus the regenerated tail may be considered as a mimic of the true tail. Some lizards are capable of repeatedly shedding their tails, but at each subsequent shedding, the separation must occur at a point closer to the body than the previous one. Such lizards cannot, therefore, afford to lose too much tail at each shedding.

W.E. Cooper & Vitt (1991) discussed the fact that autotomiseable body parts, such as lizard tails, are often brightly coloured. The function of this colouration after the autotomised organ has been shed presumably helps to increase the pursuer's interest in the shed part and thus aid in the escape of the animal itself. However, there must be a complex trade-off happening, because presumably the brightly coloured part may also serve to attract predators to the animal in the first place. W.E. Cooper & Vitt (1991) proposed a simple model for the conditions that would allow an animal to evolve an attack deflector which also increased the probability that it will be detected in the first place. First, they define the probability that a palatable prey animal will be attacked (and presumably killed), call it P_k, in terms of two other probabilities, P_d and P_e, which are respectively the probability that the animal will be detected and the probability that it will escape an attack. Thus,

$$P_k = P_d(1 - P_e)$$

If having an attack deflector renders them more likely to be detected, say by an amount α, there must be at least some balancing increase in the probability that they will escape; call that increment β. Evolution of a deflector at each stage will therefore depend on the overall chance of the animal escaping ($P_{k'}$) being increased, i.e.

$$P_{k'} = (P_d + \alpha)(1 - P_e - \beta)$$

must be less than P_k. Rearranging this equation shows that evolution of the deflector depends on how large the increment in escape probability, β, is. Specifically

$$\beta > \frac{\alpha(1 - P_e)}{\alpha + P_d}$$

This has several fairly intuitive implications. Obviously, if the increase in the risk of detection (α) is very small, then almost any small increase in the probability of escape will be favoured (just replace α by zero, and the right-hand side of the equation becomes zero). But the numerator will also be small if P_e is close to unity, i.e. the prey is good at escaping.

Fig. 8.20 Two birds performing broken wing displays to distract perceived predator threat to their eggs or nestlings, known as Aristotelian mimicry. (a) Killdeer, *Charadrius vociferus*; (b) ringed plover, *C. hiaticula*. (Source: a, © iStock.com/MikeLane45; b, © iStock.com/Orchidpoet.)

It is therefore a balancing act; deflectors that do not greatly increase risk of detection are more likely to evolve and, as Ruxton et al. (2004a) conclude, if they do render a prey more conspicuous they are most likely to evolve if that prey already has a good escape strategy. The way most blue butterflies waggle their false antennae is certainly going to draw attention to them, but perhaps not from any considerable distance, so the butterfly may already have had a high chance of detection. Tail-shedding in lizards, one might

suppose, evolved first and the added conspicuous colouration of the tail in some species may have evolved later.

FLASH AND STARTLE COLOURATION

A number of cryptic animals, particularly among the winged insects, employ a defensive strategy of suddenly revealing otherwise hidden brightly coloured areas. This is

Fig. 8.21 Examples of catocaline and other moths with cryptic fore wings and brightly coloured hind wings that have been proposed to serve as startle colouration and diversity of which may help reduce habituation by predators. (a) The French red underwing, *Catocala elocala* (Erebidae); (b) *C. conversa*; (c) white underwing moth, *Catephia alchymista* (Erebidae); (d) *Blenina chrysochlora* or close (not fully matching) (Nolidae); (e) unidentified moth from Oriental region; (f) female castor semi-looper, *Achaea janata* (Noctuidae). (Identifications thanks to Alberto Zilli.)

often the hind wings that are concealed or folded underneath cryptic fore wings[5] (Figs 8.21 and 8.22a–d,f–h). In some stick insects, such as many diaphomerids (Fig. 8.22e), the fore wings are reduced, but the anterior part of the large hind wings is cryptically coloured and at rest they are folded

so as to conceal the large, often brightly coloured posterior hind wing membrane.

Underwing moths, especially members of the large Holarctic genus *Catocala* (Noctuidae) (Fig. 8.21a–c), have been the topic of extensive investigations of flash colouration, especially by Sargent and co-workers at the University of Massachusetts (1969–78). These moths have cryptically coloured, bark-like forewings that when folded at rest completely hide the hind wings which, in many species, are brightly coloured and patterned with pink, red, blue or

---

5. The toughened fore wings of Orthoptera, such as grasshoppers, mantids and stick insects, are usually referred to as tegmina (singular tegmen) by entomologists.

Fig. 8.22 Examples of hind wing flash colours (and an eyespot-like pattern) in various Orthoptera, Phasmatodea and Hemiptera. (a) Giant purple grasshopper, *Titanacris albipes* (Romaleidae), north-eastern neotropics; (b) *Phymateus saxosus*, Madagascar; (c,d) male and female, *Ommatoptera pictifolia* (Tettigoniidae), respectively, Brazil; (e) female yellow umbrella stick insect, *Tagesoidea nigrofasciata* (Diapheromeridae), Malaysia; (f) cicada from South-East Asia, probably *Platypleura* sp. (Cicadidae); (g) *Pyrops astarte* (Fulgoridae), Malaysia; (h) *P. pyrorhyncha*, Malaysia.

yellow (chromatic ones) and in others largely black (achromatic ones) (Fig. 8.21) (Sargent 1973, 1977, Sargent & Owen 1975). Frequently there may be several (up to 40) different species of *Catocala* at any given site. One surprising finding of Sargent and colleagues' long-term light-trapping survey was that, despite fluctuation in the abundances of particular species, the proportion of chromatic to achromatic individuals at sites remained remarkably constant. Sargent's (1978) interpretation of this was that the diversity in moth hind wing colouration was being maintained

by selection favouring anomalous startle effects. When a bird encounters a *Catocala* with a particular flash pattern for the first few times, the bird will be startled and the moth probably has a better chance of escaping, but the birds are likely to start to learn that despite the flash colouration, no harm comes to them and so would be expected to habituate and start actually predating on the moths. Furthermore, moths with different hind wing flash colours would appear almost equally startling and so also experience greater protection. The range of different hind wing colours has become known as 'aspect diversity'. These conclusions are not consistently supported, and while Ricklefs & O'Rourke's (1975) data were consistent with Sargent's (1978), Ricklefs (2009), with more samples from a range of latitudes from Ecuador to Canada, failed to find any relationship between aspect diversity and either numbers of species in the samples or with latitude.

D.L. Evans (1983) surveyed the responses of a diverse range of resting moths and found that when their dorsal wing surfaces were touched, a large proportion responded with some sort of defensive display ranging from flight, falling to the ground with associated thanatosis (see Chapter 3, section *Thanatosis (death feigning)*) or flashing coloured hind wings or some combination. A larger proportion of palatable species exhibited such reactions than did unpalatable ones (78–100% versus 10–52%) and the stimulus intensity thresholds were higher among the unpalatable taxa.

Intimidating displays and bizarre mimicries

In some cases, the human observer may be less than certain what a particular display represents. The European nymphalid butterfly, the peacock, *Inachis io* (Nymphalidae), is famous for having a conspicuous eyespot towards the outer angle of its fore wings (see Fig. 8.1c). Vallin et al. (2005) presented evidence that the peacock butterfly's eyespots may be intimidating to predators. Vallin et al. (2006) compared survivorship of three Western European nymphalid butterflies – the peacock, the comma, *Polygonia c-album*, and the small tortoiseshell, *Aglais urticae* – that have boldly patterned fore wings but cryptic hind wings and which

hibernate as adults, often against a background of dead leaves. In trials with blue tits as predators, peacocks took significantly longer, on average, to be discovered than commas (Tukey's HSD: $p = 0.04$) though there was no difference between small tortoiseshells and the others. Furthermore, both peacocks and commas survived significantly better than small tortoiseshells (Fisher's exact tests: $p < 0.001$) (Fig. 8.23). When only those data are presented for butterflies that were discovered by the blue tits, the peacocks survived significantly better than the commas or small tortoiseshells (Fisher's exact tests: $p < 0.001$, $p = 0.002$, respectively) (Fig. 8.23). Vallin et al. (2006) attribute this to the startle effect of the wing 'flicking' by the butterflies, which the peacocks initiated while the birds were at a greater distance away than when the small tortoiseshells did. However, the peacock is also a larger species than the others and it is possible that size could have played a role.

A subsequent comparison of the effects of eyespots and deimatic displays by peacock butterflies and eyed hawkmoths, *Smerinthus ocellatus*, gave rather mixed results and sample sizes were rather small, though peacocks survived bird attacks significantly better than the hawkmoths (Vallin et al. 2007).

Merilaita et al. (2011) experimentally manipulated peacock butterfly wings to discover the importance of their four large eyespots, obliterating either just the fore wing pair or all four. Naïve pied flycatchers, *Ficedula hypoleuca*, the experimental predators, as found in previous studies, showed a significant and innate reluctance to attack insects with visible eyespots, but Merilaita et al. (2011) found no difference in attack response to butterflies with four compared to just two eyespots, and therefore concluded that conspicuousness

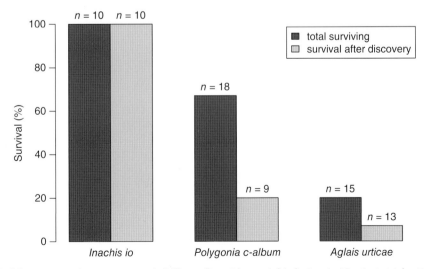

Fig. 8.23 Survival of three species of European nymphalid butterflies with cryptic hind wings in 40-minute trials with blue tits as predators. (Source: Adapted from Vallin et al. 2006 with permission from Springer.)

alone might not be the whole story behind the butterflies' pattern. However, again the results strictly only pertain to the one species of predator used.

Olofsson et al. (2012) investigated the startle display of adults of the European swallowtail butterfly, *Papilio machaon*, against attacks by a small insectivorous bird, the great tit, *Parus major*. When disturbed, resting swallowtails quickly open their wings, displaying their brightly coloured upper surfaces, and if the disturbance persists they start rhythmically jerking their wings. By presenting great tits with both living and dead (freeze-killed and freshly thawed) swallowtails in a two-choice experiment, they found that the startle display was quite effective in frightening the birds, often causing the bird to flee, and overall it took the birds four times longer to attack the living butterflies.[6] Twenty-four out of 27 of their live butterfly subjects survived their 45-minute exposures to the predator, which suggests that their deimatic display formed an effective defence against a small passerine predator. However, interpreting why the birds should be so affected by the butterfly's behaviour is not totally straightforward. The black and yellow colour combination on the butterfly's wings is typical of many warningly coloured animals, and the butterflies' larvae feed upon fennel, *Foeniculum vulgare* (Apiaceae), which contains various flavonoids and the glucoside foeniculoside I, but Olofsson et al. (2012) showed that the birds readily ate the freshly killed ones even though they had *ad libitum* access to other palatable energy-rich food (suet). The wing-flicking might be an honest signal that the butterfly was a difficult prey to handle but since, in their experiment, the great tits readily attacked and ate mealworms that were attached directly below the live butterflies, that might not be the case. Therefore Olofsson et al. (2012) conclude that the birds probably have an innate defensive response to any signal that might indicate that they were in some real danger, and that the swallowtail's wing-flicking display successfully tapped into such an inbuilt self-protection system. Such successful bluff systems will therefore depend upon three things: the predator must also have its own predators that it will have evolved to be wary of, encounters between the bluffing prey and any individual predator must be rare so that habituation to the deimatic display does not occur, and, correlated with the former, the predator must be a generalist.

Other studies of startle display include Vaughan (1983) and Schlenof (1985), who both worked with bluejays, *Cyanocitta cristata*, as predators. In the first paper, the birds were trained to push aside a flap and to pick out a coloured disc underneath the flap to find a mealworm reward. Birds showed startle responses when the discs were either novel or rare, thus providing evidence that sudden appearance of a novel colour could provide additional time for a prey to escape. Schlenoff's (1985) study was aimed to test the deimatic flash displays of *Catocala* (Noctuidae) underwing moths, and used baited moth models with cardboard fore wings that the birds had to move aside to obtain the reward; in so doing they revealed the patterned 'hind wings'. She found that "Jays which had been trained on models with grey hind wings exhibited a startle response[7] when they were exposed to *Catocala*-patterned hind wings. In contrast to this, subjects trained on *Catocala* models did not startle to a novel grey hind wing" and also that the startle effect of a novel pattern lasted several days. I think it is particularly interesting that the birds displayed beak-wiping in the absence of any aversive chemical taste or odour (see also D.L. Evans & Waldbauer 1982).

Also within this broad category should be mentioned strange, possibly coincidental, but sometimes convincing resemblances. Hinton (1974) drew attention to the remarkably mammal-like appearances of some butterfly pupae of the blues family Lycaenidae, and Janzen et al. (2010) also illustrate quite a number of Costa Rican Lepidoptera pupae with eyespots of varyingly convincing complexity. The pupa of the nymphalid butterfly *Dynastor darius* is also remarkably like the head of a snake (see figure 2 in Pough 1988b). Cloudsley-Thompson (1980, 1981) illustrates and refers to the likeness of the head of a lanternfly (Hemiptera: Fulgoridae) to the head of an alligator. Both cases might seem absurd considering the size differences, but Hinton (1974) explains that this might not be a problem that needs to be overcome if one considers the 'rapid peering' strategy of small gleaning birds. They might quickly turn a corner, so to speak, and see something that has a resemblance to an image innately associated with danger. What should the bird do? – clearly not remain in full sight and advance closer to check it out. Better by far to scurry away alive and search for a meal elsewhere.

SCHOOLING, FLOCKING AND PREDATOR CONFUSION

Vane-Wright (1976), in his classification of mimicry systems, noted the similarity between the members of insect swarms, fish schools, bird flocks and mammal herds, which may be considered bipolar S1 + S2 systems (see Chapter 1,

---

6. Olofsson et al.'s (2012) article also includes online supplementary videos of the birds' responses to live and dead swallowtails.

7. This includes dropping the food item, raising its crest, flying away rapidly to the side of the cage, emitting alarm calls and beak wiping.

section *Vane-Wright's system*). These large assemblages of individuals serve a protective purpose because many predators are unable successfully to target an individual prey organism among a large moving mass of similar organisms. Thus it may be considered that the similarity is a form of mimicry; certainly it is a form of adaptive resemblance, in that any individual that differs markedly from the remainder of the group will be a comparatively easy target for a predator. Vane-Wright (1976) refers to the fact that a predatory aquatic *Chaoborus* fly larvae will starve to death if presented with very large numbers of *Daphnia* prey, because it is unable to track and capture individual prey among the crowd. This also provides an explanation as to why butterflies that aggregate around mud puddles in the tropics keep in more-or-less single-species clusters (see cover illustration) (van Sommeren & Jackson 1959).

Among fish and birds, schools and flocks are not infrequently made up of members of more than one species (Randall 2005, Cipresso Pereira et al. 2011), but there is seldom a great deal of difference in appearance between the taxa that routinely tend to associate together in such groups. Ehrlich & Ehrlich (1973) noted that heterotypic schools of tropical reef fish, that is schools comprising more than one species, are often composed of fish of similar size and colouration (Fig. 8.24). Sometimes essentially solitary species, at least ones that do not form schools of their own, nevertheless routinely swim among schools of another species that they resemble. Dafni & Diamant (1984), who describe a case involving a fangblenny, *Meiacanthus nigrolineatus* (Blenniidae), refer to this as "school-orientated mimicry". Ehrlich & Ehrlich (1973) likened heterotypic schooling to Müllerian mimicry because the more individuals in a school, the greater the protection, but I do not think it is helpful to use that term for all mutualist mimicries.

It would not be surprising if schooling fish species had instinctive behaviours that caused them to school with similar-looking individuals, after all recognition of conspecifics is an essential component of schooling behaviour. Randall & Guézé (1980) described a new species of goatfish, *Mulloidichthys mimicus*, that mimics the colour pattern of the blue-striped snapper, *Lutjanus kasmira*, and suggested that this may reduce predation risk when the two species shoal together. In the Indian Ocean, near Kenya and Tanzania and off the coast of Sri Lanka, a blue-striped colour variety of *M. vanicolensis* occurs that looks very similar to their new species and *M. vanicolensis* also forms schools with *L. kasmira*, and the individuals that do so have the more 'typical' colouration. The authors therefore suggested that perhaps *M. vanicolensis* may be capable of changing its pattern quite quickly between mimic and normal forms depending upon which species it associates with. There

Fig. 8.24 Mixed-species shoals of visibly similar fish. (a) Bigeye snapper, *Lutjanus lutjanus*, schooling with bluestripe snapper, *Lutjanus kasmira*, Maldives; (b) yellowfin goatfish, *Mulloidichthys vanicolensis*, schooling with gold-lined sea-bream, *Gnathodentex aureolineatus*, Borneo. (Source: © Linda Pitkin Underwater Photography. Reproduced with permission from Linda Pitkin.)

seems to be no direct evidence of this, however, and I think it is unlikely except in an evolutionary sense with different selection pressures acting upon different populations.

J.H. Brown et al. (1973) compared 16 species of reef-dwelling damsel fish (Pomatocentridae) and found they could be divided into solitary, pairing and social species, and that there was a strong associated colour signal, with six of the eight social species being patterned with black (or brown) vertical stripes on a white or yellow background, whereas all of the solitary or pairing species were quite different. It seems likely that the stripes either play a role in group cohesion, or help protect groups from predation – perhaps through flicker-fusion effects.

Pielowski (1959, 1961) was one of the first people to draw importance to protective nature of the similarity between members of a flock. During his observations of predation on flocks of wild pigeons, *Columba livia*, by gosshawks, *Accipiter gentilis*, he noticed that the hawks nearly always attacked and captured those pigeons that differed most from the rest of the flock. Similarly, Mueller (1977) found the same selection of minority forms by predatory American kestrels, *Falco sparverius*. Mueller (1971a) conducted a large series of experiments (1600 in all) with two species of hawk to determine the most important aspects of prey that led the birds to select them. Odd prey, i.e. ones that differed from the majority, were selected against. He also found evidence that the birds continued to prefer a particular sort of prey (as if they had a specific search image for it) even if it was just as common as, or even somewhat rarer than, the alternative.

As with hawks, predatory fish also find it hard to successfully attack individuals in schools if they all look the same, and they are also somewhat easier to conduct experiments with than raptors. Ohguchi (1978) found the same results as Pielowski (1959, 1961) and Mueller (1971a, 1977) in experiments with three-spined sticklebacks, *Gasterosteus aculeatus*, attacking normal and dyed waterfleas, *Daphnia*, as did Landeau & Terborgh (1986), who carried out experiments with groups of minnows, *Hybognathus nuchalis*, and month-old largemouth bass, *Micropterus salmoides*, predators. Landeau & Terborgh (1986) used groups of eight prey, some of normal colour and some dyed blue to render them distinctive, to assess the effect on survival of not standing out from the crowd. Overall capture rate was greatly reduced when the minnow colour types approached 50:50 (Fig. 8.25) and the kill rate massively decreased as school size increased, with only about one in eight attacks being successful when schools comprised 15 prey (Fig. 8.26). Thus the phenomenon is taxonomically widespread among vertebrates but I cannot find any examples involving errant, visually hunting invertebrates that might encounter similar situations – such as dragonflies and insect swarms.

Disrupting the cohesiveness of schooling is therefore likely to be a good strategy for a predator and this appears to be the strategy being employed by *Spheniscus* penguins and some other conspicuously black and white marked marine piscivores (R.P. Wilson et al. 1987).

Several similar instances of likely aggressive mimicry involving coral reef goatfish (Mullidae) as 'nuclear species' and co-schooling 'attendant species' have been documented (e.g. Sikkel & Hardison 1992). Krajewski et al. (2004) describe how off the coast of Brazil, the similar yellow goatfish, *Mulloidichthys martinicus*, and the smallmouth grunt, *Haemulon chrysargyreum*, school together. When threatened, small groups of the goatfish join the grunts and change their swimming behaviour to resemble the latter, though both species no doubt benefit from the safety-in-numbers principle.

'Social' mimicry in birds and fish

Many people have noticed that vertebrate species that frequently associate with one another often have similar appearances. Among birds, members of mixed-species flocks often have similar wing bars and tail flashes. Such similarities are often referred to as 'social' mimicry (Moynihan 1968, 1981, C.J. Barnard 1979, 1982). Several possible explanations for this have been put forward and discussed, including the suggestion that some of it might be straightforward Batesian and/or Müllerian mimicry. Burtt & Gatz Jr (1982) suggest that while mimicry might be involved, similarities in habitat or flight leading to feather abrasion or damage due to turbulence might also be responsible for convergence on similar patterns. E.O. Willis (1976b, 1989), followed by Sazima (2010), considered the visual similarity between the co-flocking cloud forest birds *Philydor*

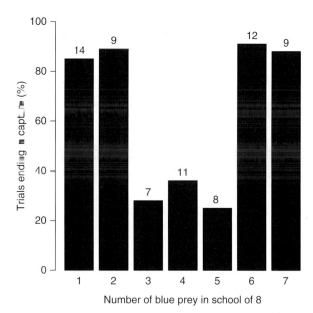

Fig. 8.25 Individuals in a mixed school of ordinary-coloured and blue-dyed silvery minnows (*Hybognathus nuchalis*) are more at risk of predation when the school contains a small number of odd-coloured individuals. Number of trials given above the bars. (Source: Adapted from Landeau & Terborgh 1986 with permission from Elsevier.)

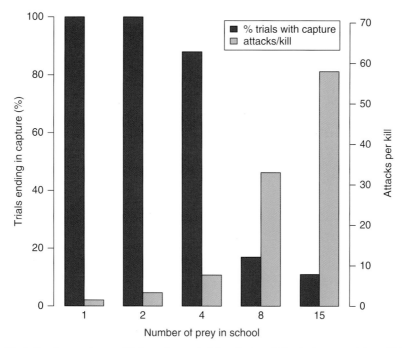

Fig. 8.26 Protection afforded to silvery minnows (*Hybognathus nuchalis*) against predation by largemouth bass (*Micropterus salmoides*) increases with number of minnows in a school. (Source: Data from Landeau & Terborgh 1986.)

rufus, *Orchesticus abeillei*, *Pachyramphus castaneus*, and other species in south-eastern Brazil as having a primarily anti-predator role as with mixed-species fish schools, possibly enhanced by the difference in escape strategies displayed by some of the species.

In addition to visual 'social' mimicry, it may be that some birds imitate the calls of species with which they usually flock. E. Goodale et al. (2014) suggest that in the greater racket-tailed drongo, *Dicrurus paradiseus lophorhinus*, mimicry of non-alarm calls made by other bird species with which it normally associates may be used to get the other birds to coalesce into a group which is beneficial to the drongos. Furthermore, mimicry of other species' 'mobbing calls' might also serve to bring together other species to join in mobbing a potential predator and therefore reduce the risk of the activity to the drongo.

Alarm call mimicry for protection

Igic & Magrath (2013) and Igic et al. (2015) compared sonograms of calls, including mimetic ones, made by the African brown thornbill, *Acanthiza pusilla*, and other bird species that co-occurred with it. In particular, it was found to mimic the aerial alarm calls of a range of other bird species. Igic & Magrath (2014) showed that its use of mimic alarm calls was context-dependent, with birds

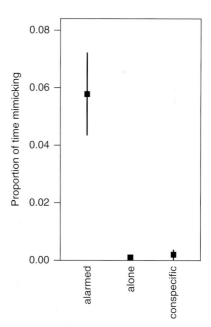

Fig. 8.27 Time spent performing mimetic calls by male spotted bowerbirds, *Chlamydera maculatus*, is significantly higher when their bowers are under threat (e.g. by a human intruder) than when they are alone or being visited by a female conspecific. (Source: Adapted from Kelley & Healey 2012 with permission from John Wiley & Sons.)

most likely to use them either when under aerial threat, in which case they mimicked aerial alarm calls which caused nearby birds to flee, or when their nests were under threat, in which case they mimicked mobbing alarm calls which might get other birds – heterospecifics – to help mob and drive away a potential nest predator. Similarly, male spotted bowerbirds, *Chlamydera maculatus* (Ptilonorhynchidae), produce mimetic heterospecific alarm calls predominantly when they perceive that their bowers are under threat (Fig. 8.27), and this suggests that the mimicry may

be involved in confusing or deterring potential predators, which might thus behave as if their presence has been detected by numerous other birds (L.A. Kelley & Healy 2012). Use of mimicked alarm calls to recruit heterospecific birds to help in deterring predators has been proposed in a number of other cases but there is little hard evidence (Morton 1976, Chu 2001, E. Goodale & Kotagama 2006). Whether this is true of other bowerbird species that mimic the calls of other species (e.g. Frith et al. 1996) is not known.

Chapter 9

ANTI-HERBIVORY DECEPTIONS

For millions of years flowers have been producing thorns. For millions of years sheep have been eating them all the same. And it's not serious, trying to understand why flowers go to such trouble to produce thorns that are good for nothing? It's not important, the war between the sheep and the flowers?

The Little Prince, a novella by Antoine de Saint-Exupéry

Mimicry, Crypsis, Masquerade and other Adaptive Resemblances, First Edition. Donald L. J. Quicke.
© 2017 John Wiley & Sons Ltd. Published 2017 by John Wiley & Sons Ltd.

INTRODUCTION

With the exception of mimicries involved in pollination (see Chapter 12, section *Pollinator deception*) and seed dispersal (see Chapter 14, section *Fruit and seed dispersal by birds*), there has been remarkably little work undertaken on defensive deception in plants. It is possible, of course, that for various reasons there are not so many as in animals or that we are missing them because of our own perceptions. Certainly plants are generally constrained to be green, and larger ones are able to tolerate some degree of herbivory, whereas most animals die if a large chunk of them is bitten off. This second point might mean that selection pressure to protect themselves is, on average, lower than among animals. Yet plants collectively show enormous variety in growth form, leaf shape, longevity, size and toxicity. Leaf shape, for example, may be determined not only by mechanical and physiological factors but also potentially by mimicry of leaves of other species or of already partially consumed leaves, or masquerade of inanimate objects. Intraspecific variation (between individuals, within plants, between young and old plants) as well as such factors as leaf subdivision might also reduce recognition by herbivores (V.K. Brown et al. 1991).

In Chapter 12 I discuss a lot of examples of pollination by deception, many cases of which involve duping male bees and other hymenopteran insects that the flowers, predominantly of orchids, are conspecific females. These almost always involve a combination of visual as well as olfactory mimicry. Lev-Yadun (2014a) suggests the possibility that when the visual aspect is particularly accurate it might have a secondary benefit to the plant in deterring herbivory of the blossom, which might be interpreted as having a stinging bee or wasp on it.

CRYPSIS AS PROTECTION IN PLANTS

Plants may defend themselves from herbivores by being difficult to locate. Schaefer & Ruxton (2009) discuss much about mimicry among plants, in particular how it may result from the 'exploitation of perceptual biases' (EPBs) in a receiver – i.e. how innate biases in a receiver's sensory or information processing system may lead them to perform particular behaviours. The EPB model they describe is really the sum of the ideas incorporated in sensory exploitation and sensory trap models because it includes everything from signal detection to higher cognitive processing. Many mammalian herbivores, for example, lack red colour vision, and therefore plants that use reds or other colours to signal young foliage may be perceived as just dark leaf tissue. Cognitive issues may be important too; to a human, leaf variegation may appear attractive but to a herbivore it

might suggest prior damage or it may make leaves cryptic against a complex background (Givnish 1990, Campitelli et al. 2008, Soltau et al. 2009, Lev-Yadun 2014a). Unlike mimicry, EPB by plants is not constrained by such things as temporal synchrony, as with many mimetic pollinator deceptions (see Chapter 12, section *Pollinator deception*), though it may be the precursor of mimicry.

K.C. Burns (2010) provided accounts of several likely cases of plant crypsis based on the flora of New Zealand, and outlined a number of conditions that need to be satisfied for a plant's colour/growth form to be considered cryptic. Notably:

1. they need to be eaten by visually orientated herbivores,
2. they need to be habitat specialists, such that they grow in environments that provide a fairly consistent visual background,
3. normally they would be short-statured, such that they grow in close physical proximity to the visual background generated by their preferred habitat, and
4. the background against which they are cryptic should be unpalatable, whether it is inanimate or composed of other plants.

While most anti-predator mimicries appear to be found among animals, there are nevertheless a number of fairly well-documented, and a few more tentatively suggested, cases known among plants too (Barthlott 1995). Indeed, this has been recognised since the early days of mimicry theory (Wallace 1878, Rothrock 1888). One of the first to be discussed was that of arid land succulent plants bearing a close resemblance to stones. Pre-eminent among these are the aptly called stone plants of South Africa, belonging to the genus *Lithops* (Aizoaceae), though stone mimicry also occurs in more than 30 other genera in at least seven families (Wiens 1978). Sir Alfred Russel Wallace, while noting that plants are generally green due to chlorophyll and mainly obtain protection by virtue of toughness, spines or toxins, relates that *Lithops* and various other succulent plants in arid places such as South Africa's Karoo probably avoid detection by herbivores through their very impressive resemblances to the surrounding soil and pebbles (Wallace 1878). In these cases it is probably not just their potential value as food that makes them desirable but the fact that they are succulent means that they contain that valuable desert resource, water. Similar background matching occurs in many plants. The New Zealand penwiper plant, *Notothlaspi rosulatum* (Brassicaceae), for example, with its slaty grey leaves, is particularly difficult to find on the scree slopes where it occurs unless it is in bloom (K.C. Burns 2010). It is well known that many plants that live close to the sea or among sand dunes are often whitish due to their possession of white trichomes, sticky glandular trichomes, or because of light-coloured waxes, and thus their surfaces may become partially self-camouflaging because grains of sand or clay particles tend to adhere to them (Lev-Yadun 2006). In addition to stone masquerade, various arid land plants also mimic leafless sticks, dead grass and possibly even faeces (Wiens 1978).

Corydalis benecincta (Papaveraceae) is a small annual alpine plant that lives on scree slopes in China. It is dimorphic for leaf colouration – some plants have grey leaves and some green – and therefore provides an opportunity to test the crypsis hypothesis. Those with grey leaves are clearly more cryptic to human vision against their normal background. Niu et al. (2014) tested the premise that herbivores (*Parnassius* butterflies) cannot distinguish grey leaves plants from a scree background using spectrography and a butterfly colour vision model, and they also compared photosynthetic activity of the two morphs. Grey-leaved plants had similar photosynthetic activity to green ones but experienced lower levels of herbivory, suggesting that they were indeed cryptic for some herbivores. This was supported by the vision model, which showed that the green morphs would be much more easily distinguished from the scree by *Parnassius*.

Leaf mottling and variegation for crypsis

It can often be difficult to know what is the most important aspect of leaf colouration. For example, Lev-Yadun (2014b) considers that the highly conspicuous green and white variegation of the European milk thistle, *Silybum marianum* (Asteraceae) (Fig. 9.1), could have evolved through a dazzle effect, as a deterrent to insect landing, or as mimicry to

Fig. 9.1 Leaf variegation in European milk thistle, *Silybum marianum* (Asteraceae), the white markings of which result from sub-epidermal air spaces.

tunnelling. As the plant is spiny and also can be toxic to ruminants, it could be aposematism.

Many plants, but especially low-growing, understorey herbs in temperate and tropical forests, have variegated leaves or glossy ones with strong three-dimensional surface shaping. Variegation carries a cost in that the white and darkened parts have reduced photosynthetic capability. Furthermore, in numerous species there is an apparently stable balance between individuals with variegated and unvariegated leaves, suggesting a trade-off. Indeed, most gardeners will be aware that cherished variegated plants often revert to all-green forms.

Givnish (1990) highlights that some groups of small plants might be especially profitable for herbivores to eat for various reasons – spring ephemerals, spring leaves of summer species because of their relatively high protein (leaf nitrogen) content, species that have green leaves while most other plants have lost theirs, such as evergreen species, wintergreen species, and winter leaves of dimorphic species. In addition, leaf value might be especially high for some groups of plants in which the cost of leaf replacement is high, such as those living on nutrient-poor soils or in low light conditions, or both, such as many understorey plants in tropical forests (Fig. 9.2). In a survey of eastern North American plants, Givnish (1990) found that plants in virtually all of these categories had disproportionately high levels of leaf mottling compared to other categories of plants.

A.P. Smith (1986) proposed seven hypotheses to explain the occurrence of persistent variegation:

1. *Temporal Heterogeneity Hypothesis* – different leaf morphs are adaptations to changing light conditions throughout the seasons,
2. *Spatial Heterogeneity Hypothesis* – production of different ecotypes adapted to sites of different canopy openness has an evolutionary advantage,
3. *Leaf Apparency (crypsis)* – variegated leaves might better evade herbivores than plain leaves do,
4. *Frequency* or *Density-Dependent Herbivory Hypothesis* – the morph that is dominant at any given moment experiences the most intense herbivore pressure,
5. *Mimicry Hypothesis* – variegated morphs mimic herbivory damage,
6. *Aposematic Colouration Hypothesis* – distinct leaf colour patterns are linked to certain chemical defence, and
7. *Neutral Hypothesis* – colour polymorphism is not in itself adaptive but might persist as a relic trait.

So far, only a few studies have tested these hypotheses.

Givnish (1990) nevertheless rejects most of A.P. Smith's (1986) hypotheses, at least when it comes to the vast majority of cases, which he suggests are simply acting to disrupt the outline of the leaf shape and thus deceive colour-blind mammalian herbivores, but it should be noted that many

Fig. 9.2 Variegated and dazzling plants (left-hand images) with associated versions lacking red signal to illustrate how they will appear visually to mammalian herbivores which lack red colour vision (right-hand images). (a–d) Understorey assemblages of Cat Tien National Park, Vietnam; (e,f) *Pellionia repens* (Urticaceae), an understorey plant from Thailand that shows marked changes in leaf appearance through development – note the difference in contrast between leaves and reddish soil.

mammals may have some blue-green dichromatism, so there is still potential for variegation to help conceal leaf shape. As Soltau et al. (2009) point out, the mimicry hypothesis has not been subject to much rigorous study.

Mistletoes and lianas

One of the best-documented cases of probable Batesian mimicry in plants concerns various Australian mistletoes and their host plants, to which they show a considerable resemblance (B.A. Barlow & Wiens 1977). *Amyema biniflora*, for example, is parasitic on several *Eucalyptus* species, and from a distance is often quite hard to spot because of its similar colouration and leaf size (Fig. 9.3a). On close inspection

it can be seen that its leaves have almost the same shape and texture as that of its host (Fig. 9.3b). The host/hemiparasite pair *A. cambagei* and *Casuarina cunninghamiana* similarly share very narrow green stems that can be difficult to distinguish unless the mistletoe is in flower. Canyon & Hill (1997) showed that the mistletoes have a higher nitrogen content than their host plants and interpreted this as indicating a higher protein content and therefore greater nutritional value of the mistletoes compared with their host plants. K.C. Burns (2010) noted a similar likely mistletoe mimicry in the New Zealand species *Korthasella salicornioides* Santalaceae, formerly (Viscaceae), which infects and closely resembles the chemically defended shrub-like tree *Leptospermum scoparium* (Myrtaceae). However, in New Zealand, mistletoes of the Loricanthaceae look quite

Fig. 9.3 The Australian mistletoe, *Amyaena biniflora* (Loranthaceae), a mimic of its host tree. (a) *A. biniflora* in *Eucalyptus clarksoniana*, Queensland, Australia; (b) detail of *A. biniflora* plant showing similarity of leaf shape and colouration to that of its host. (Source: Russell Cumming. Reproduced with permission from Russell Cumming.)

different from their hosts. A possible explanation for this suggested by Burns (2010) is that the two plant families, Loricanthaceae and Santalaceae, have different host plant height preferences and these would tend to have different herbivores. In New Zealand, selection for host mimicry might have been acting upon the lower growing hosts of parasitic Santalaceae, and while the herbivores responsible are not known, extinct moas are a definite possibility.

In tropical America there are numerous species of vine whose leaf shapes resemble that of the tree species through which they routinely climb (L.E. Gilbert 1975, Williamson 1982). If the vine leaves are mimetic it implies that they are mistaken by a herbivore for the host plant's leaves and are avoided because the herbivore would be avoiding those of the host plant as a result of past experience. However, if the vine's leaves are cryptic, the herbivore merely fails to detect the vine's leaves against the general background of leaves.

In South American *Passiflora* species, young leaves are often of a very different shape to mature leaves, which may confuse ovipositing *Heliconius* butterflies. This could be particularly beneficial for small plants which might suffer proportionately greater damage from the feeding caterpillar. Closely related sympatric *Passiflora* species also often have very different leaf morphologies (L.E. Gilbert 1982) and therefore it is harder for herbivores that find them generally palatable to recognise them as a group.

Gianoli & Carrasco-Urra (2014) have shown that the mimetic Neotropical vine, *Boquila trifoliolata* (Lardizabalaceae), gains protection from herbivory by resembling the leaves of its host trees in numerous features (size, shape, colour, orientation, petiole length and/or tip spininess). Leaves of *B. trifoliolata* that are on plants climbing through trees with leaves differ both from those of unsupported vines and those climbing through leafless trees. Unlike many similar instances of vines mimicking just one tree host species, of *B. trifoliolata* can mimic several species.

Fruit masquerade by leaves

Hakea trifurcata, an Australian proteaceous plant, provides a probable example of self-crypsis that helps to protect its fruit (strictly its follicles). In this species, there are two types of leaves: needle-like ones and some broad ones (Fig. 9.4). The broad leaves rather closely resemble the green but mature seed follicles, but they are only produced by mature plants, and they grow close to the follicles. Groom et al. (1994) propose that as the broad leaves provide no reward for potential follicle predators such as black cockatoos, they effectively produce an uninteresting background against which the rewarding follicles are cryptic. If this interpretation is

Fig. 9.4 Photograph of *Hakea trifurcata* (Proteaceae), showing that the majority of leaves are narrow, round in section and more or less thorn-like, but a proportion are obovate and greatly resemble the mature fruit. (Source: K. J. & S. J. Hall. Reproduced with permission from K. J. Hall.)

correct, then this provides a non-human example of an intraspecific (conjunct) synergic mimicry of Vane-Wright (1976) (see Table 1.4).

PROTECTIVE BATESIAN AND MÜLLERIAN MIMICRY IN PLANTS

In addition to crypsis, some plants display mimetic features that render them less likely to be palatable to herbivores (Augner & Bernays 1998, Lev-Yadun 2009a). Despite some mathematical modelling indicating that plants should also be able to engage in Batesian and Müllerian mimicry, there have been only a few experimental studies of true Batesian and Müllerian mimicry in plants, but there seems to be no fundamental reason why an unpalatable plant should not evolve aposematic traits to signal its unpalatability to potential predators (e.g. grazing mammals) and once warningly advertised plants have evolved they could surely act as models for Batesian mimics or become members of Müllerian complexes. In Europe, a classic example is the similarity in leaf shape and habitat between the stinging nettle, *Urtica dioica*, and the white dead-nettle, *Lamium alba*, which lacks stinging hairs and is relatively palatable when young (Fig. 9.5a). Children, including myself when I was young,

frequently play tricks on others with it, and both large and small herbivorous mammals quickly learn to avoid eating stinging nettles (Rothschild 1964b).

Lubbock (1905) notes two other, little discussed, possible examples of palatable plants closely resembling others with a bitter taste that are apparently shunned by herbivores. German chamomile, *Matricaria chamomilla*, he suggests as a mimic of the poisonous Roman chamomile, *Anthemis nobilis*, but both plants belong to the subfamily Anthemideae of the Asteraceae and so some of their resemblance is likely due to this, however they are not sister groups (Oberprieler & Vogt 2006). His other suggested example is the resemblance between *Ajuga chamaepitys* (Lamiaceae) and cypress spurge, *Euphorbia cyparissias* (Euphorbiaceae), which often grow together. Wickler (1968) also mentions the similarity of fool's parsley, *Aethusa cynapium*, to hemlock, *Conium maculatum* (both Apiaceae), but without evidence that the latter is avoided.

Several plants are cyanogenic, most work having been done on various Fabaceae such as white clover, *Trifolium repens*, and bird's-foot trefoil, *Lotus corniculatus* (D.A. Jones 1962, Till 1987) (see section *Experimental evidence for plant aposematism and Batesian mimetic potential in plants*). The ability to produce cyanide is controlled by two gene loci and is polymorphic in both species. However, expression of the genes is also

Fig. 9.5 Examples of leaf shapes that are possibly mimetic. (a) A naturally occurring patch with stinging nettles, *Urtica dioica*, and its leaf mimic (flowering in this picture), the white dead-nettle, *Lamium album*. (*continued*)

Fig. 9.5 (Continued) (b–e) Putative resemblances to leaves that have already been partly eaten by some animal and therefore are likely to contain higher levels of induced toxic chemicals: (b) young plant of *Monstera adansonii* (Araceae) from South America; (c) *Pterocymbium diversifolium* (Malvaceae), Thailand; (d) unidentified plant, Thailand; (e) *Artocarpus* sp. (Moraceae), Thailand. (Identification: c,e, Pitoon Kongnoo.) (Source: a, Steve Cook, 2014. Reproduced under the terms of the Creative Commons Attribution Share-Alike Licence CC BY-SA 2; Tim a n mpol pompholj ncoma)

influenced by environmental factors and it has been found that within a single plant some leaves may be cyanogenic and others not. Cyanogenesis in these has a substantial feeding deterrent effect on a wide range of herbivores, so an acyanogenic leaf on a plant that also has cyanogenic ones (comprising a separate ramet) is effectively a Batesian mimic. Larger herbivores will generally recognise the whole plant but they cannot distinguish externally which leaves are potentially edible and which are toxic. Small herbivores, such as slugs, that are usually dietary generalists, will be able to sample the foliage around them on a more leaf-by-leaf basis.

To try to understand why in some plant species there is both intra-plant and inter-plant variation in cyanogenesis Till-Bottraud & Gouyon (1992) proposed an evolutionarily stable strategy (ESS) model with cyanogenesis being costly as well as protective, the former being responsible for maintaining the acyanogenic phenotype, otherwise all plants would evolve to be cyanogenic throughout. Their model shows that the optimal frequency of cyanogenic leaves is determined by the amount of herbivory and, furthermore, that the ESS is lower than this optimum when there are more generalist herbivores present.

False indicators of damage or likely future damage

Herbivores may also assess the suitability of food/host plants in terms of how healthy they appear and whether or not they have already been eaten by slugs or insect larvae. Such assessment may have been responsible for the evolution of certain leaf forms and colour patterns that make the plant look as though it has already been eaten or infected by a pathogen. Furthermore, insect herbivory, both by chewers and suckers, is well known to induce plants to produce defensive chemicals that can have a negative impact on the original causer of the damage and other subsequent herbivores (e.g. Duffey & Stout 1996, Inbar et al. 1999), and this opens up the possibility that plants might evolve to appear to have feeding damage in order to reduce the probability of other herbivores attacking them. A number of plants in various families produce leaves that are especially ragged and, at a casual glance, give the appearance that they have been partially eaten (Fig. 9.5b–e). Niemelä & Tuomi (1987) first proposed that the leaf shapes of various Moraceae (e.g. *Broussonetia* and *Morus* species) may serve this function and it seems entirely reasonable to me that if the main herbivores, vertebrate or invertebrate, are deterred from eating or ovipositing on what are likely to be particularly toxin-laden leaves and that vision plays a role in the decision, then selection would favour plants whose leaves mimicked such damage. However, there have been no experimental studies that I can locate about whether such leaf shapes provide significant protection. Niemelä & Tuomi (1987) also suggest that apparently damaged leaves might attract insect parasitoids or insect predators, but I suspect that at least the former is unlikely to be important in this case as parasitoids of folivores tend to be specialists and rely largely on volatiles to locate the correct species of host.

Conspicuousness of leafmines

Members of four orders of insect (Diptera, Lepidoptera, Coleoptera and Hymenoptera), comprising thousands of species, feed as juveniles between the upper and lower surfaces of leaves, consuming nutritious plant tissue and almost invariably producing conspicuous leafmines. Yamazaki (2010) asks the question why the mines of these insects are so conspicuous. It could be, of course, that by eating the chlorophyll-containing, protein-rich cells they inevitably remove the green colouration. Many insect mines are highly apparent to humans from some distance, and in some plants, such as holly (*Ilex* spp.), birds selectively predate on the associated holly leafminer and almost certainly do so using visual cues. But holly shrubs have robust, which facilitates searching by the birds, which can hold on the leaves and twigs, whereas in other plants, the more mobile, soft leaves and shoots are not so suitable for gleaning and so the leafminers are primarily mainly subject to parasitisation and to accidental herbivory. Yamazaki (2010) postulates that the mines themselves may resemble, and even have evolved to mimic, fungal infection, animal excrement or necrotic plant tissues, all things that might deter a herbivore from consuming the mined leaf and so protect the miner. An arms race exists, with the plant evolving to defend itself from the miners and the miners to be able to complete their development on the plant, and thus, at least in some cases and probably the majority, the feeding activity of the miner induces local chemical defence in the plant with unconsumed parts of mined leaves potentially having higher concentrations of toxins and lower relative amounts of highly nutritious tissue.

While it is not entirely clear by what means it occurs (and there may be several), leaf variegation and mottling appear generally to be evolved mechanisms that lead to reduced herbivory (Cahn & Harper 1976, Lev-Yadun 2003a, 2006). Some workers have also considered that conspicuous leafmines resemble leaf mottling. Thus conspicuous leafmines might be mimicking variegation. On the other hand, it has also been argued that some plant variegation is mimetic of leafmines (Wiens 1978, A.P. Smith 1986, Soltau et al. 2009), which would make sense if mammalian or other herbivores avoided leaf-mined vegetation for whatever reason.

The only experimental study examining this I can find is that of Soltau et al. (2009), who investigated insect mines and variegation in the arum *Caladium steudneriifolium* in Ecuador. This plant produces both unvariegated (about 33%, Fig. 9.6a) and variegated (with whitish spots) (about 66%; Fig. 9.6c) leaves. They noticed that the plain leaves were especially prone to attack by an unidentified leafmining moth and that the mines (Fig. 9.6b) bore a resemblance to the spots on the variegated leaves, suggesting that the spots were leafmine mimics which might deter oviposition by adult female moths 'wanting' to avoid laying on already mined leaves. To test this they carried out a manipulation experiment in which they artificially variegated plain green leaves using Tipp-Ex correction fluid (Fig. 9.6d) as well as doing a control manipulation, and placed these out along with unmanipulated plain and variegated leaves and allowed the moths to oviposit on them for three months. The results were quite clear (Table 9.1), infestation rates were reduced by a factor of four when the leaves bore white spots. Similarly, Campitelli et al. (2008) found that leaf variegation in *Hydrophyllum virginianum* was associated with reduced herbivore damage.

Dark central florets in some Apiaceae

As noted by Darwin (1888), there is variation among wild carrots, *Daucus carota* (Apiaceae), in whether the central floret[1] of their umbels has a dark central spot. It has been

---

1. In this species this is actually an umbellet that is reduced to a single but conspicuous dark floret.

Fig. 9.6 (a) Plain, (b) leaf-mined, (c) variegated and (d) experimentally modified leaves of *Caladium steudneriifolium* (Araceae) used to test the potential protective role of variegation against leafmining insects. (Source: Soltau et al. 2009. Reproduced with permission from Springer Science + Business Media.)

Table 9.1 Infestation rates (%) by a leafmining moth on leaves of *Caladium steudneriifolium* (Araceae) depends on their variegation, and provides evidence that the white spots are leafmine mimics. (Soltau et al. 2009. Reproduced with permission from Springer.)

	Plain leaves	Leaves with white spots
Unmanipulated plants	7.88	1.61
Manipulated plants	9.12	0.41

postulated that when present this is mimetic of an insect (Eisikowitch 1980) and thus may attract other pollinators. Alternatively, it has been postulated that it might resemble the central parts of normal, or possibly composite, flowers, thus serving to attract more potential pollinators or to orientate them. By analysing video recordings of pollinator landings and visit durations, Polte & Reinhold (2013) found no evidence of any difference in pollinator activity between spotted and plain umbels. They did find, however, that umbels with a dark central floret were significantly less parasitised by the cecidomyiid gall fly, *Kiefferia pericarpiicola*, and therefore they interpret the dark floret as a probable mimic of an already existing *Kiefferia* gall, which might deter the fly from attacking that inflorescence. Thus the fact that the putatively mimetic mark was associated with a flower led people to assume that its function, if any, was connected directly with plant reproduction, whereas it now seems likely to be a mimetic defence against phytophagy.

Mimicry of silk or fungal hyphae

Some plants have a very dense covering of long trichomes (white, hair-like cuticular structures), which give the appearance that the plant is either infected heavily by fungus, which might deter mammalian herbivores, or has an infestation of silk-producing spiders, mites or caterpillars (Yamazaki & Lev-Yadun 2015). As examples, they show photographs of new buds of *Onopordum* and *Carthamus* spp., flower heads of *Arctium tomentosum* (Asteraceae), a small leaf of coltsfoot, *Tussilago farfara* (all Asteraceae) and new fronds of the fern, *Osmunda japonica*. While accepting that such structures may have other roles, Yamazaki & Lev-Yadun propose that they may often also constitute a case of mimicry, with the model being the same species of plant but rendered less palatable by some other species.

Insect egg mimics

Two small and distantly related groups of plants have evolved so as to exploit egg-laying adaptations of particular butterflies. Numerous butterflies are known to deposit their eggs singly and widely dispersed on their larval food plants. This behaviour may be associated with maximising food resources per larva, and reducing risks of parasitism and avian predation. Some butterflies not only regulate the dispersion of their own eggs but also assess the number of eggs already laid on a given potential food plant before deciding whether or not to lay on it (see Chapter 4, section *Egg load assessment*). In the case of *Heliconius* butterflies in South America, this may also be due to the fact that many *Heliconius* caterpillars are, at least to some extent, cannibalistic (Rausher 1979). This egg load assessment is apparently largely a visual phenomenon, and is therefore relatively vulnerable to deception. At least two species of South American passionflower (*Passiflora*) develop false egg-like structures (vine eggs) on the ends of their elongate stipules (Fig. 9.7a), which serve to deceive egg-laying females of *Heliconius* butterflies (Benson et al. 1975, L.E. Gilbert 1975, K.S. Williams & Gilbert 1981). In this case, the location of the egg mimic is probably crucial in that the species of butterfly only oviposits on the tips of very young shoots. Other *Passiflora* species have other probable egg mimic structures or marks. Some have the swollen extrafloral nectaries on the leaf petiole, some have swollen egg dummies on the edge of their leaves (e.g. *P. oerstedii*), some have pale yellow spots on their leaves, which may be either symmetrically (Fig. 9.7b) or asymmetrically arranged, and according to species may be slightly raised as in *P. candollei* (L.E. Gilbert 1982, MacDougal 2003). Pasteur (1982) termed this S1R/S2 system Gilbertian mimicry. Rothschild (1974) additionally proposed that in some species of *Passiflora*, modified stipules give the plants the appearance of already having a (horned) caterpillar feeding upon them.

Similarly, Shapiro (1981a, 1981b) reported that small orange 'callosities' on the tips of the erect cauline leaves of the North American crucifers *Streptanthus glandulosus* and *S. breweri* are mimics of the orange eggs of the white butterfly, *Pieris sisymbrii* (Pieridae). The mimicry hypothesis was tested experimentally by removing the egg mimics from some *Streptanthus* plants and comparing the number of *P. sisymbrii* eggs laid on these with those laid on untreated controls. In this way Shapiro (1981a, 1981b) showed that the presence of egg mimics significantly reduced the likelihood of a plant being oviposited on by *P. sisymbrii* (Fig. 9.8). The egg mimics described by Shapiro (1981a, 1981b) are not widespread, however, and are restricted to just these two species of *Streptanthus*. They are even absent from some

Fig. 9.7 Butterfly egg mimicry by *Passiflora* vines. (a) Young shoot of *P. laurifoliae* showing (at left) single egg of a *Heliconius* butterfly, (near centre) the plant's egg-dummy, a swollen yellow tip to a stipule, and (at right) pair of yellow swollen extrafloral nectaries on leaf petiole; (b) *P. biflora*, food plant of *Heliconius cluonnumun* from Costa Rica, showing egg-mimics on the leaves. (Source: a, Professor James Mallet. Reproduced with permission from Professor Mallet; b, Kenji Nishida. Reproduced with permission from Kenji Nishida.)

S. glandulosus populations that occur on non-serpentine soils, indicating that the selection pressure exerted by *P. sisymbrii* must be significantly higher in the poor serpentine slopes investigated. Furthermore, it should be noted that this particular species of butterfly preferentially consumes flowers and seeds of its host plant and thus, as the *Streptanthus* species are annuals, the selection pressures on the plant to avoid caterpillar damage are likely to be considerable. Conversely, the reason why *P. sisymbrii* is an egg load-assessing species whereas many other pierids are not, is likely to be associated with the fact that a single host plant is only likely to provide enough food for the complete development of a single caterpillar and the fact that cannibalism

is frequent. The production of conspicuous red or orange rather than white, yellowish or green eggs by some pierids (the pierid red egg syndrome) is therefore probably the result of the same selection pressures, i.e. utilising small dispersed foodplants (Shapiro 1981a). In both these cases there is a relatively high degree of dietary dependence between the butterfly and the host plant and, perhaps more importantly, the butterfly constitutes a major component of herbivory on its particular host plant.

Defensive aphid and caterpillar mimicry in plants

There is considerable evidence that some macro-herbivores selectively avoid eating plants that are infested with various phytophagous invertebrates. Thus it would not be too surprising if some plants had been selected to possess markings or structures that give the impression that they have invertebrates feeding upon them. Lev-Yadun & Inbar (2002) postulate that the dark anthers of *Paspalum paspaloides* (Poaceae) and dark flecks on the stems of *Alcea setosa* (Malvaceae) both resemble dark aphids that they may be mimicking.

A second-hand (multitrophic) aspect of this was postulated by Yamazaki (2016), who noted that various Eurasian and East Palaearctic plant galls caused by certain aphids, jumping plantlice, thrips and gall midges bear a considerable resemblance to caterpillars. While this might attract some insectivores, the notion put forward was that it might also deter herbivory of the gall and also of adjacent foliage (see section *Conspicuousness of leafmines* earlier).

Aphid deterrence by alarm pheromone mimicry

It is hard to know how commonly plants may deter attack by aphids through the use of mimicry because so few systems have been investigated. However, R.W. Gibson & Pickett (1983) explored wild potatoes, *Solanum berthaultii*, for the mechanisms that they may use to deter herbivores and found that the plants leaves had a repellent effect on the major crop pest aphid, *Myzus persicae* (Hemiptera: Aphidae), due to their release of (E)-β-farnesene, the aphid's alarm pheromone. It seems quite possible that many other temperate[2] plants might use the same strategy, but most work on plant volatiles has concentrated on crop plants.

---

2. Aphids are a predominantly temperate group of insects with rather few tropical species, or at least abundant ones.

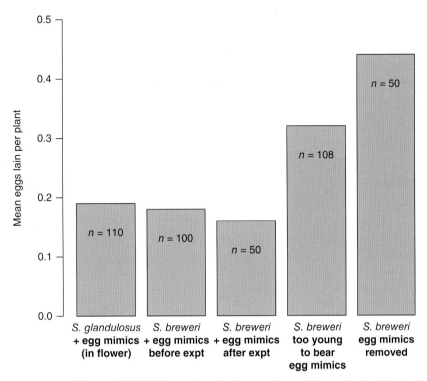

Fig. 9.8 Graphical representation of Shapiro's (1981b) data on the effectiveness of egg mimics on the plant *Streptanthus breweri* (Brassicaceae) in deterring oviposition on plants by the butterfly *Pieris sisymbrii* (Pieridae).

Ant mimicry in plants

It also seems likely that some plants may mimic being covered with ants to deter potential macro-herbivores (Lev-Yadun & Inbar 2002) but why this is not more widespread is a bit of a mystery. Perhaps the revulsion that largely hairless humans have to arboreal ants and their nests is not shared to such an extent by many large, hairy herbivores. *Xanthium strumarium* (Asteraceae), shown in Fig. 7.12f, is a candidate for such ant mimicry. It has numerous dark epidermal markings that give the impression that small dark ants, such as formicines, may be swarming over it. Lev-Yadun & Inbar (2002) also suggest that similar marks on the petiole and inflorescence stems of *Arisarum vulgarum* may serve the same purpose. Some *Hydrangea* species also have very similar markings.

A quite different form of ant mimicry is utilised by the South American ant plant, *Cordia nodosa*, which forms domatia in which its symbiotic *Allomerus* ants rear their broods. The ants patrol the *Cordia* plant for prey and protect them from all sorts of herbivores. D.P. Edwards et al. (2007) discovered that new shoots of the *Cordia* plant, which are the parts most vulnerable to herbivory, release odours that mimic those of the ant broods, so attracting patrolling *Allomerus* towards the new leaves. This is not a particularly easy system to interpret since the ants probably catch more insect prey around the new growth compared with around older tougher leaves, so they may be being deceived in one sense, but benefitting in another.

Of orchids and bees

Lev-Yadun & Ne'eman (2012) posed the question of whether bee or wasp mimicry by orchid flowers might additionally deter herbivores, because grazing mammals would probably avoid taking a mouthful of flowers if they thought they were covered in stinging aculeate Hymenoptera, or perhaps they might even avoid foraging on whole patches of flowering plants if they were associated with the likelihood of getting stung. Of course there may be a dose–size effect; a bee sting that is painful to a small human may be far less noticeable to a large cow. Also, the mimicries of Hymenoptera by flowers are very largely based on pheromones so the visual similarity of the flower to a bee or wasp may be far from perfect, at least from a human perspective (see Fig. 12.3).

Carrion mimicry as defence

Lev-Yadun & Gutman (2013) make an interesting inference from the avoidance behaviour of grazing cattle towards places where dead cattle had been dumped. It makes evolutionary good sense for animals to avoid coming into contact with dead individuals of their own, and possibly related species, because the carcasses will be infected with various bacteria that may themselves be harmful, or the dead individual may have died recently from a communicable disease. The avoidance shown by the cattle meant that in areas where dead cattle had been left there was a consequent increased plant biomass at the end of the season. They therefore suggest that plants that smell of carrion, something that has probably evolved to deceive pollinators (see Chapter 12, section *Brood site/oviposition site deception*), might also gain protection by deterring grazing by large herbivores.

Algae and corals

Although not strictly plants, mimicry of corals has also been described in two species of red algae – a *Rhodogorgon* species mimics gorgonian corals (J.N. Norris & Bucher 1989) and *Eucheuma arnoldii* mimics *Acropora* staghorn coral (Kraft 1972). Both presumably deceive visually browsing alga-feeding coral reef fish.

Plant galls

Many insects and some other arthropods, as well as some nematode worms, induce galls in plants, and it has long been noted that these are often quite conspicuous. The question that was usually debated was whether it was the gall-former or the plant that was in control of the gall's external appearance. Most people held the view that the gall-former was in control (Price et al. 1987), partly because detailed gall morphology is highly dependent on the species of gall-inducer, and if the gall-inducer can cause the plant to make highly nutritious gall tissue, why not also be able to control other features of the structure? The second of these hypotheses appears to be supported by the fact that galls that are attacked most by parasitoids are, on average, the ones that are most apparent (B.A. Hawkins & Gagné 1989), so perhaps conspicuousness has been a selected defensive feature of the host plant that overall reduces the number of gall-formers in the vicinity in subsequent years through increased parasitism rates. Inbar et al. (2010) put forward an alternative idea, that the conspicuousness of galls was

an aposematic signal under control of the gall-former, and while they presented no data they made a number (six) of testable predictions that could serve to determine whether aposematism was involved. Of these, I think that the following four are likely to be most easily tested and also the most telling:

1. only chemically defended galls should be brightly coloured,
2. bright colours are more likely in species whose galls are common and/or aggregated to enable predators to learn,
3. large galls (or aggregations), being more readily detected by predators, are therefore expected to be more chemically protected and more colourful, and
4. chemically protected galls are expected to also advertise their unpalatability through odours that would inform a potential predator of toxicity before they received much, or any, damage.

Of course, further investigation should also take into account possible multiple instances that are due to close phylogenetic relatedness of different species, but with both gall-former and host plant involved it is not immediately obvious how this ought to be done.

Experimental evidence for plant aposematism and Batesian mimetic potential in plants

There is very little by way of experimental evidence that plants can be Batesian mimics. In an interesting set of experiments, Launchbaugh & Provenza (1993) demonstrated that sheep could be trained to preferentially avoid cinnamon flavoured wheat if their previous experience of consuming cinnamon-flavoured ground rice had been associated with the intra-oesophageal administration of lithium chloride (LiCl) in a gelatin capsule using a balling gun.[3] These experiments demonstrated not only that sheep, and therefore presumably many other large herbivores, can learn to avoid particular foods that have unpleasant associations, which is not surprising, but also that they can generalise a particular flavour across other different types of food, in this case rice and wheat. In a follow-up set of trials using lambs and LiCl-induced toxicity, Provenza et al. (2000) explored the animal's associative learning with another novel odour (coconut) and the reputedly toxic, and certainly

---

3. Lithium chloride administered this way reliably causes food aversion in sheep.

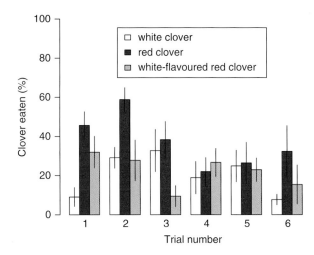

Fig. 9.9 Results of six trials showing that consumption of clover by rabbits over a 24-hour period could be affected by the clover's odour.[4] Rabbits preferentially ate red clover (*Trifolium pratense*) compared to white clover (*T. repens* var. Aran), which contains high concentrations of (non-volatile) cyanogenic glycosides, but avoided it when it was sprayed with volatile organic compounds from *T. repens*. (Source: Adapted from Massei et al. 2007. Reproduced with permission from John Wiley & Sons.)

noxious, odoriferous plant *Astragalus bisulcatus*, which gives off sulphurous organic voliatiles.[5] The lambs showed no innate aversion to either coconut or *Astragalus* odour, but when each was associated with LiCl administration they fed less on barley-straw with the relevant odour compared to controls. These results therefore clearly show the potential for Batesian and Müllerian resemblances to evolve among the plants.

Massei et al. (2007) carried out an experiment with two clover species and wild-caught rabbits to see whether the smell of unpalatable *Trifolium* would act as a feeding deterrent, even in the absence of toxicity (Fig. 9.9), the likely active components being 2-butanone and 2-propanone which the highly unpalatable 'Aran' variety of white clover releases even when undamaged. Over a series of trials they found that although rabbits preferred to eat palatable red clover, *T. pretense*, they avoided it when it was made to smell like the unpalatable white clover, *T. repens* var. Aran.

---

4. Red and white control plants had both been sprayed with volume of corn oil equivalent to the volume of white clover extract used on the treated red clover plants.

---

5. The authors used the following artificial solution designed to mimic the natural odour profile of *Astragalus bisulcatus* (all in ppm): ethanol (10.5), acetaldehyde (10.1), pentanal (0.03), hexanal (0.03), octanal (0.03), heptanal (0.02), 2-hexenal (*trans*) 0.008, methyl sulphide (0.008), methyl selenide (0.008), dimethyl disulphide (0.008), dimethyl diselenide (0.008).

Chapter 10

AGGRESSIVE DECEPTIONS

Macbeth shall never vanquished be until
Great Birnam Wood to high Dunsinane Hill
Shall come against him.

William Shakespeare, from *Macbeth*

INTRODUCTION

This chapter covers 'aggressive mimicry' in its most commonly employed way, i.e. those systems in which the aim of mimicry is primarily to obtain food and mostly concerned with capturing prey. Schuett et al. (1984) suggested that this ought to be renamed 'feeding mimicry' because it is a disjunct or a bipartite system (see Chapter 1, section *Wickler's system*),

$$S1 + R - S2$$

while Wittenberger (1981) considered aggressive mimicry in the context of intraspecific interactions. This latter usage, although logical, is such a rare thing that I prefer to consider the intraspecific cases as forms of social mimicry (see Chapter 8, section '*Social' mimicry in birds and fish*).

Often a mimicry system can be shown to be simultaneously protective and aggressive to different dupes. Many ambush predators (Fig. 10.1) and ones that use lures (see Fig. 10.7) are successful because they are inconspicuous to their own predators. A different sort of example involves the widespread Kermadec petrel, *Pterodroma neglecta*, and Herald petrel, *P. arminjoniana*, which resemble large aggressive skuas in profile, colouration and flight behaviour (Spear & Ainsley 1993, J.M. Diamond 1994). These birds seem to be attacked significantly less often by skuas than the more typically coloured and similarly sized petrels. However, the mimicry also functions in the Kermadec's feeding behaviour since it also dupes other petrels into dropping or regurgitating their meals, something they normally do in response to, or to avoid, the onslaught of large and ferocious jaegers or skuas. In this system, the nature of the mimicry varies according to

Fig. 10.1 Spiders and bugs that employ crypsis to facilitate prey capture. (a) Green lynx spider, *Peucetia viridans*, in North America, having captured a wasp, cryptic among *Baccharis* sp.; (b) crab spider, *Thomisus labefactus*; (c) the European bug, *Phymata crassipes* (Hemiptera: Reduviidae), an ambush bug that, along with congeners, frequently awaits prey on flowers. (Source: b, Itchydogimages. Reproduced with permission from John Horstman.; c, © Evan Sowder 2013, via Flickr. Reproduced with permission from Evan Sowder.)

what the petrel is doing at any one time. Which role was more important in its original evolution cannot be said with any certainty but it seems most plausible that protection may have been its first role and that the responses of other petrels subsequently enabled the Kermadec to feast more easily. Vane-Wright (1976) adds another nuance to the case of the petrels and skuas, noting that skuas themselves have a rather hawk-like appearance, which might further encourage their intended prey to rid themselves of their latest food item.

Cryptic versus alluring features

Zabka & Tembrock (1986) noted that predators have two principal deceptive ("mimetic" in their sense) strategies to facilitate prey capture: (1) to remain invisible to the prey as long as possible and (2) to be alluring to them (see Table 1.7). It is not always easy to discern what the model might be in the latter case (X.J. Nelson 2014).

Dual-purpose aggressive protective mimicry is probably best known in sit-and-wait predatory insects and spiders, which sit on or among flowers that they resemble or simply appear like flowers, and so may avoid being recognised as potential prey by birds and simultaneously attract their prey. Gaining evidence to demonstrate both roles is not necessarily easy though (see later).

Most free-living, errant predators employ one or more modalities to remain undetected by prey as they approach them, including colour, morphology and various stealthy behaviours, such as silence and slow approach. These may include chemical neutrality or acquired similarity. Thus, stealthy approach by predators is also a form of crypsis; the stealthy behaviour is to deceive the prey that the predator is not there, or is not moving towards them, or is not a threat – until it is too late. Strategies may also include masquerade, such as in the carnivorous caterpillars of some Hawaiian geometrid moths, *Eupithecia* spp., that resemble a twig, but which can rapidly bend their body to grab a nearby insect.

Aggressive allurement includes devices such as the food-like (vermiform) lures of angler fish (Pietsch & Grobecker 1978), the flower-like form of the orchid mantis (O'Hanlon et al. 2014a), the prey sex pheromone mimicry of bolas (Eberhard 1977) and probably other spiders, the mimicry of auditory or bioluminescent sexual signals of some insects, or mimicking the group calls of primates by some big cats that cause their monkey prey to simply come and investigate until they get too close for their own good. I discuss many of these below, as well as a relatively small amount of experimental work that has been conducted on aggressive mimicry.

CRYPSIS AND MASQUERADE BY PREDATORS

There are countless examples of sit-and-wait predators that employ camouflage of various types to avoid detection by their prey, and in most cases this also serves as a protection against their own predators (Fig. 10.1). The Neotropical leaffish, *Monocirrhus polyacanthus* (Polycentridae), provides a nice and perhaps less familiar example. Popular in the aquarium fish trade, this Amazonian species drifts lifelessly in murky waters, showing a very strong resemblance to a dead leaf. This probably accounts for why 63% of its diet is comprised of other fish, including fast-moving ones, that have blundered too close (Catarino & Zuanon 2010).

Stealth

It may seem obvious, but if animals are hunting prey that can hear or see them they will generally have evolved to do so as quietly and slowly as possible. If the prey can smell, then evolution will have either led to reduced predator odour or behaviours to reduce its detection by prey. Anyone who has been close to a large cat in a zoo or in the wild will know that they can be quite malodorous. Such large mammalian predators tend to approach their prey from downwind, reducing the chance that the prey will be alerted by odour.

The apices of the flight feathers of owls are well known to have a soft fringe that reduces the formation of noise-producing vortices during flight and, not surprisingly, the detailed mechanisms have received considerable attention from researchers in military and aeronautical fields (Graham 1934, Lilley 1998). The morphological differences between owl wing feathers and those of more noisy fliers are a little more sophisticated than just having a fringe (Bachmann et al. 2007). The best studied of the owls in this respect is the crepuscular-hunting barn owl, *Tyto alba*, but all owls investigated have the same set of adaptations to silent flight: the leading edge of the outer feather vane has comb-like serrations whose function is not fully understood but is likely to be involved in reducing air turbulence. The inner vane's trailing edge has a fringe resembling the plumaceous barbs found at the bases of other birds' feathers, and the dorsal surfaces of the inner vanes have a velvety texture. In addition, barn owl flight feathers are relatively large and this leads to lower wing loading and hence the ability to fly more slowly to facilitate hunting. Barn owls use a combination of sight and bi-aural hearing to locate prey such as voles, and their silent flight not only helps to prevent prey from detecting the owl and taking evasive action, but also helps the owl to hear noises made by its prey. Voles have excellent hearing between 2 and 20 kHz, but the adaptations of the owl's feathers and slow-gliding flight means that any noise that is produced is at frequencies lower than 2 kHz and is therefore largely outside the vole's hearing range (Lilley 1998).

There are a set of trade-offs when stalking prey. The closer a predator can get before making the final dash, jump or strike, the higher the chance of a successful capture, but the longer the time spent in approaching the prey, the more likely it is to move on or detect the predator, or the hunt may be interrupted by competitors or other random events. The chance of a predator being detected as it approaches will depend on how visible it is. The salticid spider, *Plexippus paykulli*, modifies its prey-stalking behaviour according to whether it is camouflaged or not, stalking more slowly when camouflaged (Bear & Hasson 1997), and another salticid, *Portia fimbriata*, selectively changes its stalking strategy depending on prey species (Harland & Jackson 2001). Many mantids and chameleons that stalk their prey move in a quite distinctive way, not just moving slowly but also swaying from side to side or slowly rocking backward and forward but moving slightly more forward than backward each time. *P. fimbriata* is well known for its plasticity of prey capture behaviour (see section *Other prey and food mimicry* later). It primarily hunts and feeds on other salticid spiders, its commonest prey being *Jacksonoides queenslandicus*. In hunting them it uses a 'cryptic-stalking' strategy before getting close enough to jump on the prey, salticids being capable of quick, jumping escape. In cryptic stalking, the *Portia* holds its palps back next to its chelicerae in a typical resting posture, and responds to the target spider by stopping its approach when the prey is looking towards it. But when approaching other non-salticid,[1] prey spiders, it does not retract its palps and does not freeze. Salticids have exceptional visual capabilities and will quickly move away if they detect a threat, whereas other spiders are far less visually orientated, meaning that *Portia* does not need to take such precautions in its approach towards them.

Personal experience shows that swaying motion most likely circumvents some part of the target's perception of a potential threat approaching. Watanabe & Yano (2013) examined the prey capture success rate of the cryptic Japanese giant mantid, *Tenodera aridifolia*, in still and windy conditions and found that it both adjusted its approach

strategy towards prey depending on the wind conditions and managed to approach more successfully (i.e. came closer before detection) and had a significantly higher prey capture rate under windy conditions. When the wind ceased, the *Tenodera* would stop swaying and its approach towards prey was slowed. A similar use of wind to help disguise their approach towards prey has been observed in snakes (Fleishman 1985) and in *Portia* spiders that hunt web-building spider prey (Wilcox et al. 1996). In the laboratory, *Portia* used windy conditions (caused by a fan or simulated by mechanically vibrating the prey spider's web) as a 'smokescreen', apparently taking advantage of the fact that the movement caused by the wind reduced the prey's ability to detect and evade the approaching *Portia*. Furthermore, *Portia* only speeds up its approach in windy conditions when attacking spiders and not, for example, insects already trapped in the prey spider's web. I think this use of a 'smokescreen' could be termed stealth-by-proxy.

Shadowing

It may seem obvious that predators may try to avoid being detected by the prey they are stalking until the last possible moment. Thus lions and other large cats adopt a special low stalking posture, and chameleons and praying mantids move slowly and deliberately as they approach potential prey to within striking distance. Srinivasan & Davey (1995) have suggested that the actual path taken by the approaching predator as it tracks and homes in on a mobile prey may also be important. Specifically, they proposed that a predator will increase the effectiveness of its camouflage by following a path which keeps its image constant relative to the background from the point of view of the prey, the assumption being that if the optic flow of the predator resembles that which would be produced by a stationary object (save for a gradual increase in size), then it is less likely to be detected as a potential threat by the quarry. No empirical data that stalking predators actually do follow the calculated optimal trajectories was presented by Srinivasan & Davey (1995), but it ought to be relatively simple to test their proposal using examples such as hunting by big cats or aerial predators such as robberflies (Diptera: Asilidae) or dragonflies (Odonata).

Seasonal polymorphisms in predators

I have discussed seasonal colour polymorphism in detail in terms of its protective role in Chapter 2 (section *Winter pelage: pelts and plumage*), however, it can obviously also be

---

1. This also applies to when it was hunting individuals of the ant-mimicking salticid genus *Myrmarachne*, which it discriminated from other salticids.

important to predators. In many parts of the world the season has a dramatic effect on the background environment; for example, in rainy seasons vegetation is typically lush and green while in dry seasons, yellows and browns may dominate. In the north, snowfall can change the landscape to white for much of the year and so too predators' pelages.

Why seabirds are black and white (and grey)

One of the most widespread features of bird colouration which most people will have noticed is that seabirds, which collectively belong to several families, are almost entirely patterned with some combination of white, grey and black, and usually any patterns are typically fairly simple (G. Phillips 1964, K.E.L. Simmons 1972, Bretagnolle 1993). Brightly coloured feathers are essentially absent. Colouration among the albatrosses and petrels (Procellariiformes) correlates with foraging technique and foraging group size (Bretagnolle 1993). It is well known that most seabirds are predators of aquatic animals such as fish, squid or krill and they generally attack their prey from above by diving from the air or swimming underwater. Similar largely grey or white plumage is typical of a large number of either piscivorous or amphibian-eating wading birds (waders). The refraction of light across the surface of water means that when the surface is viewed from below, there is only a small circular zone – called Snell's window – through which the fish can see the sky. Lateral to this the aquatic animal will see reflections of things from below the surface (Fig. 10.2). There can be little doubt that in most cases, seabird and wader colouration facilitates prey capture because when viewed from below – as a prey animal would do – their white undersides would appear much like the sky above. The typically pale undersides of these birds camouflages them against the background of the sky. This hypothesis was tested by Phillips (1964), who photographed model seabirds (one half of underside black and one white) from under water, at various angles and under different surface conditions. When the model was close to the water surface (15 cm) and the camera moderately deep (1.22 m), the model was highly visible under all water conditions if it was directly overhead, but if it was near the periphery of Snell's window, the white half was virtually invisible. If the water surface has the slightest agitation, as is usually the case, birds near the window's periphery would be virtually undetectable to a fish below, which would probably have very little time to react to an attack from above the water. Some features of seabird colouration, however, such as the widespread occurrence of black wing tips, might be purely

an adaptation to reduce damage[2] due to the strong turbulence that the wing tips experience (Averill 1923), especially as many such birds rarely settle.

Bretagnolle (1993) used correspondence analysis to examine the relative importance of various foraging parameters on the colouration of 105 species of albatross and petrel. Feeding technique and foraging group size were shown to be the two most important determinants, but phylogeny, as suggested by generic placement, also had an effect.

Cairns (1986) noted that all-dark diving seabirds tend to be bottom-feeders, whereas those that are dorsally black and ventrally white tend to be midwater feeders, suggesting that colouration is primarily concerned with reducing conspicuousness to prey. However, not all seabird colouration may be for such camouflage purposes, and R.P. Wilson et al. (1987) have provided some evidence that the extra lateral dark stripes of some *Spheniscus* penguins may serve to render the birds more conspicuous, thereby causing their prey fish schools to become disrupted and so facilitate prey capture. In general, R.P. Wilson et al. (1987) found that most highly conspicuously patterned marine piscivores fed predominantly on pelagic fish that form schools, whereas these were seldom the prey of uniformly or countershaded species (Fig. 10.3).

Chemical crypsis by a predatory fish

Resetarits & Binckley (2013) describe a likely case in the predatory North American pirate perch, *Aphredoderus sayanus*, which appears to be chemically camouflaged to two groups of prey: water beetles and amphibian (such as frogs) larvae. The latter are seldom common in the diets of predatory fish, largely because ovipositing frogs actively avoid patches of water in which such fish occur. Resetarits & Binckley (2013) assessed the oviposition behaviour of three species of *Hyla* treefrog in relation to the presence of each of nine species of freshwater fish, collectively representing six families in four orders. Frogs showed significant avoidance of all of them except for *Aphredoderus*. They also examined colonisation by water beetles of tanks or pools containing pirate perch, another predatory fish, the banded sunfish, *Enneacanthus obesus*, or no fish (control). Here they found no significant reduction from control values in either numbers of individual beetles (Fig. 10.4a) or species (Fig. 10.4b)

---

2. Black colouration, often associated with melanin pigment, is associated with strengthening biological materials, including keratin and chitin, though it is not the only mechanism (Burtt 1979).

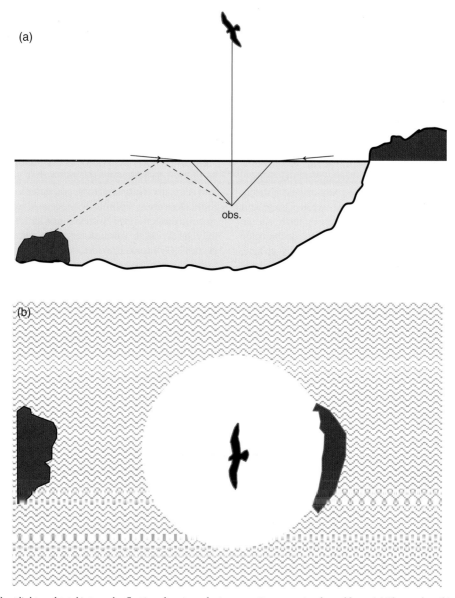

Fig. 10.2 Incident light and total internal reflection showing what an aquatic prey animal would see. (a) The angles of incident light rays; (b) view from the submerged observer, the view of the sky being Snell's window, defined by a cone from the observer with sides 49 degrees from the vertical all around. Transparent organisms with optical densities higher than that of the surrounding water, near the surface but just outside of Snell's window, may refract light such that it reaches the fish's eye, thus rendering them visible. (Source: Adapted from Phillips 1964. Reproduced with permission from *Diver*.)

when the water contained pirate perch but did when they contained the predatory sunfish, *Enneacanthus* (Centrarchidae) The authors therefore conclude that the pirate perch does not release any chemical cues into the water that either frogs or beetles may use in assessing the presence of potential predators before ovipositing.

ALLURING MIMICRIES

White & Kemp (2015) surveyed the use of colour (human visible spectrum plus UV) as lures. It has been known for some time that the brightly coloured bodies of a number of orb-weaving spiders act as general attractants for generalist

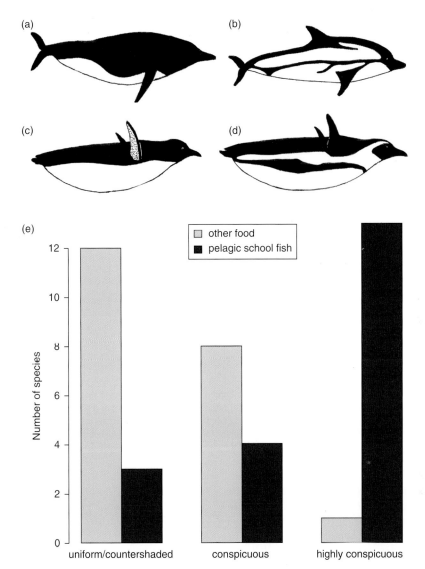

Fig. 10.3 Relationship between conspicuousness of patterning of pelagic vertebrate piscivores and whether their diet is primarily schooling pelagic fish. Although the data are not controlled for phylogenetically, each of the conspicuousness categories includes a mix of cetaceans and penguins. (a) Northern right whale dolphin, *Lissodelphis borealis*, a nocturnal squid feeder; (b) striped dolphin, *Stenella coeruleoalba*, a pelagic fish feeder; (c) Adelie penguin, *Pygoscelis adelie*, a typical non-*Spheniscus* species; (d) adult *Spheniscus* penguin; (e) bar plot of food type versus colouration. (Source: Adapted from Wilson et al. 1987. Reproduced with permission from Elsevier.)

pollinators, and thus can be viewed as employing very non-specific flower mimicry. Craig & Ebert (1994) demonstrated this by constructing a grass shield to hide either the dorsal or ventral surfaces of the orb-weaving spider *Argiope argentata* (see the similar-looking *Argiope* in Fig. 3.4), and found that prey captured from the obscured side of the spider was reduced. The UV-reflecting dorsal surface of the spider was particularly attractive during the dry season. Other diurnal (or day and night) foraging spiders for which there is evidence that bright colours are important include members of the genera *Gasteracantha*, *Nephila*, *Cyrtaphora*, *Diadea* and several others.

Perhaps surprisingly, nocturnally foraging moths may be subject to alluring visual mimicry by spiders. The orb web-weaving spider *Leucauge magnifica* has a silver, pale green and black-striped dorsum and, like many others, has strong

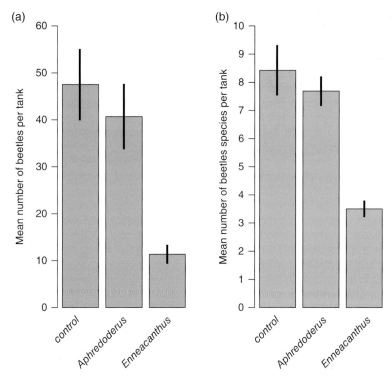

Fig. 10.4 Presence of pirate perch, *Aphredoderus sayanus*, in experimental outdoor tanks does not significantly reduce colonisation of the tanks by aquatic beetles compared to control tanks with no fish in them, whereas the presence of a banded sunfish, *Enneacanthus obesus*, in a tank reduced beetle colonisation. (a) Total numbers of colonising beetles and (b) number of colonising species in each treatment. The interpretation is that pirate perch do not give out chemical signals that would deter beetles from colonising the tanks. (Source: Adapted from Resetarits & Binckley 2013.)

longitudinal yellow and black marks on the underside of its abdomen. Large moths constitute a considerable proportion of its total prey. Tso et al. (2007) investigated the possible role of the colouration in prey capture by this species by painting either the dorsal or ventral surface of the abdomen to conceal the patterns. Their results were quite striking (Fig. 10.5), in that while concealing the dorsal markings had little effect, concealing those on the ventral side significantly reduced the spiders' success at catching moths. Furthermore, Zhang et al. (2015) have demonstrated that a prominent white stripe on the cephalothorax of the common nocturnally active huntsman spider, *Heteropoda venatoria* (Sparassidae), also acts as a visual lure predominantly attracting moths.

Larvae of the cave-dwelling antipodean fly genus *Arachnocampa* (Keroplatidae) are quite remarkable in that they hang down from the roofs of their caves on long silk threads and they bioluminesce, the light being conducted along the threads. Many larvae do this, forming a glowing curtain of strands that in some places are a tourist attraction (see figure 26 in Wickler 1968, as *Bolitophila*). The light they produce serves to attract a variety of different prey (Diptera, Isopoda, Hymenoptera, Myriapoda, Gastropoda) (Broadley & Stringer 2001, R.E. Willis et al. 2011). Why some nocturnal insects and other animals are attracted to lights is a different matter.

Flower mimicry

Several large mantids appear to mimic flowers. For example, the African devil's flower mantis, *Idolomantis* (=*Idolum*) *diabolicum* (see figure 32 in Wickler 1968), and the South-East Asian orchid mantis, *Hymenopus coronatus* (Fig. 10.6a–c), and it has often been assumed that their form, colour and display serve to attract pollinating insects, their principal prey. Recently, quite a lot of experimental work has been carried out on the orchid mantis and supports the flower mimicry hypotheses. The juvenile has an enlarged pronotum and enlarged pink femoral lobes on its middle and hind legs

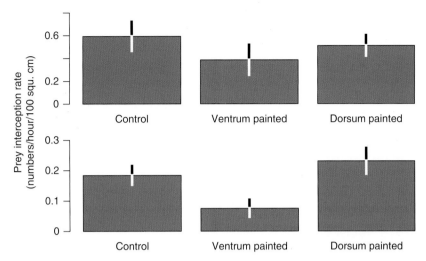

Fig. 10.5 Demonstration of the importance of ventral markings on the brightly coloured web-building orchid spider, *Leucauge magnifica*, in prey capture success. The upper bar chart shows interception rates (mean ± S.E.) for all insects, the lower one, rates for moths only. (Source: Adapted from Tso et al. 2007. Reproduced with permission from Elsevier.)

and the rest of the body is largely white, the overall effect being that of a flower's large petals (Fig. 10.6a). Under UV light, the only part that shows any marked reflectance is the area of the wing buds, which are located more or less in the middle of the body and this could be mimetic of the central part of a flower (O'Hanlon et al. 2013). O'Hanlon et al. (2014a, 2014b) found that the colouration both of the mantids as would be perceived by pollinating insects and, in particular, of their femoral lobes matched that of numerous sympatric flowers and, using geometric morphometric analysis (O'Hanlon et al. 2014b), that they were within the range of variation shown by many sympatric flower petals. However, the mantids were not mimics of any particular species. Rather they appear to be using a generalised food-deception strategy as do many non-rewarding flowers (see Chapter 12, section *Pollinator deception*). Investigating the prey capture success of *H. coronatus*, O'Hanlon et al. (2015) found that it made no difference whether they were resting on or near flowers or just sitting on foliage, but they were significantly more successful when in locations with many flowers, which supports the 'magnet species hypothesis', i.e. that the presence of real flowers leads to a higher density of potential pollinator prey rather than the alternative 'remote habitats hypothesis', which would suggest that a high density of flowers would act as competitors for the mantid thus reducing its capture success. However, the orchid mantids showed no preference for either flower-rich or plain vegetation foraging sites. Interestingly, Mizuno et al. (2014) found that juvenile females had a higher success rate at capturing oriental honey bees, *Apis cerana*, than adult females and discovered that

the juvenile mandibular glands release 3-hydroxyoctanoic acid (3HOA) and 10-hydroxy-(*E*)-2-decenoic acid (10HDA), which are both components of the pheromone communication system of the oriental honey bee. Furthermore, the juvenile mantids only released significant amounts of 3HOA when they were actively trying to capture prey. Bioassays using dummies impregnated with appropriate ratios of 3HOA and 10HDA showed that the dummies were attractive to the bees, and thus it seems likely that at least the juveniles are also using a deceptive chemical lure.

Because of the frequent presence of ambush predators, flower visiting by potential pollinators is a risky business and some pollinators are wary of the risks. Dukas & Morse (2003, 2005) found that honey bees visit flowers with crab spiders (Thomisidae) less frequently, and avoid flowers where they have experienced a (failed) ambush. Hoverflies of the genus *Sphaerophoria* (Syrphidae) hesitate before landing on a flower to check whether there is an ambush predator there, almost never alighting on a flower if a cryptic thomisid spider, *Thomisus labefactus*,[3] is on top of it and alighting significantly less even if there is a spider below a flower compared with when no spider is present at all (Yokoi & Fujisaki 2008).

Larvae of the hoverfly *Ocyptamus* sp. (Syrphidae) in Costa Rica live among colonies of whitefly but, unlike most related species, they do not feed on the whiteflies themselves.

---

3. Experiments conducted with thomisid corpses arranged in a life-like fashion.

Fig. 10.6 Flower mimicry and crypsis in praying mantids. (a–c) The South-East Asian orchid mantis, *Hymenopus coronatus*: (a) juvenile, (b) intermediate instar, (c) adult male; (d) flower mantid, *Pseudocreobotra wahlbergi*, on an inflorescence of *Coptosperma littorale* (Rubiaceae), Mozambique; (e) the Chinese mantid, *Creobroter gemmatus*, which can be cryptic among foliage and some flowers. (Source: a, Raffi Kojian (http://Gardenology.org) 2011. Reproduced under the terms of the Creative Commons Attribution Share-Alike Licence, CC BY-SA 3.0, via Wikimedia Commons; b, Ettore Balocchi, Hectonichus 2009. Reproduced under the terms of the Creative Commons Attribution Share-Alike Licence CC BY-SA 3.0, via Wikimedia Commons; c, Sander van der Wel 2010. Reproduced under the terms of the Creative Commons Attribution Share-Alike Licence CC BY-SA 2.0, via Flickr; d, Antonius Rulkens. Reproduced with permission from Antonius Rulkens; e, Itchydogimages. Reproduced with permission from John Horstman.)

Instead they wait until they are probed at by adult flies of other species visiting to feed on the sugary honeydew produced by the surrounding whiteflies. When a prey comes near, the *Ocyptamus* lashes out at the fly, biting, subduing and consuming it (Ureña & Hanson 2010). There is thought to be some chemical mimicry involved.

Rain mimicry

A rarely discussed mimicry is commonly observed among worm-eating birds such as gulls (Laridae), predominantly European species, which display a characteristic paddling behaviour on the ground, attracting earthworms to the soil surface. It has been suggested that such paddling mimics rain, which is a signal to worms that they may surface to forage for leaves, although N. Tinbergen (1962) suggests that the vibrations produced may imitate those of the subterranean earthworm predator the mole, *Talpa europaea*. I can see no reason why both might not be true. It can be a very successful strategy and, indeed, there is a curious and competitive pastime in the UK called worm-charming where contestants use the same sort of deception to lure worms to the surface – the charmer with the biggest catch wins. This human pursuit could be regarded as a case of mimicking mimicry. In North American gulls there are very few observations of terrestrial foot-paddling for worms, but O'Donnell (2008) reports one case which is interesting because the worms being duped in his study are the result of accidental human introduction, since endemic large lumbricid earthworms are absent from most of the gull's range. It is not clear whether the bird learnt the behaviour itself, or from watching another, but it is not a common behaviour in the species. In the Eurasian black-headed gull, *Chroicocephalus* (previously *Larus*) *ridibundus*, the behaviour is known to be innate (Rothschild 1962). Apart from gulls, there is also a record of foot-paddling, or perhaps better called foot-stomping, for worms by the wood turtle (Kaufmann 1986).

Physical lures

Lures that in some way resemble a prey species' own prey have evolved in a large number of different predators. Some dozen bird species in six families, though mostly herons (Ardeidae), actually use a bait in their fishing (Ruxton & Hansell 2011). These birds select a buoyant object and toss it onto the water surface within reach and if it attracts an attack from a fish, the bird has a good chance of getting a meal. Most other lures are specific physical entities combined with appropriate manipulation, but purely behavioural luring of prey also occurs widely, e.g. in some spiders and even big cats.

Angling fish

Some of the best-known and most-convincing examples of lures are the escas on the ends of the elongated anterior rays of the dorsal fins in various anglerfish (Fig. 10.7), including members of the Antennaerridae (Pietsch & Grobecker 1978), Lophiidae, Ceratiidae (Gudger 1946), Melanocetidae, etc., and even a shark (Widder 1998). Further examples include the angler catfish, *Chaca chaca* (Chacidae), which is reported to use its maxillary barbels the same way (Roberts 1982), though not everyone agrees on this point, the batfish, *Ogcocephalus vespertilio*, from Brazil, which has an illicium that it uses while walking on its fins during the early morning (Gibran & Castro 1999) and the flounder, *Asterorhombus fijiensis*, which uses a dorsal fin ray with a membranous distal structure which the fish vibrates near its mouth (Amaoka et al. 1994). The American alligator snapping turtle, *Macroclemys temminckii*, has a worm-like protuberance at the base of its mouth (Fig. 10.8). In the turtle and some of the angler fish the lure is muscularised, enabling the owner to wiggle it, thus enhancing the deception. In some of the frogfish, the lure is a multi-tasseled structure that is gently wafted in the water, with the tassels moving in the eddies.

Members of two genera of South American characid fish, *Corynopoma* and *Pterobrycon*, utilise very different lures to attract females (see Chapter 11, section *Food dummies and sex*). Grafen (1990) points out that with such systems there is no opportunity for an arms race, i.e. for the model worms to become less worm-like, and that the anglers are almost inevitably going to be far rarer than the worms that their lures mimic. Therefore, despite death being the consequence to a small vermivorous fish mistaking a lure for the real thing, their need to feed on worms and similar wriggling prey that will have diverse attributes means that it is unlikely that they could find a good clue in the lure itself to distinguish it from a worm. Nevertheless, the proximity of the anglerfish itself certainly would be a give-away so they often display spectacular camouflage or disruptive colouration and certainly do not have the typical appearance of a predatory fish. Furthermore, many anglerfish are quite variable in their colouration. The skins of some shallow water anglerfish resemble a bed of bryozoans or sponges in colour and texture (Y. Leroy 1974; Fig. 10.7c,d).

In the case of several of the deep-sea anglerfish, the lure is bioluminescent, which presumably makes it more attractive to many small vermivorous predators. For example, the lure of the deep-sea anglerfish *Himantolophus groenlandicus* (Himantolophidae) has a central bioluminescent 'bulb' as well as surrounding tassels. Bioluminescence may also play a role in luring prey in some lantern-eye or flashlight fish (Anomalopidae), which have oval cavities beneath each eye

Fig. 10.7 Various 'angler' fish with lures. (a,b) Warty frogfish, *Antennarius maculatus* (Antennariidae), from Sempini and Ambon, respectively; (c,d) unidentified frogfish, the second one possibly being *Antennarius pictus*; (e,f) striate anglerfish, *Antennarius striatus*; (f) detail of worm-like muscular lure. (Source: a,b, © Linda Pitkin Underwater Photography. Reproduced with permission from Linda Pitkin.; c–f, Denis Riek. Reproduced with permission from Denis Riek.)

that contain light-emitting bacterial colonies, although at present this alleged function is no more than a guess. Bioluminescent luring is also reported by Haddock et al. (2005) in a deep-sea siphonophore, which twitches the lures to attract fish. Interestingly, in this case the wavelength of the bioluminescence is modified by the presence of a material that fluoresces red, and thus prey are lured by a red rather than a typical green fluorescence. This is probably because red light plays a more important role in deep-sea communication.

Fig. 10.8 Pink lure in the bottom of the mouth of an alligator snapping turtle, *Macroclemys temminckii*. (Source: a, L.A. Dawson 2007. Reproduced under the terms of the Creative Commons Attribution Licence, CC BY-SA 2.5; b, Benny Mazur 2009. Reproduced under the terms of the Creative Commons Attribution Licence, CC BY 3.0, via The Biomimicry Institute, www.asknature.org.)

Caudal (and tongue) lures in reptiles

Many snakes and some lizards, especially juveniles, have been observed to move the tips of their tails in a worm-like manner while the rest of the body remains still, and it has long been assumed that such movements are alluring to potential worm predators which are also snake prey, such as frogs or lizards (Neill 1960, Heatwole & Davidson 1976, Leal & Thomas 1994, Hagman et al. 2008, Reiserer & Schuett 2008). Evidence that in the case of juveniles of the rattlesnake, *Sistrurus catenatus*, caudal luring, as it is called, is attractive to frogs (*Rana* spp.) was provided by Schuett

et al. (1984). They noted that the snake's tail movements became more pronounced when the prey was moving, and that at these times the snakes also refrained from tongue flicking, which might give their true identity away. Finally the frogs snapped at the alluring tail tips only to be struck, killed and consumed by the snake. Schuett et al. (1984) further suggest that caudal luring behaviour might have led to the evolution of the rattlesnake's rattle, with the incipient rattle serving to enhance the luring effect of the tail movements and perhaps also to protect the tip of the snake's tail against the actions of the lured victims, because during the early stages of its evolution the rattle would not have been able to produce noise. Caudal luring by snakes has been described in members of several families, including Boidae (J.B. Murphy et al. 1978), Colubridae (Sazima & Puorto 1993), Elapidae (Chiszar et al. 1990, Shine 1980), Pygopodidae (Murray et al. 1991) and various other Viperidae (H.W. Greene & Campbell 1972, Heatwole & Davidson 1976, C.C. Carpenter et al. 1978, Sazima 1991, Parellada & Santos 2002) and has been reviewed by Heatwole & Davidson (1976) and Pough (1988a).

The only experimental data that I have been able to find on the effectiveness of caudal lures in snakes is the study by Farrell et al. (2011), which involved manipulation of the colour of the snake's tail. In this study they captured, modified and released 169 neonate pygmy rattlesnakes, *Sistrurus miliarius*, and compared them upon recapture over the following three months. Snakes were treated by colouring their yellow tails either with black pigment similar to their cryptic body colour or with yellow ink as a control, but no differences were detected between treatments in terms of either growth rate or presence of palpable food in their digestive tracts. However, there can be no doubt that lures do work and one most convincing example is provided by *Pseudocerastes urarachnoides* (Viperidae), an endangered endemic snake from the mountains of western Iran, described recently by Bostanchi et al. (2006). It has a rather short tail, even for a snake, the distal supracaudal scales forming a knob-like structure and just anterior to that, the lateral dorsal caudal scales forming elongate projections, the result of which resembles the body and legs of a large mygalomorph spider (Fig. 10.9). The snake has aptly been given the English common name the spider-tailed horned viper.

Some aquatic, mangrove and saltmarsh snakes use their tongues as lures to attract their fish prey (Welsh Jr & Lind 2000, Hansknecht 2008). Snakes often use their forked tongues to 'taste' the air – scents adhere to the fork and are transferred to receptors in the roof of the mouth, and the air-tasting flicks are usually quite brief. However, in the mangrove saltmarsh watersnake *Nerodia clarkii compressi-cauda* (Colubridae), Hansknecht (2008) has demonstrated

Fig. 10.9 The Iranian viper, *Pseudocerastes urarachnoides*, showing its highly specialised caudal lure that resembles a spider. (Source: Omid Mozaffari. Reproduced with permission from Omid Mozaffari.)

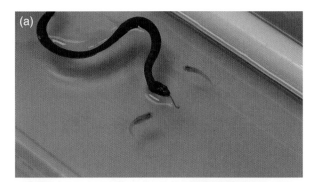

Fig. 10.10 Lingual-luring by the mangrove saltmarsh watersnake *Nerodia clarkii compressicauda* (Colubridae) being used to attract prey fish. Note the loop made by the apical forks. (a) Attracting guppies, *Poecilia reticulata*; (b) attracting *Xiphophorus ?variatus*. (Source: Dr Kerry A. Hansknecht. Reproduced with permission from Dr Hansknecht.)

the effectiveness of lingual-luring, in which the tongue remains protracted for a longer time with the tips curved over themselves in a very different manner from normal air tasting (Fig. 10.10). Some arboreal snakes also protract their tongues for longer than the normal quick flicks used to 'taste' the air, but Keiser (1975) provided evidence that in at least one such case the reason may be to do with crypsis rather than luring as had previously been suggested.

Caudal lure in a dragonfly

Edgehouse & Brown (2014) have recently described the first case of caudal, strictly abdominal, luring in an insect. The large aquatic larvae of the dragonfly, *Aeshna palmata*, upon seeing the smaller larva of the damselfly, *Argia vivida*, adopted a laterally curved body posture and started rapid abdominal lateral movements. These attracted the attention of the damselfly larva which then re-orientated itself to face the *Aeshna* abdomen, at which point the dragonfly larva struck and consumed it. The authors also observed one unsuccessful but similar luring attempt by the smaller species.

Death feigning as a lure

It seems surprising to me that there are so few reported cases of death feigning being used to attract insect prey by animals, though it is well known in plant pollination deceptions. Forsyth & Alcock (1990a) describe how the Neotropical rove beetle, *Leistotrophus versicolor* (Staphylinidae), deposits a droplet of a foul-smelling material from its abdomen onto the substrate. This acts as a lure to attract small flies (Drosophilidae and Phoridae) that either forage or oviposit on rotting material; thus lured, they become the beetle's prey. It is not known for certain whether the foul smell of some bird dropping mimicking spiders (see Chapter 3, section *Bird dropping resemblances*) is also an aggressive mimicry for luring coprophagous insects but it seems likely.

McKaye (1981) describes an aggressive corpse mimicry, often referred to as thanatosis, in cichlids of the genus *Nimbochromis* from Lake Malawi. *Nimbochromis* are large predatory fish which will rest on the lake bottom, immobile on their sides and develop a more blotchy colouration such that they resemble the body of a dead fish. This attracts the attention of various scavenger species that then become their prey. Similar behaviour has also been observed in *Lamprologus lemairii* from Lake Tanganyika (Lucanus 1998) and the Central American yellowjacket cichlid *Parachromis friedrichsthalii* (Tobler 2005).

Other prey and food mimicry

Mimicking a potential prey animal to attract a predator might seem dangerous but it has evolved on several occasions. Wignall & Taylor (2011) describe how the assassin bug, *Stenolemus bituberus* (Reduviidae), which is a specialist predator of web-building spiders, enters the spider's web and plucks its silk lines in a manner that mimics another insect becoming entangled. The spider duly rushes to attack the 'prey' only to be killed by the assassin. In this case, the vibrations produced by the bug were recorded and shown to closely resemble those caused by a subset of the spider's normal prey and to differ significantly from the similarly attractive vibrations produced by courting male spiders to allure the female. Similar behaviour occurs in *Argyrodes*, though these spiders are normally kleptoparasitic within host webs and tend to only attack the host when alternate prey are rare (Whitehouse 1986, Elgar 1993).

Several spider-eating spiders employ the same strategy (R.R. Jackson 1995a, 1995b). Members of the aranaeophagic jumping spider genus *Portia* (Salticidae) imitate the vibrations that prey cause when they become ensnared by a web-building spider's web, but otherwise often resemble dead leaves (e.g. *P. fimbriata*) that may have fallen into the web, though web-building spiders typically have rather poor eyesight so this resemblance is perhaps unlikely to be part of prey deception. In response to the vibrations caused by the *Portia*, the web-builder rushes out to attack/wrap the imitated prey, but *Portia*, with its acute vision, is careful when attacking its potentially dangerous prey and will often take detours to break visual contact with its intended victim and leap on it from behind, thus minimising risk to itself. Some other aranaeophagic salticids also invade webs and use aggressive mimicry, and they may modify their precise behaviour in a trial-and-error way until they hit on a winning strategy (R.R. Jackson 2002). The same strategy is employed by various other spiders, such as the European spiders *Pholcus phalangioides* (Pholcidae) and *Ero furcata* (Mimetidae). Both of these invade other spiders' webs and vibrate them so as to mimic a prey insect (Gerhardt 1926, Bristowe 1941, Jackson & Brassington 1987) or, in the second case, to mimic a courtship signal (Czajka 1963) until the resident spider approaches close enough to be attacked. Very little is known about the hunting strategies of other members of the Mimetidae, but similar aggressive mimicry has been described for a few species of *Mimetus* (R.R. Jackson & Whitehouse 1986, Kloock 2012), so aggressive mimicry might be the norm as the family's name suggests.

Portia spiders are particularly versatile in their mode of prey capture (R.R. Jackson & Wilcox 1993). One species from the Philippines often dupes a particularly dangerous spitting spider (Scytodidae), which is also a specialised aranaeophage, in this case making only faint vibrations in its prey's web that cause approach but not full-scale attack, and then pounces on the spider from behind, a behaviour that appears to be innate in this species (R.R. Jackson et al. 1998). *Portia* also do a lot of trial-and-error learning about how to deal successfully with different species of prey spider. The Australian species, *P. fimbriata*, instead of mimicking a web-builder's prey, imitates the courtship vibration 'song' of the *Euryattus* spiders (R.R. Jackson & Wilcox 1990).

The case of the German cockroach

Eliyahu et al. (2009) have described a very interesting case of conjunct mimicry by nymphs of the widespread anthrophilic cockroach *Blattella germanica*, which release a mimic of female sex pheromone. In this species adult males respond to nearby receptive females by raising their wings, thus exposing abdominal tergal glands whose reservoirs contain phagostimulatory substances. The females respond to this by mounting the male and consuming the nuptial food gift and thus attain a position more or less correct for copulation to proceed. However, antennal secretions from juveniles can also elicit the sexual response of an adult male and the young take advantage of the tergal gland's food source. Of course, the young might be relatives of the adult but in most cases this is unlikely and so the adult males are probably duped into feeding without gaining any reward and potentially with lost fertilisation opportunities. Eliyahu et al. (2009) identified some of the adult female's courtship pheromones (four out of six compounds) in secretions from penultimate instar female nymphs, but the courtship stimulatory compound of other nymphs appears to be a novel compound not present in the normal adult female pheromone blend.

WOLVES IN SHEEPS' CLOTHING

This term is applied to a class of aggressive mimicries in which predators or spies appear more like something innocuous and thereby deceive prey about their true threat. Wickler (1968), for example, cites instances where wartime spies pretended to be members of the Red Cross.

Vulture-like hawks

Many ornithologists have noted that the North American zone-tailed hawk, *Buteo albonotatus*, bears a striking resemblance to the turkey vulture, *Cathartes aura*, though, unlike the scavenger vulture, it is an active predator.[4] Vultures are ubiquitous and common and most animals presumably get used to seeing them regularly and so, being habituated to the vultures, might not recognise the threat posed by the zone-tailed hawk until it is too late (E.O. Willis 1963). However, Mueller (1971b) suggested that, indeed as Willis had mentioned, aerodynamic effects might be the driving force for the resemblance, and noted that the same general wing shape was displayed by other raptors that habitually soar near the ground, such as harriers, *Circus* spp. Observations on hunting by the hawk and also of the mobbing on one by other birds were presented by Zimmerman (1976), who noted that the *Buteo* was on the receiving end of less mobbing than were other raptors. The hawk is generally a rare bird that forages far from the nest and thus potential prey such as ground squirrels are relatively unlikely to encounter many of them in their lives. The flight pattern of the bird as it approached the ground was confirmed as being similar to that of vultures. Thus it only essentially gives away that it is a predator when it makes a dive in an attempt to catch a prey, whereafter other birds and mammals would be alerted and sometimes mob it. Thus far no one appears to have tested these alternatives but it seems likely to me that resemblance could indeed be coincidental but nevertheless might now lead to enhanced prey capture.

Negro (2008) comments on the remarkable visual similarity between two aberrant snake-eating eagles and bird-eating raptors. This, like many other similarities, may have multiple functions and three plausible possibilities are suggested. The similarity might:

- deceive their snake prey which might not flee from bird-eating raptors,
- lower predation or harassment by the model or other predators, and/or
- reduce mobbing by small birds that normally avoid doing so to bird-eating raptors because of the innate risks.

E.O. Willis (1976a) postulated a rather different explanation for the resemblance between two raptors with very different diets – the insect-eating rufous-thighed kite, *Harpagus diodon*, and the bird-eating bicoloured hawk, *Accipiter bicolor*. In this case, it was postulated that the hawk model scares away birds because of the danger it poses to them and also causes them to keep quiet; by achieving the same effect, the mimic kite is left alone, more easily to find and capture its cicada prey.

Sazima (2010) is somewhat more skeptical because of differences in the ranges of the birds and because no evidence

is provided that the kite uses sound to locate its prey, but the possibility exists. Sazima (2010) is also fairly skeptical about there being mimicry involved in the resemblance between the snail-eating hook-billed kite, *Chondrohierax uncinatus*, and various other small Neotropical hawks and falcons. Edelstam (2005) had proposed that this might be an instance of Batesian mimicry, as the kite is slower flying and has less-sharp talons than the other raptors, but Sazima (2010) suggests that it might just possibly be an aggressive mimicry on the part of the other species since their small mammal and bird prey would be less likely to be concerned about a snail-eating bird. Nevertheless, it all seems a little tenuous.

Cleaner fish and their mimics

Temporary associations exist in both marine and freshwater environments between different species of fish where individuals of one species, the cleaner, sets up a station which is visited by various larger fish, even sometimes predatory ones, that present their mouths, gills and bodies so that the cleaner can feed on ectoparasites and bits of dead skin (Losey 1971, Sulak 1975). The best-known and most-studied examples occur on coral reefs and the cleaner fish are typically wrasses of the genus *Labroides*, though the behaviour is also shown by a number of very similarly coloured gobies, such as *Gobiosoma*. In addition to particular colour patterns that customers recognise, they also often have particular swimming dances (see figure 38 in Wickler 1968). Although other cleaner associations are known, these do not seem to have associated colour convergence or involve mimicry (see Sulak 1975, Brockmann & Hailman 1976).

Cleaner fish and their mimics on tropical reefs have been known about for a long time. The first description of the similarity between the widely distributed bluestreak cleaner wrasse, *Labroides dimidiatus* (Labridae), and the fangblenny, *Aspidontus taeniatus* (Fig. 10.11), was by K.H. Barnard (1927) based on his fieldwork in the Gilbert Islands located in the Pacific Ocean. His observations revealed most of the essential features of the cleaning symbiosis and its mimicry. He noted that the cleaner wrasse had an unusual mode of swimming, oscillating its posterior in a swimming dance (see figure 38e in Wickler 1968), and was feeding on ectoparasitic copepods and isopods which it picked off the bodies of other fish. He concluded that the fangblenny mimicked the cleaner wrasse to gain protection from the other predatory fish. In fact, the fangblennies take further advantage of their duped customers' obliging tranquility by taking bites out of them. The cleaner wrasse and its mimics show colour polymorphism in parts of their range (Randall 2005), but it is not known whether this particular fangblenny can

---

4. See Sazima (2010) for nice colour photographs illustrating this example.

Fig. 10.11 Cleaner fish and their mimic species and their forms. (a) Cleaner wrasse tending a 'customer'; (b) interaction between individuals of bluestreak cleaner wrasse, *Labroides dimidiatus*, blue-striped fangblenny, *Plagiotremus rhinorhynchos*, and fangblenny, *Aspidontus taeniatus*; (c) interaction between juvenile model *L. dimidiatus* and its mimic, *P. rhinorhynchos*; (d) *P. rhinorhynchos*; (e) non-mimetic form of *P. rhinorhynchos*, with two prominent lateral blue stripes; (f) non-mimetic orange form of *P. rhinorhynchos*. (Source: a, © Linda Pitkin Underwater Photography. Reproduced with permission from Linda Pitkin.; b–f, Dr Even Moland. Reproduced with permission from Dr Moland.)

match its model's colour through active colour change. Interestingly, whether *A. taeniatus* makes use of aggressive mimicry is geographically variable, with populations in the Red Sea, in Indonesia and on the Great Barrier Reef rarely using this strategy and instead feeding primarily on other items, while in Polynesia the fish is mainly an aggressive mimic. Thus its resemblance to cleaner wrasse in some localities may be primarily protective (Cheney et al. 2014a).

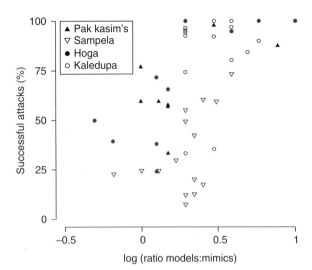

Fig. 10.12 Successful attacks made by cleaner fish mimicking blue-striped fangblennies, *Plagiotremus rhinorhynchos*, at four sites near Hoga Island, in the Wakatobi National Marine Park, southeast Sulawesi, Indonesia, depend on the relative numbers of models and mimics in the local population. (Source: Adapted from Cheney & Côté 2005. Reproduced with permission from The Royal Society.)

Study of the cleaner fish mimicry systems has been hugely aided by the invention and popularisation of subaqua equipment (Limbaugh 1961) which has allowed experimental work to be carried out in the field, leading to the elucidation of further subtleties. Cheney & Côté (2005) showed that the attack success frequency of another aggressive cleaner wrasse mimic, the blue-striped fangblenny, *Plagiotremus rhinorhynchos* (Fig.10.11d–f), is positively related to the relative abundance of models (Fig. 10.12).

The aggressive mimicry between *P. rhinorhynchos* and *Labroides dimidiatus* is facultative (Côté & Cheney 2005) and spectral analysis has shown that it is also very accurate (Cheney et al. 2008). Moland & Jones (2004) demonstrated experimentally that *P. rhinorhynchos* is an aggressive mimic of *L. dimidiatus*. Some juveniles of the former often closely resemble the latter, but only when they are of a similar size to a cleaner and also in close proximity to one. The two species showed a close spatial relationship with about 50% of juvenile fangblennies being within 1 m of a cleaner wrasse. Moland & Jones (2004) found that when cleaner wrasse were experimentally removed, the fangblennies changed colour to a general non-mimetic one and subsequently suffered an approximately 20% reduction in foraging success. The mimetic fangblennies also received less aggression from their dupes even after taking a feed from them (Côté & Cheney 2007) compared with when the dupes were attacked by the closely related but non-mimetic *Plagiotremus*

tapeinosoma, indicating that they still partly associate the colour with a helpful species. Perhaps the bitten fish are not certain (speaking anthropomorphically) that the fish that looks like a harmless cleaner really bit them; evolutionarily, perhaps the cost of losing the services of real cleaners outweighs the relatively rare nips from the fangblennies. By artificially increasing the parasite load on staghorn damselfish, *Amblyglyphidodon curacao*, Cheney & Côté (2007) were able to show that the need for cleaning affected its level of discrimination such that fangblennies were able to make more successful attacks. Thus there is apparently a trade-off for the damselfish between the risk of permitting a mimic too close and the need to rid itself of a parasite load.

In at least some cleaner fish systems, the resemblance to the model cleaner fish not only facilitates feeding but also leads to fewer attacks by predators. For example, Cheney (2010, 2013) demonstrated this for the bicolour fangblenny, *Plagiotremus laudandus*, and suggested that their mimicry might be best considered as a combined Batesian–aggressive mimicry system.

While the resemblance between *Aspidontus* and their cleaner fish models (*Labroides*, *Gobiosoma* and *Plagiotremus*) is clearly aggressive, perhaps a separate term ought to be applied to the non-deceptive convergence in behaviour and colouration among the different species – cleaner fish syndrome. By no means all cleaner fish associations involve mimicry though (e.g. Wyman & Ward 1972).

Mingling with an innocuous crowd

Boileau et al. (2015) have described a very similar system involving a scale-eating cichlid fish, *Plecodus straeleni*, from Lake Tanganyika, which closely resembles two other species, *Neolamprologus sexfasciatus* and *Cyphotilapia gibberosa*. For some while it had been assumed that this was an aggressive mimicry that probably allowed the scale-eater to get close to its presumed models. DNA barcoding of the stomach contents of the *Plecodus*, however, revealed that they attacked numerous other fish. In fact, scales from their harmless, non-scale-eating, and far more common models comprised only 0.9% of their diet. The mimetic resemblance appears, therefore, to be of the 'wolf in sheep's clothing' type.

Sazima et al. (2005) describe an instance of mixed-species schools in two fish that probably has dual functions, one being the protective effect of looking like other members of the school, but also an aggressive mimicry, though not aimed at other members of the school. These authors describe a neat mimicry complex they discovered off the coast of northern Brazil involving the naturally schooling (protection just from confusion effect) fish, the brown chromis, *Chromis multilineata*, and young of a coney fish,

Cephalopholis fulva, which co-shoals with them. The second species is piscovorous whereas the first is not. Other fish, potential prey of the coney, are not threatened by the *Chromis* school and thus do not try to avoid it, therefore allowing the coney to get close enough to them to make an attack. A number of other similar examples are known (e.g. Russell et al. 1976, Ormond 1980, Hori & Watanabe 2000, Sazima 2002a, 2002b, Schelly et al. 2007).

Many cases of conspicuous resemblance among coral reef fish may not actually be mimetic. For example, juveniles of the black snapper fish, *Apsilus dentatus* (Lutjanidae), and various hamlets, *Hypoplectrus* spp., are quite similar in colour to the blue chromis, *Chromis cyanea*, and other pomacentrids (Thresher 1978, Bunkley-Williams & Williams 2000) and of these at least eight *Hypoplectrus* have been suggested as aggressive mimics. Thresher (1978) suggested that they use the resemblance to get close to crustacean prey, but it is questionable whether many of the latter have sufficiently acute vision and/or cognitive abilities to recognise fish species by sight. Robertson (2013), in a critical appraisal of the evidence, concludes that:

> while some aspects of the interspecific associations involving *H. indigo* and, particularly, *H. unicolor* are consistent with mimicry, the supporting evidence is suggestive rather than decisive, and other evidence (e.g. the visual incompetence of crustacean prey) is counter-indicative.

Russell et al. (1976), Randall & Kuiter (1989) and Moland et al. (2005) survey a lot of what is known about mimicry among fish and described an example of juvenile groupers, *Anyperodon leucogrammicus*, from the Indo-Pacific, which mimic the wrasse *Halichoeres purpurascens*. Losey (1972) suggests that the tropical reef-dwelling blenny *Plagiotremus laudandus* may be an aggressive mimic of the poison-fang blenny, *Meiacanthus atrodorsalis*, as well as benefitting from Batesian mimicry of it, because it swims among groups of the latter which are ignored by most other reef fish, and can then dash out to attack them. Springer & Smith-Vaniz (1972) studied another *Meiacanthus* species, *M. nigrolineatus*, and two other blennies in the same mimicry complex, *Ecsenius gravieri* and *Plagiotremus townsendi*, and similarly concluded that the mimetic relationships were complicated: *Ecsenius* was proposed to be a Batesian mimic of the other two, and *Plagiotremus* an aggressive mimic of the other two as well as being in a quasi-Batesian relationship with *Meiacanthus*.

Rashed & Sherratt (2006) considered the mimicry of wasps by hoverflies (Syrphidae) in this context also. Both groups of insect spend large amounts of time feeding on nectar/pollen on flowers, often both being present together on the same inflorescence. Wasps (Vespidae) are therefore hoverfly competitors and also potentially dangerous to other flower visitors because they are also carnivorous, eating many other insects and spiders, and so might be expected to deter other insects from visiting the same flowerhead they are feeding on. Rashed & Sherratt tested this by examining visits to flowers that had pinned wasps or hoverfly specimens. Their results provided no evidence that insects avoided flowers with hoverflies but they did find that other insects avoided flowers with wasps more than they did flowers with non-mimetic insects (see also Chapter 9, section *Of orchids and bees*). I think further experiments will be needed to test this fully.

DUPING BY MIMICRY OF COMPETITORS

Rainey & Grether (2007) coined the collective term "competitor mimicry" to encompass three distinct sorts of relationships:

- mimicry of a non-competitor (of the duped species),
- mimicry of a competitor, and
- mimicry of a competitor's predator.

It seems to me that these are not only different in nuance but quite different in nature from one another and I find it hard to see how this lumping can be justified. Rainey & Grether (2007) recognise six different types of competitor "which are defined by different sets of relationships between receiver and mimic, mimic and receiver and model and receiver mimicry" (Table 10.1). As an example they cite the similarities between juvenile surgeonfish, *Acanthurus pyroferus*, and various species of Pacific Ocean *Centropyge* angelfish. For example, juvenile *A. pyroferus* in the seas around Papua New Guinea closely resemble the angelfish *Centropyge vrolikii*, and this appears then to allow the former to forage freely in damselfish, *Plectroglyphidodon lacrymatus*, territories, whereas the adults of *A. pyroferus* are vigorously repelled. Furthermore, *C. vrolikii* is probably tolerated among the damselfish because its diet does not overlap much with theirs, but the diet of *A. pyroferus* means it may be a significant competitor of the damselfish (Eagle & Jones 2004). In Table 10.2 I have superimposed Rainey & Grether's (2007) system on Vane-Wright's (1976) system and examples.

Among the predatory medium-sized (*Leopardus*, *Panthera* and *Puma*[5] spp.) and big cats (*Panthera*) that often feed on primates, there are numerous reports of the cats imitating the vocalisations of various primate prey species. Although

---

5. As might be expected from its generic name, *Puma yaguaroundi* is otherwise known as the jaguarondi and is closely related to the puma, cougar or mountain lion (*Puma concolor*), which is also widely reported to imitate primate vocalisations.

Table 10.1 Forms of competitive mimicry, with classical forms of protective mimicry included for comparison. (Adapted from Rainey & Grether 2007 with permission from John Wiley & Sons.)

Competitive mimicry types	Category	Receiver's relationship to mimic	Mimic's relationship to receiver	Model's relationship to receiver
A	Deceptive mimicry of non-competitor (unidirectional)	Competitor	Competitor	Non-competitor
B	Deceptive mimicry of non-competitor (reciprocal)	Competitor	Competitor	Non-competitor
C	Honest mimicry of non-competitor	Non-competitor	Non-competitor	Non-competitor
D	Mimicry of dominant competitor	Dominant competitor	Subordinate competitor	Dominant competitor/identical
E	Mimicry of equal competitor	Equal competitor	Equal competitor	Equal competitor/identical
F	Mimicry of competitor's predator	Competitor	Competitor	Predator

Table 10.2 Examples of competitor mimicry types (see Table 10.1) superimposed on Vane-Wright's (1976) classification of mimicry systems. (Adapted from Rainey & Grether 2007 with permission from John Wiley & Sons.)

	Synergic		Antergic
	I – warning	**III – defensive**	**VI – defensive**
Disjunct		Type F (burrowing owls)	Type A (*Acanthurus pyroferus*)
Bipolar S2/S1 + R	Type C (*Pomacanthus*) Type E (mockingbirds)		Type D (song sparrows)
Bipolar S1/S2 + R	Type C (*Pomacanthus*)		Type B (*Pomacanthus*)
Conjunct	Type E (song sharing)		Type A (sexual mimicry) Type D (indigo buntings)

many of the reports are anecdotal and often by indigenous inhabitants, there are some scientific reports (de Oliveira Calleia et al. 2009). Such vocalisations probably attract their primate prey because the latter are interested in what other conspecifics or groups of conspecifics might be nearby, possibly encroaching on their territories. E.C. Atkinson (1997) suggested that northern shrikes, *Lanius excubitor* (Laniidae), imitate the calls of various small songbirds during winter, thereby causing them to come and investigate possible territorial intruders or potential mates, and so bringing them close enough to be caught and eaten. In Vane-Wright's (1976) terminology these would be a disjunct aggressive synergic system, just like angling.

SEEMING TO BE CONSPECIFIC

Within this category there are a range of reasons why seeming to be a member of the same species as the prey will benefit the mimic. Terrestrial vertebrates are sometimes fooled because they have territories or form social groups, both of which are involved in defending resources such as mates or food, and therefore respond in often predictable ways to potential intruders. Among invertebrates, the most usual deceptions involve mimicry of a potential mate, either pheromonally or behaviourally, such as the flashing patterns of fireflies or the courtship songs and responses of crickets and cicadas.

Getting close

Puebla et al. (2007) show that coral reef fish of the genus *Hypoplectrus* (Serranidae) have diverse morphs and that those of predatory species often mimic the fish that they prey upon, the resemblance presumably resulting in them being mistaken for a harmless conspecific. In a similar case, McKaye & Kocher (1983) describe how three species of African cichlid fish of the genus *Cyrtocara* feed on the young of other mouth-brooding *Cyrtocara* species by aggressively head-ramming them. *Cyrtocara orthognathus* is able to change colour quite quickly and is therefore able to mimic females of both *C. eucinostomus* and *C. pleurotaenia*, the former being entirely silver and the latter having vertical stripes of silver.

Appearing to be a potential mate

The females of many species of firefly (Coleoptera: Lampyridae) produce sequences of bioluminescent flashes to attract mates, the timing and pattern of the flashes being highly species-specific (Lewis & Cratsley 2008) and flash communication has evolved on several independent occasions within the family. J.E. Lloyd (1975, 1984) showed that females of the predatory genus *Photuris*, in addition to their own courtship 'visual songs', will also mimic those of females of other genera such as *Photinus*, *Pyractonema* and *Aspisoma* species (Fig. 10.13), and thus attract heterospecific males which they then eat (Fig. 10.14). Appropriately, Lloyd (1975) termed these beetles "*femmes fatales*". While aggressive mimicry and predation of other firefly species is the norm in *Photuris*, it appears to be absent in the *P. congener* species group. Aggressive mimicry of fireflies also occurs in fireflies of the genus *Bicellychonia*. The situation in the *Photuris/Photinus* system is more complicated than simple aggressive mimicry: the former additionally obtains defensive steroids called lucibufagins from the males they consume (Eisner et al. 1997). Males of some *Photuris* species also mimic flashes of other species, but this is a form of sexual mimicry (see Chapter 11, section *Food dummies and sex*).

An analogous case of predatory courtship song mimicry occurs in the Australian bush cricket *Chlorobalius leucoviridis*

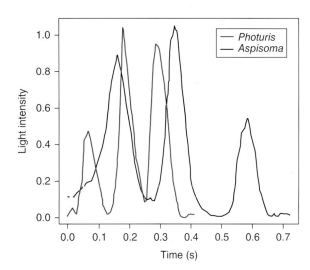

Fig. 10.13 Field-recorded bioluminescent flash patterns of the predatory firefly *Photuris trivittata* (Lampyridae) and one of its prey firefly species, *Aspisoma* sp. Traces show modulated flashes (about 8.3 Hz) of *Photuris* and of co-active *Aspisoma* sp. with form similar to that of certain flashes emitted by *P. trivittata*, but at a much slower modulation rate (mean 4.8 Hz). (Source: Adapted from Lloyd & Ballantyne 2003.)

Fig. 10.14 Two stages in the consumption of a presumed allospecific male by a female *Photuris* firefly. *Photuris* females dupe males to come to them by imitating the flash patterns of females of other species. (Source: Eisner et al. 1997. Reproduced with permission from PNAS.)

(Tettigoniidae), females of which mimic the species-specific clicks made by female cicadas to attract courting males (D.C. Marshall & Hill 2009). The bush cricket clicks are produced by a tegminal mechanism, whereas the female cicadas make their click replies by flicking their wings. Furthermore, the deception is enhanced when a male cicada approaches, by the bush cricket performing body jerks synchronised with the reply clicks that resemble the model's wing flicks. One of the most remarkable things about the bush cricket studied by D.C. Marshall & Hill (2009) is that it can mimic the specific replies of several species of cicada that occur sympatrically (Fig. 10.15) though they do sometimes respond inappropriately by clicking at the wrong time (Fig. 10.15 m,n). A similar but reversed aggressive mimicry, described by Koeniger et al. (1994), involves the Asian hornet, *Vespa affinis*. This flies among honey bee, *Apis mellifera*, drone aggregations and attracts the attention of a drone, which then follows the hornet to a perch, whereupon the hornet swiftly pounces upon it and eats it.

At a more proximal level, Waloff (1961) described what appears to be courtship signal mimicry by the parasitoid braconid wasp *Perilitus dubius*, which oviposits into adults of a chrysomelid leaf beetle. The female wasp faces the beetle head-on and "palpates its head violently", which is similar to what the beetles do in courtship. Sometimes, but not always, the beetle responds by extending its head forwards, exposing the membrane between head and thorax, as it would do with a suitable mate, and the wasp then swiftly oviposits into it through the thin exposed membrane.

Pheromone lures

American bolas spiders, *Mastophora*[6] species, are known for their unusual prey capture method, in which they swirl a sticky ball on the end of a silk thread ("the bolus") to capture their moth prey. They have been shown to enhance their prey capture rate through mimicry of female moth sex pheromones, and consequently their prey consists almost entirely of male insects (Eberhard 1977, Yeargan 1994). Three moth sex pheromone compounds, (Z)-9-tetradecenyl acetate, (Z)-9-tetradecenal and (Z)-11-hexadecenal, have been identified in volatile substances emitted by adult female *Mastophora cornigera* while they are hunting, and these attract some of the spider's moth prey species (Stowe et al.

---

6. There are approximately 50 known species in the genus and, while predominantly Neotropical, it is distributed from Minnesota in the north to southern Chile.

1987). Gemeno et al. (2000) have also worked out the chemistry of the mimetic system of another species, *M. hutchinsoni*, which primarily attracts males of the bristly cutworm, *Lacinipolia renigera* (Noctuidae), and appears to produce an allomone blend very similar to the two-compound pheromone of the moth. However, *M. hutchinsoni* also attracts males of the erebid *Tetanolita mynesalis*, whose sex pheromone blend is quite different from that of *L. renigera*. The two moth species are active at different times of night, *L. renigera* being an early evening species, the other being active later. Haynes et al. (2002) found that the spider's pheromone mimicry attracted both species, even when their photophases were adjusted to correspond to the other, thus demonstrating that the bolas spider is producing an adequate but suboptimal chemical blend throughout the evening, rather than adjusting the blend according to the time of night. The question should then arise as to how juvenile *Mastophora* feed, because spiderlings clearly cannot prey on far larger moths. The answer is that they also use a chemical attractant to lure prey that are appropriate for their size, specifically small flies of the family Psychodidae, otherwise, but purely coincidentally, often called 'mothflies' (Yeargan & Quate 1996). Haynes et al. (2001) found that female bolas spiders release their mimetic sex pheromones (allomones) before making a bolus, and they wait until they detect the wing vibrations of their duped prey, and only then construct the bolas and use it aggressively.

Similar use of a bolas also occurs in the Australian spiders *Ordgarius* (=*Dicrostichus*) (Longman 1922), *Cladomelea* from Africa (J.-M. Leroy et al. 1998) and *Exechocentrus* from Madagascar (Scharff & Hormiga 2012). Of all the known cases, *Mastophora* shows the most sophisticated behaviour (Wickler 1968). Only in *Exechocentrus* does there appear to have been any study of their possible use of pheromonal mimicry, but it seems highly likely to occur in all of them.

The bolas-wielding spiders may not be the only ones that produce pheromone-mimicking allomones, and evidence is growing that some orb web-weaving spiders do so too. Horton (1979) showed an apparently attractive effect on saturniid moths of the genus *Hemileuca* by the araneid spider *Argiope trifasciata*, and Jennings & Houseweart (1989) noted that nine web-spinner spider species captured significantly more male than female spruce budworm moths, *Choristoneura fumiferana*, in their webs (G-tests, $p < 0.05$), though the latter authors were not able to exclude the possibility that the males were more active, more numerous or concentrated nearer the webs for other reasons. More recently, further work on the *Argiope/Hemileuca* system has confirmed chemical mimicry (Warren 2014) but details of the allomones have not yet been published.

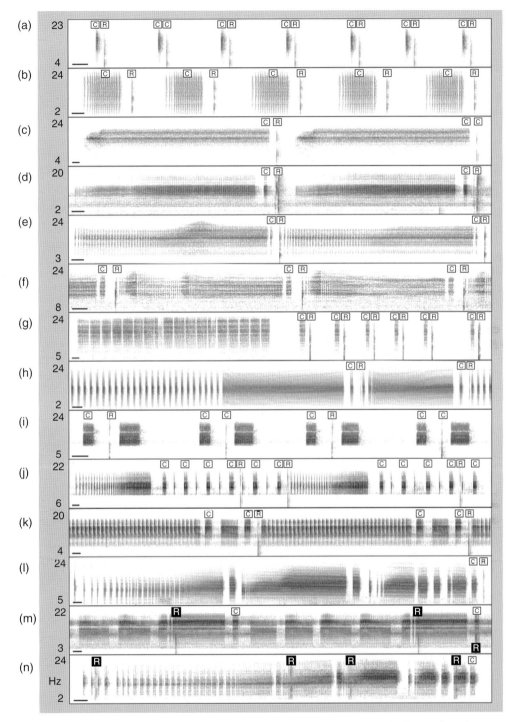

Fig. 10.15 Sonograms of the acoustic response clicks made by the bush cricket (katydid) *Chlorobalius leucoviridis* to the songs of 14 different species of cicada (tribe Cicadettini) belonging to at least nine different genera. The cicada songs are ended by a cue for a response from a female (marked with a 'C' and the *C. leucoviridis* click responses in each case are marked with an 'R'). (a) *Urabunana marshalli* – Australia (AUS); (b) undescribed genus, sp. "Nullarbor wingbanger" – AUS; (c) *Cicadetta calliope* – USA; (d) *Maoricicada campbelli* – New Zealand (NZ); (e) undescribed genus, sp. "Kynuna" – AUS; (f) undescribed genus, sp. "pale grass cicada" – AUS; (g) *Cicadetta viridis* – AUS; (h) *Pauropsalta* sp. "Sandstone" – AUS; (i) *Kikihia* sp. "tuta" – NZ; (j) *Kikihia* sp. "nelsonensis"; (k) *Kikihia subalpina* – NZ; (l) undescribed genus, sp. "swinging tigris" – AUS; (m) *Kikihia scutellaris* – NZ; (n) undescribed genus, sp. "troublesome tigris" – AUS. The white 'R' in a black box in sonograms (m) and (n) indicates an incorrect reply, all other katydid replies are correctly placed. Scale on left of each sonogram shows frequency range in kHz. (Source: Marshall et al. 2009. Reproduced under the terms of the Creative Commons Attribution Licence CC BY, via PLoS ONE.)

MIMICKING DANGER AS A FLUSHING DEVICE

Several reptiles use distractive tail movements while foraging for prey, and not simply as lures. Catania (2009) showed how the fish-eating tentacle snake, *Erpeton tentacularis*, feigns a danger sign by moving the side of its body into a more pronounced C-shape once a prey fish has swum into the concavity between its head and body. This causes the prey to turn to try to avoid the perceived threat but instead ends up swimming towards the snake's bite.

The ant-mimicking spider *Myrmarachne melanotarsa* (Salticidae) from East Africa, which feeds on the broods of other spiders, makes use of the fact that many spiders in the tropics keep well clear of potentially dangerous ants. When either aggressive *Crematogaster* ants or *M. melanotarsa* come close to females of the spider *Menemerus* sp. that are guarding their eggs or spiderlings, the *Menemerus* usually leave their broods unguarded to avoid the risk of being captured by the ants, whereas they generally remain with their broods in response to other spiders or insects (Fig. 10.16). Ants cannot usually attack the *Menemerus* broods because they cannot negotiate the silk threads, but *Myrmarachne*

has no such problem and so can feed on the duped spider's brood (X.J. Nelson & Jackson 2009b).

Human use of aggressive mimicry

Humans have long made use of deception in hunting prey, including the use of lures such as fishing flies, and disguise. North American plains Indians used to drape buffalo skins over themselves and their horses and then, riding with both their own and their horses' heads held low, they were able to approach and mingle with buffalo herds until they got close enough to make their attack. Similarly, men of the San people of South Africa would disguise themselves as ostriches in order to approach herds of game.

These tribal instances are effectively the same as 'western' human uses of duck calls and duck decoys to attract waterfowl for the pot. Willughby (1678) described the alluring use of a live tame duck in the Netherlands at the entrance to a trap, but duck call use by humans probably far predates that and may have originated in the Far East. Similarly, fly fishers are variously expert at making fishing 'flies' to resemble numerous forms of insect that fish might feed upon.

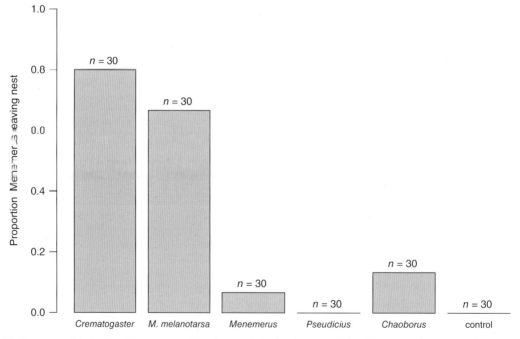

Fig. 10.16 Frequency of test spider (*Menemerus* sp.) females remaining inside nest with hatchlings or vacating nest when surrounded by groups of 20 individual ants (*Crematogaster* sp.), ant-like salticids (*Myrmarachne melanotarsa*), non-ant-like salticids (*Menemerus* sp. and *Pseudicius* sp.), midges (*Chaoborus* sp.) and control empty stimulus chambers. Responses to ants and *M. melanotarsa* do not differ significantly from one another, but both differ significantly from all others (χ^2 tests, *p* values <0.0001). (Source: Adapted from Nelson & Jackson 2009b.)

Quite the opposite strategy was employed by North American plains Indians who sometimes dressed in wolf skins in order to be able to herd bison to their place of slaughter more effectively, as illustrated by American artist George Catlin (c. 1833) in his painting 'Buffalo Hunt under the Wolf-skin Mask'.[7]

CUCKOLDRY, INQUILINES AND BROOD PARASITISM

Brood parasitism almost invariably involves some sort of mimicry. The most well-known cases are those of a considerable number of bird species that lay their eggs in the nests of other birds, and those of various insects that usurp the brood-caring behaviours of other, usually social insect, species such as bees, wasps and ants. Although there are broad similarities in that the parasitic juveniles usually resemble those of their hosts, visually, chemically or behaviourally, there are also pronounced differences. In the case of brood-parasitic birds the adults often do not resemble their hosts but in some brood-parasitic insects they do. Therefore I shall deal with these different systems largely separately. One of the big problems in dealing with cuckoldry systems, as with many other apparent cases of mimicry, is that there are often several distinct possible explanations for various aspects of the resemblance, and little or no experimental evidence in favour of any of them. A typical case is that of various species of robber fly (Diptera: Asilidae) that mimic carpenter bees (Hymenoptera: Apidae: Xylocopini) in the Old World tropics. In several instances, the mimicry is so perfect that series of bees in poorly curated collections have been found to conceal specimens of their robber fly mimics. However, the situation may be more complicated because, at least in some instances, the larvae of the robber flies are predatory on those of the wood-boring bees. The question arises, therefore, as to whether the resemblance between the flies and bees enables the flies to approach the nesting sites of the bees without detection, or whether it permits the newly emerged and rather vulnerable flies to escape from the area of the bees' nests more safely, or whether the mimicry is purely a Batesian one with no aggressive overtones. Both Rettenmeyer (1970) and H.E. Evans & Eberhardt (1970, pp. 227–28) concluded that the mimicry was probably purely Batesian in function (L.P. Brower et al. 1960).

The term 'cuckoldry' has been widely applied both to cuckoos and to a wide range of animals with similar reproductive behaviours and, of course, the term is also applied to

people when a husband's wife commits adultery, perhaps even leading to the husband thinking that any resulting child is his own and investing effort of some form in rearing it. The analogy to cuckoos and other non-human cases is that the male may be mistaken about the parentage of any offspring. Females of many pair-breeding birds often indulge in extra-pair copulations, as do males, causing cases of cuckoldry closer to the human situation.

In spite of the great phylogenetic distance between avian cuckoos and insect examples, the systems show many evolutionary similarities. Kilner & Langmore (2011), in reviewing what is known, identified at least five common evolutionary features:

1. directional coevolution of weaponry and armoury,
2. furtiveness in the parasite countered by strategies in the host to expose the parasite,
3. specialist parasites mimicking hosts who escape by diversifying their genetic signatures (which may be seen as two separate trends),
4. generalist parasites mimicking hosts who escape by favouring signatures that force specialisation in the parasite,
5. parasites using crypsis to evade recognition by hosts who then simplify their signatures to make the parasite more detectable

Of course, we can only observe and study those relationships in which the brood parasite is successful. No one can say for sure whether a bird or bee species that lacks 'cuckoos' has won a past arms race or whether the 'cuckoos' never evolved to attempt to parasitise them. However, Marchetti's (2000) study of egg discrimination by the yellow-browed leaf warbler, *Phylloscopus humei*, which most probably has been subject to cuckoo parasitism during its evolutionary history though not any longer, shows that females have a very strong ability to discriminate against eggs in their nests based on relative size. In this case, the female will sometimes reject one of her own eggs if its size does not match that of the others in her brood.

Cuckoldry in birds

Apart from the common Eurasian cuckoo, *Cuculus canorus*, and other true cuckoos (Cuculidae[8]), other brood-parasitic birds include indigobirds and whydahs (*Vidua*: Viduidae) (Fig. 10.17a,b), honeyguides (Indicatoridae) (Fig. 10.17c,d), cowbirds (Icteridae), and some finches (Fig. 10.18a–c) and

---

7. Painting held in the Smithsonian American Art Museum collection.

8. Approximately 45 out of its 127 or so species are nest parasites of other birds.

Fig. 10.17 Cuckoldry in birds and associated mimicry. (a) Chick of an African host, the melba finch, *Pytilia melba*, showing the ornate and brightly coloured gape that is mimicked by its brood-parasitic *Vidua* species; (b) purple indigobird, *Vidua purpurascens*, chick, showing similar ornate gape; (c) African greater honeyguide, *Indicator indicator*, chick, just a couple of days old, attacking the egg of its bee-eater host, which is facilitated by a sharp apical tooth on its bill; (d) honeyguide chick attacking bee-eater chick; (e) honeyguide chick holding onto finger, showing the power of the grip of its bill. (Source: Dr Claire Spottiswoode. Reproduced with permission from Dr Spottiswoode.)

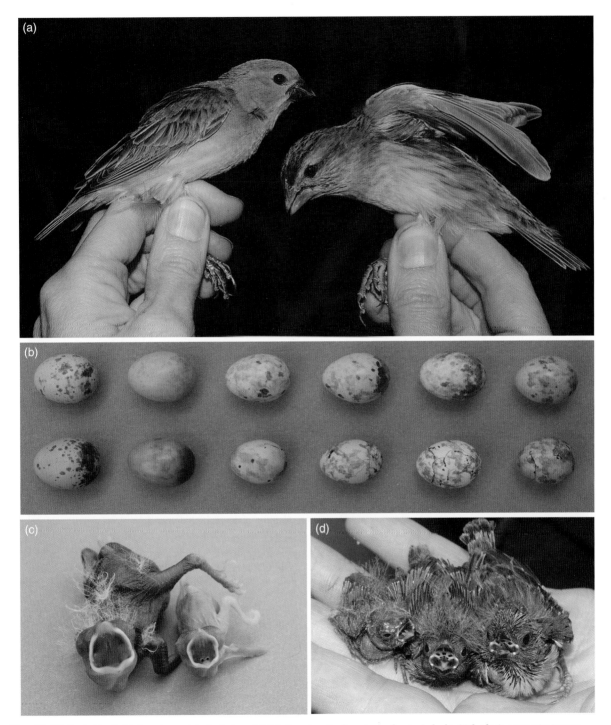

Fig. 10.18 Cuckoldry in birds and associated mimicry. (a) Adult cuckoo finches, *Anomalospiza imberbis* (Viduidae), composite image; (b) eggs of cuckoo finches and those of their associated hosts; (c) chicks of cuckoo finch (left) and its associated host, the African tawny-flanked prinia, *Prinia subflava* (Cisticolidae) (right); (d) indigobird, *Vidua* (Viduidae) chick (right) and those of its host firefinch (two on the left) showing disparity in size. (Source: Dr Claire Spottiswoode. Reproduced with permission from Dr Spottiswoode.)

even some ducks. Indeed, nearly 1% of all bird species perform some sort of cuckoldry on other species (Payne 1977, Stevens 2013b) and the vast majority involve some sort of mimetic deception (Davies 2000, 2011) as well as many other adaptations. Females of some species will even destroy host nests that already contain host chicks so that the host bird may start nesting again and the cuckoo can then parasitise them (Davies & Brooke 1998, Davies 2000, Langmore 2013).

Mimicry in bird brood parasites is complex in several respects. At least for female cuckoo gentes, it is one of rather few one-to-one species associations in terrestrial vertebrates. While the concept of cuckoldry may seem straightforward enough, whereby one individual or species dupes another into rearing its young for it, the term conceals a wealth of variation. For example, the cuckolded host may be either the same or a different species as its brood parasite, and both male and female hosts may be deceived. When it comes to mimetic aspects of cuckoldry, the situation becomes more complex as the cuckold may mimic its dupe, the eggs may mimic those of the dupe and/or the offspring may be mimics of the dupe's offspring.

Grim (2005) stresses that simple similarity between the eggs and/or nestlings (which he refers to as propagules) of cuckoos and those of their host species does not necessarily mean that it is due to mimicry, and that proper experimental tests are required. Alternative hypotheses may be sufficient to explain some features (Table 10.3).

Gentes and 'cuckoo' eggs

While it has been known for a long time that Eurasian cuckoos, *Cuculus canorus*, have different races, called gentes (or sometimes 'gens'), each associated with different host

Table 10.3 Possible explanations for similarities between cuckoo eggs and chicks and those of their hosts. (Grim 2005. Reproduced with permission from John Wiley & Sons.)

Explanation	Eggs	Nestlings
Phylogenetic constraints	+	+
Random matching	+	+
Spatial autocorrelation	+	–
Nest predation	+	+
Egg replacement by cuckoos	+	–
Host discrimination	+	+
Non-random matching	–	+
Pre-existing preferences	–	+
Vocal imitation	–	+

species, and that the cuckoo's eggs closely resemble those of their given hosts, the underlying mechanism was unknown, though many suspected, erroneously as it turns out, that the nestling cuckoos imprinted on their host species and so chose to select these as their own hosts. Brooke & Davies (1991) in a 2-year experiment involving cross-fostering between reed warbler, *Acrocephalus scirpaceus*, and European robin, *Erithacus rubecula*, hosts, demonstrated that imprinting did not occur, which means that host selection must also be genetically controlled.

The specific gene complex controlling a cuckoo's egg mimicry is located on the bird's W chromosome. Unlike in mammals, birds and butterflies have heterogametic females. Male birds are ZZ and females are ZW. The X and Y chromosomes of mammals are not homologous to those of birds and they share no genes. This form of karyotypic sex determination means three things. First, the W chromosome genes can only be expressed in females, second, they are never heterozygotic, thus third, they cannot recombine. In many ways this makes them like the purely maternally inherited mitochondrial genome. It also poses some problems, in that the expression of Z chromosome genes does not appear to show dosage compensation, whereas it does for X chromosome ones. Chandra (1991) suggested that this may be offset, at least in some groups, by the large maternal legacy of RNA in the very large eggs of birds and reptiles.

Although in early DNA studies, Gibbs et al. (1996) and Marchetti et al. (1998) failed to find fixed differences between common cuckoo gentes, Gibbs et al. (2000) subsequently did, though they noted that it was possible that some gentes might have had multiple origins and that host switching (colonisation of new hosts by female lineages) has been fairly frequent over evolutionary history. While females of a given gente show a high level of host specificity, males are not at all selective about which females they mate with (Marchetti et al. 1998), and therefore the female preference must be carried by genes that are sex-linked. Mitochondrial DNA data indicate that the various British cuckoo gentes evolved from a common ancestor about 80,000 years ago (Gibbs et al. 2000). In contrast, Spottiswoode et al. (2011), based on estimated rates of mitochondrial sequence substitutions, found that the two female gentes of the African honeyguide, *Indicator indicator*, one of which parasitises cavity-nesting birds such as hoopoes and woodpeckers, the other parasitising ground hole-nesting birds such as bee-eaters (see Fig. 10.17c–e), diverged approximately three million years ago, although as the nuclear DNA showed no such differentiation, the females of the gentes are clearly mating with males at random with respect to host type.

Kelly (1987) published an evolutionary model for the rate of spread of egg mimicry and egg rejection by host birds, but at that time it was not known that cuckoo egg colour was governed entirely by maternally inherited genes so his model assumed that both host and cuckoo characters could have intermediate heterozygous states. His model showed both cuckoo and host genotypes transitioning from being nearly monomorphic for one state to monomorphic for the mimicking and rejecting states via a period when heterozygotes were most abundant. He also found that while mimicry in the cuckoos did evolve to fixation, a substantial proportion of heterozygous, poor rejecting hosts persisted for a long time. In Fig. 10.19 I present results from a simulation model based on discrete host–parasitoid biology (Beddington et al. 1975) modified to allow survival of adults whose nests were parasitised, and with overlapping generations such that a proportion of the previous season's adults breed again in successive years. The equations I derived are very similar to

those of May & Robinson (1985) and had similar outputs, though they only discussed what would most likely happen in the case of host acceptor–rejector polymorphism. In my model, cuckoos only have two genotypes (controlling either mimetic or non-mimetic eggs) and hosts have three (rejecting discriminators, poor discriminators and universal acceptors). In the model, host panmixis is assumed such that at each season gene frequencies are in Hardy–Weinberg equilibrium. A cost factor for the ability to reject cuckoos was included but had little effect unless set quite high. The outcome of the simulations depended a great deal on cuckoo nest parasitisation rate. At low rates the mimicry genotype remained at low frequency (Fig. 10.19a) or went extinct, and the rejector allele increased to a high frequency (Fig. 10.19b) but with cuckoos persisting at low population

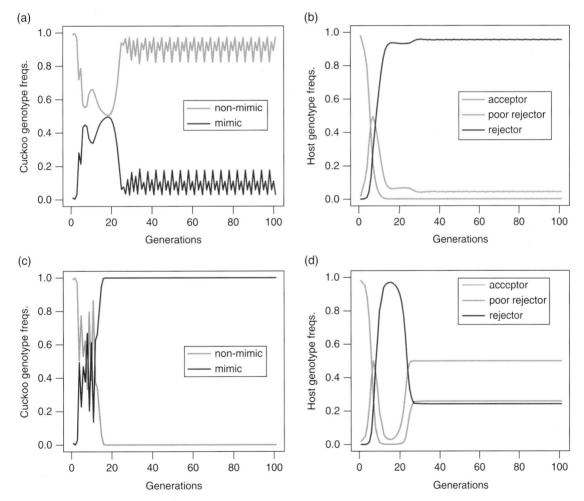

Fig. 10.19 Simulations of cuckoo mimetic haplotype and host acceptor–rejector alleles. Evolutionary dynamics for two sets of parameter combinations: (a,b) with low host nest parasitism rate; (c,d) with high nest parasitism rate.

density through being able to attack intermediate rejectors and acceptors. At high parasitisation rates the mimicry rapidly evolved to fixation (Fig. 10.19c) whereas hosts reached a Hardy–Weinberg stable equilibrium with most being intermediate rejectors (Fig. 10.19d). A variety of other dynamics can be recovered with various other parameter combinations.

Servedio & Lande (2003) applied a more parameter-rich quantitative genetic model to the evolution of egg similarity between cuckoo and host which assumed that the phenotypic variation had a complicated genetic basis, and also allowed for recognition errors by the host. Recognition errors were treated as an emergent property of the discrimination rule rather than being set at some arbitrary level. Their results show that recognition errors played a crucial role, and that while their model did allow the evolution of discrimination, at equilibrium this might still be rather weak. As with the simulations (see earlier), their model predicted both lags and equilibrium, and also that, after initial colonisation of a genetically naïve host species/population, both host and cuckoo populations went through crashes before reaching stable equilibria.

Moksnes et al. (1993) found that cuckoo egg (or model cuckoo egg) rejection or nest desertion by meadow pipits, *Anthus pratensis*, varied depending on whether the adult bird was aware of nearby cuckoos. Thus when a dummy cuckoo was mounted near to a pipit nest, the pipits significantly increased their ability to recognise the presence of a cuckoo egg. Davies et al. (1996) determined that the probability of reed warblers rejecting a cuckoo egg depended in part on their perception or history of the risk of brood parasitism. Avilés et al. (2007) found that reed warbler and cuckoo egg phenotype was influenced by local climatic conditions. Avilés et al. (2011a) found a lack of local adaptation of the cuckoo to its hosts, but Avilés et al. (2011b) did find that egg matching correlated with local egg rejection rate once spring climate effects had been factored out of the analysis. Across host species M.C. Stoddard & Stevens (2010) have found a significant positive correlation between the number of matching attributes between cuckoo and host eggs and the mean rejection rate of cuckoo eggs by hosts (Fig. 10.20).

In some species, the differentiation of gentes is not immediately apparent to human inspection as it may either involve wavelengths outside of our visual range, or more subtle features of hue that humans do not readily appreciate (e.g. Takasu et al. 2009). Starling et al. (2006) studied the eggs of the pallid cuckoo, *Cuculus pallidus*, and, using reflectance spectrophotometry across the 300–700 nm range relevant to bird vision, found that its eggs did indeed more closely resemble those of their four respective different host species. Their findings could indicate that gentes are

more prevalent among cuckolding bird species than previously reported.

Cuckoldry by birds should select for mechanisms that reduce intra-brood egg variation but increase differences between individual host females (Stokke et al. 2002). Birds can potentially protect themselves from brood parasitism by recognising cuckoo eggs and abandoning their parasitised nests to start afresh. Furthermore, they can potentially maximise the chance of being able to discriminate eggs, by different females laying eggs with different colours, thereby making it harder for their cuckoo to find nests that contain eggs appropriate to their gentes. The African tawny-flanked prinia, *Prinia subflava*, lays many different coloured eggs (Fig. 10.18b) and appears to be capable of rapid evolution, as evidenced by changes in its eggs and its brood parasite, the cuckoo finch, *Anomalospiza imberbis*, over the past 30 or so years (Spottiswoode & Stevens 2012). Table 10.4 shows changes in egg appearance, as would be seen by the host bird using an avian visual model, from samples collected recently (2007–2009) and older specimens (mainly 20–30 years old) collected at a site in Zambia. Both the host prinia eggs and the cuckoo finch eggs have increased in phenotypic diversity in this evolutionarily very short time interval, suggesting that there is strong selection pressure in the evolutionary arms race between host and parasite. Current-day cuckoo finch eggs are significantly more similar to current-day prinia eggs than the historical cuckoo finch eggs are to current-day prinia ones and, conversely, historical host eggs are more different to current-day parasite ones than they were to historical parasite ones, showing that there has been a tracking of egg phenotypes over the period (Table 10.5). Interestingly, Spottiswoode & Stevens (2011) found that the host species with the most polymorphic eggs also has the crudest discriminatory ability against the cuckoos, whereas the least polymorphic one was the most discriminating, but although this could be interpreted in several ways, I suspect that cognitive limitations play a role. Of course, just because the prinia and cuckoo finch both evolved egg colour variation rapidly does not mean that other, unrelated species have the same genetic potential.

P. Mason & Rothstein (1987) investigated whether shiny cowbird, *Molothrus bonariensis*, eggs were truly mimetic in all cases or whether both their own and their host eggs were simply cryptic. Individuals of this cowbird lay plain (immaculate) or spotted eggs. Transfer experiments showed that their common host, the rufous-collared sparrow, *Zonotrichia capensis*, which lays speckled eggs in an open cup nest, shows no egg discrimination at all, so matching in this case is presumed to be due to mutual selection for laying cryptic eggs. They therefore conclude that the variation between cowbirds must be a result of selection for mimicry by other common host species. Similarly, the white eggs of the large

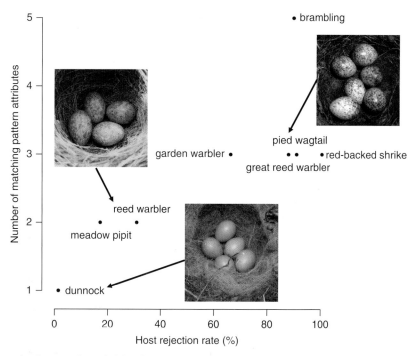

Fig. 10.20 The relationship between the probability that a species of host bird will reject a common cuckoo egg and the number of colour pattern attributes the egg of the gente shares with those of the host bird. [Scientific names of species: brambling (*Fringilla montifringilla*), dunnock (*Prunella modularis*), garden warbler (*Sylvia borin*), great reed warbler (*Acrocephalus arundinaceus*), meadow pipit (*Anthus pratensis*), pied wagtail (*Motacilla alba*), red-backed shrike (*Lanius collurio*), reed warbler (*Acrocephalus scirpaceus*).] (Source: Adapted from Stoddard & Stevens 2010 with permission from The Royal Society; image of *A. scirpaceus* nest, NottsExMiner 2013. Reproduced under the terms of the Creative Commons Attribution Share-Alike Licence CC BY-SA 2.0, via Flickr; image of *P. modularis* nest, NottsExMiner 2011. Reproduced under the terms of the Creative Commons Attribution Share-Alike Licence CC BY-SA 2.0, via Flickr; image of *M. alba* nest, Walcoford 2014. Reproduced under the terms of the Creative Commons Attribution Share-Alike Licence CC BY-SA 3.0, via Wikimedia Commons.)

Table 10.4 Rapid evolutionary change in the appearance of host bird eggs, African prinia, and its brood-parasitic cuckoo finch, in a population in Zambia provides evidence for an arms race and strong selection. Variation among both host and brood parasite eggs has increased significantly over approximately 30 years. (Data from Spottiswoode & Stevens 2012.)

Species	Time period	Mean colour difference in jnds (range)	*p*-value
Prinia	Historical	4.8 (3.8–9.1)	*p* < 0.001
	Present day	6.1 (4.85–12.88)	
Cuckoo finch	Historical	3.87 (3.25–8.17)	*p* < 0.001
	Present day	4.82 (3.63–10.82)	

jnd, just noticeable difference.

hawk-cuckoo, *Hierococcyx sparverioides*, do not closely resemble the blue eggs of its host the Chinese babax, *Babax lanceolatus*, a fact which Yang et al. (2012) take to indicate that their association is of recent origin.

Cues for egg rejection

The actual features of cuckoo eggs leading to rejection are not yet well understood, but Honza et al. (2007) found that song thrushes, *Turdus philomelos*, utilised both UV reflectance and green visible light reflectance in their decision to reject or not and, because of this, rejected some model eggs that were intended to be mimetic. M.C. Stoddard & Stevens (2011) found that when eggs of passerine hosts were considered from the point of view of an appropriate avian visual model, the cuckoo's matching was highly accurate, and better in those species that were good rejectors.

Study of reed warbler/European cuckoo associations have revealed many interesting trade-offs. Cuckoo eggs are

Table 10.5 Time-shift analyses of phenotypic distance between host and parasite eggs; comparisons of historical and current-day hosts and cuckoos has increased over a time period of about 30 years. The upper two rows show that historical host eggs were significantly more similar to historical parasite eggs than to current-day parasite eggs; the lower two rows show that current-day parasite eggs are more similar to current-day host eggs than they are to historical host eggs. See legend to Table 10.4 for further details. (Data from Spottiswoode & Stevens 2012.)

Comparison	Mean phenotypic difference in jnds (range)	Probability
Historical host vs. historical cuckoo finch	4.55 (3.1–8.4)	$p < 0.001$
Historical host vs. current-day cuckoo finch	5.25 (3.45–12.12)	
Current-day host vs. historical cuckoo finch	6.1 (3.1–13.3)	$p = 0.048$
Current-day host vs. current-day cuckoo finch	5.6 (3.4–14.1)	

jnd, just noticeable difference.

usually slightly larger than those of their hosts (E.C.S. Baker 1942, Lack 1968, Alvarez 1994), but are smaller than would be expected given the cuckoo's adult size (Lack 1968). This is also likely to be an adaptation that allows the cuckoo to lay relatively quickly, as well as reducing the size differential between their eggs and those of the host. Size would seem to be a potentially easy way that host birds might be able to discriminate between their own and the parasite's eggs. Davies & Brooke (1988) inserted large model cuckoo eggs (similar in size to those of non-parasitic cuckoo species), coloured like the appropriate gente, into reed warblers nests and found that these were more likely to be rejected than model mimetic eggs of normal cuckoo-size. However, this is not the case with all host species, and Alvarez (2000) found that the bigger the cuckoo egg model, the less likely that rufous bush chats, *Cercotrichas galactotes*, would reject them – the large eggs apparently acting as somewhat of a supernormal stimulus. Ability to recognise a size difference, and possibly other differences, almost certainly requires that there is a cuckoo and at least one host egg in the nest at the same time so that the host bird can

notice the difference (Moskát & Hauber 2007) and thus, not surprisingly, egg rejection rate is not fixed but varies according to the timing of nest parasitism relative to the hosts' laying period.

Rothstein (1982) used the North American robin, *Turdus migratorius* (Turdidae), which is a common host of the brown-headed cowbird, *Molothrus ater* (Icteridae) to test what influenced rejection of cowbird eggs. Compared to cowbird eggs, those of *Turdus* are larger, blue and unicolourous. By experimentally placing egg models that differed in all combinations of these three parameters Rothstein found that significant levels of egg rejection never occurred if the model egg differed in a single parameter, but those that differed in two or three parameters were usually rejected. The results of this and other experiments suggest that the host bird has evolved to allow some degree of tolerance to eggs that differ somewhat from the norm, and parasitism by cowbirds has not led to the evolution of a specific cowbird egg recognition system. In another investigation of which features hosts use to discriminate between their own eggs and those of cuckoos, Spottiswoode & Stevens (2010) used avian visual models to compare eggs of the African cuckoo finch, *Anomalospiza imberbis*, and those of its host, the tawny-flanked prinia, *Prinia subflava* (see Fig. 10.18b). They quantified colour, luminance ("perceived lightness") and pattern information (dispersion of markings, principal marking size and variability in marking size) and found that the features used by the hosts were exactly the same as the features revealed to be most discriminatory in independent analysis, i.e. the hosts are using the best cues available to them.

Lotem et al. (1995) have shown, in Japan, that the imperfection with which great reed warbler *Acrocephalus arundinaceus*, hosts of the Eurasian cuckoo, can recognise their own eggs provides a constraint on the level of discrimination that can evolve. In fact, in their study area only 61.5% of cuckoo eggs were rejected. Young host birds in particular lay more variably patterned eggs than do older birds, and so if they reject eggs that differ only slightly from the others in their nest, they risk throwing out their own offspring along with cuckoos. Further, Lotem et al. (1995) provide evidence that the ability of the host to discriminate between their own and cuckoo eggs depends on a learning process akin to imprinting, in which the bird learns to recognise its own eggs better as its brood increases.

Blue-green bird eggs (see dunnock in Fig. 10.20) are now generally believed to be expensive to produce, and Soler et al. (2012) note that this also has consequences for the cuckoos that parasitise nests of birds that lay blue-green eggs. They considered three different aspects of this, none of which was mutually exclusive. Hosts that lay blue-green eggs may have strong sensory biases favouring this colour, as it indicates that the resulting chicks may be fit in this same sense and therefore it might be less likely that they would evolve discriminatory ability against similarly coloured, mimetic cuckoo eggs. But if blue-green eggs are costly it might be especially difficult for cuckoos to mimic them because cuckoos lay many more eggs per bird than do their hosts, since they do not have to feed the chicks, i.e. the total cost to the cuckoo of making such expensive mimic eggs is higher than any one host's cost, and therefore attacking such hosts must be more than offset by the various other missed costs such as chick-rearing. Third, they suggest that if production of blue-green eggs by hosts is an indicator of likely parental quality, then such hosts should be selected preferentially for parasitism because they could be expected to be better 'parents' to the cuckoo's chick.

However, some host birds, such as the dunnock, *Prunella modularis*, a host of the common cuckoo, do not discriminate at all between their own pale blue eggs and the brown speckled ones of their cuckoo gente (Brooke & Davies 1988, Davies 2002). In the absence of discrimination by the dunnock, their cuckoo parasites have been under no selection pressure to evolve mimetic eggs. Lotem et al. (1995) experimentally inserted real, model and simulated cuckoo eggs in great reed warbler, *Acrocephalus arundinaceus*, nests and found that only 6% of host individuals would accept highly dissimilar eggs, suggesting that true 'universal acceptor' genotypes were rare, at least in the study population, though acceptance of dissimilar eggs was higher among birds breeding later in the year, which they attributed to the higher proportion of young – and presumably less experienced – female birds breeding in the mid-season.

Hole-nesting Eurasian redstarts, *Phoenicurus phoenicurus*, find it difficult to recognise highly similar cuckoo eggs and never eject them when they are detected, but instead may desert their nest and start nesting afresh (Avilés et al. 2005). Because cuckoos find parasitism of redstart nests comparatively difficult and use them less than the exposed nests of other species, the redstarts suffer a lower overall level of parasitism than many other hosts. The low incidence of parasitism may lead to a situation where actually taking the risk and accepting a similar-looking cuckoo egg and carrying on nesting is actually the optimal strategy (Davies & Brooke 1988, Marchetti 1992).

Moskát et al. (2008) manipulated both spotting density and colour of great reed warbler, *Acrocephalus arundinaceus*, eggs and found that spot density varying between 15 and 75% egg coverage did not have any significant influence on the normal 8–20% cuckoo egg rejection rate, but when the eggs were made 100% spotted (i.e. totally brown), rejection rate jumped sharply to 100%. Background egg colour had almost no effect. Moskát et al. (2012) compared the similarities between cuckoo eggs and great reed warbler eggs in

Hungary with the relationship between cuckoos and *A. orientalis*, a very closely related species, in Japan. In the Hungarian system there was high intraspecific and intragente variation in egg colour which might make it relatively more difficult for cuckoos to achieve good matches there; reflectance differences were also higher than in Japan. Their results suggest that different features have, by chance, become important in the cuckoo host arms race in the two systems.

Instead of varying a particular egg parameter, Moskát et al. (2014), again with the great reed warblers/cuckoo system, varied the variance of the features of the eggs in a clutch. This enabled them to discriminate between specific eggshell cues that might trigger rejection by the host and a host's response to differences rather than absolutes. Their results suggested that the hosts did indeed examine their clutches for discordance as an indicator that their clutches had been parasitised.

Moksnes et al. (1990) experimentally tested acceptance rates of artificial, non-mimetic eggs across a large range of potential host species, ranging from commonly used, often used, occasionally used and never (and could never be) used species and found that the dummies were accepted most by common cuckoo hosts and least by least-used hosts, suggesting that cuckoos are utilising, in general, the most easily parasitised species. However, those species that are presumed never to have experienced cuckoldry for various biological reasons, accepted 100% of the dummy eggs, presumably because there has never been any selection to discriminate parasite eggs (within evolutionary memory).

Horsfield's bronze cuckoo, *Chrysococcyx* (previously *Chalcites*) *basalis*, an Australian species that parasitises the nests of various fairy-wrens, *Malurus* spp., lays eggs that quite closely resemble those of the host. Langmore & Kilner (2009) experimentally parasitised host nests with painted non-mimetic eggs to test whether the cuckoo preferentially removed an odd-looking egg from the clutch, as would be expected if it had evolved to remove eggs of other cuckoos which might not be particularly good matches, but no such effect was apparent. However, that does not necessarily mean that it was not an initial driver of selection for the egg mimicry in this cuckoo, because now that the mimicry is very precise, selection may no longer favour maintenance of an ability to discriminate. It might also be that because only about 2.5% of parasitised host nests are doubly parasitised, selection for discrimination might not be sufficiently strong, especially if there is some trade-off involved. Horsfield's bronze cuckoo is also interesting in that it utilises 17 different fairy-wren host species and approximately 27 other species, but although the eggs of the different hosts do differ from one another, albeit in rather subtle ways, the cuckoo does not have host-related gentes, and instead its eggs appear to be rather a compromise, being sufficiently similar to hosts to enable it to be a generalist (Feeney et al. 2014). Some other bronze cuckoos, *Chrysococcyx* (previously *Chalcites*) spp., that parasitise eggs of birds such as large-billed gerygones, *Gerygone magnirostris*, lay dark olive-green coloured cryptic eggs that do not resemble the pale spotted ones of their hosts (Gloag et al. 2014, see also figure 7 in Langmore 2013). It is suspected that rather than simply being due to a lack of visual discrimination by the host, they are cryptic against conspecific competitor cuckoos, since cuckoos will often remove an egg from a host clutch before laying one of their own. The nests of gerygones are domed and their interiors are dark, and visual modelling shows that the low luminance of the cuckoo egg would render them very difficult to see against the very similarly coloured nest lining. Therefore if a second bronze cuckoo comes to parasitise the same nest, it is far more likely to remove an egg that it can see easily, i.e. a host egg, than one of the original cuckoo's cryptic ones

In the case of the common cuckoo, *Cuculus canorus*, and the diederik cuckoo, *Chrysococcyx caprius*, Spottiswoode (2010) has shown that selection has also led to a positive correlation between parasite eggshell thickness, and thus presumably eggshell strength, and the probability of rejection by a given host species, as would be expected if greater strength discouraged rejection. This is the first example of a non-visual or textural association in cuckoo egg morphology.

Having a defensive adaptation, such as a host bird's ability to distinguish between its own eggs and those of a cuckoo, say, nearly always comes with some costs, and in this case it could be time expenditure or occasional errors where a host rejects one of its own eggs by mistake. Detecting such costs may be possible when a species occurs alone in some parts of its range. Individual females of the African village weaver, *Ploceus cucullatus* (Ploceidae), lay clutches of rather similar eggs but different individuals lay quite different ones, variation that is thought to have evolved to make it harder for *Vidua* spp. to parasitise their nests. Any given *Vidua* would only produce one egg type and that would be unlikely to match that of any given weaver female. Indeed, nesting birds reject experimentally introduced dissimilar eggs, but Cruz & Wiley (1989) describe how descendants from birds originally introduced in the eighteenth century to the island of Hispaniola, where there were no endemic cuckoos, no longer show any egg discrimination. However, with the subsequent introduction of the shiny cowbird, *Molothrus bonariensis* (Icteridae), in the 1970s, the weavers are now subject to brood parasitism, which is having a substantial effect on their breeding success. It is thought that during the years without parasitism, the lack of selection pressure meant that egg discrimination behaviour was lost, presumably because of associated costs. It will be interesting to know whether it will re-evolve in due course, and if so, how long it takes.

While the most obvious interpretation of the resemblance of cuckoo eggs to that of their host birds is to reduce the risk of the latter rejecting the cuckoo's egg, Brooker et al. (1990) also note that other cuckoos might be the dupes because cuckoos finding a nest that has already been parasitised and in which they can distinguish the pre-existing cuckoo egg from the host's clutch, may selectively remove the first cuckoo's egg and replace it with their own. In fact, using a mathematical population genetics model they show that the model in which cuckoos recognised and rejected cuckoo eggs would lead to far quicker evolution of mimesis than host egg/nest rejection by the host bird would. Evidence that this may be the case was obtained by Spottiswoode (2013), who showed that laying female honeyguides, *Indicator indicator*, which parasitise nests made in dark holes, preferentially punctured experimental eggs resembling those of other honeyguides more than they did host or control eggs. Since honeyguide chicks kill all other chicks in the nest, and hosts are poor discriminators, an ability to detect other honeyguide eggs is clearly important.

Mimicry by chicks – genetic and substantive differences

The colour and texture of a cuckoo's eggs depends on the genotype of the adult female that manufactures and lays them, and so her genes directly affect the survivorship of her eggs in the nest of the host bird. However, once the cuckoo chick has hatched, it is the similarity of the chick itself to its foster parents' own nestlings that will be critical, and this depends on the genes of the chick which come from both parents, though the mother might still have influence through epigenetic effects.

Cuckoo chick appearance

Different systems exhibit different degrees of resemblance between cuckoo chicks and those of their hosts. These differences may reflect sensory limitations of hosts or they may indicate more complicated evolutionary chases.

A classic example of mimicry between cuckoo chicks and their hosts is provided by African whydahs and indigobirds (Viduidae) where host estrildid finch chicks (Fig. 10.17a) and their brood parasites (Fig. 10.17b) both have elaborately coloured gapes that appear to mimic one another (see also Fig. 10.18c,d and figure 42 in Wickler 1968). The standard explanation for this similarity is that it is the result of an evolutionary chase between the species, based on the parent birds selecting for features that enabled them to discriminate their own chicks from the cuckoo and the cuckoo evolving to minimise the parent bird's ability to do so. Hauber & Kilner (2007), however, found that recent work only provides moderate support for this and is also largely consistent with a simple stimulus strength model, with nestlings competing for provisioning by the parent. If so, this puts a completely different slant on the issue because it would be the host chicks that were being selected to look like the 'cuckoo' and therefore able to compete better with the larger parasite for food.

In North America, the colour of the gape of chicks of the brown-headed cowbird, *Molothrus ater*, varies with locality. Those from the southwest have yellow rictal flanges, while birds from the rest of the continent have white ones. Furthermore, cowbird chicks raised in sympatric nests of the yellow warbler, *Setophaga petechia*, and song sparrow, *Melospiza melodia*, in California display significant host-related flange colours (Croston et al. 2012), but it is not known whether these differences are due to preferential host nest selection by female cowbirds or some developmental effect on the chicks themselves, possibly as a result of the diet fed to them. Rothstein (1978) suggested that while the host birds would feed the parasitic cowbird nestlings irrespective of their gape colour, those chicks whose rictal flanges most closely matched the host's own young might receive higher quality rearing. Similarly, Langmore & Kilner (2010) describe how rejection of the Australian Horsfield's bronze cuckoo, *Chalcites basalis*, chicks by their host superb fairy-wrens, *Malurus cyaneus*, has resulted in an arms race similar to those postulated for the African whydahs and indigobirds. Thus, the chicks of different species of bronze cuckoo often closely resemble their hosts in overall colouration, some being all black with conspicuous white feather tufts, some all yellow, some all pink and some drab (see figure 1 in Langmore et al. 2010, and figure 9 in Langmore 2013), though the degree of similarity depends on the degree of host specificity for each species.

In contrast to the *Vidua/Prinia* and *Chalcites/Malurus* systems above, many hosts of the common cuckoo, *Cuculus canorus*, are quite good at rejecting cuckoo eggs, although they are not good at discriminating between their own and cuckoo nestlings, which seems paradoxical, especially since human observers have no such difficulty (Alvarez et al. 1976, Davies & Brooke 1988, 1989). Consider the widely known photographs of a small host bird feeding a very large cuckoo chick that has grown beyond all possibility of being one of their own brood (see Fig. 10.18d). Grim (2006) carried out a meta-analysis of known cases but unfortunately it did not really yield any firm conclusions. It appears that there would be no cognitive limitation on the host bird's ability to discriminate cuckoo chicks, especially when they are older, so we are left with the selection pressure to do so necessarily being rather low. Nestling discrimination might also be expected to be more developed against cuckoos whose eggs were very good mimics, such that the host bird cannot discriminate between them at that stage, but there is no evidence that this is the case.

Some authors also considered that the bright gape colouration of various cuckoos probably acted as a supernormal stimulus to the host bird, such that they would preferentially provision the cuckoo chick over their own (N. Tinbergen 1951, R. Dawkins & Krebs 1979). Over the years, this view generally lost support, and Noble et al. (1999) found no evidence for it in an experimental test. However, these results may have been due to a failure to take into account the host bird's visual spectrum and sensitivity and to consider the likely lighting conditions in the nest. Tanaka et al. (2011) took these factors into account when studying the gapes of Horsfield's hawk-cuckoo, *Cuculus fugax*, and that of its host, the red-flanked bluetail, *Tarsiger cyanurus*, and concluded that, within the nest situation, the cuckoo chick's colour traits (which include also wing features) could indeed act as supernormal stimuli.

A final complication to the nestling discrimination issue is created by the habit of common cuckoos predating on nestling birds. Guilford & Read (1990) discuss this and pose the interesting conundrum that if cuckoos only predated nestlings in nests containing a cuckoo nestling as well as host ones, then they would be preferentially selecting against hosts that were acceptors, and that, in the long run, might increase selection for discrimination. However, in the short term, such behaviour might reduce competition for food for the young cuckoo and thus aid its development. Alternatively, if a cuckoo has temporarily run out of eggs, it might benefit by predating host chicks, forcing the hosts to start nesting and laying again, thus providing the cuckoo with renewed opportunity to parasitise nests. Finally, if cuckoos selectively predated nests of discriminating hosts they might both reduce their fitness and drive them away.

Begging calls

There are also numerous reports that the calls emitted by cuckoo chicks mimic the begging calls of their hosts' chicks, though quite a lot of these just involved human qualitative appraisal rather than sonograms (McLean & Waas 1987, Davies et al. 1998, Redondo & Arias de Reyna 1988, Langmore et al. 2003). Nevertheless, there is good evidence that chicks of the greater spotted cuckoo, *Clamator glandarius*, do this (Mundy 1973, Redondo & Arias de Reyna 1988) and also the specialist New Zealand shining cuckoo, *Chrysococcyx lucidus*, whose host is the grey warbler, *Gerygone igata* (M.G. Anderson et al. 2009). Phenetic analyses of chick begging calls show that those of the shining cuckoo are more similar to its host than to those of a range of other local birds (Fig. 10.21). Using cross-fostering of chicks, Langmore & Kilner (2010) showed that Horsfield's bronze cuckoo chicks learn to mimic the calls of their different hosts; chicks from fairy-wren nests transferred to thornbill (*Acanthiza*) broods quickly learned

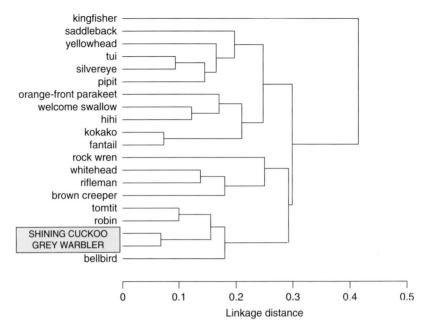

Fig. 10.21 Phenetic similarity between begging calls of chicks of various New Zealand passerine birds and the shining cuckoo, *Chrysococcyx lucidus*, which is a specialist brood parasite of the grey warbler, *Gerygone igata*, showing that the cuckoo's calls are most similar to those of its host. (Source: Data from M.G. Anderson et al. 2009.)

to imitate the long, rasping begging calls of their new brood-mates, possibly partly as a result of reinforcement by the host adult's provisioning behaviour.

Cuckoo chicks may also exaggerate the begging calls of the host chicks so as to maximise their share of the food being delivered by the host parents. Pagnucco et al. (2008) found that in the cowbird/song sparrow system, the exaggerated calling by the cowbird chick leads to the host chicks also changing their calls (increased amplitude and higher frequency), thus making them more similar to those of the intruder. This change in the direction of the mimicry has thus gained the catch phrase, a sheep in wolves' clothing.

Cuckoo and host coevolution

Klein & Payne (1998) used molecular sequence data to examine the evolutionary relationships between parasitic *Vidua* species (both indigobirds and paradise whydahs) and their estrildid finch hosts and found that the current host associations were unlikely to be the result of parallel-speciation events. In particular, they found that in general the *Vidua* species within a geographic region were more closely related to each other than they were to allopatric species attacking the same hosts.

Mimicry between adult cuckoos and their hosts

The cuckolding African paradise whydahs, *Vidua* spp., mimic the songs of their estrildid hosts (*Pytilia*) (Payne 1973a, 1973b, Payne & Payne 1994). Several possible functions have been proposed for this, but playback experiments (DaCosta & Sorenson 2014) involving male mimetic and non mimetic songs seem to show that rather than being deceptive, it provides an important intraspecific recognition signal in these territorial birds. This is likely because cuckoo chicks do not have the ready opportunity to learn their own species-specific songs from their parents and it may indeed be a reason why the call of the common (onomatopoeic) cuckoo is so simple – because it has to be genetically pre-programmed. The same problem applies also to imprinting, which juvenile cuckoos of all types must lack, at least as regards finding mates.

Hawk mimicry by adult cuckoos

Adults of many cuckoos, including the common cuckoo, *C. canorus*, are fairly similar to raptors (birds of prey) in appearance, with barred breast stripes and long tails, a resemblance that has long been recognised

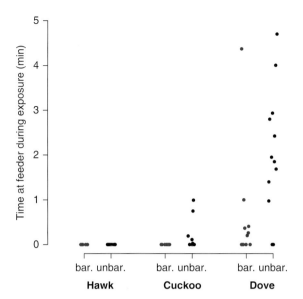

Fig. 10.22 Deterrence effect of natural and manipulated stuffed birds (hawk, cuckoo and dove) placed 50 cm from a bird-feeder for 5 minutes on the feeding behaviour of resident blue and great tits. Both cuckoos and hawks deterred birds, but the effect was reduced if the breast barring was obliterated, while the dove had little deterrent effect though somewhat more when its breast was artificially barred. (Source: Adapted from Davies & Walbergen 2008 under the terms of the Creative Commons Attribution Licence CC BY, via Proceedings of the Royal Society of London, Series B: Biological Sciences.)

(see Fig. 10.23a,b). The Indian hawk-cuckoo, *Hierococcyx varius*, bears a particularly close resemblance to the shikra, *Accipiter badius*. It was often suspected that this similarity was a true mimicry as many of the host birds would mob passing hawks, thus allowing the female cuckoo more easily to gain access to an unguarded host nest. Alternatively, the similarity could also simply be convergent, as in some sort of crypsis, or might provide the cuckoos with some protection from bird-eating birds of prey.

Davies & Welbergen (2008) investigated the effect of presenting mounted specimens of four similarly sized birds, common cuckoos, hawks, collard doves and teal, the latter two as controls, in the vicinity of a bird-feeder, and recorded the responses of non-host birds, blue tits and great tits. The tits responded equally to the unmodified hawk and cuckoo specimens, but not to the dove (Fig. 10.22). Then they manipulated the appearance of specimens, and found that barring in the stuffed cuckoos was important for the tits to respond, but that they would respond to the hawk even if the barring pattern were obliterated, while still not responding to the dove even when it was manipulated to be barred. Thus, these naïve birds recognised hawks on the basis of

some features other than barring, but the cuckoo with barring was clearly sufficiently hawk-like for it to elicit a response even though its other features were not a perfect hawk-like match. A similar study to assess the effectiveness of apparent hawk mimicry by the common cuckoo was carried out by Trnka & Prokop (2012), who presented various pairwise combinations of taxidermy mounts of cuckoo, sparrowhawk and turtle dove (a harmless control) in the vicinity of great reed warblers, *Acrocephalus arundinaceus*, nests, a well-known aggressive host. Both cuckoo and sparrowhawk mounts elicited a lot of aggression, but the cuckoo received the most attacks. They concluded that this particular host shows a strong generalised nest defence behaviour to all potential threats but is able to discriminate between the threats. The risk from the cuckoo, of course, is losing a brood, whereas the risk from the sparrowhawk is death.

In addition to these experimental tests, M.C. Stoddard (2012) used an avian visual model to compare how similar the common cuckoo was to the sparrowhawk, its presumed model. Consistent with the mimicry hypothesis, she found that female plumage was a markedly closer match to the hawks (10/17 closely matching body parts against equivalent parts of both male and female sparrowhawks) than that of the males (only 5/16 closely matching body parts against equivalent parts of both male and female sparrowhawks). Stoddard's (2012) approach further illuminates which body regions might be most important in terms of deceiving host birds of the cuckoo's hawkishness, and specifically the crown, rump, back and belly appeared to be the main contributors to mimetic resemblance. This might then lend itself to further experimental testing.

In a large survey of cuckoos and hawks, Gluckman & Mundy (2013) found that barred breast plumage was far more prevalent among parasitic cuckoos (35 out of 58 species) than non-parasitic ones (4 out of 83 species), strongly suggesting that it plays a role in their biology. Although the study was unable to take into account cuckoo phylogeny, other evidence indicates that barring in the five genera of parasitic cuckoos evolved after parasitism. Using digital image analysis, they compared the barring patterns of many cuckoo species with those of sympatric and allopatric raptors and found that the overall barring pattern of cuckoos more closely resembled those of sympatric hawks rather than allopatric ones, which again was supportive of the mimetic hypothesis.

Some cuckoos are polymorphic, with two adult colour patterns, and it turns out that polymorphism is far more prevalent among species with a hawk-like appearance (28%) than those without (8%) (Thorogood & Davies 2013). Using Sorenson & Payne's (2005) independent molecular phylogeny of cuckoos, it was shown that these two features are probably functionally related.

Host birds, such as barn swallows, *Hirundo rustica*, may show considerable aggression towards cuckoos and can be successful in driving them away from their nests. Liang & Møller (2014) hypothesised that the response of host birds towards hawk-like cuckoos might depend on the prevalence of hawks in the local area. Using dummies, they tested whether barn swallows may have evolved local adaptations, such that in areas where there are high densities of *Accipiter* hawks they might show a lower level of aggression against cuckoos. At sites in Denmark and China the authors placed dummies of cuckoo, *C. canorus*, sparrowhawk, *A. nisus*, and, as a control, Oriental turtledove, *Streptopelia orientalis*, at the nest sites of breeding barn swallows. Contrary to expectations, since hawks are relatively far more abundant in Denmark, the Danish swallows were far more aggressive towards both sparrowhawk and cuckoo dummies than their Chinese counterparts, but they also distinguished between the two types of threat. Thus, some local adaptation seems to have occurred but in a way contrary to the initial hypothesis, so some other factors have no doubt played a role, but with investigations only at two sites it is not possible to suggest what these might be.

Møller et al. (2015) found further evidence that hosts of common cuckoos perceive the cuckoos as hawks by investigating flight initiation distance, i.e. how close a cuckoo or sparrowhawk can get before the birds take flight. Flight initiation distance (adjusted for body mass) was negatively related to sparrowhawk predation rate but positively related to cuckoo nest parasitism rate. Therefore birds that flee from potential hawk danger earlier leave their nests more vulnerable to colonisation by cuckoos.

Alfred Russel Wallace (1889) suggested that hawk mimicry could be protective against attacks by real hawks, as cuckoos themselves are relatively defenseless and spend much time on exposed perches looking for potential hosts. He also suggested that the resemblance of drongo cuckoos, *Surniculus* spp. (Fig. 10.23c) to drongos, *Dicrurus* spp. (Fig. 10.23d), could be because the latter are very aggressive towards hawks and other large birds (Davies & Welbergen 2008).

Mimicry of harmless birds by adult cuckoos

As an alternative to mimicry of a predatory bird, i.e. hawks, females of the African cuckoo finch, *Anomalospiza imberbis*, appear to be mimics of harmless *Euplectes* weaver birds that occur abundantly near where their *Prinia* hosts occur (Feeney et al. 2015). The plumage of the cuckoo finch resembles that of the supposed models more closely than it does that of close relatives such as *Vidua* species. However, how well the deception works seems unclear in that host females were more likely to reject eggs after seeing either a cuckoo finch or bishop bird, *E. orix*, near to their nests.

Brood parasitism and inquilinism in social insects

Social insects offer potentially large rewards to predators that can gain access to their colonies and avoid attack. Members of many groups of insects are inquilines within the nests of ants for part or most of their life cycle, and collectively these play a range of roles from predators to mutualists and most, if not all, of these no doubt have to mimic ants to some extent, at least chemically, in order to avoid being attacked and eaten by their hosts.

While the vast majority of animal species that live within ant nests are other insects, there are also a few specialised vertebrates. The West African savanna frog, *Phrynomantis microps* (Microhylidae), lives in the subterranean nests of aggressive ponerine ants, *Paltothyreus tarsatus*, and does not elicit aggression from them. Rödel et al. (2013) found that treating termites and mealworms with the frog's skin secretions also protected these from attack, and chemical analysis subsequently showed that the frog secreted two peptides that are interpreted as being mimics of ant appeasement compounds. The ant's natural appeasement pheromones, as with *Polygergus*, are likely hydrocarbon compounds (e.g. Mori et al. 2000), so the substances produced by the frog most probably have similar epitopes rather than being chemically related.

Cuckoo bees and cuckoo wasps

Brood parasitism and social parasitism are widespread in the Hymenoptera, in fact one whole family have been given the name, cuckoo wasps (Chrysididae). Brood parasites such as chrysidid wasps, enter host nests and lay an egg in each of the host's brood cells where, after hatching, the parasite kills and eats any host egg or early stage larva and then proceeds to complete development, consuming the host egg/larva and brood chamber's food contents. Social parasites, such as cuckoo bumblebees, *Bombus* ('*Psithyrus*') spp.,[9] enter a social wasp or bee colony, kill the resident queen and start laying eggs in her place. These eggs are then tended by the original queen's worker daughters.

---

9. The cuckoo bumblebees were for a long time considered to be a separate genus (*Psithyrus*) from the social bumblebees (*Bombus*), however morphological, molecular and combined phylogenetic studies show that they are just a species group within *Bombus* (P.H. Williams 1994, Kawakita et al. 2004). Some workers retained it as a subgenus, but this would necessitate creating many more, rather useless subgenera, so I therefore place '*Psithyrus*' in inverted commas to indicate that it is just a convenient handle.

Fig. 10.23 Adult cuckoo mimicry of hawks and drongo. (a) Eurasian cuckoo, *Cuculus canorus* (Cuculidae); (b) female sparrowhawk, *Accipiter nisus* (Accipitridae); (c) drongo cuckoo, *Surniculus lugubris* (Cuculidae) from Thailand – its mimicry probably serves as protection from hawks since the drongo models are very aggressive; (d) black drongo, *Dicrurus macrocercus* (Dicruridae). (Source: a, Locaguapa 2010. Reproduced under the terms of the Creative Commons Attribution Share-Alike Licence CC BY-SA 3.0, via Wikimedia Commons; b, © Walter Baxter 2016. Reproduced under the terms of the Creative Commons Attribution Licence, CC BY-SA 2.0, via Geograph; d, Dr Raju Kasambe 2012. Reproduced under the terms of the Creative Commons Attribution Share-Alike Licence CC BY-SA 3.0, via Wikimedia Commons.)

Cuckoo bumblebees often show a marked resemblance to their host bumblebees (*Bombus* spp.). Whether such resemblances are mimetic in any way other than a Müllerian sense is debatable. It has been argued that since most of the interaction between a '*Psithyrus*' and its *Bombus* host take place within the darkness of the nest, the visible resemblance between them must be coincidental. However, it could be that it also plays a part in allowing the cuckoo bumblebee to approach its host's nest without provoking an attack, or perhaps more importantly, it may allow a recently emerged and therefore not fully hardened '*Psithyrus*' to escape from the vicinity of its host's nest in safety. Figure 1.5b summarises the potential mimetic links, in this case using semiotic notation. In comparison with their *Bombus* hosts, '*Psithyrus*' species are particularly well armoured and so too are cuckoo wasps, so part of their defence relies on physical resilience. However, olfactory crypsis and mimicry become important within the host's nest.

S.J. Martin et al. (2010) confirmed many people's suspicions that the '*Psithyrus*' cuckoo bees use chemical mimicry as part of their strategy, and found that host *Bombus* had species-specific alkene positional isomer profiles that are stable over large parts of their geographical ranges and that these were mimicked by three host-specific '*Psithyrus*' species. However, in some cases the chemical mimicry was less accurate, possibly due to the associations having had more recent evolutionary origins, but the '*Psithyrus*' species in these cases had a different chemical trick in that they produced dodecyl acetate,[10] which is known to be repellent to their hosts (Zimma et al. 2003), and probably helps them to avoid potentially dangerous close encounters when they enter the host nests. Not only do adult '*Psithyrus*' females have to enter host nests, but newly emerging males and females have to be able to leave them intact once they complete development. Lhomme et al. (2012) showed that emerging males of *B.* '*Psithyrus*' *vestalis* release secretions from their head region that also reduce aggression in host *B. terrestris* workers. The overall levels of aggression shown by workers to lures coated with various extracts as well as controls were markedly higher than in nests that were parasitised by the cuckoo bee, whether or not they contained emerging males of the '*Psithyrus*', which suggests that perhaps the female '*Psithyrus*' is releasing aggression suppressors within the colony. A further complexity in the systems is that, having usurped the original *Bombus* queen, the cuckoo bee female releases chemical odour signals that mimic the ones released by the host queen *Bombus* she has killed. These signal to the workers that she is still reproductively fit, so the workers will carry on caring for the brood, which now includes those of the usurping '*Psithyrus*' (Kreuter et al. 2011).

Not only are numerous bee species brood parasites, but so too are a good number of vespid wasps, including various species of *Dolichovespula* and *Vespula*. In most cases nothing is known about their molecular camouflage or mimicry. Bagnères et al. (1996) have studied the usurpation of host *Polistes* species nests by the obligate social parasite *Sulcopolistes* (as *Polistes*) *atrimandibularis*, and found that the colony-usurping female can switch off a whole family of cuticular hydrocarbons to facilitate her taking over a host colony. In another social parasite *Polistes* species, *P. sulcifer*, Sledge et al. (2001) found that 'queens' acquire cuticular hydrocarbon profiles that are characteristic of the individual host nest they have usurped and so are recognised as foreign by host *P. dominula* individuals from other colonies. Uboni et al. (2012) compared cuticular hydrocarbon profiles of *S. atrimandibularis* with those of the kleptoparasitic velvet ant, *Mutilla europaea* (Mutillidae), both of which attack colonies of *Polistes buglumis*. Both of the invasive species had fewer recognition signalling hydrocarbons but only the social parasite adopted a mimetic strategy within the host *Polistes* nests, and this is probably because they need to manipulate the host colony's social structure, whereas the velvet ant does not.

Cuckoo wasps of the family Chrysididae, while still aculeate hymenopterans, belong to a separate superfamily from the ants, vespid or sphecoid (including bees) wasps which collectively comprise the Vespoidea and Apoidea. They are some of the most startlingly coloured insects, usually brightly coloured in metallic iridescent greens, reds and blues, and have an exceptionally hard exoskeleton and an ability to roll up into a tight ball to protect themselves from attack. Nevertheless, they have to enter the nest burrows of their solitary aculeate wasp hosts and at least some employ chemical mimicry to facilitate this. The European cuckoo wasp, *Hedychrum rutilans*, is a specialist brood parasite of the beewolf, *Philanthus triangulum*, which in turn is a specialist predator of the honey bee, *Apis mellifera*, with which it stocks its brood cells. Beewolves will attack *H. rutilans* outside of their nests, presumably because they recognise them visually, but they do not molest them once in the dark inside of the nest (Strohm et al. 2008). Unexpectedly, female beewolves were found to be dimorphic in their hydrocarbon profiles, one morph having (Z)-9-C25:1 and the other (Z)-9-C27:1 as the major component. Phenetic clustering of gas chromatography–mass spectrometry (GC-MS) profiles of cuticular hydrocarbons from female *H. rutilans* were more similar to those of the (Z)-9-C27:1 chemical 'morphotype' of the female beewolves and generally more similar to beewolves than to either congeneric *H. nobile* females or the non-host sphecid, *Cerceris arenaria*, females (Fig. 10.24).

---

10. Also known as lauryl acetate.

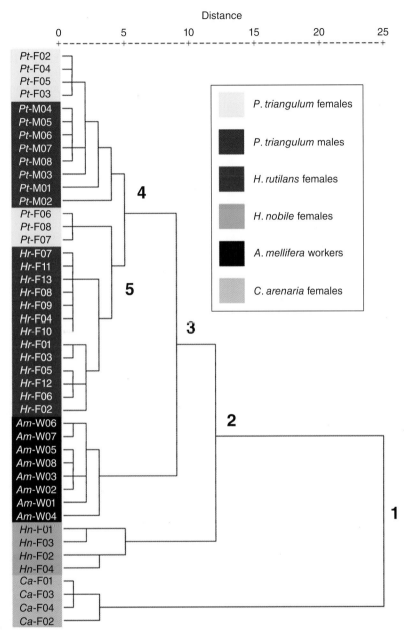

Fig. 10.24 Cluster analysis of the cuticular hydrocarbon compounds revealed by GC-MS of beewolf, *Philanthus triangulum*, females and males (Pt-F and Pt-M, respectively) a predator of the honey bee, brood parasite cuckoo wasp, *Hedychrum rutilans* (females, Hr-F) which is a parasite and kleptoparasite of *P. triangulum*, honey bee, *Apis mellifera* workers (Am-W), and biologically unrelated *Cerceris arenaria* females (Ca-F), and *H. nobile* females (Hn-F). Results show the considerable similarity in chemical profiles between females of the *H. rutilans* brood parasite and females of one of the chemo-morphs of its beewolf host, which enables the brood parasite to remain unharmed within its host's nests. Numbers after the species label indicate the different individuals. (Source: Strohm et al. 2008. Reproduced under the terms of the Creative Commons Attribution Licence, CC BY 2.0, via *Frontiers in Zoology*.)

Kleptoparasites of bees

Members of bee genus *Nomada* are kleptoparasites and are very host-specific, with each species being associated with a single *Andrena*, *Mellita* or similar, solitary bee species (Tengö & Bergström 1976). The female enters a host nest and lays her own eggs in host brood cells that are still open and being provisioned, and her larvae then develop, consuming the pollen and nectar provided by the host. Female *Nomada* obtain their chemical camouflage when they mate, as the males synthesise the mimetic compounds in their mandibular glands and spray them onto their mates (Tengö & Bergström 1977). In the pair *Andrena haemorrhoa*/*Nomada bifida*, the host and the male *Nomada* bee heads contain all-*trans* farnesyl hexanoate, which is the overwhelmingly abundant compound, whereas in the pairs *Andrena helvola*/*Nomada panzeri* and *Andrena clarkella*/*Nomada leucophthalma* the major component is geranyl octanoate.

Another group of bee kleptoparasites are the death's head hawkmoths, *Acherontia* species, whose adults enter the nests of various honey bee species and feed on their stored honey and nectar. Their thick cuticle may help protect them from bee stings, and they appear to be at least partially immune to bee venom, but they are nevertheless seldom attacked, even though honey bee colonies have efficient defence systems. Moritz et al. (1991) showed that hexane extracts of the European death's head hawkmoth, *Acherontia atropos*, contain only a very small number of compounds in any quantity and these are also abundant compounds in honey bee extracts, but they lack many of the other cuticular hydrocarbons of honey bees that the bees use to distinguish non-nest mates or even distinguish between individuals with differing degrees of relatedness.

Myrmecophily

A number of groups of beetles are particularly associated with ants, notably paussine carabids and at least a dozen lineages of staphylinid rove beetles. Many of the latter are morphologically specialised and all have a waist or short stem separating the thorax from the abdomen. In the case of the subfamily Clavigerinae, the adults have greatly reduced mouthparts because they dupe their host ants into feeding them through trophallaxis. The clavigerine association with modern ants has been shown to go back to the early Eocene, which is when modern ants started to become a dominant part of the insect fossil record (J. Parker & Grimaldi 2014). The degree of morphological convergence depends to some extent on the ant's eyesight: those inquilines of species that rely more heavily on vision, e.g. *Eciton* army ants, tend to show greater resemblance in colour and form than do ones associated with blind hosts (Seevers 1965).

Beetles of the carabid subfamily Paussinae are obligate myrmecophiles and feed on the broods of their predominantly myrmecine and formicine ants (Geiselhardt et al. 2007). Unlike many myrmecophiles, at least some paussines are not completely host-specific. Some species rely heavily on armour and shorter appendages to facilitate access, whereas others use a more chemical-based strategy, though the details have not been worked out, and may involve both mimicry and the secretion of compounds that may 'appease' the host ants. W.F. Kirby was quoted in *Transactions of the Entomological Society of London* (1883) as suggesting that some ants might tolerate paussines in their colonies because they are capable of producing chemical explosions when threatened, giving them the common name flanged bombardier beetles.

An extension to Wasmannian myrmecophilic mimicries was described by Lohman et al. (2006) who examined the cuticular hydrocarbon profiles of the woolly alder aphid, *Prociphilus tessellatus* (Hemiptera: Aphididae), its predators *Feniseca tarquinius* (Lepidoptera: Lycaenidae), *Chrysopa slossonae* (Neuroptera: Chrysopidae) and *Syrphus ribesii* (Diptera: Syrphidae) and the ants that tend them mutualistically. Each species of predator had a distinct cuticular hydrocarbon profile and, interestingly, each of these showed a greater similarity to that of the aphids rather than to that of the ants that tend the aphid. This implies that they have not evolved to resemble the ants *per se* as in Wasmannian mimicry, but rather that the ants are recognising the aphids on the basis of their hydrocarbons and so most likely are misinterpreting the signals from their competing aphid predator species as being those of their honeydew source.

Acquired chemical mimicry in social parasites and inquilines

As Dettner & Liepert (1994) point out, there are two routes by which chemical camouflage may be obtained – they can either be synthesised *de novo* or acquired from the hosts, and in the latter case the inquilines must usually have evolved specialised behaviours to ensure they remain safe while being in close enough physical contact with the host to gain enough cuticular hydrocarbons from them that they can then pass among them unrecognised. On the other hand, acquisition of colony identification compounds from hosts might allow the inquiline to utilise multiple host colonies or species, as seems to be the case for the myrmecophilous cricket *Myrmecophilus* (Akino et al. 1996).

R.W. Howard et al. (1990) interpreted chemical mimicry as meaning that the mimic synthesises their own mimetic compounds, and chemical camouflage if they acquired their deceptive chemicals passively, via contact with their hosts. However, that goes against the main usage of the terms because whether something is mimicking or camouflaging are defined by their effect on the signal receiver (Dettner & Liepert 1994).

Most mimics that synthesise their own deceptive chemicals probably do so by up- or downregulating synthetic pathways that they already have and then slowly may evolve some additional innovative biosynthetic pathways to improve the match. Many members of the myrmecophilic hoverfly genus *Microdon* (Diptera: Syrphidae) have larvae that develop as inquilines within the nests of various species of *Formica*, *Myrmica* and *Camponotus* and are obligate predators of the brood (larvae and pupae) of their host ants. The adult flies hatch from their puparia in the ant colony, mate outside and the females lay their eggs either on or near host ant nests. *Microdon piperi*, which are associated with *Campanotus* ants, are particularly unusual because they synthesise monomethyl- and dimethyl-(Z)-4-alkenes, which are very unusual insect compounds apart from among ants (R.W. Howard 1990, R.W. Howard et al. 1990, Dettner & Liepert 1994). Various *Microdon* species have been the subject of detailed investigation by R.W. Akre and colleagues (Akre et al. 1973, Garnett et al. 1985). During the *Microdon* second instar stage, which most strongly resemble the ants' cocoons, worker ants carry them around just like they do their true cocoons, so the fly larvae are presumably not only using chemical crypsis or resemblance to the ants, but also various tactile and possibly visual cues as well. By use of radioisotope labelling techniques, R.W. Howard et al. (1990) were able to show that at least in the case of *M. albicomatus*, a specialised associate of the ant *Myrmica incompleta*, the mimetic cuticular hydrocarbon mixture was synthesised *de novo* by the fly larvae, and not simply acquired by absorption from their ant larvae prey. Whereas most *Microdon* species are only associated with one ant species, a few are polyphagous, and this raises the question of whether these can adjust the hydrocarbon content of their cuticle to suit a particular host, which seems unlikely but perhaps not impossible, or whether there are distinct fly races that specialise on particular hosts as appears to be the case with the Eurasian cuckoo. It is also important to demonstrate that the cuticular hydrocarbon similarities observed in these and other cases actually represent adaptive convergence rather than being just by chance or the result of independent selective factors. R.W. Howard (1992) provided some evidence in this respect by comparing the cuticular hydrocarbon profiles of two

parasitic bethylid wasps and their beetle larvae hosts. In these solitary species he found virtually no similarity between parasitoid and host, thus showing that in the absence of any strong selection pressure for convergence in cuticular chemistry, associated species do not share chemical similarity but remain independent.

Von Beeren et al. (2012) suggested that each of their classes of chemical adaptive resemblance could be qualified by the terms "innate" or "acquired" to indicate the source of the chemical involved.

Witte et al. (2009) compared behaviour and chemical profiles of two very different myrmecophiles of ponerine army ant, *Leptogenys* spp. – the silverfish *Malayatelura ponerophila* (Thysanura) and the spider *Gamasomorpha maschwitzi*. Neither was a particularly good chemical mimic of the host, the spider being less convincing than the silverfish. Survival from ant predation in the silverfish relied largely on its ability to escape when detected, whereas the spider does not and remains in contact with the potential aggressor until aggression ceases. Perhaps the aggressing ant habituates or the spider absorbs some of the aggressor's hydrocarbons. Observations on ant predating larvae of the *Thermophilum* species (Carabidae) in North Africa also strongly suggest rapidly acquired chemical camouflage. Eggs are laid outside of the host ant's nest and the newly hatched larvae locate a nest entrance by following the ant chemical trails and enter it, rapidly pursued by attacking ants. However, within a short time they can move around freely within the nest, consuming the hosts' brood (Dinter et al. 2002).

In the case of the oriental social parasitic hornet *Vespa dybalskii* (Vespidae), S.J. Martin et al. (2008) found that its eggs did not possess hydrocarbon profiles similar to those of its two host species (*V. simillima* and *V. crabro*) though they were not lacking in total hydrocarbon content. The main difference between their profiles and those of their hosts and other related wasps was that they only contained a very low percentage (<1%) of methyl-branched compounds, which are the compounds postulated as being important in species recognition. Thus, rather than being concealed through chemical similarity, they may be chemically "transparent", as the authors term it.

The ectoparasitic mite *Varroa jacobsoni* is a major pest of honey bees, *Apis mellifera*. Bees are, of course, hugely economically important and also fastidious under normal circumstances in maintaining the cleanliness of the colony and, not surprisingly, *Varroa* has evolved chemical mimicry that minimises its detection by the host bees (Nation et al. 1992). What is particularly interesting about this example is that, although not perfectly similar, the cuticular hydrocarbon profiles of the mites varied with host developmental stage and tracked the changing host profiles, thus presumably reducing the likelihood of being detected and killed as the host develops (C. Martin et al. 2001). In addition, Kather et al. (2015) found that *Varroa* also mimics host colony-specific aspects of the chemical deception, especially in terms of *n*-alkanes and *n*-alkenes.

Brood-parasitic and slave-making ants

Various ant species use one of two strategies to make use of the workers of another ant species; these are called brood parasitism and slave-making. In brood parasitism, a new fertilised queen enters the nest of a host species, kills the resident queen (regicide) and then lays her own eggs, which are then tended by the host queen's surviving workers, exactly as in some of the bee and wasp cases discussed earlier. Slave-makers,[11] on the other hand, capture the brood of another ant species and transport them to the slave-maker's nest, where they perform all the worker ant duties that they would have in their original colonies. In the case of *Polyergus rufescens*, a new queen has to penetrate a host colony, kill its queen and let the previous owner's workers rear her brood, and at this stage may be considered a social parasite. Because the host workers are not being replenished, the *Polyergus* workers then raid other nests of the slave species and carry back their larvae and pupae to the originally usurped nest. The *Polyergus* workers cannot feed themselves and carry out no nest duties other than raiding other host species' nests. Approximately 2% of ant species are social parasites or slave-makers that, usually following regicide, usurp the foraging and nest-upkeep behaviour of a host ant species.

Cuticular hydrocarbon and volatile organic compound similarities between slave-maker ants and their hosts have been found in a number of systems (G. Bergström & Löfqvist 1968). The Palaearctic slave-making ant *Polyergus rufescens* synthesises a cuticular hydrocarbon blend that contains many of the compounds of its *Formica rufibarbis* and *F. cunicularia* slaves, but no unique ones of its own (Bonavita-Cougourdan et al. 1996). Furthermore, having usurped a host nest, the cuticular hydrocarbon pattern of the *Polyergus* adapted more closely to resemble the pattern of the given host species or colony, suggesting that natural selection has led to the evolution of a feedback mechanism to adjust the proportions of particular components synthesised in accordance with the detected profiles of their hosts. Mori et al. (1996) found that the two *Formica* slave species did not distinguish the *Polyergus* cocoons from their own cocoons,

nor did they differ in how they reared their own or *Polyergus* broods, but *F. lugubris*, a species which is never enslaved, readily and always destroyed cocoons of the *Polyergus*.

Because ants rely heavily on sophisticated chemical communication and recognition systems, the slave-makers nearly always have to rely on some form of chemical subterfuge to gain access to the host nest and stay alive within it. Interestingly, among ants, there is often a close phylogenetic relationship between slave-making species and their hosts (e.g. Hasegawa et al. 2002) and this might have made evolution of similar cuticular hydrocarbon profiles or mimetic pheromone compounds/blends easier. However, it seems that crypsis is far more common, if not the only mechanism (Lenoir et al. 2001), and typically progresses from chemical inertness to acquired mimicry by absorption of host odours, sometimes with associated behaviours that maximise odour acquisition (Lenoir et al. 1997). D'Ettorre & Heinze (2001) suggest that this is inevitable because each individual host ant colony will have its own particular odour characteristics because of environmental influences (or intraspecific genetic variation), which makes it virtually impossible for a slave-maker or inquiline to mimic a future host's smell precisely. Thus, in ants, mimicry is not involved so much in smelling like a host, but in producing mimetic pheromones that can change the behaviour of host workers to the mimic's advantage.

Queens, but not the workers, of *Bothriomyrmex syrius* produce large quantities of the alarm substance of their host ants *Tapinoma* spp. (6-methyl-5-hepten-2-one), and this chemical mimicry is thought to facilitate penetration of the *Bothriomyrmex* queen into the host colony (H. Lloyd et al. 1986). Although little is known of *Bothriomyrmex*'s biology, at least some species are temporary social parasites, with the queen of *B. syrius* allowing herself to be dragged into a *Tapinoma* nest by the *Tapinoma* workers and then decapitating the *Tapinoma* queen before laying her own eggs.

The compounds released by a usurping hymenopterans are not always clearly mimetic. D'Ettorre et al. (2000) showed that the slave-making ant *Polyergus rufescens* uses a dual strategy to usurp the nests of its host, *Formica cunicularia*. The mated *Polyergus* queen may successfully sneak into the host ant's nest to kill the host queen in mortal combat, first by being chemically inert. However, some host workers are liable to show persistent aggression towards her, and her second strategy is to release a mixture of butanoic and acetic acid esters, of which decyl butanoate comprises over 80%, from her Dufour's gland. These have a repellent effect upon host workers, despite the fact that the compound does not appear to be utilised in the chemical communication system of the host ant, though it is in related species. It is possible that the *Polyergus* is tapping into an evolutionary left-over response pathway.

---

11. I use this term though some workers consider it "culturally insensitive", "offensive" or inaccurate, and prefer terms such as cleptergic.

The Mesoamerican ant *Ectatomma ruidum* (Formicidae) is not a parasite or slave-maker, but it is kleptobiotic in that workers sometimes steal food from conspecifics from different colonies which they locate by following their trails (Perfecto & Vandermeer 1993). Breed et al. (1992) described how individual ants were able to acquire the colony-recognition labels (cuticular hydrocarbons) from conspecific individuals from different colonies, and thus were able to steal from them. Different individuals apparently only acquire labels from a single other colony and they specialise in robbing from members of that particular colony.

Chemical mimicry and ant and termite inquilines

Myrmecophiles may avoid recognition as nest intruders by various strategies, including behavioural mimicry, release of appeasement chemicals that reduce ant aggression, inherited chemical mimicry, acquired chemical mimicry and probably chemical crypsis (Kistner 1979). Many staphylinid beetles live in association with ants (Akre & Rettenmeyer 1966, Chopard 1967) and these include species with various degrees of myrmecomorphy, though in the case of army ant inquilines, Akre & Rettenmeyer (1966) note that those that are morphologically most similar to the ants tend to be rare, live in the centres of the colonies migrating with the brood, and are generally more integrated into colony life, whereas the least myrmecomorphic ones tended to be more peripheral and more often attacked by the ants. Rarity is relative, and ants normally out-number their inquilines by a ratio of thousands to one. Collectively, these staphylinids prey variously on the ants' prey, the ants' brood and, in some species, even on adult ants. Given how important organic volatiles are in ant communication, colony recognition, etc., it is certain that all these commensals must smell like the host ants. Some species regularly make physical contact with their hosts, even stridulating them and gaining food by trophallaxis, and this may also serve to help them acquire host ant hydrocarbons, thus increasing their resemblance to the ant colony.

Some of the myrmecophilic staphylinids are likely Batesian mimics of the ants (Taniguchi et al. 2005). *Drusilla inflatae* is associated with *Crematogaster* ants both morphologically and in colour pattern, which it is believed to be a Batesian mimic of, foraging among and scavenging on dead hosts, avoiding contact with them and probably producing a repellent secretion (Maruyama et al. 2003). Members of the myrmecophilous staphylinid genus *Pella* live along the foraging trunk routes of the ant *Lasius fuliginosus*, and prey on living ants and scavenge on dead ones in the ants' garbage dumps. These beetles avoid host ant aggression by mimicking the ants' alarm pheromones (Kistner & Blum 1971, Steidle & Dettner 1993). The two main compounds

produced are undecane and sulcatone (6-methyl-5-hepten-2-one), which seem to be alarm and panic pheromones, respectively (Stoeffler et al. 2007). The undecane alone elicits aggression from the ant, but when sulcatone is also present the ants avoid the area and make fewer attacks. Most chemical mimicry among myrmecophiles involves cuticular hydrocarbons, but the mimetic pheromones of the *Pella* beetles are produced by specialised tergal glands. The tergal gland secretions of two other myrmecophilic staphylinids, *Zyras collaris* and *Z. haworthi*, are very unlike those of related staphylinids that live with ants, instead comprising α-pinene, β-pinene, myrcene and limonene. For these, Stoeffler et al. (2013) suggest that their secretions mimic the odours of aphids that the host ants tend in return for honeydew (see section *Ants and aphid trophallaxis* later)

Another staphylinid beetle, *Trichopsenius frosti*, which is an inquiline of eastern subterranean termite, *Reticulitermes flavipes*, colonies in North America, synthesises precisely the same pattern of cuticular hydrocarbons as does its host (R.W. Howard et al. 1980).

The aphodiid beetle *Myrmecaphodius excavaticollis* is an inquiline within colonies of *Solenopsis* fire ants and can integrate itself into different colonies and several species. By itself it has no innate chemical mimicry, though it may be rather chemically neutral, but once inside a colony it picks up that colony's odour (Vander Meer & Wojcik 1982).

A brood-parasitic aphid

Salazar et al. (2015) recently described a most remarkable case of polymorphism in an ant-associated aphid. Aphids that inhabit ant nests feed on sap from plant roots and are tolerated or even farmed by their hosts, who receive sugar-rich aphid honeydew in return through trophallaxis. The aphid *Paracletus cimiciformis*, which inhabits *Tetramorium* ant nests, has two morphs, one with typical mutualistic trophobiont biology, but the other an aggressive mimic of the host. The aggressive morph has a cuticular hydrocarbon profile that more closely resembles that of the host ant's larvae and dupes the ants into transporting it to the brood chamber, where it actively predates (sucks the haemolymph from) the host ant's larvae.

Ants and aphid trophallaxis

Kloft (1959) and Way (1963) observed the way ants that tend (farm) aphid colonies stroke the hind legs of the aphids to elicit the aphid releasing a drop of honeydew for them to feed on. Both considered that the way the ants do this begging is identical to the way that ants elicit feeding from other conspecifics (trophallaxis), and if you take an ant's-eye view of the posterior of an aphid there is a definite resemblance in profile (see figures 21 and 22 in Wickler 1968), which led to the hypothesis that the ants were simply mistaking the aphid for one of their own. That the aphids supply the ants with food while the ants protect the aphids from natural enemies makes this resemblance truly mutualistic. The view that this is a form of mimicry was supported by Wickler (1968) and by Fowler et al. (1991), as well as by Kloft's (1959) observations that ants also sometimes attempt to feed the aphids' posteriors with regurgitated food – obviously in vain. Further evidence in support of this mimetic hypothesis would be to see whether the physical similarity, perhaps through having longer siphuncles, was less in those aphids that are not farmed by ants. However, as the siphuncles produce defensive compounds, one would expect them to be better developed in species that had to rely more on their own defences rather than having ants do the job.

Aphidiine parasitoids of ant-attended aphids

Many aphids are regularly protected from parasitoid wasps of the braconid subfamily Aphidiinae by ants in exchange for honeydew. Nevertheless, aphidiines can often be observed ovipositing into aphids while ants are present, seemingly without attracting any attention from the ants. The explanation appears to be that the aphidiines possess a mix of cuticular hydrocarbons that closely resemble those of their aphid hosts (Völkl & Mackauer 1993, Liepert & Dettner 1996, Dieckhoff & Heimpel 2010), thus this host mimicry works by virtue of deceiving a mutualist of the parasitoid's hosts.

Many species of aphid are tended symbiotically by ants, the ants gaining access to honeydew while the aphids gain protection from predators and parasitoids. Liepert & Dettner (1996) showed that *Lasius niger* ants, tending colonies of the aphid *Aphis fabae*, ignored individuals of the aphid parasitoid *Lysiphlebus cardui* (Hymenoptera; Braconidae), but attacked another related parasitoid, *Trioxys angelicae*. Hexane-washed *T. angelicae* and *L. cardui* were both ignored by the ants, and *T. angelicae* treated with *L. cardui* hexane extract were similarly ignored. But, if *L. cardui* were treated with hexane washings of *T. angelicae*, they were attacked by the ants, indicating that *T. angelicae* cuticle contains hydrocarbons that elicit attack by the ants, but that these stimulants were absent from the *Lysiphlebus*. Given that ants will attack a wide range of aphid predators and other insects, it seems likely that the absence of aggression-provoking compounds in the *Lysiphlebus* cuticle may represent an adaptation towards parasitising ant-tended aphids. If this is the case then this situation could be interpreted as an example of olfactory crypsis. *Lysiphlebus fabarum* might actually be unique among the ant-mimicking aphidiines because it

additionally solicits honeydew secretion from its aphid host, *Aphis fabae*, by antennating them in a manner that is very similar to the way their ant-attenders do (Rasekh et al. 2010).

DOES AGGRESSIVE MIMICRY OCCUR IN PLANTS?

Carnivorous (mostly insectivorous) plants mostly frequent bright, wet habitats with very poor nutrient availability and thus insectivory is a means of obtaining minerals, nitrogen, and perhaps some energy (Givnish et al. 1984). They often have traps that resemble flowers in a general sort of way and therefore they may attract visiting insects. Some of them have traps with bold UV absorbance patterns, as do many flowers (Joel et al. 1985). The insect traps of carnivorous plants are also commonly red in colour, and this has also been hypothesised to be part of a deception, either by resembling flowers or increasing contrast (F.E. Lloyd 1942, Jaffé et al. 1992, Ichiishi et al. 1999, Schaefer & Ruxton 2008). However, this is not always the case, as Foot et al. (2014) have shown that, at least in the sundew, *Drosera rotundifolia* (Droseraceae), the red colouration does not serve any attractive function, and these authors also comment that some previous experimental evidence seeming to

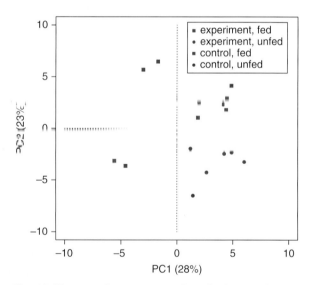

Fig. 10.25 Principal components analysis plot showing change in profile of volatile organic compounds produced by the Venus fly trap, *Dionaea muscipula* (Droseraceae), in response to having been fed and thus receiving a nitrogen fix. (Source: Adapted from Kreuzwieser et al. 2014. Reproduced under the terms of the Creative Commons Attribution Share-Alike Licence CC BY 3.0, via the *Journal of Experimental Botany*.)

show the importance of the red trap colour in attracting prey may well be flawed, for example through lack of ecological relevance or other methodological issues.

The well-known pitcher plant *Saracenia flava* from the eastern USA produces an enol diacetal monoterpene called sarracenin in quantities (Miles et al. 1976) and Jaffé et al (1995) similarly found this to be an important component of the volatiles emitted by the spoon-shaped pitcher leaves of the confamilial *Heliamphora heterodoxa* and *H. tatei* at the time they attract insect prey, as well as various phenolics and methyl esters. Interestingly, in addition to attracting insects that fall into their traps, these plants may also reduce the chance of them subsequently escaping by releasing paralytic compounds into their traps (Mody et al. 1976). Recently, Kreuzwieser et al. (2014) found that the Venus fly trap, *Dionaea muscipula* (Droseraceae), also emits an insect-attracting blend of volatile organic compounds (VOC) with an odour resembling "the bouquet of fruit and plant flowers". The authors found that when plants were fed with *Drosophila*, thereby receiving nitrogen supplement, their VOC emission profile changed significantly (Fig. 10.25) with a reduction in the terpene component of their emissions. However, this odour change did not appear to affect their attractiveness to other *Drosophila*. Floral scents are also employed by the insectivorous pitcher plant *Nepenthes rafflesiana* (Nepenthaceae) (Di Giusto et al. 2010), and the pitchers, which are insect traps, also provide a nectar reward.

As pointed out by Williamson (1982), these, and of course similar animal examples, offer little chance of learning since duped insects are eliminated. Pitcher plants of the Sarracenaceae attract a wide variety of insects to their trap leaves, though ants are particularly important for some species. Ruxton & Schaefer (2011) do not consider that the *Nepenthes* traps are in any way examples of mimicry, and that instead they are simply exploiting perceptual biases in the insects they capture. Joel (1988) also suggested there was no evidence that any insects visit a pitcher because they have been duped. In no cases have any specific flower models been identified, and the pitchers provide a nectar reward that insects can exploit if they do not get trapped first time around. Furthermore, there should not be any selection pressure for pitchers to remain rare, relative to flowers, because, overall, flowers do not limit the numbers of insects that the pitchers might trap. Thus Ruxton & Schaeffer (2011) conclude that "there is also no evidence that mimicry (in any sensory modality) plays any part in carnivory by plants". Schaefer & Ruxton (2014) discuss another taxonomically widespread feature of pitcher plants that has been suggested to be deceptive – fenestration, i.e. transparent patches near the upper parts of the pitchers that might deceive visiting insects about the location of the exit from the pitcher. Their experimental test did not support this idea, but did indicate that fenestration might make the pitchers more attractive from a distance.

Chapter 11

SEXUAL MIMICRIES IN ANIMALS (INCLUDING HUMANS)

Never attempt to win by force what can be won by deception.

Niccolò Machiavelli, *The Prince*

Mimicry, Crypsis, Masquerade and other Adaptive Resemblances, First Edition. Donald L. J. Quicke.
© 2017 John Wiley & Sons Ltd. Published 2017 by John Wiley & Sons Ltd.

INTRODUCTION

The term sexual mimicry is used here for those adaptive resemblances that increase an individual's chance of successful mating or, more generally, that increase the chance of obtaining fertilisations in some other way. In most cases this involves individuals of one sex mimicking conspecifics of the other, though in a few cases it involves mimicry of other species or, in some plants, the mimicry may be of different parts of the same individual (see Chapter 12). The term has also been used by various authors for deceptions involving mimicry of one sex by another that are not directly concerned with gaining parenthood (Table 11.1) and these cases are mostly dealt with in other chapters. In particular, Zabka & Tembrock (1986) used the term "reproductive mimicry" as a catch-all and a quick inspection of their table will show that such a broad concept is of very little value or meaning (see Table 1.9).

MIMICKING THE OPPOSITE SEX

Female mimicry by males

Males of externally fertilising species, such as the majority of fish, seem likely to be especially vulnerable to losing fertilisations to other sneaky males who might get close enough to a mating pair. Not surprisingly, many males are therefore vigilant of other males and chase them away or attack them. In a number of nest-building fish species, some males do not develop to typical male size or colouration but instead mimic conspecific females. This polymorphism has been best documented in the bluegill sunfish, *Lepomis macrochirus* (Centrarchidae) (Gross 1979, 1982, Dominey 1980), though it also occurs in a number of other species of fish (Table 11.2) and some other groups of organisms (see later) (Weldon & Burghardt 1984). By analysis of otolith annuli in the bluegill sunfish, Dominey (1980) was able to show that female-mimicking males had a similar age distribution to typical, nest-building males, and also that nest-building males had not put on a sudden burst of growth late in life. It was therefore concluded that the mimic males do not represent a developmental stage and may in fact be the result of a genetic polymorphism.

In the sexually dimorphic North Atlantic corkwing wrasse, *Symphodus* (*Crenilabrus*) *melops*, some 20% of males so closely resemble females in their secondary sexual characters that they cannot be distinguished from each other except by their gonads (Dipper 1981, Uglem et al. 2000). Another wrasse, *S. ocellatus*, has males with four distinct types of reproductive strategy: brightly coloured territory holders, sneakers and two classes of satellite males, one

tolerated and one not (Taborsky et al. 1987). Intermediate morphologies in at least some wrasse may be due to their ability to change sex. As with the sunfish, the corkwing wrasse is a nest-building species. In the reef fish *Pomatocentrus amboinensis*, juveniles sport a black spot on the dorsal fin (see Chapter 8, section *Eyespots and fish*). Although the spot usually fades away completely by the time the fish reach sexual maturity, a small proportion of sexually mature males retain the mark (Gagliano & Depczynski 2013). These males are also untypical in another way, their body shape more closely resembling that of immature females rather than dominant mature males. It is not yet known whether these female-like males use the spot and shape deceptively, though it seems likely that they do.

Macías-Garcia & Valero (2001) describe a case in which female mimicry is affected by a male's performance in male–male contests. Subordinate males of the goodeid (splitfin), *Girardinichthys multiradiatus*, an endemic freshwater fish from Mexico, develop a darker black spot around the vent that resembles the 'pregnancy spot' of females. The authors suggest that this facultative change helps these less-dominant males deceive dominant ones into being less aggressive towards them, possibly allowing them to get sneaky fertilisations.

In species other than fish, this sort of cuckoldry is also known in two cuttlefish, *Sepia apama* and *S. plangon* (Norman et al. 1999, Hanlon et al. 2005), a crab (Laufer & Ahl 1995), a bird (Jukema & Piersma 2006), a marine isopod (Shuster & Wade 1991) and in some insects, such as the rove beetle *Leistotrophus versicolor* (Forsyth & Alcock 1990b) and crickets (W. Bailey et al. 2006). The first cuttlefish example (Norman et al. 1999) differs from the others in that because of their ability to undergo rapid colour change (see Chapter 2, section *Cephalopod chromatophores*) the female mimicry is very transient and is also facilitated because, although cuttlefish have acute vision, they have poor social recognition ability (Boal 1996), i.e. they cannot readily recognise that a duping she-male is a male they have encountered previously. Males of the second cuttlefish species, *S. plangon*, use their ability to mimic female signals in an altogether more sophisticated way to deceive other males (Hanlon et al. 2005). This species has the ability to display male courtship patterns on one side of its body, while appearing female-like on the other. Thus when a male is courting a female in the presence of a rival, he displays male signals to her and female signals to the other male (C. Brown et al. 2012).

Mank & Avise (2006) plotted the occurrence of various types of male alternative reproductive tactics (what they call MARTs) on a supertree phylogeny of fish, and found evidence for an evolutionary progression from male monopolisation to sneaking and to full female mimicry.

Table 11.1 Summary and examples of intraspecific mimicry of one sex by another.

	Gain fertilisations/matings	Avoid/minimise sperm competition	Avoid unwanted courtship attempts	Avoid aggression/increase social bonding
Females mimic males	Homosexual pseudocopulation *Ecballium* flowers (Dukas 1987)		*Drosophila* (D. Scott 1986) Various damsel & dragonflies (Fincke et al. 2005)	Spotted hyaena, *Crocuta crocuta* (Racey & Skinner 1979, East et al. 1993)
Males mimic females	See Table 11.2 for fish examples Giant Australian cuttlefish, *Sepia apama* (Hanlon et al. 2005); *S. plangon* (C. Brown et al. 2012) Garter snakes (R.T. Mason & Crews 1985) Marine isopod, *Paracerceis sculpta* (Shuster & Wade 1991) Rove beetle, *Leistotrophus versicolor* (Forsyth & Alcock 1990b) Parasitoid wasp, *Cotesia rubecula* (Field and Keller 1993)	Salamanders (Arnold 1976)		*Cardiocondyla* ant (Cremer et al. 2002) Augrabies flat lizard, *Platysaurus broadleyi* (Whiting et al. 2009)
Males cause females to mimic males		*Drosophila* (D. Scott 1986)		

Table 11.2 Selection of fish species in which males are known to mimic females to gain sneaky matings.

Common name	Species	Family	Type of water	References
Bluegill sunfish	*Lepomis macrochirus*	Centrarchidae	Fresh	Gross 1979, 1982, Dominey 1980
Nine-spined stickleback*	*Pungitius pungitius*	Gasterosteidae	Fresh	Morris 1950
Fourspine stickleback	*Apeltes quadracus*	Gasterosteidae	Fresh, brackish and salt	H.E. Willmott & Foster 1995
Dark-edged splitfin	*Girardinichthys multiradiatus*	Goodeiidae	Fresh	Macías-Garcia & Valero 2001
Leaf fish	*Polycentrus schomburgkii*	Polycentridae	Fresh	G.W. Barlow 1967
Corkwing wrasse	*Symphodus (Crenilabrus) melops*	Labriidae	Salt	Dipper 1981
	S. ocellatus	Labriidae	Salt	Taborsky et al. 1987
	Tripterygion	Tripterygiidae	Salt	Wirtz 1978, de Jonge & Videler 1989, de Jonge et al. 1989
Rusty blenny	*Parablenius sanguinolentus*	Blenniidae	Salt	R.S. Santos 1985, Ruchon et al. 1995, R.F. Oliveira et al. 2001
Peacock blenny	*Salaria pavo*	Blenniidae	Salt	Gonçalves et al. 1996
	Rhinogobius	Gobiidae	Fresh	Okuda et al. 2003
Red-spotted masu salmon	*Oncorhynchus masou*	Salmonidae	Fresh	Kano et al. 2006

* Also called the ten-spined stickleback.

Both obtaining fertilisations by sneaking without mimicry and female mimicry appear to be tactics to usurp male monopolisation of mates.

In the case of the swordtail *Xiphophorus nezahualcoyotl* (Poeciliidae), mimicry of females by a small percentage of males which display a false brood spot does not seem to be connected to sneaky mating. These males, however, also have an even longer sword (on the tail) than males without a false brood spot. This is a character selected for by females. Other males seem to show the mimics less aggression, thus allowing them to spend more time foraging (Rios-Cardenas et al. 2010). The low incidence of mimetic males means that the presence of a brood spot is still a largely reliable signal to other males that individuals with brood spots are females.

A rather different situation has been described by R.T. Mason & Crews (1985) for North American red-sided garter snakes, *Thamnophis sirtalis parietalis*. These, like other snakes, are internal fertilisers, yet a small proportion of males display female characters and release pheromones that attract other males as do normal females. Obviously, such 'she-males', as they have been termed, cannot get sneaky fertilisations, unlike the externally fertilising 'transvestite' fish described above. Nevertheless, she-male garter snakes were shown to outperform normal males in mating trials. An important aspect of garter snake mating strategy that seems to have allowed the evolution of the she-male strategy is that mating normally occurs in mating aggregations in the form of an intertwined ball of competing 10–100 normal males around a single female. She-males were also observed to attract groups of 5–17 males around themselves. R.T. Mason & Crews (1985) concluded that the she-males benefit from their pheromonal mimicry of normal females by causing normal males in mating aggregations to ignore the true female in favour of the she-male and thus to allow the she-male to manoeuvre to a more advantageous position in the mating ball than its normal counterparts. Shine et al. (2012) further demonstrated that the she-males do not waste effort in attracting other males to court them unless it is useful and, in this case, this also means they are cold and newly emerged from hibernation, since the courting behaviour and tangle of other males helps them to increase their body warmth and thus their potential speed in the event of predator attacks.

Unfortunately for some of the she-male snakes, "in extreme cases, the courted male may be forcibly copulated (Pfrender et al. 2001) or suffocated" (Shine et al. 2000).

S.J. Arnold (1976) described a behaviour in two North American species of salamander, *Ambystoma tigrinum* and *Plethodon jordani*, in which males mimic females and thereby attract other males to court with them, leading ultimately to other males depositing their spermatophore inappropriately. Thus the mimic male dupes the other into wasting its spermatophore and this is assumed to increase the chance of the mimic successfully using its own retained spermatophore. This behaviour no doubt evolved in *P. jordani* because in this species, the male has a complex courtship behaviour and usually only deposits a single spermatophore. Another species studied by Arnold (1976), *A. maculatum*, is very different in that its courtship is very simple and it deposits many spermatophores, though it has a low success rate with each of these. In this species, female mimicry is not present and in any case it would have little effect as any duped male would still have reserves of many more spermatophores. *A. tigrinum* is intermediate in its courtship behaviour, though the fact that some males mimic females would indicate that this strategy does increase the number of successful fertilisations by reducing competition from other males with significant sperm reserves.

Males of most cicadas emit courtship songs to attract females, the sound being produced by their tymbal organs. Receptive females often respond by producing a sound using wing stridulation. Males of *Subpsaltria yangi* are unusual in that they also have a well-developed stridulatory organ and use it to make a sound closely resembling a female's response call (Luo & Wei 2015). They seem to carry out this mimicry so that receptive females will act as if there is a female competitor nearby, perhaps increasing the likelihood that the deceived female will produce response sounds. As males rely largely on female responses to locate them, this behaviour increases the male mimic's chances of locating a mate.

In Augrabies flat lizards, *Platysaurus broadleyi*, from South Africa, Whiting et al. (2008) found that the visual mimicry of she-males was effective against other males at a distance, but they also found, by using hexane-washing and relabelling of she-males, that he-males could distinguish them from females by their smell. Thus, in this species, the deception does not work at close quarters and the she-males in the wild keep themselves at a distance from he-males to avoid detection.

Among invertebrates there are some similar examples. Thornhill (1979) described the female-mimicking behaviour in a nuptial gift-giving scorpionfly, *Hylobittacus apicalis* (Mecoptera). In this case the reward is either a 'free meal' or a ticket to reproductive success, since the dupe gives away the nuptial prey gift to the female-mimicking male rather than to a potential mate. Furthermore, Thornhill (1979) notes that the males that obtain gifts by acts of piracy both copulate more frequently than duped ones and probably also avoid additional dangers, such as predation, associated with having to search for and capture their own living nuptial gift prey directly.

Avoiding aggression from competing males

Avoiding other male aggression is important in many instances. Cremer et al. (2002) describe a morphological, behavioural and chemical polymorphism in the tropical tramp ant species *Cardiocondyla obscurior* (Formicidae), in which virgin queens are mated within their natal nests. Males of this species are either wingless and aggressive or winged and docile, and the latter would be expected to lose in aggressive competition with wingless males. However, their odour closely resembles that of virgin queens and so, rather than being attacked by the aggressive morph, the latter often attempt to mate with them. Nevertheless, the winged morph have good success in terms of fertilisations gained, and Cremer et al. (2002) point out that this example seems to differ from many other cases of she-males gaining sneaky matings, in that these winged individuals are not conspicuously weaker than the wingless aggressive ones.

Mate guarding through distracting other males

Having successfully copulated with a female, a male's chance of fathering her offspring will almost inevitably be reduced if she subsequently goes on to mate with other males, and in many animals, strategies have evolved in males to minimise the risk of this. Some insects have evolved ways of plugging their mate's reproductive orifice, sometimes with elaborate structures, and in some extreme cases, males may castrate themselves by leaving their genitalia in the female. In some species, post-mated males start to mimic females and in that way distract other males from mating with the same female.

Field & Keller (1993) report an instance in the solitary parasitic braconid wasp, *Cotesia rubecula*, an insect in which several males typically compete to mate with a single female. Males may court a female directly with their own full courtship display, attempt sneaky mating of a female that is being courted by another male, or mimic female behaviour, thus distracting an already courting male away from his originally intended mate, and then mate with her himself. However, although females normally only mate once, there is a short window of opportunity after her first mating when she might still be receptive, and thus other males might get an opportunity to inseminate the female too. Thus males that have just mated can also start mimicking a female wasp, and so distract other males away from his mate during this window of opportunity.

Androchromatism and male mimicry by females

In several families of damselflies and dragonflies (Odonata), males and females are markedly sexually dimorphic in their colouration, females typically being drabber or less boldly marked. However, in some species, a small proportion of females display male-type patterns (Fincke et al. 2005). This is particularly common in the coenagrionid genera, *Ceriagrion*, *Enallagma* (Fig. 11.1) and *Ischnura*, but also occurs in some dragonflies, such as *Aeshna* (Aeshnidae) (Hilton 1987). Such individuals have been referred to under a variety of terms in the past, including heterochromatics, homochromes, isochromatic, isomorphic, etc. As Hilton (1987) pointed out, many of these are misnomers, and a more appropriate term is androchromatism, which has largely been adopted in the research literature on the subject.

The function of androchromatism in the Odonata remains somewhat of a mystery and could involve female choice, mate location or avoiding harassment, thus increasing her fitness (Sherratt 2001). That it truly is a mimicry was nicely demonstrated by Iserbyt et al. (2011), who found that, as expected in mimicry systems, the fidelity of the androchromatism in the North American sedge sprite, *Nehalennia irene*, was highest in populations where the proportion of androchromic females to males was highest, i.e. where there would be greater selection pressure by males to distinguish between them. Bots et al. (2009) showed that androchromatic females had proportionately higher protein contents the greater the operational sex ratio, the latter being a proxy variable for likely levels of harassment by males. The function is likely to be complicated and understanding it is made harder by different and often a limited number of life history parameters measured in any particular study. Van Gossum et al. (2005) tried to overcome this by using a number of manipulated insectary populations of *Ischnura elegans*, varying in numbers of androchromes and overall density, and found, contrary to expectation, that androchromes did not do better than gynochromes at high density, indeed the opposite. Similarly, Iserbyt & Van Gossum (2011), who examined this phenomenon in the coenagrionid damselfly *Nehalennia irene*, by manipulating the appearance of individuals, found that males made more courtship attempts to the male-like (andromorphic) females than to other real males and, surprisingly, more than the typical (gynomorph) female forms, and therefore concluded that colour alone was not the only cue affecting male interactions. Likewise, Schultz & Fincke (2013), who tethered females of both morphs amidst vegetation, found that males located andromorphs more easily. These results suggest the possibility of negative frequency-dependent selection in maintaining the female polymorphism depending on the abundance of males, i.e. when the andromorph is rare it has a selective advantage until it becomes the commoner, and then the gynomorph has the advantage. Takahashi et al. (2010), studying a population of *Ischnura senegalensis*, found precisely this for each of the female morphs and observed that their relative frequencies oscillated on a 2-year cycle. The two morphs also appear to display different reproductive strategies, with them having different egg sizes, such that the andromorphs are apparently an r-selected morph and the gynomorphs are K-selected (Takahashi & Watanabe 2010). Fincke (2015) tested whether the female colour polymorphism in *Enallagma hageni* is a mixed evolutionarily stable straegy (ESS), with the morphs obtaining protection from harassment differently, one via crypsis and one through mimicry. Some trade-offs were found that were consistent with the ESS hypothesis, but they were dependent on the context, i.e. availability of suitable crypsis-providing background and on whether other males were present near the andromorph such that they acted as distractors.

A very different proposal to all of these was put forward by Sherratt & Forbes (2001). In their scenario, if males can use colour as a clue to sexual identity of another individual, and scramble competition for mates in which females have little choice is normal, and if male:male interactions, as they inevitably would, incur a cost (wasted time or energy), then female dimorphism should be expected to evolve. Furthermore, males should then be selected for having even brighter colouration that is not easy for females to evolve. In this intriguing model, the male's bright colours can be interpreted as aposematic, warning themselves to other males as 'unprofitable mates'. In this scenario, the female polymorphism of the coenagrionid damselflies could be the first intraspecific example of aposematism.

The genetics of androchromatism has been investigated in *Ischnura graellsii* (Cordero 1990) and in *I. elegans* (Sánchez-Gullén et al. 2005). In both cases, a single autosomal locus with female-limited expression is involved, and its three alleles have a hierarchical dominance pattern.

Among butterflies with female sex-limited mimicry, there is often a small proportion of females that do not display mimicry patterns, but resemble the males. S.E. Cook et al. (1994) showed that in the case of the mocker swallowtail, *Papilio dardanus*, these females experience less harassment by males and thus male-like pattern retention in females may be part of a trade-off between protection from predation and less energy expenditure.

Fig. 11.1 Heterochromatism (sexual dimorphism) and androchomatism in three species of *Enallagma* damselflies (Coenagrionidae). In each pair of photographs a normal heterochromatic pairing is shown on the left and an androchromatic pair on the right. The male clasps the female behind the head and the female collects his sperm from his secondary genitalia on his anteroventral abdominal segments. (a,b) *E. hageni*; (c,d) *E. civile*; (e,f) *E. geminatum*. (Source: Professor Thomas Schultz. Reproduced with permission from Professor Schultz.)

Egg dummies on fish

In many fish, it is the male that cares for the brood rather than the female. This different asymmetry was explained by R. Dawkins & Carlisle (1976), who postulated that in external fertilisers such as most fish, the male has to wait before releasing his sperm until the female has laid her eggs, and therefore the female has the opportunity to desert first. One consequence of this is that the female can desert before the male has completed his job of fertilising her eggs and thus evolution has led to the males in such species looking after the brood.

In some species of fish with paternal care, females have been shown to prefer males that are already caring for a brood, even though these will belong to a different female. Such a preference, whatever its origin, leaves the female open to deception by males. Thus, Page & Swofford (1984) proposed that egg-like modifications of the distal margin of the dorsal fin in several species of darter, *Etheostoma* (*Catonotus*) (Percidae), have evolved for this reason. In fact, darters have quite a wealth of potential egg-mimicking structures (Page 1989) and Page & Swofford (1984) proposed that the males initially evolved fleshy masses on the tips of their otherwise sharp dorsal spines because it reduced the risk of rupturing the eggs they were guarding, and then through pleiotropic effects some came to display structures on the tips of their soft dorsal fin rays. These blobs are then hypothesised to have been selected to appear more egg-like, stimulating the females to spawn. In the subgenus *Boleosoma*, egg dummies are present on the pelvic and pectoral fins, though in these it is unlikely that any initial evolution was associated with egg protection. Page & Swofford's (1984) hypothesis was subsequently tested by Knapp & Sargent (1989), who studied the North American fan-tailed darter *E. flabellare*, and compared both the preference of females for males already guarding eggs, and the success of males with an intact line of fleshy egg-like swellings on the dorsal fin, with males in which these structures had been removed. Female darters were found to prefer males with eggs to those without, and males with egg spots to those without, suggesting that the female preference for males that were already guarding eggs probably led to the evolution of the males' egg spot mimicry. Interestingly, Knapp & Sargent (1989) also noted that a significant proportion of males from one site had missing egg-mimics, which they attributed to either agonistic interactions or attempted predation. It would be interesting to investigate whether males in the wild might attempt to remove other males' egg-mimics in order to increase their own chances of obtaining mates.

Better known than the darters are the endemic African haplochromine cichlid fish, in which males possess egg-like markings on their anal fins (Fig. 11.2a–c). These markings, which are most prominent in sexually active males, are generally called egg spots or egg dummies. Haplochromines are maternal mouth brooders and the males play no part in brood care. Wickler (1962a, 1962b) described the role of the egg dummies in the cichlid *Haplochromis wingatii*. The male lures a receptive female to a nest pit he has dug, and the female starts laying batches of eggs but then very quickly turns and takes them into her mouth, not giving the male any chance to fertilise them. The male then moves over the spot where the eggs have been laid and spreads his anal fin to reveal the strongly contrasting yellow anal ocelli over the dark background, and simultaneously releases sperm. The female, seeing the fin markings, turns around to try to take them into her mouth as if they were more eggs, but instead only succeeds in taking the male's sperm into her mouth, where her eggs get fertilised. As noted by Wickler (1962a, 1962b), the markings on the male's anal fin are not eye-like in that they do not have a dark centre, which eye-mimicking markings typically do.

In another group of African mouth-brooding cichlids, members of the *Oreochromis* (as *Tilapia* in Wickler 1962a, 1962b), the problem of getting the female to take the male's sperm into her mouth is achieved by his possession of bright-coloured genital papillae (Fig. 11.2d). In some species, these are long, semitransparent and "festooned with bright orange blobs of tissue", which resemble the female's eggs and which she likewise nibbles at or tries to take up as if they were real eggs. Wickler (1962a, 1962b) refers to these as egg dummies.

While in many species, the egg spots on the males' anal fins are of a similar size to the female's real eggs, and in these there are usually several spots on the fin (Fig. 11.2a) a few species have only a single egg spot that is markedly larger than the real eggs (Fig. 11.2b). It has been postulated that these create a supernormal stimulus for the female because of the way that her neuronal processing is locked into just the key features of an egg spot.

Hert (1989) compared the reproductive success of *Astatotilapia burtoni* (Fig. 11.2c) with and without egg spots and found that males in which the spots had been artificially removed produced significantly fewer clutches (Fig. 11.3). In contrast, Theis et al. (2012) carried out further female choice experiments with the same species using males that had been manipulated to have various numbers of egg spots or no egg spots at all. Females were found generally to prefer males without egg spots (experiment 1: GLMM, $n = 21$, $z = 21.897$, $p = 0.058$; experiment 2: Binomial test, $n = 23$, $p = 0.001$), but males without egg spots were subject to significantly higher levels of aggression than males with spots (GLMM, $n = 13$, $z = 22.218$, $p = 0.027$). Experiments with other species, e.g. *Pseudotropheus aurora* (Hert 1991),

Fig. 11.2 Egg spots on anal fins of four cichlid species. (a) *Labidochromis* sp. 'Hongi', with four black-margined yellow spots; (b) *Astatotilapia latifasciata*, with single large egg spot; (c) series of images showing variation on egg spots on anal fin of *A. burtoni*; (d) *Oreochromis squamipennis* male, showing conspicuous, tassel-like egg dummies. (Source: a, Gerard Delaney 2006. Reproduced under the terms of the Creative Commons Attribution Share-Alike Licence CC BY-SA 2.5, via Wikipedia; b, Dr Jessica Drake 2007. Reproduced under the terms of the Creative Commons Attribution Share-Alike Licence CC BY-SA 2.5, via Wikipedia; c, Theis et al. 2012. Reproduced under the terms of the Creative Commons Attribution Licence CC BY, via PLoS ONE; d, Haplochromis 2007. Reproduced under the terms of the Creative Commons Attribution Share-Alike Licence CC BY-SA 3.0, via Wikimedia Commons.)

have generally shown female preference for mating with males with egg spots, and Couldridge (2002) found that females of another *Pseudotropheus* species prefer males with a single large egg spot.

Since fertilisation rate does not seem to be influenced by male egg spot number, the initial proposal that the egg spots deceive a female into behaving as if they are eggs to be taken up into their mouths for brooding but instead take up sperm to fertilise the real eggs in her mouth, does not seem to hold. In at least some species, therefore, rather than increasing fertilisation rates within a clutch, egg spots may simply increase mating success.

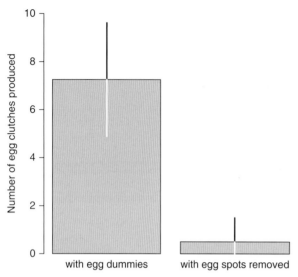

Fig. 11.3 Mean numbers of egg clutches fertilised by males with and without egg spots (dummies) on their caudal fins from four trials each with 14 female *Astatotilapia elegans* (Cichlidae: Haplochrominae). Egg spots had been removed previously by freeze-branding using dry ice. (Source: Adapted from Hert 1989. Reproduced with permission from Elsevier.)

Wickler (1962a, 1962b) postulated that egg spot markings were derived originally from the patterns formed by dark cross-bands between the male's anal fin rays, with evolution leading to a slight central narrowing of the cross-bars, giving the pale central spots a more circular appearance, and then subsequent enlargement and enhancement of a few spots and loss of other cross-bands, leaving just a few egg-like circles. T. Goldschmidt & de Visser (1990) have extended the potential significance of cichlid egg dummies by suggesting that they may also be important in speciation of the Cichlidae in the African lakes, where they are extremely specious. They suggested two possible mechanisms by which egg dummies could be involved in forming effective reproductive barriers between incipient species, leading to complete speciation. The first scenario is based on the assumption that egg mimicry is precisely maintained within a species, i.e. the size and colour of the egg dummies accurately matches that of the species' own eggs. If two isolates of such a species were to be subject to independent selection for different-sized eggs (perhaps through r & K selection factors), then selection would also quickly result in the egg dummies of these two isolates becoming closely matched with their respective models. Upon subsequent recontact between the two demes, females, which play an important role in mate choice in haplochromines, would prefer members of their own deme, egg dummy recognition by females thus forming an isolating mechanism. In T.

Goldschmidt & de Visser's (1990) second scenario, egg dummies themselves would be subject to different selection pressures in two temporarily isolated demes and would thus come to look different. Again, female preference for males with a particular type of egg dummy would, upon subsequent recontact between the demes, continue to prefer mates with appropriate egg dummies. In an attempt to examine these possibilities, T. Goldschmidt & de Visser (1990) compared egg dummy size with egg size but found no correlation between the two and therefore the first scenario could seemingly be discounted, at least for the majority of species. Their second model would certainly be consistent with the explosive speciation observed among the haplochromine cichlids.

Lehtonen & Meyer (2011) found that egg spot number in *Astatotilapia burtoni* was a heritable trait, and F. Henning & Meyer (2012) found that it was positively correlated with body size and that in artificial selection experiments over two generations, selection for fewer egg spots had been successful but it was not possible to increase the number. The underlying genetic mechanisms of egg spot formation have now been at least partially understood using modern genomic techniques (Salzburger et al. 2007, M.E. Santos et al. 2014). M.E. Santos et al. (2014) identified two novel pigmentation genes, *fh12a* and *fh12b*, and showed that the b-paralogue in particular, which is more rapidly evolving, is associated with egg spot formation, but only if there is a particular transposable genetic element within the gene's *cis*-regulatory region. Cichlid species that lack the transposable element in *fh12b* do not have egg spots. By creating transgenic zebra fish, *Danio rerio*, the *fh12b* gene with the transposable element was shown to have effects directly on the iridophore pigment cells, such as are present in the haplochromine egg spots.

Food dummies and sex

Two genera of South American characid fish, *Corynopoma* (sometimes referred to as *Stevardia*) and *Pterobrycon*, utilise very different lures to attract females. In *Corynopoma* the operculum is developed into a long thin projection with a water flea-like lure at its apex, and it can be wafted out to attract a female's attention. Kolm et al. (2012) surveyed wild populations of the swordtail characin (*C. riisei*) in Trinidad and found that the morphological similarity of the male's opercular flag ornament to ants was significantly correlated with the proportion of ants in the fishes' diet in the various streams ($r = 0.60$, d.f. = 17, $p = 0.011$) (Fig. 11.4). They then went on to conduct two experiments in which females were trained on either ants as food or on two alternatives – *Drosophila* or flakes of fish food – and

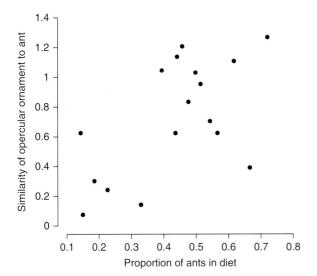

Fig. 11.4 Relationship between the proportion of ants in the diet of swordtail characin fish, *Corynopoma riisei*, and the morphological similarity of the male's opercular food-mimic flag ornament to ants. (Source: Adapted from Kolm et al. 2012. Reproduced with permission from Elsevier.)

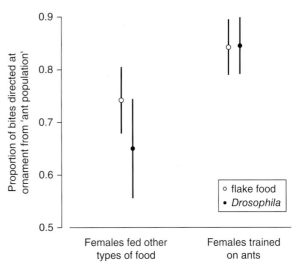

Fig. 11.5 Effect of female training diet in swordtail characin, *Corynopoma riisei*, on their preference for biting at the opercular ornaments of males from populations with higher proportions of ants in their natural diets and which had more ant-like ornaments (see Fig. 11.4). (Source: Adapted from Kolm et al. 2012. Reproduced with permission from Elsevier.)

showed that after training on ants the females showed a higher preference for biting at the ornaments of males that came from populations with a high proportion of ants in their diet (Fig. 11.5).

Although mimetic flashing in some fireflies is predominantly carried out by females as an aggressive mimicry (q.v.), males of *Photuris* species also mimic flashes but rather than those of conspecifics, the models are males of various *Photinus* and *Pyractomena* species (J.E. Lloyd 1980). Some species can only imitate the flash pattern of one other sympatric species, while others can mimic at least two species. However, the reason behind the mimicry is not predation as it is with female *Photuris*, and Lloyd (1980) suggests that they are using it to locate their own conspecific females, duping them into responding with their aggressive flash mimicry. As he points out, this could be the reason that *Photuris* is taxonomically such an exceedingly difficult genus, since in other, non-mimetic genera, workers often rely on flash patterns to enable them to distinguish between the morphologically very similar (cryptic) species.

Mimicry by sperm-dependent all-female lineages

An interesting interspecific sexual mimicry has been described by Lima et al. (1996) and involves an all-female species of poeciliid fish. As in many all-female vertebrate species (and also some plants), the wild-occurring hybrid

Poeciliopsis monacha-lucida requires its eggs to undergo the process of fertilisation before they can develop – a tricky issue in an all-female species. The evolutionary solution has been that they have to mate with males of the sexual *P. lucida*. But natural selection will strongly favour the latter, discriminating between females of their own sexual species which they can inseminate and thus get genes into the next generation, and those of the asexual *P. monacha-lucida*, which do not use the male's sperm genes at all, and therefore waste their time, energy and resources. Equally, selection is acting on the asexual females to resemble more closely those of the sexual species, and in some populations where there are exceptional mimics of *lucida* females, genetic evidence shows that the *monacha* genome controlling such things as genital pigmentation is only weakly expressed. Given that unisexual *P. monacha-lucida* hybrids probably arise not infrequently, the resemblance is likely to reflect clonal selection, with only those lines with more *lucida*-like genital appearance being successful.

Female genital mimicry in a female

In Chapter 13 I describe several examples in mammals in which male mimicry of female genitalia and female mimicry of male genitalia have evolved to reduce aggression and/or modify status in a social hierarchy, but not to increase mating success directly. An example where genital mimicry in a primate may be involved in increasing the probability of mating is described by Morris (1967). Female gelada baboons, *Theropithecus gelada*, have a largely hairless red patch on their chests that closely resembles their genitalia and surrounding red peri-genital skin. Interestingly, the intensity of the red colour varies through the female's menstrual cycle, and it seems likely that this serves to emphasise to males when she is maximally fertile.

Energy-saving cheating for sex

Photeros annecohenae, formerly *Vargula annecohenae*, is a marine ostracod crustacean, males of which release pulses of luciferin and luciferase into the sea, producing spectacular trains of bioluminescent flashes to attract mates. Very much as in fireflies, different species have characteristic flash patterns. Some male bioluminescent ostracods 'cheat' by swimming in another male's flashing light but not flashing themselves, and thus stealing his mate. Although the female crustacean is deceived in that her eggs will be fertilised by a different male than the one that produced the flash to attract her, this is not really a deception as many species of animal obtain sneaky matings when they can. Some other males are called entrainers, in that, upon seeing a nearby male starting a display, they swim in parallel nearby, producing their own competing, similar luminescent pattern in loose synchrony with the initiator. They are thus potentially parasitising the first male's photic 'song' (Morin 1986, Rivers & Morin 2009). Single males can swap readily between initiating their own train of light pulses, entraining and sneaking. In the case of the sneaker males, there is no doubt that they are saving considerable energy but they probably have a lower mating success rate.

BEHAVIOURAL DECEPTIONS IN HIGHER VERTEBRATES

It may come as a surprise to many that humans are not the only animals to regularly deceive their mates – either male or female. While not falling as clearly in the realm of mimicry as some deceptions, behavioural deceptions are nevertheless adaptive and involve resemblance in a very general sense and thus warrant at least a brief mention here.

Polygynous birds

In many bird species, males are known to be polygynous, even though they still help with rearing some of their offspring, usually those that are conceived with their preferred

mate. Thus one might ask why females, which will have a lower reproductive success if they mate with a male who has already paired with another female, should accept these matings. Two possibilities exist: either they cannot detect that the male has already mated and paired with another female and therefore are deceived, or they can detect this but, for other reasons, choose to accept him anyway.

Males of the hole-nesting Tengmalm's owl, *Aegolius funereus*, attract mates by singing at potential nest holes and, after mating, the male helps to provision its mate and subsequently the nestlings. Having attracted one mate, however, many males continue to sing but always from a hole further than 300 m away from the one where his first mate is nesting (Carlsson 1991). This continued singing is frequently successful in attracting secondary females, which the male also provides with provisions, but to a significantly lesser extent than for the primary female and, consequently, secondary broods have a far lower fledging success rate than do primaries. Thus, in Tengmalm's owl, secondary females are disadvantaged compared with primary ones, and the fact that the male sings from a site well removed from his primary nest hole suggests that his behaviour has evolved so as to conceal the fact that he is attempting to be polygynous. Perhaps an even stronger example of behavioural deception in mate attraction has been described in the polygynous northern harrier, *Circus cyaneus*, of North America. R.E. Simmons (1988) showed that a female's mate choice is affected almost entirely by a potential mate's ability to provision them with small mammals, mostly meadow voles. However, while deceitful males continue to provision their first mate (alpha female) and clutch, they provision secondary females with large amounts of courtship food to achieve a mating, but they always abandon them once they have been provided with enough food to lay a large batch of eggs. As a result, the secondary female harriers on average only successfully rear about one third as many chicks as do the alpha females. Thus mate choice is in many respects honest, the females are preferring males that can do energy-expensive aerobatic displays and also provide lots of food – but only up to a point.

Dale & Slagsvold (1994) have showed that females accept already-mated males in the pied flycatcher, *Ficedula hypoleuca*, and concluded that while at any one time mated and unmated males behave similarly, females prospecting for mates may nevertheless distinguish between them because mated males spend less time singing and more time away from their territory. Assessment by the female therefore requires her to make repeated visits to the potential mate's territory to assess the amount of time he spends there. On the basis of a mathematical model developed, it was shown that females may reject a high proportion of already-mated males by visiting their singing territory at the frequency that wild females are observed to do. However, many females accept mated males, and Slagsvold & Dale (1994) suggest that these secondary females have to take into account the cost of time spent searching for an unmated male. Thus, the apparent deception by the males, although potentially counterable, does work, for if they were immediately identifiable as being already mated, the cost experienced by a female as a result of testing a larger number of potential mates in order to find an unmated one would be reduced sufficiently to make this a preferable alternative.

In a similar study, Temrin & Stenius (1994) also showed that females of the similarly polyterritorial wood warbler, *Phylloscopus sibilatrix*, could assess the mating status of singing males by repeated visits, but that the situation was complicated by changes in the males' behaviour according to both time of day and season.

Some evidence that already-mated male pied flycatchers actually succeed in deceiving females was provided by Rätti (1994), who experimentally, temporarily removed males from a territory soon after they had been accepted by a female, so as to 'mimic' the behaviour of a previously mated male who would spend time with the primary female. The majority of the recently settled females stayed with the experimental male in spite of its imposed absence, indicating perhaps that they were not assessing its absence as a sign that they may in fact be secondary females and therefore at a disadvantage.

Deceptive use of alarm calls and paternity protection

Møller (1990) reported an unusual instance of deception in the common swallow, *Hirundo rustica*, that helps to maintain an individual's paternity of his mate's offspring. Having mated, some males will issue alarm calls upon detecting other males nearby, in the experimental case a mounted specimen of a male, if other males had been seen during the pre-laying period, and they would also issue alarm calls when another male was attempting to secure an extra-pair copulation. These alarm calls would often cause all nearby birds to take flight and thus were able to successfully interrupt some extra-pair mating.

Similarly to the bird examples, Bro-Jørgensen & Pangle (2010) report that males of the African topi antelope or sassaby, *Damaliscus lunatus*, when they perceive that oestrous females in their vicinity may be about to wander away, utter false alarm snorts that delay or deter their departure from the group where presumably there is safety in numbers. The males issue significantly more such false calls when oestrous females are attempting to depart than when

Table 11.3 Frequency of male topi antelope, *Damaliscus lunatus*, issuing false alarm snorts depends on the oestrous state and behaviour of accompanying females. (Source: Data from Bro-Jørgensen & Pangle 2010.)

Alarm call type	Female response	Mean male alarm calls per hour (± S.E.M.)
False	Visited by oestrous female attempting to walk away	0.8 ± 0.065
False	Visited by oestrous female not attempting to walk away	0.15 ± 0.014
False	Visited by non-oestrous females	0.008 ± 0.008
False	Solitary	0
True		0.035 ± 0.018

they are not showing signs of leaving (Table 11.3) and they hardly ever issue false alarm calls if females with them are not in oestrus. From their responses, females appear to be unable to distinguish the male's false alarm snorts from ones indicating genuine danger (Fig. 11.6).

Female–female mounting behaviour in mammals and birds

Many species display what is often reported in the literature as homosexual behaviour, in which a male or more often a female mounts and goes through copulatory-like movements with another individual of the same sex. Such mounting behaviour has been most extensively studied in primates but is widely found in mammals as diverse as domestic dogs and cattle (Sommer & Vasey 2006). In cattle, the behaviour is so common it is referred to as 'bulling' and is a key indicator to farmers that a female is in heat (oestrus), as during this time they are more likely both to allow others to mount them and to mount others. Why then do animals show these behaviours? Almost all examples of this type of behaviour occur in mammals (Beach 1968) and also in some birds that either live in groups all the time or, at least, come together in numbers at breeding time. G.A. Parker & Pearson (1976) suggested that, in the case of females mounting other females, the activity may serve to attract the attention of males which, in the case of herd and harem

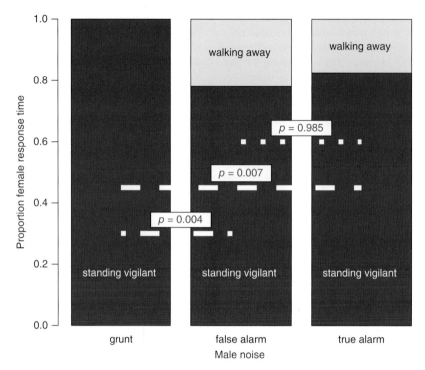

Fig. 11.6 Responses of female topi antelopes to playbacks of male false alarm snorts, genuine male alarm snorts and control male grunts. The *p*-values are for Tukey's honest significant difference test post hoc comparisons. Females do not respond significantly differently to true or false male alarm calls. (Source: Adapted from Bro-Jørgensen & Pangle 2010.)

species, are on continuous lookout for intruders that may be about to steal fertilisations, and, as the authors put it, "should act as a powerful attractor of the dominant herd bull". In this case, as would be expected, the females that are indulging in the homosexual pseudocopulation may be in heat and therefore likely to benefit if they can attract the dominant male to mate with them. Sexual performance[1] in male goats has been shown to be stimulated by the sight of female–female mounting (Shearer & Katz 2006, Katz 2007).

In those species that exhibit female–female mounting, it is also not uncommon for an anoestrous female to mount a female that is in oestrus, and G.A. Parker & Pearson (1976) propose that this might be considered as reciprocal altruism and possibly have been selected in part due to kin selection. In the more common case of an oestrous female mounting another, it is the mountee that is behaving altruistically.

It also appears that female–female mounting acts to attract male attention in some insects. Harari & Brockmann (1999) describe the behaviour in the Caribbean weevil, *Diaprepes abbreviatus* (Curculionidae), a pest of citrus trees. They also found that small males were not attracted by seeing female–female mounting, whereas larger males were. This suggests that it enables the females to select the largest, and presumably fittest, mates for copulation.

In the above-mentioned bovids and beetle, the function of female–female mounting is to do with mate attraction, but in primates mounting is usually considered to be a socio-sexual behaviour. It has the outward appearance of sex but serves other social functions, i.e. the behaviour is mimetic but not to attract mates. It occurs most often in troop-living species and is probably best studied in Japanese macaques, *Macaca fuscata*. In this species there is strong evidence that female–female mounting is directly to do with gaining sexual physical stimulation. Female pairs are long lasting, individuals take turns in mounting, they even have uterine contractions characteristic of orgasm (Vasey & Duckworth 2006), and it tends to occur in macaques that have separated themselves from the rest of the group. This was also

---

1. Increased ejaculation frequency, and decreased latency to first mount and ejaculation, post-ejaculatory interval, and the interval between ejaculations.

noted in the Hanuman langurs, *Presbytis entellas*, investigated by Srivastava et al. (1991). Furthermore, male macaques generally appear to be uninterested in females when they are performing the act. Whether this is a by-product of some previous social function or a *de novo* behavioural act due to the discovery of the physical stimulation is unclear.

Mimicry in humans

I do not wish to go into any detail on the huge volume of work on unconscious and conscious imitation of other individuals in voice, posture, mannerisms, etc. of human beings – there are many books on the topic. The observation that humans involuntarily start mimicking aspects of other people they encounter is not only widely known but also constitutes a large field in human psychology. It is not only humans that show this behaviour. It has been noted, not surprisingly, in chimpanzees, and a similar, non-vocal mimicry of movements has even been described in a grey parrot that had bonded with humans (B.R. Moore 1992). Tooke & Camire (1991) provide a nice overview of different types of human–human deceptions to do with 'mating' and this list is very long.

Pasteur (1982) did not consider aspects of behavioural copying in birds or higher primates as mimicry at all because they do not necessarily involve deception. He commented:

> It is unfortunate that certain phenomena that do not give rise to mimicry systems, and therefore are not mimicry in Bates's sense, are frequently called "mimicry." These include vocal imitations by some birds, especially Mimidae and *Psittacus*, as well as vocal and gesturing imitations by dolphins and nonhuman primates. Human mimicry is one step further removed, since it is not only conscious, but also conscious that it is conscious. Still, human mimicry, being aimed at deceit, gives rise to mimicry systems comparable to aggressive and protective biological ones.

It is now well demonstrated that many higher primates and other mammals and birds carry out behavioural deceptions, indicating a considerable degree of cognition of their situations (Byrne & Whiten 1990), though they also have the cognitive ability to recognise when they have been deceived and may retaliate (Hauser 1992). Involuntary mimicry by people is a major part of the ways humans interact socially to appease potential threats and to gain acceptance, often by not standing out from the crowd, as in picking up local accents. Pasteur would probably not class that as mimicry, yet it is a form of natural deception and, I would say, more akin to passive chemical camouflage than to outright lying.

Make-up, clothes and silicone

Many aspects of human dress and make-up can be seen as involving mimicry. Wearing clothes with horizontal hoops has often been thought to render the wearer seemingly more solidly built, which is a desirable feature in some societies and social groups, whereas the opposite effect may be achieved by means of vertical stripes (Fig. 11.7). This suggestion is based on a well-known optical illusion first described by v. Helmholtz (1867) and has been illustrated to show the veracity of the idea. However, recent work has actually cast considerable doubt on this assumption of western female (in particular) fashion. P. Thompson & Mikellidou (2011) have found quite the opposite, in fact.

Fig. 11.7 The Helmholtz illusion. When a flat figure of a person (or any other flat shape for that matter) is patterned with horizontal stripes they appear fatter to most people than they do when the stripes are vertical.

Although not involving optical illusions in the same sense, wearing padded bras or even having breast-enhancement surgery may well serve to attract males but, of course, the resulting deceptions, if initially successful, will almost certainly be found out in the end! Human females are, in fact, unique among apes and probably other primates in that their breasts develop before pregnancy and remain enlarged throughout life. Furthermore, in many individuals they are far larger than need be to produce adequate milk for their offspring – the extra size being due to adipose tissue rather than mammary glands. These may well be the result of sexual selection by males, though other alternatives include the possibility that they facilitate nursing at the hip, are involved in disguising the oestrous state or are a non-adaptive by-product of selection for general increase in fat deposits (see Mascia-Lees et al. 1986, C.M. Anderson & Bielert 1994, Arieli 2004, and Caro 1987 for a review, and section *Cryptic oestrus in humans* later). If they are releasers for male sexual interest, then padded bras will quite possibly help gain initial interest.

High beehive hairstyles and high-heeled shoes can give the impression that someone is taller than they are. Perhaps, likewise, the cod-pieces of old suits of armour may have served more than simple protective function. In terms of human attraction it seems that it is often just the initial stage that can be important. Rouge on the cheeks, lipstick on the lips, eye shadow and many other aspects of make-up all serve to create features that approach more closely ideals as perceived by various societies and thus ultimately to attract the highest possible quality mates. Some make-up is aimed at hiding blemishes, such as pimples, thus hiding potential signs of less than immaculate health. In the past in Europe, syphilis took its fair (or unfair) toll of noses and ears, and replacement parts were made to redress the loss, though given that some of the wealthy sufferers chose to wear gold prostheses, the replacement part could hardly be said to be deceptive in the way that toupees, wigs, false teeth or various modern prostheses usually are.

Human behavioural sexual deceptions are not only limited to visual displays. Perfumes, for example, may deceive by masking other, less pleasant odours or perhaps they may allure mates through the incorporation of musky-smelling essences that were once obtained from the caudal glands of the musk deer, *Moschus* spp.[2] (Moschidae) and civetone from the African civet, *Civetticus civetta* (Viverridae). Some plants or their seeds also have a strong musky odour, such as angelica, *Angelica archangelica* (Apiaceae), abelmosk,

Abelmoschus moschatus (Malvaceae), muskflower, *Mimulus moschatus* (Phrymaceae), and muskwood, *Olearia argophylla* (Asteraceae). Some of these are still used in perfumery although synthetic compounds are more commonly used nowadays because of relatively lower costs.

Cryptic oestrus in humans

One of the most interesting features of human female sexual cycles compared with those of other primates is that human females do not go through obvious changes in appearance during their fertile periods (Dixson 2012). This has been interpreted as mimicry, in that a woman in her fertile period is mimicking one who is not, that is, her ovulation is concealed. It has often been postulated that this presumed loss of obvious visual (and perhaps other) signals during ovulation was associated with life in multi-male societies, in which there was no need to attract the attention of a single dominant individual. Concealed fertility in primates is not completely limited to humans and occurs in a dozen or so other species, including a langur and a macaque (Heistermann et al. 2001, Fürtbauer et al. 2011). Schröder (1993) has argued that if, as seems likely, early hominid evolution passed through a phase of social groups with a single dominant male, the concealment of ovulation could have provided females with a greater choice of mates, giving them a selective advantage. Sillén-Tullberg & Møller (1993) used phylogenetic contrasts to determine whether mating system was a likely correlated factor and showed that ovulation signalling was lost 0–1 times among monogamous species and between 8 and 11 times in non-monogamous systems. Therefore they conclude that "the lack of ovulatory signs is more likely to promote monogamy than monogamy is to promote a lack of ovulatory signs". Pawłowski (1999), however, is more doubtful about the role of sexual selection in humans leading to concealment of oestrus. It should also be noted that although human females do not advertise their ovulation by developing a bright sex skin, there are still potentially detectable changes through the menstrual cycle, for example, in body odour (Havlíček et al. 2006). Whether most modern men are aware of this – perhaps subconsciously – I am not sure.

Flirting in humans

I thank Ruud van der Weele for drawing my attention to an article by Gersick & Kurzban (2014) highlighting why humans are different from [all] other animals and one of the features mentioned is flirting. But what is the evidence that other animals do not flirt? Well, there are several issues

---

2. Musk deer are now protected under the Convention on International Trade in Endangered Species of Wild Fauna and Flora (CITES) but a lot of illegal poaching still goes on.

here, and I do not think that flirting is at all restricted to humans. Is that not more or less what is going on when female birds, such as black-cocks or birds-of-paradise explore and examine males in a lek or a bower, but then perhaps decide to go and see what other males have to offer? A member of either sex, though perhaps most likely females, might obtain more interest from a potential mate if they advertise that they are potentially interested in mating. Would one then describe flirting as a sort of mimicry – where a potential mate that is eventually rejected has wasted a lot of time and effort on an individual without reward? In human society, especially since humans are capable of rape, which no other animals are, this can lead to very terrible results.

Chapter 12

REPRODUCTIVE MIMICRIES IN PLANTS

Everyone sees what you appear to be, few experience what you really are.

Niccolò Machiavelli, *The Prince*

Mimicry, Crypsis, Masquerade and other Adaptive Resemblances, First Edition. Donald L. J. Quicke.
© 2017 John Wiley & Sons Ltd. Published 2017 by John Wiley & Sons Ltd.

INTRODUCTION

The vast majority of flowering plants (angiosperms) rely on other organisms for transferring pollen between individuals, which is no doubt the reason why conspicuous flowers first evolved. The vast majority of pollinators are insects, though members of various other animal groups are involved and loss of animal pollination by, for example, wind or water dispersal of pollen, or selfing, are derived states that have evolved on numerous independent occasions among the flowering plants (Vereecken & McNeil 2010). Why do insects, etc., visit flowers? The main reason is that the flower is offering a reward, maybe nectar, maybe pollen itself which is protein rich, often both, and sometimes something else, such as oils and perfumes. Rewards have a cost to the plant and it is hardly surprising therefore, that on thousands of occasions, evolution has led to deception of pollinators, sometimes within a plant, sometimes within a species and, probably most interestingly, between species. Renner (2006) presents a large list of rewardless angiosperm plants collectively representing some 32 separate families while noting, however, that it is often difficult to be certain that a plant is rewardless because rewards vary so much depending on the pollinator. They may also be transient and therefore easy to overlook.

POLLINATOR DECEPTION

While deceptive pollination syndromes are recognised in a wide range of angiosperms (flowering plants) they are most prevalent, proportionately and numerically, in the huge orchid family Orchidaceae[1] (Jersáková et al 2009). Most orchids produce distinct packets of pollen, called pollinia, which become attached to potential pollinating agents by special morphological adaptations of the floral mechanism. The species that do not produce discrete pollinia nevertheless have very clumped, sticky pollen rather than the dusty pollen produced by most plant species. This use of discrete, or relatively discrete, gamete packages is closely associated with their use of pollination deceptions. If they transmitted thousands of separate pollen grains onto a pollinator, it might well visit any number of other plants, diluting the pollen load, before it encountered another plant of the same orchid species.

While the majority of orchids species provide some form of reward to at least some of their pollinators, including stigma exudates (Dafni & Woodell 1986, Papadulos et al. 2013), approximately 33% of species achieve pollination through deception, which is a much higher proportion than in any other large group of plants (Schiestl 2005). Furthermore, orchids are the only plants that use sexual deception of male insects to achieve pollination (Peakall & Beattie 1996). Since the Orchidaceae is one of the largest families of flowering plant, with an estimated 25,000–30,000 species, it follows that it includes a very large number of species that obtain pollination by deception (Ackerman 1986, Nilsson 1992). Two main hypotheses have been put forward to explain the evolution of floral deceptions (Jersáková et al. 2006): the plants may benefit by diverting expensive pollinator rewards to other functions or they may achieve a higher level of cross-pollination as a result.

Ackerman (1986) suggested that food-deceptive pollination falls into two modes, one relying on mimicry of another species, be it another plant or an insect, the other relying on naïve pollinators which might be migrants or just recently emerged. The efficiency of the former would be expected to be negatively density-dependent with respect to the orchid itself. Another factor that might make reward deception in orchids advantageous is that at the very low population densities that some display, they are unlikely to be able to rely on regular reward pollination, as the pollinator will be unlikely to encounter another orchid of the same species without the latter using a lure of a high potential value, such as mimetic pollinator sex pheromones.

Floral mimicries have only recently started to be the subject of experimental investigation. In fact they have lagged behind their animal counterparts in this respect. However, there are a few notable exceptions to this generalisation and the last few years have seen a great increase in studies in this area. Although the existence of sexual chemical deceit has been appreciated for some time, it required advances in techniques of chemical analysis to reveal the details. And as with much chemical research, the early work of groups such as Schiestl's in Austria required amassing large amounts of samples for analysis, something that modern gas chromatography–mass spectrometry (GC-MS) has greatly reduced the need for. In Table 12.1, I present a short and far from comprehensive summary of the uses of chemical deception in plants.

The fact that some orchid flowers do not produce nectar has been known for a long time (Sprengel 1793), even though Darwin (1885) subsequently considered as untenable the idea that *Orchis morio*, for example, with its obviously insect-attracting flowers, did not provide its pollinating insects with a reward (Nilsson 1984). Indeed, Sprengel (1793) coined the

---

1. Here the Orchidaceae are meant in the traditional sense, even though in some recent studies it has been treated as a monophyletic group of several separate families.
2. The name 'orchid' is derived from ancient Greek *orchis*, meaning testicle, because of the similarity of the tubers to mammalian descended testicles.

Table 12.1 Some examples of deceptions employed by various plants to attract pollinators.

Expected reward	Pollinating agent (dupe)	Examples	Family	References
Nectar	Bumblebees (*Bombus* spp)	*Calypso bulbosa*, *Dactylorhiza sambucina*	Orchidaceae	Ackerman 1981, 1986, Boyden 1982, Gigord et al. 2002
Oil	*Centris* bees	*Oncidium lucayanum*, *Cyrtopodium* spp.	Orchidaceae	Nierenberg 1972, Vale et al. 2011
Mate	Bees, wasps, ichneumonids	Numerous orchids	Orchidaceae	See section *Pollinator sex pheromone mimicry*
Aphids	Female aphidophagous hoverflies	*Epipactis consimilis*	Orchidaceae	Ivri & Dafni 1977
Fungus for oviposition	Fungus gnats (Mycetophilidae); *Zygothrica* spp. (Drosophilidae)	*Asarum* Corybas, *Cypreoidium*, *Dracula*, *Lepanthes*	Araceae Orchidaceae	Vogel 1978, Ackerman 1986, Blanco & Barboza 2005, Endara et al. 2010
Honey bee prey	*Vespa bicolor*	*Dendrobium sirense*	Orchidaceae	Brodmann et al. 2009
Carrion, dung, decaying matter for oviposition	Carrion flies; *Scaptodrosophila bangi* (Drosophilidae)	Stapelia, Orbea, Huernia, Edithcolea Rafflesia Various Araceae Satyrium pumilum, Gastrodia similis Eucomis spp.	Apocynaceae Rafflesiaceae Araceae Orchidaceae Hyacinthaceae	Meve & Liede 1994, Kite 1995, Kite & Hetterschieide 1997, S.D. Johnson 1997, T.S. Woodcock et al. 2014, S.D. Johnson et al. 2007; Martos et al. 2015; Shuttleworth & Johnson 2010
Sleeping site	Male bees	*Serapias vomeracea*	Orchidaceae	Dafni et al.1981
Oil-flowers	*Centris* bees	*Tolumnia guibertiana*	Orchidaceae	Van der Cingel 2001; Vale et al. 2011
Green-leaf volatiles	Prey-hunting *Vespula* wasps	*Epipactis helleborine, E. purpurata*	Orchidaceae	Brodmann et al. 2008
Food source mimic	Euglossine bees	*Cochleanthes lipscombiae*	Orchidaceae	Ackerman 1983
Drive away competing males	*Centris* bees	Oncidium (Orchidaceae)	Orchidaceae	Dodson & Frymire 1961

term "Scheinsaftblumen" for such flowers, which means "sham nectar flowers" (Dafni & Woodell 1986).

Within the Orchidaceae there are two major types of pollinator deception employed: (i) food deception, in which the plant mimics the nectar-rewarding flowers of another species of plant, either specific species or just generally attractive flower signals (Peter & Johnson 2013), and (ii) sexual deception, in which the orchid flower mimics the sexual attraction signals of a female insect (Schiestl et al. 1999, Schiestl 2005, Peakall et al. 2010). It has been suggested that the enormous species-richness of the family may be largely due to the widespread use of pollinator deceptions of various sorts, but the evidence for this is mixed, with Tremblay et al. (2005) and Cozzolino & Widmer (2005) lending support, but Givnish et al. (2015) showing that deception does not accelerate the net rate of diversification.

Generalised floral mimics would be expected to do better in areas where there are numerous other similar but rewarding flowers due to two effects: (i) the mimic effect, whereby local pollinators have become conditioned to the rewarding flowers and therefore seek out similar-looking flowers, and (ii) the magnet effect, with areas rich in rewarding flowers generally attracting large numbers of potential pollinators (S.D. Johnson et al. 2003a, Peter & Johnson 2008) (see also Chapter 10, section *Flower mimicry*). In the early-flowering, rewardless North American orchid *Calypso bulbosa*, pollination seems to rely on deception of naïve, recently emerged bumblebees that very quickly learn to avoid the flowers, but although pollination success is therefore often rather low, the orchid, when successfully fertilised, produces a very large number of seeds (Boyden 1982). Interestingly, when *C. bulbosa* is pollinated, the flowers senesce quite quickly compared to when a pollinia is removed, and H.C. Proctor & Harder (1995) suggest that this is especially likely to be true of sexually deceptive, unrewarding species because the pollinator has to be duped twice in order to successfully pollinate the flower.

Steve D. Johnston and collaborators work on pollination in various South African plants, but their research on *Disa* orchids in particular has provided a lot of new insights into the evolution of floral deception types and the multiple re-emergence of nectar production (e.g. S.D. Johnson 1992, 1994, 1997, 2000). S.D. Johnson et al. (1998) used a molecular phylogeny to explore the evolutionary trends in *Disa*'s pollination biology. This genus, with some 169 known species ranging from Yemen to South Africa, is believed ancestrally to have been a pollinator deceiver (S.D. Johnson et al. 2013). Combining the *Disa* phylogeny with detailed microscopic study of nectaries in those species that provide a floral reward has shown that nectaries have evolved in the genus three times, both through *de novo* epidermal specialisation and recapitulation of the nectary type present in its sister genus (Hobbhahn et al. 2013). The phylogeny

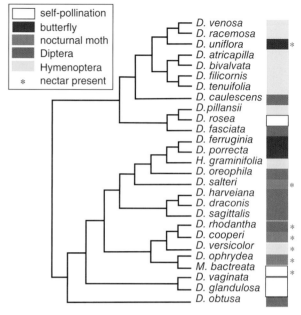

Fig. 12.1 Major pollinator groups and whether the plants are rewarding in terms of nectar production indicated on morphological cladogram of South African *Disa* orchids showing multiple shifts between major pollinator groups. (Source: Adapted from Johnson et al. 1998. Reproduced with permission from the *American Journal of Botany*.)

(Fig. 12.1) also illustrates a balance between pollinator group conservatism and with multiple shifts between groups in both major clades. Members of the *Disa draconis* complex are particularly interesting, in that they are pollinated by long tongued flies (various Tabanidae and Nemestrinidae) and these use their extremely long proboscises (up to 60 mm) to extract nectar from other deep-spurred flowers. They are attracted to the orchid but can only pollinate it if they can probe into a sufficiently long floral spur, even though these *Disa* orchids provide no reward (S.D. Johnson & Steiner 1997). If the flowers' spurs are artificially shortened, flies carrying pollinia near to the base of the proboscis cannot penetrate deeply enough into the spur to deliver the pollinia to the stigma.

The South African plant *Gorteria diffusa* (Asteraceae) appears to use a purely visual deception to attract males of the bombylid fly *Megapalpus nitidus* (S.D. Johnson & Midgley 1997). Some of its petals have black raised spots, which somewhat resemble other flies on the flower. Removing the petals with the spots reduced visitations by males of the bee-fly, but creating artificial spots with ink also reduced visitations, and it appears from electron microscopic study of the spots that their epidermal sculpturing and possibly UV reflectance may also be important.

One obvious potential advantage to be gained by floral mimicry is a reduction in energy costs through avoiding having to produce nectar. However, it must be doubtful whether this is a sufficient reason, as the main energy components of nectar are sugars, and plants are seldom in short supply of these. Furthermore, for the mimicry to be effective there may have to be substantial investment by the plant in terms of alluring scent production or morphological adaptation. Dafni (1983, 1984) notes that many floral mimics reduce the chance that a pollinator will learn to avoid them by showing considerable flower colour polymorphism (see also Heinrich 1975, 1979).

Cozzolino & Widmer (2005), using allozyme data published by Forrest et al. (2004), found, rather surprisingly, that gene flow between deceptive orchids was significantly greater than that between rewarding ones, as evidenced by smaller overall genetic distances between populations (Fig. 12.2a). Further analysis of the deceptive species showed that there was a marked and highly significant difference among deceptive species depending upon whether they used sexual deception or food deception, e.g. mimicking another nectariferous species of plant (Fig. 12.2b). Members of the

sexually deceptive clade examined showed far smaller pairwise genetic distances, indicating closer relatedness between species. Interpreting this is not straightforward since it could mean that sexually deceptive species are speciating more rapidly or that there is more gene flow between species due to errors which may lead to hybrids. Both, of course, may be true. However, Stökl et al. (2009) found that selection by the *Andrena* bee pollinators of *Ophrys* species serves to maintain discrete floral odours of three species, even though genetic markers show that some hydridisation does occur between *O. bilunulata* and both *O. lupercalis* and *O. fabrella*. The differences necessary for maintaining pollinator specificity at the molecular level can be quite small. Sadeek et al. (2016) demonstrated that in *O. exaltata*, the production of mimetic 7-alkenes to attract and elicit copulation attempts by its pollinator bee, *Colletes cunicularius*, involves a homologue, SAD5, of a stearoyl-acyl-carrier-protein desaturase. Comparison of the sequence of the pheromone-producing homologue with a reconstructed ancestral version shows that changes in just two amino acids at the bottom of the enzyme's substrate-binding cavity are responsible for the double-bond position in the alkene

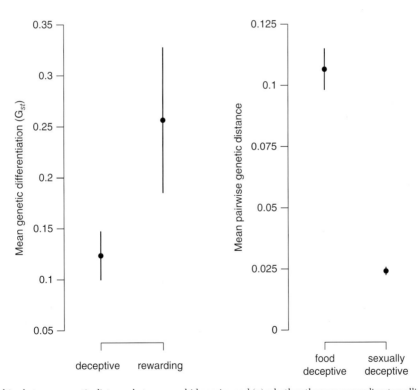

Fig. 12.2 Relationships between genetic distance between orchid species and (a) whether they are rewarding to pollinators or not, and (b) if they are unrewarding, whether they use food or sexual deception to dupe their pollinators (mean ± S.E.). (Source: Adapted from Cozzolino & Widmer 2005. Reproduced with permission from Elsevier.)

that allows pollinator specificity and reproductive isolation from closely related *Ophrys sphegodes*.

Scopece et al. (2010) collated information about features of reproductive success across a large sample of orchid species for which data were available and found some very interesting results. Although the analyses were not phylogenetically controlled, the lability of pollination systems in the family suggests that the observed trends are probably not artefacts. In general, food-rewarding species had significantly higher pollination efficiency, and male and female reproductive success, than either food-deceptive or sexually deceptive ones. Male reproductive success was also significantly greater in food-deceptive than in sexually deceptive species, whereas overall pollination efficiency was significantly lower. Deceptive species also had significantly less interpopulation genetic variation (G_{ST}) than rewarding species, suggesting that deception increased gene flow between populations.

Pollinator sex pheromone mimicry

Many genera and species of orchid succeed in deceiving males of their pollinator insect species by mimicking their females, usually by combined chemical, physical and visual means. The Eurasian genus *Ophrys* includes several species whose common English names give a good indication of what their deceptive flowers might be duping, for example the bee orchid (*O. apifera*) (Fig. 12.3a), bumblebee orchid (*O. bombyliflora*), fly orchid (*O. insectifera*) (Fig. 12.3b), early spider[3] orchid (*O. sphegodes*), etc. Interestingly, the vast majority (>91%) of sexually deceived orchid pollinators are male solitary haplodiploid hymenopterans, with just 2.7% social hymenopterans and 5.2% other diploid insects, such as a few cases involving male flies (van der Pijl & Dodson 1966, Blanco & Barboza 2005, Gaskett et al. 2008). Proving that an orchid is using sexual deception to achieve pollination might seem straightforward, but it does require a fairly specific set of criteria to be met (R.D. Phillips et al. 2014), specifically:

1. the pollinators should only be males,
2. only one (normally) species of pollinator should be involved,
3. there should be no reward,
4. the flowers should (normally) be insectiform,
5. there should be chemical attractants, though it is, of course, theoretically possible that visual mimicry alone might be sufficient.

---

3. The so-called spider orchids are not pollinated by spiders, however, nor man orchids by man.

Confirmation might be through direct observation of attempted copulation or through evidence of ejaculation. Rather few studies satisfy all these criteria though I do not doubt that the great majority of purported cases do nevertheless involve sexual deception. The diversity of sexually attractive volatiles emitted by orchid species and the fact that in some *Ophrys* species the biologically active floral odour constituents also happen to be widespread, if not ubiquitous, plant cuticular hydrocarbons, suggested to Schiestl (2005) that pheromone mimicry must also be supported by other cues for the hymenopteran pollinators, and that particular orchids have evolved to exploit specific pollinator behavioural traits.

The deceived male hymenopterans, often bees but frequently other sorts of wasps such as tiphiids (Alcock 2000), scoliids, sphecids, ichneumonids (Coleman 1928, 1938) and even sawflies and ants (Peakall 1990), alight on the orchid flowers and usually attempt to mate with them in a behaviour often referred to as pseudocopulation. This sexual deception, frequently referred to as Pouyannian[4] mimicry, even occasionally leads to ejaculation; nearly 75% of *Lissopimpla* males visiting *Chiloglottis* orchid flowers for the first time are duped so well that they ended up depositing some sperm (Gaskett et al. 2008). Gaskett et al. recognised four different levels of sexual interaction between pollinating bee/wasp species and their duping orchid flowers:

1. Pseudocopulate with ejaculation, e.g. *Cryptostylis* (Fig. 12.3c,d).
2. Pseudocopulate without ejaculation, e.g. *Ophrys*, *Geoblasta*.
3. Grip the orchid's hinged major petal, i.e. *Caladenia*, *Chiloglottis*, *Drakaea* (Fig. 12.3e,f).
4. Briefly trapped before collecting the pollination departure, i.e. *Pterostylis*.

Using these levels they found a positive correlation with the degree to which the duped hymenopteran is induced to try to copulate with the flowers and pollination success.

This type of sexual mimicry appears to be particularly well developed in Australia, where more than 70 orchid species rely on this method to achieve pollination, although a few splendid, if less well-studied, examples occur in other places, e.g. *Bipinnula penicillata* in South America (Ciotek et al. 2006). Many Australian thynnine tiphiid wasp males seek out their respective pheromone-emitting wingless

---

4. Maurice-Alexandre Pouyanne, after whom Pasteur (1982) coined the term, was a naturalist as well as being president of the Court of Appeal in Sidi-Bel-Abbes, Algeria. He described in great detail the pollination activities of various bees and the *Ophrys* orchids in the region where he lived.

Fig. 12.3 Sexually deceptive orchids that mimic their dupe hymenopterans both chemically and, at least partially, in morphology and colouration. (a) The bee orchid, *Ophrys apifera*; (b) fly orchid, *O. insectifera*; (c) bonnet orchid, *Cryptostylis erecta*: note the narrow double band of pale and dark spots up the middle of the labellum; (d) male *Lissopimpla* ichneumonid wasp visiting *Cryptostylis* flower: note pollinia attached to the apex of its metasoma; (e) two *Drakaea micrantha* flower spikes showing the contrasting female tiphiid-mimicking labella; (f) *D. glyptodon* flower on left and wingless female tiphiid wasp, *Zaspilothynnus trilobatus*, on adjacent stalk on right. (Source: a,b, Chris Raper. Reproduced with permission from Chris Raper; c, David Lochlin 2012. Reproduced under the terms of the Creative Commons Attribution Licence CC BY 2.0, via Flickr; d–f, Professor Rod Peakall. Reproduced with permission from Professor Peakall.)

females and carry them away *in copulo*, a behaviour that has been particularly subject to orchid deceptions. Peakall (1990) showed experimentally that males of *Zaspilothynnus trilobatus* were quickly able to locate flowers of the orchid *Drakaea glyptodon* because of their appearance (Fig. 12.3f) and, probably more importantly, their scent, which mimics that of the wasp female's sex pheromones. In the field. 21% of male wasp/flower interactions resulted in the wasp grasping the flower in a way that was sufficient for the orchid's pollinia to be attached to it.

Peakall's (1990) study, which involved marking and observing individual male tiphiids, is particularly significant in that it showed that after one encounter with an orchid flower, the wasp would immediately fly off a long way and only return to the same area after a prolonged period. Thus the behaviour of these deceived male wasps might be expected to contribute significantly to outbreeding. As Dafni (1984) amusingly noted, "except for the rare occurrence of ejaculation during pseudocopulation, sexual attraction is based on lack of satisfaction". However, from

an evolutionary point of view, the time, energy and sperm spent in such events, indeed in all sexually duped pollinators, is almost certain to reduce their fecundity (see later).

In the case of the Australian orchid *Chiloglottis trapeziformis*, there is a double negative effect on its thynnine wasp pollinator, *Neozeleboria cryptoides* (Wong & Schiestl 2002). Although males are unable to distinguish the orchid's mimetic pheromones from those of conspecific virgin females, they do learn that they are being duped and tend therefore to leave areas where the orchid is common to seek out mates elsewhere, thereby increasing their chances of parenthood. However, a corollary of this is that females in areas where orchids are abundant are likely to have far lower mating success both because of the 'competition' from orchid flowers and a reduced prevalence of males. Field observations further showed that females in orchid-rich sites not only received fewer visits by males but actually received no copulation attempts. Molecular phylogenies of *Chiloglottis* and pollinating *Neozeleboria* species show highly significant congruence (Fig. 12.4), though this does not

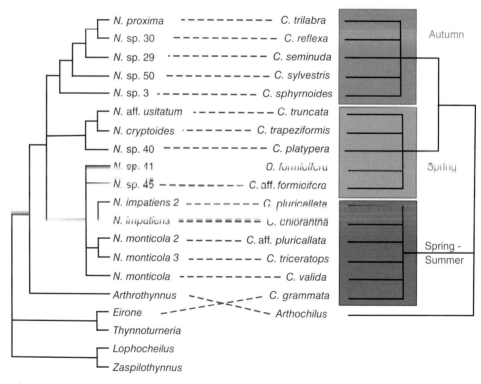

Fig. 12.4 Largely congruent maximum-parsimony molecular phylogenies for sexually deceptive *Chiloglottis* orchids and their thynnine wasp pollinators (*Neozeleboria* species), indicating that flowering season is fairly strongly conserved but could alternatively reflect phylogenetic pattern in the pollinator sex pheromones. Broken blue lines between orchid species on the right and pollinator on the left indicate phylogenetically congruent associations, whereas the three broken red lines indicate incongruent ones. (Source: Adapted from Mant et al. 2002. Reproduced with permission from John Wiley & Sons.)

necessarily mean that they are co-speciating (Mant et al. 2002). Rather, it may reflect constraints on the speciation of the orchids such that the daughter species still flower at the same time of year, perhaps because of the similarities in the pheromones of the thynnines at a given time of year.

Borg-Karlson (1990) provided an extensive review of pollination in members of the non-rewarding orchid genus *Ophrys*. Chemicals emitted by *Ophrys* flowers were shown to be highly attractive to males of various Hymenoptera. Two main classes of flower volatiles – aliphatics and terpenoids – were detected. The orchid flower volatiles (particularly for the *Ophrys fuscii-luteae* section) have been shown to resemble those produced by the mandibular glands of their male pollinating bees, *Andrena* species. Furthermore, female bee Dufour's gland components were also present in the flower emissions. The main attractive compounds identified were aliphatic 1-alcohols and 2-alcohols and various terpenes, notably geraniol, geranial, linalool and *E,E*-farnesol. Studies on the sexually deceptive chemicals have been carried out now for several orchid species (Table 12.2). Borg-Karlson et al. (1993) found that different subspecies

of the fly orchid, *O. insectifera* (Fig. 12.3b), produced different sex pheromone-mimicking odour blends and attracted quite different hymenopteranan pollinators. *O. i. insectifera* attracted males of two species of the sphecid wasp genus *Argogorytes*, while *O. i. aymoninii* attracted the bee *Andrena combinata*. It is suggested that this variation might be a possible route to sympatric speciation because of the high pollinator specificity linked to their floral odours (Schiestl 2005). M.R. Whitehead & Peakall's (2014) study of two closely related and sympatric *Chiloglottis* orchids in Australia with distinct floral odour blends confirmed that they specifically attract, and are pollinated by, two different *Neozeleboria* typhoid wasp species. While it was possible to create interspecific hybrids, that even showed hybrid vigour, by hand fertilisation, there was no evidence of any introgression of genes from one orchid to the other in the wild. Pollinator specificity in most sexually deceptive orchids is likely to be important both in maintaining species integrity and probably also in promoting speciation (see Schiestl & Ayasse 2002, Stökl et al. 2009, Jersáková et al. 2009 and Xu et al. 2012 for a review).

Table 12.2 Research studies on the sexually mimetic volatiles of various orchids.

Orchid	Reference	Notes
Caladenia barbarossa	Bohman et al. 2013	Pyrazines detected, role in pollination not fully confirmed; possible chemical convergence
Chiloglottis trapeziformis	Schiestl et al. 2003	Chiloglottone
Chiloglottis valida	Schiestl & Peakall 2005	Same as *C. trapeziformis*
Various *Chiloglottis* spp.	Peakall et al. 2010	Phylogenetic implications, electroantennagrams, comparative chemistry
Cryptostylis	Schiestl et al. 2004	
Drakaea livida	Bohman & Peakall 2014	Pheromone contains pyrazines
Mormolyca ringens	Flach et al. 2006	Including non-volatile virgin queen cuticular wax mimics
Ophrys exaltata	Vereecken & Schiestl 2008	Orchid odour significantly more attractive to male bee pollinator than virgin female bees
Ophrys fusca group	Stökl et al. 2005	Orchid species sharing same pollinator rely on same mimetic odour compounds; phylogeny; electroantennograms
Ophrys speculum	Ayasse et al. 2003	Trace amounts of chemicals that are novel to plants: (omega-1)-hydroxy and (omega-1)-oxo acids; blend and enantiomers important
Ophrys insectifera	Borg-Karlson et al. 1987, 1993	Different forms attract different pollinators
Ophrys iricolor	Stökl et al. 2007	Comparison with pollinator sex pheromone
Ophrys lutea	Borg-Karlson & Tengö 1986	
Ophrys speculum	Ayasse et al. 2003	Enantiomer specificity
Ophrys sphegodes	Schiestl et al. 2000; Ayasse et al. 2000	Variation between allopatric subspecies with different pollinators; individual flower variation and bee learning
Ophrys iricolor	Stökl et al. 2007	Comparison with bee model, and use of electrophysiology

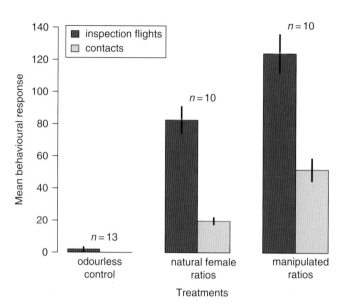

Fig. 12.5 Attractiveness of manipulated odour bouquets to patrolling males of the bee, *Colletes cunicularius*, the pollinator of the southern European orchid *Ophrys exaltata*. Data show the responses of male bees patrolling for virgin females at a site in France to black cylindrical plastic beads, 4–5 mm, mounted on insect pins set out in the field, treated with either hexane (control), extract of virgin females, or extracts of virgin female bees with increased ratios of the key odour compounds [(Z)-7-C21, (Z)-7-C23, and (Z)-7-C25]. By increasing the ratio of the key odour compounds, the dummies were made significantly more attractive than unmanipulated odour mix (Mann–Whitney U test: $p < 0.05$). Means ± S.E. number of inspecting flights and contacts of the male bees (as determined by microphone near dummy recording wing noise). (Source: Adapted from Vereecken & Schiestl 2008 with permission from National Academy of Sciences, U.S.A.)

While some orchid flowers do bear some visual resemblance to their sexually deceived insect pollinators (Fig. 12.3a,b), the resemblance is not always that convincing, but as humans, we do not know what the specific clues are that males of the pollinators are specifically evolved to recognise and respond to. Gaskett (2012) investigated the importance of floral shape. In members of the Australian orchid genus *Cryptostylis* (Fig. 12.3f–g), which has several species, all pollinated by the parasitic ichneumonid wasp *Lissopimpla excelsa* (Fig. 12.3e) (Coleman 1928, 1938). Furthermore, the volatile chemical mixtures given off by *Ophrys* orchids are not always that accurate a representation of the virgin pollinator bee female, although, rather surprisingly, Vereecken & Schiestl (2008) found that the orchid's perfume was more attractive to its male pollinators than their own virgin queens. By adding more of the compounds 'over-expressed' by the orchid, they found they could significantly increase male bee visitations and attempted copulations with impregnated dummies (Fig. 12.5). In this case, therefore, the orchid is actually benefitting by being an imperfect mimic.

Many sexually deceptive orchids have, to the human eye, rather attractive flowers, seemingly sometimes far more elaborate and colourful than would be needed to imitate the key features of the female insect that they are supposedly mimicking (e.g. Fig. 12.3a). Spaethe et al. (2007, 2010) attempted to understand why this should be especially since brightly coloured flowers might attract non-specific pollinators that could accidentally dislodge and remove the flower's pollinia with virtually no prospect of transferring them to another flower of the same species. The Mediterranean orchid *Ophrys heldreichii*, for example, has a deep pink perianth, whereas *O. cephalonica* from the same general region has a green perianth. The first is pollinated by male *Eucera* (*Tetralonia*) *berlandi*, the second by *Andrena* spp. Male *Euceros*, however, as their name suggests, have particularly long antennae, which, of course, could indicate that they rely more on olfaction in mate-finding. Streinzer et al. (2009) video-recorded bees approaching unmodified flowers and ones from which the conspicuous perianth had been removed and found that although orchid location from a distance involved olfaction, at close range (<30 cm whereby the flower would subtend an angle of 5 degrees or more to the bee's eye) orientation in this species was purely visual. A similar conclusion was reached by Spaethe et al. (2007) using dual-choice tests of flowers with and without perianth and also photographs of flowers with and without perianth. A survey of known pollinators of *Ophrys* orchids[5] by Spaethe et al. (2010) found a strong correlation between whether flower perianths were green or conspicuous (white or pink) and their pollinator species, albeit without phylogenetic control. In total, 83% of species with brightly coloured flowers ($n = 40$) were pollinated by eucerotine bees (i.e. *Eucera* and relatives, all with males with especially well-developed antennae) whereas only 9% of those with greenish perianths were ($n = 70$), the latter mostly being pollinated by *Andrena* species.

Steiner et al. (1994) studied the specificity of two species of non-rewarding orchids in South Africa. *Disa atricapilla* is pollinated by the sphecid wasp *Podalonia canescens*, while the very similar sister species, *Disa bivalvata*, is pollinated by the pompilid (spider-hunting) wasp *Hemipepsis hilaris*. The fidelity of the two pollinators is so high that the occasional occurrence of hybrids between the sibling orchid species could be attributed to the activities of secondary beetle pollinators.

The orchid *Trigonidium obtusum* is pollinated by sexually deceiving males of the bee *Plebeia droryana*, and attachment of the pollinia is achieved by the bee slipping on the waxy perianth and then becoming trapped in the tube-like flower (Singer 2002). This orchid is apparently self-compatible, but the risk of selfing is reduced by the pollinia being initially too large for a male bee carrying it to get trapped in another flower soon after, but after 40 minutes the pollinia have typically dehydrated and shrunk sufficiently for the male to pollinate another flower. By then it has almost certainly moved on to a different plant (Singer 2002). The European orchid *O. sphegodes* appears to minimise the risk of self-pollination by a chemical means. Ayasse et al. (2000) found that each plant produces a distinctly different mix of volatiles and that male bees, once duped, learned to recognise the scent of a particular flower and avoid attempting to mate with it a second time.

The response of some wasps to orchid flowers is, in fact, so dependable that Handel & Peakall (1993) were able to employ flowers of the Australian orchid *Chiloglottis reflexa* to investigate mate-seeking behaviour of another thynnine wasp, *Neozeloboria* sp., much more easily than could be done using the females, which spend much of their time at or below ground level. In this way they showed that the wasps prefer flowers that are closer to the ground and, of course, this has implications also for the success of orchids of different flowering heights. In the same system, Wong et al. (2004) found that the orchid deception had a somewhat negative effect on the male wasps' potential fecundity, but

males quickly habituated to patches where there were more flowers of the orchid. Conversely, the mating success of female wasps increased with their distance from the nearest orchid flower, apparently because her sex pheromones are not being masked by the chemically mimetic orchid emissions. Female thynnines, which are flightless, might therefore be able to increase their mating success by walking away from *Chiloglottis* patches, but whether they do so is not yet known. *Chiloglottis* is not the only Australian orchid to rely on thynnines for pollination; so too do the rare hammer orchids, *Drakaea* spp. These have a dark purple and black, insect-like labellum, contrasting with the pale green and pinkish other parts of the flower, and are fairly accurate mimics of the wingless females of the model wasp (Peakall 1990, Menz et al. 2013).

In the *Lissopimpla excelsa/Cryptostylis ovata* system, the ichneumonid male pollinator fairly quickly habituated to the sexually deceptive orchid flowers with reduced frequency of attempted copulations over successive visits (Weinstein et al. 2016). Mark–release–recapture experiments showed that individual ichneumonid wasps ranged over considerable distances (maximum observed being 625 m) but, nevertheless, a high percentage revisited the same inflorescences over a period of a day or week.

Other important factors are likely to be concerned with achieving fertilisation in generally rare plant species where pollinator fidelity is even more important (Bobisud & Neuhaus 1975). This is especially true for orchids in which the entire pollen load of a flower is in the pollinia and, therefore, once removed, the flower is unable to fertilise another. Odours can be detected at great distances as receivers may have highly sensitive receptors, far more sensitive than visual or auditory ones (Dettner & Liepert 1994). Successful pollination in orchids is considerably higher in nectariferous orchids in both temperate and tropical regions, with nectar production approximately doubling fruit-set overall (Neiland & Wilcock 1998). However, fruit-set in tropical orchids is markedly less frequent than in temperate orchid floras, which may be due to lower per species abundance, or perhaps higher diversity of the local flora, making it less probable that a pollinator will return to the same species. Only approximately 14% of non-nectar-producing orchid species can be regarded as being highly successful (abundant), and Neiland & Wilcock (1998) suggest that multiple evolutions of nectar production in the family has occurred in response to low pollinator visitation frequency. I see this in a somewhat different light, in that mimicry that attracts a particular pollinator, e.g. by mimicking sex pheromones, by many species of orchid is an adaptation to maximise fertilisation, while otherwise being constrained to be at low population density. It should be emphasised that species cannot be selected to be rare (see

---

5. The genus *Ophrys* contains approximately 300 described species.

for example Schiestl 2005), but rare species will be selected to have adaptations that enable them to persist when otherwise constrained to be rare.

Food deception

Mimicry of nectivorous plants by non-nectivorous ones, usually referred to as floral mimicry, has evolved on numerous occasions. Beardsell et al. (1986), in common with many other botanists, discuss these as being examples of either Müllerian or Batesian mimicry (B.A. Roy & Widmer 1999), whereas such floral mimicries are neither, or at least not in the same sense of protection in which the terms are used in describing animal cases. In botanical parlance, Batesian floral mimicry involves pollinator deception with one species or sex of flower offering no reward, and Müllerian floral mimicry covers convergence in floral form in rewarding species that benefit mutually by acting as pollinator magnets for one another (M.C.F. Proctor & Yeo 1973, Dafni 1984).

Female pollen-eating hoverflies, *Allograpta javana* and *Episyrphus balteatus* (Syrphidae), are the pollinators of the Chinese slipper orchid *Paphiopedilum barbigerum*, and in this case at least part of the deception appears to involve an innate colour preference (yellow, 500–560 nm wavelength) of the flies (Shi et al. 2008), leading them to settle on the bright yellow staminode. GC-MS detected no scent molecules so the attraction appears to be purely visual. Bänziger (1996) describes attraction of pollenivorous hoverflies to *Paphiopedilum villosum*, which also has a bright yellow staminode which glistens as if covered in droplets of honeydew[6] or water. Next to this is a slippery wart which is a deceptive perch such that when the hoverfly settles, it cannot grip on and falls into the orchid's pouch. In these and other slipper orchids, escape from the pouch forces the fly to pass next to the stigma, where any of the orchid's sticky pollen would be deposited, and then close to the sticky pollen clusters, which will adhere to the fly and remain viable for weeks.

Food-mimic orchids generally produce fewer flowers per plant than rewarding ones (Ren et al. 2014). Shi et al. (2008) contrast *P. barbigerum* to *P. dianthum*, which is a hoverfly brood-site mimic (Shi et al. 2007; see later) and which has a higher rate of fruit-set, inferring, albeit on the basis of just one species pair, that brood-site mimicry is a more effective strategy than weak pollen food mimicry.

Specific floral mimicry

Pasteur (1982) referred to this type of relationship as Dodsonian mimicry and cites the following as a prime example. Boyden (1980) described similarity of the nectarless orchid *Epidendrum radicans* (as *ibaguensis radicans*) (Fig. 12.6a) to those of the verbenaceous shrub *Lantana camara* (Fig. 12.6b) and the milkweed *Asclepias curassavica* (Fig. 12.6c), both of which produce abundant nectar, and suggested that they form a mimicry complex. In Costa Rica and elsewhere in Central America, *E. radicans* commonly occurs together with *Lantana* and *Asclepias* along roadside verges, and they are all pollinated by the same butterflies, ostensibly the last two reinforcing pollination in one another and the orchid obtaining pollination by deceit. Kjellsson et al. (1985) describe a very similar situation involving the orchid *Dendrobium infundibulum* and two co-flowering rewarding plants in Thailand, and Schemske (1981) provided evidence that two species of *Costus* (Costaceae) in Panama have also converged to share the same bee pollinator. However, Bierzychudek (1981) investigated this suggestion in relation to the very similar *E. radicans* but found that the frequencies at which the species are visited by potential pollinating insects within stands, comprising varying proportions of the three species, failed to support either mimicry hypothesis. Instead it was suggested that these floral resemblances, at least between *A. curassavica* and *L. camara*, might be protective rather than reproductive, since both these species are toxic to vertebrate herbivores, which might recognise the plants, in part at least, on the basis of their flowers. Papadopulos et al. (2013) used molecular phylogenetics in combination with measurements of floral reflectance to show that flowers of Neotropical orchids of the Oncidiinae (e.g. *Trichocentrum ascendens* and *Oncidium nebulosum*) are more similar to various rewarding species of the distantly related Malpighiaceae than would be expected by chance. In this case, instead of just one model species, each orchid has multiple models, and the mimetic resemblance has evolved at least 14 times within the Oncidiinae, suggesting that the mimicry has played a role in the diversification of the orchids. Nevertheless, seed-set in *Oncidium* orchids generally appear to be pollinator limited (Ackerman 1986).

In contrast to Bierzychudek's (1981) study, Nilsson (1983) provided evidence that the helleborine orchid *Cephalanthera rubra*, of southern and central Europe, mimics two blue-flowered *Campanula* species, namely *C. persicifolia* and *C. rotundifolia*. The orchid, which possesses purple (Fig. 12.7a) to brilliant red flowers, provides no food reward for flower-visiting insects, but nevertheless it attracts males of two species of megachilid bee – *Chelostoma fuliginosum* and *C. campanularum* – which are its principal pollinating agents. Females of these two megachilids provision their

---

6. The significance of any resemblance to honeydew is likely to be that this alone is a sufficient oviposition stimulus for some aphidophagous syrphids (Budenberg & Powell 1992).

Fig. 12.6 Mimicry complex of three butterfly pollinated plants. (a) Nectarless *Epidendrum radicans* (Orchidaceae); (b) nectariferous *Lantana camara* (Verbenaceae); (c) nectariferous *Asclepias curassavica* (Apocynaceae). (Source: a, Yricordel 2012. Reproduced under the terms of the Creative Commons Attribution Share-Alike Licence CC BY-SA 3.0, via Wikimedia Commons; b, Mercewiki 2007. Reproduced under the terms of the Creative Commons Attribution Share-Alike Licence CC BY-SA 3.0, via Wikimedia Commons; c, Guérin Nicolas 2008. Reproduced under the terms of the Creative Commons Attribution Share-Alike Licence CC BY-SA 3.0, via Wikimedia Commons.)

broods almost entirely on pollen from these *Campanula* species, but to the human eye they appear quite blue (Fig. 12.7b), while males of these bees frequent the same plants in search of food and mates. Colourimetric analysis demonstrated that the reflectance of the orchid and *Campanula* flowers are very similar over the range of wavelengths that are visually detectable by bees, with both models and mimics having strong absorbances around 420 nm,

Fig. 12.7 Flowers of (a) the non-rewarding orchid *Cephalanthera rubra*, (b) its rewarding model, *Campanula persicifolia*, and (c) their visual reflectance spectra showing that they differ markedly towards the long wavelength red end of the spectrum that humans can see, but are very similar over the range, extending into the UV, that bees can see. (Source: a, Stefan.lefnaer 2012. Reproduced under the terms of the Creative Commons Attribution Share-Alike Licence CC BY-SA 3.0, via Wikimedia Commons; b, Andrea Moro. Reproduced under the terms of the Creative Commons Attribution Share-Alike Licence CC BY-SA, via Università di Trieste, Dip. di Scienze della Vita; c, adapted from Nilsson 1983 with permission from Macmillan.)

even though to a human observer they appear quite different due to the orchid reflecting strongly between 600 and 650 nm (Fig. 12.7c). The flowering season of the orchid peaks about two weeks before that of the *Campanula* species, in fact at the peak of orchid flowering fewer than 10% of the *Campanula* plants were in bloom. However, as in many Hymenoptera, males of *Chelostoma* usually emerge before

the females and thus the orchid's flowering phenology corresponds well with that of the mate-seeking male bees.

Finally, Nilsson (1983) was able to show that seed-set in the orchid was highly dependent on its proximity to campanulas and consequently to patrolling males of *Chelostoma* bees. As females of *Chelostoma* do not visit the orchid, it seems likely that their attraction to *Campanula* flowers involves other signals. Interestingly, the *Cephalanthera rubra* mimicry is dependent not only on sympatry with the campanulas and the bee but also on the occurrence of wood that has been bored by various appropriately sized beetles, forming suitable nesting sites for the bees.

A similar mimicry complex involving the African orchid, *Disa ferruginea*, has been subject to investigation by S.D. Johnson (1994). The orchid, whose flowers lack nectar, has two colour forms and is pollinated solely by the nymphalid butterfly, *Aeropetes* (=*Meneris*) *tulbaghia*. In the Southwest Cape area, a red-flowered form appears to mimic the iris *Tritoniopsis triticea*, while in the Langeberg Mountains, an orange-flowered form mimics the red hot poker, *Kniphofia uvaria*. As in the case of *Cephalanthera* and its models, the reflectance spectra of each *Disa* colour form and their respective models are very similar. S.D. Johnson (1994) showed that pollination and seed-set was greater in *D. ferruginea* occurring in mixed stands together with the putative models, than in *D. ferruginea* in stands by itself. E. Newman et al. (2012) carried out reciprocal transplantation experiments as well as setting up areas with artificial flowers and found that the pollinator preferred the red phenotype and red artificial flowers in the west of the orchid's range, and correspondingly orange flowers in the east, thus indicating that the pollinator had local blossom colour preferences corresponding to the local rewarding model flowers. It is not known, however, whether there is a genetic component to this.

Quite often the relationship between model and mimic is rather precise (Table 12.3). As long ago as 1877, Sir Alfred Russel Wallace reported the resemblance between a South African orchid and the labiate plant *Ajuga ophrydis*, which bears a very striking resemblance to some *Ophrys* orchids (Wallace 1877). In this case the *Ajuga* is suggested to be the mimic, but this does not seem to have been investigated further, save for other people commenting on the resemblance. The Australian orchid *Diurus maculata* mimics members of a small group of native legumes (Fabaceae) (Beardsell et al. 1986). Wiens (1978, p. 393) mentions a number of additional suggested instances but evidence for these is scant at best.

In South Africa, the non-rewarding orchid *Disa pulchra* is pollinated by a tabanid fly, *Philoliche aethiopica*, which also pollinates the pink–flowered rewarding *Watsonia lepida* (Iridaceae), which is believed to be the orchid's model in the

Table 12.3 Some examples of specific floral mimicries between nectivorous and non-nectivorous flowers and the pollinating agent.

Mimic	Model	Dupe	Reference
Euphrasia (Scrophulariaceae)	*Calluna* (Ericaceae)	Various bees	Yeo 1968
Anacamptis (as *Orchis*) *israelitica*	*Bellevalia flexuosa* (Liliaceae)	*Eucera clypeata, Bombylius* sp. and possibly *Anthophora* sp.	Dafni & Ivri 1981
Cephalanthera rubra (Orchidaceae)	*Campanula persicifolia, C. rotundifolia*	*Chelostoma* spp (Megachilidae)	Nilsson 1983
Cypripedium macranthos var. *rebunense* (Orchidaceae)	*Pedicularis schistostegia* (Scophulariaceae)	*Bombus* spp. (Apidae)	Sugiura et al. (2002)
Disa cephalotes	*Scabiosa columbaria* (Dipsacaceae), *Brownleea galpinii* (Orchidaceae)	Nemestrinidae	S.D. Johnson et al. 2003b
Disa draconis (Orchidaceae)	*Pelargonium*	*Moegistorynchus longirostris*	S.D. Johnson & Steiner 1997, Jersáková et al. 2009
Disa ferruginea (Orchidaceae)	*Kniphofia uvaria* (Asphondelaceae), *Tritoniopsis triticea* (Iridaceae)	*Aeropetes tulbagha* (Nymphalidae)	S.D. Johnson 1994
Disa harveiana	?	*Philoliche rostrata* (Tabanidae)	S.D. Johnson & Steiner 1997
—	*Tritoniopsis triticea* (Iridaceae)	*Aeropetes tulbagha* (Nymphalidae)	S.D. Johnson 1994
Disa karooica	*Pelargonium stipulaceum*	*Philoliche gulosa*	Combs & Pauw 2009
Diurus maculata (Orchidaceae)	*Daviesia* spp, *Pultenaea scabra* (Fabaceae)	Various bees, mostly Colletinae	Beardsell et al. 1986
Diurus nervosa (Orchidaceae)	*Watsonia densiflora* (Iridaceae)	Tabanidae	S.D. Johnson & Morita (2006)
Diurus nivea (Orchidaceae)	*Zaluzyanskia microsiphon* (Scophulariaceae)	Nemestrinidae	Anderson et al. 2005
Disa pulchra	*Watsonia lepida* (Iridaceae)	*Philoliche aethiopica* (Tabanidae)	S.D. Johnson 2000
Disa oreophila erecta	*Brownleea macroceras* (Orchidaceae)	*Prosoeca ganglebaueri* (Nemestrinidae)	S.D. Johnson & Steiner 1995
Epidendrum ibaguense (Orchidaceae)	*Asclepias curassavica, Lantana camara* (Verbenaceae)	Many insects, especially butterflies	Boyden 1980
Epidendrum radicans (Orchidaceae)	*Asclepias curassavica, Lantana camara* (Verbenaceae)	Many insects, especially butterflies	Bierzychudek 1981
Eulophia zeyheriana (Orchidaceae)	*Wahlenbergia cuspidata* (Campanulaceae)	Halictid bees	Peter & Johnson 2008
Oncidium cosymbephorum (Orchidaceae)	*Malphigia glabra* (Malphigiaceae)	Anthophoridae	Carmona-Díaz 2001
Various *Oncidium* and relatives (Orchidaceae)	Malphigiaceae	?	Papadopulos et al. 2013
Ophrys spp.	*Ajuga ophrydis* (Lamiaceae)	?	Wallace 1878

same area (S.D. Johnson & Morita 2006), and *D. karooica* and its mimic *Pelargonium stipulaceum* are both pollinated by the same tabanid fly, *Philoliche gulosa* (Combs & Pauw 2009). Jersáková et al. (2012) investigated what features of the orchid flower the flies were responding to by presenting a range of plastic flowers in the field that differed in colour, brightness, shape, inflorescence architecture and the presence or absence of nectar guide patterns. Flies visited plastic flowers readily if their colour (as would be seen according to a fly visual model) was indistinguishable from the *Watsonia*, whereas intensity of colour and inflorescence architecture had little effect. After colour, the next most important visual signals to the flies were flower shape and nectar guides. This is, of course, just one pollinator and it would be interesting to know how much variation there was among different species and different higher taxa of pollinators, and whether any generalisations were possible.

Advergent evolution of *Turnera sidoides* ssp. *pinnatifida* (Turneraceae) to resemble co-flowering mallows (Malvaceae) in Argentina is described by Benitez-Vieyra et al. (2007). In different regions there are different mallow species with different flower colours, which the *Turnera* locally resembles. Both are rewarding but the principal mallow pollinator is primarily after the mallow pollen, but does not distinguish the two plant species. Thus the *Turnera* benefits from the pollinator magnet effect of the mallows. Pellmyr (1986) and Laverty (1992) describe similar 'commensal' situations for rewardless North American *Cimifuga* (Ranunculaceae) species and *Podophyllum peltatum* (Berberidaceae), respectively, and the highly rewarding plants they are usually associated with. In the second case, although queen bumblebee pollinator visits are few, the plant flowers early in the season when there are few other rewarding flowers in bloom. Ehrlén (1993) suggests that the non-fruiting flowers of the northern European legume *Lathyrus vernus* may be acting as intraplant pollinator magnets, enhancing the attractiveness of the individual plants.

Generalised floral mimicry

Orchids that do not provide rewards for pollinators and do not appear to have any specific model often form aggregations and also often display polymorphism in their flower colour. Some are associated in a vague way with local rewarding flowers of the same colour (e.g. Dafni 1983, 1986, 1987, Dafni & Calder 1987, Burns-Balogh et al. 1987). Such trends may be interpreted in terms of deceiving naïve generalist pollinators such as bumblebees. This view is supported by the flowering phenologies of such orchids, which tend to bloom early in the season when there are many recently emerged and inexperienced

potential pollinators (Nilsson 1992). A similar example involving the South African orchid *Disa tenuifolia* was recently described by S.D. Johnson & Steiner (1994). *D. tenuifolia* blooms *en masse* following wildfires and its non-rewarding flowers, which do not appear to mimic any other particular plant, are pollinated by various megachilid bees. S.D. Johnson & Steiner (1994) presumed that the orchid takes advantage of the general instinct of these bees to visit bright flowers. In the temperate region, orchids that rely on generalised floral mimicry are typically gregarious and flower early in the spring when there are numerous naïve pollinators, such as newly emerged queen bumblebees (Heinrich 1975, Jersáková et al. 2006, Peter & Johnson 2008, 2013). They are also quite commonly polymorphic for flower colour, which has been proposed to reduce pollinator learning that the flowers provide no reward so that pollinators carrying pollinia will still visit other conspecific *Disa* flowers, or it could simply indicate relaxed selection pressure for maintaining a particular flower colour (Jersáková et al. 2009).

The non-rewarding European globe orchid, *Traunsteinera globosa*, has a quite atypical, domed pink to bluish inflorescence, which closely resembles the flower heads of various fly-pollinated plants such as various sympatric *Knautia* and *Scabiosa* (Dipsacaceae), and *Valeriana* (Caprifoliaceae). Jersáková et al. (2016) showed that its flowers were visited at essentially the same rate by potential pollinators, as were those of these putative rewarding models, whether they were naïve or experienced with the orchid. This applied especially to Diptera, which frequently picked up pollinaria. Furthermore, the orchid's floral colour overlaps several of these putative rewarding models though the flower scents of nearly all the species were fairly distinct. This suggests that as the pollinators had no reason to discriminate between the various rewarding species in the complex on the basis of scent, they pay little attention to it, whereas the similarity in inflorescence colour and morphology are likely to be most important to them.

J. Kunze & Gumbert (2001) used an artificial flower system to investigate the roles that flower colour and odour play in food-deceptive pollinator attraction. They used the common European bumblebee, *Bombus terrestris* (Apidae), and artificial rewarding and deceptive flowers. In one experiment, rewarding and non-rewarding flowers were presented simultaneously and the bees learned to discriminate between mimic and models more quickly, either if there was a difference in scent or if the mimic was scentless while the model had a scent. This implies that odour mimicry would be important for food-deceptive flowers for at least this pollinator species, although the authors note that there is, as yet, no evidence of this being the case in natural systems. When the model and mimic flowers were presented successively, the bees learned to avoid the non-rewarding flowers quickest if they had a scent rather than if they did not. Overall, it appears that scent enhances the learning of colour cues. Whether this is due to enhanced learning or better memory retrieval is not known.

Gigord et al. (2002) also investigated foraging by bumble-bees, in their case using the rewardless elder-flowered orchid, *Dactylorhiza sambucina*, which has two colour forms (purple or cream) throughout most of its range, and reward-ing models of various species with approximately the same flower colour as one colour form of the orchid, but not the other. Naïve bumblebees visited both colour forms of the *Dactylorhiza* at random, but after experiencing the reward-ing flowers, they preferentially visited those of the similarly coloured *Dactylorhiza* form. Furthermore, the bees would still visit the similarly coloured orchid blossoms even after the rewarding plants had been removed, thus demonstrat-ing that, under appropriate circumstances, selection could lead to general floral mimicry by rewardless plants. In west-ern Germany only the cream-blossomed form occurs and yet they still achieve pollination success comparable to that achieved by the species elsewhere (Kropf & Renner 2005).

In a laboratory experiment with rewarding and non-rewarding artificial flowers, Smithson & Macnair (1997) found that initial experience of pollinators with a non-rewarding 'flower' colour that was different from any rewarding flowers in the background experienced negative frequency-dependent selection, in that the bumblebees vis-ited rare non-rewarding morphs more than expected by chance and moved more between different rare unreward-ing forms, suggesting that they were using a probability-based sampling strategy. This pollinator behaviour would therefore favour the evolution of polymorphism. In con-trast, Levin (1972) found that rare flower morphs were selected against (positive frequency-dependent selection) in a rewarding *Phlox* species.

In an elegant experimental study, Kunin (1993) tested the effect of spacing between plants (nearest-neighbour dis-tance) and background on pollinator visitation, pollinator constancy and seed-set in the self-incompatible, yellow-flowered cruciferous plant *Brassica kaber*. Plants of *B. kaber* were grown in plots by themselves, i.e. with no alternative flowers present for pollinators to visit, or together with the florally similar *Brassica hirta*, in which case pollinators were found to act as generalists, or they were planted among dis-similar flowered species (*Fagopyrum esculentum* or *Lolium multiflorum*), in which case the pollinators again behaved as specialists. Kunin (1993) found that both pollinator visita-tion and seed-set in *B. kaber* were greatly influenced by spacing and by the background planting. In all treatments, the greater the nearest-neighbour distance between *B. kaber* plants, the lower the seed-set, but the decline was far more rapid if the plants were growing among a background of similar rather than dissimilar flowers or no flowers at all.

The highest numbers of pollinator visits, particularly to plants with larger nearest-neighbour distances, occurred where *B. kaber* were interplanted with the similarly yellow-flowered *B. hirta* (Fig. 12.8) and in the other treatments, pollinator visitation to *B. kaber* flowers rapidly declined with nearest-neighbour distance. However, while pollinator visitation was enhanced by the presence of florally similar *B. hirta*, seed-set was not. Thus, when the pollinators were acting as generalists, the bulk of the pollen they deposited on the *B. kaber* flowers would have been from the other spe-cies, namely *B. hirta*. These experiments, showing that seed-set can be strongly influenced by pollen quality rather than just the number of pollinator visits, have considerable impli-cations for plant mimicry theory. For example, this finding may help to explain why non-rewarding floral mimics are so much more abundant among the Orchidaceae compared with other groups of plants, since in the orchid family, pol-len is packaged within pollinia and so an orchid's pollen will not become mixed or diluted and thereby diminish in qual-ity, even if there are long intervals between the pollinator visiting orchid flowers. Furthermore, however, it casts some doubt on whether, in the absence of such specialised mech-anisms, floral mutualistic mimicry will be advantageous.

Another *Disa* example is that of the rare South African *D. nivea*, which only occurs in places where there are sub-stantial numbers of its nectar-producing model, the mountain drumstick, *Zaluzianskya microsiphon* (Scrophu-lariaceae), which has a specific pollinator, the tangle-veined fly, *Prosoeca ganglbaueri* (Nemestrinidae), a species with an exceptionally long proboscis (B.C. Anderson et al. 2005). By manipulating the relative densities of the two plant species, B.C. Anderson & Johnson (2006) showed that pol-lination of the *Disa* was reduced when it was relatively abundant compared to the model, and this appeared to be due to the pollinating fly avoiding nearby flowers after hav-ing encountered a rewardless orchid. However, in this case the authors did not detect any negative effect of the orchid on pollination of the model.

It has been suggested that floral mimicry may achieve a higher frequency of outbreeding than that achievable by plants providing nectar or other rewards (Dressler 1981, L.A Nilsson 1983) but whether it does this, or conversely leads to high levels of inbreeding, will depend on the type of mimicry and the behaviour of the pollinator. Some pollina-tors, such as many bees which are seeking nectar rewards, will sample a few plants at a site and if they are not rewarded they quickly fly off considerable distances to explore new areas (Heinrich 1979, L.A. Nilsson 1984). On the other hand, if pollination is effected by deceit of male insects that are duped into mistaking the orchid blossom for a female,

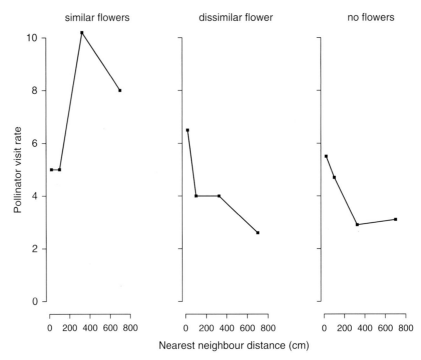

Fig. 12.8 Effect of nearest-neighbour flower similarity on pollinator visits to focal plant showing that when a flowering plant is surrounded by similarly coloured neighbouring blossoms it receives more pollinator visits. (Source: Adapted from Kunin 1993 with permission from John Wiley & Sons.)

then the same males will often tend to remain in the vicinity of their first floral encounters, especially if the species is one which tends to hold territories.

Pellissier et al. (2010) have shown that food-deceptive orchids generally occur at lower altitudes and bloom earlier than non-deceptive ones, though this could be a simple function of earlier flowering in general at lower altitudes because of temperature, and the presence of more non-duping flowers there.

Mimicry of a fungus-infected plant

The Chinese slipper orchid, *Cypripedium fargesii*, is pollinated by the syrphid fly *Cheilosia lucida*, misidentified in the original paper as a member of the Platypezidae (Ren et al. 2011, 2012). The orchid leaves have dark setose spots similar to *Cladosporium* fungus-infected leaves, and the flowers emit some *Cladosporium* volatiles. Flies visiting the plants normally feed as adults on *Cladosporium*, and when they get caught in the orchid flowers they are guided to escape in a way that leads to them picking up the orchid's pollen. Flies visiting the flowers were also found to be carrying *Cladosporium* spores, but these appear unable to infect the

orchid. Presumably if the orchid was susceptible to the fungus such a pollination system would not have evolved.

Brood-site/oviposition-site deception

Numerous flowers and inflorescences belonging to plants in many families produce odours that smell like decaying flesh or fungi. These attract insects, predominantly flies (Diptera), whose larvae feed on such substances and thus the plants are mimicking their pollinator's oviposition sites. Often referred to as oviposition-site mimicry or brood-site deception, this system occurs in many species of Araceae, but has evolved independently in the Aristolochiaceae, Asclepiadaceae, Iridaceae, Orchidaceae and Solanaceae (Atwood 1985, Burgess et al. 2004, Goldblatt et al. 2009, Urru et al. 2011, Moré et al. 2013, Jin et al. 2014). In brood-site deception it is common for the flowers to include some sort of trap that maximises the chance of the pollinator getting pollen onto it (Vogel & Martens 2000).

Brood-site mimicry is often thought to be rather unspecialised, but in fact there are many exceptions in which various sorts of filters are employed that narrow the range of attracted insects. Another interesting feature is that it is frequently associated with floral gigantism (Jürgens & Shuttleworth 2015). Probably the most extreme deception known to date is that of the South African plant *Ceropegia sandersonii* (Apocynaceae), the flowers of which emit volatiles that mimic those of honey bees that have been attacked, and therefore, whose bodies/corpses would be a suitable brood site for milichid flies of the genus *Desmometopa* which scavenge dead insects such as bees (Heiduk et al. 2106). Unlike *Stapelias*, for example, this plant has trap-type flowers that retain their deceived pollinators for a while, maximising the chance of successful pollen transfer.

Jürgens et al. (2013) carried out a phylogenetically informed survey of such strategies and found that release of blends of oligosulphides, which are universal infochemicals used by carrion-attracted flies and beetles, has evolved independently in at least five plant families: Annonaceae, Apocynaceae, Araceae, Orchidaceae and Rafflesiaceae. Furthermore, they found that the volatile profiles fell into four distinct clusters in chemical phenotype space, with plants mimicking animal carrion, decaying plant material, herbivore dung and omnivore/carnivore faeces. Some 11% of orchid genera include species that use brood-site mimicry, and thus this is also one of the most widespread deceptive pollination syndromes in that family.

Orchids of the South American genus *Dracula* are normally pollinated by fungivorous flies, predominantly Drosophilidae, and in addition to mimetic organic volatiles some species have the labellum closely resembling the cap and gills pattern of small mushrooms (Fig. 12.9) (Kaiser 2006, Dentinger & Roy 2010). In a clever application of new technology, Policha et al. (2016) investigated what aspects of these orchids' floral morphology and odour were responsible for attracting the flies. Using three-dimensional printing they made accurate moulds of flower parts from scent-free surgical silicone, and combined them with either the calyx or labellum of real *Dracula* flowers to make chimeras. In field tests, drosophilid flies landed on the presented 'flowers' in the order: real flowers ($n=151$) > chimeras with labellum ($n=127$) > chimeras with calyx ($n=54$) > totally artificial flowers ($n=28$). The flies were attracted most by both the labellum's volatile and visual clues, and also by the coloured calyx. They also showed that the volatile mix released from the orchid flowers showed a "remarkable overlap" with the volatiles from co-occurring fungal fruiting bodies.

It had long been supposed that mycophilous flies pollinated flowers while laying eggs on them (Vogel 1978), but Endara et al. (2010), who studied pollination of *Dracula lafleurii* and *D. felix* in Ecuadorian cloud forest, found that while drosophilid flies of the genus *Zygothrica* courted and mated in the orchid flowers, there was no attempt at oviposition, at least in the case of these two species. Fungus gnats and drosophilids are also the pollinators for various other orchids, particularly members of the Pleurothallidinae, including *Octomeria*, *Pleurothallis* and *Stellis*.

A particularly interesting deception by an orchid involves the marsh helleborine, *Epipactis veratrifolia*, a terrestrial orchid with a rather unobtrusive flower that emits α- and β-pinene, β-myrcene and β-phellandrene and is pollinated by aphidophagous hoverfly females that often attempt oviposition on it. This organic volatile mix is remarkably similar to the alarm pheromone mixes of various aphids, and thus it seems likely that the plant is duping the hoverflies by making itself appear to be a suitable oviposition site (Stökl et al. 2011). Since the orchid's flowers are also often heavily infested with aphids while they are still in bud, the hoverflies may oviposit on the aphid colony where their larvae develop, thus being partially rewarded (Jin et al. 2014). Then the *Epipactis* orchids could therefore be partly on the way to true pollinator brood-site deception but are currently hitch-hiking on part of their own defensive mechanism by tri-trophically signalling to aphidophagous hoverflies for the mutualistic reasons of reducing aphid-induced damage. Mapping reconstructed ancestral character states onto a phylogenetic tree of *Epipactis* (Fig. 12.10) suggests that pollination by hoverflies is actually plesiomorphic for the genus. It would be interesting to know whether the more derived, non-hoverfly-pollinated taxa are as susceptible to attack by aphids or whether they have evolved alternative defences against them.

Flowers of the orchid *Dendrobium sinense*, which is endemic to the island of Hainan off the coast of mainland China, mimic the alarm pheromone of the European and Asian honey bees, *Apis mellifera* and *A. cerana*, which also attracts the bee-predating hornet, *Vespa bicolor*, which is believed to be the orchid's normal pollinator (Brodmann et al. 2009). The active compound in the orchid's emission, (Z)-11-eicosen-1-ol, is also a major component of the bee alarm pheromone mixes.

Arum lily pollination is probably entirely through deception of flies, or less commonly beetles, through mimicry of odours of their oviposition sites, which are normally decaying organic matter, including rotting fruit or dung (Kite 1995, Gibernau et al. 2004, T.S. Woodcock et al. 2014). Care has to be taken in interpreting what the real pollinators are as many non-pollinators or only occasional ones may visit the flowers and get caught in the traps. Flowers pollinated by saprophilous or dung flies are typically rather dark red, brown or black, though of course flies cannot see red and the relative significance of odour versus colour is not really understood (T.S. Woodcock et al. 2014). Flies such as blow fly greenbottles (*Lucilia*) and bluebottles (*Calliphora*) and the flesh flies (*Sarcophaga*) have been found to have different colour preferences depending on whether they are seeking food or oviposition sites (Kugler 1956).

Fig. 12.9 Flowers of epiphytic *Dracula* orchids that resemble fungi both visually and odoriferously, and are pollinated by 'mycophilous flies', mainly drosophilids. (a) *Dracula bella*, detail of labellum showing irregular pattern of ridges resembling the gills of a small fungus; (b,c) *D. morleyi in situ* in the field in Ecuador, (b) in the correct dangling orientation showing resemblance of underside of the labellum to a mushroom cap; (d) flower of *D. roezlii*, an orchid from moist high-altitude forest of Colombia. (Source: Bryn Dentinger. Reproduced with permission from Bryn Dentinger.)

Flowers of the Solomon's lily, *Arum palaestinum* (Araceae), mimic the smell of yeast fermenting a substrate in order to attract its pollinators, which are drosophilid flies that normally lay their eggs in fermenting substrates (Stökl et al. 2010) but which might also be places that the adult flies simply go to feed (Gibernau et al. 2004). Flowers of some other tropical Araceae are well known for their unpleasant smell to humans, and large *Amorphophallus titanum*, sometimes called the 'corpse flower', in botanical gardens will attract public visitors just for the experience. These plants are producing odour cocktails that mimic the smell of dung (Kite & Hetterscheide 1997, Kite et al. 1998) and consequently also attract visits from various dungflies and beetles. B.N. Smith & Meeuse (1966) found that several arums release skatole, a common component of faeces odour, as well as the polyamines putrescine and cadaverine. These polyamines, as well as spermidine, have very little odour themselves but are easily oxidized to form 1-pyrroline which is the primary odour of semen[7]. Not surprisingly therefore, some sapromyophilous flowers smell strongly of semen due to the presence of considerable quantities of 1-pyrroline in their floral scents (Raguso et al. 2007, Chen et al. 2015, Shuttleworth 2016). Stensmyr et al. (2002) compared the odour mix of the aptly named dead-horse arum, *Helicodiceros muscivorus*, during anthesis with that of rotting flesh and found an excellent match.

---

7. Some people are specifically unable to detect the smell of 1-pyrroline or semen (Amoore et al. 1975).

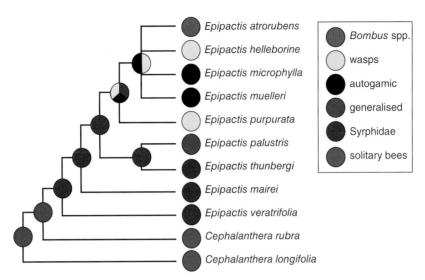

Fig. 12.10 Reconstructed phylogeny and ancestral pollinator use for members of the orchid genus *Epipactis* indicating that aphidopha-gous hoverflies (Syrphidae) are probably the ancestral pollinator group for the genus, at least one member of which, *E. veratrifolia*, releases aphid alarm pheromones to attract its pollinators. Only the maximum-parsimony reconstruction is shown; the maximum-likelihood one is very similar but a little more complex. (Source: Adapted from Jin et al. 2012. Reproduced under the terms of the Creative Commons Attribution Licence CC BY 2.0, via BMC Plant Biology.)

Members of the South African genus *Stapelia*[8] (Apocynaceae) typically have somewhat hairy flowers that visually may resemble carrion, and their odours have earned them the common epiphet 'carrion flowers'. These are especially attractive to flies of various groups (Fig. 12.11) and Jürgens et al. (2006) have shown that while sulphur compounds, benzenoids, fatty acid derivatives or nitrogen-containing compounds dominate the odours of the species in general, different species produce quite different odours, and these could be tentatively classified into four groups with likely different models (Table 12.4). Another fly-polli-nated South African carrion-mimic flower is the terrestrial orchid *Satyrium pumilum*, which has dark red-flecked flow-ers formed very close to the ground (S.D. Johnson et al. 2007). van der Niet et al. (2011) compared its pollinating fly assemblage to those of real animal carcasses and found a marked difference. Real carcasses were visited predomi-nantly by blow flies (Calliphoridae), house flies (Muscidae) and flesh flies (Sarcophagidae), but only the last of these visited the orchid flowers. However, the difference did not appear to be due to the odoriferous compounds in the orchid's scent, as these (oligosulphides, 2-heptanone, *p*-cresol and indole) were essentially similar to those given off by carcasses. Instead, it appeared that the relatively

Fig. 12.11 Carrion flies attracted to the odour of decaying flesh emanating from the flower of a *Stapelia gigantea* (identification thanks to David Goyder.)

small quantities of the carrion smell-mimicking blend com-ing from the orchids was only sufficient to attract Sarcophagidae, and this might relate to the fact that they will lay on small animal carcasses, such as those of rodents and even dead snails and insects. Interestingly, sarcophagids are larviparous, depositing larvae on food sources rather than egg masses, which means that smaller resources are potentially just as viable for their offspring. By staining

---

8. *Stapelia*, once a large genus, has been substantially split in recent revision, so quite a few species are now placed in other genera.

Table 12.4 The chemical compositions of four odour groups of stapeliad (Apocynaceae) species. (Data from Jürgens et al. 2006.)

Chemical profile	Model
High p-cresol content but low amounts of polysulphides	Herbivore faeces
Mainly polysulphides with low amounts of p-cresol	Carnivore/omnivore faeces or carcasses
High amounts of heptanal and octanal	Carnivore/omnivore faeces or carcasses
Hexanoic acid	Mammal urine

pollinia, it was possible to show that the flies frequently deposited pollinia on the same plant (self-pollination by proxy), so it seems that carrion mimicry, at least in this species, did not evolve primarily to achieve out-crossing.

Other foul, carrion- or excrement-smelling flowers include the South-East Asian *Rafflesia kerri*[9] (Rafflesiaceae/Euphorbiaceae), which is notable for having the largest single flower of any living plant (Bänziger 1991), *Hydnora africana* (Hydnoraceae), *Sterculia foetida* (Sterculiaceae) and various Burmanniaceae, Rhamnaceae, Taccaceae and a few Solanaceae (Wiens 1978, Alves et al. 2005, Woodward et al. 2007, Bolin et al. 2009, Teichert et al. 2012, Moré et al. 2013). Most *Aristolochia* species also seem to use floral odour deception to attract pollinating saprophagous flies, which are usually trapped in the flower during its initial female phase and released the following day after pollen has been released onto them, but two *Mononeuria* species studied by Sakai (2002) are honest in that they lack the traps and instead shed their large flowers which, having fallen on to the forest floor, rot and so provide larval food for their associated saprophagous flies.

The deceptive strategy recently determined for another *Aristolochia* species, *A. rotunda* from the Mediterranean region, is quite different from the above and might serve as a warning that not all plants assumed to be pollinated by saprophilous flies might be so. It turns out that the pollinators of *A. rotunda* are female flies of the family Chloropidae. These are normally quite specific food thieves (kleptophages/kleptoparasites) which are attracted to mired bugs (Heteroptera) that are being consumed by some predatory arthropods such as spiders (Sivinski et al. 1999). Chemical analyses and bioassays showed that it is the specific smell of a freshly killed and wounded mirid bug that the *Aristolochia*

flowers mimic, specifically hexyl butyrate and (*E*)-2-hexenyl butyrate (Oelschlägel et al. 2015, see also commentary by F. González & Pabón-Mora 2015). I have recently seen several chloropids as the sole insects visiting the spathes of various species of Thai Araceae, and wondered at the time what the attraction might be, but now I have a clue.

Shelter mimicry

A rather less well-known set of floral mimicries are those in which the flower appears to offer a nesting site or roosting shelter for its pollinator. Vereecken et al. (2013) show fairly conclusively, using a bee visual model and chemical analyses of floral scent, that the tubular flowers of *Iris atropurpurea*[10] (Iridaceae) and relatives resemble shelters to male eucerotine bees, which duly pollinate them, despite the plant providing no nectar. Flowers that look red to humans are typically not attractive to bees or other insects because they cannot see red, so unless they reflect strongly in the UV they would appear black and unattractive to most. To a bee, the iris flower would resemble a large, achromatic protective shelter, and the authors found no evidence that the male pollinators were particularly attracted to the relatively large amounts of *n*-alkenes in the flower's scent. This particular example is perhaps on the borderline of adaptive resemblance since the male bees are not deceived, the plant is providing a roosting tunnel either during overcast weather or overnight, but the bees would presumably not normally (in the absence of the *Iris*) be seeking a flower for this purpose, so the *Iris* flower is gaining pollination by resembling a different sort of roosting bee hole, such as "rock crevices, under flat stones or in hollow wood stems".

Vereecken et al. (2012) present observations of both hymenopteran roosting behaviour and phylogenetic reconstructions. They show not only that there are various other cases of shelter mimicry involving both orchids and irises, but that these are often closely associated phylogenetically with sexual deceptive species and the evolution can lead to transitions in both directions between these two pollination categories.

FLOWER SIMILARITY OVER TIME

As Thomson (1980) put it, "a species may be effectively familiar if pollinators have been accustomed to visit an earlier-blooming species very much like it [at the same place]". Thus, in its simplest terms, if a plant that blooms later in the

---

9. Also sometimes confusingly called 'corpse flower' because of its odour.

10. Other members of the section *Oncocyclus* of the genus *Iris* similarly provide male bees with shelter.

season than a species whose pollinators have already discovered and learned about its type of signal from earlier flowering species, the later-appearing species might benefit from resembling the earlier one but not having to train a completely new set of pollinators. Thomson's hypothesis and data also suggest that when an unfamiliar type of flower blooms, the pattern of flower opening ought to be positively skewed such that most open after pollinators have had a chance to learn that they are rewarding.

FLOWER AUTOMIMICRY – INTRASPECIFIC FOOD DECEPTION (BAKERIAN MIMICRY)

Some plants have separate male and female flowers (monoecious plants) and in some cases different individuals produce only one sex of flower, and are called dioecious. In many monoecious and dioecious angiosperms flowers of the different sexes offer different levels of rewards to pollinators, and sometimes only the flowers of one sex are rewarding at all. Willson & Ågren (1989) found that male flowers tend to offer more rewards than female ones, but if the reward is only nectar sometimes the female offers the larger reward. Also, it appears correspondingly that female flowers are usually the mimics rather than the models. A consequence of this type of mimicry noted by Willson & Ågren (1989) is that it is likely to reduce incidence of intrasexual flower polymorphism because that defeats the mimicry, and naïve pollinators are likely to be relatively unimportant as pollinating agents.

H.G. Baker (1976) drew attention to this in relation to the papaya family of plants, the Caricaceae, and hence when deception is involved this has often been referred to as Bakerian mimicry. The papaya, *Carica papaya*, is dioecious and the male and female flowers are quite different in appearance to a human observer, which begs the question, why does an insect which has visited rewarding female flowers (Fig. 12.12a) start also visiting the non-rewarding male flowers, which are white and tubular (Fig. 12.12b)? What is likely critical, though, is how the main pollinating insects perceive the flowers; maybe aroma and reflectance spectra are important but not shape. Intersexual automimicry has also been proposed for some Caricaceae (e.g. Bawa 1980).

In *Begonia involucrata*, male flowers have nectaries but the similar-looking female ones do not and therefore the latter are mimics of the former. Schemske & Ågren (1995) used artificial female flowers of this *Begonia* to test whether female flower size was constrained by stabilising selection to be similar to the mean size of the rewarding male flowers. Their results (Fig. 12.13) clearly demonstrated that this was not the case, as larger flowers were more attractive to potential pollinators in terms of both approaches to flowers and visits. Interpretation is complicated, however, because a plant's ability to produce large flowers may be a trade-off with the number of flowers, and a pollinator in nature may respond to both size and number. Schemske et al. (1996) described the pollination biology of another sexually deceptive species, *Begonia oaxacana*, with very similar results. In this species the mimetic female flowers face downwards, as do the newly opened male flowers, which are the ones that offer the highest reward and are most preferred by pollinating bumblebees. A very similar situation occurs in the monoecious plant *Ecballium elaterium* (Cucurbitaceae), with relatively rare, non-rewarding female flowers being mimics of the commoner, nectiferous male ones (Dukas 1987). In *Clusia nemorosa* (Clusiaceae) the idea of deception is less

Fig. 12.12 Bakerian mimicry in papaya, *Carica papaya* (Caricaceae). (a) Rewarding female flowers (note the branched stigma); (b) unrewarding male flowers (note typical long tube).

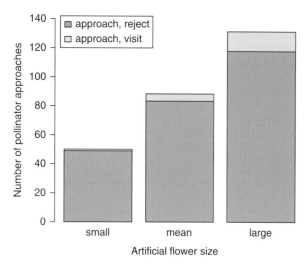

Fig. 12.13 Results of presenting artificial non-rewarding female flowers of different sizes to pollinators, with the medium-sized category representing the mean size of the rewarding, model male flowers of *Begonia involucrata*. (Source: Adapted from Schemske & Ågren 1995. Reproduced with permission from John Wiley & Sons.)

obvious, because male and female flowers offer different rewards – males pollen and the mimetic female flowers resin (Vlasáková & Jarau 2011).

Wickler (1968) notes that some plants whose pollinators are primarily interested in the pollen itself as a food, though almost unavoidably will transmit some between plants, might be partially deceived by the plants producing both fertile pollen and a still rewarding (though perhaps less so) variant that is sterile. The exact interpretation of whether the sterile pollen is deceptive or not depends on how important the nucleic acid component of the 'pollen' might be to the pollen-feeding pollinator. Another of his examples involves the flowers of the Parnassus lily, *Parnassia palustris* (Celastraceae), which has three finger-like processes, each ending in a glistening button arising from the bases of its petals, though these provide no reward but do appear to be responsible for attracting various larger flies which are its pollinators. The petals themselves, however, do provide some nectar and therefore the flies are only deceived by the initial visual lure and presumably usually get a real reward. These glistening dummies are called pseudonectaries (van der Pijl & Dodson 1966). In some members of the South American cloud forest orchid genus *Stelis*, brightly reflecting and refractile clusters of calcium oxalate crystals[11] are located at the tips

of the petals and on the lip (Chase & Peacor 1987). Some *Stelis* species produce nectar and those ones lack the crystals, whereas all the ones with crystals are nectarless, strongly suggesting that they are nectar mimics. Perhaps it is less costly to produce calcium oxalate, which does not need replenishing, than continuously making nectar.

I have to say that I think the term 'mimicry' in some of these cases is being rather stretched because although deception is involved, one might presume that in an ancestral plant species with bisexual, rewarding flowers, all flowers look more or less the same, certainly within an individual plant. Therefore, when evolution led to a separation of sexes, either between flowers on a plant or separate plants, it seems most likely to me that both had similar-appearing flowers. Any marked differences might therefore be expected to reduce the likelihood of a pollen-bearing pollinator visiting a female. Thus, after separation of flower sexes, evolution is likely to act to maintain sufficient similarity for pollinators to visit both. That male flowers are usually the rewarding ones makes sense, as it would be unprofitable for the female flowers to reward pollinators that then did not keep visiting unrewarding males, because they would then be rather poor pollinators.

MATHEMATICAL MODELLING OF SEXUAL DECEPTION BY PLANTS

Lehtonen & Whitehead (2014) modelled sexual deception and examined the effects of the cost of mimicry, the cost to receiver for being fooled, the density of mimics and the relative magnitude of a mimicry-independent component of fitness. Their modelled situation assumed a range of acceptable variation in a given trait (from $-y$ to $+y$) in the real model which overlaps with some of the variation in the mimic (Fig. 12.14a). Within the acceptable zone, potential mates can be choosy about how close the model is to the ideal. Depending on the costliness of the mimicry, the model predicts evolutionarily stable strategies (ESSs) with choosiness at equilibrium increasing with costliness (Fig. 12.14b–d), though with increasing cost the zone of attraction decreases a little.

POLLINATOR GUILD SYNDROMES

Bird-pollinated systems

A number of large-flowered tropical and subtropical shrubs, trees and some herbs are pollinated primarily by birds. In the Old World, these are mainly sunbirds (Nectariniidae),

---

11. These are not raphides in the strict sense because they are extracellular, whereas calcium oxalate raphides are intracellular.

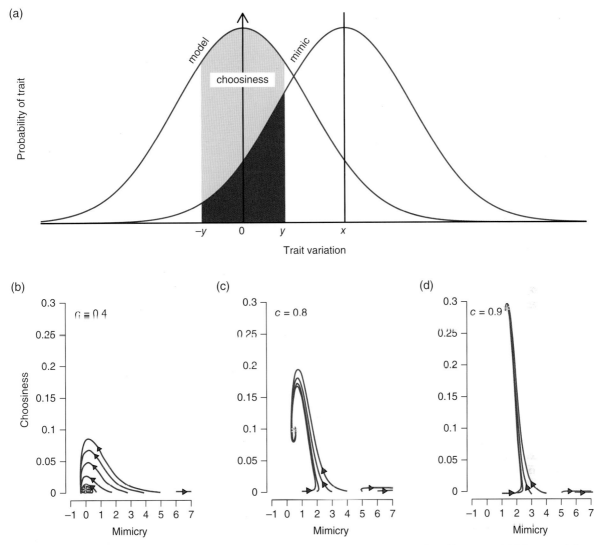

Fig. 12.14 Coevolution between a sexually deceptive mimic and its model can be an evolutionarily stable strategy (ESS) because of the trade-off between the cost of discriminating the real signal from the mimic's and the cost of falsely rejecting a mating opportunity. (a) The model framework with the differing sexual signals of model and mimic, and a zone of acceptable variation in the real model (i.e. a real virgin female bee) and its overlap with part of the mimic's variation; (b–d) model outputs showing the evolutionary trajectories leading to an ESS (indicated by the pale blue asterisk) for increasingly costly mimicry (c). (Source: Adapted from Lehtonen & Whitehead 2014. Reproduced with permission from Oxford University Press.)

whereas in the Neotropics, it is hummingbirds (Trochilidae) that are normally involved, however there are other families too. Some 500 plant genera include bird-pollinated species (Renner 2006). These diverse plants have brightly coloured, very often red, flowers (K.G. Grant 1966), and in the more specialised ones the flowers are tubular (Fig. 12.15b), though in others they are very open, with long stamens, often in clusters (J.H. Brown & Kodric-Brown 1979,

Rodríguez-Gironés & Santamaría 2004) (Fig. 12.15a). Given the size and energy demands of their pollinators, these flowers provide a substantial reward, which is usually prolific nectar, but in some cases water, which may be a very valuable resource for the pollinators at hot, dry times of day. The flowers of the African tulip tree, *Spathodea campanulata* (Bignoniaceae) (Fig. 12.15c), for example, are boat-shaped and provide water as well as nectar for their visitors

Fig. 12.15 Red flowers are characteristic of many bird-pollinated plants, while other bird- and bat-pollinated ones, such as many bananas (*Musa* spp.), may have red bracts acting as flags. (a) Mainly bird-pollinated *Combretum constrictum* (Combretaceae) from South America has red flowers with very exposed nectar sources; (b) *Odontonema strictum* (Acanthaceae) is hummingbird-pollinated and originally from South America, but now widely cultivated throughout the tropics; (c) *Spathodea campanulata* (Bignoniaceae), native to Africa, holds rainwater as well as nectar in its large boat-shaped flowers; (d) a banana, *Musa ?velutina* (Musaceae) from Thailand. Species of banana with erect flowers are bird-pollinated and those with pendant flowers are bat-pollinated.

(Rangaiah et al. 2006). The combination of floral characters in this species suggest that it may be undergoing a pollinator shift from bats to diurnal birds. The largely 'red-flowered' bananas, *Musa* spp. (Musaceae) (Fig. 12.15d), are variously pollinated by birds and bats (Itino et al. 1991), showing the relative ease with which this transition between pollinator guilds can evolve. In bananas the flowers themselves may be yellow but the bracts are usually red or purple.

J. H. Brown & Kodric-Brown (1979) studied a community of red tubular hummingbird-pollinated flowers in the White Mountains of Arizona, and found that hummingbirds did not visit particular species preferentially and often carried pollen from four or more plant species on their bills. Although most of the nine plant species secreted nectar at comparable rates, *Lobelia cardinalis* is nectarless, though still visited by the hummingbirds, and therefore achieves pollination by deception. Unlike many species, it does so through membership of a general colour guild with no specific model.

Rose & Barthlott (1994) noted that there was a strong correlation between pollen colour and pollinating agent among some 360 species of cacti distributed among 71 genera, with those pollinated by hummingbirds having dark red or brown pollen, in contrast to the paler coloured pollen of insect-pollinated species. The association was interpreted as a form of crypsis, with the darker pollen being less conspicuous against the dark bills of the pollinators and therefore running less risk of causing the bird to groom it off before it could deposit it.

Chapter 13

INTRA- AND INTERSPECIFIC COOPERATION, COMPETITION AND HIERARCHIES

As each man fell, throughout that long and awful day, he had propped him up, wounded or dead, set the rifle in its place, fired it, and bluffed the Arabs that every wall and every embrasure and loophole of every wall was fully manned.

P.C. Wren, *Beau Geste*, John Murray (as quoted by John R. Krebs 1977)

Mimicry, Crypsis, Masquerade and other Adaptive Resemblances, First Edition. Donald L. J. Quicke.
© 2017 John Wiley & Sons Ltd. Published 2017 by John Wiley & Sons Ltd.

INTRODUCTION

Here I deal with a rather diverse set of resemblances that are concerned in some way with duping conspecifics into being less aggressive towards the mimic, or allowing them access to food resources, territory or social status. For example, although it is not always the primary function, many instances in which males either resemble females or younger males probably function to reduce aggression from other males and, sometimes, appeasement is the only apparent function.

C.J. Barnard (1979, 1982) used the term social mimicry in a more precise way, meaning visual and behavioural similarities in mixed-species flocks of birds or schools of fish, and I deal with that topic in Chapter 8 (section *Schooling, flocking and predator confusion*), though it is likely that the similarities also maintain flock cohesion and may have other functions, such as making it harder for predators to single out odd-looking individuals.

REMAINING LOOKING YOUNG

Juveniles of many vertebrate species differ in appearance from older individuals, even if they have reached, or almost reached, adult size and sexual maturity. There are various reasons for this: a juvenile appearance may reduce intraspecific aggression and facilitate access to resources, or it may be involved in sexual deceptions to gain sneaky matings (see Chapter 11). It is possible that in some cases the retained juvenile appearance serves both functions.

Delayed plumage maturation

Adult males of approximately 30% of sexually dichromic bird species delay acquisition of the adult male plumage and often resemble females (Rohwer 1978, Foster 1987, Rohwer & Niles 1979, Lyon & Montgomerie 1986) though only in some cases is sexual mimicry involved. This delay is especially common among passerine birds but also occurs in other groups (e.g. Vergara & Fargallo 2007). Two hypotheses have been widely discussed to explain this. The status-signalling hypothesis postulates that the sex of the juvenile male is still apparent to the older ones, but that its colouration has evolved to indicate lower status and thus elicit less aggression, whereas the female mimicry hypothesis proposes that older males cannot distinguish them from females but because of their resemblance to females they avoid aggression and may be allowed to forage within the territory of an older male, which may increase their chance of breeding with older females. Hakkarainen et al. (1993) provide evidence using caged European kestrels, *Falco tinnunculus*, and how they responded to other individuals presented to them, that, at least in this species, the older males were unable to distinguish between females and female-like younger males. This may be the general situation because a similar finding for pied flycatchers, *Ficedula hypoleuca*, was described by Slagsvold & Sætre (1991).

It seems likely that in most cases this is connected with avoiding aggression from conspecific older males, which might be deceived into assuming they are females and therefore potential mates. One reason for this, if it is a deception, could be that by avoiding agonistic interactions with nearby fully adult males they might acquire information that will be useful in future reproductive years and/or gain direct access to resources by being tolerated within a territory. As suggested by Lack (1954), selection would favour delaying maturation if there was a high risk of mortality as a result of breeding during the first season and if potential fecundity increased with age. Furthermore, because females of birds are generally choosy about mates and have selected for characteristic adult male plumage, these young males are unlikely to get many matings, even if they tried, and as passerines lack a penis, rape is not a real option.

Rohwer et al. (1980) suggest that having female-like plumage gives young males the best opportunity for establishing a territory. Small birds that breed in temperate regions usually have rather short adult lives, and so territory acquisition and expansion is likely to be facilitated by the fairly frequent deaths of others, whereas in the tropics mortality prior to establishing a territory seems to be far higher, but once gained, the males often live for 10 or more years (Wiersma et al. 2007). Apart from territory establishment, being allowed to live within the established territories of older males might enable direct access to food resources. Rohwer (1978) tested this hypothesis by placing specimens of adult males, and the studliest and least studliest subadult male red-winged blackbirds, *Agelaius phoeniceus*, in the territories of fully adult males. Although the subadult red-winged blackbirds are not particularly good female mimics, the adult males responded most aggressively to mounted specimens of adult males and least aggressively to the least studliest subadult specimens, which seems to be a clear indication that they were deceived, though it is always possible that they did not think that the subadults were actually females. Procter-Gray & Holmes (1981) also tried to test Rohwer's (1978) hypothesis by examining reproductive success of young and older males of the American redstart, *Setophaga ruticilla*. In this species, although some first-year males remained unmated, those that did manage to pair actually produced slightly more fledglings than the older males. However, the young paired males tended to occupy different parts of the forest than older males and, therefore,

Procter-Gray & Holmes' (1981) results are not consistent with either the territory acquisition or the reproductive success hypotheses, and they propose instead that the less conspicuous female-like plumage of young males might be retained as it reduces predation, whereas older males benefit from having brighter colours by achieving higher mating success. Rohwer (1983) proposed a number of ways in which conflicting hypotheses about the function of female mimicry in birds might be tested.

Rohwer (1977) experimentally dyed the crown and chest feathers of subordinate males of the North American Harris sparrow, *Zonotrichia querula*, so as to resemble those of dominant males, to see whether it would alter their position in the pecking order; however it did not, and instead they suffered greater persecution. Two possible reasons were discussed for this failure in deception – that the dyed bird could not cheat because the other birds recognised who it was despite the colour change, i.e. hierarchy is socially controlled, or that despite appearing colourwise as a dominant the subordinant did not behave like a dominant bird, and this was called the incongruence hypothesis. Rohwer & Rohwer (1978) went on to test the two hypotheses by both dying the subordinates and administering them with testosterone, such that their behaviour was shifted to that of a more dominant type. The dramatic increase in social status with these doubly modified birds supported the incongruence hypothesis, and therefore in this species it appears that throat and chest and crown colour are an honest and uncheatable status signal.

Interspecific social dominance mimicry

The social mimicry of multi-species bird flocks described in Chapter 8 (section *Schooling, flocking and predator confusion*) may have another twist. Alfred Russel Wallace (1863, 1869) suggested that the smaller subordinate species in the complex may deceive other smaller and less competitive species that they are members of the dominant species, and that this would allow the mimics greater access to resources. This has become known as the interspecific social dominance mimicry hypothesis. Prum (2014) reviewed and illustrates a large number[1] of proposed instances of putative mimetic resemblances that do not involve either cuckoldry or aposematism and found that in general the evidence supported the hypothesis. The putative mimic species, on average, had mean body masses 56–58% of those of the models and, while this may seem a lot, it corresponds to a rather smaller (17%) difference

in linear dimensions, and Prum (2014) suggests that viewing birds from different distances could easily produce the same effect in species with poor distance judgement. Prum & Samuelson (2016) went on to explore the evolution of interspecific social dominance mimicry in cases with two and three species, noting that the mimicry would result in the model evolving always to separate itself from the ever-evolving mimic, or evolutionary traps "in which a subdominant species is evolutionarily constrained from evading mimicry by a third, subordinate mimic species".

A good example of this type of mimicry is provided by two species of woodpecker, the hairy and downy woodpeckers (Picidae), *Picoides villosus* and *P. pubescens*, respectively, which independent molecular work (Weibel & Moore 2005) has shown to be due to convergence. The downy, despite being very hard to distinguish from the hairy in the field, has less than half its body mass and would clearly lose out in any direct conflict. The resemblance is therefore interpreted as most likely being to allow the smaller downy to split food resources with the dominant, but duped, hairy woodpecker. Prum & Samuelson (2012) presented a game theory model to explain how such mimicry can come about which they named after these two birds, the Hairy-Downy game – in essence a variant on hawk–dove games.

A very similar mimetic relationship was noted by Wallace (1869), which involved pugnacious and relatively large friarbirds (*Philemon* spp.: Meliphagidae) and rather smaller orioles (*Oriolus*: Oreolidae) as mimics. These two birds, whose taxonomic status is still somewhat confusing, showed marked parallel geographic variation in plumage across the islands of the Indonesian archipelago, which confirmed that the resemblance was almost certainly mimetic rather than just coincidental. The mimetic relationship between orioles and friarbirds is, in fact, more complicated, with greater visual mimicry existing between those pairs where the size discrepancy between them is greatest. The orioles only mimicked the largest of the friarbird species on those islands where there was more than one species (J.M. Diamond 1982). Thus the bigger and therefore potentially more aggressive/dangerous the friarbird, the more suitable it was to be a model. As J.M. Diamond (1994) suggested, many other systems may need to be examined more carefully for instances of multipurpose, or a "two-faced" mimicry, as he calls it, that may have been overlooked in the past.

Bird song and alarm call mimicry – deceptive acquisition of resources

Birds may imitate (mimic) both specific calls of other species and also whole courtship or territorial songs, and nearly 14% of species are known to do so (Baylis 1982,

---

1. Fifty phylogenetically independent examples involving 60 model and 93 mimic species, subspecies and morphs from 30 families.

L.A. Kelley et al. 2008). Vocal mimicry by birds can be both extensive and highly accurate. Some birds, such as Indian hill mynas, *Gracula religiosa* (Sturnidae) and various parrots, perhaps most notably the African grey parrot, *Psittacus erithacus*, are frequently kept as pets because of their ability to learn to 'talk' and have sometimes been thought to do so in an intelligent and helpful way. Some larger birds and others that do not adapt so well to captivity are also especially good vocal mimics, e.g. common Eurasian starlings, *Sturnus vulgaris*, the American mockingbird, *Mimus polyglottos* (R.D. Howard 1974) and Australian lyrebirds, *Menura novaehollandiae* (Dalziell & Magrath 2012). Pasteur (1982) considered it unfortunate that the vocal imitation by parrots was called mimicry, as he did not consider that the behaviour itself might be selected because it could deceive another operator. Dalziel et al. (2015) suggested a modification to Vane-Wright's (1980) definition of mimicry to enable bird vocal mimicry to be separated from other forms of vocal imitation that many birds exhibit but which do not appear to be targeted at a given receiver in a way that changes their behaviour (see Table 1.4).

There has been much speculation over the years as to the function of song mimicry but very little by way of experiment or analysis (Baylis 1982, L.A. Kelley et al. 2008). Although several theories have been proposed to explain this phenomenon, there is no strong evidence to favour any of the so-called functional explanations, i.e. ones that suggest that selection has favoured individuals that imitate the songs of other species, but no single explanation is likely to fit all cases. L.A. Kelley et al. (2008) conclude that in many cases birds are simply making mistakes, as they are evolutionarily programmed to learn and repeat sounds they hear around them over a certain period of their development. There seems, in general, to be little evidence for Rechten's (1978) suggestion, based on John Krebs' (1977) earlier exploration of intraspecific song repertoire mimicry (see section *Beau Geste and seeming to be more than you are*), that heterospecific song mimicry could be generally advantageous in deterring potential competitors, because the individuals likely to be competing most with you would be expected to be conspecifics. However, large repertoires, including heterospecific songs or song elements, could still possibly deter heterospecific competitors (R.D. Howard 1974) as there is often considerable dietary overlap between species in a community, as is clearly shown by the diversity of songbirds that will visit a feeding table in winter.

Superb lyrebirds, *Menura novaehollandiae*, are interesting in that females are very good mimics of heterospecific song as well as alarm calls (Dalziell & Wellbergen 2016). Importantly, the types of call they produce is context-dependent: females are silent during courtship at male leks,

and employ most types of mimetic songs and calls during nest defence, while during foraging they mimic predator vocalisations and non-alarm vocalisations of heterospecifics. Female lyrebirds defend territories and males do not play any role in brood care, therefore it seems likely that mimetic vocalisations made during foraging have a functional role, possibly as part of female–female competition for breeding territories.

Young male American song sparrows, *Melospiza melodia* (Melospizidae), develop their call repertoire by learning songs from more mature males in surrounding territories, preferentially selecting songs that different neighbouring males shared (Beecher et al. 1994). By doing so, they suggest that the new young bird, which has not yet gained a territory, is positioning itself optimally for when it can set up its own territory, for example by replacing one of its tutor males which may have died, by seeming to be like the familiar neighbour that other older males recognise.

'Mimicry' of songs and calls does not necessarily have to be concerned with deception but might sometimes be a result of convergence. However, in some cases the call and song copying does appear to have a deceptive role. American song sparrow songs are also imitated by sympatric white-crowned sparrows, *Zonotrichia leucophrys*, and vice versa (Catchpole & Baptista 1988, Baptista & Catchpole 1989). In this case it is fairly well demonstrated that the role is associated with interspecific competition for what are probably broadly overlapping food resources, so each species is attempting to persuade members of the other species to avoid its territory, though sometimes aggressive encounters occur as a result.

Møller (1988) described how great tits, *Parus major*, frequently make alarm calls during the winter when they are typically members of mixed flocks, even though there are no predators around. Instead the calls, which can be directed to both conspecifics and heterospecifics, were made when there was strong competition between individuals for a concentrated food resource, and especially when there was the greatest need for food, such as in very cold weather. Similar deceptive use of alarm calls to minimise competition for food has been described in drongos (Dicruridae) (Flower 2011, E. Goodale et al. 2014). However, the drongo investigated in detail by E. Goodale et al. (2014) actually produces quite a few different alarm calls that may well serve different functions. These are danger calls, which includes their species-typical alarm vocalisations, imitations of calls made by predators themselves, and imitations of mobbing vocalisations made by other, non-predator species. Use of these different danger calls may well be context-specific. Flower et al. (2014) found that the dupes of the false alarm calls of fork-tailed drongos, *Dicrurus adsimilis*, get accustomed to and stop responding to the alarms, but the drongos monitor the dupe responses and change to mimicking the calls of a different species when original dupes are no longer behaving as 'desired'. Fork-tailed drongos even mimic the alarm calls of meerkats, *Suricata suricata*, causing them to abandon prey, which allows the drongo to steal it (Flower 2011).

WICKLERIAN MIMICRY – MIMICRY OF OPPOSITE SEX TO REDUCE AGGRESSION

Pasteur (1982) coined the name Wicklerian mimicry for cases in which a feature of one sex (usually males), be it behavioural, morphological or both, which acts as a submissive signal to the opposite sex, is mimicked by members of the opposite sex so as to avoid or minimise conflict. Why this should not also include cases where the same aim is achieved by juveniles or subadults of the same sex is unclear to me as I think they are almost the same thing, except that adult mimics might be potential threats to paternity.

Female resemblance in male primates

The primary example cited by Pasteur (1982) and perhaps the most stunning, involves the brilliant swollen red buttocks (Fig. 13.1a) of male hamadryas baboons, *Papio hamadryas*, which closely resemble the sexual skin of females in oestrus (Fig. 13.1b). When subordinate males are threatened, they present their red rears to the dominant, threatening male, which signals their subordinate position, and the dominant male may even mount them for homosexual copulation, thus reinforcing his dominance (Wickler 1967). While the buttocks of the hamadryas baboon are by far the most spectacular example of this type of mimicry among the primates, males of several other species show lesser degrees of female genital mimicry. Wickler (1967), whose study was based to a large extent on published anatomical descriptions, was unable to be totally sure of all possible cases, since the literature often omitted to describe the colour of the male rump patches, even if they were in fact conspicuously red just as in the females. Wickler's (1967) own study of the vervet monkey, *Cercopithecus pygerythrus*, in Tanzania showed that females undergo a cyclical change in genital colouration associated with oestrus and that the colouration of males closely matches that of the females on heat. In oestrous females the perineum is deep red, the protruding vulval edge is bright blue and the clitoris which is bright red, is clearly visible, whereas at other times the females' genitalia are far paler and the clitoris and vulval edge are hardly visible. In males, the perineum is

Fig. 13.1 Wicklerian mimicry: hamadryas baboon, *Papio hamadryas*, male with brightly coloured buttocks (a) that closely resemble the swollen sex skin of a female in oestrus (b), and in doing so appease more dominant and potentially aggressive males. (Source: a, © iStock.com/maratr; b, © iStock.com/Jean-nicolas Nault.)

similarly red but the blue of the female's vulva is replaced by the blue scrotum; the male's penis is bright red, corresponding with the female's clitoris. Such exact correspondence and the occurrence of bright blue, a rare colour in animals, suggests that this is another example of Wicklerian mimicry, though behavioural observations to confirm the function of the resemblance still remain to be carried out.

In *Colobus verus*, the olive colobus monkey, juvenile males are also frequently mistaken for females because they exhibit a pronounced circumanal swelling that resembles a female's sex skin, even to the extent that it has a pseudovaginal orifice (Dixson 1983, Napier & Napier 1985). Mimicry is again strongly implicated because the structures involved in the resemblance are non-homologous. For instance, the pseudovaginal orifice or pseudovulva results from the presence in males of a sulcus across the swollen area, and the mimicry is further enhanced by the inconspicuous penis and reduced scrotum in juvenile males. Wickler (1967) proposed that such a resemblance reduces harassment by dominant males.

Social appeasement by female mimicry in an insect

Peschke (1987) demonstrates that young males of the rove beetle, *Aleochara curtula*, release female sex pheromone as a means of avoiding aggression from other male conspecifics. Males of this species are especially aggressive towards one another, biting at limbs of competitors as well as trying to smear opponents with abdominal defensive secretions. The female pheromone mimic blend produced by young males before they reach sexual maturity immediately suppresses the aggressive responses of older males. However, there is a trade-off because females will not accept males producing female pheromones as mates and therefore pheromone production must cease prior to courtship and mating.

HYPERFEMININITY IN PREREPRODUCTIVE ADOLESCENT PRIMATES

In a considerable number of Old World monkeys (catarrhines), adolescent females display exaggerated versions of the signals displayed by adult females in oestrus. It has been proposed that these may constitute super-normal sexual stimuli for males and they "stimulate the most enthusiastic mating responses from adult males" (C.M. Anderson & Bielert 1994). Whether deception is involved is a different matter – in some cases the

adolescent's signal is almost certainly associated with reducing aggression and/or enhancing competitive ability, but as adolescent females are relatively unlikely to become pregnant, the males are certainly being deceived in the sense that they are not going to achieve immediate fertilisations. However, if the bond continues until the female has reached sexual maturity then the males may be compensated.

MIMICRY OF MALE GENITALIA BY FEMALES

The case of the spotted hyaena

Not only do males mimic females in the context of submission in social interactions, but in spotted hyaenas, *Crocuta crocuta*, females possess a highly erectile pseudopenis derived from the clitoris, complete with surrounding prepuce, and a pseudoscrotum derived from the fused labiae with an adipose tissue-filled swelling (Fig. 13.2a,b) (Racey & Skinner 1979, Frank et al. 1990, 1995, Frank & Glickman 1994, Glickman et al. 1998, 2005, Cunha et al. 2003, Place & Glickman 2004). The possible significance of this genitalic mimicry has been pondered for a long time, even since Aristotle in the fourth century BC (Wickler 1968), but only recently has its function as a submissive signal been clearly demonstrated (East et al. 1993), and it is indeed often displayed by very young animals (Fig. 13.2c). The genitalia of female spotted hyaenas are, in fact, rather more modified than it may at first appear, with the urogenital opening being forced anterior to the base of the pseudopenis by the fused labiae. This has two important consequences for reproduction. First, males are only able to mate with the female's full cooperation but, perhaps more importantly, the birth of the female's first litter involves a large rupturing of the pseudopenis and frequently massive mortality of the pups. Such high costs signify that there must also be considerable fitness advantages associated with the male genital mimicry in spotted hyaenas.

Mimicry of male genitalia in other mammals

Possible mimicry of male genitalia also occurs in a number of other mammals, notably platyrrhine monkeys from South America, in which females frequently have such greatly elongated clitorises and pseudoscrota formed from the labia minora that the sexes may be difficult to tell apart in the field (Wickler 1967). As far as is known, the pseudopenis cannot be erected in any of

Fig. 13.2 Penis mimicry in the spotted hyaena, *Crocuta crocuta*. (a) Erect penis of subdominant male; (b) erect clitoris of adult female; (c) young, zoo-bred individuals showing erected genitalia, the individuals being thought (based on penis/clitoris morphology) to be, from left to right, male, female and male. (Source: a,b, Adapted from Frank et al. 1990 with permission from John Wiley & Sons; c, Glickman et al. 2005. Reproduced with permission from Elsevier and S. Glickman.)

these in the same way that it can in the spotted hyaena (see earlier), though since there is often a pseudoscrotum it seems likely that they have evolved as part of some form of mimicry.

Some female genital mimicry of males is also found in the fossa, *Cryptoprocta ferox*, a Madagascan viverrid carnivore, though it varies between individuals and is transient (C.E. Hawkins et al. 2002). In this species, young females have a very high testosterone titre and the associated aggressiveness may serve to help them compete with their brothers. It seems, in this case therefore, that the visual resemblance may be a hormonal by-product (see Clutton-Brock 2009), the main or only selective advantage arising from their behaviour.

Phallic mimicry by males

Without doubt the most obvious case of automimicry is that of the mandrill's brightly coloured muzzle, with its blue skin with a bright red median strip, and bright flaxen beard. These closely resemble in shape, proportion and colour the male's scrotum, penis and surrounding hairs, though they are a little larger (Fig. 13.3a). Desmond Morris (1967) comments that:

> When the male mandrill approaches another animal, its genital display tends to be concealed by its body posture, but it can still apparently transmit the vital messages by using its phallic face.

Pasteur (1982) coined the term Wicklerian–Guthrian mimicry for this one particular example, and suggested that

Fig. 13.3 Body self-mimicry. (a) Male mandrill, *Mandrillus sphinx*, a forest baboon, whose muzzle is believed (convincingly to me) to be mimetic of his genitalia and probably plays a role socially in emphasising his dominance; (b) male proboscis monkey, *Nasalis larvatus*, showing modified nose that has been suggested to be either a penis mimic or a mimic of the large nose of a human, or even [presumed] *Homo erectus*, because of the potential danger that a tool-using intelligent ape might be to a predator. (Source: a, © iStock.com/Gleb_ Ivanov; b, Dr Frank van Veen. Reproduced with permission from Dr van Veen.)

because it is used in dominance it should be considered as a conjunct aggressive mimicry. Renoult et al. (2011) show that colours of the snout maximise contrast against background vegetation and that, in terms of dominance signal, the key factor is the red–blue contrast.

The large nose of the South-East Asian proboscis monkey, *Nasalis larvatus* (Fig. 13.3b), may be a penis mimic though there are numerous other mimetic or functional interpretations. The nose is considerably larger in adult males than in females, though the latter's nose is still very large in comparison to other monkeys. Interestingly, unlike in most other colobine monkeys, male proboscis monkeys frequently perform a display with the penis fully erect, with legs apart, perhaps indicating the importance of this organ as a signal (Yeager 1992).

APPETITIVE (FORAGING) MIMICRY

Sick (1997) discusses what, despite them being congeners, is a probable case of mimicry between two Neotropical toucans, the white-throated toucan, *Ramphastos tucanus cuvieri*, and the smaller channel-billed toucan, *Ramphastos vitellinus culminata*. These two species have a very similar colour pattern whether viewed from above or below (see figures 9 and 10 in Sazima 2010), and

both forage together in the same trees. It is proposed that the smaller bird deceives the larger into accepting it as a conspecific and thus allowing it to compete with them in food trees, whereas if it were distinguishable, as for example other toucan species are, the white-throated toucan would expel them.

Appetitive mimicry and deceptive use of alarm calls

Tramer (1994) described an interesting example in which the North American white-breasted nuthatch, *Sitta carolinensis* (Sittidae), deceives conspecific individuals by producing alarm calls. The other birds fly off in response and so allow the deceiver to gain easier access to bird-feeders. South American capuchin monkeys do much the same thing (B.C. Wheeler 2009). Similarly, the South American tanager, *Lanio versicolor*, and the antshrike, *Thamnomanes schistogynus*, both of which lead mixed-species foraging flocks, appear to use their alarm calls deceptively to distract the attention of other flock members, increasing their own chances of capturing flushed prey (Munn 1986), but no doubt because these species also act as sentinel birds that give true warnings of approaching hawks, they can get away with some degree of deception.

Fig. 13.4 Weapon automimicry. (a) Grant's gazelle, *Nanger granti*; (b) Beisa oryx, *Oryx gazelle beisa*; (c) head of male pronghorn, *Antilocapra americana*, showing the dark lines that extend anteriorly in front of the horns so exaggerating the apparent length of the horns; (d) two hornless females showing their dark and hook-tipped ears. (Source: a, Robbert van der Steeg 2008. Reproduced under the terms of the Creative Commons Attribution Share-Alike Licence CC BY-SA 2, via Flickr; b, ChrisHodgesUK 2007. Reproduced under the terms of the Creative Commons Attribution Share-Alike Licence CC BY-SA 3.0, via Wikimedia Commons; c, MONGO 2006, via Wikimedia Commons; d, Jollymama 2012, via pixabay.)

Beau Geste and seeming to be more than you are

J.R. Krebs (1977), who was researching nesting and territoriality in great tits, *Parus major*, in Wytham Wood near Oxford, sought to understand why males, which are not territorial in this species, had huge song repertoires. He suggested that the birds were using conspecific call diversity as a way of assessing the density of potential competitors in a local piece of woodland. On hearing a high diversity of songs, new birds are less likely to stay because they will perceive that the area is already densely occupied, and therefore individual resident male birds may reduce the likelihood of new competitors entering their zone by 'pretending' to be more than one individual. This might, of course, have led to an arms race, so instead of just singing a couple of songs they have been selected to sing increasingly larger repertoires. The "Beau Geste hypothesis" as Krebs (1977) had dubbed it (see quotation at beginning of this chapter), has been tested in several species with different, though not generally supportive, results (R.D. Howard 1974, Dawson & Jenkins 1983, Yasukawa & Searcy 1985), and other explanations may have been proposed, especially where repertoires include imitations of songs or song features of neighbours.

Appearing older than you are

Several authors have suggested that young animals, or generally less fit individuals, may benefit by mimicking older or fitter ones, in other words 'age class mimicry' (Weldon & Burghardt 1984). Such deceptions may facilitate foraging or mate acquisition. Wynn et al. (1998) suggest that some juveniles (approximately one quarter) of the African penguin, *Spheniscus demersus*, moult into adult plumage (see Fig. 10.3d) to reduce aggression from adults and to be permitted to join cooperative feeding groups. The aggression by adults may be viewed as them maximising their own fitness by excluding inexperienced hunters from their communal feeding groups. However, the authors note that probably only the fittest juveniles moult into adult-like plumage so, although they are inexperienced at cooperative hunting, the level of deception is possibly less than if less fit youngsters achieved the same deception.

WEAPON AUTOMIMICRY

Many large mammals use their ears for intraspecific signalling, and in the artiodactyls it may be no coincidence that the ears are anatomically close to their horns. Their ears are usually boldly marked and their hair fringes often increase their similarity to horns (Guthrie & Petocz 1970). The horns themselves may be accentuated by facial markings which seem to extend their length, for example in the Grant's gazelle, *Nanger granti* (Fig. 13.4a), oryx, *Oryx* spp. (Fig. 13.4b) and sable antelope, *Hippotragus niger*. Warthogs, *Phacochoerus aethiopicus*, have white whisker tufts behind their tusks (elongated canine teeth) which may effectively exaggerate the animal's potential to do harm.

In the North American pronghorn antelope, *Antilocapra americana*, the males have distinctive horns with an anterior prong and dark lines on the head and muzzle that extend their apparent length, especially in a front-on view if its head is lowered (Fig. 13.4c). Female pronghorns may be horned, but when horned they lack the anterior prong, some only have short horns and some remain completely hornless. However, the ears of the females are particularly pointed, with the tip hooked such that from a distance they look like they possess male horns (Fig. 13.4d). Male Chinese water deer, *Hydropotes inermis*, have well-developed protruding white canines, whereas females do not, but older females develop a white hairy tuft in the same position with a similar appearance to male tusks.

The last example I want to discuss in this section is one that I find very hard to classify. Backwell et al. (2000) describe a case in a fiddler crab, *Uca annulipes*, that involves individuals with an inferior weapon compensating by making it bigger. In fiddler crabs, males have a small claw used for feeding and a large one used in agonistic displays against other males and in courtship. The size of the larger claw might therefore, to be an honest signal of the male's potential. It turns out, however, that males that have lost their claw in some misadventure can regenerate it over successive moults, and although the replacement is structurally inferior to the lost original, it is bigger. Because the animals assess the potential of a male by the size of its larger claw these regenerators have evolved a way to cheat, because their claws are no longer honest signals. Indeed, the authors found that more than 40% of males had regenerated their large claw and yet the size deception is apparently still effective.

Chapter 14

ADAPTIVE RESEMBLANCES AND DISPERSAL: SEEDS, SPORES AND EGGS

I value my garden more for being full of blackbirds than of cherries, and very frankly give them fruit for their songs.

Joseph Addison, *The Spectator*, 1712

INTRODUCTION

Whether various mimicries that are involved in propagule dispersal should be regarded as sexual mimicries or aggressive ones is debatable Roper (1993), so I treat this as a separate category, though it seems that quite a few such mimicries are more complicated than they might at first appear. Mimicries that have evolved to achieve the dispersal of propagules (as opposed to pollen grains which are gametes) are mostly confined to plants and fungi, and possibly bacteria, and are similar in many ways to those involved in achieving pollination through deception, some even involving pollinators.

FRUIT AND SEED DISPERSAL BY BIRDS

In this section I discuss seed dispersal syndromes. It may be questioned whether they clearly fall under the remit of this book's title, specifically 'adaptive resemblance', because it may be that their appearances are all individually adaptive and not benefitting through similarity to those of other species. On the other hand, one could lump most of them under the adaptive resemblance notion that they have all evolved the trait of not looking like their leafy or unripe fruit backgrounds. The fact that there are significant interspecific correlations in fruit colour does suggest that, in at least some cases, the similarity is due to mutualism (Schaefer et al. 2014). Whatever the reasons, it is an interesting topic with vastly more work needed to understand it and thus far there has been very little experimental work, in particular on the UV aspect (but see Schaefer et al. 2007). Indeed, the evolutionarily important seed dispersers of many plants are unknown, and although it might seem reasonable to suggest that if species X eats a lot of fruit they must be the important ones, that might not necessarily be so, as what counts is whether dropped seeds reach maturity and result in fit plants. However, if some important seed dispersers have innate colour preferences, or in a region they develop a colour preference, then there may be selection pressure on other plants to conform to that colour until disperser capabilities become saturated.

Birds are, in general, the most important agents in seed dispersal, and plants that rely on birds for this purpose usually provide the latter with a reward, i.e. the fruit. The seeds within fruits are typically highly resistant to digestion by the natural frugivorous dispersal agent and necessarily remain viable after their brief passage through the gut. Indeed, a good number of seeds will not germinate until they have been through the digestive process.

The majority of fleshy fruit become contrastingly coloured when ripe, often orange, red, purple, black, or sometimes yellow, as in bananas, although overall, fruit have achieved only half the range of colour variation that flowers have (Stournaras et al. 2013). Those that do not appear to contrast to the human eye may have other mechanisms to attract dispersers, such as pure olfactory signals or maybe UV reflectance, but very little seems to be known about this area. Using an avian visual model, Schaefer et al. (2007) found that, in general, fruit/seed contrast to the background was greater in the UV (at least in bright light conditions) and, as with intersexual signalling (epigamic signalling), this might give passerines an advantage over other groups of bird. However, many small dark fruit may not be that conspicuous from a distance against a complex background of foliage (Schaefer 2011). Furthermore, it is not necessarily just the pulp of the fruit that matters, the pigment of the skin itself may also be important. Schaefer (2011) shows that fruit anthocyanin pigments are important to some dispersers such as the European blackcap, *Sylvia atricapilla*, since they function as antioxidants and increase the bird's immune response. It is therefore likely that the birds are deliberately selecting for fruit with higher anthocyanin content, although birds other than the seed dispersers may parasitise the system by also eating the fruit with more anthocyanin.

Willson & Whelan (1990) discuss the various hypotheses that have been put forward to explain why most fleshy fruit are brightly or conspicuously coloured and the evidence for and against each. The hypotheses can be summarised (abbreviated and slightly modified) as:

- fruit-eating, seed-dispersing birds have innate colour preferences,
- fruit colours have been selected to be conspicuous from long distances,
- fruit colours indicate fruit maturity (ripeness), thus concentrating frugivory on those whose seeds are capable of germination,
- fruit colours facilitate the birds' recognition and learning of appropriate resources,
- fruit providing only low levels of reward may mimic ones providing seed dispersers with higher nutritional reward,
- red fruit are inconspicuous to visually foraging arthropods that lack red spectral sensitivity,
- fruit colour may signal unsuitability to various vertebrate foragers that would be unsuitable dispersers, perhaps indicating poisonous properties,
- colours, or rather the chemicals responsible for them, have other protective functions against, for example, plant pathogens, or may be an unimportant by-product of such defences,
- dark-coloured fruit may absorb more light energy, becoming warmer and so ripening faster, thus reducing exposure to unsuitable dispersers or arthropod attack,

- colour diversity may limit competition between different species of seed disperser,
- particular pigments may themselves be nutrients important for particular dispersers, resulting in special mutualisms.

It is obvious that any or all of these might be true in particular cases. That the pigments are sometimes restricted to the epidermis and in other cases distributed throughout the pulp may indicate different roles or costs.

Lomáscolo & Schaefer (2010) performed a large analysis of fruit colours using an avian visual model and both dichromatic and trichromatic primate visual models. They found that all seed dispersers were able to distinguish reliably those fruits that they are the normal dispersers of, and also that fruit colouration was explained better by the type of seed disperser than by plant phylogeny. Thus their data broadly support the signal convergence hypothesis, which states that fruit colour should evolve to maximise the probability that they are dispersed by the appropriate type of animal. Although it was suggested for a long time that primates and birds dispersed the same coloured fruits, Lomáscolo & Schaefer (2010) found that primates dispersed more green fruit and birds dispersed more blue ones. The general assumption from workers in the north temperate region that bird-dispersed fruit are red does not hold up in the Neotropics at least, where the majority of ripe fruit are black (Wheelwright & Janson 1985). Interestingly, while red may be a widespread warning signal when it concerns insect prey, the opposite might be true when it involves fruit.

In seasonal habitats the time of year that particular coloured fruits are most abundant is also highly seasonal (Camargo et al. 2013). K.C. Burns & Dalen (2002) surveyed North American bird-dispersed fruit and found that red fruit predominated during the earlier part of the year when leaves are green, and black fruits were mostly available when the leaves were turning yellow or red. With manipulation experiments and artificial backgrounds they found that the high contrast of red against green led to high fruit removal, with a lesser effect of black against red-orange backgrounds. While each species of plant is therefore maximising its own potential for seed dispersal, it is also likely that with multiple plants in an area, either of the same or different species but with the same signal, they might be achieving a better seed dispersal because of a magnet effect on seed-disperser birds and also any learning that birds might do to find fruit more easily.

V. Schmidt et al. (2004) note that there is still no clear explanation for the strong and cosmopolitan prevalence of red and black fruits and seeds. One might ask why bright blue, for example, or yellow are nowhere near so common yet both would be quite conspicuous – indeed, blue is a rather rare colour in nature. V. Schmidt et al. (2004)

speculate that the prevalence of red and black colours in fruit has nothing to do *per se* with the colours themselves but instead with creating a contrast such that the seed disperser can more easily locate them. Red and black fruits were shown to have the strongest contrasts against foliage under natural conditions and ambient lighting. In their experimental tests using four bird species they found that they strongly preferred contrasting red-green or black-green fruit displays over uni-coloured red, green or black ones, and the birds showed no preference for any particular hues. Their measurements of fruit reflectance showed that most species had little UV 'colour' but some black fruit did have a marked UV reflectance component so these might be termed 'UV coloured'. White fruit reflected all light visible to humans plus UV light more or less equally but, in contrast, different natural plant backgrounds reflected little UV.

The temperate blackcap, *Sylvia atricapilla*, is a small omnivorous Eurasian bird that frequently feeds on small berries. Adults show no particular colour preference, but V. Schmidt & Schaefer (2004) demonstrated that hand-reared birds showed a significant preference for red artificial fruit over blue, green, yellow and white. Perhaps significantly, in the area from which the birds originated, 53% of appropriate (i.e. bird-dispersed) edible fruit were red during the summer when the fledgling birds would be foraging on their own. Thus their innate preference would seem likely to increase their initial foraging success until they had gained more experience, but because older birds had lost the preference it is unlikely that it would, at least in this species, act as a strong selection pressure on fruit colour. Older blackcaps in the Mediterranean area, however, learn to preferentially select less brightly coloured fruit (less chromatic ones) because in that region dark fruit is generally associated with a higher lipid content, e.g. olives (Schaefer et al. 2014).

Many plants that have black ripe fruits or seeds seem to advertise them by having surrounding objects that are bright red, or the seeds themselves may be bright red (Wheelwright & Janson 1985). The fruit of many species that are black when ripe pass through an intermediate red stage and frequently the individual fruit (called a drupe) mature at a different rate so the black ripe ones may be conspicuous against a background of red ones, the so-called 'flag' hypothesis. In the genus *Ochna* (Ochnaceae), a small cluster of large black fruit are arranged around an enlarged and vivid red receptacle, but in this genus the fruit transform more or less directly from green to black (Fig. 14.1a) so there can be little doubt about the receptacle forming a flag. The same effect is achieved by many other species but often in different ways (Fig. 14.1b). In many plants glossy black, exposed seeds are rendered tempting and conspicuous by contrasting orange/red edible arils that only partially conceal the seed, e.g. South-East Asian tindalo tree, *Afzelia rhomboidea* (Fabaceae), several *Paeonia* (e.g. *P. clusii*), various

Fig. 14.1 Contrasting colours, flags and seed maturation. (a) Mickey-mouse plant, *Ochna integerrima* (Ochnaceae), drupelet showing the bright red open receptacle acting as a flag to advertise the seeds and the direct transition from green to black at maturation of the seeds themselves; (b) lomentaceous fruit of *Clerodendrum* sp. (Lamiaceae), with black seeds contrasting against disarticulated pericarp; (c) *Alpinia* sp. (Zingiberaceae), Vietnam; (d) unidentified Vittaceae from Thailand showing red stalks leading to black berries; (e) *Amischotolype* sp. (Commelinaceae), showing orange-red aril of seed protruding from purple inflorescence, Thailand; (f) open seed pod of *Sterculia quadrifida* (Sterculiaceae) from Australia; (g) coral vine, *Abrus precatorius* (Fabaceae), showing contrasting black base to bright red, and very hard seeds without an aril reward. (Identifications of c and e by Tim Utteridge, RBG Kew.) (Source: f, Mark Marathon 2012. Reproduced under the terms of the Creative Commons Attribution Share-Alike Licence CC BY-SA 3.0, via Wikimedia Commons.)

Fig. 14.1 (Continued)

Clerodendrums, some *Acacias*, etc. Some plants, such as the vine shown in Fig. 14.1d and European elderberry, *Sambucus nigra*, have vivid red stalks (peduncles) leading to their black berries, the colour being most intense when the fruits contain their highest sugar concentration. In the latter species, the red anthocyanin pigments have been suggested to have more of a protective than advertising role (Cooney et al. 2015), but this was not the conclusion of Z. Wang & Schaefer (2014), who found that captive blackcaps, *Sylvia atricapilla*, selected infructescences on the basis of peduncle colour rather than fruit colour. The birds were, however, unable to detect artificially modified infructescences with bright red peduncles but low-sugar berries because they consume the berries whole and therefore cannot detect their sugar content by taste.

The scarlet bean, *Archidendron ramiflorum* (Fabaceae), goes one better in that the outside of the pod is bright red but, when it opens, the black seeds are contrasted even more strongly with the exterior red, and the peanut tree, *Sterculia quadrifida* (Sterculiaceae), has a red pericarp but the black seeds are contrasted even more against a yellow line (Fig. 14.1f), while the African mgambo tree, *Majidea zangueberica* (Sapindaceae), has the inside of the pericarp red but the outside whitish. There are numerous other examples, such is the frequency of evolution of this black and red signalling system.

Wheelwright & Janson (1985) note that in the Neotropics fruit that are black, brown, blue or green when ripe tended to be associated with contrastingly coloured accessory structures (bracts, peduncles, persistent calyces). In contrast, red, orange, white and yellow fruit tend to occur alone, without contrastingly coloured accessory structures. It may be that much of the black pigment formation has an intermediate reddish phase, but in some plants this is only transient such that effectively there may only be green and black fruit present in an infructescence most of the time. In the common European blackberries, *Rubus* spp., fruit maturation passes through a prolonged red phase before the ripe black fruit is ready to be consumed. When less than 100% of the fruit on a bush are black, the black fruit present are nearly 100% fully ripe, indicating that in this situation, birds quickly consume the fruit as they become ripe, but when nearly 100% of the fruit on a plant were black, and so birds had not been eating them, many were deteriorated, withered and infected by pathogens (Greig-Smith 1986). These data are concordant with the flag hypothesis.

Willson & Melampy (1983) tested the flag hypothesis of mixed red and black fruiting displays with manipulative experiments using wild black cherry, *Prunus serotina*, which has infructescences with red unripe and black ripe fruit, and pokeweed, *Phytolacca americana*, which has ripe black fruit against red stems. In both species, fruit removal was greatest when there was a red-black colour contrast, especially in forest gap situations where there was more ambient light. Greig-Smith (1986) used Willson & Melampy's (1983) data to support his flag hypothesis, but re-analysis by Janson (1987) showed that the data did not provide unequivocal evidence that birds actually prefer bicoloured fruit displays because of the potentially higher mean quality of the fruit pulp they provide. In a further test of the flag hypothesis, Fuentes (1995) carried out experiments with the rather unusually coloured infructescences of the Mediterranean shrub *Pistacia terebinthus* (Anacardiaceae), which have bright red unripe fruit and green ripe ones. The bright red-coloured unripe fruit would be expected to act as a flag to seed dispersers and, indeed, in a comparison of infructescences with only ripe fruit and ones with ripe and unripe ones, wild birds removed more fruit from those with both ripe and unripe fruit, supporting the flag hypothesis. However, in a second experiment comparing fruit removal from a mixed colour infructescence and ones from which all unripe fruit had been removed, Fuentes (1995) demonstrated that a higher proportion of ripe fruit were taken in the latter case, indicating that the presence of unripe fruit can also act as a handicap making it harder for birds to eat the ripe ones. I think, surprisingly, Schaefer et al. (2007) found that fruit colour in 130 species of bird-dispersed fruit were no more visually contrasting with their backgrounds than the fruit would be against other general plant backgrounds, therefore they showed no sign of having been selected for maximising their visual contrast to attract dispersers.

Willson & Thompson (1982) hypothesised that small fruit (fruit displays) should be more conspicuous colour-wise, than large ones and that particularly conspicuous displays (in the temperate region) ought to be more common when seed-dispersing birds are fewer in number, because of the increased selection to attract dispersers. They found no evidence that smaller displays were more conspicuous than larger ones, but did find strong support for the displays being more conspicuous when dispersers were least abundant. It has also been hypothesised that plants, particularly temperate deciduous species, may have evolved various foliar flags, and specifically early autumn leaf colours, to signal the presence of fruit. However, in a survey by Willson & Hoppes (1986) of fruiting species in a deciduous forest in Illinois, USA, no evidence was found of either a particular annual succession in fruit colours or of early autumn leaf colour being associated with fruit ripeness. Manipulation experiments in which they artificially 'flagged' some individuals of three plant species showed no difference in fruit removal between treatments and controls. Another aspect of the fruit conspicuousness discussion that is far less researched, especially in natural communities, is that conspicuous colour could also attract non-dispersing, seed predator insects.

It is not deceptive, of course, but from the point of view of seed dispersal it does not matter to the plant that its seed disperser is also dispersing seeds of other species that might be in the vicinity at the time. Evolution will select always to favour those traits that lead to the seed's and seedling's highest chance of reaching reproductive maturity.

Warningly coloured fruit

Lev-Yadun et al. (2009) suggests that in those cases in which fruit progress through a red phase before reaching a different colour when they are fully ripe, the red colouration could be a warning signal. Some fruit, including walnut, *Juglans regia*, olives, *Olea europaea*, quinces, *Cydonia oblonga*, and several species of fig, *Ficus* spp., have conspicuous white markings when they are unripe and Lev-Yadun (2013) similarly postulates that these constitute a warning signal. My own experience of touching unripe walnut flesh to my lips would strongly support this, in that it was very painful, whereas quince was just incredibly sour, and olives, of course, need considerable processing before they become edible, at least to humans.

Fruit mimicry by seeds

A few, mostly tropical, plants, for example *Erythrina*, *Ormosia* and *Abrus* species (Fabaceae), produce brightly coloured red or black-and-red seeds that lack a fleshy (aril) reward (Ridley 1930, Williamson 1982, Peres & van Roosmalen 1996, Foster & Delay 1998). In A. *precatorius* (Fig. 14.1g), for example, the seed bicoloured red with an extremely contrasting black base. Two hypotheses have been put forward to explain this: (1) that they are simple mimics that are deceiving avian dispersers about their reward, and (2) that some of these seeds are exceptionally hard (e.g. those of *Ormosia*) and they act as grit to help break up other food in the birds' crops. If the latter is true, then the birds benefit in a different way and the seeds themselves get partially abraded within the bird's crop which improves their germination ability, i.e. the relationship is mutualistic rather than deceptive. Foster & Delay (1998) examined seed dispersal in *Ormosia isthamensis* and *O. macrocalyx* and concluded that both these species deceived arboreally foraging frugivores, though seed dispersal rates were low, perhaps because birds quickly learned of the deception. However, seeds on the ground under the *Ormosia* trees were also dispersed by other ground-feeding birds, which was consistent with the mutualistic grit hypothesis.

Seed dispersal by humans, arable weeds and Vavilovian mimicry

Human activity has not only been involved in the largely deliberate selective breeding of various plants and animals but has also had marked, but completely unintentional, selective effects, many of which have been far from beneficial. For example, the activity of mowing lawns in Europe incidentally selected for low-flowering, dwarf dandelions. One of the most interesting types of inadvertent selection due to human action involves the evolution of specific forms of various weed species that have regularly been associated with arable crops, a process first recognised by Vavilov (1922).

It is also widely believed that rye, the grain used in the famous German rye bread pumpernickel, originated through Vavilovian mimicry of wheat, with those individuals bearing a closer resemblance to the latter being less likely to be weeded out (Vavilov 1922) and thereby continuing to exist. Another example noticed by the Russian scientists was a variety of the false flax weed, *Camelina sativa* (named, appropriately, var. *linicola*) (Brassicaceae), which occurs in flax, *Linum usitatissimum* (Linaceae), fields. Natural selection appears to have acted to create the overall etiolated plant phenotype through competition among the dense shading of the flax plants, but in this case human action may have led to its seeds evolving to resemble those of the flax because weed seeds are usually picked out of the harvested flax crop by hand. Incidentally, *C. sativa* is also cultivated in its own right as an oil-seed crop. Similarly, an agro-ecotype of common vetch, *Vicia sativa*, possesses rather flattened, lens-shaped seeds that mimic those of lentils, *Lens esculenta* (both Fabaceae), and thus may be confused with it (Rowlands 1959). This mutation is apparently rather recent, having first been noted in the early twentieth century, and is due to a single recessive mutation.

S.C.H. Barrett (1983, 1987) recognises that selection can act both on the vegetative parts and on the seeds. Domesticated rice, *Oryza sativa*, is apparently mimicked by members of the *Echinochloa crus-galli* grass complex. This grass has two varieties, *Echinochloa crus-galli* var. *crus-galli* and var. *oryzicola*, the second being the crop mimic. A few weeks after germination, model and mimic appear remarkably similar and a farmer would be hard-pressed to weed out the *Echinochloa*. Similarly in Africa (Swaziland), S.C.H. Barrett (1987) reported how two different wild rice species, *O. rufipogon* and *O. punctata*, had successfully infiltrated the domesticated rice crop because of the difficulty of distinguishing them in the vegetative phase. Subsequently, the job became even more difficult because *O. rufipogon* started to hybridise with the domesticated varieties, producing more intermediate plants. A similar situation occurs in other parts of Africa with various varieties of sorghum and other, predominantly graminaceous, crops. In some cases selection favouring mimetic resemblances is facilitated by gene flow when the mimic is a close relative of the true crop plant (Harlan et al. 1973).

Seed elaiosomes and their insect mimics

The seeds of a wide range of plants,[1] particularly among shrubs in dry habitats in Africa and Australia and among herb-layer plants in mesic forests, possess lipid-rich appendages called elaiosomes at one end that attract ants, for which they form an important food source (Beattie 1985). The ants (e.g. *Campanotus femoratus* and *Crematogaster* species) carry the seeds to their nests and remove and consume the elaiosomes, while the remainder of the seed often

germinates there. Thus the ants are largely responsible for seed dispersal in these plants. In itself, this does not constitute a case of adaptive resemblance, and the elaiosomes could simply have resulted from convergent evolution for ant-mediated seed dispersal acting independently on each plant. However, that other factors may play a role is indicated by the fact that many of the ants involved with these myrmechores are normally insectivorous. The seed surfaces of many of these plants, as well as their interiors, have been found to contain 6-methylsalicylate, which is also a component of *C. femoratus* mandibular gland secretions. At low concentrations 6-methylsalicylate stimulates excitement, seed handling and sometimes seed transport, whereas at higher concentrations it has a repellent effect on the ants (Davidson et al. 1990). Thus the plants may be using chemical mimicry to modify ant behaviour in a way that increases their seed dispersal.

The elaiosome story gets more complicated because some stick insects (Phasmatodea) lay eggs that bear a close resemblance to the elaiosome-bearing seeds of ant-dispersed plants such as the castor oil plant, *Ricinus communis* (Euphorbiaceae). Ants do not appear to distinguish between the castor oil plant seeds and morphologically similar stick insect eggs, which have a swollen structure called a capitulum attached to the operculum which opens, allowing the first instar stick insect to hatch (Hughes & Westoby 1992, P.D. Moore 1993). Moore (1993) commented that if the ants transported phasmid eggs into their nests, the emerging nymph might then be attacked by the ants, but Compton & Ware (1991) found that the nymphs are not attacked, and thus are presumably chemically cryptic. Suggestions that birds might also be responsible for dispersing seed-like phasmid eggs now appear to be disproven since Shelomi (2011) showed that they do not survive passage through the digestive systems of two granivores – chickens and quail.

MIMICRY BY PARASITES TO FACILITATE HOST FINDING

The trematode and the snail

Members of the Holarctic trematode (Platyhelminthes) fluke genus *Leucochloridium* have a complex life cycle, with their primary or definitive hosts being various birds, although their juvenile development takes place within snails of the family Succineidae. Species of the fluke are hard to differentiate and the banding pattern of the snail's infected tentacles is a useful guide (Mönnig 1922). Generally, the snails become infected by consuming eggs of the trematode that have been released into the bird's droppings by the adult worms, which are endoparasitic within the bird's rectums. The consumed egg(s) hatch to give rise to a miracidia larval stage which

---

1. Notable examples include Araceae, Bromeliaceae and Cactaceae.

Fig. 14.2 *Succinea* snails infected with the trematode *Leucochloridium paradoxum*, showing the brightly banded broodsac(s) of the trematode's sporocyst, which are filled with the cercaria larval stage of the worm and, in life, pulsate, making the snail, which is still fully mobile, far more conspicuous to avian predators. (a) Snail with only one sporocyst thus far entered into its tentacle; (b) succineid snail with both tentacles now occupied by *Leucochloridium* sporocyst sacs. (Source: a, Thomas Hahmann 2009. Reproduced under the terms of the Creative Commons Attribution Share-Alike Licence CC BY-SA 3.0, via Wikimedia Commons; b, Gilles San Martin 2012. CC BY-SA 3.0 (http://creativecommons.org/licenses/by-sa/3.0), via Wikimedia Commons.)

completes its development within the snail's gut and transforms into another larval stage, called a sporocyst, which is mobile within the snail and which produces, by asexual reproduction, a very large number of infective cercaria stage larvae (Fig. 14.2). The tri-lobed, banded sporocysts have pulsating broodsacs, and when one or two enter the snail's tentacles, they give increased prominence and vaguely resemble a caterpillar or perhaps some other interesting dietary morsel for birds such as crows, sparrows or finches. These birds are then likely to eat the snail and thus become infected with the worm's cercaria larvae which subsequently metamorphose again into the adult. The model is very non-specific bird food, and this mimicry could also be construed as a case of aggressive mimicry, especially since the adult worms primarily absorb nutrients from the food passing along the bird's intestinal tract.

A particularly interesting aspect of this mimicry is that whereas *Succinea* snails and their relatives are normally negatively phototaxic, remaining in darker places where they presumably are at less risk of both predation and dehydration, individuals harbouring sporocysts are significantly more likely to remain out in bright light, which increases the sporocysts' chances of detection and being consumed by a bird (Wesołowska & Wesołowski 2014). Furthermore, E.J. Robinson Jr (1947) found that the pulsations of the bands in the sporocyst are light-dependent and when the snail is in the dark, the pulsating movements cease.

The trematode and the fish

The giant, free-living cercaria of the trematode *Cercaria* (*Azygia*) *mirabilis*,[2] commonly called 'anchor-tails', are approximately 6–7 mm long and move like small fish or mosquito larvae and so attract the attention of predatory fish that will be the final host (see figure 28 in Wickler 1968). Since these do not need to penetrate hard ectoderm they have lost the normal anterior hooks that most cercaria have.

Pocketbook clams and fish

The North American pocketbook clam, *Lampsilis ovata* (see figure 30 in Wickler 1968), *L. fasciola* and the Higgins' eye pearly mussel, *L. higginsii* (Fig. 14.3), use their lures indirectly, not to obtain their own food, but to secure a host for their parasitic glochidia larvae. In these species, the edge of

2. There seems to be a good deal of taxonomic confusion regarding this animal. Frolova & Shcherbina (1975) even described a new species of *Azygia* from a perch host, *Essox lucius*, and stated that the larva was *Cercaria mirabilis*, while earlier, L. Szidat (1932) had shown that it was the larva of *A. lucii*. The problem is that the cercaria lack much in the way of morphological characters and have often been described independently of the adult fluke stages.

the mantle (pallium) resembles a small fish and thus attracts predatory fish that come to snap at them. The illusion is enhanced by the clam being orientated such that the passing water currents make the mantel lobes undulate rhythmically. When a predator fish comes to snap at the fish mimic, the clam releases its glochidia larvae into the fish's mouth and these then attach to the gills and act as parasites until they metamorphose to a free-living juvenile stage. Wickler (1968) supposes that the protruding mantel folds probably initially evolved as an aid to respiration and then became elaborated as the benefit of attracting hosts of their larvae to them became realisable. Perhaps even more remarkable are some other species of *Lampsilis* release many glochidia as a discrete mass, called a superconglutinate, which itself resembles a small fish in both shape and colouration, and thus attract their predatory host fish directly. Mimetic superconglutinates are known to occur in *L. perovalis* and *L. subangulata* (O'Brien & Brim-Box 1999, Haag et al. 1995).

'Termite balls'

Inside the nests of several *Reticulitermes* termite species it is frequently possible to find so-called 'termite balls', which are actually a fungus, and although many *Reticulitermes* cultivate associated fungi, the 'termite ball' fungi is a different species and can also be found among the eggs of *R. okinawensis*, which does not have a fungal association (Matsuura et al. 2000). By treating appropriately sized glass

beads with egg extract, Matsuura (2006) found that only coated dummy beads of almost precisely the size of the termite's own eggs, and also the size of the 'termite balls', were gathered and transported by termite workers.

Pseudoflowers, pseudo-anthers and pseudo-pollen

A number of floral pathogens of plants produce pseudo-flowers that may at least visually attract pollinating insects and, by coating the insect with their spores, they are dispersed. Pseudoflowers caused by the parasitic rust fungus *Puccinia monoica*, as well as various other species (Fig. 14.4a), do not necessarily have any particular model flower and may just be creating a generalised flower-like structure, and some flower visitors may actually visit them deliberately to obtain spores as food (B.A. Roy 1993, 1994).

Fig. 14.4 Examples of plant-pathogenic fungi that deceive pollinators into visiting them thus enabling their dispersal from plant to plant. (a) *Puccinia urticata* on stinging nettle, *Urtica* sp.; (b) mummy-berry, *Monilinia vaccinii-corymbosi* pseudo-fruit on a blueberry plant, *Vaccinium* sp. (Source: a, Rambatino 2010. Reproduced under the terms of the Creative Commons Attribution Share-Alike Licence CC BY-SA 2.5, via Wikipedia.; b, University of Georgia Plant Pathology Archive, University of Georgia, 2007. Reproduced under the terms of the Creative Commons Attribution Share-Alike Licence CC BY 3.0, via www.bugwood.org.)

Fig. 14.3 The rare unionid bivalve mollusc, *Lampsila higginsii*, from the Mississippi, commonly called the Higgins' eye pearly mussel, displaying its mantle flap lure that attracts its larval host fish which include largemouth bass (*Micropterus salmoides*), smallmouth bass (*M. dolomieu*), walleye (*Sander vitreus*) and yellow perch (*Perca flavescens*). (Source: Photograph by Mark Hove.)

Figure 4G in Vereecken & McNeil (2010) illustrates a nice *Puccinia*-infected plant. When *Puccinia*-induced pseudoflowers on *Arabis* spp. coexist with flowering buttercups, *Ranunculus inamoenus*, there may be some negative effect on the latter if the predominant pollinators of the buttercup are flies, since these may prefer to visit the fungus. B.A. Roy (1996) experimented with the system in which the rust fungus *P. monoica* produces quite 'showy' pseudoflowers on *Arabis holboellii* (a small Brassicaceae) and commonly co-occurs with blooming pasque flowers (*Pulsatilla patens*, as *Anemone*) in Colorado, USA. He found that the presence of pseudoflowers enhanced overall pollinator visits to the true flowers over a range of densities, whereas visitations to the pseudoflowers was not affected by the density of true pasque flowers. Nevertheless, the sticky nature of the fungal pseudoflowers might result in removal of pollen from pollinators and the deposition of fungal spores on the pasque flowers might interfere with pollination and reduce seed set. B.A. Roy & Raguso (1997) and Raguso & Roy (1998) went on to show that the pseudoflowers actually release a blend of volatiles that mimics floral odours. Other related species that induce pseudoflowers include *Puccinia arrhenatheri* on *Berberis vulgaris* (Berberidaceae), *P. punctiformis* on the thistle, *Cirsium arvense*, and *Uromyces pisi* on cypress spurge, *Euphorbia cyparissias* (Euphorbiaceae). However, it could be argued that all the yellow and reddish discolourations caused on plant foliage by rust fungi make the infected leaves visually conspicuous such they attract insects, and therefore potential spore dispersers, to investigate them.

When fungi of the genus *Monilinia* infect leaves or shoots of blueberries (*Vaccinium* spp.) and huckleberries (*Gaylussacia* sp.) they become UV-reflective, fragrant and start to secrete sugars from induced lesions. The normal pollinators of these plants are also attracted to the discoloured leaves where they feed on the secreted sugars, and transmit fungal conidia to their flowers (Batra & Batra 1985), where the fungus then attacks the fruit as they develop, filling them with fungal mycelia (Fig. 14.4b). The infected fruit, known as mummyberries, shrivel and drop and in commercial terms can be very costly. The *Monilinia* thus induces floral mimicry based partly on odour and partly on colour, but the pollinators are not in a sense duped because they do get a sugar reward, albeit not nectar. However, the mimicry by the fungus is somewhat more complex than that because its conidia also resemble the host plants' pollen grains, both physically and physiologically, facilitating the fungus' access to the ovary by growing down from the stigma along the inside of the style (Ngugi & Scherm 2004). Thus the system has two distinct mimetic phases, as illustrated diagrammatically by Ngugi & Scherm (2006): the pollinator is deceived by the fungus-infected leaves, but does not suffer much because the infected leaves provide sugar though not pollen (Fig. 14.5a), and the

plant is deceived and suffers both by the fungus' leaf-infecting stage and its conidia (Fig. 14.5b).

Some members of the rust fungus genus *Ravenelia*, which are parasitic on *Acacia* species (Fabaceae), produce very unusual teliospores which resemble, and are believed to mimic, the pollen of their host plants. Various bees that collect *Acacia* pollen also collect these spores (Savile 1976, 1980) and will transport them from tree to tree.

A.R. Diamond Jr et al. (2006) describe a less spectacular but still probably mimetic system from the southern USA. The fungus *Fusarium semitectum* infects the flowering heads of the rare composite *Rudbeckia auriculata* (Asteraceae) and renders them sterile. Its spores are orangey or pinkish and so, on the central cluster of flowers, look quite similar to the natural pollen, and the infected *Rudbeckia* flowers are still visited by its specialist pollinating bee, *Andrena aliciae*, and so transferred to other plants. Luckily, the plant is perennial and mostly reproduces asexually via its stolons, so at the sites where it occurs the fungus is probably not threatening its existence, at least in the short term. In the same way that *F. semitectum* renders the flowers that it infects sterile, so too does the basidiomycete anther smut fungus, *Microbotryum violaceum*, an obligate parasite of many species of carnation (Caryophyllaceae). This affects the plant's actual flowers and they become sterile, with their anthers bearing the fungal spores rather than pollen but otherwise still resembling true anthers, though how much appearance matters in this case is less certain (Alexander 1990, Antonovics & Alexander 1992).

Truffles

Another case of mimicry involving fungi involves the highly prized subterranean truffle fungi. These ascomycete fungi belong to the genus *Tuber* and have a pleasant musky odour, which is due to the production of the slightly volatile steroid androstenone (5α-androst-16-en-3-one) (Claus et al. 1981). This molecule has no androgenic activity but is present, for example, in the axillary (underarm) sweat of men. This steroid is also synthesised in the testes of pigs and wild boar and transported to the salivary glands, where it is released and causes receptive females to assume a standing posture (lordosis) suitable for mating. Thus it is used by pig-breeders (marketed as 'Boarmate'®) as a way of detecting which females are in oestrus. However, in the case of truffles, the fruiting body is produced underground, sometimes up to a metre below the surface, and the production of odoriferous steroid serves to attract wild pigs that will dig for them and, in the process, bring them to the surface whereby the truffle spores can be spread. Many sex shops and similar trade outlets even market sprays that contain androstenone or androstenol, or both, that are supposed to attract women.

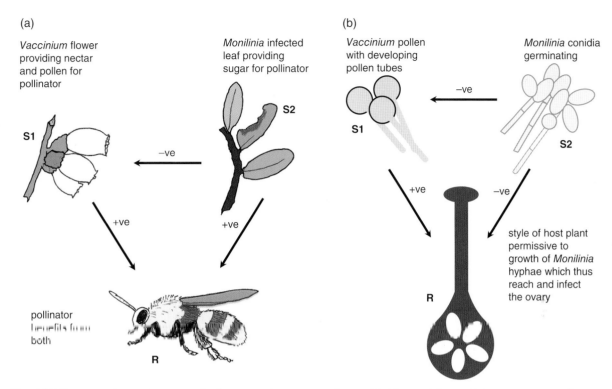

Fig. 14.5 Diagrams of aggressive mimicry by the mummy-berry-causing fungus, *Monilinia vaccinii-corymbosi*, gaining access to the ovary of its blueberry host plant, *Vaccinium* spp. (Ericaceae). (a) Pollinator is partially duped into visiting fungus-infected leaves as well as flowers, though there is some reward; (b) fungus spores mimic host pollen at both physical and molecular levels, allowing them to grow hyphae down the host plant style and then infect the ovules. (Source: Adapted from Ngugi & Scherm 2006. Reproduced with permission from Oxford University Press.)

Mimicry of dead flesh by fungi and mosses

The Eurasian stinkhorn fungus, *Phallus impudicus*, is renowned for its strong fetid odour which, while repugnant to humans, is clearly attractive to a range of carrion-feeding flies (K.G.V. Smith 1956, Wickler 1968). It is by no means the only such species. *Clathrus archeri*, the so-called octopus stinkhorn, smells strongly of rotting meat and, as with many arum plants (Araceae), gives off a mix of oligo-sulphide compounds (S.D. Johnson & Jürgens 2010). Spores the pan-tropical *P. (=Dictyophora) indusiatus*,[3] and the related Hawaiian *Ithyphallus coralloides* are known to be able to pass through the digestive tracts of their disperser flies unharmed, thus facilitating dispersal (Cobb 1906, Tuno 1998).

Several mosses of the family Splachnaceae, commonly called stink mosses, such as *Tetraplodon mnioides*, the black fruited stink moss, attract carrion and dung flies which then get coated in their sticky spores (Bequaert 1921, Erlanson 1930, Wiens 1978, Marino et al. 2009) and subsequently disperse them.

Deception of dung beetles by fruit

Seeds of the African spineless monkey orange trees (*Strychnos madagascariensis*) mimic the smell of dung thus duping the dung beetle, *Pachylomerus* into dispersing them by rolling them (Burger & Petersen 1991, Midgley et al. 2015). Oddly, *Pachylomerus* are not dung-ball rollers.

---

3. The smell of this species is reportedly able to induce spontaneous orgasms in some women (Holliday & Soule 2001).

Chapter 15

MOLECULAR MIMICRY: PARASITES, PATHOGENS AND PLANTS

Perfect replication is the enemy of any robust system... Lacking a central nervous system – much less a brain – the parasite is a simple system designed to compromise a very specific target host. The more uniform the host, the more effective the infestation.

Daniel Suarez (2006) *Daemon*. Dutton, New York

INTRODUCTION

The term 'molecular mimicry' was coined by Damian (1964) to describe the shared antigenic properties of hosts and their parasites. Von Beeren et al. (2012) clarified the use of the term 'adaptive resemblance' as used in the literature on chemical ecology which has not always been consistent (see Table 1.2), and suggested three definitions based on their usage in whole-organism situations:

1. chemical crypsis – the operator does not detect the mimic as a discrete entity (background matching),
2. chemical masquerade – the operator detects the mimic but misidentifies it as an uninteresting entity, and
3. chemical mimicry – the operator detects the organism as an interesting entity.

Pasteur (1982) considered molecular mimicries as fundamentally different from all the other types discussed previously in this book, in that "no sensory perception" is involved, but from an evolutionary point of view molecular mimicry is the same: a feature of the mimic is selected because it deceives some system within the receiver about its true identity and therefore I include a brief overview here, though not in as much detail as I have for organismal examples. Starrett (1993) also includes most of these cases within his concept of adaptive resemblance, though points out that the last two cases that I cover in this chapter (concerning animal hormone mimicries by plants) are perhaps more borderline. These could also be considered as a sort of delayed-action toxin or secondary defences, as nearly all toxins and venoms after all, act because some part of them 'mimics' some part of a receptor agonist or enzyme substrate.

MACRO-ANIMAL SYSTEMS

Anemone fish

Anemome fish, also called clown fish, are a group of marine fish belonging to the Pomacentridae (Amphiprioninae), most of which spend much of their time swimming among the stinging tentacles of their various sea anemone hosts with which they have a symbiotic, mutualistic relationship. They occur exclusively in the Indian and Pacific Oceans. Schlichter (1976) investigated how Clark's anemone fish, *Amphiprion clarkii*, often misspelled as *clarki*, gains protection from the anemones among whose tentacles they dwell. Anemone nematocysts (cnidocytes) are stimulated to discharge by contact with foreign stimulating substances, such as food items, but are inhibited from discharging by substances produced by the anemone itself so as to prevent self-stinging. It appears that the fish accumulate anemone-produced glycoproteins on their bodies which inhibit the firing of the anemone's stinging cells. Isolation of anemone fish from their habitual hosts lead to a loss of these protective substances such that, when returned to an anemone, they get stung. This acquired chemical camouflage resembles that of schistosomes, which incorporate host antigens into their integuments, thus rendering them 'invisible' to the host's immune system (Smithers 1972).

Parasitic helminthes

Helminth parasites that live inside a host organism's body have to avoid or circumvent the host's immune defence mechanisms. In many cases this does not just mean a single host, as many parasitic 'worms' have evolved complex life cycles with one or more intermediate hosts before their final host, in which they become sexually mature. For example, the blackfly (Simulidae) parasitic nematode *Onchocerca volvulus* alternates between human and fly hosts, sexual reproduction occurring in the human and the larval stages in the fly. An even more complex example is that of the nematode *Anisakis* spp., whose final hosts are dolphins (humans can also become infected, suffering anisakiasis as a result, though it is a dead end for the parasite[1]). Before reaching adulthood, the *Anisakis* eggs excreted by the dolphins must first hatch to develop into free-swimming larvae, which are then consumed by various crustacea which are, in turn, consumed by various squid and fish, and maybe these are further consumed by other fish, each time transferring the *Anisakis* larvae until a fish (or squid) is again consumed by a dolphin, in which the larvae complete development and reach sexual maturity. It should be clear from this that the various larval and adult stages of the *Anisakis* must be able to protect themselves from a variety of different types of immune system.

Most work on how helminths evade host immune defences concerns the final host, and is really only well understood in a few cases of medical or veterinary importance. There is very little information on the biochemical or molecular strategies employed by the vast majority of species, including most of those of medical importance.

Platyhelminthes (Trematoda)

Liver flukes are just one group (the class Trematoda) of the phylum Platyhelminthes that are obligate parasites of other animals. The helminth defence molecule (HDM) secreted by

1. In humans, *Anisakis* larvae fail to penetrate through the intestinal wall, die and cause an immune reaction causing the condition known as anisakiasis, which can be very unpleasant, mimicking Crohn's disease, with vomiting and other symptoms.

the liver fluke, *Fasciola hepatica*, in this case the form called FhHDM-1, and a post-translation terminal peptide that is cut off from it by the enzyme cathepsin, also produced by the fluke, show remarkable molecular similarity to host mammal antimicrobial peptides, especially one called LL-37 (M.W. Robinson et al. 2011). Binding of the HDM peptide to the lipopolysaccharides (LPS) prevents them from binding to the LPS-binding protein receptors on the host's macrophage blood cells. In this way, the fluke is effectively inactivating part of the host's normal macrophage-mediated response to parasites and pathogens. Because of their action, helminth HDM and the released terminal peptide, or at least synthetic analogues thereof, have potential use in anti-inflammatory therapies and the prevention of associated sepsis.

Schistosomes are another group of parasitic trematode, otherwise known as blood flukes, because the adults live within a vertebrate body inside capillary vessels associated with the intestinal mesenteries or bladder, and they are responsible for a number of serious diseases in humans (e.g. schistosomiasis, bilharzia) and other animals. They have complex life cycles: eggs are released into the environment through the host's urine or faeces, depending on the species, and their parasitic larval stages infect various snails. The adults of several species are now known to protect themselves from host immune defences by means of chemical mimicry or crypsis (Lumsden 1975, Salzet et al. 2000).

Bayne & Stephens (1983) raised antibodies in rabbits to sporocysts of the human blood fluke, *Schistosoma mansoni*, and found that these were reactive to the haemocytes of its intermediary host, the aquatic snail *Biomphalaria glabrata* (Planorbidae), indicating that the sporocyst surface possesses the same immunoreactive epitopes as the snail, strongly suggesting that the schistosome escapes the snail's lectin based immune system. Again with rabbits, Yoshino & Bayne (1983) generated antibodies against two different strains of the snail, one susceptible to the schistosome and one resistant to infection by it. Both sets of antibodies were reactive against the surface epithelium, the ciliary membranes and the surface membranes of intercellular ridges of the schistosome miracidial stage, thus demonstrating that the features of the schistosome surface chemistry had high similarity to basic features of the snail's haemocytes, suggesting molecular crypsis. By rearing schistosome miracidia and sporocysts in media devoid of snail host materials and demonstrating that they were still immunoreactive, Yoshino & Bayne showed that the immunoreactive epitopes are encoded by the parasite genome and not the result of binding of snail factors to the parasite's surface. Yoshino et al. (2012) provided evidence that inducible, highly diverse, lectin-like recognition molecules in snail haemocytes mediate their

response to *S. mansoni* larvae, which may help explain why some snail biotypes are resistant and others are not.

Tapeworms (Platyhelminthes: Cestoda)

As with flukes (Trematoda), tapeworms (Cestoda) also have complex life cycles with intermediate hosts, though probably less is known about their defences against primary and intermediate host immune systems. In some cases, the adults seem to release inhibitors, compounds that interact with and block host defensive enzymes (Lumsden 1975). However, Hammerschmidt & Kurtz (2005) studied *Schistocephalus solidus*, which has two intermediate hosts, a freshwater copepod crustacean, *Macrocyclops albidus*, and the three-spined stickleback fish, *Gasterosteus aculeatus*, before entering its definitive host, a fish-eating bird. The cestode appears to show genetically determined variation in the sugar groups conjugated to their surface (glycans), and by manipulating these with specific sugar-binding proteins called lectins, they showed that the particular sugar groups affected its fitness and that the tapeworm juveniles can change their surface according to the intermediate invertebrate or vertebrate host. Another complex aspect of helminth biology is that they migrate through different host tissues during their development, which means that they are exposed to a variety of different potential immune challenges even within a single host. How they overcome these is still not well understood. Glycans (complex polysaccharides) probably play an important role in enabling helminth parasites to evade their host's immune system and helminths have a large diversity of these molecules (van Die & Cummings 2010). Interaction of these with host glycan-receptor proteins is likely to be an important part of how they help defend the worms, rather than direct mimicry of host glycan epitopes. T.O. Hebert et al. (2015) in the *Gasterosteus/Schistocephalus* system have more recently identified 94 tapeworm proteins with close local sequence homology to stickleback proteins and found that four of them are secreted membrane proteins which are most likely either averting host immune response or interfering with host physiology.

Parasitic nematodes

The phylum Nematoda is huge, with many free-living as well as parasitic species, and they occupy virtually all habitats on the planet. They are commonly called roundworms, but that epithet is only really a good description of a subset of the group. Rather little seems to be known about how the many species of parasitic nematode worms avoid host immune defences. The roundworm *Onchocerca lienalis* attacks cattle as definitive hosts, but like its relative, *O. volvulus*, the causative agent of human river blindness, it has a complex life cycle that involves a blackfly (Simulidae) intermediate host. *O. lienalis* is thought to mimic host sugars on its surface so as not be recognised as foreign, at least by the invertebrate host's immune system (Jacobson & Doyle 1996).

Approximately two billion humans worldwide are infected with the large roundworm, *Ascaris lumbricoides*, which passes through the gut wall as a larva, enters the bloodstream, moves to the lungs, and is finally coughed up and swallowed to finish up in the intestine where it releases its eggs. Avoiding host immunity during its early stages appears to be achieved by surface absorption of host ABO blood group antigens on to its surface (Ponce de León & Valverde 2003).

Parasitoid wasp eggs

The eggs and larvae of endoparasitic parasitoid wasps may overcome host immunity either actively through venoms and other systems that target the host's immune system, or passively by their surface not evoking an immune response (i.e. haemocyte adhesion followed by encapsulation and other defence mechanisms). In a number of cases, the outer layer of the wasp egg is known to have a fibrous texture and to become covered in a layer of immune-neutral particles, or possibly glycoproteins, as it passes along the wasp's oviduct on the way to being laid (Quicke 2015).

PATHOGENIC FUNGI

Blastomyces dermatitidis is a yeast endemic to North America where it normally dwells in the soil, but is highly pathogenic to humans, dogs and a range of other mammals, causing a type of pneumonia which can be lethal if not treated. It can also cause serious skin lesions. Sterkel et al. (2016) describe how the fungus produces a dipeptidyl-peptidase IVA (DppIVA) that closely mimics the ectopeptidase CD26 of its mammalian host. CD26, also known as dipeptidyl peptidase-4, is expressed on the surface of most mammalian cell types and plays an important role in regulating immune response. Its fungal mimic cleaves and inactivates host cytokines and chemokines, blocks the recruitment and differentiation of host monocytes, and prevents activation of host phagocytes, thus preventing the host from mounting an immune response against it. That this fungal enzyme was the principal, and possibly only, agent responsible for the yeast's invasive pathogenicity was demonstrated by pharmacologically blocking DppIVA.

PROTISTA

Chagas' disease

Chagas' disease is an unpleasant vector-borne infection that is widespread in Central and South America, and is widely suspected as having been the cause of Charles Darwin's ill-health in later life. The infective agent in Chagas' disease, *Trypanosoma cruzi*, which is transmitted between hosts by the blood-sucking, anthrophilic, 'kissing bugs', such as *Triatoma infestans* and *Rhodnius prolixus* (Reduviidae: Triatominae), has evolved molecular mimicry (Gironès et al. 2005). The disease is unusual in that during its chronic phase, which in many patients (30%) involves severe heart pathology, the actual parasite is quite scarce, and the symptoms experienced are actually an autoimmune response caused by similarity of, for example, a trypanosome 160-kDa flagellum-associated surface protein (Fl-160) to epitopes in the host nervous tissue and similarity of the immunodominant *T. cruzi* B13 protein to host (human) heart myosin (Avila 1992, Christen 2014).

MICROBIAL SYSTEMS

Bacterial chemical mimicry and autoimmune responses

Autoimmune responses due to pathogen molecular mimicries of specific host molecules (antigenic determinants) are quite widespread (Damian 1964, 1987), and include Lyme's disease, Guillain–Barré syndrome, myasthenia gravis, leprosy, listeria and typhoid disease (Cossart et al. 2003, Kirvan et al. 2003, Cusick et al. 2012, Singh et al. 2015). Antigenic cross-reactivity may also be important in the pathogenic effects of some parasites (Rowley & Jenkin 1962). Here I will only mention a couple of important human-related cases but there can be little doubt that similar effects are to be found throughout the higher Caudata, with their sophisticated immune systems. In some cases the mimetic microbial molecules are homologous to those of their host, possibly obtained through horizontal gene transfer, whereas in others they are the result of convergent molecular evolution (Stebbins & Galán 2001). Modern genomics and protein-structural approaches are rapidly shedding new light on how this class of mimicries work at the molecular level (Ludin et al. 2011) and it seems highly likely that far more examples will be discovered in the near future. Damian (1997) suggests that when an epitope of a parasite or pathogen leads to autoimmune disease in the host, it ought to be termed an 'autoimmunogenic epitope'.

Helicobacter pylori

Helicobacter pylori is a major cause of human gastroenteritis and is associated with various illnesses such as the development of ulcers and sometimes the development of cancers. The bacterium has surface antigens that are similar to Lewis blood group antigens, and as these antigens are also expressed by the surface membranes of human intestinal epithelial cells, where they are associated with a gastric proton pump, it seems likely that these features of the bacterium are involved in molecular crypsis from the host's immune system. But when the bacterial infection is severe or chronic and causing damage to systems, this is likely to be, in part, due to the patient raising antibodies against that particular bacterial epitope that subsequently go on to attack the patient's own gastrointestinal tissue.

Campylobacter jejuni

Campylobacter jejuni, another major cause of gastroenteritis, is the most common infection preceding the onset of Guillain–Barré syndrome (GBS), an inflammatory neuropathy (Moran & Prendergast 2001). The surface of *C. jejuni* strains that are associated with the onset of the neuropathic autoimmune disease share immunoreactive features with human gangliosides. It is likely that the bacterium causes an immune response in the host and the antibodies produced, in addition to attacking the bacteria, also attack the patient's nervous tissue because of this similarity.

Mimicry by plant-pathogenic bacteria

The bacterium *Xanthomonas axonopodis* is the causative agent of citrus canker and produces lesions on the host tree leaves with wet edges, which is different from those produced by close relatives. This wetness is no doubt beneficial to the bacteria and might even facilitate transmission. DNA sequence data revealed that the bacterium's genome encodes a protein that is very similar to the natriuretic peptides of plants that are involved in many physiological processes, including water uptake (Nembaware et al. 2004). The degree of sequence homology observed could indicate that the bacterial gene is truly homologous to that of the host plant and may have originated through horizontal gene transfer.

VIRUSES

Viruses have also evolved to maximise their reproduction and potential to spread from one host to another, and clearly this is dependent on successful counteraction or evasion of host immune defences. In vertebrates, white blood cells respond to small signal proteins called chemokines (chemotactic cytokines) secreted by cells and are therefore guided to the site of infection. In normal function, the chemokines react with membrane-bound G protein-coupled receptors (GPCRs) in the white blood cells. Various virus genomes also encode both chemokines and chemokine receptors (P.M. Murphy 2001, Alcami 2003). Burg et al. (2015) have shown that a GPCR encoded by human *Cytomegalovirus*, a herpes-type virus that is normally prevalent[2] in the salivary gland and is asymptomatic in people with active immune systems, binds to human chemokines through molecular similarity to human GPCR and therefore diverts host chemokines away from the target blood cells, preventing the inflammatory process.

Molecular mimicry is also demonstrated or invoked in a wide range of other virus-associated autoimmune responses, and Oldstone (1989) includes a relatively early review. A few more recent examples include: Herpes Simplex Virus (Zhao et al. 1998), dengue (Lin et al. 2011) and HIV-1 (P.M. Murphy 2001, Tsiakalos et al. 2011).

PLANTS

Sugar, toxin and satiation mimicry

Here I discuss three types of protective device that rely on deceiving a herbivore that it does not want to eat a plant, either because it mistakenly regards it as toxic or it is deceived into behaving as if it has had enough of the nutrient it needs at that time.

First let us consider the well-known artificial sweeteners, i.e. sugar mimics, such as aspartame and saccharin. The former is an ester of a dipeptide (N-(L-α-aspartyl)-L-phenylalanine,1-methyl ester) and the latter a heterocyclic molecule ($2H$-$1\lambda^6$,2-benzothiazol-1,1,3-trione). Neither of them sugars, yet both are potent agonists of human sugar receptors. On the other hand, many chemical sugars are not sweet at all. All that is required is that somewhere on the molecule's surface is the appropriate three-dimensional configuration of charges and hydrophobicity to be able to interact with the receptor. In addition to these artificial sweeteners, as well as many other unrelated artificial sweeteners, there is a growing list of potent natural sugar receptor agonists. The list includes moneti (3000× sweeter than sugar) from *Sclerochiton ilicifolius* (Acanthaceae), thaumatin (2000×), a protein from *Thaumatococcus daniellii* (Marantaceae), brazzein (1000–2000×), a protein from *Pentadiplandra brazzeana* (Pentadiplandraceae), monellin (1500×) from

---

2. Between 50 and 80% of the general human population carry this particular cytomegalovirus.

Dioscoreophyllum volkensii (Menispermaceae), rebaudioside, a glycoside (200×) from *Stevia rebaudiana* (Asteraceae), and luo han guo (300×) from *Siraitia grosvenorii* (Cucurbitaceae). Whether any of these are mimetic of sugars to frugivores or herbivores does not seem to be known.

Non-sugar artificial and natural sweeteners are, in some ways, super-stimuli because, due to their affinity for the sugar receptor, only very small amounts are required for a large response. Saccharin, for example is 300× sweeter than sucrose. Evolution could have led to sugar receptors that would respond to true sugars in the same way they do to aspartame or saccharin, but that would have been both unnecessary and even harmful to the animal, since sugars are required for energy and only sources that contain considerable quantities of them are going to be worthwhile consuming.

Humans only have five types of taste receptor and a lot of the flavour of our food depends on volatiles that are detected in the nose. But many toxins are not volatile, or only slightly so. It is well known that humans, and indeed all mammals, find that capsaicin,[3] the main active ingredient of chilli peppers (*Solanum fructicosa* and relatives) can cause extreme pain, but it is, except at very high dosage, not dangerous. Why then is this compound present in rather large concentrations, predominantly in the placental and other fleshy tissue of chilli plant pods which advertise their ripeness to potential seed dispersers by turning bright red? The answer is simply that their seed dispersers are birds and not mammals, and birds are not affected by capsaicin (Tewksbury & Nabhan 2001). Thus, the capsaicin of the chillies is duping the mammals that the plant is toxic (Ruxton & Kennedy 2006). The reason such deceit can occur rests on the nature of the receptors of the herbivore. These will be selected to provide information about the presence both of nutritious foods and of toxins in a potential food. Given enough time, no doubt selection could lead to the possession of many different, highly specialist toxin receptors, but as Ruxton & Kennedy (2006) point out, there would normally be little to be gained by a predator or herbivore actually being able to identify the precise toxin, all that is necessary is to be able to know that the food should not be eaten. Therefore, selection is more likely to lead to more generalist receptors, and that means

that they will be triggered by some non-toxins as well as toxins. It is no doubt through this route that chilli peppers have managed to evolve a compound that selectively triggers receptors only in mammals.

A corollary of the predator/herbivore's generalist versus specialist toxin receptors is whether it is better for their prey to be 'honest' producers of real toxins or to evolve non-toxins that mimic their consumers' toxin receptors. Importantly, real toxins are generally expensive to synthesise and/or sequester and, most likely, also incur a fitness cost through the prey having to have evolved a modified physiology that is less susceptible, if not totally immune, to them. Such modified physiology is likely to be less efficient than an unconstrained one.[4] Therefore, it is generally going to be less costly to synthesise/store aversive non-toxins than the real thing, though this will inevitably lead to an arms race with predators that can distinguish the truly toxic from its chemical mimic being at a selective advantage over those that cannot. Making such a distinction might not be too difficult in principle, since non-toxic chemical mimics are often not at all chemically similar to their models.

The situation in plants is likely to be somewhat different on several counts. Apart from small plants and annuals/biennials in particular, herbivory often leaves the individual plant to recover, whereas, apart from browsing on some colonial animals such as corals, predation on animals is usually fatal. Second, the physiological pathways of plants are far more different from animals than they are between animals, and vice versa, so it is easier for plants to evolve toxins that may be effective against a range of animal herbivores while requiring less modification to their own system to accommodate them. Third, plants rely heavily on inducible defences (Tollrian & Harvell 1999) which animals only do against pathogens and parasites.

Another possible chemical mimicry by plants is provided by the mammalian appetite suppressor in the South African plant *Hoodia gordonii* (Apocynaceae) (Van Huesen & Minsky 2007) though there is some debate about whether it does indeed lead to false satiation, at least in humans. The structure of the putative active ingredient was patented as P57 (an oxypregnane steroidal glycoside) but the chemical structure is so complex that commercial synthesis is not economically viable.

---

3. Capsaicin and related capsaicinoids belong to the vanilloid family of chemicals. They bind to vanilloid subtype 1 receptors (TRPV1) which control a type of cation channel in sensory neurons. These receptors are normally triggered by heat, acids and physical abrasion and hence capsaicin is often described as causing a burning sensation.

4. Many studies on other systems show that there are evolutionary costs associated with being able to defend oneself against pathogens, parasites/parasitoids or insecticides, even in the absence of the danger, which is why, once some pesticide such as DDT stops being used, alleles for resistance to DDT rapidly get selected against.

Phytoecdysteroids – plant chemicals that mimic insect moulting hormone

In the mid-1960s, Czech entomologist Karel Sláma made an interesting accidental discovery while trying to culture the plant bug *Pyrrhocoris apterus*. For some reason, he could not get them to complete development when their rearing containers were lined with paper from the USA, but they grew unimpeded if the paper was from Europe or Japan (Sláma & Williams 1965, C.M. Williams 1970). The reason turned out to be the wood pulp that the American paper was made from, which came from balsam fir, *Abies balsamea*, and this tree produces a triterpenoid derivative that is a molecular mimic of insect moulting hormone, ecdysone (Sláma 1993). Since then, the pharmacological actions of these compounds has even gained some reputed value in human medicine.

Plant oestrogens – phyto-contraceptives

Oestrogen-mimicking molecules are produced by a number of pasture plants, notably various Fabaceae (e.g. *Trifolium repens*, *T. subterraneum* and *Medicago sativa*) but also a wide range of other plant families. These can cause reproductive disorders including infertility in grazing animals, such as rabbits, horses and sheep (Shutt 1976) as well as primates. Sheep can be particularly strongly affected. From an evolutionary point of view it must be a possibility that in long-lived, clonal or kin-structured groups of plants, especially ones rich in foliar nitrogen such as legumes, they have been selected for because over a medium term of a few months (rabbits) or years (sheep) they could reduce the number of herbivores present. They may, however, have other roles, such as antimicrobial activity.

EXTENDED GLOSSARY

For ease of reference, the three components of a mimicry system, the model, mimic and dupe, are referred to below as S1, S2 and R, respectively, in accordance with Wickler's (1965) terminology which has become widely adopted.

actual model the model in a mimetic system belongs to an exact species.

aculeate in reference to the insect order Hymenoptera, members of the clade Aculeata, in which the ovipositor is no longer used for laying eggs (which emerge at its base) but has evolved into a stinger used for defence or paralysing prey. The Aculeata includes ants, bees, vespid wasps (Vespidae, Sphecidae), spider-hunting wasps (Pompilidae), mutillids, bethylids and chrysidids, mentioned herein.

adaline an alkaloid constituent of the defensive secretion (reflex blood) of the ladybird beetle, *Adalia bipunctata* (Coccinellidae) (de Jong et al. 1991).

additive mimicry *see* **arithmetic mimicry**.

adventitious mimicry the use of objects from the environment deliberately added to an animal's exterior that help to conceal it, such as the use of fragments of seaweed etc. by decorator crabs.

advergence when only one member of a pair of species is evolving towards the other (L.P. Brower & Brower 1972).

aerial crypsis phenomenon that at low light intensities such as at dusk, darker insects are less visible during flight than pale-coloured ones (Kettlewell 1961).

aestivation period of dormancy similar to hibernation but taking place during the summer.

aggressive mimicry in animals, mimicry in which a predator or parasite avoids detection by its prey or host until it is close enough to make a successful capture and may involve devices to lure a prey closer; in plants term sometimes used for pollinator deception, because of its negative impact on the dupe.

aide mémoire mimicry signals, such as behaviours or odours, that remind a would-be predator of an unpleasant or unsuccessful experience they had in a previous attempt at capturing or ingesting a given prey type (Rothschild 1984). See Chapter 4.

alkaloid a member of a diverse group of chemicals synthesised by predominantly short-lived plants for defence against herbivory.

allaesthetic any displayed feature of an animal, behavioural or structural, that has an effect on the behaviour of other individuals of its own or another species (Chance & Russell 1959).

Allee effect extinction resulting from a sexually reproducing species' density in a habitat getting so low that individuals are no longer likely to encounter a member of the opposite sex. In terms of mimicry, it may have particular relevance to the deceit pollination of rare plants such as some orchids (Duffy et al. 2013).

alleles different versions of a given gene.

allopatric living in different, non-overlapping places.

anachoresis hiding from view, for example spiders hiding in crevasses or inside curled leaves.

anamnesis term used by R.E. Owen & Owen (1984) for memory.

androchromatism or **androchromatypic** term proposed by Hilton (1987) for females whose colour patterns mimic those of males (cf. gynochromatypic).

androgenic of a hormone that promotes the development (or maintenance) of male characteristics.

andromimesis general term for female special resemblance to males.

andromorphy evolved similarity of female shape to that typical of males (Hilton 1987)

anisakiasis disease in humans caused by parasitic nematodes of the genus *Anisakis*.

antergic cases in which the mimicry or resemblance is harmful to the signal receiver (Vane-Wright 1976) (cf. **synergic**).

anthesis the time that a flower opens and is fully functional in terms of pollen production; also can mean the whole period that a flower is open and fully functional.

anthocyanin water-soluble red, purple or blue flavonoid pigments produced by plants.

anthrophilic associatied with humans or human habitation.

antiapostatic selection Differential predation on two forms in which the rarer form is preferred and so suffers disproportionately more mortality than does the common form.

anticryptic resemblance term usually applied to alluring, aggressive mimicries though I do not think that it is a helpful term (Poulton 1890, Starrett 1993).

antigenic pertaining to antigens, which are molecules or parts of molecules (*see* **epitope**) that stimulates the production of antibodies in vertebrates

apatetic colouration colouration that serves as camouflage (van Sommeren & Jackson 1959).

Apocrita Large monophyletic group of Hymenoptera (Insecta) recognisable by their narrow waist (separating thorax plus first abdominal segment from remainder of abdomen) which affords great body flexibility for oviposition and stinging. It includes the Aculeata (q.v.).

apomorphy a derived character state, relative to the one which is ancestral (*see also* **plesiomorphy**, **synapomorphy**, **symplesiomorphic**).

aposematism term originally coined by Poulton (1890) for a warning signal that is conspicuously different from background, usually referring to bright and/or contrasting colour pattern but could also be a smell, sound or bioluminescence.

apostasy term used for polymorphism or variation that thwarts a search image or search image-like behaviour (Endler 1981; *see also* **reflexive selection**).

apostate the rarer form of a species.

apostatic selection differential predation on two forms in which the commoner form is preferred and so suffers disproportionately more mortality than does the rarer form.

aril an outgrowth, often fleshy, of a seed or its attachment point (arillode), and surrounds it creating a fruit-like structure, though technically these are 'false fruit' (*see also* **pericarp**).

Aristotelian mimicry injury feigning by nesting birds in the presence of a potential predator that serves to distract the latter from the bird's nest or nestlings.

arithmetic mimicry a non-deceitful system in which two or more edible species appear similar thus 'sharing' predation by any predator that comes to be able to recognise them and possibly benefitting from confusion caused by having large numbers of similar-looking individuals (van Sommeren & Jackson 1959, Holling 1965) and assumes some limitation to predator attack rate; in so doing it also prevents rare species suffering from **antiapostatic selection**.

aspect diversity numbers of different colour patterns; applies particularly to either fore wings of cryptic moths or hind wing flash colours of moths and other insects.

associative learning learning in which a response becomes linked to a particular stimulus.

asymptotic attack probability the rate at which an experienced predator attacks the members of a Batesian–Müllerian mimicry complex if it cannot distinguish them, and is some function of its attack probabilities for each of the prey types alone (see Mallet & Joron 1999).

autoimmune disease class of diseases caused by a vertebrate's immune system recognising some of the animal's own tissue as foreign and so causing immune defence reactions against them.

automimicry generally used to mean Batesian mimicry within a given species, as is the case in this book and, for example, L.P. Brower et al. (1967b) and Ruxton et al. (2004a), but can be confusing as it covers various types from appeasement, bluff, some reproductive mimicry (but *see also* **weapon automimicry**). Sometimes also used for **self mimicry** (q.v.).

autotomy voluntary shedding of one or more appendages by an animal.

avoidance learning an active process in which an animal develops a new behaviour that reduces its exposure to noxious stimuli rather than simply reducing an activity, so for a predator with noxious prey, it means active avoidance of the noxious forms and not a reduction in overall predation effort (see Roper & Wistow 1986, Balogh et al. 2008).

Bakerian mimicry term coined by Pasteur (1982) for mimicry by non-nectar producing (female) flowers of **nectariferous** (male) ones of the same species, after a paper by H.G. Baker (1976) (S1S2/R).

Batesian mimicry (1) most commonly used, especially for animal examples, for anti-predator mimicry of an unpalatable potential prey by a more palatable one; (2) in botanical literature, in particular, the term has been extensively [mis]used to include all mimicries where the model is disadvantaged by the mimic and so, for example, may include floral mimicry aimed at increasing pollination success, and some authors have even employed the term to cover masquerade, where a species has evolved to resemble a harmless item of its general background such as a leaf or twig (Speed 2014). I prefer only to include anti-predator mimicries where the model is considered a potential prey by a predator.

Batesian–Poultonian mimicry term coined by Pasteur (1982) for S1R/S2 systems in which the model is agreeable to the dupes, as when a prey mimics a potential predator.

Batesian–Wallacian mimicry term coined by Pasteur (1982) for aggressive mimicry where the model and the dupe belong to the same species and the dupe preyed upon by the mimic (S1R/S2). Examples include prey sex pheromone mimicry by bolas spiders (see Eberhard 1977), prey bioluminescent or auditory mimicry (J.L. Lloyd 1984, D. C. Marshall & Hill 2009), and a dolichopodid fly from Papua New Guinea that imitates a sexually receptive female of a psychodid fly to attract and then eat males of the psychodid (Hampton Carson cited as pers. comm. in Pasteur 1982).

Beau Geste a deception whereby an animal makes it appear that there are more individuals in a territory than there are, most often applied to song birds that produce multiple types of calls: named after the character Michael "Beau" Geste in the P.C. Wren novel, and subsequent film adaptations, who arranged dead foreign legionnaires at a fort as if they were still alive and thus deceived the enemy into believing that there was still strong opposition.

biogenic amines organic compounds with two or more amine groups, including several that have signal function in the nervous system or inflammatory pathways and various other physiological activities. Examples include histamine, serotonin, dopamine and adrenaline (=epinephrine).

bioluminescence production of visible light by enzymatic action in an organism, most commonly by the action of luciferase on luciferin.

bipolar of a mimetic system in which there are only two species, i.e. one species acts as both model and mimic (S1 and S2), model and dupe (S1 and R) or mimic and dupe (S2 and R) (Pasteur 1982).

Bitrex® trademark name for phenylmethyl-[2-[(2,6-dimethylphenyl)amino]-2-oxoethyl]-diethylammonium benzoate (or denatonium benzoate for short), an extremely bitter-tasting compound often added to dangerous products to prevent accidental ingestion.

bizarre forms inexplicable complex structures exhibited by some insects and spiders believed to be possibly mimetic but without obvious (to the human observer!) models; for example, complex thoracic structures of some membracid bugs or the 'lanterns' of fulgorid bugs.

brood spot virtually feather-free, highly vascularised patch of skin on a bird used for incubating eggs, induced by prolactin in the breeding season.

Browerian mimicry term coined by Pasteur (1982) for an intraspecific Batesian **automimicry** as occurs when some individuals of an **aposematic** species are greatly less palatable than others (S1S2/R).

cadaverine pentane-1,5-diamine, a foul-smelling compound produced by protein breakdown during putrescence and, together with putrecine, largely responsible for the smell of rotting flesh; produced by various fly-pollinated flowers.

cantharidine a terpenoid that is highly toxic to many (but not all) animals and, in humans, is a systemic poison with an LD_{50} of approximately 0.5 mg/kg. At [far] lower doses it has been used for millennia as an aphrodisiac (Spanish fly) through its irritant effects, especially on the bladder. It is principally produced by beetles of the families Meloidae and Oedemeridae and other species actively seek out these insects to eat and sequester their cantharidin for secondary protection.

cantharophilic pollinated by beetles.

cardenolide the steroidal part of a cardiac glycoside molecule that can cause cardiac arrest in vertebrates.

cardiac glycosides glycosides (q.v.) in which the non-sugar part is the nucleus of a steroid molecule, such as in digitoxin. They are major components of the protective chemicals sequestered by insects from plants such as milkweeds (Asclepiadoideae), foxgloves (*Digitalis:* Plantaginaceae), and various Crassulaceae, Ruscaceae and Hyacinthaceae.

carrying capacity the maximum population size of a species that its environment can support indefinitely.

catalepsis *see* **thanatosis**.

catalpol an iridoid glucoside produced by various plants, notably in Bignoniaceae (*Catalpa* spp.), Plantaginaceae, Laminaceae and Scrophulariaceae and often sequestered by insects that feed upon them.

caudal concerned with the tail of an animal.

caudal autotomy automatic shedding of the tail by some lizards when they are seized. The shed part, which remains twitching or convulsing on the ground, often distracts the predator's attention. The fracture line is normally through the middle of a vertebra along a narrow, unossified zone. Also called tail-shedding or **urotomy**.

cerata fleshy, often brightly- coloured protuberances on the dorsal surface of many sea-slugs (opisthobranch molluscs) that aid in respiration and may contain nematocytes obtained from their cnidarian food.

cercaria free-living motile larval stage of flukes (parasitic trematode flatworms).

chromatophore a cell or an organ of colour change. In cephalopods they are under direct nervous, rather than hormonal, control, which is normal for most other groups, and comprise an elastic pigment cell surrounded by radial muscles.

circumocular around the eyes.

***cis*-region** non-coding parts of the genome that regulate the expression of nearby genes; also called *cis*-regulatory elements (CREs).

cleptergic stealing work; used in literature on slave-making ants.

cnidocyte *see* **nematocyte**.

cnidosacs structures at the ends of the **cerata** of aeolid sea-slugs which contain undischarged nematocysts from their prey that are transported through their guts.

coccineline an alkaloid secreted by various ladybirds (Coleoptera: Coccinellidae).

collective mimicry mimicry in which the mimetic resemblance is achieved by a group of conspecific individuals and cannot be achieved by single individuals of the mimic species (Haas 1945, Pasteur 1982, Fricke 1970).

commensal relationship between two organisms in which one benefits from the other without affecting it adversely.

competitive mimicry systems in which mimicry plays a role in **interference competition** (Rashed & Sherratt 2006, Rainey & Grether 2007).

compound mimicry term employed by X.J. Nelson & Jackson (2009a) for when the mimicry is not of a single individual model but of a model in combination with some other feature, such as an ant carrying a food item.

conditional probability the probability of an event occurring if (and only if) another probabilistic criterion has been satisfied.

congener member of the same genus.

conidia asexual spores produced at the tips of specialised hyphae by some fungi.

conjunct mimicry systems in which the three components (S1, S2, R) are all members of the same species.

conspecifics individuals belonging to the same species.

correlational selection selection favouring combinations of characters such that there can be multiple local fitness optima.

countershading the common phenomenon whereby the side of an animal that normally faces uppermost (towards the sky) is darker than its lower parts. Its role may be to cancel out the three dimensional effect caused by its own shadow on the lower parts (see **self-shadow concealment**) or it may simply be involved in making the animal maximally cryptic from above and below to different observers (see Ruxton et al. 2004a, 2004b).

crepuscular active around the time of dawn or dusk.

crypsis type of camouflage in which the resemblance is to the general background in terms of things like general colour, shading, texture (**eucrypsis**), or to common abundant elements of the background that nevertheless a dupe would not be interested in (**cryptic mimesis**) (Pasteur (1982).

cryptic mimesis term coined by Pasteur (1982) for resemblance to common, even dominant elements of an animal's surroundings such as sticks, flower buds, flowers, bark, leaves, pebbles, soil, etc. that the dupe does not react to.

cuckoldry term with precise usage depending on context. In humans and various other animals it refers to action by which a female gets fertilised by a male other than her husband or regular partner; the latter then is duped into helping to rear the child of the other male. In cuckoos and similar birds, as well as various insect social parasites, it can be both males and females or just one sex that is deceived.

cursorial capable of or adapted for running.

decoy a device that attracts attention of a dupe rather than them noticing the real thing. Coy ducks was the old name for caged, noisy ducks that were used to attract waterfowl to traps.

deimatic display a threat display such as a moth suddenly revealing eyespots, or a mantid or large spider rearing up and splaying front legs to either reveal bright pattern or chelicerae, or the rattling noise of a rattlesnake, or the rearing and spreading of the hood of a cobra (M. Edmunds 1974a, Cloudsley-Thompson 1981).

deme an interbreeding local population of a species which has a gene pool distinct from other populations.

density-dependence the magnitude of an effect is causally related to the density of the organism, either positively or negatively. Protection of **aposematic**, defended prey is classically positive density-dependent.

diaposematism term used by G.A.K. Marshall (1908) for a concept proposed by Dixey (1894) of reciprocal mimicry, i.e. that two or more unrelated distasteful species may evolve to look like one another through both simultaneously homing in on the same appearance (Müllerian mimicry) rather than the perceived wisdom at the time that one species will remain static while others evolve to look like it (see also Dixey 1909).

dichromat an animal with two types of colour receptor, each maximally sensitive to a different wavelength of light.

dioecious plants that have either male or female reproductive modes in a given individual, but never both.

disjunct mimicry system in which the model, mimic and signal receiver are each different species (see Vane-Wright 1976).

dispersal mimicry term first introduced by Wiens (1978) for mimicries which have evolved specifically to enhance or enable the dispersal of **propagules**.

disruptive colouration features of pattern that serve to make detection of an animal's (or plant's) outline hard to identify.

Dodsonian mimicry a deceptive pollination mimicry in which a non-rewarding species of plant mimics a co-occurring rewarding one (cf. **Bakerian mimicry**).

drupe a fleshy fruit containing a 'stone', i.e. hardened endocarp enclosing the seed.

dual signals concept first expounded by Miriam Rothschild that the same feature might have two functions at different distances, for example colouration might aid crypsis at a distance but be **aposematic** once the organism has been detected.

Dufour's gland an unpaired, tubular gland associated with the female ovipositor or stinger in the Hymenoptera which secretes predominantly volatile fatty compounds used in host marking (by parasitoid species) and intraspecific communication, such as colony recognition signals and sex pheromones.

dulosis a term used for the slave-making strategy of some ant species which capture workers of their host species and bring them to their own nest to act as workers there.

dupe term for the individual or species which is deceived by an adaptive resemblance; the signal receiver.

eatability same as palatability; suitable to be eaten.

Ecdysozoa a monophyletic group of invertebrate phyla dominated by Arthropoda and Nematoda characterised by having a cuticle that is moulted periodically enabling the animal to grow.

egg dummies markings, or sometimes three-dimensional structures, on the anal fin of males of some mouth-brooding cichlid fish that are believed by some to mimic lain eggs, such that the female will be attracted to take them into her mouth. As she does so, the male releases sperm that she also takes up, thus fertilising the eggs.

egg spots circular markings as for **egg dummies**, but only when they are simple colour markings.

elaiosome lipid- and protein-rich fleshy appendages attached to the seeds of many angiosperm plants that attract seed-dispersers, principally ants.

empathic learning *see* **observational learning**.

Emsleyan mimicry a disjunct protective system applied almost entirely to mimicry of highly venomous coral snakes and some pit vipers, which were thought likely to kill any predator rather than the predator having a chance to avoid them on subsequent occasions. In this type of mimicry the supposed model instead is not deadly and only painful, and therefore can be mimicked both by less toxic species as well as by the deadly ones (Emsley 1966, Pasteur 1982). Same as **Mertensian mimicry**, as defined by Wickler (1968).

endogenous camouflage camouflage that is brought about by the animal's own body colour, shape, texture, smell, etc.

endoparasitic living within the body of a host organism.

epigamic signalling features associated with providing information to a potential mate, such as courtship displays.

epigenetic heritable moderation of gene expression by external not genomic (DNA) factors.

episematic resemblance similarity in markings that helps individuals to recognise others of the same species, usually in the context of keeping them together as in herding, flocking or schooling species; Starrett (1993) gives it a somewhat broader meaning of any alluring or pacifying resemblance that allows or encourages approach by a signal receiver or mimic for trophic, defensive or reproductive purposes.

epitope that part of an antigen that is recognised by the immune system; another name for **antigenic determinant**.

esca the fleshy or frilly tip of the lure of an anglerfish; *see also* **illicium**.

escape mimicry Batesian and Müllerian systems in which predators reject the model or co-model on sight because they have failed so often in the past to successfully capture them.

ESS *see* **evolutionarily stable strategy**.

etiolated of plants that grow tall and thin (and often) pale due to lack of light.

eucrypsis a general term for simple crypsis with respect to the non-biological aspects of the background. It includes **countershading**, general **homochromy**, and **disruptive colouration** (M.H. Robinson 1969), where the model is of no interest to the dupe. Also called **mimesis**.

evasive mimicry *see* **escape mimicry**.

evolutionarily stable strategy (ESS) a strategy that a population of species can adopt (under a given set of conditions) that cannot be taken over by a different strategy such as might occur in some (rare) individuals through, for example, mutation or recombination.

exogenous camouflage use of particles or other things from the environment to decorate an animal (e.g. decorator crabs) or plant (sticky succulents that soil/sand adhere to) to help conceal it (cf. **endogenous camouflage**).

facultative crypsis the ability of some colour-changing prey to produce different camouflage types in response to detecting different types of predator (see Stuart-Fox & Moussalli 2009).

feeding mimicry term proposed by Schuett et al. (1984) for **aggressive mimicry** q.v. because these mimicries involve predators and prey and are therefore conceptually different from aggressive interactions, which usually involve conspecifics or competitors but do not involve prey.

female mimicry condition-dependent mimicry of females by some conspecific males that may enable them to gain sneaky matings, reduce aggression by other males, or reduce risk of predation and/or may possibly have physiological benefits.

flicker-fusion effect of an intermittent light stimulus (e.g. flashes, moving stripes) appearing to be completely steady or a constant intermediate shade to the observer (typically an average human).

frequency-dependent selection fitness/survivorship of a species or form is a function of that animal's abundance – it may be a negative or positive relationship.

fruit-set production of fruit as a result of fertilisation of an ovule or ovules.

functional response the relationship between prey abundance and a predator's rate of prey capture.

gaster the part of the abdomen of an hymenopteran insect posterior to the narrow waist or waists.

gentes 'races' of cuckoos that lay differently marked/coloured eggs and favour different brood host species; gente traits are inherited/expressed only matrilineally.

***Gestalt* perception** perception of a pattern as a whole, rather than responding only to single elements of it.

Gilbertian mimicry term coined by Pasteur (1982) for instances in which a potential prey or host (or part of it) mimics its potential predator, parasite or herbivore to reduce probability of attack. Examples include butterfly egg dummies on *Passiflora* and *Streptanthus* plants that reduce oviposition by butterflies, or spider mimicry by various insects (S1R/S2).

glochidia the microscopic larval stage of some freshwater bivalve molluscs.

glucoside a **glycoside** (q.v.) in which the sugar group is a glucose molecule.

glucosinolates class of pungent chemical compounds including sinigrin that are characteristic of Brassicaceae and often sequestered by insects that feed on them. Their chemical structure includes a sugar ring with a sulphur, nitrogen and oxygen-containing side chain.

glycan either a polysaccharide formed from many mono-saccharide units (sugars) linked by glycosidic bonds or the sugar subunits of glycoproteins or glycolipids.

glycoside an organic molecule comprising a sugar group and some other functional group linked by a glycosidic bond.

gorgonians marine, sessile Cnidaria of the subclass Octocorallia, order Gorgonacea, primarily from tropical and subtropical shallow waters with branched structures that form the substrate for many associated cryptic animals.

green beard selection as formulated by Guilford (1988), an alternative to **kin selection** as an explanation for the evolution of **aposematism** in unpalatable prey. In his formulation, the **aposematism** is uncheatable, like a 'green beard' indicating an altruistic individual. It differs from **kin selection** in that the individuals with the 'green beard' character share a common phenotype but are not necessarily close kin and are not gaining benefit because of kinship, only their phenotype. *See also* **synergistic selection** and S.B. Malcolm (1990).

group mimicry special term used to describe a few rare cases of (mostly protective) mimicries in which the resemblance to the model depends on there being a group of mimics arranged or interacting in a particular way. Examples include mimicry of sea anemones by groups of worms or fish (Wickler 1968) and mimicry of the nest of polistine wasps by caterpillars (Millot 1946; but see Chapter 7, section *Wasp (and bee) mimicry*).

G$_{ST}$ standardised genetic variation, a measure used to express how much of the total genetic variation of a species/sample is due to variation between populations. Low values (near zero) indicate that little of the variation is due to differences between populations, whereas high values denote that a large amount of the variation is due to inter-population differences.

gynochromatypic term proposed by Hilton (1987) for females with typical female colour patterns – *see* **female mimicry** (cf. **androchromatypic**)

handicap principle concept introduced by Zahavi (1977) that if an individual can survive to reproduction while expressing some feature that is unnecessary and costly for survival (the handicap) their other genes must surely be of high quality to compensate, e.g. the large and heavy tails of male peacocks.

haplochromines members of the cichlid fish tribe Haplochromini.

Hardy–Weinberg equilibrium in large populations, in the absence of selection and random mating, the expected frequencies of the genotypes of individuals with two alleles having frequencies p and q ($q = 1 - p$) will be p^2, $2pq$ and q^2.

hemiparasite a plant that gains nourishment through its own photosynthesis as well as parasitically from another plant.

heterochromy having two or more different colour forms.

heterochromatypic term sometimes used for females with typical female colour patterns, but this usage is confusing and originates from a misuse of the prefix hetero- (Hilton 1987) (cf. **homomorphic, androchromatypic**).

heterozygote an individual of a diploid species that has two different alleles at a given gene locus, one from each of its parents (cf. **homozygote**).

homeochromy *see* **homochromy**.

homochemy having the same chemical appearance.

homochromy (1) term used by Pasteur (1982) for resemblances only based on colour; (2) sometimes used to indicate mimicry of males by females but this usage is confusing and originates from a misuse of the prefix 'homo-' (Hilton 1987).

homoelectry term coined by Pasteur (1982) for, as yet unknown though potentially possible cases of resemblance of an electrical signal.

homokinemy term coined by Pasteur (1982) for resemblances involving mode of movement, such as ant mimics might display.

homomorphy term coined by Pasteur (1982) for resemblances involving body shape or outline.

homophony term coined by Pasteur (1982) for resemblances involving sound.

homophoty term coined by Pasteur (1982) for resemblances involving bioluminescence.

homothermy term normally used in connection with warm-blooded animals that maintain a more or less constant body temperature. Rarely used to refer to situations in which the mimic is at the same temperature as the model and the dupe can detect temperature.

homotypy broadly, a term coined by Pasteur (1982) for resemblance to another living organism that causes a reaction in the signal receiver, but he gives particular attention to its usefulness in cases of **arithmetic** and **Müllerian mimicry** which do not involve any deceit of a predator. Can be divided into 'concrete homotypy', in which the model is an identifiable species or group of species, and 'abstract homotypy', in which the model is some general category of organism, e.g. a snake or a wasp.

homozygote an individual of a diploid species that has two effectively identical alleles at a given gene locus, one from each of its parents (cf. **heterozygote**)

hypertelic crypsis originally used in various publications by the Austrian orthopterist Brunner von Wattenwyl for cases of crypsis or masquerade which in his view were far more than was necessary to provide protection from predators. Given the spectacular mimicries of some orthopteroids (see Figs 3.14c,d and 8.2c,d) it is easy to understand his thoughts, but natural selection led to the evolution of these minute details so it cannot possibly have been excessive.

illicium the combined elongate fin ray and its distal lure of certain 'anglerfish'; *see also* **esca**.

industrial melanism evolved melanism (q.v.) as a result of industrial pollution.

infructescence a cluster of fruit derived from the multiple separate flowers of an inflorescence.

inquiline an animal (often an insect) that lives or develops in a nest or gall made by a different species.

instar each separate growth stage of development of an insect larva separated by moults.

interference competition competition for a scarce resource in which foragers interact (fight) with each other directly.

introgression the transfer of gene alleles from one species to another by occasional hybridisation events and subsequent crossing of the hybrid with members of one of the parental species.

iridoid glycosides compounds formed from an iridoid monoterpene and a sugar group, which is the form in which many plants store their iridoid chemicals, for example, catalpol. Main plant sources include Ericaceae, Loganiaceae, Gentianaceae, Rubiaceae, Verbenaceae, Lamiaceae, Oleaceae, Plantaginaceae, Scrophulariaceae, Valerianaceae and Menyanthaceae, but they are also synthesised by some ants, notably *Iridomyrmex*.

iridophores type of chromatophore cells that contain white, reflective or iridescent granules.

isochromy sometimes used to indicate mimicry of males by females (Hilton 1987).

isomorphy *see* **isochromy**.

kin selection a mechanism in natural selection whereby the frequency of an allele (or other genetic change) can evolve through the net advantage it bestows on a kin group even if some individuals with the genetic feature may suffer mortality as a result (Maynard Smith 1964, Guilford 1985). See R. Dawkins (1979) for common misuses of the term.

Kirbyan mimicry term proposed by Pasteur (1982) for mimicry by brood parasites such as cuckoo eggs and nestlings. The Reverend William Kirby had initially described apparent host mimicry by hoverflies and robberflies associated with aculeate hymenopteran colonies (Kirby & Spence 1823). Quicke et al. (1992) discussed the lack of precise definition of this (and some other) term(s) and apparent discordance between some of the examples used in their original definitions.

kleptoparasitic obtaining food by stealing it from another animal that has already caught it.

lectins tetrameric proteins that bind and aggregate sugar residues (moieties) on cell surface membranes; includes phytohaemagglutinins.

linamarin a cyanogenic glucoside that produces hydrocyanic acid upon hydrolysis.

linkage disequilibrium genetic term for when the frequencies of allele combinations at two polymorphic loci are not independent. For example, with alleles A and a at locus one, and B and b at locus two, the frequencies of individuals with any combination, i.e. AB, Ab, aB or ab, departs significantly from random expectations, and may result from selective mortality of some combinations or genetic linkage on the same chromosome.

lithium chloride inorganic salt that is an emetic for mammals, usually causing vomiting within 6 hours of consumption. It has a distinctive taste that mammals, such as ungulates, can learn to associate with its ill-effects.

locomotor mimicry similarity in swimming, walking or flying behaviours of a mimic that have evolved to reduce the ability of a predator to differentiate between them and the model on the basis of their movement (Srygley 1994).

lotaustralin a cyanogenic glucoside that produces hydrocyanic acid upon hydrolysis.

masquerade this term, introduced by Endler (1981), is now well-established to specify those types of camouflage where the thing being mimicked is clearly defined (identifiable) but not of potential interest to a predator or prey, e.g. twig-mimicking caterpillars or stick insects, potoos (Nyctibiidae) and frogmouths (Podargidae) resembling broken stumps, various fish and insects resembling dead leaves, etc.

melanin a black or dark brown pigment found widely in metazoans and generally located in specialised cells called melanocytes. It (or more strictly, when a pigment, the form called eumelanin) is formed by the action of tyrosinase on tyrosine, leading to the production of DOPA and then L-dopaquinone; the last molecule then undergoes ring closure to form L-leucodopachrome which is oxidised to form L-dopachrome which is the monomeric unit of the melanin polymer chain. As well as a pigment function, melanised structures are frequently also physically harder than non-melanised ones (see Bonser 1995).

melanism occurrence of higher proportion than normal of a very dark (often nearly entirely black) form of a colour-polymorphic species.

melanophore a dermal cell containing multiple vesicles filled with the dark pigment, **melanin**, which can change overall appearance between light and dark depending on whether the **melanosomes** are aggregated into a single small dark patch or broadly spread through the cell's cytoplasm.

melanosome a **melanin**-rich intracellular particle/vesicle in a dermal **melanophore** cell.

Mertensian mimicry *see* **Emsleyan mimicry**.

mesic of habitats with a moderate amount of available moisture/water.

mesosoma the middle part of the body of a hymenopteran insect (ant, bee, wasp) which comprises the three thoracic segments and the first abdominal one.

metasoma the posterior-most part of the body of a hymenopteran insect (ant, bee, wasp) which comprises the abdominal segments excluding the first one.

methiocarb a carbamate chemical (3,5-dimethyl-4-(methylthio)phenyl methylcarbamate) used as a bird repellent as well as a pesticide.

methyl anthranilate compound distasteful to birds and used to flavour food in some prey-aversion learning experiments.

mimesis essentially almost the same as **masquerade**, the model is defined as uninteresting to the dupe, in shape as well as colour and sometimes movement. Pasteur (1982) subdivides this into cryptic mimesis and **phaneric mimesis** (q.v.).

miracidia free-living, mobile larval stage of a fluke (Trematoda) that give rise to the sporocyst stage when they infect an intermediate host.

monochromatic being of only one colour though usually of varying intensities.

monoecious (1) in a broad sense, plants that have male and female reproductive modes in the same individual (cf. **dioecious**); (2) in a narrow sense, as in (1), but male and female flowers are separate, as opposed to each flower being bisexual.

monomorphic having only one form.

motion camouflage moving in a way that reduces the probability of the movement being detected.

motion dazzle where markings make estimation of speed and direction of movement difficult for an observer to determine.

Müllerian mimicry/convergence (1) resemblance between equally or near equally unpalatable, **aposematic** prey which dilutes the cost of predator learning across two or more species. It is unlikely that it is ever precisely achieved in nature but approximately similar unpalatabilities will have the same effect if there are alternative prey. (2) In plants, the term is sometimes, and I believe quite inappropriately, used for similarity between co-occurring flowers of different plant species, which resemble one another and all offer some degree of reward (e.g. Dafni 1984).

multimodal model concept – not well-supported – that poor mimics have evolved to resemble an average appearance of a range of different potential model species rather than being a good mimic of any one of them (M. Edmunds 2000).

mutualists different species of organism whose actions benefit each other.

mycelia masses of fungal hyphae.

myrmechory being carried and dispersed by ants, such as various plant seeds and putatively mimetic phasmid eggs.

myrmecomorphy structural resemblance to an ant as shown, for example, by numerous species of spider (P.O. Oliveira & Sazima 1984), bugs and other insects; see McIver & Stonedahl (1993) for a review, and Chapter 7.

myrmecophile an organism that lives in close association with ants, such as many staphylinid beetles that inhabit ants nests or aphids that are regularly attended by ants.

nectariferous producing nectar.

nematocytes cells of cnidarians that contain on a large organelle (the nematocyst) which explosively fires a venom-containing thread when triggered (also called cnidocyte/cnidocyst).

neophobia literally, fear of the new; useful term, often thought to be a contributing factor in the evolution of **aposematism** associated with unpalatability, in that a mutation leading to increased conspicuousness might add to the attack latency or avoidance of a potential predator.

Nicolaian mimicry named in reference to a paper by Nicolai (1964) concerns bipolar S1/S2R systems in which a female will only mate with a male that mimics her host or prey. The original system involves brood-parasitic widow birds in which, to gain a mating, the male must mimic the estrildid finch host's vocalisations. Pasteur (1982) points out that mating in some *Photuris* and *Photinus* fireflies follows the same system.

novel worlds concerning experiments where prey signals would be completely unfamiliar to predators (e.g. Fig. 4.8) so as to avoid biases due to innate or previously learned signals (see LG Morrell & Turner 1970, Lea & Turner 1972 for early examples)

numerical mimicry *see* **arithmetic mimicry**.

obligate necessary.

obliterative shading an aspect of **countershading** (q.v.) that makes the three-dimensional nature of an object hard to detect.

observational learning when an animal sees the effect of a behaviour (for example eating a toxic prey or being bitten by a snake) on another individual (usually a conspecific) and learns to avoid these prey because of that rather than firsthand experience.

ommochromes a family of heterocyclic pigments derived from tryptophan found in invertebrate eyes and cephalopod chromatophores.

operator another name used for **signal receiver** or **dupe**, specifically the species in a mimicry system that brings about or maintains the mimicry through being deceived.

opportunity cost a particular appearance, for example being cryptic against a particular sort of background, may limit an individual's potential to forage on different types of background because they will no longer be as cryptic and so will experience increased predation; introduced by Speed & Ruxton (2005a) to help explain why some species evolve **aposematism**.

osmophores scent glands, usually referring to those on plants that secrete scent compounds associated with flowering.

otilith annuli growth rings in the calcium carbonate inner ear bones of fish. Denser zones are formed when growth is slowest and typically indicate winters, in which case the number of rings corresponds to age in years.

Oudemans' phenomenon when a pattern on two or more separate parts of a body match up when the animal is in its normal posture, such as the striped pattern that extends across the back, thigh and shin of many raniid frogs when they are sitting at rest, or across the wings of resting moths (Oudemans 1903; see also Cott 1940, Komárek 2003).

Owen–Owen effect when **aposematic** species are not equally unpalatable, small relative numbers of the more palatable species can lead to increased predation on the mimicry complex as a whole (Speed 1999a).

palatability general term indicating the desirability of an organism as a food resource for a given predator, cf. **unpalatability**.

panmixis when all individuals in a sexual species are equally likely to mate with all others.

parasitoid an animal, usually an insect, that feeds on or in another organism, its host, for a period but ultimately kills it.

Pavlovian predator a predator that learns about the potential reward of a prey from associative learning, as with the salivation in response to hearing a bell associated with feeding time in Pavlov's experimental dogs (Speed 1999b, Honma et al. 2008). Such predators are assumed (1) to have sufficient alternative prey that they have a constant hunger level; (2) to be naïve predators with some neophobia so attack novel prey with an intermediate probability; (3) to show reinforced learning up to some asymptotic value; (4) to forget over time (Speed 1993a).

Peckhamian mimicry term coined by Wickler (1968) for any **aggressive mimicry**. However, Pasteur (1982) considered the term unacceptable because there are several different sorts of **aggressive mimicry**, and because the concept was well understood before Peckham's work. Nevertheless, Peckham (1889) and Peckham & Peckham (1892) only deal with the specific example of spiders (and by implication others) that mimic a group-foraging animal, such as ants, such that they can infiltrate among them to facilitate attacking them. Endler (1981) notes that it might include three separate types, depending on whether the model is just the background, a specific part of the background, or an animal or reproductive part of a plant.

perianth that part of the flower comprising the calyx (sepals) and corolla (petals) that surround the reproductive parts.

pericarp the outer fleshy layer of a fruit which develops from the ovary wall and surrounds the seeds (*see also* **aril**).

pericaudal near the tail or tail end of an animal.

phagostimulatory compound or compounds that stimulate feeding.

phaneric mimesis term introduced by Pasteur (1982) for **masquerade** cases in which the model is a relatively rare and inanimate part of the environmental background; his examples are mimics of animal droppings, pebbles, etc..

phenotype physical appearance and general physiological capability of an organism resulting from the interaction between its genotype and its environment.

pheromone a chemical compound released by one individual of a species and received by another of the same species, whose behaviour and/or physiology they affect.

phospholipase enzyme commonly present in vespid wasp venoms that break cell membranes. There are two major groups of these 'A' and 'B'.

phragmosis protection of a burrowing animal which blocks and sometimes camouflages the entrance to its burrow with part of its own body (e.g. D.E. Wheeler & Holldobler 1985).

phytoecdysteroid triterpenoid chemical mimics of insect moulting hormone (ecdysone) synthesised by various plants, which, when consumed by insects, lead to premature moulting, weight loss and often death (Dinan 2001).

phytomimesis animal mimetic resemblance to plants (Wickler 1968).

piperidine alkaloids major venom constituents of fire ants (*Solenopsis* spp.) and, responsible for a lot of the pain they cause.

pleiotropy effects of a particular allele other than the main effect that is normally considered; for example, an allele that causes **melanism** (q.v.) might also affect cuticle hardness, resistance to disease or fertility (or all three for that matter). Probably the majority of genes have some pleiotropic effects.

plesiomorphy the ancestral character state (*see also* **apomorphic**, **synapomorphy**, **symplesiomorphy**).

plurivoltine having more than one generation per year.

poikilothermic organisms whose body temperature is highly variable and largely controlled by their environmental temperature.

polarotactic organisms whose movements are influenced by their detection of polarised light.

pollinia (pollinaria) the coherent bags of pollen made by many orchid flowers and also some milkweeds (Asclepiadoideae) that detach from the anther as they attach to the potential pollinator as a single unit.

polygynous of males that have more than one mate.

Pouyannian mimicry term coined by Pasteur (1982) after the paper by Pouyanne (1917) for cases in which (normally) plants obtain pollination through deceitful induction of pseudocopulation by an insect – typically a hymenopteran (S1R/S2).

predator mimicry term originally coined to cover a particular case of presumed protective mimicry in some cannibalistic cichlid fish, in which resemblance to a predator is believed to deter conspecifics from attempting cannibalism (Zaret 1977, Wiens 1978); however, other instances in which potential predators are mimicked include situations in which the predator is the **signal receiver** ('sheep in wolves clothing') and situations which can appear to be essentially either **Batesian** or **Müllerian** or both.

prepuce the foreskin of a penis or the hood of the clitoris.

primary defence an aspect of an organism's phenotype that stops a predator or herbivore initiating an attack, either by circumventing detection by hiding, being camouflaged or giving signals, such as warning colouration, on which a predator bases a decision not to attack (cf. **secondary defence**).

procrypsis term originally coined by Poulton (1890) for protective resemblance to the surroundings – no longer used much (see also Starrett 1993).

proteaceous belonging to the plant family Proteaceae.

protean display highly unpredictable behaviour of prey animal that makes it impossible for a predator to predict exactly where it will move at any given moment.

proxy variable a variable that is correlated with the variable of interest, the latter usually being difficult or impossible to measure directly.

pseudo-aposematic sometimes used to mean the same as **Batesian** mimicry (Poulton 1890; see Starrett 1993), but more precisely means conspicuous colouration of a palatable animal, for example the brightly- coloured hind wings of many cryptic moths that are used in **startle displays** (Komárek 2003).

pseudo-episematic a character that leads to an animal making an incorrect identification of the species displaying it.

pseudocopulation in which a sexually deceptive mimetic plant, such as many orchids, dupe males of their pollinator species such that they attempt to copulate with the flower as if it was a conspecific female. Also called **Pouyannian mimicry**.

pseudoflowers flower-like structures that attract pollinators, but which are caused by plant-pathogenic fungi as a result of natural selection towards a means of spore dispersal utilising the host plant's normal pollinators.

pseudonectaries glistening structures on plants that appear to potential pollinators as droplets of nectar but which are not.

pseudopenis a structure resembling a penis, usually erect.

pseudopheromone term sometimes used for the insect sex pheromone-mimicking chemical blends given off by sexually deceptive orchids or bolas spiders.

pseudoreplication a problem in the statistical analysis of experimental data due to replicating samples but not replicating treatments, for example, the same experimental animal may be used in a series of replicates of a test, but all that can be inferred from the results is the behaviour of that single individual. Similarly, you might be trying to test whether pollination of a target species is affected by the number of other flowers open at the same time and make measurements every month but only at one site (repeated measures). Any 'significant' effect could be due to completely extrinsic factors and/or be site-specific. To make a valid inference, the measurements must each come from different (independent) sites.

pseudosting a sting- or tail-like structure at the end of the abdomen of a non-stinging animal. Rothschild (1984) illustrates one in the zygaenid moth, *Cyclosia papilionaris*. Males of several Hymenoptera have the aedeagus and parameres of their genitalia strongly produced and jab these in their **pseudostinging behaviour** (q.v.) into potential predators. *See also* **aide mémoire mimicry**.

pseudostinging behaviour usually by insects, the probing of the tip of the abdomen when an individual is captured in a way that more or less resembles the stinging motion made by captured aculeate Hymenoptera females. Some hymenopteran taxa such as the typhiid wasp, *Myzinum*, have the male genitalia protruding quite visibly, reminiscent of a female's true 'stinger' (Fig. 7.11).

putrecine tetramethylenediamine, a foul-smelling compound produced by amino acid breakdown during putrescence and, together with **cadaverine**, largely responsible for the smell of rotting flesh; produced by various fly-pollinated flowers.

pyrazines heterocyclic aromatic organic compounds based on the ring structure ($C_4H_4N_2$; 1,4-diazabenzene). 2-Methoxy-3-alkylpyrazines are odour components associated with various **aposematic** and distasteful insects, including *Danaus plexippus*, *Zygaena lonicerae* and various arctiine moths such as *Amata* species (see Fig. 4.3). There is evidence that they have a warning odour function (Rothschild et al. 1984b, Guilford et al. 1987). Some insect pheromone blends also contain pyrazines and those of a thynnine wasp are mimicked by the sexually deceptive orchid, *Drakaea livida* (Bohman & Peakall 2014).

pyrrolizidine alkaloids (PAs) various dicyclic alkaloids of plant origin that generally act as toxins against herbivores, but which are also sequestered by certain insect species as a protection against predation, though not always via their larval diet (Nishida 1995). They are mostly known to protect against predation by invertebrate predators (K.S. Brown 1984a, A.R. Masters 1990), but there is some evidence that vertebrates will also avoid prey containing them (Cardoso 1997). They are mostly associated with Apocynaceae, Asteraceae, Boraginaceae, Convolvulaceae, Fabaceae, Orchidaceae and Poaceae. PAs are highly toxic to mammals but their effects are not immediate and cause, among other things, liver failure and carcinomas.

quasi-Batesian mimicry anti-predator Müllerian mimicry system in which some unpalatable species (individuals) are less unpalatable than others (Speed 1993a) and which can result in a mutualistic relationship between model and mimic (Balogh et al. 2008) or the more unpalatable model can suffer because of the presence of more palatable mimics (MacDougall & Dawkins 1998).

quasi-Müllerian mimicry situation in which reduced discrimination error by a predator causes mimicry by a much more palatable prey to actually increase the fitness of the model (MacDougall & Dawkins 1998, Kokko et al. 2003).

quinine, quinine dihydrochloride a very bitter compound often used as the harmless distasteful agent in experimental mimicry systems.

ramet individual plant in a clonally produced colony.

raphids plant internal needle-shaped rods of calcium oxalate that are sharp at both ends and produced in special parenchymal cells in plants called idioblasts. When the tissue is damaged they disperse and may penetrate a herbivore's gut wall.

receiver the organism in a mimicry system which is deceived and whose responses exert the selection pressure to create and maintain the resemblances.

receiver psychology term used to encompass various basic aspects of signal receivers, predominantly in the context of **aposematism** and protective colouration, that affect their response to various signals, such as what features they find most memorable, how they react to novel stimuli, how ready they are to 'experiment', etc. (Guilford & Dawkins 1991, 1993a).

red-green colour blind condition in humans in which an individual has a defective allele coding for one or other of the long-wavelength (red) or medium-wavelength (green) opsins present in colour-sensing retinal cone cells; the defective alleles are recessive and carried on the X chromosome so red-green colour blindness is approximately 80 times more prevalent in males than in females.

reflexive selection term coined by Moment (1962a) for situations in which selection has led to extreme diversity of colour and/or form so as to reduce effectiveness of a predator's search image capabilities.

regicide in brood-parasitic Hymenoptera, the killing of the queen of the colony.

reproductive mimicry term first introduced by Wiens (1978) for those cases in which mimicry has evolved to facilitate or enable fertilisation, such as the many pollination deceptions of flowering plants.

rictal flanges fleshy, often brightly coloured tissue near the base of a juvenile bird's beak that is exposed when the gape is open.

salicin an alcoholic β-glucoside, chemically related to aspirin, that can be toxic to some animals especially in high concentration. It is sequestered by various *Salix*-feeding insects.

sapromyiophilous name for flowers such as *Stapelias* that attract carrion flies and other insects that are normally associated with rotting environments.

sarmentosin a cyanogenic glucoside sequestered by some zygaenid, geometrid and yponomeutid larvae that feed on Crassulaceae and Celastraceae species (Nishida 1995).

satyric mimicry an explanation of the success of apparently imperfect mimicry (especially **Batesian mimicry**) that involves a predator having an incomplete/imperfect cognition of the pattern, which leads to a small delay in signal processing (prey recognition), that might therefore give an imperfect mimic the additional time needed to escape (see Howse & Allen 1994, Howse 2013).

school-orientated mimicry term proposed by Dafni & Diamant (1984) for the benefit obtained by different species resembling one another schooling together, thus increasing overall school size and making it harder for predators to attack individuals successfully.

secondary defence a protective aspect of an individual that makes it unprofitable for a predator or herbivore to continue pursuing an attack, having already detected and started to attack them. The category includes aggression, i.e. posing a danger to the predator, speed of escape, spines, toxins, warning odours, resilience to the attack such as an armadillo rolling into a ball, etc. (see M.H. Robinson 1969, M. Edmunds 1974a, Ruxton et al. 2004a).

seed-set production of seeds after successful fertilisation of one or more ovules.

selective agent the organism whose response leads to, or maintains, the mimetic resemblance.

self mimesis another term for **thanatosis** (Pasteur 1982).

self mimicry another name for a category of **automimicry** (q.v.).

self shadow concealment one of the functions of **countershading** (q.v.) in many organisms whereby the shadow a body casts on itself is counterbalanced by that shadowed part being paler coloured.

senesce deterioration as a result of age.

serial mimicry phenomenon described by Plowright & Owen (1980) in relation to the mimetic resemblances of sympatric bumblebee and cuckoo bumblebee species in which the mimic species emerges later in the season than its model, affording predators first to experience the unpalatable type. In reality, predators might have learned from the previous year, or survival from attack might be high enough to make timing less important (also see Mallet 2001).

sex-limited mimicry cases of mimicry in which members of only one sex are mimics or in which the two sexes have different models.

signal receiver the individual (or species) that detects and responds (or does not respond) to a deceptive signal.

siphuncles pair of cuticular tubes protruding from postero-dorsal part of an aphid's abdomen that release pheromones and defensive secretions.

skatole 3-methylindole, a compound produced naturally in faeces from tryptophan degradation. It has a strong faecal odour and is a constituent of the volatiles given off by several dung-mimicking Araceae (arums) at **anthesis**, i.e. the time the flower becomes fully functional.

Snell's window the zone in which a submerged object viewed through the water surface from above is visible, because light from the object hitting the water surface at angles of 49 degrees or more are reflected back under water (see Fig. 10.2).

social mimicry (1) in non-humans usually referring refers to the similarity between species that associate together in schools or flocks; (2) in humans term is generally applied for human–human interactions in which a person from one social category (consciously) mimics someone from another in the hope of gaining some advantage. It is also relevant to various non-human primates.

somatolysis disruptive cryptic colouration or pattern which makes an animal seem to be two or more separate entities.

special protective resemblance or **special resemblance** term coined by Cott (1940) to cover **masquerade**, i.e. similarity to definable elements of the environment that are of no interest to a potential predator; for example mimicry of twigs, leaves, thorns, bird- droppings, pebbles, bubbles, etc.

sporocyst an immobile, sack-like larval stage of a fluke (Trematoda) in which germ cells divide to form clusters that will each develop into the next stage, the mobile redia larvae.

stabilimentum (-a) dense and highly visible patch(es) of silk made by some orb web-weaving spiders of uncertain

and probably mixed function, some of which may be deceptive. Sometimes they involve addition of plant debris or egg sacs (see Eberhard 2003).

staminode a rudimentary or aborted stamen that does not produce pollen.

startle displays a sudden behaviour, often associated with flash colouration, that may, at least temporarily deter an attack.

stigma exudates secretions of a plant's stigma that may reward pollinators.

stolons horizontal stems, usually just under the soil surface, connecting two plants.

stridulation sound production through rubbing a peg or ridge over a corrugated surface.

studliness term employed by Rohwer (1975) to describe the extent of bright and contrasting colouration of male individuals within bird flocks (Balph et al. 1979).

sulcus technical term for a groove.

superconglutinate a cohesive mass of parasitic **glochidia** larvae of some freshwater bivalve molluscs such as some *Lampsilis* species.

super-Müllerian mimicry system in which the benefits of resemblance between two **aposematic unpalatable** species exceed that of traditional **Müllerian mimicry** (Balogh et al. 2008).

super stimulus situation in which a presented stimulus that is larger than that which would normally occur naturally, because of the way an organism's brain processes the key signal features, causes a greater than normal response and may override other conflicting signals.

switching when a predator's preference for a commoner type of prey reduces its population such that it is no longer the commonest and so preference shifts to a formerly rarer type; *see* **apostatic selection**.

sympatric of two species (or sometimes races) that co-exist at the same locality.

symphile an organism (usually referring to insects) that live within the colony of a social insect (usually ants or termites) and which are fed by the host insect, for example the larvae of some lycaenid butterflies.

symplesiomorphy an underived (primitive) character state found in multiple species as a result of it having been present in their common ancestor but subsequently lost in other members of the group and thus provides no information about their relationship to each other.

synapomorphy a derived character state shared by two or more species as a result of it having evolved in their common ancestor and so possession of the state indicates that the species form a monophyletic group.

synaposematic more or less the same as being a **Batesian mimic** of a warningly coloured, defended species.

syncryptic similarity between organisms that results from independent selection on them to resemble their environment.

synechthran an insect that is an unwelcome 'guest' within the nest of a social insect.

synergistic selection selection favouring only those individuals that share a trait as in **green beard selection** (q.v.) (cf. **kin selection**).

synoekete an insect that lives within the colony of a social insect without being attacked.

teliospore thick-walled dormant stage of some smut and rust fungi.

termitophile an animal that habitually and usually obligately lives in association with termite colonies.

thanatosis death feigning, i.e. mimicry of itself in a dead state; derived from *thanatos*, the Greek word for death.

titre concentration of a chemical or antibody, originally as determined by titration.

tonic immobility *see* **thanatosis**.

trade-off in biology, the idea that expression of a character (i.e. of a gene) improves one aspect of fitness but necessarily reduces another.

transcription factor a protein that binds to a specific DNA sequence and then controls the rate at which messenger RNA is produced through transcription of the DNA.

transformational mimicry the mimetic tracking of multiple, differently sized models by a mimic as it grows, for example, various ant-mimicking Hemiptera and spiders (e.g. J.F. Jackson & Drummond 1974).

transposable element a DNA region that can insert itself in other parts of the genome, sometimes through duplication; also called 'jumping genes' or transposons.

trichomes small hair-like cells protruding from the surface of a plant.

trichromat an animal with three types of colour receptor, each maximally sensitive to a different wavelength of light.

triungulin a specialised first instar larval stage of beetles of the family Meloidae, that is highly scleritised and mobile and involved in host location (a subclass of planidium).

trophallaxis direct transfer of food from one individual to another, either mouth to mouth or anus to mouth.

trophobiont species (or individual) that provides food for another, usually in the form of a secretion.

übernormale Attrappen of a mimetic feature that is far larger or bolder than the thing being mimicked and which has a stronger effect on the receiver.

umbel an inflorescence in which the branching stalks leading to the individual flowers arise from a common point and can result in a rather flat surface of flowers, typical of the Apiaceae, or nearly spherical flower arrangements, depending on branch lengths.

univoltine having only one generation per year.

unpalatable general term applied to any feature that makes an organism, or part there-of, something a predator or herbivore does not want to eat. It could be flavour, poison, texture, thorns, odour or lack of digestibility.

urotomy shedding of the tail, as exhibited by some snakes and lizards as a means of escaping capture and often, as the discarded part may carry on wriggling, as a distraction to their pursuer (*see also* **caudal autotomy**).

Vavilovian mimicry also called **weed mimicry** and named after Nikolai Vavilov because of his 1922 paper. Vavilov was a famous Russian botanist and geneticist who worked on improving cereal crops and identifying their regions of origin. The mimicry evolved primarily through the action of humans who, by separating out weed seeds or killing weed plants among crops before they set seed, effectively selected for ones whose mature form and seeds were more similar to those of the crop plant itself.

vermiform worm-like in shape.

virtual model in those mimetic cases where the model is not a particular (definable) species.

Wasmannian mimicry as usually defined, this is mimicry by **myrmecophilous** arthropods of the body form and texture of their ant host or associated ants (Wasmann 1925, McIver & Stonedahl 1993). Pasteur (1982) additionally included chemical mimicry of ants. **Mutualist** and **commensal** organisms in ants' nests or with army ants are included, as well as **trophobiont** aphids. The term thus covers several distinct types of mimetic relationship.

weapon automimicry similarity between a non-dangerous part of an animal and its own true weapons through exaggeration, or similarity to weapons of other individuals of its own species (or of another similar one), e.g. similarity of female pronghorn ear markings to male horns (Guthrie & Petocz 1970).

weed mimicry *see* **Vavilovian mimicry**.

Wicklerian–Barlowian mimicry term introduced by Pasteur (1982) for a class of mimicry in which females resemble **conspecific** males, i.e. **androchromatism**.

Wicklerian–Eisnerian mimicry term introduced by Pasteur (1982) for a **disjunct aggressive mimicry**, as with the sabre-toothed blenny/cleaner-fish system.

Wicklerian–Guthrian mimicry a termed coined by Pasteur (1982) for the **automimicry** displayed by male mandrills (*Mandrillus sphinx*), whose face mimics their genitals.

Wicklerian mimicry term created by Pasteur (1982) to cover **conspecific** mimicries by the opposite sex that function to minimise conflict/aggression, e.g. the red buttocks of subordinate male hamadryas baboons.

xanthophore chromatophore cell containing yellow pigment.

zoomimesis animal mimetic resemblance to animal or animal artefact (Wickler 1968).

REFERENCES

Abbott CE (1926) Death feigning in *Anax junius* and *Aeschna* sp. *Psyche*, **33**, 8–10.

Abramowitz AA (1937) The chromatophorotropic hormone of the Crustacea: standardization, properties and physiology of the eye-stalk glands. *Biological Bulletin*, **72**, 344–65. doi:10.2307/1537694

Acheampong A, Mitchell BK (1997) Quiescence in the Colorado potato beetle, *Leptinotarsa decemlineata*. *Entomologia Experimentalis et Applicata*, **82**, 83–9.

Ackerman JD (1981) Pollination biology of *Calypso bulbosa* var. *occidentalis* (Orchidaceae): a food-deception system. *Madroño*, **28**, 101–10.

Ackerman JD (1983) Euglossine bee pollination of the orchid *Cochleanthes lipscombiae*: a food source mimic. *American Journal of Botany*, **70**, 830–4.

Ackerman JD (1986) Mechanisms and evolution of food-deceptive pollination systems in orchids. *Lindleyana*, **1**, 108–13.

Adams JM, Kang C, June Wells M (2014) Are tropical butterflies more colorful? *Ecological Research*, **29**, 685–91.

Aiello A, Brown KS Jr (1988) Mimicry by illusion in a sexually dimorphic, day-flying moth, *Dysschema jansonis* (Lepidoptera: Arctiidae: Pericopinae). *Journal of Research on the Lepidoptera*, **26**, 173–6.

Akcali CK, Pfennig DW (2014) Rapid evolution of mimicry following local model extinction. *Biology Letters*, **10**: 20140304.

Akino T (2005) Chemical and behavioral study on the phytomimetic giant geometer *Biston robustum* Butler (Lepidoptera: Geometridae). *Applied Entomology and Zoology*, **40**, 497–505.

Akino T, Yamaoka R (1998) Chemical mimicry in the root aphid parasitoid *Paralipsis eikoae* Yasumatsu (Hymenoptera: Aphidiidae) of the aphid-attending ant *Lasius sakagamii* Yamauchi & Hayashida (Hymenoptera: Formicidae). *Chemoecology*, **8**, 153–61.

Akino T, Mochizuki R, Morimoto M, Yamaoka R (1996) Chemical camouflage of myrmecophilous cricket, *Myrmecophilus* sp., to be integrated with several ant species. *Japanese Journal of Applied Entomology and Zoology*, **40**, 39–46. [Japanese with English summary.]

Akino T, Knapp JJ, Thomas JA, Elmes GW (1999) Chemical mimicry and host specificity in the butterfly *Maculinea rebeli*, a social parasite of *Myrmica* ant colonies. *Proceedings of the Royal Society of London, Series B: Biological Sciences*, **266**, 1419–26.

Akino T, Nakamura K, Wakamura S (2004) Diet-induced chemical phytomimesis by twig-like caterpillars of *Biston robustum* Butler (Lepidoptera: Geometridae). *Chemoecology*, **14**, 165–74.

Akre RD, Rettenmeyer CW (1966) Behaviour of Staphylinidae associated with army ants (Formicidae: Ecitonini). *Journal of the Kansas Entomological Society*, **39**, 745–82.

Akre RD, Alpert G, Alpert T (1973) Life cycle and behaviour of *Microdon cothurnatus* in Washington (Diptera: Syrphidae). *Journal of the Kansas Entomological Society*, **46**, 827–38.

Alatalo RV, Mappes J (1996) Tracking the evolution of warning signals. *Nature*, **382**, 708–10.

Alcami A (2003) Viral mimicry of cytokines, chemokines and their receptors. *Nature Reviews Immunology*, **3**, 36–50.

Alcock J (1971) Interspecific differences in avian feeding behaviour and the evolution of Batesian mimicry. *Behaviour*, **40**, 1–9.

Alcock J (1973) The feeding response of hand-reared red-winged blackbirds (*Agelaius phoeniceus*) to a stink bug (*Euschistus conspersus*). *American Midland Naturalist*, **89**, 307–13.

Alcock J (2000) Interactions between the sexually deceptive orchid *Spiculaea ciliata* and its wasp pollinator *Thynnoturneria* sp. (Hymenoptera: Thynninae). *Journal of Natural History*, **34**, 629–36.

Aldrich JR, Blum MS (1978) Aposematic aggregation of a bug (Hemiptera: Coreidae): defensive display and formation of aggregations. *Biotropica*, **10**, 58–61.

Alexander HM (1990) Epidemiology of anther-smut infection in *Silene alba* caused by *Ustilago violacea*: patterns of spore deposition and disease incidence. *Journal of Ecology*, **78**, 166–79.

Alexandrou MA, Oliveira C, Maillard M, McGill RAR, Newton J, Creer S, Taylor MI (2011) Competition and phylogeny determine community structure in Müllerian co-mimics. *Nature*, **469**, 84–8.

Allan RA, Elgar MA (2001) Exploitation of the green tree ant, *Oecophylla smaragdina*, by the salticid spider, *Cosmophasis bitaeniata*. *Australian Journal of Zoology*, **49**, 129–37.

Allan RA, Capon RJ, Brown WV, Elgar MA (2002) Mimicry of host cuticular hydrocarbons by salticid spider *Cosmophasis bitaeniata* that preys on larvae of tree ants *Oecophylla smaragdina*. *Journal of Chemical Ecology*, **28**, 835–48.

Allen AA (1990) Death-feigning in *Exochomus quadripustulatus* L. (Col.:Coccinellidae). *Entomological Record*, **102**, 23.

Allen JA (1972) Evidence for stabilizing and apostatic selection by wild blackbirds. *Nature*, **237**, 348–9.

Allen JA (1988) Reflexive selection is apostatic selection. *Oikos*, **51**, 251–3.

Allen JA (1989) Searching for search image. *Trends in Ecology and Evolution*, **4**, 361.

Allen JA, Anderson KP (1984) Selection by passerine birds is anti-apostatic at high prey density. *Biological Journal of the Linnean Society*, **23**, 237–46.

Allen JA, Clarke B (1968) Evidence for apostatic selection by wild passerines. *Nature*, **220**, 501–2.

Allen JA, Cooper JM (1985) Crypsis and masquerade. *Journal of Biological Education*, **19**, 268–70.

Allen JA, Raymond DL, Geburtig MA (1988) Wild birds prefer the familiar morph when feeding on pastry-filled shells of the land-snail *Cepaea hortensis* (Muell.). *Biological Journal of the Linnean Society*, **33**, 395–401.

Allen JJ, Mäthger LM, Barbosa A, Buresch KC, Sogin E, Schwartz J, Chubb C, Hanlon RT (2010) Cuttlefish dynamic camouflage: responses to substrate choice and integration of multiple visual cues. *Proceedings of the Royal Society of London, Series B: Biological Sciences*, **277**, 1031–9.

Allen JJ, Bell GRR, Kuzirian AM, Velankar SS, Hanlon RT (2014) Comparative morphology of changeable skin papillae in octopus and cuttlefish. *Journal of Morphology*, **275**, 371–90.

Allen JJ, Akkaynak D, Sugden A, Hanlon RT (2015) Adaptive body patterning, three-dimensional skin morphology and camouflage measures of the slender filefish *Monacanthus tuckeri* on a Caribbean coral reef. *Biological Journal of the Linnean Society*, **116**, 377–96.

Allen WL, Baddeley R, Scott-Samuel NE, Cuthill IC (2013) The evolution and function of pattern diversity in snakes. *Behavioural Ecology*, **24**, 1237–50.

Alonso-Mejía A, Brower LP (1994) From model to mimic: age-dependent unpalatability in monarch butterflies. *Experientia*, **50**, 176–81.

Alvarez F (1994) A gens of cuckoo *Cuculus canorus* parasitizing rufous bush chat *Cercotrichas galactotes*. *Journal of Avian Biology*, **25**, 239–43.

Alvarez F (2000) Response to common cuckoo *Cuculus canorus* model egg size by a parasitized population of rufous bush chat *Cercotrichas galactotes*. *Ibis*, **142**, 683–6.

Alvarez F, Arias de Reyna L, Segura M (1976) Experimental brood parasitism of the magpie, *Pica pica*. *Animal Behaviour*, **24**, 907–16.

Alves RJV, Pinto AC, Da Costa AVM, Rezende CM (2005) *Zizyphus mauritiana* Lam. (Rhamnaceae) and the chemical composition of its floral fecal odor. *Journal of the Brazilian Chemical Society*, **16**, 654–6.

Amaoka K, Senou H, Ono A (1994) Record of the bothid flounder *Asterorhombus fijiensis* from the western Pacific, with observations on the use of the 1st dorsal fin ray as a lure. *Japanese Journal of Ichthyology*, **41**, 23–8.

Amoore JE, Forrester LJ, Buttery RG (1975) Specific anosmia to 1-pyrroline: The spermous primary odor. *Journal of Chemical Ecology*, **1**, 299–310.

Anderson BC, Johnson SD (2006) The effects of floral mimics and models on each others' fitness. *Proceedings of the Royal Society of London, Series B: Biological Sciences*, **273**, 969–74.

Anderson BC, Johnson SD, Carbutt C (2005) Exploitation of a specialized mutualism by a deceptive orchid. *American Journal of Botany*, **92**, 1342–9.

Anderson CM, Bielert CF (1994) Adolescent exaggeration in female catarrhine primates. *Primates*, **35**, 283–300.

Anderson MG, Ross HA, Brunton DH, Hauber ME (2009) Begging call matching between a specialist brood parasite and its host: a comparative approach to detect coevolution. *Biological Journal of the Linnean Society*, **98**, 208–16.

Andersson M (1976) *Lemmus lemmus*: a possible case of aposematic coloration and behavior. *Journal of Mammalogy*, **57**, 461–9.

Andersson M (2015) Aposematism and crypsis in a rodent: anti-predator defence of the Norwegian lemming. *Behavioral Ecology and Sociobiology*, **69**, 571–81.

Andrén C, Nilson G (1981) Reproductive success and risk of predation in normal and melanistic colour morphs of the adder, *Vipera berus*. *Biological Journal of the Linnean Society*, **15**, 235–46.

Antonovics J, Alexander HM (1992) Epidemiology of anther-smut infection in *Silene alba* caused by *Ustilago violacea*: patterns of spore deposition in experimental populations. *Proceedings of the Royal Society of London, Series B: Biological Sciences*, **250**, 157–63.

Aplin RT, D'Arcy Ward R, Rothschild M (1975) Examination of large and small white butterflies (*Pieris* spp) for presence of mustard oils and mustard oil glycosides. *Entomology (A)*, **50**, 73–8.

Arenas LM, Troscianko J, Stevens M (2014) Colour contrasts and stability as key elements for effective warning signals. *Frontiers in Ecology and Evolution*, **2**:25.

Ari C (2014) Rapid coloration changes of manta rays (Mobulidae). *Biological Journal of the Linnean Society*, **113**, 180–93.

Arieli R (2004) Breasts, buttocks and the camel hump. *Israel Journal of Zoology*, **50**, 87–91.

Armstrong EA (1954) The ecology of distraction display. *Animal Behaviour*, **2**, 111–17.

Arnold EN (1984) Evolutionary aspects of tail shedding in lizards and their relatives. *Journal of Natural History*, **18**, 127–69.

Arnold EN (1988) Caudal autotomy as a defence. In C Gans and R B Huey (eds), *Biology of the Reptilia*, Liss, New York, pp. 235–73.

Arnold SJ (1976) Sexual behavior, sexual interference and sexual defense in the salamanders *Ambystoma maculatum*, *Ambystoma tigrinum* and *Plethodon jordani*. *Zeitschrift für Tierpsychologie*, **42**, 247–300.

Arnold SJ (1977) Polymorphism and geographic variation in the feeding behavior of the garter snake, *Thamnophis elegans*. *Science*, **197**, 676–8.

Aronsson M, Gamberale-Stille G (2008) Domestic chicks primarily attend to colour, not pattern, when learning an aposematic colouration. *Animal Behaviour*, **75**, 417–23.

Aronsson M, Gamberale-Stille G (2009) Importance of internal pattern contrast and contrast against the background in aposematic signals. *Behavioral Ecology*, **20**, 1356–62.

Atkins D (1926) On nocturnal colour change in the pea-crab (*Pinnotheres veterum*). *Nature*, **117**, 415–6.

Atkins EL (1948) Mimicry between the drone fly, *Eristalis tenax* and the honeybee, *Apis mellifera*. Its significance in ancient mythology and present-day thought. *Annals of the Entomological Society of America*, **41**, 387–92.

Atkinson EC (1997) Singing for your supper: acoustical luring of avian prey by northern shrikes. *The Condor*, **99**, 203–6.

Atkinson WD, Warwick T (1983) The role of selection in the colour polymorphism of *Littorina rudis* Maton and *Littorina arcana* Hannaford-Ellis (Prosobranchia: Littorinidae). *Biological Journal of the Linnean Society*, **20**, 137–51.

Atwood JT (1985) Pollination of *Paphiopedilum rothschildianum*: brood-site imitation. *National Geographic Research*, **1**, 247–54.

Aubret F, Mangin A (2014) The snake hiss: potential acoustic mimicry in a viper–colubrid complex. *Biological Journal of the Linnean Society*, **113**, 1107–14.

Augner M, Bernays E (1998) Plant defence signals and Batesian mimicry. *Evolutionary Ecology*, **12**, 667–79.

Auko TH, Trad BM, Silvestre R (2015) Bird dropping masquerading of the nest by the potter wasp *Minixi suffusum* (Fox, 1899) (Hymenoptera: Vespidae: Eumeninae). *Tropical Zoology*, **28**, 56–65.

Averill CK (1923) Black wing tips. *The Condor*, **25**, 57–9.

Aviezer I, Lev-Yadun S (2015) Pod and seed defensive coloration (camouflage and mimicry) in the genus *Pisum*. *Israel Journal of Plant Sciences*, **62**, 39–51.

Avila IL (1992) Molecular mimicry between *Trypanosoma cruzi* and host nervous tissues. *Acta Científica Venezolana*, **43**, 330–40.

Avilés JM, Rutila J, Møller AP (2005) Should the redstart *Phoenicurus phoenicurus* accept or reject cuckoo *Cuculus canorus* eggs? *Behavioral Ecology & Sociobiology*, **58**, 608–17.

Avilés JM, Stokke BG, Moksnes A, Røskaft E, Møller AP (2007) Environmental conditions influence egg color of reed warblers *Acrocephalus scirpaceus* and their parasite, the common cuckoo *Cuculus canorus*. *Behavioral Ecology & Sociobiology*, **61**, 475–85.

Avilés JM, Vikan JR, Fossøy F, Antonov A, Moksnes A, Røskaft E, Shykoff JA, Møller AP, Jensen H, Procházka P, Stokke BG (2011a) The common cuckoo *Cuculus canorus* is not locally adapted to its reed warbler *Acrocephalus scirpaceus* host. *Journal of Evolutionary Biology*, **24**, 314–25.

Avilés JM, Vikan JR, Fossøy F, Antonov A, Moksnes A, Røskaft E, Shykoff JA, Møller AP, Stokke BG (2011b) Egg phenotype matching by cuckoos in relation to discrimination by hosts and climatic conditions. *Proceedings of the Royal Society of London, Series B: Biological Sciences*, **279**, 1967–76.

Ayasse M, Schiestl FP, Paulus HF, Löfstedt C, Hannson W, Ibarra F, Francke W (2000) Evolution of reproductive strategies in the sexually deceptive orchid *Ophrys sphegodes*: how does flower-specific variation of odor signals influence reproductive success? *Evolution*, **54**, 1995–2006.

Ayasse M, Schiestl FP, Paulus HF, Ibarra F, Francke W (2003) Pollinator attraction in a sexually deceptive orchid by means of unconventional chemicals. *Proceedings of the Royal Society of London, Series B: Biological Sciences*, **270**, 517–22.

Bachmann T, Klän S, Baumgartner W, Klaas M, Schröder W, Wagner H (2007) Morphometric characterisation of wing feathers of the barn owl *Tyto alba pratincola* and the pigeon *Columba livia*. *Frontiers in Zoology*, **4**: 23.

Backwell PRY, Christy JH, Telford SR, Jennions MD, Passmore NI (2000) Dishonest signalling in a fiddler crab. *Proceedings of the Royal Society of London, Series B: Biological Sciences*, **267**, 719–24.

Badura LL, Goldman BD (1992) Prolactin-dependent seasonal changes in pelage: role of the pineal gland and dopamine. *Journal of Experimental Zoology*, **261**, 27–33.

Bagnara JT, Hadley ME (1973) *Chromatophores and Colour Change: The Comparative Physiology of Animal Pigmentation*. Prentice Hall, Englewood Cliffs, NJ.

Bagnères AG, Lorenzi MC, Dusticier G, Turillazzi S, Clément J-L (1996) Chemical usurpation of a nest by paper wasp parasites. *Science*, **272**, 889–92.

Bailey W, Macleay C, Gordon T (2006) Acoustic mimicry and disruptive alternative calling tactics in an Australian bushcricket (*Caedicia*; Phaneropterinae; Tettigoniidae; Orthoptera): does mating influence male calling tactic? *Physiological Entomology*, **31**, 201–10.

Bain RS, Rashed A, Cowper VJ, Gilbert FS, Sherratt TN (2007) The key mimetic features of hoverflies through avian eyes. *Proceedings of the Royal Society of London, Series B: Biological Sciences*, **274**, 1949–54.

Baker ECS (1942) *Cuckoo Problems*. H. H. & G. Witherby Ltd., London

Baker HG (1976) "Mistake pollination" as a reproductive system with special reference to the Caricaceae. In J Burley and BT Styles (eds), *Tropical Trees: Variation, Breeding and Conservation*. Academic Press, London. pp. 161–9.

Baker RR (1985) Bird coloration: in defence of unprofitable prey. *Animal Behaviour*, **33**, 1387–8.

Baker RR, Hounsome MV (1983) Bird coloration: unprofitable prey model supported by ringing data. *Animal Behaviour*, **31**, 614–5.

Baker RR, Parker GA (1979) The evolution of bird coloration. *Philosophical Transactions of the Royal Society London B: Biological Sciences*, **287**, 63–130.

Balgooyen TG (1997) Evasive mimicry involving a butterfly model and grasshopper mimic. *American Midland Naturalist*, **137**, 183–7.

Balleto E, Arillo A, Mensi P, Churchi MA (1987) Zygaenid moths, cyanide and thanatosis. *Bolletino di Zoologia*, **54**, 59–63.

Balogh ACV, Leimar O (2005) Mullerian mimicry: an examination of Fisher's theory of gradual evolutionary change. *Proceedings of the Royal Society of London, Series B: Biological Sciences*, **272**, 2269–75.

Balogh ACV, Gamberale-Stille G, Leimar O (2008) Learning and the mimicry spectrum: from quasi-Bates to super-Müller. *Animal Behaviour*, **76**, 1591–9.

Balogh ACV, Gamberale-Stille G, Sillén-Tullberg B, Leimar O (2010) Feature theory and the two-step hypothesis of Müllerian mimicry evolution. *Evolution*, **64**, 810–22.

Balph MH, Balph DF, Romesburg HC (1979) Social status signalling in winter flocking birds: an examination of a current hypothesis. *The Auk*, **96**, 78–93.

Bantock CR, Harvey PH (1974) Colour polymorphism and selective experiments. *Journal of Biological Education*, **8**, 323–9.

Bänziger H (1991) Stench and fragrance: unique pollination lure of Thailand's largest flower, *Rafflesia kerrii* Meijer. *History Bulletin of the Siam Society*, **39**, 19–52.

Bänziger H (1996) The mesmerizing wart: the pollination strategy of epiphytic lady slipper orchid *Paphiopedilum villosum* (Lindl.) Stein (Orchidaceae). *Botanical Journal of the Linnean Society*, **121**, 59–90.

Baptista LF, Catchpole CK (1989) Vocal mimicry and interspecific aggression in songbirds: experiments using white-crowned sparrow imitation of song sparrow song. *Behaviour*, **109**, 247–57.

Barber JR, Leavell BC, Keener AL, Breinholt JW, Chadwell BA, McClure CJW, Hill GM, Kawaharab AY (2015) Moth tails divert bat attack: evolution of acoustic deflection. *Proceedings of the National Academy of Sciences, U.S.A.*, **112**, 2812–6.

Barbosa A, Mäthger LM, Buresch KC, Kelly J, Chubb C, Chiao C-C, Hanlon RT (2008) Cuttlefish camouflage: the effects of substrate contrast and size in evoking uniform, mottle or disruptive body patterns. *Vision Research*, **48**, 1242–53.

Barlow BA, Wiens D (1977) Host-parasite resemblance in Australian mistletoes: the case for cryptic mimicry. *Evolution*, **31**, 69–84.

Barlow GW (1967) Social Behavior of a South American leaf fish, *Polycentrus schomburgkii*, with an account of recurring pseudofemale behavior. *American Midland Naturalist*, **78**, 215.

Barlow GW (1972) The attitude of fish eye-lines in relation to body shape and to stripes and bars. *Copeia*, **1972**, 4–12.

Barnard CJ (1979) Predation and the evolution of social mimicry in birds. *The American Naturalist*, **113**, 613–18.

Barnard CJ (1982) Social mimicry and interspecific exploitation. *The American Naturalist*, **120**, 411–15.

Barnard KH (1927) A monograph of the marine fishes of South Africa. *Annals of the South African Museum*, **21**, v + 419–1065.

Barnett CA, Bateson M, Rowe C (2007) State-dependent decision making: educated predators strategically trade off the costs and benefits of consuming aposematic prey. *Behavioral Ecology*, **18**, 645–51.

Barrett JA (1976) The maintenance of non-mimetic forms in a dimorphic Batesian mimic species. *Evolution*, **30**, 82–5.

Barrett SCH (1983) Crop mimicry in weeds. *Economic Botany*, **37**, 255–82.

Barrett SCH (1987) Mimicry in plants. *Scientific American*, **255**, 76–83.

Barthlott W (1995) Mimikry Nachahmung und Täuschung im Pflanzenreich. *Biologie in unserer Zeit*, **25**, 74–82.

Bates HW (1862) Contributions to an insect fauna of the Amazon valley. Lepidoptera: Heliconidae. *Transactions of the Linnean Society*, **23**, 495–566.

Bates HW (1864) *A Naturalist on the River Amazons: A Record of Adventures, Habits of Animals, Sketches of Brazilian and Indian Life, and Aspects of Nature under the Equator, during Eleven Years of Travel.* John Murray, London.

Bates HW (1981) Contributions to an insect fauna of the Amazon valley. Lepidoptera: Heliconidae. *Biological Journal of the Linnean Society*, **16**, 41–54. [Abridged version of his 1862 paper.]

Bateson W (1889) Notes on the senses and habits of some Crustacea. *Journal of the Marine Biological Association of the United Kingdom*, **1**, 211–14.

Batra LR, Batra S (1985) Floral mimicry induced by mummy-berry fungus exploits host's pollinators as vectors. *Science*, **228**, 1011–13.

Bawa KS (1980) Mimicry of male by female flowers and intrasexual competition for pollinators in *Jacaratia dolichaula* (D. Smith) Woodson (Caricaceae). *Evolution*, **34**, 467–74.

Baxter SW, Papa R, Chamberlain N, Humphray SJ, Joron M, Morrison C, ffrench-Constant RH, McMillan WO, Jiggins CD (2008) Convergent evolution in the genetic basis of Müllerian mimicry in *Heliconius* butterflies. *Genetics*, **180**, 1567–77.

Baxter SW, Nadeau NJ, Maroja LS, Wilkinson P, Counterman BA, Dawson A, Beltran M, Perez-Espona S, Chamberlain N, Ferguson L, Clark R, Davidson C, Glithero R, Mallet J, McMillan WO, Kronforst MR, Joron M, ffrench-Constant RH, Jiggins CD (2010) Genomic hotspots for adaptation: the population genetics of Müllerian mimicry in the *Heliconius melpomene* clade. *PLoS Genetics*, **6**(2): e1000794.

Baylis JR (1982) Avian vocal mimicry: its function and evolution. In DE Kroodsma and EH Miller (eds), *Acoustic Communication in Birds.* Academic Press, New York, pp. 51–80.

Bayne CJ, Stephens JA (1983) *Schistosoma mansoni* and *Biomphalaria glabrata* share epitopes: antibodies to sporocysts bind host snail hemocytes. *Journal of Invertebrate Pathology*, **42**, 221–3.

Beach SA (1968) Factors involved in the control of mounting behaviour by female mammals. In M Diamond (ed.), *Perspectives in Reproduction and Sexual Behaviour.* Indiana University Press, Bloomington and London, pp. 83–131.

Bear A, Hasson O (1997) The predatory response of a stalking spider, *Plexippus paykulli*, to camouflage and prey type. *Animal Behaviour*, **54**, 993–8.

Beardsell DV, Clements MA, Hutchinson JF, Williams EG (1986) Pollination of *Diuris maculata* R Br (Orchidaceae) by floral mimicry of the native legumes *Daviesia* spp and *Pultenaea scabra* R Br. *Australian Journal of Botany*, **34**, 165–73.

Beatty CD, Beirinckx K, Sherratt TN (2004) The evolution of Müllerian mimicry in multispecies communities. *Nature*, **431**, 63–6.

Beccaloni GW (1997a) Vertical stratification of ithomiine butterfly (Nymphalidae: Ithomiinae) mimicry complexes: the relationship between adult flight height and larval host-plant height. *Biological Journal of the Linnean Society*, **62**, 313–41.

Beccaloni GW (1997b) Ecology, natural history and behaviour of ithomiine butterflies and their mimics in Ecuador (Lepidoptera: Nymphalidae: Ithomiinae). *Tropical Lepidoptera*, **8**, 103–24.

Beck J, Fiedler K (2009) Adult life spans of butterflies (Lepidoptera: Papilionoidea + Hesperioidea): broadscale contingencies with adult and larval traits in multi-species comparisons. *Biological Journal of the Linnean Society*, **96**, 166–84.

Beckers GJL, Leenders TAAM (1996) Coral snake mimicry: live snakes not avoided by a mammalian predator. *Oecologia (Berlin)*, **106**, 461–3.

Beddington JR, Free CA, Lawton JH (1975) Dynamic and complexity in predator–prey models framed in difference equations. *Nature*, **255**, 58–60.

Beebe W (1955) Polymorphism in reared broods of *Heliconius* butterflies from Surinam and Trinidad. *Zoologica, New York*, **40**, 139–43, plates I-VI.

Beecher MD, Campbell SE, Stoddard PK (1994) Correlation of song learning and territory establishment strategies in the song sparrow. *Proceedings of the National Academy of Sciences, U.S.A.,* **91**, 1450–4.

Belk MC, Smith MH (1996) Pelage coloration in Oldfield mice (*Peromyscus polionotus*) – antipredator adaptation. *Journal of Mammology*, **77**, 882–90.

Belt T (1874) *The Naturalist in Nicaragua*. Murray, London, UK.

Benitez-Vieyra S, de Ibarra NH, Wertlen AM, Coccuci AA (2007) How to look like a mallow: evidence of floral mimicry between Turneraceae and Malvaceae. *Proceedings of the Royal Society of London, Series B: Biological Sciences*, **274**, 2239–48.

Benjamin DR (1995) *Mushrooms Poisons and Panaceas: A Handbook for Naturalists, Mycologists, and Physicians*. W.H. Freeman, San Francisco, CA.

Benson WW (1971) Evidence for the evolution of unpalatability through kin selection in the Heliconiinae. *The American Naturalist*, **105**, 213–26.

Benson WW (1972) Natural selection for Müllerian mimicry in *Heliconius erato* in Costa Rica. *Science*, **176**, 936–9.

Benson WW (1977) On the supposed spectrum between Batesian and Müllerian mimicry. *Evolution*, **31**, 454–5.

Benson WW, Brown KS Jr, Gilbert LE (1975) Coevolution of plants and herbivores: passion flower butterflies. *Evolution*, **29**, 659–80.

Bent AC (1938) *Life histories of North American Birds of Prey, Part 2, Orders Falconiformes and Strigiformes*. United States National Museum Bulletin No. 170.

Bequaert J (1921) On the dispersal by flies of the spores of certain mosses of the family Splachnaceae. *Bryologist*, **14**, 1–4.

Berenbaum MR (1995) Aposematism and mimicry in caterpillars. *Journal of the Lepidopterists' Society*, **49**, 386–96.

Bergström G, Löfqvist J (1968) Odour similarities between the slave-keeping ants *Formica sanguinea* and *Polyergus rufescens* and their slaves *Formica fusca* and *Formica rufibarbis*. *Journal of Insect Physiology*, **14**, 995–1011.

Bergstrom PW (1988) Breeding displays and vocalizations of Wilson's Plovers. *Wilson Bulletin*, **100**, 36–49.

Bernays E, Edgar JA, Rothschild M (1977) Pyrrolizidine alkaloids sequestered and stored by the aposematic grasshopper, *Zonocerus variegatus*. *Zoological Journal of the Linnean Society*, **182**, 85–7.

Berry RJ (1990) Industrial melanism and peppered moths (*Biston betularia* (L.)). *Biological Journal of the Linnean Society*, **39**, 301–22.

Bezzerides AL, McGraw KJ, Parker RS, Husseini J (2007) Elytra color as a signal of chemical defense in the Asian ladybird beetle *Harmonia axyridis*. *Behavioral Ecology & Sociobiology*, **61**, 1401–8.

Bierzychudek P (1981) *Asclepias, Lantana* and *Epidendrum*: a floral mimicry complex? *Biotropica*, **13**, 54–8.

Bilde T, Tuni C, Elsayed R, Pekár S, Toft S (2006) Death feigning in the face of sexual cannibalism. *Biology Letters*, **2**, 23–5.

Blahó M, Egri Á, Száz D, Kriska G, Åkesson S, Horváth G (2013) Stripes disrupt odour attractiveness to biting horseflies: battle between ammonia, CO_2, and colour pattern for dominance in the sensory systems of host-seeking tabanids. *Physiology & Behavior*, **119**, 168–74.

Blakiston T, Alexander T (1884) Protection by mimicry – a problem of mathematical zoology. *Nature*, **29**, 405–6.

Blanco MA, Barboza G (2005) Pseudocopulatory pollination in *Lepanthes* (Orchidaceae: Pleurothallidinae) by fungus gnats. *Annals of Botany*, **95**, 763–72.

Blanco MA, Sherman PW (2005) Maximum longevities of chemically protected and non-protected fishes, reptiles and amphibians support evolutionary hypothesis of aging. *Mechanisms of Ageing and Development*, **126**, 794–803.

Blest AD (1957a) The evolution of protective displays in the Saturnioidea and Sphingidae (Lepidoptera). *Behaviour*, **11**, 257–309.

Blest AD (1957b) The function of eye-spot patterns in the Lepidoptera. *Behaviour*, **11**, 209–56.

Blest AD (1963) Longevity, palatability and natural selection in five species of New World saturniid moth. *Nature*, **197**, 1183–6.

Blest AD (1964) Protective display and sound production in some new world arctiid and ctenuchid moths. *Zoologica, New York*, **49**, 161–81.

Blest AD, Collett TS, Pye JD (1963) The generation of ultrasonic signals by a New World arctiid moth. *Proceedings of the Royal Society of London, Series B: Biological Sciences*, **158**, 196–207.

Blount JD, Speed MP, Ruxton GD, Stephens PA (2009) Warning displays may function as honest signals of toxicity. *Proceedings of the Royal Society of London, Series B: Biological Sciences*, **276**, 871–7.

Blount JD, Rowland HM, Drijfhout FP, Endler JA, Inger R, Hoggard JJ, Hurst GDD, Hodgson DJ, Speed MP (2012) How the ladybird got its spots: effects of resource limitation on the honesty of aposematic signals. *Functional Ecology*, **26**, 334–42.

Blum MS, Sannasi A (1974) Reflex bleeding in the lampyrid *Photinus pyralis*: defensive function. *Journal of Insect Physiology*, **20**, 451–60.

Blut C, Wilbrandt J, Fels D, Girgel EI, Lunau K (2012) The 'sparkle' in fake eyes: the protective effect of mimic eyespots in Lepidoptera. *Entomologia Experimentalis et Applicata*, **143**, 231–44.

Boal JG (1996) Absence of social recognition in laboratory-reared cuttlefish, *Sepia officinalis* L. (Mollusca: Cephalopoda). *Animal Behaviour*, **52**, 529–37.

Boardman M, Askew RR, Cook LM (1974) Experiments on resting site selection by nocturnal moths. *Journal of Zoology, London*, **172**, 343–55.

Bobisud LE (1978) Optimal time of appearance of mimics. *The American Naturalist*, **112**, 962–5.

Bobisud LE, Neuhaus RJ (1975) Pollinator constancy and survival of rare species. *Oecologia (Berlin)*, **21**, 263–72.

Bobisud LE, Potratz CJ (1976) One-trial versus multi-trial learning for a predator encountering a model-mimic system. *American Naturalist*, **110**, 121–8.

Bocak L, Yagi T (2010) Evolution of mimicry patterns in *Metriorrhynchus* (Coleoptera: Lycidae): the history of dispersal and speciation in Southeast Asia. *Evolution*, **64**, 39–52.

Bohl E (1982) Food supply and prey selection in planktivorous Cyprinidae. *Oecologia (Berlin)*, **53**, 134–8.

Bohman B, Peakall R (2014) Pyrazines attract *Catocheilus* thynnine wasps. *Insects*, **5**, 474–87.

Bohman B, Phillips RD, Flematti G, Peakall R, Barrow RA (2013) Sharing of pyrazine semiochemicals between genera of sexually deceptive orchids. *Natural Product Communications*, **8**, 701–2.

Boileau N, Cortesi F, Egger B, Muschick M, Indermaur A, Theis A, Büscher HH, Salzburger W (2015) A complex mode of aggressive mimicry in a scale-eating cichlid fish. *Biology Letters*, **11**: 20150521.

Bolin JF, Maass E, Musselman LJ (2009) Pollination biology of *Hydnora africana* Thunb. (Hydnoraceae) in Namibia: brood-site mimicry with insect imprisonment. *International Journal of Plant Science*, **170**, 157–63.

Bonavita-Cougourdan A, Riviére G, Provost E, Bagnéres A-G, Roux M, Dusticier G, Clément J-L (1996) Selective adaptation of the cuticular hydrocarbon profiles of the slave-making ants *Polyergus rufescens* Latr. and their *Formica rufibarbis* Fab. and *F. cunicularia* Latr. slaves. *Comparative Biochemistry and Physiology Part B: Biochemistry and Molecular Biology*, **113**, 313–29.

Bond AB (1983) Visual search and selection of natural stimuli in the pigeon: the attention threshold hypothesis. *Journal of Experimental Psychology: Animal Behavior Processes*, **9**, 292–306.

Bond AB (2007) The evolution of color polymorphism: crypticity, searching images, and apostatic selection. *Annual Review of Ecology, Evolution and Systematics*, **38**, 489–514.

Bond AB, Kamil AC (1998) Apostatic selection by blue jays produces balanced polymorphism in virtual prey. *Nature*, **395**, 594–6.

Bond AB, Kamil AC (2002) Visual predators select for crypticity and polymorphism in virtual prey. *Nature*, **415**, 609–13.

Bonser RHC (1995) Melanin and the abrasion resistance of feathers. *The Condor*, **97**, 590–1.

Booth CL (1990) Evolutionary significance of ontogenetic colour change in animals. *Biological Journal of the Linnean Society*, **40**, 125–63.

Boppré M (1978) Chemical communication, plant relationships, and mimicry in the evolution of danaid butterflies. *Entomologia Experimentalis et Applicata*, **24**, 264–77.

Boppré M (1990) Lepidoptera and pyrrolizidine alkaloids. Exemplification of complexity in chemical ecology. *Journal of Chemical Ecology*, **16**, 165–85.

Boppré M, Vane-Wright RI, Wickler W (2016) A hypothesis to explain accuracy of wasp resemblances. *Ecology and Evolution*, **2016**: 1–9.

Borg-Karlson A-K (1990) Chemical and ethological studies of pollination in the genus *Ophrys* (Orchidaceae). *Phytochemistry*, **29**, 1359–87.

Borg-Karlson A-K, Tengö J (1986) Odor mimetism? Key substances in *Ophrys lutea-Andrena* pollination relationship (Orchidaceae: Andrenidae). *Journal of Chemical Ecology*, **12**, 1927–41.

Borg-Karlson A-K, Bergström G, Kullenberg B (1987) Chemical basis for the relationship between *Ophrys* orchids and their pollinators. 2. Volatile compounds of *Ophrys insectifera* and *O. speculum* as insect mimetic attractants/excitants. *Chemica Scripta*, **27**, 303–11.

Borg-Karlson A-K, Groth I, Ågren L, Kullenberg B (1993) Form-specific fragances from *Ophrys insectifera* L. (Orchidaceae) attract species of different pollinator genera. Evidence of sympatric speciation? *Chemoecology*, **4**, 39–45.

Bostanchi H, Anderson SC, Gholi Kami H, Papenfuss TJ (2006) A new species of *Pseudocerastes* with elaborate tail ornamentation from western Iran (Squamata: Viperidae). *Proceedings of the California Academy of Science, (Series 4)*, **57**, 443–50.

Bots J, de Bruyn L, Van Dongen S, Smolders R, Van Gossum H (2009) Female polymorphism, condition differences, and variation in male harassment and ambient temperature. *Biological Journal of the Linnean Society*, **97**, 545–54.

Bowden SR (1952) Pupal colour and diapause in *Pieris napi* L. *Entomologist*, **85**, 175–8.

Bowers MD (1980) Unpalatability as a defense strategy of *Euphydrym phaeton* Drury (Lepidoptera: Nymphalidae). *Evolution*, **34**, 586–600.

Bowers MD (1983) Mimicry in North American checkerspot butterflies: *Euphydryas phaeton* and *Chlosyne harrisii* (Nymphalidae). *Ecological Entomology*, **8**, 1–8.

Bowers MD (1988) Chemistry and coevolution: iridoid glycoside, plants and herbivorous insects. In K Spencer (ed.), *Chemical Mediation of Coevolution*. Academic Press, New York, pp. 133–54.

Bowers MD (1993) Aposematic caterpillars: life-styles of the warningly colored and unpalatable. In NE Stamp and TM Casey (eds), *Caterpillars. Ecological and Evolutionary Constraints on Foraging.* Chapman and Hall, New York, pp. 331–71.

Bowers MD, Larin Z (1989) Acquired chemical defense in the lycaenid butterfly, *Eumaeus atala*. *Journal of Chemical Ecology*, **15**, 1133–46.

Bowers MD, Wiernasz DC (1979) Avian predation on the palatable butterfly, *Cercyonis pegala* (Satyridae). *Ecological Entomology*, **4**, 205–9.

Bowers MD, Williams EH (1995) Variable chemical defence in the checkerspot butterfly *Euphydryas gillettii* (Lepidoptera: Nymphalidae). *Ecological Entomology*, **20**, 208–12.

Bowers MD, Brown IL, Wheye D (1985) Bird predation as a selective agent in a butterfly population. *Evolution*, **39**, 93–103.

Boyden TC (1976) Butterfly palatability and mimicry: experiments with *Ameiva* lizards. *Evolution*, **30**, 73–81.

Boyden TC (1980) Floral mimicry by *Epidendrum ibaguense* (Orchidaceae) in Panama. *Evolution*, **34**, 135–6.

Boyden TC (1982) The pollination biology of *Calypso bulbosa* var. *americana* (Orchidaceae): initial deception of bumblebee visitors. *Oecologia (Berlin)*, **55**, 178–84.

Brach V (1978) *Brachynemurus nebulosus* (Neuroptera: Myrmeleontidae): a possible Batesian mimic of Florida mutillid wasps (Hymenoptera: Mutillidae). *Entomological News*, **89**, 153–6.

Brady J, Shereni W (1988) Landing responses of the tsetse fly *Glossina morsitans morsitans* Westwood and the stable fly *Stomoxys calcitrans* (L.) (Diptera: Glossinidae & Muscidae) to black-and-white patterns: a laboratory study. *Bulletin of Entomological Research*, **78**, 301–11.

Brady PC, Gilerson AA, Kattawar GW, Sullivan JM, Twardowski MS, Dierssen HM, Gao M, Travis K, Etheredge RI, Tonizzo A, Ibrahim A, Carrizo C, Gu Y, Russell BJ, Mislinski K, Zhao S, Cummings ME (2015) Open-ocean fish reveal an omnidirectional solution to camouflage in polarized environments. *Science*, **350**, 965–9.

Brakefield PM (1985) Polymorphic Müllerian mimicry and interactions with thermal melanism in ladybirds and a soldier beetle a hypothesis. *Biological Journal of the Linnean Society*, **26**, 243–67.

Brakefield PM (2003) The power of evo-devo to explore evolutionary constraints: experiments with butterfly eyespots. *Zoology (Jena)*, **106**, 283–90.

Brakefield PM, French V (1993) Butterfly wing patterns – developmental mechanisms and evolutionary change. *Acta Biotheoretica*, **41**, 447–68.

Brakefield PM, Larsen TB (1984) The evolutionary significance of dry and wet season forms in some tropical butterflies. *Biological Journal of the Linnean Society*, **22**, 1–12.

Brakefield PM, Reitsma N (1991) Phenotypic plasticity, seasonal climate and the population biology of *Bicyclus* butterflies. *Ecological Entomology*, **16**, 291–303.

Brandão RA, Motta PC (2005) Circumstantial evidences for mimicry of scorpions by the neotropical gecko *Coleodactylus brachystoma* (Squamata, Gekkonidae) in the Cerrados of central Brazil. *Phyllomedusa*, **4**, 139–45.

Brandon RA, Labanick GM, Huheey JE (1979a) Learned avoidance of brown efts, *Nofophthalmus viridescens louisianensis* (Amphibia, Urodela, Salamandridae), by chickens. *Journal of Herpetology*, **13**, 171–6.

Brandon RA, Labanick GM, Huheey JE (1979b) Relative palatability, defensive behavior, and mimetic relationships of red salamanders (*Pseudotriton ruber*), mud salamanders (*Pseudotriton montanus*), and red efts (*Notophthalmus viridescens*). *Herpetologica*, **35**, 289–303.

Brattstrom BH (1955) The coral snake "mimic" problem and protective coloration. *Evolution*, **9**, 217–9.

Breed MD, Snyder LE, Lynn TL, Morhart JA (1992) Acquired chemical camouflage in a tropical ant. *Animal Behaviour*, **44**, 519–23.

Bretagnolle V (1993) Adaptive significance of seabird coloration: the case of Procellariiformes. *The American Naturalist*, **142**, 141–73.

Brilot BO, Normandale CL, Parkin A, Bateson M (2009) Can we use starlings' aversion to eyespots as the basis for a novel 'cognitive bias' task? *Applied Animal Behaviour Science*, **118**, 182–90.

Brinton BA, Curran MC (2015) The effects of the parasite *Probopyrus pandalicola* (Packard, 1879) Isopoda, Bopyridae) on the behavior, transparent camouflage, and predators of *Palaemonetes pugio* Holthuis, 1949 (Decapoda, Palaemonidae). *Crustaceana*, **88**, 1265–1281.

Bristowe WS (1941) *The Comity of Spiders*, Vol. II. Ray Society, London.

Broadley RA, Stringer IA (2001) Prey attraction by larvae of the New Zealand glowworm, *Arachnocampa luminosa* (Diptera: Mycetophilidae). *Invertebrate Biology*, **120**, 170–7.

Brockmann HJ, Hailman JP (1976) Fish cleaning symbiosis: notes on juvenile angelfishes (*Pomacanthus*, Chaetondontidae) and comparisons with other species. *Zeitschrift für Tierpsychologie*, **42**, 129–38.

Brodie III ED (1992) Correlational selection for color pattern and antipredator behaviour in the garter snake *Thamnophis ordinoides*. *Evolution*, **46**, 1284–98.

Brodie III ED (1993) Differential avoidance of coral snake banded patterns by free-ranging avian predators in Costa Rica. *Evolution*, **47**, 227–35.

Brodie III ED, Howard RR (1973) Experimental study of Batesian mimicry in the salamanders *Plethodon jordani* and *Desmognathus ochrophaeus*. *The American Naturalist*, **90**, 38–46.

Brodie III ED, Janzen FJ (1995) Experimental studies of coral snake mimicry – generalized avoidance of ringed snake patterns by free-ranging predators. *Functional Ecology*, **9**, 186–190.

Brodie III ED, Moore AJ (1995) Experimental studies of coral snake mimicry: do snakes mimic millipedes? *Animal Behaviour*, **49**, 534–6.

Brodie III ED, Moore AJ, Janzen FJ (1995) Visualizing and quantifying natural selection. *Trends in Ecology & Evolution*, **10**, 313–8.

Brodmann J, Twele R, Francke W, Hölzler G, Zhang Q-H, Ayasse M (2008) Orchids mimic green-leaf volatiles to attract prey-hunting wasps for pollination. *Current Biology*, **18**, 740–4.

Brodmann J, Twele R, Francke W, Yi-bo L, Xi-qiang S, Ayasse M (2009) Orchid mimics honey bee alarm pheromone in order to attract hornets for pollination. *Current Biology*, **19**, 1368–72.

Bro-Jørgensen J, Pangle WM (2010) Male topi antelopes alarm snort deceptively to retain females for mating. *The American Naturalist*, **176**, E33–E39.

Brooke M de L, Davies NB (1988) Egg mimicry by cuckoos *Cuculus canorus* in relation to discrimination by hosts. *Nature*, **335**, 630–2.

Brooke M de L, Davies NB (1991) A failure to demonstrate host imprinting in the cuckoo (*Cuculus canorus*) an alternative hypothesis for the maintenance of egg mimicry. *Ethology*, **89**, 154–66.

Brooker LC, Brooker MG, Brooker AMH (1990) An alternative population/genetics model for the evolution of egg mimesis and egg crypsis in cuckoos. *Journal of Theoretical Biology*, **146**, 123–43.

Broom M, Speed MP, Ruxton GD (2005) Evolutionarily stable investment in secondary defences. *Functional Ecology*, **19**, 836–43.

Broom M, Speed MP, Ruxton GD (2006) Evolutionarily stable defence and signalling of that defence. *Journal of Theoretical Biology*, **242**, 32–43.

Brower AVZ (1994) Rapid morphological radiation and convergence among races of the butterfly *Heliconius erato* inferred from patterns of mitochondrial DNA evolution. *Proceedings of the National Academy of Sciences, U.S.A.*, **91**, 6491–5.

Brower AVZ (1995) Locomotor mimicry in butterflies? A critical review of the evidence. *Philosophical Transactions of the Royal Society of London B: Biological Sciences*, **347**, 413–25.

Brower AVZ (1996) Parallel race formation and the evolution of mimicry in *Heliconius* butterflies: a phylogenetic hypothesis from mitochondrial DNA sequences. *Evolution*, **50**, 195–221.

Brower AVZ, Sime KR (1998) A reconsideration of mimicry and aposematism in caterpillars in the *Papilio machaon* group. *Journal of the Lepidopterists' Society*, **52**, 206–12.

Brower JVZ (1958a) Experimental studies of mimicry in some North American butterflies. Part I. The monarch *Danaus plexippus*, and the viceroy, *Limenitis archippus archippus*. *Evolution*, **12**, 32–47.

Brower JVZ (1958b) Experimental studies of mimicry in some North American butterflies. Part II. *Battus philenor* and *Papilio troilus*, *P. polyxenes* and *P. glaucus*. *Evolution*, **12**, 123–36.

Brower JVZ (1958c) Experimental studies of mimicry in some North American butterflies. Part III. *Danaus gilippus berenice* and *Limenitis archippus floridensis*. *Evolution*, **12**, 273–85.

Brower JVZ (1960) Experimental studies of mimicry. IV. The reactions of starlings to different proportions of models and mimics. *The American Naturalist*, **94**, 271–82.

Brower JVZ (1963) Experimental studies and new evidence on the evolution of mimicry in butterflies. *Proceedings of the 16th International Congress of Zoology, Washington*, **4**, 156–61.

Brower JVZ, Brower LP (1961) Palatability of North American model and mimic butterflies to caged mice. *Journal of the Lepidopterists' Society*, **15**, 23–4.

Brower JVZ, Brower LP (1962) Experimental studies of mimicry. 6. The reaction of toads (*Bufo terrestris*) to honey bees (*Apis mellifera*) and their dronefly mimics (*Eristalis vinetorum*). *The American Naturalist*, **96**, 297–308.

Brower LP (1963) The evolution of sex-limited mimicry in butterflies. *Proceedings of the 16th International Congress of Zoology, Washington*, **4**, 173–9.

Brower LP (1971) Prey coloration and predator behavior. In V Dethier (ed.), *The BIO Source Book, Section 6, Animal Behaviour*. Harper & Row Publishers, New York, pp. 360–70.

Brower LP (1984) Chemical defense in butterflies. In RI Vane-Wright and PR Ackery (eds), *The Biology of Butterflies*. Academic Press, London, pp. 109–34.

Brower LP, Brower JVZ (1962) The relative abundance of the model and mimic butterflies in natural populations of the *Battus philenor* mimicry complex. *Ecology*, **43**, 154–8.

Brower LP, Brower JVZ (1972) Parallelism, convergence, divergence and the new concept of advergence in the evolution of mimicry. *Transactions of the Connecticut Academy of Arts and Sciences*, **44**, 59–67.

Brower LP, Glazier SC (1975) Localization of heart poisons in the monarch butterfly. *Science*, **188**, 19–25.

Brower LP, Brower JVZ, Westcott PW (1960) Experimental studies of mimicry. 5. The reactions of toads (*Bufo terrestris*) to bumblebees (*Bombus americanorum*) and their robberfly mimics (*Mallophora bomboides*) with discussion of aggressive mimicry. *The American Naturalist* **94**, 343–55.

Brower LP, Brower JVZ, Collins CT (1963) Experimental studies of mimicry. 7. Relative palatability and Müllerian mimicry among Neotropical butterflies of the subfamily Heliconiinae. *Zoologica, New York*, **48**, 65–84 + 1 plate.

Brower LP, Brower JVZ, Stiles FG, Croze HJ, Hower AS (1964) Mimicry: differential advantage of color patterns in the natural environment. *Science*, **144**, 183–5.

Brower LP, Brower JVZ, Cranston F (1965) Courtship behavior of the queen butter-fly, *Danaus gilippus berenice* (Cramer). *Zoologica, New York*, **50**, 1–40.

Brower LP, Cook LM, Croze HJ (1967a) Predator responses to artificial Batesian mimics released in a Neotropical environment. *Evolution*, **21**, 11–23.

Brower LP, Brower JVZ, Corvino JM (1967b) Plant poisons in a terrestrial food chain. *Proceedings of the National Academy of Sciences, U.S.A.*, **57**, 893–8.

Brower LP, Ryerson WN, Coppinger LI, Glazier SC (1968) Ecological chemistry and the palatability spectrum. *Science*, **161**, 1342–81.

Brower LP, Pough FH, Meck HR (1970) Theoretical investigations of automimicry. I. Single trial learning. *Proceedings of the National Academy of Sciences, U.S.A.*, **66**, 1059–66.

Brower LP, Alcock J, Brower JVZ (1971) Avian feeding behaviour and the selective advantage of incipient mimicry. In R Creed (ed.), *Ecological Genetics and Evolution*. Oxford, Blackwell, pp. 261–74.

Brower LP, Edmunds M, Moffitt CM (1975) Cardenolide content and palatability in a population of *Danaus chrysippus* butterflies from West Africa. *Journal of Entomology (A)*, **49**, 183–96.

Brower LP, Gibson DO, Moffitt CM, Panchen AL (1978) Cardenolide content of *Danaus chrysippus* butterflies from three areas of East Africa. *Biological Journal of the Linnean Society*, **10**, 251–73.

Brown C, Garwood MP, Williamson JE (2012) It pays to cheat: tactical deception in a cephalopod social signaling system. *Biology Letters*, **8**, 729–32.

Brown JH, Kodric-Brown A (1979) Convergence, competition, and mimicry in a temperate community of hummingbird-pollinated flowers. *Ecology*, **60**, 1022–35.

Brown JH, Cantrell MA, Evans SM (1973) Observations on the behaviour and coloration of some coral reef fish (Family: Pomacentridae). *Marine Behaviour and Physiology*, **2**, 63–71.

Brown KS Jr (1979) *Ecologia geografica e evolucao nas florestas neotropicas*. Campinas, Brasil. Universidade Estadual de Campinas, Livre de Docencia.

Brown KS Jr (1981) The biology of *Heliconius* and related genera. *Annual Review of Entomology*, **26**, 427–56.

Brown KS Jr (1982) Paleoecology and regional patterns of evolution in neotropical forest butterflies. In GT Prance (ed.), *Biological Diversification in the Tropics*. Columbia University Press, New York, pp. 255–308.

Brown KS Jr (1984a) Adult-obtained pyrrolizidine alkaloids defend ithomiine butterflies against a spider predator. *Nature*, **309**, 707–9.

Brown KS Jr (1984b) Chemical ecology of dehydropyrrolizidine alkaloids in adult Ithomiinae (Lepidoptera: Nymphalidae). *Revista Brasileira de Biologia*, **44**, 435–60.

Brown KS Jr (1988) Mimicry, aposematism and crypsis in neotropical Lepidoptera: the importance of dual signals. *Bulletin de la Société zoologique de France*, **113**, 83–101.

Brown KS, Benson WW (1974) Adaptive polymorphism associated with multiple Müllerian mimicry in *Heliconius numata* (Lepid.: Nymph.). *Biotropica*, **6**, 205–28.

Brown KS Jr, Francini RB (1990) Evolutionary strategies of chemical defense in aposematic butterflies: cyanogenesis in Asteraceae-feeding American Acraeinae. *Chemoecology*, **1**, 52–6.

Brown KS Jr, Trigo JR (1995) Multi-level complexity in the use of plant allelochemicals by aposematic insects. *Chemoecology*, 5/6, 119–26.

Brown KS Jr, Vasconcellos-Neto J (1976) Predation on aposematic ithomiine butterflies by tanagers (*Pipraeidea melanonota*). *Biotropica*, **8**, 136–41.

Brown KS Jr, Sheppard PM, Turner JRG (1974) Quaternary refugia in tropical America: evidence from race formation in *Heliconius* butterflies. *Proceedings of the Royal Society of London, Series B: Biological Sciences*, **187**, 369–78.

Brown KS Jr, Trigo JR, Francini RB, Barros de Morais AB, Motta PC (1991) Aposematic insects on toxic host plants: coevolution, colonization, and chemical emancipation. In PW Price, TM Lewinsohn, GW Fernandes and WW Benson (eds), *Plant-Animal*

Interactions: Evolutionary Ecology in Tropical and Temperate Regions. John Wiley, New York, pp. 375–402.

Brown RM (2006) A case of suspected coral snake (*Hemibungarus calligaster*) mimicry by lepidopteran larvae (*Bracca* sp.) from Luzon Island, Philippines. *The Raffles Bulletin of Zoology*, **54**, 225–7.

Brown VK, Lawton JH, Grubb PJ (1991) Herbivory and the evolution of leaf size and shape [and discussion]. *Philosophical Transactions of the Royal Society of London, Series B, Biological Sciences*, **333**, 265–72.

Bückmann D, Maisch A (1987) Extraction and partial purification of the pupal melanization reducing factor (PMRF) from *Inachis io* (Lepidoptera). *Insect Biochemistry*, **17**, 841–4.

Budenberg WJ, Powell W (1992) The role of honeydew as an ovipositional stimulant for two species of syrphids. *Entomologia Experimentalis et Applicata*, **64**, 57–61.

Bullini L, Sbordoni V (1971) Ricerche sperimentali sul valore mimetico delle forme efialtoidi rosse de *Zygaena ephialtes* (Lepidoptera, Zygaenidae). *Bolletino di Zoologia*, **38**, 502.

Bunkley-Williams L, Williams EH (2000) Juvenile black snapper, *Apsilus dentatus* (Lutjanidae), mimic blue chromis, *Chromis cyanea* (Pomacentridae). *Copeia*, **2000**, 579–81.

Burd M (1994) Butterfly wing colour patterns and flying heights in the seasonally wet forest of Barro Colorado Island, Panama. *Journal of Tropical Ecology*, **10**, 601–10.

Burg JS, Ingram JR, Venkatakrishnan AJ, Jude KM, Dukkipati A, Feinberg EN, Angelini A, Waghray D, Dror RO, Ploegh HL, Garcia KC (2015) Structural basis for chemokine recognition and activation of a viral G protein-coupled receptor. *Science*, **347**, 1113–17.

Burger BV, Petersen WGB (1991) Semiochemicals of the Scarabaeinae, III: Identification of the attractant for the dung beetle *Pachylomerus femoralis* in the fruit of the spineless monkey orange tree, *Strychnos madagascariensis*. *Zeitschrift fur Naturforschung*, **46**, 1073–9.

Burger J, Gochfeld M (2001) Smooth-billed ani (*Crotophaga ani*) predation on butterflies in Mato Grosso, Brazil: risk decreases with increased group size. *Behavioral Ecology and Sociobiology*, **49**, 482–92.

Burgess KS, Singheld J, Melendez V, Kevan PG (2004) Pollination biology of *Aristolochia grandiflora* (Aristolochiaceae) in Veracruz, Mexico. *Annals of the Missouri Botanical Garden*, **91**, 346–56.

Burghardt GM (1975) Chemical prey preference polymorphism in newborn garter snakes *Thamnophis sirtalis*. *Behaviour*, **52**, 202–25.

Burghardt GM, Greene HW (1988) Predator simulation and duration of death feigning in neonate hognose snakes. *Animal Behaviour*, **36**, 842–44.

Burns JM (1966) Preferential mating versus mimicry: disruptive selection and sex-limited dimorphism in *Papilio glaucus*. *Science*, **153**, 551–3.

Burns KC (2010) Is crypsis a common defensive strategy in plants. Speculation on signal deception in the New Zealand flora. *Plant Signaling & Behavior*, **5**, 9–13.

Burns KC, Dalen JL (2002) Foliage color contrasts and adaptive fruit color variation in a bird-dispersed plant community. *Oikos*, **96**, 463–9.

Burns-Balogh P, Szlachetko DL, Dafni A (1987) Evolution, pollination, and systematics of the tribe Neottieae (Orchidaceae). *Plant Systematics and Evolution*, **156**, 91–115.

Burtt EH (1979) Tips on wings and other things. In EH Burtt, Jr (ed.) *The Behavioral Significance of Color*. Garland, New York, pp. 75–125.

Burtt EH, Gatz Jr AJ (1982) Color convergence: is it only mimicry? *The American Naturalist*, **119**, 738–40.

Byrkjedal I (1989) Nest defense behavior of lesser golden-plovers. *Wilson Bulletin*, **101**, 579–90.

Byrne R, Whiten A (1990) Tactical deception in primates: the 1990 database. *Primate Report*, **27**, 1–101.

Cahn MG, Harper JL (1976) The biology of the leaf mark polymorphism in *Trifolium repens* L. 2. Evidence for the selection of leaf marks by rumen fistulated sheep. *Heredity*, **37**, 327–33.

Cain AJ, Sheppard PM (1950) Selection in the polymorphic land snail *Cepaea nemoralis* L. *Heredity*, **4**, 275–94.

Cain AJ, Sheppard PM (1954) Natural selection in *Cepaea*. *Genetics*, **39**, 86–116.

Cain AJ, Sheppard PM, King JMB (1968) Studies on *Cepaea*. I. The genetics of some morphs and varieties of *Cepaea nemoralis*. *Proceedings of the Royal Society of London, Series B: Biological Sciences*, **253**, 383–396.

Cairns DK (1986) Plumage colour in pursuit-diving seabirds: why do penguins wear tuxedos? *Bird Behaviour*, **6**, 58–65.

Caldwell GS, Rubinoff RW (1983) Avoidance of venomous sea snakes by naive herons and egrets. *The Auk*, **100**, 195–8.

Caley MJ, Schluter D (2003) Predators favour mimicry in a tropical reef fish. *Proceedings of the Royal Society of London, Series B: Biological Sciences*, **270**, 667–72.

Camara MD (1997) Physiological mechanisms underlying the costs of chemical defence in *Junonia coenia* Hubner (Nymphalidae): a gravimetric and quantitative genetic analysis. *Evolutionary Ecology*, **11**, 451–69.

Camargo MGG, Cazetta E, Schaefer HM, Morellato LPC (2013) Fruit color and contrast in seasonal habitats – a case study from a cerrado savanna. *Oikos*, **122**, 1335–42.

Camazine SM (1985) Olfactory aposematism – association of food toxicity with naturally-occurring odours. *Journal of Chemical Ecology*, **11**, 1289–95.

Camazine SM, Resch JF, Eisner T, Meinwald J (1983) Mushroom chemical defense: pungent sesquiterpenoid dialdehyde antifeedant to opossum. *Journal of Chemical Ecology*, **9**, 1439–47.

Campbell JA, Lamar WW (1989) *The venomous reptiles of Latin America*. Cornell University Press, Ithaca, New York.

Campitelli BE, Steglik J, Stinchcombe JR (2008) Leaf variegation is associated with reduced herbivore damage in *Hydrophyllum virginianum*. *Botany*, **86**, 306–13.

Canfield MR, Pierce NE (2010) Facultative mimicry? The evolutionary significance of seasonal forms in several Indo-Australian butterflies in the family Pieridae. *Tropical Lepidoptera Research*, **20**, 1–7.

Canyon DV, Hill CJ (1997) Mistletoe host-resemblance: a study of herbivory, nitrogen and moisture in two Australian mistletoes and their host trees. *Australian Journal of Ecology*, **22**, 395–403.

Cardoso MZ (1997) Testing chemical defence based on pyrrolizidine alkaloids. *Animal Behaviour*, **54**, 985–91.

Carlberg U (1980) Defensive behaviour in adult female *Extatosoma tiaratum* (Macleay) (Phasmidae). *Entomologist's Monthly Magazine*, **116**, 133–8.

Carlberg U (1981a) Defensive behaviour in females of the stick insect *Sipyloides sipylus* (Westwood) (Phasmida). *Zoologischer Anzeiger*, **207**, 177–80.

Carlberg U (1981b) An analysis of the secondary defence reactions of stick insects (Phasmida). *Biologisches Zentralblatt*, **100**, 295–303.

Carlberg U (1983) Diversity in the genus *Baculum* Saussure (Insecta: Phasmida). *Zoologische Jahrbuecher Systematik*, **110**, 127–40.

Carlberg U (1985a) Chemical defence in *Extatosoma tiaratum* (Macleay) (Insecta: Phasmida). *Zoologischer Anzeiger*, **214**, 185–92.

Carlberg U (1985b) Chemical defence in *Anisomorpha bubrestoides* (Houtluyn in Stoll) (Insecta: Phasmida). *Zoologischer Anzeiger*, **215**, 177–88.

Carlberg U (1985c) Evolutionary and ecological aspects of hatching time and defensive behaviour of Phasmida (Insecta). *Biologisches Zentralblatt*, **104**, 529–37.

Carlberg U (1986a) Chemical defence in *Sipyloides sipylus* (Westwood) (Insecta: Phasmida). *Zoologischer Anzeiger*, **217**, 31–8.

Carlberg U (1986b) Thanatosis and automimicry as defence in *Baculum* sp. 1 (Insecta: Phasmida). *Zoologischer Anzeiger*, **217**, 39–53.

Carlson SP (1935) The color changes in *Uca pugilator*. *Proceedings of the National Academy of Sciences, U.S.A.*, **21**, 549–51.

Carlsson B-G (1991) Recruitment of mates and deceptive behaviour by male Tengmalm's owls. *Behavioral Ecology and Sociobiology*, **28**, 321–8.

Carmona-Díaz G (2001) Mimetismo floral entre *Oncidium cosymbephorum* Morren (Orchidaceae) y *Malpighia glabra* L. (Malpighiaceae). Maestro en Neuroetologia. Universidad Veracruzana, Xalapa, Veracruz.

Caro T (1987) Human breasts: unsupported hypotheses reviewed. *Human Evolution*, **2**, 271–82.

Caro T (2009) Contrasting coloration in terrestrial mammals. *Philosophical Transactions of the Royal Society of London, Series B, Biological Sciences*, **364**, 537–48.

Caro T (2011) The functions of black-and-white coloration in mammals: review and synthesis. In M Stevens and S Merilaita (eds), *Animal Camouflage: Mechanisms and Function*. Cambridge University Press, Cambridge, pp. 298–329.

Caro T (2014) Antipredator deception in terrestrial vertebrates. *Current Zoology*, **60**, 16–25.

Carpenter CC, Murphy JB, Carpenter GC (1978) Tail luring in the death adder, Acanthophis antarcticus (Reptilia, Serpentes, Elapidae). *Journal of Herpetology*, **12**, 574–7.

Carpenter GDH (1921) Experiments on the relative edibility of insects, with special reference to their coloration. *Transactions of the Royal Entomological Society of London*, **54**, 1–105.

Carpenter GDH (1933) Gregarious roosting habits of aposematic butterflies. *Proceedings of the Entomological Society of London*, **8**, 110–1.

Carpenter GDH (1936) *Pseudacraea eurytus* (L.) and its models in the Budongo Forest, Bunyoro, Western Uganda (Lepidoptera). *Proceedings of the Entomological Society of London*, **11**, 22–8.

Carpenter GDH (1937) Further evidence that birds do attack and eat butterflies. *Proceedings of the Zoological Society of London*, **107**, 223–47.

Carpenter GDH (1938) Audible emission of defensive froth by insects with an appendix on the anatomical structures concerned in a moth by H. Eltringham. *Proceedings of the Zoological Society of London*, **108**, 243–52.

Carpenter GDH (1939) Birds as enemies of butterflies, with special reference to mimicry. *Proceedings of the VII International Kongress of Entomology, Berlin*, **1938**, 1061–74.

Carpenter GDH (1941) The relative frequency of beak marks on butterflies of different edibility to birds. *Proceedings of the Zoological Society of London (Series A)*, **3**, 223–31.

Carpenter GDH (1942) Observations and experiments in Africa by the late C. F. M. Swynnerton on wild birds eating butterflies and the preference shown. *Proceedings of the Linnean Society of London*, **154**, 10–46.

Carpenter GDH (1949) *Pseudacraea eurytus* (L.) (Lep. Nymphalidae): a study of a polymorphic mimic in various degrees of speciation. *Transactions of the Royal Entomological Society of London*, **100**, 71–133

Carpenter GDH, Ford EB (1933) *Mimicry*. Methuen, London.

Carrick R (1936) Experiments to test the efficiency of protective adaptations in insects. *Transactions of the Royal Entomological Society of London*, **85**, 131–40.

Carroll J, Korshikov E, Sherratt TN (2011) Post-reproductive senescence in moths as a consequence of kin selection: Blest's theory revisited. *Biological Journal of the Linnean Society*, **104**, 633–41.

Case JF, Warner JA, Barnes AT, Lowenstine M (1977) Bioluminescence of lantern fish (Myctophidae) in response to changes in the light intensity. *Nature*, **265**, 179–81.

Catania KC (2009) Tentacled snakes turn C-starts to their advantage and predict future prey behavior. *Proceedings of the National Academy of Sciences, U.S.A.*, **106**, 11183–7.

Catarino MF, Zuanon J (2010) Feeding ecology of the leaf fish *Monocirrhus polyacanthus* (Perciformes: Polycentridae) in a terra firme stream in the Brazilian Amazon. *Neotropical Ichthyology*, **8**, 183–6.

Catchpole CK, Baptista LF (1988) A test of the competition hypothesis of vocal mimicry, using song sparrow imitation of white-crowned sparrow song. *Behaviour*, **106**, 119–28.

Ceccarelli FS, Crozier RH (2007) Dynamics of the evolution of Batesian mimicry: molecular phylogenetic analysis of ant-mimicking *Myrmarachne* (Araneae: Salticidae) species and their ant models. *Journal of Evolutionary Biology*, **20**, 286–95.

Chai P (1996) Butterfly visual characteristics and ontogeny of responses to butterflies by a specialized bird. *Biological Journal of the Linnean Society*, **59**, 37–67.

Chai P, Srygley RB (1990) Predation and the flight, morphology, and temperature of neotropical rain-forest butterflies. *The American Naturalist*, **135**, 748–65.

Chance MRA, Russell WMS (1959) Protean displays: a form of allaesthetic behaviour. *Journal of Zoology*, **132**, 65–70.

Chandra HS (1991) How do heterogametic females survive without gene dosage compensation? *Journal of Genetics*, **70**, 137–46.

Charlesworth B (1994) The genetics of adaptation: lessons from mimicry. *The American Naturalist*, **144**, 839–47.

Charlesworth B, Lande R, Slatkin M (1982) A new-Darwinian commentary on macro-evolution. *Evolution*, **36**, 474–98.

Charlesworth D, Charlesworth B (1975) Theoretical genetics of Batesian mimicry I. Single-locus models. *Journal of Theoretical Biology*, **55**, 283–303.

Charlesworth D, Charlesworth B (1976a) Theoretical genetics of Batesian mimicry. II. Evolution of supergenes. *Journal of Theoretical Biology*, **55**, 305–24.

Charlesworth D, Charlesworth B (1976b) Theoretical genetics of Batesian mimicry III. Evolution of dominance. *Journal of Theoretical Biology*, **55**, 325–37.

Chase MW, Peacor CR (1987) Crystals of calcium oxalate hydrate on the perianth of *Stelis* Sw. *Lindleyana*, **2**, 91–4.

Chazot N, Willmott KR, Santacruz Endara PG, Toporov A, Hill RI, Jiggins CD, Elias M (2014) Mutualistic mimicry and filtering by altitude shape the structure of Andean butterfly communities. *The American Naturalist*, **183**, 26–39.

Chemsak JA, Linsley EG (1970) Death-feigning among North American Cerambycidae (Coleoptera). *Pan-Pacific Entomologist*, **46**, 305–7.

Chen G, Jürgens A, Shao L, Liu Y, Sun W, Xia C (2007) Semen-like floral scents and pollination biology of a sapromyophilous plant *Stemona japonica* (Stemonaceae). *Journal of Chemical Ecology*, **11**, 244–52.

Cheney KL (2010) Multiple selective pressures apply to a coral reef fish mimic: a case of Batesian–aggressive mimicry. *Proceedings of the Royal Society of London, Series B: Biological Sciences*, **277**, 1849–55.

Cheney KL (2013) Cleaner fish coloration decreases predation risk in aggressive fangblenny mimics. *Behavioral Ecology*, **24**,1161–5.

Cheney KL, Côté IM (2005) Frequency-dependent success of aggressive mimics in a cleaning symbiosis. *Proceedings of the Royal Society of London, Series B: Biological Sciences*, **272**, 2635–9.

Cheney KL, Côté IM (2007) Aggressive mimics profit from a model-signal receiver mutualism. *Proceedings of the Royal Society of London, Series B: Biological Sciences*, **274**, 2087–91.

Cheney KL, Grutter AS, Marshall JN (2008) Facultative mimicry: cues for colour change and colour accuracy in a coral reef fish. *Proceedings of the Royal Society of London, Series B: Biological Sciences*, **275**, 117–22.

Cheney KL, Grutter AS, Bshary R (2014a) Geographical variation in the benefits obtained by a coral reef fish mimic. *Animal Behaviour*, **88**, 85–90.

Cheney KL, Cortes F, How MJ, Wilson NG, Blomberg SP, Winters AE, Umanzör S, Marshall NJ (2014b) Conspicuous visual signals do not coevolve with increased body size in marine sea slugs. *Journal of Evolutionary Biology*, **27**, 676–87.

Chiao C-C, Wickiser KJ, Allen JJ, Genter B, Hanlon RT (2011) Hyperspectral imaging of cuttlefish camouflage indicates good color match in the eyes of fish predators. *Proceedings of the National Academy of Sciences, U.S.A.*, **108**, 9148–53.

China WE (1929) Historical survey of notes on the "protective" resemblance to a spike of blossom borne by clusters of an African homopteran of the genus *Ityraea*. *Annals and Magazine of Natural History*, **10**, 347–54.

Chiszar D, Boyer D, Lee R, Murphy JB, Radcliffe CW (1990) Caudal luring in the southern death adder *Acanthophis antarcticus*. *Journal of Herpetology*, **24**, 253–60.

Chittka L (2001) Camouflage of predator crab spiders on flowers and the colour perception of bees (Aranida: Thomisidae/ Hymenoptera: Apidae). *Entomologia Generalis*, **25**, 181–7.

Chittka L, Osorio D (2007) Cognitive dimensions of predator responses to imperfect mimicry. *PLoS Biology*, **5**: e339.

Chopard L (1967) Les staphylins commensaux des terribles fourmis legionnaires et leur extraordinaire mimetisme. *Nature, Paris*, **3384**, 146–9.

Chouteau M, Angers B (2012) Wright's shifting balance theory and the diversification of aposematic signals. *PLoS ONE*, **7**: e34028.

Chouteau M, Arias M, Joron M (2016) Warning signals are under positive frequency-dependent selection in nature. *Proceedings of the National Academy of Sciences, U.S.A.*, **113**, 2164–2169.

Christen U (2014) Molecular mimicry. In Y Shoenfeld, PL Meroni and ME Gershwin (eds), *Autoantibodies*. 3rd edn. Elsevier, New York, pp. 35–42.

Chu M (2001) Vocal mimicry in distress calls of phainopeplas. *The Condor*, **103**, 389–95.

Ciotek L, Giorgis P, Benitez-Vieyra S, Cocucci AA (2006) First confirmed case of pseudocopulation in terrestrial orchids of South America: pollination of *Geoblasta pennicillata* (Orchidaceae) by *Campsomeris bistrimacula* (Hymenoptera, Scoliidae). *Flora*, **201**, 365–9.

Cipresso Pereira PH, Leão Feitosa JL, Padovani Ferreira B (2011) Mixed-species schooling behavior and protective mimicry involving coral reef fish from the genus *Haemulon* (Haemulidae). *Neotropical Ichthyology*, **9**, 741–6.

Claridge MF (1974) Stridulation and defensive behaviour in the ground beetle, *Cychrus caraboides* (L.). *Journal of Entomology (A)*, **49**, 7–15.

Clark R, Brown SM, Collins SC, Jiggins CD, Heckel DG, Vogler AP (2008) Colour pattern specification in the Mocker swallowtail *Papilio dardanus*: the transcription factor *invected* is a candidate for the mimicry locus H. *Proceedings of the Royal Society of London, Series B: Biological Sciences*, **275**, 1181–8.

Clarke B (1962) Natural selection in mixed populations of two polymorphic snails. *Heredity*, **17**, 319–45.

Clarke CA, Sheppard PM (1959a) The genetics of some mimetic forms of *Papilio dardanus*, Brown, and *Papilio glaucus*, Linn. *Journal of Genetics*, **56**, 236–59.

Clarke CA, Sheppard PM (1959b) The genetics of *Papilio dardanus*, Brown. I. Race *cenea* from South Africa. *Genetics*, **44**, 1347–58.

Clarke CA, Sheppard PM (1960a) The evolution of mimicry in the butterfly *Papilio dardanus*. *Heredity*, **14**, 163–73.

Clarke CA, Sheppard PM (1960b) The genetics of *Papilio dardanus*, Brown. II. Races *dardanus*, *polytrophus*, *meseres*, and *tibullus*. *Genetics*, **45**, 439–57.

Clarke CA, Sheppard PM (1960c) The genetics of *Papilio dardanus*, Brown. III. Race *antinorii* from Abyssinia and race *meriones* from Madagascar. *Genetics*, **45**, 683–98.

Clarke CA, Sheppard PM (1960d) Supergenes and mimicry. *Heredity*, **14**, 175–85.

Clarke CA, Sheppard PM (1962a) The genetics of *Papilio dardanus*, Brown. IV. Data on race *ochracea*, race *flavicornis*, and further information on races *polytrophus* and *dardanus*. *Genetics*, **47**, 909–20.

Clarke CA, Sheppard PM (1962b) The genetics of the mimetic butterfly *Papilio glaucus*. *Ecology*, **43**, 159–61.

Clarke CA, Sheppard PM (1963) Interactions between major genes and polygenes in the determination of the mimetic patterns of *Papilio dardanus*. *Evolution*, **17**, 404–13.

Clarke CA, Sheppard PM (1971) Further studies on the genetics of the mimetic butterfly *Papilio memnon*. *Proceedings and Transactions of the Royal Society of London, Series B*, **263**, 35–70.

Clarke CA, Sheppard PM (1972a) The genetics of the mimetic butterfly *Papilio polytes*. *Proceedings and Transactions of the Royal Society of London, Series B*, **263**, 431–58.

Clarke CA, Sheppard PM (1972b) Genetic and environmental factors influencing pupal colour in the swallowtail butterflies *Battus philenor* (L.) and *Papilio polytes* L. *Journal of Entomology (A)*, **46**, 123–33.

Clarke CA, Sheppard PM (1973) The genetics of four new forms of the mimetic butterfly *Papilio memnon* L. *Proceedings of the Royal Society of London, Series B: Biological Sciences*, **184**, 1–14.

Clarke CA, Sheppard PM (1975) The genetics of the mimetic butterfly *Hypolimnas bolina* (L.). *Philosophical Transactions of the Royal Society of London. Series B, Biological Sciences*, **272**, 229–65.

Clarke CA, Sheppard PM (1977) Data suggesting absence of linkage between two loci in the mimetic butterfly *Hypolimnas bolina* (Nymphalidae). *Journal of the Lepidopterists' Society*, **31**, 139–43.

Clarke CA, Sheppard PM, Thornton IWB (1968) The genetics of the mimetic butterfly *Papilio memnon*. *Philosophical Transactions of the Royal Society of London. Series B, Biological Sciences*, **254**, 37–89.

Clarke CA, Clarke FMM, Collins SC, Gill ACL, Turner JRG (1985) Male-like females, mimicry and transvestism in butterflies (Lepidoptera: Papilionidae). *Systematic Entomology*, **10**, 257–83.

Clarke CA, Clarke FMM, Gordon IJ, Marsh NA (1989) Rule-breaking mimics: palatability of the butterflies *Hypolimnas bolina* and *Hypolimnas misippus*, a sister species pair. *Biological Journal of the Linnean Society*, **37**, 359–67.

Clarke GL, Denton EJ (1962) Light and animal life. In MN Hill (ed.), *The Sea*. Vol. I. Interscience, New York, pp. 456–68.

Clarke JM, Schluter D (2011) Colour plasticity and background matching in a three spine stickleback species pair. *Biological Journal of the Linnean Society*, **102**, 902–14.

Claus R, Hoppen HO, Karg H (1981) The secret of truffles: a steroidal pheromone? *Experientia*, **37**, 1178–9.

Cloudsley-Thompson JL (1977) The genus *Cossyphus* (Col., Tenebrionidae): a striking instance of protective resemblance. *Entomologist's Monthly Magazine*, **113**, 151–2.

Cloudsley-Thompson JL (1980) *Tooth and Claw: Defensive Strategies in the Animal World*. J M Dent & Sons, London.

Cloudsley-Thompson JL (1981) Comments on the nature of deception. *Biological Journal of the Linnean Society*, **16**, 11–14.

Clucas B, Owings DH, Rowe MP (2008) Donning your enemy's cloak: ground squirrels exploit rattlesnake scent to reduce predation risk. *Proceedings of the Royal Society of London, Series B: Biological Sciences*, **275**, 847–52.

Clutton-Brock T (2009) Sexual selection in females. *Animal Behaviour*, **77**, 3–11.

Cobb NA (1906) Fungus malady of the sugar cane. *Bulletin of the Hawaiian Sugar Planters' Association. Pathological and Physiological series*, **5**, 1–254.

Cobben RH (1986) A most strikingly myrmecomorphic mirid from Africa, with some notes on ant-mimicry and chromosomes in hallodapines (Miridae, Heteroptera). *Journal of the New York Entomological Society*, **94**, 194–204.

Cody ML (1969) Convergent characteristics in sympatric species: a possible relation to interspecific competition and aggression. *The Condor*, **71**, 223–39.

Cohen JA (1985) Differences and similarities in cardenolide contents of queen and monarch butterflies in Florida and their ecological and evolutionary implications. *Journal of Chemical Ecology*, **11**, 85–103.

Coleman E (1928) Pollination of an Australian orchid by the male ichneumonid *Lissopimpla semipunctata*, Kirby. *Transactions of the Entomological Society of London*, **76**, 533–9.

Coleman E (1938) Further observations on the pseudocopulation of the male *Lissopimpla semipunctata* Kirby (Hymenoptera, Parasitica) with the Australian orchid *Cryptostylis leptochila* F.v.M. *Proceedings of the Royal Entomological Society of London*, **13**, 82–3.

Collenette CL (1935) Notes concerning attacks by British birds on butterflies. *Proceedings of the Zoological Society of London*, **1935**, 201–17.

Collinge WE (1937) Wild birds and butterflies. *Nature*, **140**, 974.

Combs JK, Pauw A (2009) Preliminary evidence that the long-proboscid fly, *Philoliche gulosa*, pollinates *Disa karooica* and its proposed Batesian model *Pelargonium stipulaceum*. *South African Journal of Botany*, **75**, 757–61.

Compton SG, Ware AB (1991) Ants disperse the elaiosome-bearing eggs of an African stick insect. *Psyche*, **98**, 207–14.

Congdon JH, Vitt LJ, King WW (1974) Geckos: adaptive significance and energetics of tail autotomy. *Science*, **184**, 1379–80.

Cook LM (2000) Changing views on melanic moths. *Biological Journal of the Linnean Society*, **69**, 431–41.

Cook LM (2003) The rise and fall of the *Carbonaria* form of the peppered moth. *Quarterly Review of Biology*, **78**, 399–417.

Cook LM, Turner JRG (2008) Decline in melanism in two British moths: spatial, temporal and interspecific variation. *Heredity*, **101**, 483–9.

Cook LM, Brower LP, Alcock J (1969) An attempt to verify mimetic advantage in a neotropical environment. *Evolution*, **23**, 339–45.

Cook LM, Grant BS, Saccheri IJ, Mallet J (2012) Selective bird predation on the peppered moth: the last experiment of Michael Majerus. *Biology Letters*, **8**, 609–12.

Cook SE, Vernon JG, Bateson M, Guilford T (1994) Mate choice in the polymorphic African swallowtail butterfly, *Papilio dardanus*: male-like females may avoid sexual harassment. *Animal Behaviour*, **47**, 389–97.

Cooney LJ, Schaefer HM, Logan BA, Cox B, Gould KS (2015) Functional significance of anthocyanins in peduncles of

Sambucus nigra. Environmental and Experimental Botany, **119**, 18–26.

Cooper JM (1984) Apostatic selection on prey that match the background. *Biological Journal of the Linnean Society,* **23**, 221–8.

Cooper JM, Allen JA (1994) Selection by wild birds on artificial dimorphic prey on varied backgrounds. *Biological Journal of the Linnean Society,* **51**, 433–46.

Cooper SM, Owen-Smith N (1986) Effects of plant spinescence on large mammalian herbivores. *Oecologia (Berlin),* **68**, 446–55.

Cooper WE Jr (1981) Mimicry and spatial occupation in the mydas fly, *Mydas clavatus. Journal of the Alabama Academy of Science,* **52**, 58–65.

Cooper WE Jr (1992) Does gregariousness reduce attacks on aposematic prey? Limitations of one experimental test. *Animal Behaviour,* **43**, 163–4.

Cooper WE Jr, Vitt LJ (1991) Influence of detectability and ability to escape on natural selection of conspicuous autotomous defenses. *Canadian Journal of Zoology,* **69**, 757–64.

Coppinger RP (1969) The effect of experience and novelty on avian feeding behaviour with reference to the evolution of warning coloration in butterflies. I. Reactions of wild-caught blue jays to novel insects. *Behaviour,* **35**, 45–60.

Coppinger RP (1970) The effects of experience and novelty on avian feeding behavior with reference to the evolution of warning coloration in butterflies: reactions of naïve birds to novel insects. *The American Naturalist,* **101**, 323–35.

Cordero A (1990) The inheritance of female polymorphism in the damselfly *Ischnura graellsii* (Rambur) (Odonata: Coenagrionidae). *Heredity,* **64**, 341–6.

Cortesi F, Cheney KL (2010) Conspicuousness is correlated with toxicity in marine opisthobranchs. *Journal of Evolutionary Biology,* **23**, 1509–18.

Cortesi F, Feeney WE, Ferrari MCO, Waldie PA, Phillips GAC, McClure EC, Sköld HN, Salzburger W, Marshall NJ, Cheney KL (2015) Phenotypic plasticity confers multiple fitness benefits to a mimic. *Current Biology,* **25**, 949–54.

Coss RG (1979) Delayed plasticity of an instinct: recognition and avoidance of 2 facing eyes by the jewel fish. *Developmental Psychobiology,* **12**, 335–45.

Coonunt D, Pinsunu, Custé J, Lrrut M (2003) Invasion of mammalian cells by *Listeria monocytogenes*: functional mimicry to subvert cellular functions. *TRENDS in Cell Biology,* **13**, 23–31.

Côté IM, Cheney KL (2005) Choosing when to be a cleaner-fish mimic. *Nature,* **433**, 211–2.

Côté IM, Cheney KL (2007) Protective function for aggressive mimicry? *Proceedings of the Royal Society of London, Series B: Biological Sciences,* **274**, 2445–8.

Cott HB (1940) *Adaptive Colouration in Animals.* Methuen, London.

Cott HB (1947) The edibility of birds: illustrated by five years' experiments and observations (1941–1946) on the food preferences of the hornet, cat and man: and considered with special reference to the theories of adaptive coloration. *Proceedings of the Zoological Society of London,* **116**, 371–524.

Couldridge VCK (2002) Experimental manipulation of male egg-spots demonstrates female preference for one large spot in *Pseudotropheus lombardoi. Journal of Fish Biology,* **60**, 726–30.

Cournoyer BL, Cohen JH (2011) Cryptic coloration as a predator avoidance strategy in seagrass arrow shrimp colormorphs. *Journal of Experimental Marine Biology and Ecology,* **402**, 27–34.

Cox AHM (1930) Nestling wood-warblers "hissing". *British Birds,* **23**, 219–20.

Cox GW (1960) A life history of the mourning warbler. *The Wilson Bulletin,* **72**, 5–28.

Cox S, Chandler S, Barron C, Work K (2009) Benthic fish exhibit more plastic crypsis than non-benthic species in a freshwater spring. *Journal of Ethology,* **27**, 497–505.

Coyne JA (1998) Not black and white. Review of 'Melanism: evolution in action' by Michael E.N. Majerus. *Nature,* **396**, 35–6.

Cozzolino S, Widmer A (2005) Orchid diversity: an evolutionary consequence of deception? *Trends in Ecology and Evolution,* **20**, 487–94.

Craig CL, Ebert K (1994) Colour and pattern in predator prey interactions: the bright body colours and patterns of a tropical orb-spinning spider attract flower-seeking prey. *Functional Ecology,* **8**, 616–20.

Cranfield MR, Chang S, Pierce NE (2009) The double cloak of invisibility: phenotypic plasticity and larval decoration in a geometrid moth, *Synchlora frondaria*, across three diet treatments. *Ecological Entomology,* **31**, 112–1.

Crawley MJ (2007) *The R Book.* John Wiley & Sons, Chichester, UK.

Creed ER (1971) Industrial melanism in the two-spot ladybird and smoke abatement. *Evolution,* **23**, 290–3.

Cremer S, Sledge MF, Heinze J (2002) Male ants disguised by the queen's bouquet. *Nature,* **419**, 897.

Cronin TW, Nadav Shashar N, Caldwell RL, Marshall J, Cheroske AG, Chiou T-H (2003) Polarization vision and its role in biological signalling. *Integrative and Comparative Biology,* **43**, 549–58.

Croston R, Tonra C, Heath SK, Hauber ME (2012) Flange color differences of brood parasitic brown-headed cowbirds from nests of two host species. *The Wilson Journal of Ornithology,* **124**, 139–45.

Croze H (1970) *Searching image in carrion crows.* Paul Parey, Berlin.

Cruz A, Wiley JW (1989) The decline of an adaptation in the absence of a presumed selection pressure. *Evolution,* **43**, 55–62.

Cunha GR, Wang Y, Place NJ, Liu W, Baskin L, Glickman SE (2003) Urogenital system of the spotted hyena (*Crocuta crocuta* Erxleben): a functional histological study. *Journal of Morphology,* **256**, 205.

Currey JD, Cain AJ (1968) Studies on Cepaea. IV. Climate and selection of banding morphs in *Cepaea* from the climatic optimum to the present day. *Philosophical Transactions of the Royal Society London B: Biological Sciences,* **253**, 483–98.

Cushing PE (1997) Myrmecomorphy and myrmecophily in spiders: a review. *Florida Entomologist,* **80**, 165–93.

Cushing PE (2012) Spider-ant associations: an updated review of myrmecomorphy, myrmecophily, and myrmecophagy in spiders. *Psyche,* **2012**: 151989.

Cusick MF, Libbey JE, Fujinami RS (2012) Molecular mimicry as a mechanism of autoimmune disease. *Clinical Reviews in Allergy & Immunology,* **42**, 102–11.

Cuthill IC, Bennett ATD (1993) Mimicry and the eye of the beholder. *Proceedings of the Royal Society of London, Series B: Biological Sciences,* **253**, 203–4.

Cuthill IC, Stevens M, Sheppard J, Maddocks T, Párraga CA, Troscianko TS (2005) Disruptive coloration and background pattern matching. *Nature*, **434**, 72–4.

Cuthill IC, Stevens M, Windsor AMM, Walker HJ (2006) The effects of pattern symmetry on detection of disruptive and background-matching coloration. *Behavioral Ecology*, **17**, 828–32.

Cuthill JH, Charleston M (2012) Phylogenetic codivergence supports coevolution of mimetic *Heliconius* butterflies. *PLoS ONE*, **7(5)**: e36464.

Cutler B (1991) Reduced predation on the antlike jumping spider *Synageles occidentalis* (Araneae: Salticidae). *Journal of Insect Behavior*, **4**, 401–7.

Czajka M (1963) Unknown facts of the biology of the spider *Ero furcata*. *Polskie Pismo Entomologiczne*, **33**, 229–31.

DaCosta JM, Sorenson MD (2014) An experimental test of host song mimicry as a species recognition cue among male brood parasitic indigobirds (*Vidua* spp.). *The Auk*, **131**, 549–58.

Dafni A (1983) Pollination of *Orchis caspia* – a nectarless plant which deceives the pollinators of nectariferous species from other plant families. *Journal of Ecology*, **71**, 467–74.

Dafni A (1984) Mimicry and deception in pollination. *Annual Review of Ecology and Systematics*, **15**, 259–78.

Dafni A (1986) Floral mimicry – mutualism and unidirectional exploitation of insects by plants. In TRF Southwood & BE Juniper (eds), *The Plant Surface and Insects*. Arnold, London, pp. 81–90.

Dafni A (1987) On the evolution from reward to deception in *Orchis* s.1. (Orchidaceae) and its related genera. In J Arditti (ed.), *Orchid Biology, Reviews and Perspectives 3*. Cornell University Press, Ithaca, NY, pp. 79–104.

Dafni A, Calder DM (1987) Pollination by deceit and floral mimesis in *Thelymitra antennifera* (Orchidaceae). *Plant Systematics and Evolution*, **158**, 11–22.

Dafni J, Diamant A (1984) School-oriented mimicry, a new type of mimicry in fishes. *Marine Ecology Progress Series*, **20**, 45–50.

Dafni A, Ivri Y (1981) Floral mimicry between *Orchis israelitica* Baumann and Dafni (Orchidaceae) and *Bellevalia flexuosa* Boiss. (Liliaceae). *Oecologia (Berlin)*, **49**, 229–32.

Dafni A, Woodell SRJ (1986) Stigmatic exudate and the pollination of *Dactylorhiza fuchsii* (Druce) Soo. *Flora*, **178**, 343–50.

Dafni A, Cohen D, Noy Meir I (1981) Life cycle variation in geophytes. *Annals of the Missouri Botanical Garden*, **68**, 652–60.

Dahlgren U (1916) Production of light by animals. *Journal of the Franklin Institute*, **181**, 525–56.

Dale S, Slagsvold T (1994) Mate choice on multiple criteria in the pied flycatcher. *Journal of Ornithology*, **135**, 119.

Daloze D, Pasteels JM (1979) Production of cardiac glycosides by chrysomelid beetles and larvae. *Journal of Chemical Ecology*, **5**, 63–77.

Daloze D, Braekman JC, Pasteels JM (1986) A toxic dipeptide from the defense glands of the Colorado beetle. *Science*, **233**, 221–3.

Daloze D, Braekman JC, Pasteels JM (1994) Ladybird defence alkaloids: structural, chemotaxonomic and biosynthetic aspects (Col.: Coccinellidae). *Chemoecology*, **5**, 173–83.

Daly JW (1998) Thirty years of discovering arthropod alkaloids in amphibian skin. *Journal of Natural Products*, **61**, 162–72.

Daly JW, Myers CW (1967) Toxicity of Panamanian poison frogs (*Dendrobates*): some biological and chemical tests. *Science*, **156**, 970–3.

Dalziell AH, Magrath RD (2012) Fooling the experts: accurate vocal mimicry in the song of the superb lyrebird, *Menura novaehollandiae*. *Animal Behaviour*, **83**, 1401–10.

Dalziell AH, Welbergen JA (2016) Elaborate mimetic vocal displays by female superb lyrebirds. *Frontiers in Ecology and Evolution*, **4**: 34.

Dalziell AH, Welbergen JA, Igic B, Magrath RD (2015) Avian vocal mimicry: a unified conceptual framework. *Biological Reviews*, **90**, 643–68.

Damian RT (1964) Molecular mimicry: antigen sharing by parasite and host and its consequences. *The American Naturalist*, **98**, 129–49.

Damian RT (1987) Molecular mimicry revisited. *Parasitology Today*, **3**, 263–6.

Damian RT (1997) Parasite immune evasion and exploitation: reflections and projections. *Parasitology*, **115**, 169–75.

Damman H (1986) The osmeterial glands of the swallowtail butterfly, *Eurytides marcellus* as a defence against natural enemies. *Ecological Entomology*, **11**, 261–5.

Dare J, Montgomery JC (in press) The conflict between feeding and camouflage in a planktivorous fish (*Trachurus novaezelandiae*). *Behavioral Ecology*.

Darlington PJ (1938) Experiments on mimicry in Cuba with suggestions for future study. *Transactions of the Royal Entomological Society of London*, **87**, 681–95.

Darnell MZ (2012) Ecological physiology of the circadian pigmentation rhythm in the fiddler crab *Uca panacea*. *Journal of Experimental Marine Biology and Ecology*, **426**, 39–47.

Darst C, Cummings ME, Cannatella DC (2006) A mechanism for diversity in warning signals: conspicuousness versus toxicity in poison frogs. *Proceedings of the National Academy of Sciences, U.S.A.*, **103**, 5852–7.

Darwin C (1863) [Review of] Contributions to an insect fauna of the Amazon Valley. By Henry Walter Bates, Esq. Transact. Linnean Soc. vol. XXIII. 1862. *Natural History Review*, **3**, 219–24.

Darwin C (1871) *The Descent of Man and Selection in Relation to Sex*. John Murray, London.

Darwin C (1885) *On the various contrivances by which orchids are fertilised by insects*. John Murray, London.

Darwin C (1888) *The Different Forms of Flowers on Plants of the Same Species*, 3rd edn. John Murray, London.

Davidson DW, Seidel JL, Epstein WW (1990) Neotropical ant gardens II. Bioassays of seed components. *Journal of Chemical Ecology*, **16**, 2993–3013.

Davies NB (2000) *Cuckoos, Cowbirds and Other Cheats*. T. & A. D. Poyser, London.

Davies NB (2002) Cuckoo tricks with eggs and chicks. *British Birds*, **95**, 101–15.

Davies NB (2011) Cuckoo adaptations: trickery and tuning. *Journal of Zoology*, **284**, 1–14.

Davies NB, Brooke M de L (1988) Cuckoos vs. reed warblers: adaptations and counteradaptations. *Animal Behaviour*, **36**, 262–84.

Davies NB, Brooke M de L (1989) An experimental study of co-evolution between the cuckoo, *Cuculus canorus*, and its hosts. I. Host egg discrimination. *Journal of Animal Ecology*, **58**, 207–24.

Davies NB, Brooke M de L (1998) Cuckoos versus hosts: experimental evidence for co-evolution. In SI Rothstein and SK Robinson (eds), *Parasitic Birds and Their Hosts: Studies in Coevolution*. Oxford University Press, Oxford, pp. 59–79.

Davies NB, Welbergen JA (2008) Cuckoo-hawk mimicry? An experimental test. *Proceedings of the Royal Society of London, Series B: Biological Sciences*, **275**, 1817–22.

Davies NB, Brooke M de L, Kacelnik A (1996) Recognition errors and probability of parasitism determine whether reed warblers should accept or reject mimetic cuckoo eggs. *Proceedings of the Royal Society of London, Series B: Biological Sciences*, **263**, 925–31.

Davies NB, Kilner RM, Noble DG (1998) Nestling cuckoos, *Cuculus canorus*, exploit hosts with begging calls that mimic a brood. *Proceedings of the Royal Society of London, Series B: Biological Sciences*, **265**, 673–8.

Davis JRA (1903) *The Natural History of Animals*. Gresham Publishing Company, London.

Davis Rabosky AR, Cox CL, Rabosky DL, Title PO, Holmes IA, Feldman A, McGuire JA (2016) Coral snakes predict the evolution of mimicry across New World snakes. Nature Communications, **7**: 11484.

Dawkins MS (1971) Perceptual changes in chicks: another look at the 'search image' concept. *Animal Behaviour*, **19**, 556–74.

Dawkins MS, Guilford T (1991) The corruption of honest signalling. *Animal Behaviour*, **41**, 865–73.

Dawkins R (1979) Twelve misunderstandings of kin selection. *Zeitschrift für Tierpsychologie*, **51**, 184–200.

Dawkins R (1999) *The Extended Phenotype: The Long Reach of the Gene*. Oxford University Press, Oxford.

Dawkins R, Carlisle TR (1976) Parental investment, mate desertion and a fallacy. *Nature*, **262**, 131–3.

Dawkins R, Krebs JR (1979) Arms races between and within species. *Proceedings of the Royal Society of London, Series B: Biological Sciences*, **205**, 489–511.

Dawson SM, Jenkins PF (1983) Chaffinch song repertoires and the Beau Geste hypothesis. *Behaviour*, **87**, 256–69.

De Bona S, Valkonen JK, López-Sepulcre A, Mappes J (2015) Predator mimicry, not conspicuousness, explains the efficacy of butterfly eyespots. *Proceedings of the Royal Society of London, Series B: Biological Sciences*, **282**: 20150202.

De Cock R, Matthysen E (2003) Glow-worm larvae bioluminescence (Coleoptera: Lampyridae) operates as an aposematic signal upon toads (*Bufo bufo*). *Behavioral Ecology*, **14**, 103–8.

Deeleman-Reinhold CL (1993) A new spider genus from Thailand with a unique ant-mimicking device, with description of some other castianeirine spiders (Araneae: Corinnidae: Catianeirinae). *The Natural History Bulletin of the Siam Society*, **40**, 167–84.

de Gunst JH (1950) Een fraai voorbeeld van "mimicry". *Entomologische Berichten, Amsterdam*, **13**, 142.

de Jong PW, Holloway GJ, Brakefield PM, Vos H (1991) Chemical defense in the ladybird beetles (Coccinellidae). II. Amount of reflex fluid, the alkaloid adaline and individual variation in defence in 2-spot ladybirds (*Adalia bipunctata*). *Chemoecology*, **2**, 15–9.

de Jonge J, Videler JJ (1989) Differences between the reproductive biologies of *Tripterygion tripteronotus* and *T. delaisei* (Pisces,

Perciformes, Tripterygiidae): the adaptive significance of an alternative mating strategy and a red instead of a yellow nuptial colour. *Marine Biology*, **100**, 431–37.

de Jonge J, de Ruiter AJH, van den Hurk R (1989) Testis-testicular gland complex of two *Tripterygion* species (Blennoidei, Telostei): differences between territorial and non-territorial males. *Journal of Fish Biology*, **35**, 497–508.

del Río M, Lanteri AA (2013) Taxonomic revision of the genus *Stenocyphus* Marshall (Coleoptera, Curculionidae) from Brazil. *ZooKeys*, **357**, 29–43.

Dentinger BTM, Roy BA (2010) A mushroom by any other name would smell as sweet: *Dracula* orchids. *McIlvainea*, **19**, 1–13.

Denton EJ (1971) Reflectors in fish. *Scientific American*, **224**, 65–72.

Denton EJ, Gilpin-Brown JB, Wright PG (1972) The angular distribution of the light produced by some mesopelagic fish in relation to their camouflage. *Proceedings of the Royal Society of London, Series B: Biological Sciences*, **182**, 145–58.

de Oliveira Calleia F, Rohe F, Gordo M (2009) Hunting strategy of the margay (*Leopardus wiedii*) to attract the wild pied tamarin (*Saguinus bicolor*). *Neotropical Primates*, **16**, 32–4.

Deppe C, Holt D, Tewksbury J, Broberg L, Peterson J, Wood K (2003) Effect of northern pygmy-owl (*Glaucidium gnoma*) eyespots on avian mobbing. *The Auk*, **120**, 765–71.

Dorns C, Pasteels JM (1977) Defensive mechanisms against predation in the Colorado beetle (*Leptinotarsa decemlineata*, Say). *Archives of Biology*, **88**, 289–304.

de Ruiter L (1952) Some experiments on the camouflage of stick caterpillars. *Behaviour*, **4**, 222–32.

de Ruiter L (1956) Countershading in caterpillars: an analysis of its adaptive significance. *Archives Neerlandaises de Zoologie*, **11**, 285–341.

de Ruiter L (1958) Some remarks on problems of the ecology and evolution of mimicry. *Archives Néerlandaises de Zoologie*, **13**, 351–68.

Dettner K (1997) Inter-and intraspecific transfer of toxic insect compound cantharidin. In K Dettner, G Bauer and W Völkl (eds), *Vertical Foodwebs Interactions*. Ecological Studies Vol. 130. Springer Verlag, Berlin, Heidelberg, pp. 115–45.

Dettner K, Liepert C (1994) Chemical mimicry and camouflage. *Annual Review of Entomology*, **39**, 129–54.

D'Ettorre P, Heinze J (2001) Sociobiology of slave-making ants. *Acta Ethologica*, **3**, 67–82.

D'Ettorre P, Errard C, Ibarra F, Francke W, Hefetz A (2000) Sneak in or repel your enemy: Dufour's gland repellent as a strategy for successful usurpation in the slave-maker *Polyergus rufescens*. *Chemoecology*, **10**, 135–42.

Detto T, Hemmi JM, Backwell PRY (2008) Colouration and colour changes of the fiddlercrab, *Uca capricornis*: a descriptive study. *PLoS ONE*, **3**: e1629.

DeVries PJ (2002) Differential wing toughness in distasteful and palatable butterflies: direct evidence supports unpalatable theory. *Biotropica*, **34**, 176–81.

DeVries PJ (2003) Tough African models and weak mimics: new horizons in the evolution of bad taste. *Journal of the Lepidopterists' Society*, **57**, 235–8.

DeVries PJ, Murray D, Lande R (1997) Species diversity in vertical, horizontal, and temporal dimensions of a fruit-feeding butterfly community in an Ecuadorian rainforest. *Biological Journal of the Linnean Society*, **62**, 343–64.

DeVries PJ, Lande R, Murray D (1999) Associations of co-mimetic ithomiine butterflies on small spatial and temporal scales in a neotropical rainforest. *Biological Journal of the Linnean Society*, **67**, 73–85.

de Wert L, Mahon K, Ruxton GD (2012) Protection by association: evidence for aposematic commensalism. *Biological Journal of the Linnean Society*, **106**, 81–9.

Diamond AR Jr, El Mayas H, Boyd RS (2006) *Rudbeckia auriculata* infected with a pollen-mimic fungus in Alabama. *Southeastern Naturalist*, **5**, 103–12.

Diamond J, Bond AB (2013) *Concealing Coloration in Animals*. The Belknap Press of Harvard University Press, Cambridge, Massachusetts.

Diamond JM (1982) Mimicry of friarbirds by orioles. *The Auk*, **99**, 187–96.

Diamond JM (1994) Two-faced mimicry. *Nature*, **367**, 683–4.

Dieckhoff C, Heimpel GE (2010) Role of cuticular hydrocarbons of aphid parasitoids in their relationship to aphid-attending ants. *Entomologia Experimentalis et Applicata*, **136**, 254–61.

Di Giusto B, Bessière J-M, Guéroult M, Lim LBL, Marshall DJ, Hossaert-McKey M, Gaume L (2010) Flower-scent mimicry masks a deadly trap in the carnivorous plant *Nepenthes rafflesiana*. *Journal of Ecology*, **98**, 845–56.

Dill LM (1975) Calculated risk-taking by predators as a factor in Batesian mimicry. *Canadian Journal of Zoology*, **53**, 1614–21.

Dimitrova M, Stobbe N, Schaefer HM, Merilaita S (2009) Concealed by conspicuousness: distractive prey markings and backgrounds. *Proceedings of the Royal Society B-Biological Sciences*, **276**, 1905–10.

Dinan L (2001) Phytoecdysteroids: biological aspects. *Phytochemistry*, **57**, 325–39.

Dinter K, Paarmann W, Peschke K, Arndt E (2002) Ecological, hydrochemical and chemical adaptations to ant predation in species of *Thermophilum* and *Graphipterus* (Coleoptera: Carabidae) in the Sahara desert. *Journal of Arid Environments*, **50**, 267–86.

Dipper FA (1981) The strange sex lives of British wrasse. *New Scientist*, **1981**, 444–5.

Ditmars RL (1937) *Reptiles of the World: The Crocodilians, Lizards, Snakes, Turtles and Tortoises of the Eastern and Western Hemispheres*. The Macmillan Company, New York, NY.

Dittrich WH, Gilbert F, Green PR, McGregor PK, Grewcock D (1993) Imperfect mimicry – a pigeon's perspective. *Proceedings of the Royal Society of London, Series B: Biological Sciences*, **251**, 195–200.

Dixey FA (1894) Development of concept of "Reciprocal mimicry". *Transactions of the Royal Entomological Society of London*, **1894**, 296.

Dixey FA (1909) On Müllerian mimicry and diaposematism. *Transactions of the Royal Entomological Society of London*, **1908**, 559–83.

Dixson AF (1983) Observations on the evolution and behavioral significance of "sexual skin" in female primates. In J Rosenblatt, R Hinde, C Beer and M-C Busnel (eds), *Advances in the Study of Behavior*, Vol. 13. Academic Press, New York, pp. 63–106.

Dixson AF (2012) *Primate Sexuality: Comparative Studies of the Prosimians, Monkeys, Apes, and Humans*, 2nd edn. Oxford University Press, New York.

Dodson CH, Frymire G (1961) Natural pollination of orchids. *Missouri Botanical Garden Bulletin*, **49**, 133–52.

Dolenská M, Nedvěd O, Veselý P, Tesařová M, Fuchs R (2009) What constitutes optical warning signals of ladybirds (Coleoptera: Coccinellidae) towards bird predators: colour, pattern or general look? *Biological Journal of the Linnean Society*, **98**, 234–42.

Dominey WJ (1980) Female mimicry in male bluegill sunfish: a genetic-polymorphism. *Nature*, **284**, 546–8.

dos Santos MB, Martins de Oliveira MCL, Verrastro L, Marques Tozetti AM (2010) Playing dead to stay alive: death-feigning in *Liolaemus occipitalis* (Squamata: Liolaemidae). *Biota Neotropical*, **10**, 361–4.

Dressler R (1981) *The Orchids – Natural History and Classification*. Harvard University Press, Cambridge, MA.

Duckworth WD, Eichlin TD (1974) Clearwing moths of Australia and New Zealand (Lepidoptera: Sessiidae). *Smithsonian Contributions to Zoology*, **180**, 1–45.

Dudgeon CL, White WT (2012) First record of potential Batesian mimicry in an elasmobranch: juvenile zebra sharks mimic banded sea snakes? *Marine and Freshwater Research*, **63**, 545–551.

Dudley R (1989) Thanatosis in the neotropical butterfly *Caligo ilioneus* (Nymphalidae: Brassolinae). *Journal of Research in Lepidoptera*, **28**, 125–6.

Duffey SS, Stout MJ (1996) Antinutritive and toxic compounds of plant defense against insects. *Archives of Insect Biochemistry and Physiology*, **32**, 3–37.

Duffy K, Patrick K, Johnson SD (2013) Does the likelihood of an Allee effect on plant fecundity depend on the type of pollinator? *Journal of Ecology*, **101**, 953–62.

Dukas R (1987) Foraging behavior of three bee species in a natural mimicry system: female flowers which mimic male flowers in *Ecballium elaterium* (Cucurbitaceae), *Oecologia (Berlin)*, **74**, 256–63.

Dukas R, Morse DH (2003) Crab spiders affect flower visitation by bees. *Oikos*, **101**, 157–63.

Dukas R, Morse DH (2005) Crab spiders show mixed effects on flower-visiting bees and no effect on plant fitness. *Ecoscience*, **12**, 244–7.

Dumbacher JP, Pruett-Jones S (1996) Avian chemical defence. In V Nolan Jr and ED Ketterson (eds), *Current Ornithology*, Vol. **13**. Plenum Press, New York, pp. 137–74.

Dumbacher JP, Beehler BM, Spande TF, Garraffo HM, Daly JW (1992) Homobatrachotoxin in the genus *Pitohui*: chemical defense in birds? *Science*, **258**, 799–801.

Dumbacher JP, Wako A, Derrickson SR, Samuelson A, Spande TF, Daly JW (2004) Melyrid beetles (*Choresine*): a putative source for the batrachotoxin alkaloids found in poison-dart frogs and passerine birds. *Proceedings of the National Academy of Sciences, U.S.A.*, **101**, 15857–60.

Duncan CJ, Sheppard PM (1965) Sensory discrimination and its role in the evolution of Batesian mimicry. *Behaviour*, **24**, 269–82.

Duncan MJ, Goldman BD (1984) Hormonal regulation of the annual pelage color cycle in the Djungarian hamster, *Phodopus sungorus*. II. Role of prolactin. *Journal of Experimental Zoology*, **230**, 97–103.

Dunlap-Pianka H, Boggs CL, Gilbert LE (1977) Ovarian dynamics in Heliconiine butterflies: programmed senescence versus eternal youth. *Science*, **197**, 487–90.

Dunn ER (1927) A new mountain race of *Desmognathus*. *Copeia*, **1927**, 84–6.

Dunn ER (1954) The coral snake "mimic" problem in Panama. *Evolution*, **8**, 97–102.

Dunning DC (1968) Warning sounds of moths. *Zeitschrift für Tierpsychologie*, **25**, 129–38.

Dunning DC, Roeder KD (1965) Moth sounds and insect catching behavior of bats. *Science*, **147**, 173–4.

Durkee CA, Weiss MR, Uma DB (2011) Ant mimicry lessens predation on a North American jumping spider by larger salticid spiders. *Environmental Entomology*, **40**, 1223–31.

Dussourd DE, Harvis CA, Meinwald J, Eisner T (1989) Paternal allocation of sequestered plant pyrrolizidine alkaloid to eggs in the danaine butterfly, *Danaus gilippus*. *Experientia*, **45**, 896–8.

Duval C, Cassey P, Lovell PG, Mikšík I, Reynolds SJ, Spencer KA (2016) Maternal influence on eggshell maculation: implications for cryptic camouflaged eggs. *Journal of Ornithology*, **157**, 303–10.

Eagle JV, Jones GP (2004) Mimicry in coral reef fishes: ecological and behavioural responses of a mimic to its model. *Journal of Zoology*, **264**, 33–43.

East ML, Hofer H, Wickler W (1993) The erect 'penis' is a flag of submission in a female-dominated society: greetings in Serengeti spotted hyenas. *Behavioral Ecology and Sociobiology*, **33**, 355.

Eaton MD (2005) Human vision fails to distinguish widespread sexual dichromatism among sexually monochromatic birds. *Proceedings of the National Academy of Sciences, U.S.A.*, **102**, 10942–6.

Eaton RL (1976) A possible case of mimicry in larger mammals. *Evolution*, **30**, 853–6.

Eberhard WG (1977) Aggressive chemical mimicry by a bolas spider. *Science*, **198**, 1173–4.

Eberhard WG (2003) Substitution of silk stabilimenta for egg sacs by *Allocyclosa bifurca* (Araneae; Araneidae) suggests that silk stabilimenta function as camouflage devices. *Behaviour*, **140**, 847–68.

Edelstam C (2005) Moult patterns, age criteria and polymorphism. In J Ferguson-Lees and DA Christie (eds), *Raptors of the World*. Christopher Helm, London, pp. 57–62.

Edgar JA, Culvenor CCJ, Pliske TE (1974) Coevolution of danaid butterflies and their host plants. *Nature*, **250**, 646–8.

Edgar JA, Cockrum PA, Frahn JL (1976) Pyrrolizidine alkaloids in *Danaus plexippus* L. and *Danaus chrysippus* L. *Experientia*, **32**, 1535–7.

Edgar JA, Boppré M, Schneider D (1979) Pyrrolizidine alkaloid storage in African and Australian danaid butterflies. *Experientia*, **35**, 1447–8.

Edgar JA, Culvenor CCJ, Cockrum PA, Smith LW, Rothschild M (1980) Callimorphine: identification and synthesis of cinnabar moth "metabolite". *Tetrahedron Letters*, **21**, 1383–4.

Edgehouse M, Brown CP (2014) Predatory luring behavior of odonates. *Journal of Insect Science*, **14**: 146.

Edmunds J (1986) The stabilimenta of *Argiope flavipalpis* and *Argiope trifasciata* in West Africa, with a discussion of the function of stabilimenta. In WG Eberhard, YD Lubin and BC Robinson (eds), *Proceedings of the Ninth International Congress of Arachnology, Panama 1983*. Smithsonian Institution Press, Washington, DC, pp. 61–72.

Edmunds J, Edmunds M (1974) Polymorphic mimicry and natural selection: a reappraisal. *Evolution*, **28**, 402–7.

Edmunds M (1969) Polymorphism in the mimetic butterfly *Hypolimnas misippus* L. in Ghana. *Heredity*, **24**, 281–302.

Edmunds M (1974a) *Defence in Animals: A Survey of Anti-Predator Defences*. Longman, Burnt Mill, Essex, UK.

Edmunds M (1974b) Significance of beak marks on butterfly wings. *Oikos*, **25**, 117–18.

Edmunds M (1981) On defining 'mimicry'. *Biological Journal of the Linnean Society*, **16**, 9–10.

Edmunds M (1987) Color in opisthobranchs. *American Malacological Bulletin*, **5**, 185–96.

Edmunds M (1990) The evolution of cryptic coloration. In DL Evans and JO Schmidt (eds), *Insect Defenses*. State University of New York Press, Albany, NY, pp. 3–21.

Edmunds M (1991) Does warning colouration occur in nudibranchs? *Malacologia*, **32**, 241–55.

Edmunds M (1993) Does mimicry of ants reduce predation by wasps on salticid spiders? *Memoirs of the Queensland Museum*, **33**, 507–12.

Edmunds M (2000) Why are there good and poor mimics? *Biological Journal of the Linnean Society*, **70**, 459–66.

Edmunds M (2006) Do Malaysian *Myrmarachne* associate with particular species of ant? *Biological Journal of the Linnean Society*, **88**, 645–53.

Edmunds M (2009) Do nematocysts sequestered by aeolid nudibranchs deter predators? – a background to the debate. *Journal of Molluscan Studies*, **75**, 203–5.

Edmunds M, Dewhirst RA (1994) The survival value of countershading with wild birds as predators. *Biological Journal of the Linnean Society*, **51**, 447–52.

Edmunds M, Grayson J (1991) Camouflage and selective predation in caterpillars of the poplar and eyed hawkmoths (*Lathoe populi* and *Smerinthus ocellata*). *Biological Journal of the Linnean Society*, **42**, 467–80.

Edwards DP, Arauco R, Hassall M, Sutherland WJ, Chamberlain K, Wadhams LJ, Yu DW (2007) Protection in an ant-plant mutualism: an adaptation or a sensory trap? *Animal Behaviour*, **74**, 377–85.

Edwards GB (1984) Mimicry of velvet ants (Hymenoptera: Mutillidae) by jumping spiders (Araneae: Salticidae). *Peckhamia*, **2**, 46–9.

Egri Á, Blahó M, Kriska G, Farkas R, Gyurkovszky M, Åkesson S, Horváth G (2012) Polarotactic tabanids find striped patterns with brightness and/or polarization modulation least attractive: an advantage of zebra stripes. *Journal of Experimental Biology*, **215**, 736–45.

Ehrlén J (1993) Ultimate functions of non-fruiting flowers in *Lathyrus vernus*. *Oikos*, **68**, 45–52.

Ehrlich PR, Ehrlich AH (1973) Coevolution: heterotypic schooling in Caribbean reef fishes. *The American Naturalist*, **107**, 157–60.

Ehrlich PR, Ehrlich AH (1982) Lizard predation on tropical butterflies. *Journal of the Lepidopterists' Society*, **36**, 148–52.

Eisenmann E (1962) Notes on nighthawks of the genus *Chordeiles* in southern middle America, with a description of a new race of *Chordeiles minor* breeding in Panama. *American Museum Novitates*, **2049**, 1–22.

Eisikowitch D (1980) The role of dark flowers in the pollination of certain Umbelliferae. *Journal of Natural History*, **14**, 737–42.

Eisner T (1970) Chemical defense against predation in arthropods. In E Sondheimer and JB Simeone (eds), *Chemical Ecology*. Academic Press, New York, pp. 157–217.

Eisner T (1985) A fly that mimics jumping spiders. *Psyche*, **92**, 103–4.

Eisner T, Grant RP (1981) Toxicity, odor aversion, and "olfactory aposematism" *Science*, **213**, 476.

Eisner T, Meinwald J (1987) Alkaloid-derived pheromones and sexual selection in Lepidoptera. In GD Prestwitch and GJ Blomquist (eds), *Pheromone Biochemistry*. Academic Press, New York, pp. 251–69.

Eisner T, Kafatos FC, Linsley EG (1962) Lycid predation by mimetic adult Cerambycidae (Coleoptera). *Evolution*, **16**, 316–24.

Eisner T, Pliske TE, Ikeda M, Owen DF, Vázquez L, Pérez H, Franclemont JG, Meinwald J (1970) Defense mechanisms of arthropods. XXVII. Osmeterial secretions of papilionid caterpillars (*Baronia, Papilio, Eurytides*). *Annals of the Entomological Society of America*, **63**, 914–5.

Eisner T, Hicks K, Eisner M, Robson DS (1978a) "Wolf-in-sheep's-clothing" strategy of a predaceous insect larva. *Science*, **199**, 790–4.

Eisner T, Wiemer DF, Haynes LW, Meinwald J (1978b) Lucibufagins: defensive steroids from the fireflies *Photinus ignitus* and *P. marginellus* (Coleoptera: Lampyridae). *Proceedings of the National Academy of Sciences, U.S.A.*, **75**, 905–8.

Eisner T, Ziegler R, McCormick JL, Eisner M, Hoebeke ER, Meinwald J (1994) Defensive use of an acquired substance (carminic acid) by predaceous insect larvae. *Experientia*, **50**, 610–5.

Eisner T, Goetz MA, Hill DE, Smedley SR, Meinwald J (1997) Firefly "femmes fatales" acquire defensive steroids (lucibufagins) from their firefly prey. *Proceedings of the National Academy of Sciences, U.S.A.*, **94**, 9723–8.

Eisner T, Schroeder FC, Snyder N, Grant JB, Aneshansley DJ, Utterback D, Meinwald J, Eisner M (2008) Defensive chemistry of lycid beetles and of mimetic cerambycid beetles that feed on them. *Chemoecology*, **18**, 109–19.

Elgar MA (1993) Inter-specific associations involving spiders: kleptoparasitism, mimicry and mutualism. *Memoirs of the Queensland Museum*, **33**, 411–30.

Elgar MA, Allan RA (2004) Predatory spider mimics acquire colony-specific cuticular hydrocarbons from their ant model prey. *Naturwissenschaften*, **91**, 143–7.

Elgar MA, Allan RA (2006) Chemical mimicry of the ant *Oecophylla smaragdina* by the myrmecophilous spider *Cosmophasis bitaeniata*: Is it colony-specific? *Journal of Ethology*, **24**, 239–46.

Eliyahu D, Nojima S, Mori K, Schal C (2009) Jail baits: how and why nymphs mimic adult females of the German cockroach, *Blattella germanica*. *Animal Behaviour*, **78**, 1097–105.

Elmes GW, Thomas JA, Wardlaw JC (1991) Larvae of *Maculinea rebeli*, a large-blue butterfly, and their *Myrmica* host ants: wild adoption and behaviour in ant-nests. *Journal of Zoology, London*, **223**, 447–60.

Emlen JM (1968) Batesian mimicry: a preliminary theoretical investigation of quantitative aspects. *The American Naturalist*, **102**, 235–41.

Emmel TC (1965) A new mimetic assemblage of lycid and cerambycid beetles in Central Chiapas, Mexico. *The Southwestern Naturalist*, **10**, 14–6.

Emsley MG (1964) The geographical distribution of the color-pattern components of *Heliconius erato* and *Heliconius melpomene* with genetical evidence for the systematic relationship between the two species. *Zoologica, New York*, **49**, 245–86.

Emsley MG (1966) The mimetic significance of *Erythrolamprus aesculapii ocellatus* Peters from Tobago. *Evolution*, **20**, 663–4.

Endara L, Grimaldi DA, Roy BA (2010) Lord of the flies: pollination of *Dracula* orchids. *Lankesteriana*, **10**, 1–11.

Endler JA (1978) A predator's view of animal color patterns. *Evolutionary Biology*, **11**, 319–64.

Endler JA (1981) An overview of the relationships between mimicry and crypsis. *Biological Journal of the Linnean Society*, **16**, 25–31.

Endler JA (1984) Progressive background [matching] in moths and a quantitative measure of crypsis. *Biological Journal of the Linnean Society*, **22**, 187–231.

Endler JA (1988) Frequency-dependent predation, crypsis and aposematic coloration. *Philosophical Transactions of the Royal Society of London, Series B*, **319**, 505–23.

Endler JA, Mappes J (2004) Predator mixes and the conspicuousness of aposematic signals. *The American Naturalist*, **163**, 532–47.

Endler JA, Rojas B (2009) The spatial pattern of natural selection when selection depends on experience. *The American Naturalist*, **173**, E62–E78.

Endress ME, Bruyns PV (2000) A revised classification of the Apocynaceae s.1. *Botanical Review*, **66**, 1–56.

Engen S, Järvi T, Wiklund C (1986) The evolution of aposematic coloration by individual selection: a life span survival model. *Oikos*, **46**, 397–403.

Erdtman G (1930) The attraction of carrion flies to Tetraplodon by an odoriferous substance in the hygroscopic hypnobium. **11**, 13–14.

Escobar-Lasso S, González-Duran GA (2012) Strategies employed by three Neotropical frogs (Amphibia: Anura) to avoid predation. *Herpetology Notes*, **5**, 79–84.

Estabrook GF, Jespersen DC (1974) Strategy for a predator encountering a model-mimic system. *The American Naturalist*, **108**, 443–57.

Estrada C, Jiggins CD (2008) Interspecific sexual attraction because of convergence in warning colouration: is there a conflict between natural and sexual selection in mimetic species? *Journal of Evolutionary Biology*, **21**, 749–60.

Eterovick PC, Côrtes Figuera JE (1997) Cryptic coloration and choice of escape microhabitats by grasshoppers (Orthoptera: Acrididae). *Biological Journal of the Linnean Society*, **61**, 485–99.

Eterovick PC, Oliveira FFR, Tattersall GJ (2010) Threatened tadpoles of *Bokermannohyla alvarengai* (Anura: Hylidae) choose backgrounds that enhance crypsis potential. *Biological Journal of the Linnean Society*, **101**, 437–46.

Evans DL (1983) Relative defensive behaviour of some moths and the implications to predator prey interactions. *Entomologia Experimentalis et Applicata*, **33**, 103–11.

Evans DL (1984) Reactions of some adult passerines to *Bombus pennsylvanicus* and its mimic, *Mallota bautias*. *Ibis*, **126**, 50–8.

Evans DL (1985) The defensive ensembles of two palatable moths. *Journal of the Lepidopterists' Society*, **39**, 43–7.

Evans DL (1987) Tough, harmless cryptics could evolve into tough, nasty aposematics: an individual selectionist model. *Oikos*, **48**, 114–15.

Evans DL, Waldbauer GP (1982) Behaviour of adult and naive birds when presented with a bumble bee and its mimic. *Zeitschrift für Tierpsychologie*, **59**, 247–59.

Evans DL, Castoriades N, Badruddine H (1986) Cardenolides in the defense of *Caenocoris nerii* (Hemiptera). *Oikos*, **46**, 325–9.

Evans HE (1968) Studies on neotropical Pompilidae (Hymenoptera). IV. Examples of dual sex-limited mimicry in *Chirodamus*. *Psyche*, **75**, 1–22.

Evans HE, Eberhardt MJW (1970) *The Wasps*. University of Michigan Press, Ann Arbor.

Ewer RF (1966) Juvenile behaviour of the African ground squirrel, *Xerus erythropus* (E. Geoff.) *Zeitschrift für Tierpsychologie*, **21**, 190–216.

Ewert J-P (1974) The neural basis of visually guided behavior. *Scientific American*, **230**, 34–42.

Exnerová A, Landová E, Štys P, Fuchs R, Prokopová M, Cehláriková P (2003) Reactions of passerine birds to aposematic and non-aposematic firebugs (*Pyrrhocoris apterus*; Heteroptera). *Biological Journal of the Linnean Society*, **78**, 517–25.

Exnerová A, Štys P, Fučicová E, Veselá S, Svádová K, Prokopová M, Jarošík V, Fuchs R, Landová E (2007) Avoidance of aposematic prey in European tits (Paridae): learned or innate? *Behavioral Ecology*, **18**, 148–56.

Exnerová A, Svádová K, Štys P, Barcalová S, Landová E, Prokopová M, Fuchs R, Socha R (2009) Importance of colour in the reaction of passerine predators to aposematic prey: experiments with mutants of *Pyrrhocoris apterus* (Heteroptera). *Biological Journal of the Linnean Society*, **88**, 143–53.

Fabricant SA, Herberstein ME (2015) Hidden in plain orange: aposematic coloration is cryptic to a colorblind insect predator. *Behavioral Ecology*, **26**, 38–44.

Fabricant SA, Exnerová A, Jezová D, Štys P (2014) Scared by shiny? The value of iridescence in aposematic signalling of the hibiscus harlequin bug. *Animal Behaviour*, **90**, 315–25.

Farrell TM, May PG, Andreadis PT (2011) Experimental manipulation of tail color does not affect foraging success in a caudal luring rattlesnake. *Journal of Herpetology*, **45**, 291–3.

Feeney WE, Stoddard MC, Kilner RM, Langmore NE (2014) "Jack-of-all-trades" egg mimicry in the brood parasitic Horsfield's bronze-cuckoo? *Behavioral Ecology*, **25**, 1365–73.

Feeney WE, Troscianko J, Langmore NE, Spottiswoode CN (2015) Evidence for aggressive mimicry in an adult brood parasitic bird, and generalised defences in its host. *Proceedings of the Royal Society of London B*, **282**: 20150798.

Feller KD, Cronin TW (2014) Hiding opaque eyes in transparent organisms: a potential role for larval eyeshine in stomatopod crustaceans. *Journal of Experimental Biology*, **217**, 3263–73.

Feltwell J, Rothschild M (1974) Carotenoids of 38 species of Lepidoptera. *Journal of Zoology*, **174**, 441–65.

Fenton A, Magoolagan L, Kennedy Z, Spencer KA (2011) Parasite-induced warning coloration: a novel form of host manipulation. *Animal Behaviour*, **81**, 417–22.

Ferguson GP, Messenger JB (1991) A countershading reflex in cephalopods. *Proceedings of the Royal Society of London, Series B: Biological Sciences*, **243**, 63–7.

Ferguson GP, Messenger JB, Budelmann BU (1994) Gravity and light influence the countershading reflexes of the cuttlefish *Sepia officinalis*. *Journal of Experimental Biology*, **191**, 247–56.

Ferreira WC Jr., Marcon D (2014) Revisiting the 1879 model for evolutionary mimicry by Fritz Müller: new mathematical approaches. *Ecological Complexity*, **18**, 25–38.

Field SA, Keller MA (1993) Alternative mating tactics and female mimicry as post-copulatory mate-guarding behaviour in the parasitic wasp *Cotesia rubecula*. *Animal Behaviour*, **46**, 1183–9.

Fields PG, McNeil JM (1988) The importance of seasonal variation in hair coloration for thermoregulation of *Ctenucha virginica* larvae (Lepidoptera: Arctidae). *Physiological Entomology*, **13**, 165–75.

Fingerman M (2013) *The Control of Chromatophores*. International Series of Monographs on Pure and Applied Biology. Elsevier, New York.

Fincke OM (2015) Trade-offs in female signal apparency to males offer alternative anti-harassment strategies for colour polymorphic females. *Journal of Evolutionary Biology*, **28**, 931–43.

Fincke OM, Jödicke R, Paulson D, Schultz TD (2005) The frequency of female-specific color polymorphisms in Holarctic Odonata: why are male-like females typically the minority? *International Journal of Odonatology*, **8**, 183–212.

Fink LS (1995) Food plant effects on colour morphs of *Eumorpha fasciata* caterpillars (Lepidoptera: Sphingidae). *Biological Journal of the Linnean Society*, **56**, 423–37.

Fisher RA (1927) On some objections to mimicry theory: statistical and genetic. *Transactions of the Royal Entomological Society of London*, **75**, 269–78.

Fisher RA (1930) *The Genetical Theory of Natural Selection*, 1st edn. Clarendon Press, Oxford.

Fisher RA (1958) *The Genetical Theory of Natural Selection*, 2nd edn. Dover, New York.

Fisher RM, Tuckerman RD (1986) Mimicry of bumble bees and cuckoo bumble bees by carrion beetles (Coleoptera; Silphidae). *Journal of the Kansas Entomological Society*, **59**, 20–5.

Fishlyn DA, Phillips DW (1980) Chemical camouflaging and behavioral defenses against a predaceous seastar by three species of gastropods from the surfgrass *Phyllospadix* community. *Biological Bulletin*, **158**, 34–48.

Fitzpatrick BM, Shook K, Izally R (2009) Frequency-dependent selection by wild birds promotes polymorphism in model salamanders. *BMC Ecology*, **9**: 12.

Fitzsimons VFM, Brain CK (1958) A short account of the reptiles of the Kalahari Gemsbok National Park. *Koedoe*, **1**, 99–104.

Flach A, Marsaioli AJ, Singer RB, Amaral MCE, Menezes C, Kerr WE, Batista-Pereira LG, Correa AG (2006) Pollination by sexual mimicry in *Mormolyca ringens*: a floral chemistry that remarkably matches the pheromones of virgin queens of *Scaptotrigona* sp. *Journal of Chemical Ecology*, **32**, 59–70.

Fleishman LJ (1985) Cryptic movements in the vine snake *Oxybelis aeneus*. *Copeia*, **1985**, 242–5.

Floren A, Otto S (2001) A tropical Derbidae (Fulgoroidea, Homoptera) that mimics a predator (Salticidae, Araneae). *Ecotropica*, **7**, 151–3.

Florey E (1969) Ultrastructure and function of cephalopod chromatophores. *The American Zoologist*, **9**, 429–42.

Flower TP (2011) Fork-tailed drongos use deceptive mimicked alarm calls to steal food. *Proceedings of the Royal Society of London, Series B: Biological Sciences*, **278**, 1548–55.

Flower TP, Gribble M, Ridley AR (2014) Deception by flexible alarm mimicry in an African bird. *Science*, **344**, 513–6.

Foot G, Rice SP, Millett J (2014) Red trap colour of the carnivorous plant *Drosera rotundifolia* does not serve a prey attraction or camouflage function. *Biology Letters*, **10**: 20131024.

Forbes P (2011) *Dazzled and Deceived: Mimicry and Camouflage*. Thistle Publishing, London.

Ford EB (1945) *Butterflies*. New Naturalist series 1. Collins, London.

Ford EB (1963) Mimicry. *Proceedings of the XVI International Congress of Zoology*, 184–6.

Ford EB (1964) *Ecological Genetics*. Chapman and Hall, London.

Ford EB (1967) *Moths*, 2nd edn. Collins, London.

Ford HA (1971) The degree of mimetic protection gained by new partial mimics. *Heredity*, **27**, 227–36.

Forrest AD, Hollingsworth ML, Hollingsworth PM, Bateman RM (2004) Population genetic structure in European populations of *Spiranthes romanzoffiana* set in the context of other genetic studies on orchids. *Heredity*, **92**, 218–27.

Forsman A, Herrström J (2004) Asymmetry in size, shape, and color impairs the protective value of conspicuous color patterns. *Behavioral Ecology*, **15**, 141–7.

Forsman A, Merilaita S (1999) Fearful symmetry: pattern size and asymmetry affects aposematic signal efficacy. *Evolutionary Ecology*, **13**, 131–40.

Forsman A, Karlsson M, Wennersten L, Johansson J, Karpestam E (2011) Rapid evolution of fire melanism in replicated populations of pygmy grasshoppers. *Evolution*, **65**, 2530–40.

Forsyth A, Alcock J (1990a) Ambushing and prey-luring as alternative foraging tactics of the fly-catching rove beetle *Leistotrophus versicolor* (Coleoptera: Staphylinidae). *Journal of Insect Behavior*, **3**, 703–18.

Forsyth A, Alcock J (1990b) Female mimicry and resource defense polygyny by males of a tropical rove beetle, *Leistotrophus versicolor* (Coleoptera: Staphylinidae). *Behavioral Ecology and Sociobiology*, **26**, 325–30.

Foster MA (1987) Delayed maturation, neoteny, and social system differentiation in two manakins of the genus *Chiroxiphia*. *Evolution*, **41**, 547–58.

Foster MS, Delay LS (1998) Dispersal of mimetic seeds of three species of *Ormosia* (Leguminosae). *Journal of Tropical Ecology*, **14**, 389–411.

Fowler HG, Forti LC, Brandão CRF, Delabie JHC, Vasconcelos HL (1991) Ecologia nutricional de formigas. In AR Panizzi and JRP Parra (eds), *Ecologia nutricional de insetos e suas implicações no manejo de pragas*. Editora Manole e CNPq, São Paulo, pp. 131–223.

Francq EN (1969) Behavioral aspects of feigned death in the opossum *Didelphis marsupialis*. *American Midland Naturalist*, **81**, 556–68.

Frank LG, Glickman SE (1994) Giving birth through a penile clitoris – parturition and dystocia in the spotted hyaena (*Crocuta crocuta*). *Journal of Zoology*, **234**, 659–65.

Frank LG, Glickman SE, Powch I (1990) Sexual dimorphism in the spotted hyaena (*Crocuta crocuta*). *Journal of Zoology, London*, **221**, 308–13.

Frank LG, Weldele ML, Glickman SE (1995) Costs of masculinization. *Nature*, **377**, 584–5.

Franks DW, Noble J (2004) Warning signals and predator-prey coevolution. *Proceedings of the Royal Society of London, Series B: Biological Sciences*, **271**, 1859–66.

Franks DW, Sherratt TN (2007) The evolution of multicomponent mimicry. *Journal of Theoretical Biology*, **244**, 631–9.

Franks DW, Ruxton GD, Sherratt TN (2009) Warning signals evolve to disengage Batesian mimics. *Evolution*, **63**, 256–67.

Frazer JFD, Rothschild M (1960) Defence mechanisms in warningly-coloured moths and other insects. In H Strouhal and M Beier (eds), *XI Internationaler Kongress Für Entomologie Vienna*, Vol. 3, pp. 249–56.

Freeland WJ, Janzen DH (1974) Strategies in herbivory by mammals: the role of plant secondary compounds. *American Naturalist*, **108**, 269–89.

Frick C, Wink M (1995) Uptake and sequestration of ouabain and other cardiac glycosides in *Danaus plexippus* (Lepidoptera: Danaidae): evidence for a carrier-mediated process. *Journal of Chemical Ecology*, **21**, 557–75.

Fricke HW (1970) Ein mimetisches Kollektiv: Beobachtungen an Fischschwarmen, die Seeigel nachahmen. *Marine Biology*, **5**, 307–14.

Frisch AJ (2006) Are juvenile coral-trouts (*Plectropomus*) mimics of poisonous pufferfishes (*Canthigaster*) on coral reefs? *Marine Ecology*, **27**, 247–52.

Frith CB, Borgia G, Frith DW (1996) Courts and courtship behavior of Archbold's bowerbird, *Archbaldia papuensis* in Papua New Guinea. *Ibis*, **138**, 204–11.

Froesch D, Messenger JB (1978) On leucophores and the chromatic unit of *Octopus vulgaris*. *Journal of Zoology, London*, **186**, 163–73.

Frolova EN, Shcherbina TV (1975) New species in genus *Azygia* Looss, 1899 (Trematoda, Azygiidae). *Parazitologiia*, **9**, 489–93.

Fuentes M (1995) The effect of unripe fruits on ripe fruit removal by birds in *Pistacia terebinthus*: flag or handicap? *Oecologia (Berlin)*, **101**, 55–8.

Fuentes M, Schupp EW (1998) Empty seeds reduce seed predation by birds in *Juniperus osteosperma*. *Evolutionary Ecology*, **12**, 823–7.

Fuhrman FA, Fuhrman GJ, DeRiemer K (1979) Toxicity and pharmacology of extracts from dorid nudibranchs. *Biological Bulletin*, **156**, 289–99.

Fujii R (2000) The regulation of motile activity in fish chromatophores. *Pigment Cell Research*, **13**, 300–19.

Fujii Y, Shimada K, Niizaki Y, Nambara T (1975) Cardenobufotoxin : novel conjugated cardenolide from japanese toad. *Tetrahedron Letters*, **16**, 3017–20.

Fulgione D, Trapanese M, Maselli V, Rippa D, Itri F, Avallone B, Van Damme R, Monti DM, Raia P (2014) Seeing through the skin: dermal light sensitivity provides cryptism in moorish gecko. *Journal of Zoology*, **294**, 122–8.

Fullard JH (1977) Phenology of sound-producing arctiid moths and the activity of insectivorous bats. *Nature*, **267**, 42–3.

Fullard JH, Fenton MB, Simmons JA (1979) Jamming bat echolocation: the clicks of arctiid moths. *Canadian Journal of Zoology*, **57**, 647–9.

Fullard JH, Simmons JA, Saillant PA (1994) Jamming bat echolocation: the dogbane tiger moth *Cycnia tenera* times its clicks to the terminal attack calls of the big brown bat *Eptesicus fuscus*. *s*, **194**, 285–98.

Fürtbauer I, Heistermann M, Schülke O, Ostner J (2011) Concealed fertility and extended female sexuality in a non-human primate (*Macaca assamensis*). *PLoS ONE* **6(8)**: e23105.

Furth DG (1981) The *Altica* of Israel (Coleoptera: Chrysomelidae: Alticinae). *Israel Journal of Entomology*, **14**, 55–66.

Furth DG (1983) Flea beetle mimicry. *Chrysomela*, **8**, 3,

Gagliano M (2008) On the spot: the absence of predators reveals eyespot plasticity in a marine fish, *Behavioral Ecology* **19** 733–9

Gagliano M, Depczynski M (2013) Spot the difference: mimicry in a coral reef fish. *PLoS ONE*, **8(2)**: e55938.

Gagliardo A, Guilford T (1993) Why do warning-coloured prey live gregariously? *Proceedings of the Royal Society of London, Series B: Biological Sciences*, **251**, 69–74.

Gahan CJ (1913) Mimicry in Coleoptera. *Proceedings of the South London Entomology and Natural History Society*, **1912–1914**, 28–39.

Gallego-Ropero MC, Feitosa RM (2014) Evidences of Batesian mimicry and parabiosis in ants of the Brazilian savanna. *Sociobiology*, **61**, 281–5.

Gamberale-Stille G (2000) Decision time and prey gregariousness influence attack probability in naïve and experienced predators. *Animal Behaviour*, **60**, 95–9.

Gamberale-Stille G, Guilford T (2003) Contrast versus colour in aposematic signals. *Animal Behaviour*, **65**, 1021–6.

Gamberale-Stille G, Guilford T (2004) Automimicry destabilizes aposematism: predator sample-and-reject behaviour may provide a solution. *Proceedings of the Royal Society of London, Series B: Biological Sciences*, **271**, 2621–5.

Gamberale-Stille G, Tullberg BS (1996) Evidence for a peak-shift in predator generalization among aposematic prey. *Proceedings of the Royal Society of London, Series B: Biological Sciences*, **263**, 1329–34.

Gamberale-Stille G, Tullberg. BS (1999) Experienced chicks show biased avoidance of stronger signals: and experiment with natural colour variation in live aposematic prey. *Evolutionary Ecology*, **13**, 579–89.

Gamberale-Stille G, Tullberg BS (2001) Fruit or aposematic insect? Context-dependent colour preferences in domestic chicks.

Proceedings of the Royal Society of London, Series B: Biological Sciences, **268**, 2525–9.

Gans C (1961) Mimicry in procryptically coloured snakes of the genus *Dasypeltis*. *Evolution*. **15**, 72–91.

Gans C (1964) Empathic learning and the mimicry of African snakes. *Evolution*, **18**, 705.

Garcia TS, Sih A (2003) Color change and color dependent behavior in response to predation risk in the salamander sister species *Ambystoma barbouri* and *Ambystoma texanum*. *Oecologia (Berlin)*, **137**, 131–9.

Garcia JE, Rohr D, Dyer AG (2013) Trade-off between camouflage and sexual dimorphism revealed by UV digital imaging: the case of Australian Mallee dragons (*Ctenophorus fordi*). *Journal of Experimental Biology*, **216**, 4290–8.

Garman S (1882) The scream of the young burrowing owl sounds like the warning of the rattlesnake. *Nature*, **27**, 174.

Garnett WB, Akre RD, Sehlke G (1985) Cocoon mimicry and predation by myrmecophilous Diptera (Diptera: Syrphidae). *Florida Entomologist*, **68**, 615–21.

Gaskett AC (2012) Floral shape mimicry and variation in sexually deceptive orchids with a shared pollinator. *Biological Journal of the Linnean Society*, **106**, 469–81.

Gaskett AC, Winnick CG, Herberstein ME (2008) Orchid sexual deceit provokes ejaculation. *The American Naturalist*, **171**, 206–12.

Gaul AT (1952) Audio mimicry: an adjunct to color mimicry. *Psyche*, **59**, 82–3.

Gavrilets S, Hastings A (1998) Coevolutionary chase in two-species systems with applications to mimicry. *Journal of Theoretical Biology*, **191**, 415–27.

Gehlbach FR (1972) Coral snake mimicry reconsidered: the strategy of self-mimicry. *Forma et Functio*, **5**, 311–20.

Geiselhardt SF, Peschke K, Nagel P (2007) A review of myrmecophily in ant nest beetles (Coleoptera: Carabidae: Paussinae): linking early observations with recent findings. *Naturwissenschaften*, **94**, 871–94.

Gemeno C, Yeargan KV, Haynes KF (2000) Aggressive chemical mimicry by the bolas spider *Mastophora hutchinsoni*: identification and quantification of a major prey's sex pheromone components in the spider's volatile emissions. *Journal of Chemical Ecology*, **26**, 1235–43.

Gendron RP (1986) Searching for cryptic prey: evidence for optimal search rates and the formation of search images in quail. *Animal Behaviour*, **34**, 898–912.

Gendron RP, Staddon JER (1983) Searching for cryptic prey: the effect of search rate. *The American Naturalist*, **121**, 172–86.

Gerald GW (2008) Feign versus flight: influences of temperature, body size and locomotor abilities on death feigning in neonate snakes. *Animal Behaviour*, **75**, 647–54.

Gerald GW, Claussen D (2006) Extrinsic and intrinsic factors influencing intraspecific variation in death feigning by newborn brown snakes (*Storeria dekayi*). *Animal Behavior Society 43rd Annual Meeting. Aug. 12–16. Snowbird, Utah.*

Gerhardt U (1926) Weitere Untersuchungen zur Biologic der Spinnen. *Zeitschrift für Morphologie und Ökologie der Tiere*, **6**, 1–77.

Gersick A, Kurzban R. (2014) Covert sexual signaling: human flirtation and implications for other social species. *Evolutionary Psychology*, **12**, 549–69.

Getty T (1985) Discriminability and the sigmoid functional response: how optimal foragers could stabilize model–mimic complexes. *The American Naturalist*, **125**, 239–56.

Getty T (1987) Crypsis, mimicry, and switching: the basic similarity of superficially different analyses. *The American Naturalist*, **130**, 793–7.

Gianoli E, Carrasco-Urra F (2014) Leaf mimicry in a climbing plant protects against herbivory. *Current Biology*, **24**, 984–7.

Giard A (1894) Sur la mimétisme parasitaire. *Annales de la Société Entomologique de France*, **63**, 124–8.

Gibb JA (1962) L. Tinbergen's hypothesis of the role of specific search images. *Ibis*, **104**, 106–11.

Gibbs HL, Brooke M de L, Davies NB (1996) Analysis of genetic differentiation of host races of the common cuckoo, *Cuculus canorus*, using mitochondrial and microsatellite DNA variation. *Proceedings of the Royal Society of London, Series B: Biological Sciences*, **263**, 89–96.

Gibbs HL, Sorenson MD, Marchetti K, Brooke M de L, Davies NB, Nakamura H (2000) Genetic evidence for female host specific races of the common cuckoo. *Nature*, **407**, 183–6.

Gibernau M, Macquart D, Przetak G (2004) Pollination in the genus *Arum* – a review. *Aroideana*, **27**, 148–66.

Gibran FZ, Castro RMC (1999) Activity, feeding behaviour and diet of *Ogcocephalus vespertilio* in southern west Atlantic. *Journal of Fish Biology*, **55**, 588–95.

Gibson DO (1974) Batesian mimicry without distastefulness? *Nature*, **250**, 77–9.

Gibson DO (1980) The role of escape in mimicry and polymorphism. I. The response of captive birds to artificial prey. *Biological Journal of the Linnean Society*, **14**, 201–14.

Gibson DO, Mani GS (1984) An experimental investigation of the effects of selective predation by birds and parasitoid attack on the butterfly *Danaus chrysippus* (L.). *Proceedings of the Royal Society of London, Series B: Biological Sciences*, **221**, 31–51.

Gibson G (1992) Do tsetse flies see zebras? A field study of the visual response of tsetse to striped targets. *Physiological Entomology*, **17**, 141–7.

Gibson RW, Pickett JA (1983) Wild potato repels aphids by release of aphid alarm pheromone. *Nature*, **302**, 608–9.

Gigord LDB, Macnair MR, Stritesky M, Smithson A (2002) The potential for floral mimicry in rewardless orchids: an experimental study. *Proceedings of the Royal Society of London, Series B: Biological Sciences*, **269**, 1389–95.

Giguère LA, Northcote TG (1987) Ingested prey increase risks of visual predation in transparent *Chaoborus* larvae. *Oecologia (Berlin)*, **73**, 48–52.

Gilbert F (2005) The evolution of imperfect mimicry in hoverflies. In MDE Fellowes, GJ Holloway and J Rolff (eds), *Insect Evolutionary Ecology*. CABI, Wallingford, UK., pp. 231–88.

Gilbert LE (1975) Ecological consequences of a coevolved mutualism between butterflies and plants. In LE Gilbert and PH Raven (eds), *Coevolution of Animals and Plants*. University of Texas Press, Austin, pp. 210–40.

Gilbert LE (1982) The coevolution of a butterfly and a vine. *Scientific American*, **247**, 110–21.

Gilbert LE (2003) Adaptive novelty through introgression in *Heliconius* wing patterns: evidence for shared genetic "toolbox" from synthetic hybrid zones and a theory of diversification. In CL Boggs, WB Watt and PR Ehrlich (eds), *Ecology and Evolution Taking Flight: Butterflies as Model Systems*. University of Chicago Press, Chicago, pp. 281–318.

Gillis JE (1982) Substrate colour-matching cues in the cryptic grasshopper *Circotettix rabula rabula* (Rehn & Hebard). *Animal Behaviour*, **30**, 113–6.

Gingerich PD (1975) Is the aardwolf a mimic of the hyaena? *Nature*, **253**, 191–2.

Gironès N, Cuervo H, Fresno M (2005) *Trypanosoma cruzi*-induced molecular mimicry and Chagas' disease. *Current Topics in Microbiology and Immunology*, **296**, 89–123.

Gittleman JL, Harvey PH (1980) Why are distasteful prey not cryptic? *Nature*, **286**, 149–50.

Gittleman JL, Harvey PH, Greenwood PJ (1980) The evolution of conspicuous coloration: some experiments in bad taste. *Animal Behaviour*, **28**, 879–99.

Givnish TJ (1990) Leaf mottling: relation to growth form and leaf phenology and possible role as camouflage. *Functional Ecology*, **4**, 463–74.

Givnish TJ, Burkhardt EL, Happel RE, Weintraub JD (1984) Carnivory in the bromeliad *Brocchinia reducta*, with a cost/benefit model for the general restriction of carnivorous plants to sunny, moist nutrient-poor habitats. *The American Naturalist*, **124**, 479–97.

Givnish TJ, Spalink D, Ames M, Lyon SP, Hunter SJ, Zuluaga A, Iles WJD, Clements MA, Arroyo MTK, Leebens-Mack J, Endara L, Kriebel R, Neubig KM, Whitten WM, Williams NH, Cameron KM (2015) Orchid phylogenomics and multiple drivers of their extraordinary diversification. *Proceedings of the Royal Society of London, Series B: Biological Sciences*, **282**: 20151553.

Glickman SE, Coscia EM, Frank LG, Licht P, Weldele ML, Drea CM (1998) Androgens and masculinization of genitalia in the spotted hyaena (*Crocuta crocuta*). 3. Effects of juvenile gonadectomy. *Journal of Reproduction & Fertility*, **113**, 129–35.

Glickman SE, Short RV, Renfree MB (2005) Sexual differentiation in three unconventional mammals: spotted hyenas, elephants and tammar wallabies. *Hormones & Behaviour*, **48**, 403–17.

Gloag R, Keller L-A, Langmore NE (2014) Cryptic cuckoo eggs hide from competing cuckoos. *Proceedings of the Royal Society of London, Series B: Biological Sciences*, **281**: 20141014.

Gluckman T-L, Cardoso GC (2010) The dual function of barred plumage in birds: camouflage and communication. *Journal of Evolutionary Biology*, **23**, 2501–6.

Gluckman T-L, Mundy NI (2013) Cuckoos in raptors' clothing: barred plumage illuminates a fundamental principle of Batesian mimicry. *Animal Behaviour*, **86**, 1165–81.

Gnezdilov VM, Viraktamath CA (2011) A new genus and new species of the tribe Caliscelini Amyot & Serville (Hemiptera, Fulgoroidea, Caliscelidae, Caliscelinae) from southern India. *Deutsche Entomologische Zeitschrift*, **58**, 235–9.

Gochfeld M (1984) Antipredator behaviour: aggressive and distraction displays of shorebirds. In J Burger and B Olla (eds), *Shorebirds. Volume 5. Breeding Behaviour and Populations*. Plenum Press, New York, pp. 289–377.

Goddard GHR (2015) Latitudinal variation in mimicry between aeolid nudibranchs and an amphipod crustacean in the northeast Pacific Ocean. *Marine Biodiversity*, **2015**, 1–3.

Godden DH (1972) The motor innervation of the leg musculature and motor output during thanatosis in the stick insect *Carausius morosus* Br. *Journal of Comparative Physiology*, **80**, 201–25.

Godden DH (1974) The physiological mechanism of catalepsy in the stick insect *Carausius morosus* Br. *Journal of Comparative Physiology*, **89**, 251–74.

Godfray HCJ (1994) *Parasitoids: Behavioral and Evolutionary Ecology*. Princeton University Press, Princeton, NJ.

Godfrey D, Lythgoe JN, Rumball DA (1987) Zebra stripes and tiger stripes: the spatial frequency distribution of the pattern compared to that of the background is significant in display and crypsis. *Biological Journal of the Linnean Society*, **32**, 427–33.

Gohli J, Högstedt G (2009) Explaining the evolution of warning coloration: secreted secondary defence chemicals may facilitate the evolution of visual aposematic signals. *PLoS ONE*, **4(6):** e5779.

Gohli J, Högstedt G (2010) Reliability in aposematic signaling – thoughts on evolution and aposematic life. *Communicative & Integrative Biology*, **3**, 9–11.

Goldblatt P, Bernhardt P, Manning JC (2009) Adaptive radiation of the putrid perianth: *Ferraria* (Iridaceae: Irideae) and its unusual pollinators. *Plant Systematics and Evolution*, **278**, 53–65.

Golding YC, Edmunds M (2000) Behavioural mimicry of honeybees (*Apis mellifera*) by droneflies (Diptera: Syrphidae: *Eristalis* spp.). *Proceedings of the Royal Society of London, Series B: Biological Sciences*, **267**, 903–9.

Golding YC, Ennos AR, Edmunds M (2001) Similarity in flight behaviour between the honeybee *Apis mellifera* (Hymenoptera: Apidae) and its presumed mimic, the dronefly, *Eristalis tenax* (Diptera: Syrphidae). *The Journal of Experimental Biology*, **204**, 139–45.

Golding YC, Ennos AR, Sullivan M, Edmunds M (2005a) Hoverfly mimicry deceives humans. *Journal of Zoology, London*, **266**, 395–99.

Golding YC, Edmunds M, Ennos AR (2005b) Flight behaviour during foraging of the social wasp *Vespula vulgaris* (Hymenoptera: Vespidae) and four mimetic hoverflies (Diptera: Syrphidae) *Sericomyia silentis*, *Myathropa florea*, *Helophilus* sp. and *Syrphus* sp. *Journal of Experimental Biology*, **208**, 4523–7.

Goldschmidt RB (1945) Mimetic polymorphism, a controversial chapter of Darwin. *Quarterly Review of Biology*, **20**, 205–30.

Goldschmidt T, de Visser J (1990) On the possible role of egg mimics in speciation. *Acta Biotheoretica*, **38**, 125–34.

Gomez D, Théry M (2007) Simultaneous crypsis and conspicuousness in colour patterns: comparative analysis of a Neotropical rainforest bird community. *The American Naturalist*, **169**, S42–S61.

Gómez JM, Zamora R (2002) Thorns as induced mechanical defense in a long-lived shrub (*Hormathophylla spinosa*, Cruciferae). *Ecology*, **83**, 885–90.

Gonçalves EJ, Almada VC, Oliveira RF, Santos AJ (1996) Female mimicry as a mating tactic in males of the blenniid fish *Salvaria pavo*. *Journal of the Marine Biological Association of the U.K.*, **76**, 529–38.

Gonçalves Rodrigues LR, Silva Absalão R (2005) Shell colour polymorphism in the chiton *Ischnochiton striolatus* (Gray, 1828) (Mollusca: Polyplacophora) and habitat heterogeneity. *Biological Journal of the Linnean Society*, **85**, 543–8.

González A, Rossini C, Eisner M, Eisner T (1999) Sexually transmitted chemical defense in a moth (*Utetheisa ornatrix*). *Proceedings of the National Academy of Sciences, U.S.A.*, **96**, 5570–4.

González A, Rossini C, Eisner T (2004) Mimicry: imitative depiction of discharged defensive secretion on carapace of an opilionid. *Chemoecology*, **14**, 5–7.

González F, Pabón-Mora N (2015) Trickery flowers: the extraordinary chemical mimicry of *Aristolochia* to accomplish deception to its pollinators. *New Phytologist*, **206**, 10–3.

Goodale E, Kotagama SW (2006) Context-dependent vocal mimicry in a passerine bird. *Proceedings of the Royal Society of London, Series B: Biological Sciences*, **273**, 875–80.

Goodale E, Ratnayake CP, Kotagama SW (2014) Vocal mimicry of alarm-associated sounds by a drongo elicits flee and mobbing responses from other species that participate in mixed-species bird flocks. *Ethology*, **120**, 266–74.

Goodale MA, Sneddon I (1977) The effect of distastefulness of the model on the predation of artificial mimics. *Animal Behaviour*, **25**, 660–5.

Goodman JD, Goodman JM (1975) Contrasting color and pattern as enticement in snakes. *Herpeologica*, **32**, 145–8.

Gordon IJ, Smith DAS (1989) Genetics of the mimetic African butterfly *Hypolimnas misippus*: hindwing polymorphism. *Heredity*, **63**, 409–25.

Gordon IJ, Smith DAS (1998) Body size and colour pattern genetics in the polymorphic mimetic butterfly *Hypolimnas misippus* (L.). *Heredity*, **80**, 62–9.

Gordon IJ, Smith DAS (1999) Diversity in mimicry. *Trends in Ecology and Evolution*, **14**, 150–1.

Gordon IJ, Edmunds M, Edgar JA, Lawrence J, Smith DAS (2010) Linkage disequilibrium and natural selection for mimicry in the Batesian mimic *Hypolimnas misippus* (L.) (Lepidoptera: Nymphalidae) in the Afrotropics. *Biological Journal of the Linnean Society*, **100**, 180–94.

Gosler AG, Connor OR, Bonser RHC (2011) Protoporphyrin and eggshell strength: preliminary findings from a passerine bird. *Avian Biology Research*, **4**, 214–23.

Götmark F (1992) Anti-predator effect of conspicuous plumage in a male bird. *Animal Behaviour*, **44**, 51–5.

Götmark F (1994) Are bright birds distasteful? A re-analysis of H. B. Cott's data on the edibility of birds. *Journal of Avian Biology*, **25**, 184–97.

Götmark F (1996) Simulating a color mutation – conspicuous red wings in the European blackbird reduce the risk of attacks by sparrowhawks. *Functional Ecology*, **10**, 355–9.

Gotthard K, Berger D, Bergman M, Merilaita S (2009) The evolution of alternative morphs: density-dependent determination of larval colour dimorphism in a butterfly. *Biological Journal of the Linnean Society*, **98**, 256–66.

Gould SJ (1977) *Ever Since Darwin: Reflections in Natural History*. W.W. Norton & Company, New York.

Gower BA, Nagy TR, Stetson MH (1993) Role of prolactin and the gonads in seasonal physiological changes in the collared lemming (*Dicrostonyx groenlandicus*). *Journal of Experimental Zoology*, **266**, 92–101.

Gower D, Garrett K, Stafford P (2012) Snakes. London: Natural History Museum.

Grafen A (1990) Biological signals as handicaps. *Journal of Theoretical Biology*, **144**, 517–46.

Graham RR (1934) The silent flight of owls. *Journal of the Royal Aeronautical Society*, **38**, 837–43.

Grant BS, Owen DF, Clarke CA (1996) Parallel rise and fall of melanic peppered moths in America and Britain. *Journal of Heredity*, **87**, 351–7.

Grant GG, Miller WE (1995) Larval images on lepidopteran wings – an unrecognized defense mechanism? *American Entomologist*, **41**, 44–8.

Grant JB (2007) Ontogenetic colour change and the evolution of aposematism: a case study in panic moth caterpillars. *Journal of Animal Ecology*, **76**, 439–47.

Grant KG (1966) A hypothesis concerning the prevalence of red coloration in California hummingbird flowers. *The American Naturalist*, **100**, 85–97.

Gray M (1991) Spiders that smell. *Australian Natural History*, **23**, 190–1.

Grayson J, Edmunds M (1989) The causes of colour and colour change in caterpillars of the poplar and eyed hawkmoths (*Laothoe populi* and *Smerinthus ocellata*). *Biological Journal of the Linnean Society*, **37**, 263–79.

Grayson J, Edmunds M, Evans EH, Britton G (1991) Carotenoids and colouration of poplar hawkmoth caterpillars (*Laothoe populi*). *Biological Journal of the Linnean Society*, **42**, 457–65.

Green PR, Gentle L, Peake TM, Scudamore RE, McGregor PK, Gilbert F, Dittrich WH (1999) Conditioning pigeons to discriminate naturally lit insect specimens. *Behavioural Processes*, **46**, 97–102.

Greene E (1989) A diet-induced developmental polymorphism in a caterpillar. *Science*, **243**, 643–6.

Greene E (1996) Effect of light quality and larval diet on morph induction in the polymorphic caterpillar *Nemoria arizonaria* (Lepidoptera: Geometridae). *Biological Journal of the Linnean Society*, **58**, 277–85.

Greene E, Orsak LJ, Whitman DW (1987) A tephritid fly mimics the territorial displays of its jumping spider predator. *Science*, **236**, 310–2.

Greene HW (1973) Defensive tail display by snakes and amphisbaenians. *Journal of Herpetology*, **7**, 143–61.

Greene HW (1977) The aardwolf as hyaena mimic: an open question. *Animal Behaviour*, **25**, 245–6.

Greene HW, Campbell JA (1972) Notes on the use of caudal lures by arboreal green pit vipers. *Herpetologica*, **28**, 32–4.

Greene HW, McDiarmid RW (1981) Coral snake mimicry: does it occur? *Science*, **213**, 1207–12.

Greene HW, McDiarmid RW (2005) Wallace and Savage: heroes, theories and venomous snake mimicry. In MA Donnelly, BI Crother, C Guyer and ME White (eds), *Evolution and Ecology in the Tropics: A Herpetological Perspective*. The University of Chicago Press, Chicago, pp. 190–208.

Greeney HF, Gelis RA, White R (2004) Notes on breeding birds from an Ecuadorian lowland forest. *Bulletin of the British Ornithologists' Club*, **124**, 28–37.

Greenwood JJD (1984) The functional basis of frequency-dependent food selection. *Biological Journal of the Linnean Society*, **23**, 177–99.

Greenwood JJD (1985) Frequency-dependent selection by seed predators. *Oikos*, **44**, 195–210.

Greenwood JJD (1986) Crypsis, mimicry, and switching by optimal foragers. *The American Naturalist*, **128**, 294–300.

Greenwood JJD, Wood EM, Batchelor S (1981) Apostatic selection of distasteful prey. *Heredity*, **47**, 27–34.

Greenwood JJD, Johnston JP, Thomas GE (1984) Mice prefer rare food. *Biological Journal of the Linnean Society*, **23**, 201–10.

Greenwood JJD, Blow NC, Thomas GE (1985) More mice prefer rare food. *Biological Journal of the Linnean Society*, **23**, 211–9.

Greenwood JJD, Cotton PA, Wilson DM (1989) Frequency-dependent selection on aposematic prey: some experiments. *Biological Journal of the Linnean Society*, **36**, 213–26.

Greenwood RG, Mariscal RN (1984) The utilization of cnidarian nematocysts by aeolid nudibranchs: nematocyst maintenance and release in *Spurilla*. *Tissue & Cell*, **16**, 719–30.

Greer AT, Woodson CB, Guigand CM, Cowen RK (2016) Larval fishes utilize Batesian mimicry as a survival strategy in the plankton. *Marine Ecology Progress Series*, **551**, 1–12.

Greig-Smith PW (1986) Bicolored fruit displays and frugivorous birds: the importance of fruit quality to dispersers and seed predators. *The American Naturalist*, **127**, 246–51.

Grenot C (1973) Sur la biologie d'un rongeur heliophile du Sahara, le "Goundi" (Ctenodactylidae). *Acta Tropica*, **30**, 237–50.

Grim T (2005) Mimicry vs. similarity: which resemblances between brood parasites and their hosts are mimetic and which are not? *Biological Journal of the Linnean Society*, **84**, 69–78.

Grim T (2006) The evolution of nestling discrimination by hosts of parasitic birds: why is rejection so rare? *Evolutionary Ecology Research*, **8**, 785–802.

Grim T (2013) Perspectives and debates: mimicry, signalling and co-evolution (commentary on Wolfgang Wickler–Understanding mimicry – with special reference to vocal mimicry). *Ethology*, **119**, 270–7.

Grober MS (1988) Brittle-star bioluminescence functions as an aposematic signal to deter crustacean predators. *Animal Behaviour*, **36**, 493–501.

Grober MS (1990) Luminescent aposematism: a reply to Guilford & Cuthill. *Animal Behaviour*, **39**, 411–3.

Groom PK, Lamont BB, Duff HC (1994) Self crypsis in *Hakea trifurcata* as an anti-herbivore mechanism. *Functional Ecology*, **8**, 110–7.

Gross MR (1979) Cuckoldry in sunfishes (*Lepomis*: Centrarchidae). *Canadian Journal of Zoology*, **57**, 1507–9.

Gross MR (1982) Sneakers, satellites and parentals: polymorphic strategies in North American sunfishes. 2. *Zeitschrift für Tierpsychologie*, **60**, 1–26.

Grüneberg H (1982) Pseudo-polymorphism in *Clithon oualaniensis*. *Proceedings of the Royal Society of London, Series B: Biological Sciences*, **216**, 147–57.

Gudger EW (1946) The angler-fishes, *Lophius piscatorius* et *americanus*, use the lure in fishing. *The American Naturalist*, **79**, 542–8.

Guilford T (1985) Is kin selection involved in the evolution of warning coloration? *Oikos*, **45**, 31–6.

Guilford T (1986) How do 'warning colours' work? Conspicuousness may reduce recognition errors in experienced predators. *Animal Behaviour*, **34**, 286–8.

Guilford T (1988) The evolution of conspicuous coloration. *The American Naturalist*, **131**, S7–S2.

Guilford T (1990a) The evolution of aposematism. In JO Schmidt and DL Evans (eds), *Insect Defenses: Adaptive mechanisms and Strategies of Prey and Predators*. State University of New York Press, New York, pp. 23–61.

Guilford T (1990b) The secrets of aposematism: unlearned responses to specific colours and patterns. *Trends in Ecology and Evolution*, **5**, 323.

Guilford T (1990c) Evolutionary pathways to aposematism. *Acta Oecologica*, **11**, 835–41.

Guilford T (1991) Is the viceroy a batesian mimic? *Nature*, **351**, 611.

Guilford T, Cuthill IC (1989) Aposematism and bioluminescence. *Animal Behaviour*, **37**, 339–41.

Guilford T, Cuthill IC (1991) The evolution of aposematism in marine gastropods. *Evolution*, **45**, 449–51.

Guilford T, Dawkins MS (1987) Search images not proven: a reappraisal of recent evidence. *Animal Behaviour*, **35**, 1838–45.

Guilford T, Dawkins MS (1989a) Search image versus search rate – a reply. *Animal Behaviour*, **37**, 160–2.

Guilford T, Dawkins MS (1989b) Search image versus search rate – 2 different ways to enhance prey capture. *Animal Behaviour*, **37**, 163–4.

Guilford T, Dawkins MS (1991) Receiver psychology and the evolution of animal signals. *Animal Behaviour*, **42**, 1–14.

Guilford T, Dawkins MS (1993a) Receiver psychology and the design of animal signals. *Trends in Neurosciences*, **16**, 430–6.

Guilford T, Dawkins MS (1993b) Are warning colors handicaps? *Evolution*, **47**, 400–16.

Guilford T, Read AF (1990) Zahavian cuckoos and the evolution of nestling discrimination by hosts. *Animal Behaviour*, **39**, 600–1.

Guilford T, Nicol C, Rothschild M, Moore BP (1987) The biological roles of pyrazines: evidence for a warning odour function. *Biological Journal of the Linnean Society*, **31**, 113–28.

Guthrie RD (1967) Fire mimicry among mammals. *American Midland Naturalist*, **77**, 227–30.

Guthrie RD, Petocz RG (1970) Weapon automimicry among mammals, *The American Naturalist*, **104**, 585–8.

Haag WR, Butler RS, Hartfield PD (1995) An extraordinary reproductive strategy in freshwater bivalves: prey mimicry to facilitate larval dispersal. *Freshwater Biology*, **34**, 471–6.

Haas F (1945) Collective mimicry. *Ecology*, **26**, 412–3.

Hacker SD, Madin LP (1991) Why habitat architecture and color are important to shrimps living in pelagic *Sargassum*: use of camouflage and plant-part mimicry. *Marine Ecology Progress Series*, **70**, 143–55.

Haddock SHD, Dunn CW, Pugh PR, Schnitzler CE (2005) Bioluminescent and red-fluorescent lures in a deep-sea siphonophore. *Science*, **309**, 263.

Hadeler KP, Demottoni P, Tesei A (1982) Mimetic gain in Batesian and Müllerian mimicry. *Oecologia (Berlin)*, **53**, 84–92.

Hafernik J, Saul-Gershenz LS (2000) Beetle larvae cooperate to mimic bees. *Nature*, **405**, 35–6.

Hagman M, Phillips BL, Shine R (2008) Tails of enticement: caudal luring by an ambush foraging snake (*Acanthophis praelongus*, Elapidae). *Functional Ecology*, **22**, 1134–9.

Hagman M, Forsman A (2003) Correlated evolution of conspicuous coloration and body size in poison frogs (Dendrobatidae). *Evolution*, **57**, 2904–10.

Hailman JP (1977) *Optical Signals: Animal Communication and Light*. Indiana University Press, Bloomington & London.

Hailman JP (1981) A test of symmetry-deception in a chaetodontid fish. *Animal Behaviour*, **29**, 1267.

Hakkarainen H, Korpimäki E, Huhta E, Palokangas P (1993) Delayed maturation in plumage colour: evidence for the female–mimicry hypothesis in the kestrel. *Behavioral Ecology and Sociobiology*, **33**, 247–51.

Halpern M, Raats D, Lev-Yadun S (2007a) Plant biological warfare: thorns inject pathogenic bacteria into herbivores. *Environmental Microbiology*, **9**, 584–92.

Halpern M, Raats D, Lev-Yadun S (2007b) The potential anti-herbivory role of microorganisms on plant thorns. *Plant Signaling & Behavior*, **2**, 503–4.

Halpin CG, Skelhorn J, Rowe C (2008) Naïve predators and selection for rare conspicuous defended prey: the initial evolution of aposematism revisited. *Animal Behaviour*, **75**, 771–81.

Halpin CG, Skelhorn J, Rowe C (2012) The relationship between sympatric defended species depends upon predators' discriminatory behaviour. *PLoS ONE*, **7(9)**: e44895.

Halpin CG, Skelhorn J, Rowe C (2014) Increased predation of nutrient-enriched aposematic prey. *Proceedings of the Royal Society of London, Series B: Biological Sciences*, **281**: 20133255.

Hammerschmidt K, Kurtz J (2005) Surface carbohydrate composition of a tapeworm in its consecutive intermediate hosts: individual variation and fitness consequences. *International Journal of Parasitology*, **35**, 1499–507.

Hamner WM (1995) Predation, cover, and convergent evolution in epipelagic oceans. *Marine and Freshwater Behaviour and Physiology*, **27**, 71–89.

Handel SN, Peakall R (1993) Thynnine wasps discriminate among heights when seeking mates: tests with a sexually deceptive orchid. *Oecologia*, **95**, 241–5.

Hanlon RT, Messenger JB (1988) Adaptive coloration in young cuttlefish (*Sepia officinalis* L.): the morphology and development of body patterns and their relation to behaviour. *Philosophical Transactions of the Royal Society of London, Series B, Biological Sciences*, **320**, 437–87.

Hanlon RT, Forsythe JW, Joneschild DE (1999) Crypsis, conspicuousness, mimicry and polyphenism as antipredator defences of foraging octopuses on Indo-Pacific coral reefs, with a method of quantifying crypsis from video tapes. *Biological Journal of the Linnean Society*, **66**, 1–22.

Hanlon RT, Naud M-J, Shaw PW, Havenhand JN (2005) Transient sexual mimicry leads to fertilization. *Nature*, **433**, 212.

Hanlon RT, Chiao C-C, Mäthger LM, Barbosa A, Buresch KC, Chubb C (2009) Cephalopod dynamic camouflage: bridging the continuum between background matching and disruptive coloration. *Philosophical Transactions of the Royal Society of London, Series B, Biological Sciences*, **364**, 429–37.

Hanlon RT, Watson AC, Barbosa A (2010) A "mimic octopus" in the Atlantic: flatfish mimicry and camouflage by *Macrotritopus defilippi*. *Biological Bulletin*, **218**, 15–24.

Hanlon RT, Chiao C-C, Mäthger LM, Buresch KC, Barbosa A, Allen JA, Siemann LA, Chubb C (2011) Rapid adaptive camouflage in cephalopods. In M Stevens and S Merilaita (eds), *Animal Camouflage*. Cambridge University Press, Cambridge, UK, pp. 145–63.

Hansell MH (1996) The function of lichen flakes and white spider cocoons on the outer surface of birds nests. *Journal of Natural History*, **30**, 303–11.

Hansen T, Haugsten Holen Ø, Mappes J (2010) Predators use environmental cues to discriminate between prey. *Behavioural Ecology and Sociobiology*, **64**, 1991–7.

Hanski I (1989) Fungivory: fungi, insects and ecology. In N Wilding, NM Collins, PM Hammond and JF Webber (eds), *Insect-Fungus Interactions*. Academic Press, London, pp. 25–68.

Hansknecht KA (2008) Lingual luring by mangrove saltmarsh snakes (*Nerodia clarkii compressicauda*). *Journal of Herpetology*, **42**, 9–15.

Hanson TC (1976) Notes on some Brazilian insects. *AES Bulletin*, **35**, 135–138.

Harari AR, Brockmann HJ (1999) Insect behaviour: Male beetles attracted by females mounting. *Nature*, **401**, 762–3.

Hare JF, Eisner T (1993) Pyrrolizidine alkaloid deters ant predators of *Utetheisa ornatrix* eggs: effects of alkaloid concentration, oxidation state, and prior exposure of ants to alkaloid-laden prey. *Oecologia*, **96**, 9–18.

Harlan JR, de Wet JMJ, Price EG (1973) Comparative evolution of cereals. *Evolution*, **27**, 311–25.

Harland DP, Jackson RR (2001) Prey classification by *Portia fimbriata*, a salticid spider that specializes at preying on other salticids: species that elicit cryptic stalking. *Journal of Zoology*, **255**, 445–60.

Harper GR, Pfennig DW (2007) Mimicry on the edge: why do mimics vary in resemblance to their model in different parts of their geographical range? *Proceedings of the Royal Society of London, Series B: Biological Sciences*, **274**, 1955–61.

Harper RD, Case JF (1999) Disruptive counterillumination and its antipredatory value in the plainfish midshipman *Porichthys notatus*. *Marine Biology*, **134**, 529–40.

Harris AC (1978) Mimicry by a longhorn beetle, *Neocalliprason elegans* (Coleoptera: Cerambycidae), of its parasitoid, *Xanthocryptus novozealandicus* (Hymenoptera: Ichneumonidae). *The New Zealand Entomologist*, **6**, 406–8.

Harris LG (1973) Nudibranch associations. In TC Cheng (ed.), *Current Topics in Comparative Pathobiology*, Vol. **2**. Elsevier, New York, pp. 213–308.

Harris RHTP (1930) *Report on the Bionomics of the Tsetse Fly*. Provincial Administration of Natal, Pietermaritzburg, South Africa.

Hart NS, Hunt DM (2007) Avian visual pigments: characteristics, spectral tuning, and evolution. *The American Naturalist*, **169**, S7–S26.

Harvey PH, Greenwood PJ (1978) Anti-predator defence strategies: some evolutionary problems. In JR Krebbs and NB Davies (eds), *Behavioural Ecology: An Evolutionary Approach*. Blackwell Science Publishers, Oxford, pp. 129–51.

Harvey PH, Paxton RJ (1981) The evolution of aposematic coloration. *Oikos*, **37**, 391–3.

Harvey PH, Bull JJ, Pemberton M, Paxton RJ (1982) The evolution of aposematic coloration in distasteful prey: a family model. *The American Naturalist*, **119**, 710–9.

Harvey PH, Bull JJ, Paxton RJ (1983) Why some insects look pretty nasty. *New Scientist*, **97**, 26–7.

Hasegawa E, Tinaut A, Ruano F (2002) Molecular phylogeny of two slave-making ants: *Rossomyrmex* and *Polyergus* (Hymenoptera: Formicidae). *Annales Zoologici Fennici*, **39**, 267–71.

Håstad O, Victorsson J, Ödeen A (2005) Differences in color vision make passerines less conspicuous in the eyes of their predators. *Proceedings of the National Academy of Sciences, U.S.A.*, **102**, 6391–4.

Hatle JD, Salazar BA, Whitman DW (2002) Survival advantage of sluggish individuals in aggregations of aposematic prey, during encounters with ambush predators. *Evolutionary Ecology*, **16**, 415–31.

Hauber ME, Kilner RM (2007) Coevolution, communication, and host-chick mimicry in parasitic finches: who mimics whom? *Behavioral Ecology and Sociobiology*, **61**, 497–503.

Hauser MD (1992) Costs of deception: cheaters are punished in rhesus monkeys (*Macaca mulatta*). *Proceedings of the National Academy of Sciences, U.S.A.*, **89**, 12137–9.

Havlíček J, Dvořáková R, Bartoš L, Flegr J (2006) Non-advertized does not mean concealed: body odour changes across the human menstrual cycle. *Ethology*, **112**, 81–90.

Hawkins BA, Gagné RJ (1989) Determinants of assemblage size for the parasitoids of Cecidomyiidae (Diptera). *Oecologia (Berlin)*, **81**, 75–88.

Hawkins CE, Dallas JF, Fowler PA, Woodroffe R, Racey PA (2002) Transient masculinization in the fossa, *Cryptoprocta ferox* (Carnivora, Viverridae). *Biology of Reproduction*, **66**, 610–5.

Hawryshyn CW (1992) Polarization vision in fish. *American Scientist*, **80**, 164–75.

Hayashi M, Nomura M, Nakamuta K (2016) Efficacy of chemical mimicry by aphid predators depends on aphid-learning by ants. *Journal of Chemical Ecology*, **42**, 236–9.

Haynes KF, Yeargan KV, Gemeno C (2001) Detection of prey by a spider that aggressively mimics pheromone blends. *Journal of Insect Behaviour*, **14**, 535–44.

Haynes KF, Gemeno C, Yeargan KV, Millar JG, Johnson KM (2002) Aggressive chemical mimicry of moth pheromones by a bolas spider: how does this specialist predator attract more than one species of prey? *Chemoecology*, **12**, 99–105.

Hazel WN (1977) The genetic basis of pupal colour dimorphism and its maintenance by natural selection in *Papilio polyxenes* (Papilionidae: Lepidoptera). *Heredity*, **38**, 227–36.

Hazel WN (1990) Has limited variability and mimicry in the swallowtail butterfly *Papilio polyxenes* Fabr. *Heredity*, **65**, 109–14.

Hazel WN, West DA (1979) Environmental control of pupal colour in swallowtail butterflies (Lepidoptera: Papilioninae): *Battus philenor* (L.) and *Papilio polyxenes* Fabr. *Ecological Entomology*, **4**, 393–400.

Hazel WN, West DA (1982) Pupal colour dimorphism in swallowtail butterflies as a threshold trait: selection in *Eurytides marcellus* (Gamer). *Heredity*, **49**, 295–301.

Heal J (1979) Colour patterns of Syrphidae: I. Genetic variation in the dronefly *Eristalis tenax*. *Heredity*, **42**, 223–36.

Heal J (1982) Colour patterns of Syrphidae: IV. Mimicry and variation in natural populations of *Eristalis tenax*. *Heredity*, **49**, 95–109.

Heatwole H, Davidson E (1976) A review of caudal luring in snakes with notes on its occurrence in the Saharan sand viper, *Cerastes vipera*. *Herpetologiea*, **32**, 332–6.

Hebert FO, Phelps L, Samonte I, Panchal M, Grambauer S, Barber I, Kalbe M, Landry CR, Aubin-Horth N (2015) Identification of candidate mimicry proteins involved in parasite-driven phenotypic changes. *Parasites & Vectors*, **8**: 225.

Hebert PDN (1974) Spittlebug morph mimics avian excrement. *Nature*, **250**, 352.

Hebert PDN (1983) Egg dispersal patterns and adult feeding behavior in the Lepidoptera. Canadian Entomologist, **115**, 1477–81.

Hecht MK, Marien D (1956) The coral snake mimic problem: a reinterpretation. *Journal of Morphology*, **98**, 335–6.

Hegna RH, Saporito RA, Gerow KG, Donnelly MA (2011) Contrasting colors of an aposematic poison frog do not affect predation. *Annales Zoologici Fennici*, **48**, 29–38.

Hegna RH, Saporito RA, Donnelly MA (2013) Not all colors are equal: predation and color polytypism in the aposematic poison frog *Oophaga pumilio*. *Evolutionary Ecology*, **27**, 831–45.

Heiduk A, Brake I, von Tschirnhaus M, Göhl M, Jürgens A, Johnson SD, Meve U, Dötterl S (2016) *Ceropegia sandersonii* mimics attacked honeybees to attract kleptoparasitic flies for pollination. *Current Biology*, **26**, 1–7.

Heiling AM, Herberstein ME, Chittka L (2003) Crab-spiders manipulate flower signals. *Nature*, **421**, 334.

Heiling AM, Chittka L, Cheng K, Herberstein ME (2005) Colouration in crab spiders: substrate choice and prey attraction. *Journal of Experimental Biology*, **208**, 1785–92.

Heinrich B (1975) Bee flowers: a hypothesis on flower variety and blooming times. *Evolution*, **29**, 325–34.

Heinrich B (1979) Foraging strategies of caterpillars: leaf damage and possible predator avoidance strategies. *Oecologia (Berlin)*, **42**, 325–37.

Heinrich B (1993) How avian predators constrain caterpillar foraging. In NE Stamp and TM Casey (eds), *Caterpillars: Ecological and Evolutionary Constraints on Foraging*. Chapman and Hall, London, pp. 224–47.

Heinrich B, Collins SL (1983) Caterpillar leaf damage, and the game of hide-and-seek with birds. *Ecology*, **64**, 592–602.

Heistermann M, Ziegler T, van Schaik CP, Launhardt K, Winkler P, Hodges JK (2001) Loss of oestrus, concealed ovulation and paternity confusion in free-ranging Hanuman langurs. *Proceedings of the Royal Society of London, Series B: Biological Sciences*, **268**, 2445–51.

Heller R (1980) Foraging on potentially harmful prey. *Journal of Theoretical Biology*, **85**, 807–13.

Helmholtz H v (1867) *Handbuch der physiologischen Optik*, 3rd edn. Voss, Leipzig. (English translation by J P C Southall (1962) for the Optical Society of America, 1925).

Henderson CL (2009) *Butterflies, Moths, and Other Invertebrates in Costa Rica*. University of Texas Press, Austin, TX.

Henning F, Meyer A (2012) Eggspot number and sexual selection in the cichlid fish *Astatotilapia burtoni*. *PLoS ONE*, **7(8):** e43695.

Henning SF (1983) Chemical communication between lycaenid larvae (Lepidoptera: Lycaenidae) and ants (Hymenoptera: Formicidae). *Journal of the Entomological Society of South Africa*, **46**, 341–66.

Henning SF (1997) Chemical communication between lycaenid larvae (Lepidoptera: Lycaenidae) and ants (Hymenoptera: Formicidae). *Metamorphosis (Supplement)*, **3**, 66–81.

Hensel JL, Brodie ED (1976) An experimental study of aposematic coloration in the salamander *Plethodon jordani*. *Copeia*, **1976**, 59–65.

Heredia MD, Alvarez-Lopez H (2004) Larval morphology and behavior of *Antirrhea weymeri* Salazar, Constantino & López, 1998 (Nymphalidae: Morphinae) in Colombia. *Journal of the Lepidopterists' Society*, **58**, 88–93.

Herrel J, Hazel WN (1995) Female-limited variability in mimicry in the swallowtail butterfly *Papilio polyxenes* Fabr. *Heredity*, **75**, 106–10.

Herrera C (1985) Aposematic insects as six-legged fruits: incidental short-circuiting of their defense by frugivorous birds. *The American Naturalist*, **126**, 286–93.

Hert E (1989) The function of egg-spots in an African mouth-brooding cichlid fish. *Animal Behaviour*, **37**, 726–32.

Hert E (1991) Female choice based on egg-spots in *Pseudotropheus aurora* Burgess 1976, a rock-dwelling cichlid of Lake Malawi, Africa. *Journal of Fish Biology*, **38**, 951–3.

Hespenheide HA (1971) Food preference and the extent of overlap in some insectivorous birds, with special reference to the Tyrannidae. *Ibis*, **113**, 59–72.

Hespenheide HA (1973) A novel mimicry complex: beetles and flies. *Journal of Entomology. Series A, General Entomology*, **48**, 49–56.

Hespenheide HA (1975) Reversed sex-linked mimicry in a beetle. *Evolution*, **29**, 780–3.

Hespenheide HA (1995) Mimicry in the Zygopinae (Coleoptera: Curculionidae). *Memoirs of the Entomological Society of Washington*, **14**, 145–54.

Hespenheide HA (1996) The role of plants in structuring communities of mimetic insectsI In AC Gibson (ed.), *Neotropical Biodiversity and Conservation*. Mildred E. Mathias Botanical Garden, University of California, Los Angeles CA, pp. 109–26.

Hespenheide HA (2012) New Mexican and Central American species of *Agrilus* Curtis (Coleoptera: Buprestidae) mimetic of flies. *Zootaxa*, **3181**: 1–27.

Hetz M, Slobodchikoff CN (1988) Predation pressure in an imperfect mimicry system. *Oecologia (Berlin)*, **76**, 570–3.

Hetz M, Slobodchikoff CN (1990) Reproduction and the energy cost of defense in a Batesian mimicry complex. *Oecologia (Berlin)*, **84**, 69–73.

Hewitson WC (1856) *Illustrations of New Species of Exotic Butterflies: Selected Chiefly from the Collections of W. Wilson Saunders and William C. Hewitson*. John Van Voorst, London.

Higginson AD, de Wert L, Rowland HM, Speed MP, Ruxton GD (2012) Masquerade is associated with polyphagy and larval overwintering in Lepidoptera. *Biological Journal of the Linnean Society*, **106**, 90–103.

Hilker M, Köpf A (1994) Evaluation of the palatability of chrysomelid larvae containing anthraquinones to birds. *Oecologia (Berlin)*, **100**, 421–9.

Hill RI (2010) Habitat segregation among mimetic ithomiine butterflies (Nymphalidae). *Evolutionary Ecology*, **24**, 273–85.

Hill RI, Vaca JF (2004) Differential wing strength in *Pierella* butterflies (Nymphalidae, Satyridae) supports the deflection hypothesis. *Biotropica*, **26**, 362–70.

Hilton DFJ (1987) A terminology for females with colour patterns that mimic males. *Entomological News*, **98**, 221–3.

Hines HM, Counterman BA, Papa R, Albuquerque de Moura P, Cardoso MZ, Linares M, Mallet J, Reed RD, Jiggins CD, Kronforst RF, McMillan WO (2011) Wing patterning gene redefines the mimetic history of *Heliconius* butterflies. *Proceedings of the National Academy of Sciences, U.S.A.*, **108**, 19666–71.

Hinman KE, Throop HL, Adams KL, Dake AJ, McLauchlan KK, McKone MJ (1997) Predation by free-ranging birds on partial coral snake mimics: the importance of ring width and color. *Evolution*, **51**, 1011–4.

Hinton HE (1974) Lycaenid pupae that mimic anthropoid heads. *Journal of Entomology (A)*, **49**, 65–9.

Hobbhahn N, Johnson SD, Bytebier B, Yeung EC, Harder LD (2013) The evolution of floral nectaries in *Disa* (Orchidaceae: Disinae): recapitulation or diversifying innovation? *Annals of Botany*, **112**, 1303–19.

Hockey PAR (1982) Adaptiveness of nest site selection and egg coloration in the African black oystercatcher *Haematopus moquini*. *Behavioral Ecology and Sociobiology*, **11**, 117–23.

Hocking B (1964) Fire melanism in some African grasshoppers. *Evolution*, **18**, 332–5.

Hoekstra HE, Drumm KE, Nachman MW (2004) Ecological genetics of adaptive color polymorphism in pocket mice: geographic variation in selected and neutral genes. *Evolution*, **58**, 1329–41.

Holen ØH (2013) Disentangling taste and toxicity in aposematic prey. *Proceedings of the Royal Society of London, Series B: Biological Sciences*, **280**: 20122588.

Holen ØH, Johnstone RA (2004) The evolution of mimicry under constraints. *The American Naturalist*, **164**, 598–613.

Holen ØH, Johnstone RA (2006) Context-dependent discrimination and the evolution of mimicry. *The American Naturalist*, **167**, 377–89.

Holliday JC, Soule N (2001) Spontaneous female orgasms triggered by smell of a newly found tropical *Dictyphora* species. *International Journal of Medicinal Mushrooms*, **3**, 162–7.

Holling CS (1959) Some characteristics of simple types of predation and parasitism. *Canadian Entomologist*, **91**, 385–98

Holling CS (1963) Mimicry and predator behaviour. *Proceedings of the 16th International Congress of Zoology, Washington*, **4**, 166–71.

Holling CS (1965) The functional response of predators to prey density and its role in mimicry and population regulation. *Memoirs of the Entomological Society of Canada*, **45**, 1–60.

Holloway GJ, de Jong PW, Brakefield PM, de Vos H (1991) Chemical defence in ladybird beetles (Coccinellidae). I. Distribution of coccinelline and individual variation in defence in 7-spot ladybirds (*Coccinella septempunctata*). *Chemoecology*, **2**, 7–14.

Holloway GJ, Brakefield PM, Kofman S (1993) The genetics of wing pattern elements in the polyphenic butterfly *Bicyclus anynana*. *Heredity*, **70**, 179–86.

Holloway GJ, Gilbert F, Brandt A (2002) The relationship between mimetic imperfection and phenotypic variation in insect colour patterns. *Proceedings of the Royal Society of London, Series B: Biological Sciences*, **269**, 411–6.

Holmes JS (1906) Death-feigning in *Ranatra*. *Journal of Comparative Neurology and Psychology*, **16**, 200–16.

Holmgren NMA, Enquist M (1999) Dynamics of mimicry evolution. *Biological Journal of the Linnean Society*, **66**, 145–58.

Holz C, Streil G, Dettner K, Dütemeyer J, Boland W (1994) Intersexual transfer of a toxic terpenoid during copulation and its paternal allocation to developmental stages: quantification of cantharidin in cantharidin-producing oedemerids (Coleoptera: Oedemeridae) and canthariphilous pyrochroids (Coleoptera: Pyrochroidae). *Zeitschrift fur Naturforschung*, **49c**, 856–64.

Holzkamp G, Nahrstedt A (1994) Biosynthesis of cyanogenic glycosides in the Lepidoptera – incorporation of [U-C-14]-2-methylpropanealdoxime, 2S-[U-C-14]-methylbutanealdoxime and D,L-[U-C-14]-N-hydroxyisoleucine into linamarin and lotaustralin by the larvae of *Zygaena trifolii*. *Insect Biochemistry and Molecular Biology*, **24**, 161–8.

Honma A, Oku S, Nishida T (2006) Adaptive significance of death feigning posture as a specialized inducible defence against gape-limited predators. *Proceedings of the Royal Society, series B, Biological Sciences*, **273**, 1631–6.

Honma A, Takakura K-i, Nishida T (2008) Optimal-foraging predator favors commensalistic Batesian mimicry. *PLoS ONE*, **3(10)**: e3411.

Honza M, Polacicová L, Procházka P (2007) Ultraviolet and green parts of the colour spectrum affect egg rejection in the song thrush (*Turdus philomelos*). *Biological Journal of the Linnean Society*, **92**, 269–76.

Hoogmoed MS, Avila-Pires TCS (2011) A case of voluntary tail autotomy in the snake *Dendrophidion dendrophis* (Schlegel 1837) (Reptilia: Squamata: Colubridae). *Boletim do Museu Paraense Emílio Goeldi. Ciências Naturais, Belém*, **6**, 113–7.

Hooper J (2002) *Of Moths and Men: The Untold Story of Science and the Peppered Moth*. WW Norton & Company.

Hopkins GHE, Buxton PA (1926) *Euploea* spp. frequenting dead twigs of *Tournefortia argentea* in Samoa & Tonga. *Proceedings of the Royal Entomological Society of London*, **1**, 35–7.

Hori M, Watanabe K (2000) Aggressive mimicry in the intra-populational color variation of the Tanganyikan scaleeater *Perissodus microlepis* (Cichlidae). *Environmental Biology of Fishes*, **59**, 111–5.

Horiger N, Zivanov D, Linde HHA, Meyer K (1970) Weitere Cardenolide aus Ch'an Su. *Helvetica Chimica Acta*, **53**, 2057.

Horsley DT, Lynch BM, Greenwood JJD, Hardman B, Mosley S (1979) Frequency-dependent selection by birds when the density of prey is high. *Journal of Animal Ecology*, **48**, 483–90.

Horton CC (1979) Apparent attraction of moths by the webs of araneid spiders. *Journal of Arachnology*, **7**, 88.

Horváth G, Blahó M, Kriska G, Hegedüs R, Gerics B, Farkas R, Åkesson S (2010) An unexpected advantage of whiteness in horses, the most horsefly-proof horse has a depolarizing white coat. *Proceedings of the Royal Society of London, Series B: Biological Sciences*, **277**, 1643–50.

Hossie TJ, Sherratt TN (2012) Eyespots interact with body colour to protect caterpillar-like prey from avian predators. *Animal Behaviour*, **84**, 167–73.

Hossie TJ, Sherratt TN (2013) Defensive posture and eyespots deter avian predators from attacking caterpillar models. *Animal Behaviour*, **86**, 383–9.

Hossie TJ, Sherratt TN (2014) Does defensive posture increase mimetic fidelity of caterpillars with eyespots to their putative snake models? *Current Zoology*, **60**, 76–89.

Hossie TJ, Sherratt TN, Janzen DH, Hallwachs W (2013) An eyespot that "blinks": an open and shut case of eye mimicry in *Eumorpha* caterpillars (Lepidoptera: Sphingidae). *Journal of Natural History*, **47**, 2915–26.

Houston AI, Stevens M, Cuthill IC (2007) Animal camouflage: compromise or specialize in a 2 patch-type environment? *Behavioral Ecology*, **18**, 769–75.

How MJ, Zanker JM (2014) Motion camouflage induced by zebra stripes. *Ethology*, **117**, 163–70.

Howard RD (1974) The influence of sexual selection and interspecific competition on mockingbird song (*Mimus polyglottos*). *Evolution*, **28**, 428–38.

Howard RW, McDaniel CA, Blomquist GJ (1980) Chemical mimicry as an integrating mechanism: cuticular hydrocarbons of a termitophile and its host. *Science*, **210**, 431–3.

Howard RW (1990) Biosynthesis and chemical mimicry of cuticular hydrocarbons from the obligate predator, *Microdon albicomatus* Novak (Diptera: Syrphidae) and its ant prey, *Myrmica incompleta* Provancher (Hymenoptera: Formicidae). *Journal of the Kansas Entomological Soiety*, **63**, 437–43.

Howard RW (1992) Comparative analysis of cuticular hydrocarbons from the ectoparasitoids *Cephalonomia waterstoni* and *Laelius utilis* (Hymenoptera: Bethylidae) and their respective hosts, *Cryptolestes ferrugineus* (Coleoptera: Cucujidae) and *Trogoderma variabile* (Coleoptera: Dermestidae). *Annals of the Entomological Society of America*, **85**, 317–25.

Howard RW, Akre RD, Garnett WB (1990) Chemical mimicry of an obligate predator of carpenter ants (Formicidae). *Annals of the Entomological Society of America*, **83**, 606–16.

Howard RW, Pérez-Lachaud G, Lachaud J-P (2001) Cuticular hydrocarbons of *Kapala sulcifacies* (Hymenoptera: Eucharitidae) and its host, the ponerine ant *Ectatomma ruidum* (Hymenoptera: Formicidae). *Annals of the Entomological Society of America*, **94**, 707–16.

Howarth B, Edmunds M (2000) The phenology of Syrphidae (Diptera): are they Batesian mimics of Hymenoptera? *Biological Journal of the Linnean Society*, **71**, 437–57.

Howarth B, Clee C, Edmunds M (2000) The mimicry between British Syrphidae (Diptera) and aculeate Hymenoptera. *British Journal of Entomology and Natural History*, **13**, 1–39.

Howarth B, Edmunds M, Gilbert F (2004) Does the abundance of hoverfly mimics (Syrphidae) depend on the numbers of the hymenopteran models? *Evolution*, **58**, 367–75.

Howse PE (2013) Lepidopteran wing patterns and the evolution of satyric mimicry. *Biological Journal of the Linnean Society*, **109**, 203–14.

Howse PE, Allen JA (1994) Satyric mimicry: the evolution of apparent imperfection. *Proceedings of the Royal Society of London, Series B: Biological Sciences*, **257**, 111–14.

Hristov NI, Conner WE (2005) Effectiveness of tiger moth (Lepidoptera, Arctiidae) chemical defenses against an insectivorous bat (Eptesicus fuscus). *Chemoecology*, **15**, 105–13.

Hsiao TH, Hsiao C, Rothschild M (1980) Characterization of a protein toxin from dried specimens of the garden tiger moth (*Arctia caja* L.). *Toxicon*, **18**, 291–9.

Hughes L, Westoby M (1992) Capitula on stick insect eggs and elaiosomes on seeds: convergent adaptations for burial by ants. *Functional Ecology*, **6**, 642–8.

Huheey JE (1960) Mimicry in the color pattern of certain Appalachian salamanders. *Journal of the Elisha Mitchell Scientific Society*, **76**, 246–51.

Huheey JE (1964) Studies of warning coloration and mimicry. IV. A mathematical model of model-mimic frequencies. Ecology, **45**, 185–8.

Huheey JE (1976) Studies of warning coloration and mimicry. VII. Evolutionary consequences of a Batesian-Müllerian spectrum: a model for Müllerian mimicry. *Evolution*, **30**, 86–93.

Huheey JE (1980a) Batesian and Müllerian mimicry: semantic and substantive differences of opinion. *Evolution*, **34**, 1212–5.

Huheey JE (1980b) The question of synchrony or "temporal sympatry" in mimicry. *Evolution*, **34**, 614–6.

Huheey JE (1980c) Studies in warning coloration and mimicry VIII. Further evidence for a frequency-dependent model of predation. *Journal of Herpetology*, **14**, 223–30.

Huheey JE (1988) Mathematical models of mimicry. *The American Naturalist*, **131**, S22–S41. (reprinted in Brower LP (1988) *Mimicry and the Evolutionary Process*. University of Chicago Press, Chicago).

Huheey JE, Brandon RA (1961) Further notes on mimicry in salamanders. *Herpetologica*, **17**, 63–4.

Huheey RB, Pianka ER (1977) Natural selection for juvenile lizards mimicking noxious beetles. *Science*, **195**, 201–3.

Hultgren KM, Stachowicz JJ (2008) Alternative camouflage strategies mediate predation risk among closely related co-occurring kelp crabs. *Oecologia (Berlin)*, **155**, 519–28.

Hultgren KM, Stachowicz JJ (2009) Evolution of decoration in majoid crabs: a comparative phylogenetic analysis of the role of body size and alternative defensive strategies. *The American Naturalist*, **173**, 566–78.

Hultgren KM, Stachowicz JJ (2010) Size-related habitat shifts facilitated by positive preference induction in a marine kelp crab. *Behavioral Ecology*, **21**, 329–36.

Huxley J (1957) *New Wine in New Bottles*. Chatto & Windus, London.

Ichiishi S, Nagamitsu T, Kondo Y, Iwashina T, Kondo K, Tagashira N (1999) Effects of macro-components and sucrose in the medium on in vitro red-color pigmentation in *Dionaea muscipula* Ellis and *Drosera spathulata* Labill. *Plant Biotechnology*, **16**, 235–8.

Igic B, Magrath RD (2013) Fidelity of vocal mimicry: identification and accuracy of mimicry of heterospecific alarm calls by the brown thornbill. *Animal Behaviour*, **85**, 593–603.

Igic B, Magrath RD (2014) A songbird mimics different heterospecific alarm calls in response to different types of threat. *Behavioral Ecology*, **25**, 538–48.

Igic B, McLachlan J, Lehtinen I, Magrath RD (2015) Crying wolf to a predator: deceptive vocal mimicry by a bird protecting young. *Proceedings of the Royal Society of London, Series B: Biological Sciences*, **282**: 1809.

Ihalainen E, Lindstedt C (2012) Do avian predators select for seasonal polyphenism in the European map butterfly *Araschnia levana* (Lepidoptera: Nymphalidae)? *Biological Journal of the Linnean Society*, **106**, 737–48.

Ihalainen E, Lindström L, Mappes J (2007) Investigating Müllerian mimicry: predator learning and variation in prey defences. *Journal of Evolutionary Biology*, **20**, 780–91.

Ihalainen E, Rowland HM, Speed MP, Ruxton GD, Mappes J (2012) Prey community structure affects how predators select for Mullerian mimicry. *Proceedings of the Royal Society of London, Series B: Biological Sciences*, **279**, 2099–105.

Ikin M, Turner JRG (1972) Experiments in mimicry: gestalt perception and the evolution of genetic linkage. *Nature*, **239**, 525–7.

Imperio S, Bionda R, Viterbi R, Provenzale A (2013) Climate change and human disturbance can lead to local extinction of alpine rock ptarmigan: new insight from the western Italian alps. *PLoS ONE*, **8(11)**: e81598.

Inbar M, Lev-Yadun S (2005) Conspicuous and aposematic spines in the animal kingdom. *Naturwissenschaft*, **92**, 170–2.

Inbar M, Doostdar H, Mayer RT (1999) Effects of sessile whitefly nymphs (Homoptera: Aleyrodidae) on leaf-chewing larvae (Lepidoptera: Noctuidae). *Environmental Entomology*, **28**, 353–57.

Inbar M, Izhaki I, Koplovich A, Lupo I, Silanikove N, Glasser T, Gerchman Y, Perevolotsky A, Lev-Yadun S (2010) Why do many galls have conspicuous colors? A new hypothesis. *Arthropod-Plant Interactions*, **4**, 1–6.

Inglis IR, Huson LW, Marshall MB, Neville PA (1983) The feeding behaviour of starlings (*Sturnus vulgaris*) in the presence of 'eyes'. *Zeitschrift für Tierpsychologie*, **62**, 181–208.

Iserbyt A, Van Gossum H (2011) Show your true colour: cues for male mate preference in an intra-specific mimicry system. *Ecological Entomology*, **36**, 544–8.

Iserbyt A, Bots J, Van Dongen S, Ting JJ, Van Gossum H, Sherratt TN (2011) Frequency-dependent variation in mimetic fidelity in an intraspecific mimicry system. *Proceedings of the Royal Society of London, Series B: Biological Sciences*, **278**, 3116–22.

Itino T, Kato M, Hotta M (1991) Pollination ecology of the two wild bananas, *Musa acuminata* subsp. *halabanensis* and *M. salaccensis*: chiropterophily and ornithophily. *Biotropica*, **23**, 151–8.

Ito F, Hashim R, Huei YS, Kaufmann E, Akino T, Billen J (2004) Spectacular Batesian mimicry in ants. *Naturwissenschaften*, **91**, 481–4.

Ivri Y, Dafni A (1977) Pollination ecology of *Epipactis consimilis* Don (Orchidaceae) in Israel. *New Phytologist*, **79**, 173–7.

Iyengar VK, Eisner T (2001) Heritability of male quality in an arctiid moth (*Utetheisa ornatrix*): hydroxydanaidal is the only criterion of choice. *Behavioral Ecology and Sociobiology*, **49**, 283–8.

Jablonski P, Stawarczyk T, Cygan JP (2006) Description of a nest of the three-striped warbler (*Basileuterus tristriatus chitrensis*) from Costa Rica. *Ornithologia Neotropical*, **17**, 593–5.

Jackson JF, Drummond BA (1974) A Batesian ant-mimicry complex from the mountain pine ridge of British Honduras, with an example of transformational mimicry. *American Midland Naturalist*, **91**, 248–251.

Jackson JF, Ingram III W, Campbell HW (1976) The dorsal pigmentation pattern of snakes as an antipredator strategy: a multivariate approach. *The American Naturalist*, **110**, 1029–53.

Jackson RR (1986) The biology of ant-like jumping spiders (Araneae, Salticidae): prey and predatory behaviour of *Myrmarachne* with particular attention to *M. lupata* from Queensland. *Zoological Journal of the Linnean Society*, **88**, 179–90.

Jackson RR (1995a) Cues for web invasion and aggressive mimicry signalling in *Portia* (Araneae, Salticidae). *Journal of Zoology, London*, **236**, 131–49.

Jackson RR (1995b) Eight-legged tricksters: spiders that specialize at catching other spiders. *BioScience*, **42**, 590–8.

Jackson RR (2002) Trial-and-error derivation of aggressive-mimicry signals by *Brettus* and *Cyrba*, spartaeine jumping spiders (Araneae: Salticidae) from Israel, Kenya, and Sri Lanka. *New Zealand Journal of Zoology*, **29**, 95–117.

Jackson RR, Brassington RJ (1987) The biology of *Pholcus phalangioides* (Araneae, Pholcidae): predatory versatility, araneophagy and aggressive mimicry. *Journal of Zoology*, **211**, 227–38.

Jackson RR, Whitehouse MEA (1986) The biology of New Zealand and Queensland pirate spiders (Araneae, Mimetidae): aggressive mimicries, araneophagy and prey specialization. *Journal of Zoology, London*, **210**, 279–303.

Jackson RR, Wilcox RS (1990) Aggressive mimicry, prey-specific predatory behavior and predator-recognition in the predatory prey interactions of *Portia fimbriata* and *Euryattus* sp., jumping spiders from Queensland. *Behavioral Ecology and Sociobiology*, **26**, 111–9.

Jackson RR, Wilcox RS (1993) Spider flexibly chooses aggressive mimicry signals for different prey by trial and error. *Behaviour*, **127**, 21–36.

Jackson RR, Li D, Fijn N, Barrion AT (1998) Predatory-prey interactions between aggressive-mimic jumping spiders (Salticidae) and araeneophagic spitting spiders (Scytodidae) from the Philippines. *Journal of Insect Behavior*, **11**, 319–42.

Jacobs GH (1993) The distribution and nature of colour vision among the mammals. *Biological Reviews*, **68**, 413–71.

Jacobs GH, Deegan JF II, Neitz J (1998) Photopigment basis for dichromatic color vision in cows, goats, and sheep. *Visual Neuroscience*, **15**, 581–4.

Jacobson RL, Doyle RJ (1996) Lectin parasite interactions. *Parasitology Today*, **12**, 55–61.

Jaffe K, Michelangeli F, Gonzalez JM, Miras B, Ruiz MC (1992) Carnivory in pitcher plants of the genus *Heliamphora* (Sarraceniaceae). *New Phytologist*, **122**, 733–44.

Jaffe K, Blum MS, Fales HM, Mason RT, Cabrera A (1995) On insect attractants from pitcher plants of the genus *Heliamphora* (Sarraceniaceae). *Journal of Chemical Ecology*, **21**, 379–84.

Janson CH (1987) Bird consumption of bicolored fruit displays. *The American Naturalist*, **130**, 788–92.

Janzen DH (1980) Two potential coral snake mimics in a tropical deciduous forest. *Biotropica*, **12**, 77–8.

Janzen DH, Hallwachs W, Burns JM (2010) A tropical horde of counterfeit predator eyes. *Proceedings of the National Academy of Sciences, U.S.A.*, **107**, 11659–65.

Järvi T, Sillén-Tullberg B, Wiklund C (1981a) The cost of being aposematic. An experimental study of predation on larvae of *Papilio machaon* by the great tit *Parus major*. *Oikos*, **36**, 267–72.

Järvi T, Sillén-Tullberg B, Wiklund C (1981b) Individual versus kin selection for aposematic coloration: a reply to Harvey and Paxton. *Oikos*, **37**, 393–5.

Jeffords MR, Sternburg JG, Waldbauer GP (1979) Batesian mimicry: field demonstration of the survival value of pipevine swallowtail and monarch color patterns. *Evolution*, **33**, 275–86.

Jeffords MR, Waldbauer GP, Sternburg JG (1980) Determination of the time of day at which diurnal moths painted to resemble butterflies are attacked by birds. *Evolution*, **34**, 1205–11.

Jefson M, Meinwald J, Nowicki S, Hicks K, Eisner T (1983) Chemical defense of a rove beetle (*Creophilus maxillosus*). *Journal of Chemical Ecology*, **9**, 159–80.

Jennings DT, Houseweart MW (1989) Sex-biased predation by web-spinning spiders (Araneae) on spruce budworm moths. *Journal of Arachnology*, **17**, 179–94.

Jersáková J, Johnson SD, Kindlmann (2006) Mechanisms and evolution of deceptive pollination in orchids. *Biological Reviews*, **81**, 219–35.

Jersáková J, Johnson SD, Jürgens A (2009) Deceptive behavior in plants. II. Food deception by plants: from generalized systems to specialized floral mimicry. In F Baluška (ed.), *Plant-Environment Interactions Signaling and Communication in Plants*. Springer, Berlin, Heidelberg, pp. 223–46.

Jersáková J, Jürgens A, Šmilauer P, Johnson SD (2012) The evolution of floral mimicry: identifying traits that visually attract pollinators. *Functional Ecology*, **26**, 1381–9.

Jersáková J, Spaethe J, Streinzer M, Neumayer J, Paulus H, Dotterl S, Johnston SD (2016) Does *Traunsteinera globosa* (the globe orchid) dupe its pollinators through generalized food deception or mimicry? *Botanical Journal of the Linnean Society*, **180**, 269–94.

Jetz W, Rowe C, Guilford T (2001) Non-warning odors trigger innate color aversions – as long as they are novel. *Behavioral Ecology*, **12**, 134–9.

Jiggins CD, McMillan WO (1997) The genetic basis of an adaptive radiation: warning colour in two *Heliconius* species. *Proceedings of the Royal Society of London, Series B: Biological Sciences*, **264**, 1167–75.

Jiggins CD, Naisbit RE, Coe RL, Mallet J (2001) Reproductive isolation caused by colour pattern mimicry. *Nature*, **411**, 302–5.

Jin X-H, Ren Z-X, Xu S-Z, Wang H, Li D-Z, Li Z-Y (2014) The evolution of floral deception in *Epipactis veratrifolia* (Orchidaceae): from indirect defense to pollination. *BMC Plant Biology*, **14**: 63.

Joel DM (1988) Mimicry and mutualism in carnivorous pitcher plants (Sarraceniaceae, Nepenthaceae, Cephalotaceae, Bromeliaceae). *Biological Journal of the Linnean Society*, **35**, 185–97.

Joel DM, Juniper BE, Dafni A (1985) Ultraviolet patterns in the traps of carnivorous plants. *New Phytologist*, **101**, 585–93.

Johnsen S (2001) Hidden in plain sight: the ecology and physiology of organismal transparency. *The Biological Bulletin*, **201**, 301–18.

Johnson SD (1992) Pollination of *Disa filicornis* (Orchidaceae) through deception of mason bees. *South African Journal of Botany*, **80**, 147–52.

Johnson SD (1994) Evidence for Batesian mimicry in a butterfly-pollinated orchid. *Biological Journal of the Linnean Society*, **53**, 91–104.

Johnson SD (1997) Insect pollination and floral mechanisms in South African species of *Satyrium* (Orchidaceae). *Plant Systematics and Evolution*, **204**, 195–206.

Johnson SD (2000) Batesian mimicry in the non-rewarding orchid *Disa pulchra*, and its consequences for pollinator behaviour. *Biological Journal of the Linnean Society*, **71**, 119–32.

Johnson SD, Jürgens A (2010) Convergent evolution of carrion and faecal scent mimicry in fly-pollinated angiosperm flowers and a stinkhorn fungus. *South African Journal of Botany*, **76**, 796–807.

Johnson SD, Midgley JJ (1997) Fly pollination of *Gorteria diffusa* (Asteraceae), and a possible mimetic function for dark spots on the capitulum. *American Journal of Botany*, **84**, 429–36.

Johnson SD, Morita S (2006) Lying to Pinocchio: floral deception in an orchid pollinated by long-proboscid flies. *Botanical Journal of the Linnean Society*, **152**, 271–8.

Johnson SD, Steiner KE (1994) Pollination by megachilid bees and determinants of fruit-set in the Cape orchid *Disa tenuifolia*. *Nordic Journal of Botany*, **14**, 481–5.

Johnson SD, Steiner KE (1995) Long-proboscid fly pollination of two orchids in the Cape Drakensberg Mountains, South Africa. *Plant Systematics and Evolution*, **195**, 169–75.

Johnson SD, Steiner KE (1997) Long-tongued fly pollination and evolution of floral spur length in the *Disa draconis* complex (Orchidaceae). *Evolution*, **51**, 45–53.

Johnson SD, Linder HP, Steiner KE (1998) Phylogeny and radiation of pollination systems in *Disa* (Orchidaceae). *American Journal of Botany*, **85**, 402–11.

Johnson SD, Peter CI, Nilsson LA, Ågren J (2003a) Pollination success in a deceptive orchid is enhanced by co-occurring rewarding "magnet" plants. *Ecology*, **84**, 2919–27.

Johnson SD, Alexandersson R, Linder HP (2003b) Experimental and phylogenetic evidence for floral mimicry in a guild of fly-pollinated plants. *Biological Journal of the Linnean Society*, **80**, 289–304.

Johnson SD, Ellis A, Dötterl S (2007) Specialization for pollination by beetles and wasps: the role of lollipop hairs and fragrance in *Satyrium microrrhynchum* (Orchidaceae). *American Journal of Botany*, **94**, 47–55.

Johnson SD, Hobbhahn N, Bytebier B (2013) Ancestral deceit and labile evolution of nectar production in the African orchid genus *Disa*. *Biology Letters*, **9**: 20130500.

Johnson TW (1974) A study of mottled duck broods in the Merritt Island National Wildlife Refuge. *The Wilson Bulletin*, **86**, 68–70.

Jones DA (1962) Selective eating of the cyanogenic form of the plant *Lotus corniculatus* L. by various animals. *Nature*, **193**, 1009–10.

Jones DA, Parsons J, Rothschild M (1962) Release of hydrocyanic acid from crushed tissues of all stages in the life cycle of species of the Zygaeninae (Lepidoptera). *Nature*, **193**, 52–63.

Jones FM (1932) Insect coloration and the relative acceptability of insects to birds. *Transactions of the Royal Entomological Society of London*, **80**, 345–85.

Jones RB (1980) Reactions of male domestic chicks to two-dimensional eye-like shapes. *Animal Behaviour*, **28**, 212–8.

Jones RS, Fenton A, Speed MP (2016) "Parasite-induced aposematism" protects entomopathogenic nematode parasites against invertebrate enemies. *Behavioral Ecology*, **27**, 645–51.

Jones TC, Akoury TS, Hauser CK, Neblett II NF, Linville BJ, Edge AE, Weber NO (2011) Octopamine and serotonin have opposite effects on antipredator behavior in the orb-weaving spider,

Larinioides cornutus. *Journal of Comparative Physiology, series A*, **197**, 819–25.

Jordan TM, Partidge JC, Roberts NW (2012) Nonpolarizing broadband multilayer reflectors in fish. *Nature Photonics*, **6**, 759–63.

Joron M (2005) Polymorphic mimicry, microhabitat use, and sex-specific behaviour. *Journal of Evolutionary Biology*, **18**, 547–56.

Joron M, Iwasa Y (2005) The evolution of a Müllerian mimic in a spatially distributed community. *Journal of Theoretical Biology*, **237**, 87–103.

Joron M, Mallet JLB (1998) Diversity in mimicry: paradox or paradigm? *Trends in Ecology and Evolution*, **13**, 461–6.

Joron M, Papa R, Beltrán M, Chamberlain N, Mavárez J, Baxter S, Abanto M, Bermingham E, Humphrey SJ, Beasley H, Barlow K, ffrench-Constant RH, Mallet J, McMillan WO, Jiggins CD (2006) A conserved supergene locus controls colour pattern diversity in *Heliconius* butterflies. *PLoS Biology*, **4**: e303.

Joron M, Frezal L, Jones RT, Chamberlain NL, Lee SF, Haag CR, Whibley A, Becuwe M, Baxter SW, Ferguson L, Wilkinson PA, Salazar C, Davidson C, Clark R, Quail MA, Beasley H, Glithero R, Lloyd C, Sims S, Jones MC, Rogers J, Jiggins CD, ffrench-Constant RH (2011) Chromosomal rearrangements maintain a polymorphic supergene controlling butterfly mimicry. *Nature*, **477**, 203–6.

Jouventin P, Pasteur G, Cambefort JP (1977) Observational learning of baboons and avoidance of mimics: exploratory tests. *Evolution*, **31**, 214–8.

Joyce FJ (1999) Dual mimicry in the dimorphic eusocial wasp *Mischocyttarus mastigophorus* Richards (Hymenoptera: Vespidae). *Biological Journal of the Linnean Society* **66**, 501–11.

Jukema J, Piersma T (2006) Permanent female mimics in a lekking shorebird. *Biology Letters*, **2**, 161–4.

Jürgens A, Shuttleworth A (2015) Carrion and dung mimicry in plants. In ME Benbow, JK Tomberlin and AM Tarone (eds), *Carrion Ecology, Evolution, and their Applications*. CRC Press, Boca Raton, Florida, pp. 361–86.

Jürgens A, Dötterl S, Meve U (2006) The chemical nature of fetid floral odours in stapeliads (Apocynaceae-Asclepiadoideae-Ceropegieae). *New Phytologist*, **172**, 452–68.

Jürgens A, Wee S-L, Shuttleworth A, Johnson SD (2013) Chemical mimicry of insect oviposition sites: a global analysis of convergence in angiosperms. *Ecology Letters*, **16**, 1157–67.

Kaiser R (2006) Flowers and fungi use scents to mimic each other. *Science*, **311**, 806–7.

Kalshoven LGE (1961) Larvae of *Homodes* mimicking the aggressive *Oecophylla* ant in Southeast Asia (Lepidoptera, Noctuidae). *Tijdschrift voor Entomologie*, **104**, 43–50.

Kamilar JM, Bradley BJ (2011) Countershading is related to positional behavior in primates. *Journal of Zoology*, **283**, 227–33.

Kang C-K, Moon J-Y, Lee S-I, Jablonski PG (2012) Camouflage through an active choice of a resting spot and body orientation in moths. *Journal of Evolutionary Biology*, **25**, 1695–702.

Kang C-K, Moon J-Y, Lee S-I, Jablonski PG (2013) Cryptically patterned moths perceive bark structure when choosing body orientations that match wing color pattern to the bark pattern. *PLoS ONE*, **8(10)**: e78117.

Kang C-K, Moon J-Y, Lee S-I, Jablonski PG (2014) Moths use multimodal sensory information to adopt adaptive resting orientations. *Biological Journal of the Linnean Society*, **111**, 900–4.

Kano Y, Shimizu Y, Kondou K (2006) Status-dependent female mimicry in landlocked red-spotted masu salmon. *Journal of Ethology*, **24**, 1–7.

Kapan DD (2001) Three-butterfly system provides a field test of Müllerian mimicry. *Nature*, **409**, 338–40.

Kapan DD, Flanagan NS, Tobler A, Papa R, Reed RD, Acevedo Gonzalez J, Ramirez Restrepo M, Martinez L, Maldonado K, Ritschoff C, Heckel DG, McMillan WO (2006) Localization of Müllerian mimicry genes on a dense linkage map of *Heliconius erato*. *Genetics*, **173**, 735–57.

Kardong KV (1980) Gopher snakes and rattlesnakes: presumptive Batesian mimicry. *Northwest Science*, **54**, 1–4.

Karpestam E, Merilaita S, Forsman A (2012) Reduced predation risk for melanistic pygmy grasshoppers in post-fire environments. *Ecology and Evolution*, **2**, 2204–12.

Karpestam E, Merilaita S, Forsman A (2014a) Natural levels of colour polymorphism reduce performance of visual predators searching for camouflaged prey. *Biological Journal of the Linnean Society*, **112**, 546–55.

Karpestam E, Merilaita S, Forsman A (2014b) Body size influences differently the detectabilities of colour morphs of cryptic prey. *Biological Journal of the Linnean Society*, **113**, 112–22.

Kasai A, Oshima N (2006) Light-sensitive motile iridophores and visual pigments in the neon tetra, *Paracheirodon innesi*. *Zoological Science*, **23**, 815–9.

Kassarov L (1999) Are birds able to taste and reject butterflies based on 'beak mark tasting'? A different point of view. *Behaviour*, **136**, 965–81.

Kassarov L (2004) A critical response to the paper "Tough African models and weak mimics: new horizons in the evolution of bad taste" by P. DeVries published in this journal, vol. 57(3) 2003. *Journal of the Lepidopterists' Society*, **59**, 169–72.

Kather R, Drijfhout FP, Martin SJ (2015) Evidence for colony-specific differences in chemical mimicry in the parasitic mite *Varroa destructor*. *Chemoecology*, **25**: 215.

Kats LB, van Dragt RG (1986) Background color-matching in the spring peeper, *Hyla crucifer*. *Copeia*, **1986**, 109–15.

Katz LS (2007) Sexual behavior of domesticated ruminants. *Hormones and Behaviour* **52**, 56–63.

Kaufmann JH (1986) Stomping for earthworms by wood turtles, *Clemmys insculpta*: a newly discovered foraging technique. *Copeia*, **1986**, 1001–4.

Kawakita A, Sota T, Ito M, Ascher JS, Tanaka H, Kato M, Roubik DW (2004) Phylogeny, historical biogeography, and character evolution in bumble bees (*Bombus*: Apidae) based on simultaneous analysis of three nuclear gene sequences. *Molecular Phylogenetics and Evolution*, **31**, 799–804.

Kayser H, Angersbach D (1974) Action spectra for light-controlled pupal pigmentation in *Pieris brassicae*: melanisation and level of bile pigment. *Journal of Insect Physiology*, **20**, 2277–85.

Kazemi B, Gamberale-Stille G, Tullberg BS, Leimar O (2014) Stimulus salience as an explanation for imperfect mimicry. *Current Biology*, **24**, 965–9.

Keeble FW, Gamble FW (1900) The colour-physiology of *Hippolyte varians*. *Proceedings of the Royal Society, Series B*, **65**, 461–8.

Keeton WT, Gould JL (1986) *Biological Science*, 4th edn. W.W. Norton & Co., New York.

Keiser ED Jr (1975) Observations on tongue extension of vine snakes (Genus *Oxybelis*) with suggested behavioral hypotheses. *Herpetologica*, **31**, 131–3.

Kelley JL, Fitzpatrick JL, Merilaita S (2013) Spots and stripes: ecology and colour pattern evolution in butterflyfishes. *Proceedings of the Royal Society of London, Series B: Biological Sciences*, **280**: 20122730.

Kelley LA, Healy SD (2012) Vocal mimicry in spotted bowerbirds is associated with an alarming context. *Journal of Avian Biology*, **43**, 525–30.

Kelley LA, Coe RL, Madden JR, Healy SD (2008) Vocal mimicry in songbirds. *Animal Behaviour*, **76**, 521–8.

Kelly C (1987) A model to explore the rate of spread of mimicry and rejection in hypothetical populations of cuckoos and their hosts. *Journal of Theoretical Biology*, **125**, 283–99.

Kelly CA, Norbutus AJ, Lagalante AF, Iyengar VK (2012) Male courtship pheromones as indicators of genetic quality in an arctiid moth (*Utetheisa ornatrix*). *Behavioral Ecology*, **23**, 1009–14.

Kelly DJ, Marples NM (2004) The effects of novel odour and colour cues on food acceptance by the zebra finch, *Taeniopygia guttata*. *Animal Behaviour*, **68**, 1049–54.

Kettlewell HBD (1955a) Recognition of appropriate backgrounds by the pale and black phases of Lepidoptera. *Nature*, **172**, 943–4.

Kettlewell HBD (1955b) How industrialization can alter species. *Discovery*, **16**, 507–11.

Kettlewell HBD (1956) Further selection experiments on industrial melanism in the Lepidoptera. *Heredity*, **10**, 287–301.

Kettlewell HBD (1958) A survey of the frequencies of *Biston betularia* (L.) (Lep.) and its melanic forms in Great Britain. *Heredity*, **12**, 51–72.

Kettlewell HBD (1959) Darwin's missing evidence. *Scientific American*, **200**, 48–53.

Kettlewell HBD (1961) The phenomenon of industrial melanism in Lepidoptera. *Annual Review of Entomology*, **6**, 245–62.

Kettlewell HBD (1973) *The Evolution of Melanism*. Clarendon Press, Oxford.

Kikuchi DW, Pfennig DW (2010) High-model abundance may permit the gradual evolution of Batesian mimicry: an experimental test. *Proceedings of the Royal Society B: Biological Sciences*, **277**, 1041–8.

Kikuchi DW, Pfennig DW (2013) Imperfect mimicry and the limits of natural selection. *The Quarterly Review of Biology*, **88**, 297–315.

Kikuchi DW, Mappes J, Sherratt TN, Valkonen JK (2016) Selection for multicomponent mimicry: equal feature salience and variation in preferred traits. *Behavioral Ecology*, **27**, 1515–1521.

Kilner RM, Langmore NE (2011) Cuckoos versus hosts in insects and birds: adaptations, counter-adaptations and outcomes. *Biological Reviews*, **86**, 836–52.

Kiltie RA (1988) Countershading: universally deceptive or deceptively universal? *Trends in Ecology & Evolution*, **3**, 21–3.

Kiltie RA (1989) Wildfire and the evolution of dorsal melanism in fox squirrels, *Sciurus niger*. *Journal of Mammalogy*, **70**, 726–39.

King BH, Leaich HR (2006) Variation in propensity to exhibit thanatosis in *Nasonia vitripennis* (Hymenoptera: Pteromalidae). *Journal of Insect Behavior*, **19**, 241–9.

Kingsland S (1978) Abbott Thayer and the protective coloration debate. *Journal of the History of Biology*, **11**, 223–44.

Kirby WE, Spence W (1823) *An Introduction to Entomology*, Vol. 2. Longman, Hurst, Rees, Orme & Brown, London.

Kiritani K, Yamashita H, Yamamura K (2013) Beak marks on butterfly wings with special reference to Japanese black swallowtail. *Population Ecology*, **55**, 451–9.

Kirkpatrick TW (1957) *Insect Life in the Tropics*. Longmans, Green. New York, NY.

Kirvan CA, Swedo SE, Heuser JS, Cunningham MW (2003) Mimicry and autoantibody-mediated neuronal cell signaling in Sydenham chorea. *Nature Medicine*, **9**, 914–20.

Kistner DH (1979) Social and evolutionary significance of social insect symbionts. In HR Hermann (ed.), *Social Insects*. Academic Press, New York, pp. 339–413.

Kistner DH, Blum MS (1971) Alarm pheromone of *Lasius* (*Dendrolasius*) *spathebus* (Hymenoptera: Formicidae) and its possible mimicry by two species of *Pella* (Coleoptera: Staphylinidae). *Annals of the Entomological Society of America*, **64**, 589–94.

Kitamura T, Imafuku I (2010) Behavioral Batesian mimicry involving intraspecific polymorphism in the butterfly *Papilio polytes*. *Zoological Science*, **27**, 217–21.

Kite GC (1995) The floral odour of *Arum maculatum*. *Biochemical Systematics and Ecology*, **23**, 343–54.

Kite GC, Hetterscheide WLA (1997) Inflorescence odours of *Amorphophallus* and *Pseudodracontium* (Araceae). *Phytochemistry*, **46**, 71–5.

Kite GC, Hetterscheide WLA, Lewis MJ, Boyce PC, Ollerton J, Cocklin E, Diaz A, Simmonds MSJ (1998) Inflorescence odours and pollinators of *Arum* and *Amorphophallus* (Araceae). In SJ Owens and PJ Rudall (eds), *Reproductive Biology*. Royal Botanic Gardens, Kew, pp. 295–315.

Kjellsson G, Rasmussen FN, Dupuy D (1985) Pollination of *Dendrobium infundibulum*, *Cymbidium insigne* (Orchidaceae) and *Rhododendron lyi* (Ericaceae) by *Bombus eximius* (Apidae) in Thailand: a possible case of floral mimicry. *Journal of Tropical Ecology*, **1**, 289–302.

Kjernsmo K, Merilaita S (2013) Eyespots divert attacks by fish. *Proceedings of the Royal Society of London, Series B: Biological Sciences*, **280**. 20131458.

Klein NK, Payne RB (1998) Evolutionary associations of parasitic finches (*Vidua*) and their host species: analyses of mitochondrial DNA. *Evolution*, **52**, 566–82.

Kleinholz LH, Welsh JH (1937) Colour Changes in *Hippolyte varians*. *Nature*, **140**, 851–2.

Klitzke CF, Brown KS Jr (2000) The occurrence of aristolochic acids in neotropical troidine swallowtails (Lepidoptera: Papilionidae). *Chemoecology*, **10**, 99–102.

Kloock CT (2012) Natural history of the pirate spider *Mimetus hesperus* (Araneae; Mimetidae) in Kern County, California. *The Southwestern Naturalist*, **57**, 417–20.

Kloft W (1959) Versuch einer Analyse der trophobiotischen Beziehungen von Ameisen zu Aphiden. *Biologisches Zentralblatt*, **78**, 863–70.

Knapp RA, Sargent RC (1989) Egg-mimicry as a mating strategy in the fantail darter, *Etheostoma flabellare*: females prefer males with eggs. *Behavioral Ecology and Sociobiology*, **25**, 321–6.

Knill R, Allen JA (1995) Does polymorphism protect? An experiment with human "predators". *Ethology*, **99**, 127–38.

Knipper H (1953) Beobachtungen an jungen *Plotosus anguillaris* (Bloch). *Veröffentlichungen* aus dem Übersee-Museum Bremen, **2**, 141–8.

Knipper H (1955) Ein Fall von "kollektiver Mimikry"? Schreckreaktionen eines Jungfischschwarmes. *Die Umschau in Wissenschaft und Technik, Frankfurt am Main*, **13**, 398–400.

Kobayashi TM, Watanabe M (1981) Snake-scent application behaviour in the Siberian chipmunk *Eutomias sibiricus asiaticus*. *Proceedings of the Japanese Academy*, **57B**, 141–5.

Kodandaramaiah U (2011) The evolutionary significance of butterfly eyespots. *Behavioral Ecology*, **22**, 1264–71.

Kodandaramaiah U, Vallin A, Wiklund C (2009) Fixed eyespot display in a butterfly thwarts attacking birds. *Animal Behaviour*, **77**, 1415–9.

Koeniger N, Koeniger G, Mardan M (1994) Mimicking a honeybee queen – *Vespa affinis indosinensis* Perez 1910 hunts drones of *Apis cerana* F 1793. *Ethology*, **98**, 149–53.

Kokko H, Mappes J, Lindström L (2003) Alternative prey can change model-mimic dynamics between parasitism and mutualism. *Ecology Letters*, **6**, 1068–76.

Koller G (1927) Über Chromatophorensystem, Farbensinn und Farbwechsel bei *Crangon vulgaris*. *Zeitschrift für vergleichende Physiologie*, **5**, 191–246.

Kolm N, Amcoff M, Mann RP, Arnqvist G (2012) Diversification of a food-mimicking male ornament via sensory drive. *Current Biology*, **22**, 1440–3.

Kolyer JM (1968) Note on damaged specimens. *Journal of Research on the Lepidoptera*, **7**, 105–11.

Komárek S (2003) *Mimicry, Aposematism, and Related Phenomena: Mimetism in Nature and the History of its Study*. Lincom Europa, München. 167pp. (originally published 1998 as: *Mimicry, Aposematism and Related Phenomena in Animals and Plants, Bibliography 1800–1990*. Vesmír, Prague. 269pp.).

Koptur S (1989) Mimicry of flowers by parasitic wasp pupae. *Biotropica*, **21**, 93–5.

Kozak KM, Wahlberg N, Neild A, Dasmahapatra KK, Mallet J, Jiggins CD (2015) Multilocus species trees show the recent adaptive radiation of the mimetic *Heliconius* butterflies. *Systematic Biology*, **64**, 505–24.

Kraft GT (1972) Preliminary studies of Philippine *Eucheuma* species (Rhodophyta). Part I. Taxonomy and ecology of *Eucheuma arnoldii* Weber-van Bosse. *Pacific Science*, **26**, 318–34.

Krajewski JP, Bonaldo RM, Sazima C, Sazima I (2004) The association of the goatfish *Mulloidichthys martinicus* with the grunt *Haemulon chrysargyreum*: an example of protective mimicry. *Biota Neotropica*, **4**, 1–4.

Krebs JR (1977) Significance of song repertoires: Beau Geste hypothesis. *Animal Behaviour*, **25**, 475–8.

Krebs RA, West DA (1988) Female mate preference and the evolution of female-limited Batesian mimicry. *Evolution*, **42**, 1101–4.

Kreuter K, Bunk E, Lückemeyer A, Twele R, Francke W, Ayasse M (2011) How the social parasitic bumblebee *Bombus bohemicus* sneaks into power of reproduction. *Behavioral Ecology & Sociobiology*, **66**, 475–86.

Kreuzwieser J, Scheerer U, Kruse J, Burzlaff T, Honsel A, Alfarraj S, Georgiev P, Schnitzler J-P, Ghirardo A, Kreuzer I, Hedrich R, Rennenberg H (2014) The Venus flytrap attracts insects by the release of volatile organic compounds. *Journal of Experimental Botany*, **65**, 755–66.

Kropf M, Renner SS (2005) Pollination success in monochromic yellow populations of the rewardless orchid *Dactylorhiza sambucina*. *Plant Systematics and Evolution*, **254**, 185–97.

Kugler H (1956) Über die optische Wirkung von Fliegenblumen auf Fliegen. *Berichte der Deutschen Botanischen Gesellschaft*, **69**, 387–98.

Kunin WE (1993) Sex and the single mustard: population density and pollinator behavior effects on seed-set. *Ecology*, **74**, 2145–60.

Kunte K (2008) Mimetic butterflies support Wallace's model of sexual, dimorphism. *Proceedings of the Royal Society of London, Series B: Biological Sciences*, **275**, 1617–24.

Kunte K (2009a) Female-limited mimetic polymorphism: a review of theories and a critique of sexual selection as balancing selection. *Animal Behaviour*, **78**, 1029–36.

Kunte K (2009b) The diversity and evolution of Batesian mimicry in *Papilio* swallowtail butterflies. *Evolution*, **63**, 2707–16.

Kunte K, Zhang W, Tenger-Trolander A, Palmer DH, Martin A, Reed RD, Mullen S P, Kronforst MR (2014) *doublesex* is a mimicry supergene. *Nature*, **507**, 229–32.

Kunze A, Witte L, Aregullin M, Rodriguez E, Proksch P (1996) Anthraquinones in the leaf beetle *Trirhabda geminata* (Chrysomelidae). *Zeitschrift für Naturforschung*, **51c**, 249–52.

Kunze J, Gumbert A (2001) The combined effect of color and odor on flower choice behavior of bumble bees in flower mimicry systems. *Behavioral Ecology*, **12**, 447–56.

Kuriwada T, Kumano N, Shiromoto K, Haraguchi D (2009) Copulation reduces the duration of death-feigning behaviour in the sweetpotato weevil, *Cylas formicarius*. *Animal Behaviour*, **78**, 1145–51.

Kyalangalilwa B, Boatwright JS, Daru BH, Maurin O, van der Bank M (2013). Phylogenetic position and revised classification of *Acacia* s.l. (Fabaceae: Mimosoideae) in Africa, including new combinations in *Vachellia* and *Senegalia*. *Botanical Journal of the Linnean Society*, **172**, 500–23.

Lack D (1954) *The Natural Regulation of Animal Numbers*. Clarendon Press, Oxford.

Lack D (1968) *Ecological Adaptations for Breeding in Birds*. Methuen, London.

Lamberts SWJ, Macleod RM (1990) Regulation of prolactin secretion at the level of the lactotroph. *Physiological Reviews*, **70**, 291–318.

Lamborn WA (1920) The attacks of birds on butterflies witnessed in Nyasaland by W. A. Lamborn. The marks of a bird's beak recognizable on rejected wings. *Proceedings of the Entomological Society of London*, **1920**, xxiv–xxix.

Landeau L, Terborgh J (1986) Oddity and the 'confusion effect' in predation. *Animal Behaviour*, **34**, 1372–80.

Lane C, Rothschild M (1965) A case of Müllerian mimicry of sound. *Proceedings of the Royal Entomological Society of London, Series A*, **40**, 156–8 + 10 Figs.

Langham GM (2004) Specialized avian predators repeatedly attack novel color morphs of *Heliconius* butterflies. *Evolution*, **58**, 2783–7.

Langham GM (2006) Rufous-tailed jacamars and aposematic butterflies: do older birds attack novel prey? *Behavioral Ecology*, **17**, 285–90.

Langmore NE (2013) Australian cuckoos and their adaptations for brood parasitism. *Chinese Birds*, **4**, 86–92.

Langmore NE, Kilner RM (2009) Why do Horsfield's bronze-cuckoo *Chalcites basalis* eggs mimic those of their hosts? *Behavioral Ecology & Sociobiology*, **63**, 1127–31.

Langmore NE, Kilner RM (2010) The coevolutionary arms race between Horsfield's bronze-cuckoos and superb fairy-wrens. *Emu*, **110**, 32–8.

Langmore NE, Hunt S, Kilner RM (2003) Escalation of a coevolutionary arms race through host rejection of brood parasitic young. *Nature*, **422**, 157–60.

Langmore NE, Stevens M, Maurer G, Heinsohn R, Hall ML, Peters A, Kilner RM (2010) Visual mimicry of host nestlings by cuckoos. *Proceedings of the Royal Society of London, Series B: Biological Sciences*, **278**, 2455–63.

Lanteri AA, Del Río MG (2005) Taxonomy of the monotypic genus *Trichaptus* Pascoe (Coleoptera: Curculionidae: Entiminae), a potential weevil mimic of Mutillidae. *Coleopterist's Bulletin*, **59**, 47–54.

Lariviere S, Messier F (1996) Aposematic behaviour in the striped skunk, *Mephitis mephitis. Ethology*, **102**, 986–92.

Larsen TB (1983) On the palatability of butterflies. *Entomologist's Record*, **95**, 66–7.

Larsen TB (1991) The art of feigning death – thanatosis in *Euploea* (Danainae) and other aposematic butterflies. *Entomologist's Record*, **103**, 263–6.

Latimer J (2001) *Deception in War*. John Murray, London.

Laufer H, Ahl JSB (1995) Mating-behavior and methyl farnesoate levels in male morphotypes of the spider crab, *Libinia emarginata* (Leach). Journal of *Experimental Marine Biology and Ecology*, **193**, 15–20.

Launchbaugh KL, Provenza FD (1993) Can plants practice mimicry to avoid grazing by mammalian herbivores? *Oikos*, **66**, 501–4.

Laverty TM (1992) Plant interactions for pollinator visits: a test of the magnet species effect. *Oecologia*, **89**, 502–8.

Lawrence ES (1985a) Evidence for search image in blackbirds *Turdus merula* L.: short-term learning. *Animal Behaviour*, **33**, 929–37.

Lawrence ES (1985b) Evidence for search image in blackbirds *Turdus merula* L.: long-term learning. *Animal Behaviour*, **33**, 1301–9.

Lawrence ES (1986) Can great tits (*Parus major*) acquire search images? *Oikos*, **47**, 3–12.

Lawrence ES (1989) Why blackbirds overlook cryptic prey: search rate or search image? *Animal Behaviour*, **37**, 157–60.

Lawrence ES, Allen JA (1983) On the term 'search image'. *Oikos*, **40**, 313–14.

Lea RG, Turner JRG (1972) Experiments on mimicry. II. The effect of a Batesian mimic on its model. *Behaviour*, **42**, 131–51.

Leal M, Thomas R (1994) Notes on the feeding behavior and caudal luring by juvenile *Alsophis portoricensis* (Serpentes: Colubridae). *Journal of Herpetology*, **28**, 126–8.

Lederhouse RC (1990) Avoiding the hunt: primary defences of lepidopteran caterpillars. In DL Evans and JO Schmidt (eds), *Insect Defences: Adaptive Mechanisms and Strategies of Prey and Predators*. State University of New York Press, Albany, NY, pp. 175–89.

Lee TL, Speed MP (2010) The effect of metapopulation dynamics on the survival and spread of a novel, conspicuous prey. *Journal of Theoretical Biology*, **267**, 319–29.

Lees DR (1975) Resting site selection in the geometrid moth *Phigulia pilosaria* (Lepidoptera: Geometridae). *Journal of Zoology, London*, **176**, 341–2.

Lehtonen TK, Meyer A (2011) Heritability and adaptive significance of the number of egg-dummies in the cichlid fish *Astatotilapia burtoni. Proceedings of the Royal Society of London, Series B: Biological Sciences*, **278**, 2318–24.

Lehtonen J, Whitehead MR (2014) Sexual deception: coevolution or inescapable exploitation? *Current Zoology*, **60**, 52–61.

Leimar O, Tuomi J (1998) Synergistic selection and graded traits. *Evolutionary Ecology*, **12**, 59–71.

Leimar O, Enquist M, Sillén-Tullberg B (1986) Evolutionary stability of aposematic coloration and prey unprofitability: a theoretical analysis. *The American Naturalist*, **128**, 469–90.

Leimar O, Tullberg BS, Mallet J (2012) Mimicry, saltational evolution, and the crossing of fitness valleys. In EI Svensson and R Calsbeek (eds), *The Adaptive Landscape in Evolutionary Biology*. Oxford University Press, Oxford, pp. 259–70.

Lenko K (1964) Sôbre o mimetismo do cerambicideo *Pertiya sericea* (Perty, 1830) com *Camponotus sericeiventris* (Guérin, 1830). *Papéis Avulsos do Departamento de Zoologia, São Paulo*, **16**, 89–95.

Lenoir A, Malosse C, Yamaoka R (1997) Chemical mimicry between parasitic ants of the genus *Formicoxenus* and their host *Myrmica* (Hymenoptera, Formicidae). *Biochemical Systematics and Ecology*, **25**, 379–89.

Lenoir A, D'Ettorre P, Errard C, Hefetz A (2001) Chemical ecology and social parasitism in ants. *Annual Review of Entomology*, **46**, 573–99.

Leong TM, D'Rozario V (2012) Mimicry of the weaver ant *Oecophylla smaragdina* by the moth caterpillar, *Homodes bracteigutta*, the crab spider, *Amyciaea lineatipes*, and the jumping spider, *Myrmarachne plataleoides. Nature in Singapore*, **5**, 39–56.

Le Poul Y, Whibley A, Chouteau M, Prunier F, Llaurens V, Joron M (2014) Evolution of dominance mechanisms at a butterfly mimicry supergene. *Nature Communications*, **5**: 5644.

Leroy J-M, Jocque R, Leroy A (1998) On the behaviour of the African bolas-spider *Cladomelea akermani* Hewitt (Araneae, Araneidae, Cyrtarachninae), with description of the male. *Annals of the Natal Museum*, **39**, 1–9.

Leroy Y (1974) Le mimetisme animal. *La Recherche*, **45**, 417–25.

Levin DA (1972) Low frequency disadvantage in the exploitation of pollinators by corolla variants in *Phlox. The American Naturalist*, **106**, 453–60.

Levin DA (1976) Alkaloid-bearing plants: an ecogeographic perspective. *The American Naturalist*, **110**, 261–84.

Lev-Yadun S (2001) Aposematic (warning) coloration associated with thorns in higher plants. *Journal of Theoretical Biology*, **210**, 385–8.

Lev-Yadun S (2003a) Why do some thorny plants resemble green zebras? *Journal of Theoretical Biology*, **244**, 483–9.

Lev-Yadun S (2003b) Weapon (thorn) automimicry and mimicry of aposematic colorful thorns in plants. *Journal of Theoretical Biology*, **244**, 183–8.

Lev-Yadun S (2006) Defensive functions of white coloration in coastal and dune plants. *Israel Journal of Plant Sciences*, **54**, 317–25.

Lev-Yadun S (2009a) Müllerian and Batesian mimicry rings of white-variegated aposematic spiny and thorny plants: a hypothesis. *Israel Journal of Plant Sciences*, **57**, 107–16.

Lev-Yadun S (2009b) Müllerian mimicry in aposematic spiny plants. *Plant Signaling & Behavior*, **4**, 482–3.

Lev-Yadun S (2013) Theoretical and functional complexity of white variegation of unripe fleshy fruits. *Plant Signaling & Behavior*, **8**: e25851.

Lev-Yadun S (2014a) Potential defence from herbivory by 'dazzle effects' and 'trickery coloration' of leaf variegation. *Biological Journal of the Linnean Society*, **111**, 692–7.

Lev-Yadun S (2014b) The proposed anti-herbivory roles of white leaf variegation. In U Lüttge and W Beyschlag (eds), *Progress in Botany*, Vol. 76. Springer, Heidelberg, pp. 241–69.

Lev-Yadun S, Gutman M (2013) Carrion odor and cattle grazing: evidence for plant defense by carrion odor. *Communicative & Integrative Biology*, **6**: e26111.

Lev-Yadun S, Halpern M (2007) Ergot (*Claviceps purpurea*) — an aposematic fungus. *Symbiosis Journal*, **43**, 105–8.

Lev-Yadun S, Halpern M (2008) External and internal spines in plants insert pathogenic microorganisms into herbivore's tissues for defense. In T Van Dijk (ed.), *Microbial Ecology Research Trends*. Nova Scientific Publishers, New York, pp. 155–68.

Lev-Yadun S, Inbar M (2002) Defensive ant, aphid and caterpillar mimicry in plants? *Biological Journal of the Linnean Society*, **77**, 393–8.

Lev-Yadun S, Mirsky N (2007) False satiation: the probable antiherbivory strategy of *Hoodia gordonii*. *Functional Plant Science and Biotechnology*, **1**, 56–7.

Lev-Yadun S, Ne'eman G (2004) When may green plants be aposematic? *Biological Journal of the Linnean Society*, **81**, 413–6.

Lev-Yadun S, Ne'eman G (2012) Does bee or wasp mimicry by orchid flowers also deter herbivory? *Arthropod–Plant Interactions*, **6**, 327–32.

Lev-Yadun S, Ne'eman G (2013) Bimodal colour pattern of individual *Pinus halepensis* Mill. seeds: a new type of crypsis. *Biological Journal of the Linnean Society*, **109**, 271–8.

Lev-Yadun S, Ne'eman G, Izhaki I (2009) Unripe red fruits may be aposematic. *Plant Signaling & Behavior*, **4**, 836–41.

Lewis SM, Cratsley CK (2008) Flash signal evolution, mate choice, and predation in fireflies. *Annual Review of Entomology*, **53**, 293–321.

Lewis SM, Kensley B (1982) Notes on the ecology and behaviour of *Pseudamphithoides incurvaria* (Just) (Crustacea, Amphipoda, Ampithoidae). *Journal of Natural History*, **16**, 267–74.

Lhomme P, Ayasse M, Valterová I, Lecocq T, Rasmont P (2012) Born in an alien nest: how do social parasite male offspring escape from host aggression? *PLoS ONE*, **7(9)**: e43053.

Li CC, Moment GB (1962) On "reflexive selection". *Science*, **136**, 1055–6.

Liang W, Møller AP (2014) Hawk mimicry in cuckoos and anti-parasitic aggressive behavior of barn swallows in Denmark and China. *Journal of Avian Biology*, **45**, 1–8.

Lichter-Marck IH, Wylde M, Aaron E, Oliver JC, Singer MS (2014) The struggle for safety: effectiveness of caterpillar defenses against bird predation. *Oikos*, **124**, 535–3.

Liepert C, Dettner K (1996) Role of cuticular hydrocarbons of aphid parasitoids in their relationship to aphid-attending ants. *Journal of Chemical Ecology*, **22**, 695–707.

Lilley GM (1998) A study of the silent flight of the owl. In *Proceedings of the 4th AIAA/CEAS Aeroacoustics Conference*, Toulouse, France. pp. 2004–86.

Lima NRW, Kobak CJ, Vrijenhoek RC (1996) Evolution of sexual mimicry in sperm-dependent all-female forms of *Poeciliopsis* (Actinopterygii: Poeciliidae). *Journal of Evolutionary Biology*, **9**, 185–203.

Limbaugh C (1961) Cleaning symbiosis. *Scientific American*, **205**, 42–9.

Lin Y-S, Yeh T-M, Lin C-F, Wan S-W, Chuang Y-C, Hsu T-K, Liu H-S, Liu C-C, Anderson R, Lei H-Y (2011) Molecular mimicry between virus and host and its implications for dengue disease pathogenesis. *Experimental Biology and Medicine*, **236**, 515–23.

Lindell LE, Forsman A (1996) Sexual dichromatism in snakes: support for the flicker-fusion hypothesis. *Canadian Journal of Zoology*, **74**, 2254–6.

Lindroth CH (1971) Disappearance as a protective factor. A supposed case of Batesian mimicry among beetles (Coleoptera: Carabidae and Chrysomelidae). *Entomologia Scandinavica*, **2**, 41–8.

Lindstedt C, Huttunen H, Kakko M, Mappes J (2011) Disentangling the evolution of weak warning signals: high detection risk and low production costs of chemical defences in gregarious pine sawfly larvae. *Evolutionary Ecology*, **25**, 1029–46.

Lindström L, Alatalo RV, Mappes J (1997) Imperfect Batesian mimicry – the effects of the frequency and the distastefulness of the model. *Proceedings of the Royal Society of London, Series B: Biological Sciences*, **264**, 149–53.

Lindström L, Alatalo RV, Mappes J, Riipi M, Vertainen L (1999a) Can aposematic signals evolve by gradual change? *Nature*, **397**, 249–51.

Lindström L, Alatalo RV, Mappes J (1999b) Reactions of hand reared and wild caught predators towards warningly colored, gregarious and conspicuous prey. *Behavioral Ecology*, **10**, 317–22.

Lindström L, Alatalo RV, Lyytinen A, Mappes J (2001a) Strong antiapostatic selection against novel rare aposematic prey. *Proceedings of the National Academy of Sciences, U.S.A.*, **98**, 9181–4.

Lindström L, Alatalo RV, Lyytinen A, Mappes J (2001b) Predator experience on cryptic prey affects the survival of conspicuous aposematic prey. *Proceedings of the Royal Society of London, Series B: Biological Sciences*, **268**, 357–61.

Lindström L, Rowe C, Guilford T (2001c) Pyrazine odour makes visually conspicuous prey aversive. *Proceedings of the Royal Society of London, Series B: Biological Sciences*, **268**, 159–62.

Lindström L, Alatalo RV, Lyytinen A, Mappes J (2004) The effects of alternative prey on the dynamics of imperfect Batesian and Müllerian mimicries. *Evolution*, **58**, 1294–1302.

Lindström L, Lyytinen A, Mappes J, Ojala K (2006) Relative importance of taste and visual appearance for predator education in Müllerian mimicry. *Animal Behaviour*, **72**, 323–33.

Linsley EG (1959) Mimetic form and coloration in the Cerambycidae (Coleoptera). *Annals of the Entomological Society of America*, **52**, 125–31.

Linsley EG (1960) Ethology of some bee-and wasp-killing robber-flies of southeastern Arizona and New Mexico. *University of California Publications in Entomology*, **16**, 357–92.

Linsley EG, Eisner T, Klots AB (1961) Mimetic assemblages of sibling species of lycid beetles. *Evolution*, **15**, 15–29.

Liu M-H, Blamires SJ, Liao C-P, Tso I-M (2014) Evidence of bird dropping masquerading by a spider to avoid predators. *Scientific Reports*, **4**: 5058.

Llaurens V, Billiard S, Joron M (2013) The effect of dominance on polymorphism in Müllerian mimic species. *Journal of Theoretical Biology*, **337**, 101–10.

Llaurens V, Joron M, Théry M (2014) Cryptic differences in colour among Müllerian mimics: how can the visual capacities of predators and prey shape the evolution of wing colours? *Journal of Evolutionary Biology*, **27**, 531–40.

Lloyd FE (1942) *The Carnivorous Plants*. The Chronica Botanica Company, Waltham, MA.

Lloyd H, Schmidt N, Hefetz A (1986) Chemistry of the anal glands of *Bothriomyrmex syrius* Forel. Olfactory mimetism and temporal social parasitism. *Comparative Biochemistry and Physiology*, **83B**, 71–3.

Lloyd JE (1975) Aggressive mimicry in *Photuris* fireflies: signal repertoires by femmes fatales. *Science*, **187**, 452–3.

Lloyd JE (1980) Male *Photuris* fireflies mimic sexual signals of their females' prey. *Science*, **210**, 669–71.

Lloyd JE (1984) Occurrence of aggressive mimicry in fireflies. *Florida Entomologist*, **67**, 368–76.

Lloyd JE, Ballantine LA (2003) Taxonomy and behavior of *Photuris trivittata* sp. n. (Coleoptera: Lampyridae: Photurinae); redescription of *Aspisoma trilineata* (Say) comb. n. (Coleoptera\Lampyridae: Lampyrinae\Cratomorphini). *Florida Entomologist*, **86**, 464–73.

Lloyd Morgan C (1896) *Habit and Instinct*. Arnold, London.

LoBue V, DeLoache JS (2008) Detecting the snake in the grass. Attention to fear-relevant stimuli by adults and young children. *Psychological Science*, **19**, 284–9.

Lohman DL, Liao Q, Pierce NE (2006) Convergence of chemical mimicry in a guild of aphid predators. *Ecological Entomology*, **31**, 41–51.

Lomáscolo SB, Schaefer HM (2010) Signal convergence in fruits: a result of selection by frugivores? *Journal of Evolutionary Biology*, **23**, 614–24.

Londoño GA, García DA, Martínez MAS (2015) Morphological and behavioral evidence of Batesian mimicry in nestlings of a lowland Amazonian bird. *The American Naturalist*, **185**, 135–41.

Long EC, Hahn TP, Shapiro AM (2014) Variation in wing pattern and palatability in a female-limited polymorphic mimicry system. *Ecology and Evolution*, **4**, 4543–52.

Longman HA (1922) The magnificent spider: *Dicrostichus magnificus* Rainbow. Notes on cocoon spinning and method of catching prey. *Proceedings of the Royal Society of Queensland*, **33**, 91–8.

Lorimer N (1979) The genetics of melanism in *Malacosoma disstria* Hübner (Lepidoptera: Lasiocampidae). *Genetics*, **92**, 555–61.

Losey GS (1971) Communication between fishes in cleaning symbiosis. In TC Cheng (ed.), *Aspects of the Biology of Symbiosis*. University Park Press, Baltimore, MA, pp. 145–176.

Losey GS (1972) Predation protection in the poison-fang blenny, *Meiacanthus atrodorsalis*, and its mimics *Ecsenius bicolor* and *Runula laudantus* (Blenniidae). *Pacific Science*, **26**, 129–39.

Lotem A, Nakamura H, Zahavi A (1995) Constraints on egg discrimination and cuckoo–host co-evolution. *Animal Behaviour*, **49**, 1185–209.

Lovell PG, Ruxton GD, Langridge KV, Spencer KA (2013) Egg-laying substrate selection for optimal camouflage by quail. *Current Biology*, **23**, 260–4.

Lubbock J (1905) *The Beauties of Nature and the Wonders of the World We Live in*. Macmillan, New York.

Lucanus O (1998) Darwin's pond: Malawi and Tanganyika. *Tropical Fish Hobbyist*, **47**, 150–4.

Ludin P, Nilsson D, Mäser P (2011) Genome-wide identification of molecular mimicry candidates in parasites. *PLoS ONE*, **6**: e17546.

Luedeman JK, McMorris FR, Warner DD (1981) Predators encountering a model–mimic system with alternative prey. *The American Naturalist*, **117**, 1040–8.

Lumsden RD (1975) Surface ultrastructure and cytochemistry of parasitic helminthes. *Experimental Parasitology*, **37**, 267–339.

Luo C, Wei C (2015) Intraspecific sexual mimicry for finding females in a cicada: males produce 'female sounds' to gain reproductive benefit. *Animal Behaviour*, **102**, 69–76.

Lynn SK (2005) Learning to avoid aposematic prey. *Animal Behaviour*, **70**, 1221–6.

Lynn SK, Cnaani J, Papaj DR (2005) Peak shift discrimination learning as a mechanism of signal evolution. *Evolution*, **59**, 1300–5.

Lyon BE, Montgomerie RD (1985) Conspicuous plumage of birds: sexual selection or unprofitable prey? *Animal Behaviour*, **33**, 1038–40.

Lyon BE, Montgomerie RD (1986) Delayed plumage maturation in passerine birds: reliable signaling by subordinate males? *Evolution*, **40**, 605–15.

Lythgoe JN, Shand J (1982) Changes in spectral reflexions from the iridophores of the neon tetra. *Journal of Physiology*, **325**, 23–34.

Lythgoe JN, Shand J, Foster RG (1984) Visual pigment in fish iridocytes. *Nature*, **308**, 83–4.

Lyytinen A, Alatalo RV, Lindström L, Mappes J (1999) Are European white butterflies aposematic? *Evolutionary Ecology*, **13**, 709–19.

Lyytinen A, Alatalo RV, Lindström L, Mappes J (2001) Can ultraviolet cues function as aposematic signals? *Behavioral Ecology*, **12**, 65–70.

Lyytinen A, Brakefield PM, Mappes J (2003) Significance of butterfly eyespots as an anti-predator device in ground-based and aerial attacks. *Oikos*, **100**, 373–9.

Lyytinen A, Brakefield PM, Lindström L, Mappes J (2004) Does predation maintain eyespot plasticity in *Bicyclus anynana*? *Proceedings of the Royal Society of London, Series B: Biological Sciences*, **271**, 279–83.

Maan ME, Cummings ME (2012) Poison frog colors are honest signals of toxicity, particularly for bird predators. *The American Naturalist*, **179**, E1–E14.

MacDougal JM (2003) *Passiflora boenderi* (Passifloraceae): a new egg mimic passionflower from Costa Rica. *Novon*, **13**, 454–8.

MacDougall A, Dawkins MS (1998) Predator discrimination error and the benefits of Müllerian mimicry. *Animal Behaviour*, **55**, 1281–8.

Macías-Garcia C, Valero A (2001) Context-dependent sexual mimicry in the viviparous fish *Girardinichthys multiradiatus*. *Ethology, Ecology and Evolution*, **13**, 331–9.

Mackie GO (1995) Defensive strategies in planktonic coelenterates. *Marine and Freshwater Behaviour and Physiology*, **26**, 119–29.

Maddison WP (1990) A method for testing the correlated evolution of two binary characters: are gains or losses concentrated on certain branches of a phylogenetic tree? *Evolution*, **44**, 539–57.

Magnus DBE (1967) Zur Deutung der Igelstellung beim Jungfischschwarm des Korallenwelses (*Plotosus anguillaris*), (Pisces, Nematognathi, Plotosidae), im Biotop. *Verhandlungen der Deutschen Zoologischen Gesellschaft*, **35**, 402–9.

Magnus DBE (1963) Sex-limited mimicry. II. Visual selection in the mate choice of butterflies. *Proceedings of the XVI International Congress of Zoology, Washington, D.C.*, **4**, 179–83.

Maier CT (1978) Evolution of Batesian mimicry in the Syrphidae (Diptera). *New York Entomological Society*, **86**, 307.

Maisch A, Bückmann D (1987) The control of cuticular melanin and lutein incorporation in the morphological adaptation of a nymphalid pupa, *Inachis io* L. *Journal of Insect Physiology*, **33**, 393–402.

Majerus MEN (1998) *Melanism: Evolution in Action*. Oxford University Press, Oxford.

Malcicka M, Bezemer TM, Visser B, Bloemberg M, Snart CJP, Hardy ICW, Harvey JA (2015) Multi-trait mimicry of ants by a parasitoid wasp. *Scientific Reports*, **5**: 8043.

Malcolm SB (1986) Aposematism in a soft-bodied insect: a case for kin selection. *Behavioral Ecology and Sociobiology*, **18**, 387–93.

Malcolm SB (1990) Mimicry: status of a classical evolutionary paradigm. *Trends in Ecology and Evolution*, **5**, 57–62.

Malcolm SB, Brower LP (1989) Evolutionary and ecological implications of cardenolide sequestration in the monarch butterfly. *Experientia*, **45**, 284–95.

Malcolm SB, Rothschild M (1983) A danaid Müllerian mimic, *Euploea core amymone* (Cramer) lacking cardenolides in the pupal and adult stages. *Biological Journal of the Linnean Society*, **19**, 27–33.

Malcolm SB, Cockrell BJ, Brower LP (1989) Cardenolide fingerprint of monarch butterflies reared on common milkweed, *Asclepias syriaca* L. *Journal of Chemical Ecology*, **5**, 819–53.

Malcolm WM, Hanks JP (1973) Landing site selection and searching behaviour in the micro-lepidopteron *Agonopteryx pulvipennella*. *Animal Behaviour*, **21**, 45–8.

Mallet J (1989) The genetics of warning colour in Peruvian hybrid zones of *Heliconius erato* and *H. melpomene*. *Proceedings of the Royal Society of London, Series B: Biological Sciences*, **236**, 163–85.

Mallet J [1999] (2001) Causes and consequences of a lack of co-evolution in Müllerian mimicry. *Evolutionary Ecology*, **13**, 777–806.

Mallet J (2014) Speciation: frog mimics prefer their own. *Current Biology*, **24**, R1094–6.

Mallet J, Barton NH (1989) Strong natural selection in a warning-color hybrid zone. *Evolution*, **43**, 421–31.

Mallet J, Gilbert LE (1995) Why are there so many mimicry rings? Correlations between habitat, behaviour and mimicry in *Heliconius* butterflies. *Biological Journal of the Linnean Society*, **55**, 159–80.

Mallet J, Joron M (1999) Evolution of diversity in warning color and mimicry: polymorphisms, shifting balance, and speciation. *Annual Review of Ecology and Systematics*, **30**, 201–33.

Mallet J, Singer MC (1987) Individual selection, kin selection, and the shifting balance in the evolution of warning colors: the evidence from butterflies. *Biological Journal of the Linnean Society*, **32**, 337–50.

Mallet J, Barton NH, Lamas G, Santisteban J, Muedas M, Eeley H (1990) Estimates of selection and gene flow from measures of cline width and linkage disequilibrium in *Heliconius* hybrid zones. *Genetics*, **124**, 921–36.

Mallet J, Jiggins CD, McMillan WO (1996) Mimicry meets the mitochondrion. *Current Biology*, **6**, 937–40.

Mänd T, Tammaru T, Mappes J (2007) Size dependent predation risk in cryptic and conspicuous insects. *Evolutionary Ecology*, **21**, 485–98.

Mank JE, Avise JC (2006) Comparative phylogenetic analysis of male alternative reproductive tactics in ray-finned fishes. *Evolution*, **60**, 1311–16.

Manley DG (2008) Defense adaptations in velvet ants (Hymenoptera: Mutillidae) and their possible selective pressures. In AD Austin and M Dowton (eds), *Hymenoptera: Evolution, Biodiversity and Biological Control*. CSIRO, Canberra, pp. 285–9.

Mant JG, Schiestl FP, Peakall R, Weston PH (2002) A phylogenetic study of pollinator conservatism among sexually deceptive orchids. *Evolution*, **56**, 888–98.

Mappes J, Alatalo RV (1997) Batesian mimicry and signal accuracy. *Evolution*, **51**, 2050–3.

Mappes J, Tuomi J, Alatalo RV (1999) Do palatable prey benefit from aposematic neighbors? *Ecoscience*, **6**, 159–62.

Mappes J, Marples NM, Endler JA (2005) The complex business of survival by aposematism. *Trends in Ecology & Evolution*, **20**, 598–603.

Maran T (2005) Mimicry. In *Semiotics Encyclopedia Online*. E.J. Pratt Library, Victoria University. http://www.semioticon.com/seo/M/mimicry.html (accessed 27 March 2015).

Maran T (2007) Semiotic interpretations of biological mimicry. *Semiotica*, **167**, 223–48.

Maran T (2010) Semiotic modeling of mimicry with reference to brood parasitism. *Sign Systems Studies*, **38**, 349–77.

Marchetti K (1992) Costs to host defence and the persistence of parasitic cuckoos. *Proceedings of the Royal Society of London, Series B: Biological Sciences*, **248**, 41–5.

Marchetti K (2000) Egg rejection in a passerine bird: size does matter. *Animal Behaviour*, **59**, 877–83.

Marchetti K, Nakamura H, Gibbs HL (1998) Host race formation in the common cuckoo. *Science*, **282**, 471–2.

Marden JH, Chai P (1991) Aerial predation and butterfly design: how palatability, mimicry, and the need for evasive flight constrain mass allocation. *The American Naturalist*, **138**, 15–36.

Marek PE, Bond JE (2009) A Müllerian mimicry ring in Appalachian millipedes. *Proceedings of the National Academy of Sciences, U.S.A.*, **106**, 9755–60.

Marek PE, Papaj D, Yeager J, Molina S, Moore W (2011) Bioluminescent aposematism in millipedes. *Current Biology*, **21**, R680–1.

Marino P, Raguso R, Goffinet B (2009) The ecology and evolution of fly dispersed dung mosses (Family Splachnaceae): manipulating insect behaviour through odour and visual cues. *Symbiosis*, **47**, 61–76.

Marples NM (1990) The influence of predation on ladybird colour patterns. Unpublished Ph.D. thesis, University of Wales, College Cardiff.

Marples NM (1993) Do wild birds use size to distinguish palatable and unpalatable prey types? *Animal Behaviour*, **46**, 347–54.

Marples NM, Roper TJ (1996) Effects of novel colour and smell on the response of naive chicks towards food and water. *Animal Behaviour*, **51**, 1417–24.

Marples NM, Brakefield PM, Cowie RJ (1989) Differences between the 7-spot and 2-spot ladybird beetles (Coccinellidae) in their toxic effects on a bird predator. *Ecological Entomology*, **14**, 79–84.

Marples NM, Kelly DJ, Thomas RJ (2005) Perspective: the evolution of warning coloration is not paradoxical. *Evolution*, **59**, 933–40.

Marsh NA, Rothschild M (1974) Aposematic and cryptic Lepidoptera tested on the mouse. *Journal of Zoology, London*, **174**, 89–122.

Marsh DC, Clarke CA, Rothschild M, Kellett DN (1977) *Hypolimnas bolina* (L.), a mimic of danaid butterflies, and its model *Euploea core* (Cram.) store cardioactive substances. *Nature*, **268**, 726–8.

Marshall DC, Hill KBR (2009) Versatile aggressive mimicry of cicadas by an Australian predatory katydid. *PLoS ONE*, **4(1):** e4185.

Marshall GAK (1902a) Experimental evidence of terror caused by the squeak of *Acherontia atrops*. *Transactions of the Entomological Society of London*, **1902**, 402.

Marshall GAK (1902b) Insect stridulation as a warning or intimidating character. *Transactions of the Entomological Society of London*, **1902**, 404.

Marshall GAK (1908) On diaposematism, with reference to some of the limitations of the Müllerian hypothesis of mimicry. *Transactions of the Entomological Society of London*, **1908**, 93–142.

Marshall GAK (1909) Birds as a factor in the production of mimetic resemblances among butterflies. *Transactions of the Entomological Society of London*, **1909**, 329–83.

Marshall GAK, Poulton EB (1902) Five years' observations and experiments (1896–1901) on the bionomics of South African insects, chiefly directed to the investigation of mimicry and warning colours. *Transactions of the Entomological Society of London*, **1902**, 287–697.

Marshall KLA, Gluckman T-L (2015) The evolution of pattern camouflage strategies in waterfowl and game birds. *Ecology and Evolution*, **5**, 1981–91.

Marshall NJ (2000) Communication and camouflage with the same bright colours in reef fishes. *Philosophical Transactions of the Royal Society, series B, Biological Sciences*, **355**, 1243–8.

Martin C, Salvy M, Provost E, Bagnères A-G, Roux M, Crauser D, Clement J-L, Le Conte Y (2001) Variations in chemical mimicry by the ectoparasitic mite *Varroa jacobsoni* according to the developmental stage of the host honeybee *Apis mellifera*. *Insect Biochemistry and Molecular Biology*, **31**, 365–79.

Martin DJ (1973) A spectrographic analysis of burrowing owl vocalizations. *The Auk*, **90**, 564–78.

Martin JW (1993) The samurai crab. *Terre*, **31**, 30–4.

Martin SJ, Carruthers JM, Williams PH, Drijfhout FP (2010) Host specific social parasites (*Psithyrus*) indicate chemical recognition system in bumblebees. *Journal of Chemical Ecology*, **36**, 855–63.

Martin SJ, Takahashi J-I, Ono M, Drijfhout FP (2008) Is the social parasite *Vespa dybowskii* using chemical transparency to get her eggs accepted? *Journal of Insect Physiology*, **54**, 700–7.

Martinet L, Allain D, Meunier M (1983) Regulation in pregnant mink (*Mustela vison*) of plasma progesterone and prolactin concentrations and regulation of onset of the spring moult by daylight ratio and melatonin injections. *Canadian Journal of Zoology*, **61**, 1959–1963.

Martinet L, Allain D, Weiner C (1984) Role of prolactin in the photoperiodic control of moulting in the mink (*Mustela vison*). *Journal of Endocrinology*, **103**, 9–15.

Martos F, Cariou M-L, Pailler T, Fournel J, Bytebier B, Johnson SD (2015) Chemical and morphological filters in a specialized floral mimicry system. *New Phytologist*, **207**, 225–34.

Maruyama M, Yek SH, Hashim R, Ito F (2003) A new myrmecophilous species of *Drusilla* (Coleoptera, Staphylinidae, Aleocharinae) from Peninsular Malaysia, a possible Batesian mimic associated with *Crematogaster inflata*. *Japanese Journal of Systematic Entomology*, **9**, 267–75.

Mascia-Lees FE, Relethford JH, Sorger T (1986) Evolutionary perspectives on permanent breast enlargement in human females. *American Anthropologist, New Series*, **88**, 423–8.

Mason JR, Clark L (1996) Grazing repellency of methyl anthranilate to snow geese is enhanced by a visual cue. *Crop Protection*, **15**, 97–100.

Mason P, Rothstein SI (1987) Crypsis versus mimicry and the colour of shiny cowbird eggs. *The American Naturalist*, **130**, 161–7.

Mason RT, Crews D (1985) Female mimicry in garter snakes. *Nature*, **316**, 59–60.

Massei G, Cotterill JV, Coats JC, Bryning G, Cowan DP (2007) Can Batesian mimicry help plants to deter herbivory? *Pest Management Science*, **63**, 559–63.

Masterman AT (1908) On a possible case of mimicry in the common sole. *Journal of the Linnean Society (Zoology)*, **30**, 239–44.

Masters AR (1990) Pyrrolizidine alkaloids in artificial nectar protect adult ithomiine butterflies from a spider predator. *Biotropica*, **22**, 298–304.

Masters WM (1979) Insect disturbance stridulation: its defensive role. *Behavioral Ecology and Sociobiology*, **5**, 187–200.

Matessi C, Cori R (1972) Models of population genetics of Batesian mimicry. *Theoretical Population Biology*, **3**, 41–68.

Mather MH, Roitberg BD (1987) A sheep in wolf's clothing: tephritid flies mimic spider predators. *Science*, **236**, 308–12.

Mathew AP (1934) The life-history of the spider (*Myrmarachne plataleoides*). *Journal of the Bombay Natural History Society*, **37**, 369–74.

Mathew AP (1935) Transformational deceptive resemblance as seen in the life history of a plant bug (*Riptortus pedestris*), and of a mantis (*Evantissa putchra*). *Journal of the Bombay Natural History Society*, **37**, 803–13.

Mäthger LM, Hanlon RT (2007) Malleable skin coloration in cephalopods: selective reflectance, transmission and absorbance of light by chromatophores and iridophores. *Cell and Tissue Research*, **329**, 179–86.

Matsuura K (2006) Termite-egg mimicry by a sclerotium-forming fungus. *Proceedings of the Royal Society of London, Series B: Biological Sciences*, **273**, 1203–9.

Matsuura K, Tanaka C, Nishida T (2000) Symbiosis of a termite and a sclerotium-forming fungus: sclerotia mimic termite eggs. *Ecological Resources*, **15**, 405–14.

Mattiello T, Fiore G, Brown ER, d'Ischia M, Palumbo A (2010) Nitric oxide mediates the glutamate-dependent pathway for neurotransmission in *Sepia officinalis* chromatophore organs. *The Journal of Biological Chemistry*, **285**, 24154–63.

Mawdsley JR (1994) Mimicry in Cleridae (Coleoptera). *Coleopterist's Bulletin*, **48**, 115–25.

May RM, Robinson SK (1985) Population dynamics of avian brood parasitism. *The American Naturalist*, **126**, 475–94.

Maynard Smith J (1964) Group selection and kin selection. *Nature*, **201**, 1145–7.

McAtee WL (1912) The experimental method of testing the efficiency of warning and cryptic coloration in protecting animals from their enemies. *Proceedings of the National Academy of Sciences of Philadelphia*, **64**, 281–364.

McCallum ML, Beharry S, Trauth SE (2008) A complex mimetic relationship between the central newt and Ozark Highlands leech. *Southeastern Naturalist*, **7**, 173–9.

McClintock JB, Swenson DP, Steinberg DK, Michaels AA (1996) Feeding-deterrent properties of common oceanic holoplankton from Bermudian waters. *Limnology and Oceanography*, **41**, 798–801.

McCosker JE (1977) Fright posture of the plesiopid fish *Calloplesiops altivelis*: an example of Batesian mimicry. *Science*, **197** 400–1

McGuire L, Van Gossum H, Beirinckx K, Sherratt TN (2006) An empirical test of signal detection theory as it applies to Batesian mimicry. *Behavioral Processes*, **73**, 299–307.

McIver JD (1987) On the myrmecomorph *Coquillettia insignis* Uhler: arthropod predators as operators in an ant-mimetic system. *Biological Journal of the Linnean Society*, **90**, 133–44.

McIver JD, Stonedahl G (1993) Myrmecomorphy: morphological and behavioural mimicry of ants. *Annual Review of Entomology*, **38**, 351–77.

McKaye KR (1981) Field observation on death feigning: a unique hunting behaviour by the predatory cichlids *Haplochromis livingstoni*, of Lake Malawi. *Environmental Biology of Fishes*, **6**, 361–5.

McKaye KR, Kocher T (1983) Head ramming behaviour by three paedophagous cichlids in Lake Malawi, Africa. *Animal Behaviour*, **31**, 206–10.

McLean IG, Waas JR (1987) Do cuckoo chicks mimic the begging calls of their hosts? *Animal Behaviour*, **35**, 1896–8.

McNamara JC, Taylor HH (1987) Ultrastructural modifications associated with pigment migration in palaemonid shrimp chromatophores (Decapoda, Palaemonidae). *Crustaceana*, **53**, 113–33.

Mebs D, Hansen M, Köhler G, Pogoda W, Kauert G (2010) Myrmecophagy and alkaloid sequestration in amphibians: a study on *Ameerega picta* (Dendrobatidae) and *Elachistocleis* sp. (Microhylidae) frogs. *Salamandra*, **46**, 11–5.

Meinwald J, Meinwald YC, Chalmers AM, Eisner T (1968) Dihydromatricaria acid: acetylenic acid secreted by soldier beetle. *Science*, **160**, 890–2.

Mellencamp K, Hass M, Werne A, Stark R, Hazel WN (2007) Role of larval stemmata in control of pupal color and pupation site preference in swallowtail butterflies *Papilio troilus*, *Papilio polyxenes*, *Eurytides marcellus*, and *Papilio glaucus* (Lepidoptera: Papilionidae). *Annals of the Entomological Society of America*, **100**, 53–8.

Menz MHM, Phillips RD, Dixon KW, Peakall R, Didham RK (2013) Mate-searching behaviour of common and rare wasps and the implications for pollen movement of the sexually deceptive orchids they pollinate. *PLoS ONE*, **8(3):** e59111.

Merilaita S (1998) Crypsis through disruptive coloration in an isopod. *Proceedings of the Royal Society of London Series B-Biological Sciences*, **265**, 1059–64.

Merilaita S (2003) Visual background complexity facilitates the evolution of camouflage. *Evolution*, **57**, 1248–54.

Merilaita S, Dimitrova M (2014) Accuracy of background matching and prey detection: predation by blue tits (*Cyanistes caeruleus*) indicates intense selection for highly matching prey colour pattern. *Functional Ecology*, **28**, 1208–15.

Merilaita S, Tullberg BS (2005) Constrained camouflage facilitates the evolution of conspicuous warning coloration. *Evolution*, **59**, 38–45.

Merilaita S, Tuomi J, Jormalainen V (1999) Optimization of cryptic coloration in heterogeneous habitats. *Biological Journal of the Linnean Society*, **67**, 151–61.

Merilaita S, Lyytinen A, Mappes J (2001) Selection for cryptic coloration in a visually heterogeneous environment. *Proceedings of the Royal Society of London, Series B: Biological Sciences*, **268**, 1925–9.

Merilaita S, Vallin A, Kodandaramaiah U, Dimitrova M, Ruuskanen S, Laaksonen T (2011) Number of eyespots and their intimidating effect on naive predators in the peacock butterfly. *Behavioral Ecology*, **22**, 1326–31.

Merilaita S, Schaefer HM, Dimitrova M (2013) What is camouflage through distractive markings? *Behavioral Ecology*, **24**: e1271–2.

Merritt DJ (2013) Standards of evidence for bioluminescence in cockroaches. *Naturwissenschaften*, **100**, 697–8.

Mertens R (1956) Das Problem der Mimikry bei Korallenschlangen. *Zoologische Jahrbücher. Abteilung für Systematik*, **84**, 541–76.

Mertens R (1960) *The World of Amphibians and Reptiles* (translated by H. W. Parker). George G. Harrap & Co. Ltd, London.

Messenger JB (1977) Evidence that *Octopus* is colour blind. *Journal of Experimental Biology*, **70**, 49–55.

Messenger JB (2001) Cephalopod chromatophores: neurobiology and natural history. *Biological Reviews*, **76**, 473–528.

Messenger JB, Cornwell CJ, Reed CM (1997) Glutamate and serotonin are endogenous in squid chromatophore nerves. *Journal of Experimental Biology*, **200**, 3043–54.

Meve U, Liede S (1994) Floral biology and pollination in stapeliads – new results and a literature review. *Plant Systematics and Evolution*, **192**, 99–116.

Meyer-Rochow VB (1974) Leptocephali and other transparent fish larvae from the south-eastern Atlantic ocean. *Zoologischer Anzeiger*, **192**, 240–51.

Michida K (2011) Combination of local selection pressures drives diversity in aposematic signals. *Evolutionary Ecology*, **25**, 1017–28.

Midgley JJ (2004) Why are spines of African *Acacia* species white? *African Journal of Range & Forage Science*, **21**, 211–2.

Midgley JJ, White JDM, Johnson SD, Bronner GN (2015) Faecal mimicry by seeds ensures dispersal by dung beetles. *Nature Plants*, **1**: 15141.

Miles DH, Kokpol U, Bhattacherya J, Attwood JL, Stone KE, Bryson TA, Wilson C (1976) Structure of sarracenin. An unusual enol diacetal monoterpene from the insectivorous plant *Sarracenia flava*. *Journal of the American Chemical Society*, **98**, 1569–73.

Milewski AV, Young TP, Madden D (1991) Thorns as induced defenses: experimental evidence. *Oecologia (Berlin)*, **86**, 70–5.

Milius S (2006) Why play dead? Rethinking what used to be obvious. *Science News*, **170**:18.

Millot J (1946) Chenille Malgache mimant un groupe de *Belonogaster* (Hymenopteres). *Bulletin de la Société Zoologique de France*, **71**, 197–8.

Milne AA (1926) *Winnie-the-Pooh*. 1st ed. Methuen & Co. Ltd., London.

Minkiewicz R (1907) Analyse experimentale de l'instinct de deguisement chez les Brachyures oxyrhynques. (English translation in *Annual Report of the Smithsonian Institution*, **1909**, 465–485).

Minno MC, Emmel TC (1992) Larval protective coloration in swallowtails from the Florida Keys (Lepidoptera: Papilionidae). *Tropical Lepidoptera*, **3**, 47–9.

Mirza ZA, Vaze VV, Sanap RV (2011) Death feigning behavior in two species of the genus *Lycodon* of Asia (Squamata: Colubridae). *Herpetology Notes*, **4**, 295–7.

Miyatake T (2001) Effects of starvation on death-feigning in adults of *Cylas formicarius* (Coleoptera: Brentidae). *Annals of the Entomological Society of America*, **94**, 612–6.

Miyatake M, Katayamay K, Takeda Y, Nakashima A, Sugita A, Mizumoto M (2004) Is death-feigning adaptive? Heritable variation in fitness difference of death-feigning behaviour. *Proceedings of the Royal Society of London, Series B: Biological Sciences*, **271**, 2293–6.

Mizuno T, Yamaguchi S, Yamamoto I, Yamaoka R, Akino T (2014) "Double-Trick" visual and chemical mimicry by the juvenile orchid mantis *Hymenopus coronatus* used in predation of the oriental honeybee *Apis cerana*. *Zoological Science*, **31**, 795–801.

Mody NV, Henson R, Hedin PA, Kokpol U, Miles DH (1976) Isolation of the insect paralyzing agent coniine from *Sarracenia flava*. *Experientia*, **32**, 829–30.

Moermond TC (1981) Cooperative feeding, defense of young, and flocking in the black-faced grosbeak. *The Condor*, **83**, 82–3.

Moksnes A, Røskaft E, Braa AT, Korsnes L, Lampe HM, Pedersen HC (1990) Behavioural responses of potential hosts towards artificial cuckoo eggs and dummies. *Behaviour*, **116**, 65–89.

Moksnes A, Røskaft E, Korsnes L (1993) Rejection of cuckoo (*Cuculus canorus*) eggs by meadow pipits (*Anthus pratensis*). *Behavioral Ecology*, **4**, 120–7.

Moland E, Jones GP (2004) Experimental confirmation of aggressive mimicry by a coral reef fish. *Oecologia (Berlin)*, **140**, 676–83.

Moland E, Eagle JV, Jones GP (2005) Ecology and evolution of mimicry in coral reef fishes. *Oceanography and Marine Biology: An Annual Review*, **43**, 455–82.

Molleman F, Whitaker MR, Carey JR (2010) Rating palatability of butterflies by measuring ant feeding behavior. *Entomologische Berichten*, **70**, 52–62.

Møller AP (1988) False alarm calls as a means of resource usurpation in the great tit *Parus major*. *Ethology*, **79**, 25–30.

Møller AP (1990) Deceptive use of alarm calls by male swallows, *Hirundo rustica*: a new paternity guard. *Behavioral Ecology*, **1**, 1–6.

Møller AP, Christiansen SS, Mousseau TA (2011) Sexual signals, risk of predation and escape behavior. *Behavioral Ecology*, **22**, 800–7.

Møller AP, Stokke BG, Samia DSM (2015) Hawk models, hawk mimics, and antipredator behavior of prey. *Behavioral Ecology*, doi: 10.1093/beheco/arv043.

Moment GB (1962a) Reflexive selection: a possible answer to an old puzzle. *Science*, **136**, 262–3.

Moment GB (1962b) On "reflexive selection". *Science*, **136**, 1056.

Mönnig HO (1922) Über *Leucochloridium macrostomum* (*Leucochloridium paradoxum* Carus). Ein Beitrag zur Histologie der Trematoden. Fischer, Jena.

Montevecchi WA (1976) Field experiments on the adaptive significance of avian eggshell pigmentation. *Behaviour*, **58**, 26–39.

Moody S (1978) Latitude, continental drift, and the percentage of alkaloid-bearing plants in floras. *The American Naturalist*, **112**, 965–8.

Moon HP (1976) *Henry Walter Bates FRS 1825–1892, Explorer, Scientist and Darwinian*. Leicestershire Museums, Leicester, UK.

Moore BP, Brown WV (1978) Precoccinelline and related alkaloids in the Australian soldier beetle, *Chauliognathus pulchellus* (Coleoptera: Cantharidae). *Insect Biochemistry*, **8**, 393–5.

Moore BP, Brown WV (1981) Identification of warning odour components, bitter principles and antifeedants in an aposematic beetle: *Metriorrhynchus rhipidius* (Coleoptera: Lycidae). *Insect Biochemistry*, **11**, 493–9.

Moore BP, Brown WV, Rothschild M (1990) Methylalkylpyrazines in aposematic insects. *Chemoecology*, **1**, 43–51.

Moore BR (1992) Avian movement imitation and a new form of mimicry – tracing the evolution of a complex form of learning. *Behaviour*, **122**, 231–63.

Moore CD, Christopher Hassall C (2016) A bee or not a bee: an experimental test of acoustic mimicry by hoverflies. *Behavioral Ecology*, in press.

Moore J (2002) *Parasites and the behavior of animals*. (Oxford series in ecology and evolution). Oxford University Press, Oxford.

Moore PD (1993) How to get carried away. *Nature*, **361**, 304–5.

Moran AP, Prendergast MM (2001) Molecular mimicry in *Campylobacter jejuni* and *Helicobacter pylori* lipopolysaccharides: contribution of gastrointestinal infections to autoimmunity. *Journal of Autoimmunity*, **16**, 241–56.

Moré M, Cocucci AA, Raguso RA (2013) The importance of oligo-sulfides in the attraction of fly pollinators to the brood-site deceptive species *Jaborosa rotacea* (Solanaceae). *International Journal of Plant Sciences*, **174**, 863–76.

Morgan MJ, Adam A, Mollon JD (1992) Dichromats detect colour-camouflaged objects that are not detected by trichromats. *Proceedings of the Royal Society of London, Series B: Biological Sciences*, **248**, 291–5.

Morgan RA, Brown JS (1996) Using giving-up densities to detect search images. *The American Naturalist*, **148**, 1059–74.

Mori A, D'Ettorre P, Lemoli F (1996) Selective acceptance of the brood of two formicine slave-making ants by host and non-host related species. *Insectes Sociaux*, **43**, 391–400.

Mori A, Visicchio R, Sledge MF, Grasso DA, Le Moli F, Turillazzi S, Spencer S, Jones GR (2000) Behavioural assays testing the appeasement allomone of *Polyergus rufescens* queens during host-colony usurpation. *Ethology, Ecology and Evolution*, **12**, 315–22.

Mori A, Burghardt GM, Savitzky AH, Roberts KA, Hutchinson DA, Goris RC (2012) Nuchal glands: a novel defensive system in snakes. *Chemoecology*, **22**, 187–98.

Morin JG (1986) 'Firefleas' of the sea: luminescent signaling in ostracode crustaceans. *Florida Entomologist*, **69**, 105–21.

Moritz RFA, Kirchner WH, Crewe RM (1991) Chemical camouflage of the death's head hawkmoth (*Acherontia atropos* L.) in honey-bee colonies. *Naturwissenschaften*, **78**, 179–182.

Morrell GM, Turner JRG (1970) Experiments on mimicry. I. The response of wild birds to artificial prey. *Behaviour*, **36**, 116–30.

Morrell R (1954) Eyes that are no eyes. *Malayan Nature Journal*, **9**, 94–7.

Morrell R (1969) Play snake for safety. *Animals (London)*, **12**, 154–5.

Morris D (1930) Homosexuality in the ten-spined stickleback (*Pygosteus pungitius* L.). *Behaviour*, **4**, 233–61.

Morris D (1967) *The Naked Ape: A Zoologist's Study of the Human Animal*. Jonathan Cape, London.

Morris RL, Reader T (2016) Do crab spiders perform Batesian mimicry in hoverflies? *Behavioral Ecology*, **27**, 920–31.

Morton ES (1976) Vocal mimicry in the thick-billed euphonia. *The Wilson Bulletin*, **88**, 485–7.

Moskát C, Hauber ME (2007) Conflict between egg recognition and egg rejection decisions in common cuckoo (*Cuculus canorus*) hosts. *Animal Cognition*, **10**, 377–86.

Moskát C, Székely T, Cuthill IC, Kisbenedek T (2008) Hosts' responses to parasitic eggs: which cues elicit hosts' egg discrimination? *Ethology*, **114**, 186–94.

Moskát C, Takasu F, Muñoz AR, Nakamura H, Bán H, Barta Z (2012) Cuckoo parasitism on two closely-related *Acrocephalus* warblers in distant areas: a case of parallel coevolution? *Chinese Birds*, **3**, 320–9.

Moskát C, Zölei A, Bán M, Elek Z, Tong L, Geltsch N, Hauber ME (2014) How to spot a stranger's egg? A mimicry-specific discordancy effect in the recognition of parasitic eggs. *Ethology*, **120**, 1–11.

Mostler G (1935) Beobachtungen zur Frage der Wespenmimikry. [Studies on the question of wasp mimicry]. *Zeitschrift für Morphologie und Ökologie der Tiere*, **29**, 381–454.

Mouritsen KN, Madsen J (1994) Toxic birds: defence against parasites? *Oikos*, **69**, 357–8.

Moynihan MH (1968) Social mimicry: character convergence versus character displacement. *Evolution*, **22**, 315–31.

Moynihan MH (1981) A coincidence of mimicries and other misleading coincidences. *The American Naturalist*, **117**, 372–8.

Mueller HC (1971a) Oddity and specific search image are more important than conspicuousness in prey selection. *Nature*, **233**, 345–6.

Mueller HC (1971b) Zone-tailed hawk and turkey vulture: mimicry or aerodynamics? *Condor*, **74**, 221–2.

Mueller HC (1977) Prey selection in the American kestrel: experiments with two species of prey. *The American Naturalist*, **111**, 25–9.

Mühlmann H (1934) Im Modellversuch künstlich erzeugte Mimikry und ihre Bedeutung für den "Nachahmer". *Zeitschrift für Morphologie und Ökologie der Tiere*, **28**, 259–96.

Mukherjee R, Kodandaramaiah U (2015) What makes eyespots intimidating – the importance of pairedness. *BMC Evolutionary Biology*, **15**: 34.

Müller F (1878) Über die Vortheile der mimicry bei schmetterlingen. *Zoologischer Anzeiger*, **1**, 54–5.

Müller F (1879) *Ituna* and *Thyridia*; a remarkable case of mimicry in butterflies. *Proceedings of the Entomological Society of London*, **1879**, 20.

Mundy PJ (1973) Vocal mimicry of their hosts by nestlings of the greater spotted and striped crested cuckoo. *Ibis*, **115**, 602–4.

Munn C (1986) Birds that 'cry wolf'. *Nature*, **319**, 143–5.

Murphy JB, Carpenter CC, Gillingham JC (1978) Caudal luring in the green tree python, *Chondropython viridis* (Reptilia, Serpentes, Boidae). *Journal of Herpetology*, **12**, 117–9.

Murphy PM (2001) Viral exploitation and subversion of the immune system through chemokine mimicry. *Nature Immunology*, **2**, 116–22.

Murray BA, Bradshaw SD, Edward DH (1991) Feeding behavior and the occurrence of caudal luring in Burton's pygopodid *Lialis burtonis* (Sauria: Pygopodidae). *Copeia*, **1991**, 509–16.

Muyshondt A, Muyshondt A Jr (1976) Is avian predation so important in keeping down butterfly populations? *Entomologist's Record and Journal of Variation*, **88**, 283–5.

Myczko Ł, Skórka P, Dylewski Ł, Sparks TH, Tryjanowski P (2015) Color mimicry of empty seeds influences the probability of pre-dation by birds. *Ecosphere*, **6**, 1–7.

Nadeau NJ, Jiggins CD (2010) A golden age for evolutionary genetics? Genomic studies of adaptation in natural popula-tions. *Trends in Genetics*, **26**, 484–92.

Nadeau NJ, Whibley A, Jones RT, Davey JW, Dasmahapatra KK, Baxter SW, Quail MA, Joron M, ffrench-Constant RH, Blaxter ML, Mallet J, Jiggins CD (2012) Genomic islands of divergence in hybridizing *Heliconius* butterflies identified by large-scale targeted sequencing. *Philosophical Transactions of the Royal Society, Series B, Biological Sciences*, **367**, 343–53.

Nadeau NJ, Martin SH, Kozak KM, Salazar C, Dasmahapatra KK, Davey JW, Baxter SW, Blaxter ML, Mallet J, Jiggins CD (2013)

Genome-wide patterns of divergence and gene flow across a butterfly radiation. *Molecular Ecology*, **22**, 814–26.

Nahrstedt A (1988) Cyanogenesis and the role of cyanogenic compounds in insects. In Ciba Foundation Symposium 140. *Cyanide Compounds in Biology*. Wiley, Chichester, UK, pp. 131–50.

Nahrstedt A, Davis RH (1983) Occurrence, variation and biosynthesis of the cyanogenic glucosides linamarin and lotaustralin in species of the Heliconiini (Insecta: Lepidoptera). *Comparative Biochemistry and Physiology*, **75B**, 65–73.

Nahrstedt A, Davis RH (1985) Biosynthesis and quantitative relationships of cyanoglucosides, linamarin and lotaustralin, in genera of the Heliconiini (Insecta: Lepidoptera). *Comparative Biochemistry and Physiology*, **82B**, 745–9.

Naisbit RE, Jiggins CD, Mallet J (2003) Mimicry: developmental genes that contribute to speciation. *Evolution and Development*, **5**, 269–80.

Nakata K (2009) To be or not to be conspicuous: the effects of prey availability and predator risk on spider's web decoration building. *Animal Behaviour*, **78**, 1255–60.

Nams VO (1991) Olfactory search images in striped skunks. *Behaviour*, **119**, 267–84.

Napier J, Napier P (1985) *The Natural History of the Primates*. The MIT Press, Cambridge, Massachusetts.

Nardi G, Bologna MA (2000) Cantharidin Attraction In *Pyrochroa* (Coleoptera: Pyrochroidae). *Entomological News*, **111**, 74–5.

Nation JL, Sanford MT, Milne K (1992) Cuticular hydrocarbons from *Varroa jacobsoni*. *Experimental and Applied Acarology*, **16**, 331–44.

Negro JJ (2008) Two aberrant serpent-eagles may be visual mimics of bird-eating raptors. *Ibis*, **150**, 307–14.

Neiland MR, Wilcock C (1998) Fruit set, nectar reward, and rarity in the Orchidaceae. *American Journal of Botany*, **85**, 1657–71.

Neill WT (1960) The caudal lure of various juvenile snakes. *Quarterly Journal of the Florida Academy of Science*, **23**, 173–200.

Nelson JW (1986) Ecological notes on male *Mydas xanthopterus* (Loew) (Diptera: Mydidae) and their interactions with *Hemipepsis ustulata* Dahlbohm (Hymenoptera: Pompilidae). *Pan-Pacific Entomologist*, **62**, 316–22.

Nelson XJ (2011) Evolutionary implications of deception in mimicry and masquerade. *Current Zoology*, **60**, 6–15.

Nelson XJ, Card A (2016) Locomotory mimicry in ant-like spiders. *Behavioral Ecology*, **27**, 700–7.

Nelson XJ, Jackson RR (2006a) Vision-based innate aversion to ants and ant mimics. *Behavioral Ecology*, **17**, 676–81.

Nelson XJ, Jackson RR (2006b) Compound mimicry and trading predators by the males of sexually dimorphic Batesian mimics. *Proceedings of the Royal Society of London, Series B: Biological Sciences*, **273**, 367–72.

Nelson XJ, Jackson RR (2009a) Collective Batesian mimicry of ant groups by aggregating spiders. *Animal Behaviour*, **78**, 123–9.

Nelson XJ, Jackson RR (2009b) Aggressive use of Batesian mimicry by an ant-like jumping spider. *Biology Letters*, **5**, 755–7.

Nelson XJ, Jackson RR, Edwards GB, Barrion AT (2005) Living with the enemy: jumping spiders that mimic weaver ants. *Journal of Arachnology*, **33**, 813–9.

Nelson XJ, Jackson RR, Li D, Barrion AT, Edwards GB (2006) Innate aversion to ants (Hymenoptera: Formicidae) and ant mimics:

experimental findings from mantises (Mantodea). *Biological Journal of the Linnean Society*, **88**, 23–32.

Nembaware V, Seoighe C, Sayed M, Gehring C (2004) A plant natriuretic peptide-like gene in the bacterial pathogen *Xanthomonas axonopodis* may induce hyper-hydration in the plant host: a hypothesis of molecular mimicry. *BMC Evolutionary Biology*, **4**: 10.

Nentwig W (1985a) A mimicry complex between mutillid wasps (Hymenoptera, Mutillidae) and spiders (Araneae). *Studies on Neotropical Fauna and Environment*, **20**, 113–6.

Nentwig W (1985b) A tropical caterpillar that mimics faeces, leaves and a snake (Lepidoptera: Oxytenidae: *Oxytenis naemia*). *Journal of Research on the Lepidoptera*, **24**, 136–41.

Neudecker S (1989) Eye camouflage and false eyespots: chaetodontid responses to predators. *Environmental Biology of Fishes*, **25**, 143–57.

Newman E, Anderson BC, Johnson SD (2012) Flower colour adaptation in a mimetic orchid. *Proceedings of the Royal Society of London, Series B: Biological Sciences*, **279**, 2309–13.

Newman L, Cannon L (2003) *Marine Flatworms. The World of Polyclads*. CSIRO Publishing, Canberra.

Ngugi HK, Scherm H (2004) Pollen mimicry during infection of blueberry flowers by conidia of *Monilinia vaccinii-corymbosi*. *Physiological and Molecular Plant Pathology*, **64**, 113–23.

Ngugi HK, Scherm H (2006) Mimicry in plant-parasitic fungi. *Federation of European Microbiological Societies, Microbiology Letters*, **257**, 171–6.

Nicholson AJ (1927) Presidential address. A new theory of mimicry in insects. *Australian Zoologist*, **5**, 10–24.

Nicholson AJ, Bailey VA (1935) The balance of animal populations. *Proceedings of the Zoological Society of London*, **3**, 551–98.

Nickle DA, Castner JL (1995) Strategies utilized by katydids (Orthoptera: Tettigoniidae) against diurnal predators in rainforests of northeastern Peru. *Journal of Orthoptera Research*, **4**, 75–88.

Nicolai J (1964) Der Brutparasitismus der Viduinae als ethologisches Problem: Prägungsphänomene als Faktoren der Rassen- und Artbildung. *Zeitschrift für Tierpsychologie*, **21**, 129–204.

Nicolaus LK (1987) Conditioned aversions in a guild of egg predators: implications for aposematism and prey defense mimicry. *American Midland Naturalist*, **117**, 405–19.

Niemelä P, Tuomi J (1987) Does the leaf morphology of some plants mimic caterpillar damage? *Oikos*, **50**, 256–7.

Nierenberg L (1972) The mechanism for the maintenance of species integrity in sympatrically occurring equitant oncidiums in the Caribbean. *American Orchid Society Bulletin*, **41**, 873–82.

Nijhout HF (1990) A comprehensive model for colour pattern formation in butterflies. *Proceedings of the Royal Society of London, Series B: Biological Sciences*, **239**, 81–113.

Nijhout HF (1991) *The Development and Evolution of Butterfly Wing Patterns*. Smithsonian Institution Press, Washington D.C.

Nijhout HF (1994) Developmental perspectives on evolution of butterfly mimicry. *BioScience*, **44**, 148–57.

Nijhout HF (2003) Polymorphic mimicry in *Papilio dardanus*: mosaic dominance, big effects and origins. *Evolution & Development*, **5**, 579–92.

Nijhout HF, Wray GA, Gilbert LE (1990) An analysis of the phenotypic effects of certain colour pattern genes in *Heliconius* (Lepidoptera: Nymphalidae). *Biological Journal of the Linnean Society*, **40**, 357–72.

Nilsson LA (1983) Mimesis of bellflower (*Campanula*) by the red helleborine orchid *Cephalanthera rubra*. *Nature*, **305**, 799–800.

Nilsson LA (1984) Anthecology of *Orchis morio* (Orchidaceae) as its outpost in the north. *Nova Acta Regiae Societatis Scientiarum Upsaliensis*, **3**, 167–79.

Nilsson LA (1992) Orchid pollination ecology. *Trends in Ecology & Evolution*, **7**, 255–9.

Nilsson M, Forsman A (2003) Evolution of conspicuous colouration, body size and gregariousness: a comparative analysis of lepidopteran larvae. *Evolutionary Ecology*, **17**, 51–66.

Nishida R (1995) Sequestration of plant secondary compounds by butterflies and moths. *Chemoecology*, **6**, 127–38.

Nishida R (2002) Sequestration of defensive substances from plants by Lepidoptera. *Annual Review of Entomology*, **47**, 57–92.

Nishikawa H, Iijima T, Kajitani R, Yamaguchi J, Ando T, Suzuki Y, Sugano S, Fujiyama A, Kosugi S, Hirakawa H, Tabata S, Ozaki K, Morimoto H, Ihara K, Obara M, Hori H, Itoh T, Fujiwara H (2015) A genetic mechanism for female-limited Batesian mimicry in *Papilio* butterfly. *Nature Genetics*, **47**, 405–9.

Nishino H, Sakai M (1996) Behaviorally significant immobile state of so-called thanatosis in the cricket *Gryllus bimaculatus* DeGeer: its characterization, sensory mechanism and function. *Journal of Comparative Physiology, series A*, **179**, 613–24.

Niu Y, Chen G, Peng D-L, Song B, Yang Y, Li Z-M, Sun H (2014) Grey leaves in an alpine plant: a cryptic colouration to avoid attack? *New Phytologist*, **203**, 953–63.

Noble DG, Davies NB, Hartley IR, McRae SB (1999) The red gape of the nestling cuckoo (*Cuculus canorus*) is not a supernormal stimulus for three common hosts. *Behaviour*, **136**, 759–77.

Nokelainen O, Hegna RH, Reudler JH, Lindstedt C, Mappes J (2012) Trade-off between warning signal efficacy and mating success in the wood tiger moth. *Proceedings of the Royal Society of London, Series B: Biological Sciences*, **279**, 257–65.

Nokelainen O, Valkonen JK, Lindstedt C, Mappes J (2013a) Changes in predator community structure shifts the efficacy of two warning signals in arctiid moths. *Journal of Animal Ecology*, **83**, 598–605.

Nokelainen O, Lindstedt C, Mappes J (2013b) Environment-mediated morph-linked immune and life-history responses in the aposematic wood tiger moth. *Journal of Animal Ecology*, **82**, 653–62.

Nomakuchi S, Masumoto T, Sawada K, Sunahra T, Itakura N, Suzuki N (2001) Possible age-dependent variation in egg-loaded host selectivity of the pierid butterfly, *Anthocharis scolymus* (Lepidoptera: Pieridae): a field observation. *Journal of Insect Behavior*, **14**, 451–8.

Nonacs P (1985) Foraging in a dynamic mimicry complex. *The American Naturalist*, **126**, 165–80.

Noor MAF, Parnell RS, Grant BS (2008) A reversible color polyphenism in American peppered moth (*Biston betularia cognataria*) caterpillars. *PLoS ONE*, **3(9):** e3142.

Norman MD, Hochberg FG (2005) The "mimic octopus" (*Thaumoctopus mimicus* n. gen. et sp.), a new octopus from the tropical Indo-West Pacific (Cephalopoda: Octopodidae). *Molluscan Research*, **25**, 57–70.

Norman MD, Finn J, Tregenza T (1999) Female impersonation as an alternative reproductive strategy in giant cuttlefish. *Proceedings of the Royal Society of London, Series B: Biological Sciences*, **266**, 1347–9.

Norman MD, Finn J, Tregenza T (2001) Dynamic mimicry in an Indo-Malay octopus. *Proceedings of the Royal Society of London, Series B: Biological Sciences*, **268**, 1755–8.

Norn S, Kruse PR (2004) Cardiac glycosides: From ancient history through Withering's foxglove to endogeneous cardiac glycosides. *Dan Medicinhist Arbog*, **2004**, 119–32 (in Danish).

Norris KS, Lowe CH (1964) An analysis of background matching in amphibians and reptiles. *Ecology*, **45**, 565–80.

Norris JN, Bucher KE (1989) *Rhodogorgon*, an anomalous new red algal genus from the Caribbean Sea. *Proceedings of the Biological Society of Washington*, **102**, 1050–66.

Nur U (1970) Evolutionary rates of models and mimics in Batesian mimicry. *The American Naturalist*, **104**, 477–86.

Nylin S, Gamberale-Stille G, Tullberg BS (2001) Ontogeny of defence and adaptive coloration in larvae of the comma butterfly, *Polygonia c-album*. *Journal of the Lepidopterists' Society*, **55**, 69–73.

Oaten A, Pearce CEM, Smyth MEB (1975) Batesian mimicry and signal detection theory. *Bulletin of Mathematical Biology*, **37**, 367–87.

Oberprieler C, Vogt R (2006) The taxonomic position of *Matricaria macrotis* (Compositae-Anthemideae). *Willdenowia*, **36**, 329–38.

O'Brien CA, Brim-Box J (1999) Reproductive biology and juvenile recruitment of the shinyrayed pocketbook, *Lampsilis subangulata* (Bivalvia: Unionidae) in the Gulf Coastal Plain. *American Midland Naturalist*, **142**, 129–40.

O'Day WT (1973) Luminescent silhouetting in stomiatoid fishes. *Contributions to Science*, **246**, 1–8.

Odenaal FJ, Rausher MD, Benrey B, Nunez Farfan J (1987) Predation by *Anolis* lizards on *Battus philenor* raises questions about butterfly mimicry systems. *Journal of the Lepidopterists' Society*, **41**, 141–4.

O'Donald P, Barrett JA (1973) Evolution of dominance in polymorphic Batesian mimicry. *Theoretical Population Biology*, **4**, 173–92.

O'Donald P, Pilecki C (1970) Polymorphic mimicry and natural selection. *Evolution*, **24**, 395–401.

O'Donnell RP (2008) Terrestrial foot-paddling by a glaucus-winged gull. *Western Birds*, **39**, 33–5.

Oelschlägel B, Nuss M, von Tschirnhaus M, Pätzold C, Neinhuis C, Dötterl S, Wanke S (2015) The betrayed thief – the extraordinary strategy of *Aristolochia rotunda* to deceive its pollinators. *New Phytologist*, **206**, 342–51.

O'Hanlon JC, Li D, Norma-Rashid Y (2013) Coloration and morphology of the orchid mantis *Hymenopus coronatus* (Mantodea: Hymenopodidae). *Journal of Orthoptera Research*, **22**, 35–44.

O'Hanlon JC, Holwell GI, Herberstein ME (2014a) Pollinator deception in the orchid mantis. *The American Naturalist*, **183**, 126–32.

O'Hanlon JC, Holwell GI, Herberstein ME (2014b) Predatory pollinator deception: does the orchid mantis resemble a model species? *Acta Zoologica Sinica*, **60**, 90–103.

O'Hanlon JC, Herberstein ME, Holwell GI (2015) Habitat selection in a deceptive predator: maximizing resource availability and signal efficacy. *Behavioral Ecology*, **26**, 194–9.

Ohara Y, Nagasaka K, Ohsaki N (1993) Warning coloration in sawfly *Athalia rosae* larva and concealing coloration in butterfly *Pieris rapae* larva feeding on the same plants evolved through individual selection. *Research in Population Ecology*, **35**, 223–30.

Ohguchi O (1978) Experiments on selection against color oddity of water fleas by 3-spined sticklebacks. *Zeitschrift für Tierpsychologie*, **47**, 254–67.

Ohsaki N (1995) Preferential predation of female butterflies and the evolution of Batesian mimicry. *Nature*, **378**, 173–5.

Ohsaki N (2005) A common mechanism explaining the evolution of female-limited and both-sex Batesian mimicry in butterflies. *Journal of Animal Ecology*, **74**, 728–34.

Okuda N, Ito S, Ewao H (2003) Female mimicry in a freshwater goby *Rhinogobius* sp. OR. *Ichthyology Research*, **50**, 198–200.

Oldstone MBA (1989) Molecular mimicry as a mechanism for the cause and as a probe uncovering etiologic agent(s) of autoimmune disease. *Current Topics in Microbiology and Immunology*, **145**, 127–35.

Olendorf R, Rodd FH, Punzalan D, Houde AE, Hurt C, Reznick DN, Hughes KA (2006) Frequency dependent survival in natural guppy populations. *Nature*, **441**, 633–6.

Oliveira PO, Sazima I (1984) The adaptive bases of ant-mimicry in a neotropical aphantochilid spider (Araneae: Aphantochilidae). *Biological Journal of the Linnean Society*, **22**, 145–55.

Oliveira RF, Carneiro LA, Canario AVM, Grober MS (2001) Effects of androgens on social behavior and morphology of alternative reproductive males of the Azorean rock-pool blenny. *Hormones and Behavior*, **39**, 157–66.

Oliver M (1996) Death-feigning observed in *Hippopsis lemniscata* (Fabricius) (Coleoptera; Cerambycidae). *Coleopterist's Bulletin*, **50**, 160–1.

Olofsson M, Vallin A, Jakobsson S, Wiklund C (2010) Marginal eyespots on butterfly wings deflect bird attacks under low light intensities with UV wavelengths. *PLoS ONE*, **5**: e10798.

Olofsson O, Eriksson S, Jakobsson S, Wiklund C (2012) Deimatic display in the European swallowtail butterfly as a secondary defence against attacks from great tits. *PLoS ONE*, **7(10):** e47092.

Olofsson M, Dimitrova M, Wiklund C (2013) The white 'comma' as a distractive mark on the wings of comma butterflies. *Animal Behaviour*, **86**, 1325–31.

Oniki Y (1979) Nest-egg combinations: possible antipredatory adaptations in Amazonian birds. *Revista Brasiliana*, **39**, 747–67.

Ormond RFG (1980) Aggressive mimicry and other interspecific feeding associations among Red Sea predators. *Journal of Zoology*, **191**, 247–62.

Orr LP (1962) Supposed mimicry in salamander. *Journal of the Ohio Herpetological Society*, **3**, 61.

Orr LP (1967) Feeding experiments with a supposed mimetic complex in salamanders. *American Midland Naturalist*, **77**, 147–55.

Orr LP, Coyne JA (1992) The genetics of adaptation: a reassessment. *The American Naturalist*, **140**, 725–42.

Oshima N (2001) Direct reception of light by chromatophores of lower vertebrates. *Pigment Cell Research*, **14**, 312–9.

Oshima N, Nakata E, Ohta M, Kamagata S (1998) Light-induced pigment aggregation in xanthophores of the medaka, *Oryzias latipes*. *Pigment Cell Research*, **11**, 362–7.

Osorio D, Srinivasan MV (1991) Camouflage by edge enhancement in animal coloration mechanisms and its implications for visual mechanisms. *Proceedings of the Royal Society of London, Series B: Biological Sciences*, **244**, 81–5.

Oudemans JT (1903) Étude sur la position du repos chez les Lépidopteres. *Verhandelingen der Koninklijke Nederlandsche Akademie van Wetenschappen. Afdeeling Natuurkunde, Sectie 2*, **10**: 1–90.

Outomuro D, Ángel-Giraldo P, Corral-Lopez A, Realpe E (2016) Multitrait aposematic signal in Batesian mimicry. *Evolution*, **70**, 1596–608.

Owada M, Thinh TH (2002) Biodiversity and mimicry complex of diurnal insects in Vietnam. In Kubodera et al. (eds) *Proceedings of the 3rd and 4th Symposia on Collection Building and Natural History Studies in Asia and the Pacific Rim. National Science Monographs* **22**, pp. 179–88.

Owen DF (1971a) Pupal color in *Papilio demodocus* (Papilionidae) in relation to the season of the year. *Journal of the Lepidopterist's Society*, **25**, 271–4.

Owen DF (1971b) Seasonal changes in the frequency of sex-limited colour forms in the butterfly, *Catopsilia florella* Fabricius. *Revue de Zoologie et de Botanique Africaines*, **83**, 317–21.

Owen DF (1980) *Camouflage and Mimicry*. Oxford University Press, Oxford.

Owen DF, Chanter DO (1972) Polymorphic mimicry in a population of the African butterfly, *Pseudacraea eurytus* (L.) (Lep. Nymphalidae). *Entomologica Scandinavica*, **3**, 258–66.

Owen DF, Whiteley DAA (1986) Reflexive selection: Moment's hypothesis resurrected. *Oikos*, **47**, 117–20.

Owen DF, Whiteley DAA (1988) The beach clams of Thessaloniki: reflexive or apostatic selection? *Oikos*, **51**, 253–5.

Owen DF, Smith DAS, Gordon IJ, Owiny AM (1994) Polymorphic Müllerian mimicry in a group of African butterflies: a reassessment of the relationship between *Danaus chrysippus*, *Acraea encedon* and *Acraea encedana* (Lepidoptera: Nymphalidae). *Journal of Zoology*, **232**, 93–108.

Owen RE, Owen ARG (1984) Mathematical paradigms for mimicry: recurrent sampling. *Journal of Theoretical Biology*, **109**, 217–47.

Page LM (1989) Egg mimics in darters (Pisces: Percidae). *Copeia*, **1989**, 514–8.

Page LM, Swofford DC (1984) Morphological correlates of ecological specialization in darters. *Environmental Biology of Fishes*, **11**, 139–59.

Pagnucco K, Zanette L, Clinchy M, Leonard ML (2008) Sheep in wolf's clothing: most nestling vocalizations resemble their cowbird competitor's. *Proceedings of the Royal Society of London, Series B: Biological Sciences*, **275**, 1061–5.

Palmerino CC, Rusiniak KW, Garcia J (1980) Flavor–illness aversions: the peculiar roles of odor and taste in memory for poison. *Science*, **208**, 753–5.

Papadopulos AST, Powell MP, Pupulin F, Warner J, Hawkins JA, Salamin N, Chittka L, Williams NH, Whitten WM, Loader D, Valente LM, Chase MW, Savolainen V (2013) Convergent evolution of floral signals underlies the success of Neotropical orchids. *Proceedings of the Royal Society of London, Series B: Biological Sciences*, **280**: 20130960.

Papageorgis C (1975) Mimicry in neotropical butterflies. Why are there so many wing-coloration complexes in one place? *American Scientist*, **63**, 522–32.

Papavero N (1964) Notes on the myrmecomimicry of *Syringogaster rufa* Cresson, 1912 (Diptera, Acalyptratae, Megamerinidae). *Papéis Avulsos do Departamento de Zoologia, Secretaria da Agricultura, São Paulo, Brazil*, **16**, 109–13.

Pardo-Diaz C, Jiggins CD (2014) Neighboring genes shaping a single adaptive mimetic trait. *Evolution & Development*, **16**, 3–12.

Parellada X, Santos X (2002) Caudal luring in free-ranging adult *Vipera latasti*. *Amphibia-Reptilia*, **23**, 343–7.

Parker GA, Pearson RG (1976) A possible origin and adaptive significance of the mounting behaviour shown by some female mammals in oestrus. *Journal of Natural History*, **10**, 241–5.

Parker J, Grimaldi DA (2014) Specialized myrmecophily at the ecological dawn of modern ants. *Current Biology*, **24**, 2428–34.

Parker WS, Pianka ER (1974) Further ecological observations on the western banded gecko, *Coleonyx variegatus*. *Copeia*, **1974**, 528–31.

Pasteels JM, Daloze D (1977) Cardiac glycosides in the defensive secretion of chrysomelid beetles: evidence for their production by the insects. *Science*, **197**, 70–2.

Pasteels JM, Daloze D, van Dorsser W, Roba J (1979) Cardiac glycosides in the defensive secretion of *Chrysolina herbacea* (Coleoptera: Chrysomelidae). Identification, biological role and pharmacological activity. *Comparative Biochemistry and Physiology*, **63C**, 117–21.

Pasteels JM, Rowell-Rahier M, Braekman J-C, Daloze D (1984) Chemical defences in leaf beetles and their larvae: the ecological, evolutionary and taxonomic significance. *Biochemical Systematics and Ecology*, **12**, 395–406.

Pasteur G (1972) *Le Mimétisme, Que sais je?*. Presses Universitaires de France, Paris (in French)

Pasteur G (1982) A classificatory review of mimicry systems. *Annual Review of Ecology and Systematics*, **13**, 169–99.

Pawlowski B (1999) Loss of oestrus and concealed ovulation in human evolution. *Current Biology*, **40**, 257–76.

Paxton CGM, Magurran AE, Zschokke S (1994) Caudal eyespots on fish predators influence the inspection behaviour of Trinidadian guppies, *Poecilia reticulata*. *Journal of Fish Biology*, **44**, 175–7.

Payne RB (1973a) Vocal mimicry of the paradise whydahs (*Vidua*) and response of female whydahs to the songs of their hosts (*Pytilia*) and their mimics. *Animal Behaviour*, **21**, 762–71.

Payne RB (1973b) Behaviour, mimetic songs and song dialects, and relationships of the parasitic indigobirds (*Vidua*) of Africa. *Ornithological Monographs*, No. 11.

Payne RB (1977) The ecology of brood parasitism in birds. *Annual Review of Ecology and Systematics*, **8**, 1–28.

Payne RB, Payne LL (1994) Song mimicry and species associations of West-African indigo birds *Vidua* with quail-finch *Ortygospiza*

atricollis, goldbreast *Amandava subflava* and brown twinspot *Clytospiza monteiri*. *Ibis*, **136**, 291–304.

Peakall R (1990) Responses of male *Zaspilothynnus trilobatus* wasps to females and the sexually deceptive orchid it pollinates. *Functional Ecology*, **4**, 159–67.

Peakall R, Beattie AJ (1996) Ecological and genetic consequences of pollination by sexual deception in the orchid *Caladenia tentaculata*. *Evolution*, **50**, 2207–20.

Peakall R, Ebert D, Poldy J, Barrow RA, Francke W, Bower CC, Schiestl FP (2010) Pollinator specificity, floral odour chemistry and the phylogeny of Australian sexually deceptive *Chiloglottis* orchids: implications for pollinator-driven speciation. *New Phytologist*, **88**, 437–50.

Peckham GW (1889) Protective resemblances of spiders. *Occasional Papers of the Natural History Society of Wisconsin*, **1**, 61–113.

Peckham GW, Peckham EG (1892) Ant-like spiders of the family Attidae. *Occasional Papers of the Natural History Society of Wisconsin*, **2**, 1–83.

Pegram KV, Lillo MJ, Rutowski RL (2013) Iridescent blue and orange components contribute to the recognition of a multicomponent warning signal. *Behaviour*, **150**, 321–36.

Pekár S (2014) Is inaccurate mimicry ancestral to accurate in myrmecomorphic spiders (Araneae)? *Biological Journal of the Linnean Society*, **113**, 97–111.

Pekár S, Jiroš P (2011) Do ant mimics imitate cuticular hydrocarbons of their models? *Animal Behaviour*, **82**, 1193–9.

Pekár S, Král J (2002) Mimicry complex in two central European zodariid spiders (Araneae: Zodariidae): how *Zodarion* deceives ants. *Biological Journal of the Linnean Society*, **75**, 517–32.

Pellissier L, Vittoz P, Internicola AI, Bienvenu Gigord LD (2010) Generalized food-deceptive orchid species flower earlier and occur at lower altitudes than rewarding ones. *Journal of Plant Ecology*, **3**, 243–50.

Pellmyr O (1986) The pollination ecology of two nectarless *Cimifuga* sp. (Ranunculaceae) in North America. *Nordic Journal of Botany*, **6**, 713–23.

Penney HD, Hassall C, Skevington JH, Abbott KR, Sherratt TN (2012) A comparative analysis of the evolution of imperfect mimicry. *Nature*, **483**, 461–4.

Penney HD, Hassall C, Skevington JH, Lamborn B, Sherratt TN (2014) The relationship between morphological and behavioral mimicry in hover flies (Diptera: Syrphidae). *The American Naturalist*, **183**, 281–9.

Peres CA, van Roosmalen MGM (1996) Avian dispersal of mimetic seeds of *Ormosia lignivalvis* by terrestrial granivores – deception or mutualism. *Oikos*, **75**, 249–58.

Perfecto I, Vandermeer JH (1993) Cleptobiosis in the ant *Ectatomma ruidum* in Nicaragua. *Insectes Sociaux*, **40**, 295–9.

Perrard A, Arca M, Rome Q, Muller F, Tan J, Bista S, Nugroho H, Baudoin R, Baylac M, Silvain J-F, Carpenter JM, Villemant C (2014) Geographic variation of melanisation patterns in a hornet species: genetic differences, climatic pressures or aposematic constraints? *PLoS ONE*, **9**: e94162.

Peschke K (1987) Male aggression, female mimicry and female choice in the rove beetle, *Aleochara curtula* (Coleoptera, Staphylinidae). *Ethology*, **75**, 265–84.

Peter CI, Johnson SD (2008) Mimics and magnets: the importance of color and ecological facilitation in floral deception. *Ecology,* **89,** 1583–95.

Peter CI, Johnson SD (2013) Generalized food-deception: colour signals and efficient pollen transfer in bee-pollinated species of *Eulophia* (Orchidaceae). *Botanical Journal of the Linnean Society,* **171,** 713–29.

Pfennig DW, Kikuchi Q (2012) Competition and the evolution of imperfect mimicry. *Current Zoology,* **58,** 608–19.

Pfennig DW, Mullen SP (2010) Mimics without models: causes and consequences of allopatry in Batesian mimicry complexes. *Proceedings of the Royal Society of London, Series B: Biological Sciences,* **277,** 2577–85.

Pfennig DW, Harcombe WR, Pfennig KS (2001) Frequency-dependent Batesian mimicry. *Nature,* **410,** 323.

Pfrender M, Mason RT, Wilmslow JT, Shine R (2001) *Thamnophis sirtalis parietalis* (red-sided gartersnake). Male-male copulation. *Herpetological Review,* **32,** 52.

Phillips G (1964) The coloration of sea birds. *Triton Magazine,* Nov.-Dec.

Phillips RD, Scaccabarozzi D, Retter BA, Hayes C, Brown GR, Dixon KW, Peakall R (2014) Caught in the act: pollination of sexually deceptive trap-flowers by fungus gnats in *Pterostylis* (Orchidaceae). *Annals of Botany,* **113,** 629–41.

Pianka ER, Vitt LJ (2003) *Lizards. Windows to the Evolution of Diversity.* University of California Press, Berkeley, CA.

Pie MR (2005) Signal evolution in prey recognition systems. *Behavioural Processes,* **68,** 47–50.

Pie MR, Del-Claro K (2002) Male-male agonistic behavior and ant-mimicry in a Neotropical richardiid (Diptera: Richardiidae). *Studies on Neotropical Fauna and Environment,* **37,** 19–22.

Pielowski Z (1959) Studies on the relationship: predator (Goshawk) and prey (pigeon). *Bulletin of the Polish Academy of Sciences Biological Sciences,* **7,** 401–3.

Pielowski Z (1961) Über Den Unifikationseinfluss der selektiven Nahrungswahl des habichts (*Accipiter gentilis* L.) auf Haustauben. *Ekologia Polska, Series A,* **9,** 183–92.

Pierce NE, Braby MF, Heath A, Lohman DJ, Mathew J, Rand DB, Travassos MA (2002) The ecology and evolution of ant association in the Lycaenidae (Lepidoptera). *Annual Review of Entomology,* **47,** 733–71.

Pietrewicz AT, Kamil AC (1977) Visual detection of cryptic prey by blue jays (*Cyanocitta cristata*). *Science,* **195,** 580–2.

Pietsch TW, Grobecker DB (1978) The compleat angler: aggressive mimicry in an antennariid anglerfish. *Science,* **201,** 369–70.

Pilecki C, O'Donald P (1971) The effects of predation on artificial mimetic polymorphisms with perfect and imperfect mimics at varying frequencies. *Evolution,* **25,** 365–70.

Pinheiro CEG (1996) Palatability and escaping ability in neotropical butterflies: tests with wild kingbirds (*Tyrannus melancholicus*). *Biological Journal of the Linnean Society,* **59,** 351–65.

Pinheiro CEG (2003) Does Müllerian mimicry work in nature? Experiments with butterflies and birds (Tyrannidae). *Biotropica,* **35,** 356–64.

Pinheiro CEG (2004) Jacamars (Aves, Galbulidae) as selective agents of mimicry in neotropical butterflies. *Ararajuba,* **12,** 137–9.

Pinheiro CEG, Freitas AVL (2014) Some possible cases of escape mimicry in Neotropical butterflies. *Neotropical Entomology,* **43,** 393–8.

Pinheiro CEG, Freitas AVL, Campos VC, Devries P, Penz C (2016) Both palatable and unpalatable butterflies use bright colors to signal difficulty of capture to predators. *Neotropical Entomology,* **45,** 107–13.

Place NJ, Glickman SE (2004) Masculinization of female mammals: lessons from nature. *Advances in Experimental Medicine and Biology,* **545,** 243–53.

Plaisted KC, Mackintosh NJ (1995) Visual search for cryptic stimuli in pigeons: implications for the search image and search rate hypotheses. *Animal Behaviour,* **50,** 1219–32.

Platt AP, Brower LP (1968) Mimetic versus disruptive coloration in intergrading populations of *Limenitis arthemis* and *astyanax* butterflies. *Evolution,* **22,** 699–718.

Platt AP, Coppinger RP, Brower LP (1971) Demonstration of the selective advantage of mimetic *Limenitis* butterflies presented to caged avian predators. *Evolution,* **25,** 692–701.

Pliske TE (1975) Pollination of pyrrolizidine alkaloid-containing plants by male Lepidoptera. *Environmental Entomology,* **4,** 474–9.

Plowright RC, Owen RE (1980) The evolutionary significance of bumble bee color patterns: a mimetic interpretation. *Evolution,* **34,** 622–37.

Pocock RI (1908) Warning coloration in the musteline carnivore. *Proceedings of the Zoological Society of London,* **1908,** 944–59.

Pocock RI (1910) Ant-mimicry by the larvae of a species of mantis. *Proceedings of the Zoological Society of London,* **1910,** 837–9.

Policha T, Davis A, Barnadas M, Dentinger BTM, Raguso RA, Roy BA (2016) Disentangling visual and olfactory signals in mushroom-mimicking *Dracula* orchids using realistic three-dimensional printed flowers. *New Phytologist,* **210,** 1058–71.

Polidori C, Nieves-Aldrey JL, Gilbert F, Rotheray GE (2014) Hidden in taxonomy: Batesian mimicry by a syrphid fly towards a Patagonian bumblebee. *Insect Conservation and Diversity,* **7,** 32–40.

Polte S, Reinhold K (2013) The function of the wild carrot's dark central floret: attract, guide or deter? *Plant Species Biology,* **28,** 81–6.

Poole RW (1970) Convergent evolution in the larvae of two penstemon-feeding geometrids (Lepidoptera: Geometridae). *Journal of the Kansas Entomological Society,* **43,** 292–7.

Ponce de León P, Valverde J (2003) ABO system: molecular mimicry of *Ascaris lumbricoides*. *Revista do Instituto de Medicina Tropical de São Paulo,* **45,** 107–8.

Popescu C, Broadhead E, Shorrocks B (1978) Industrial melanism in *Mesopsocus unipunctatus* (Mull.) (Psocoptera) in northern England. *Ecological Entomology,* **3,** 209–19.

Portugal AHA, Trigo JR (2005) Similarity of cuticular lipids between a caterpillar and its host plant: a way to make prey undetectable for predatory ants? *Journal of Chemical Ecology,* **31,** 2551–61.

Pough FH (1976) Multiple cryptic effects of crossbanded and ringed patterns of snakes. *Copeia,* **1976,** 834–6.

Pough FH (1988a) Mimicry and related phenomena. In C Gans and RB Huey (eds), *Biology of the Reptilia. Volume 16. Defence and Life History.* Alan R. Liss Inc., New York, pp. 153–234.

Pough FH (1988b) Mimicry of vertebrates: are the rules different? *The American Naturalist*, **131**(Suppl). S67–S102 (reprinted in Brower LP (1988) *Mimicry and the Evolutionary Process*. University of Chicago Press, Chicago).

Pough FH, Brower LP, Meck HR, Kessell SR (1973) Theoretical investigations of automimicry: multiple trial learning and the palatability spectrum. *Proceedings of the National Academy of Sciences, U.S.A.*, **70**, 2261–5.

Poulton EB (1885) The essential nature of the colouring of phytophagous larvae (and their pupae): with an account of some experiments upon the relation between the colour of such larvae and that of their food-plants. *Proceedings of the Royal Society of London, Series B: Biological Sciences*, **237**, 269–315.

Poulton EB (1886) A further enquiry into a special colour relation between the larva of *Smerinthus ocellatus* and its food-plants *Proceedings of the Royal Society of London, Series B: Biological Sciences*, **243**, 135–73.

Poulton EB (1887a) An inquiry into the cause and extent of a special colour relation between certain exposed lepidopterous pupae and the surfaces which immediately surround them. *Proceedings of the Royal Society of London, Series B: Biological Sciences*, **42**, 94–108.

Poulton EB (1887b) Experimental proof of value of colour in insects. *Proceedings of the Zoological Society of London*, **17**, 191–274.

Poulton EB (1888) Notes in 1887 upon lepidopterus larvae &c. *Transactions of the Royal Entomological Society of London*, **1888**, 393–8.

Poulton EB (1890) *The Colours of Animals, their Meaning and Use, Especially Considered in the Case of insects*. D. Appleton and Company, New York.

Poulton EB (1891) On an interesting example of protective mimicry discovered by Mr. W.L. Sclater in British Guiana. *Proceedings of the Zoological Society of London*, **59**, 462–4.

Poulton EB (1892) Further experiments upon the colour-relation between certain lepidopterous larvae, pupae, cocoons, and imagines and their surroundings. *Transactions of the Royal Entomological Society of London*, **1892**, 293–487.

Poulton EB (1902) The meaning of the white undersides of animals. *Nature*, **65**, 296–7.

Poulton EB (1906) Predaceous insects and their prey. Part I. Predaceous Diptera, Neuroptera, Hemiptera, Orthoptera, and Coleoptera. *Transactions of the Royal Entomological Society of London*, **1906**, 323–409.

Poulton EB (1924) Papilio dardanus. The most interesting butterfly in the world. *Journal of the East African and Ugandan Natural History Society*, **20**, 4–22.

Pouyanne A (1917) La fécondation des *Ophrys* par les insectes. *Bulletin de la Société d'histoire naturelle d'Afrique du Nord*, **8**, 6–7.

Powell RA (1982) Evolution of black-tipped tails in weasels: predator confusion. *The American Naturalist*, **119**, 126–31.

Powell S, Del-Claro K, Feitosa RM, Brandao CRF (2014) Mimicry and eavesdropping enable a new form of social parasitism in ants. *The American Naturalist*, **184**, 500–9.

Prates I, Antoniazzi MM, Sciani JM, Pimenta DC, Toledo LF, Haddad CFB, Jared C (2012) Skin glands, poison and mimicry in dendrobatid and leptodactylid amphibians. *Journal of Morphology*, **273**, 279–90.

Price PW, Fernandes WG, Waring GL (1987) Adaptive nature of insect galls. *Environmental Entomology*, **16**, 15–24.

Procter-Gray E, Holmes RT (1981) Adaptive significance of delayed attainment of plumage in male American redstarts: tests of two hypotheses. *Evolution*, **35**, 742–51.

Proctor HC, Harder LD (1995) Effect of pollination success on floral longevity in the orchid *Calypso bulbosa* (Orchidaceae). *American Journal of Botany*, **82**, 1131–6.

Proctor MCF, Yeo P (1973) *Pollination in Flowers*. Collins, London.

Prohammer LA, Wade MJ (1981) Geographic and genetic variation in death-feigning behavior in the flour beetle, *Tribolium castaneum*. *Behavior Genetics*, **11**, 395–401.

Provenza FD, Kimball BA, Villalba JJ (2000) Roles of odor, taste, and toxicity in the food preferences of lambs: implications for mimicry in plants. *Oikos*, **88**, 424–32.

Prudic KL, Oliver JC, Sperling FAH (2007a) The signal environment is more important than diet or chemical specialization in the evolution of warning coloration. *Proceedings of the National Academy of Sciences, U.S.A.*, **104**, 19381–6.

Prudic KL, Khera S, Sólyom A, Timmermann BN (2007b) Isolation, identification, and quantification of potential defensive compounds in the viceroy butterfly and its larval host-plant, Carolina willow. *Journal of Chemical Ecology*, **33**, 1149–59.

Prum RO (2014) Interspecific social dominance mimicry in birds. *Zoological Journal of the Linnean Society*, **172**, 910–41.

Prum RO, Samuelson L (2012) The hairy–downy game: a model of interspecific social dominance mimicry. *Journal of Theoretical Biology*, **313**, 42–60.

Prum RO, Samuelson L (2016) Mimicry cycles, traps, and chains: the coevolution of toucan and kiskadee mimicry. *The American Naturalist*, **187**, 753–64.

Przeczek K, Mueller C, Vamosi SM (2008) The evolution of aposematism is accompanied by increased diversification. *Integrative Zoology*, **3**, 149–56.

Puebla O, Bermingham E, Guichard F, Whiteman E (2007) Colour pattern as a single trait driving speciation in *Hypoplectrus* coral reef fishes? *Proceedings of the Royal Society B: Biological Sciences*, **274**, 1265–71.

Punnett RC (1915) *Mimicry in Butterflies*. Cambridge University Press, Cambridge, UK.

Puurtinen M, Kaitala V (2006) Conditions for the spread of conspicuous warning signals – a numerical model with novel insights. *Evolution*, **60**, 2246–56.

Quicke DLJ (1984) Are rogadine puparia (Hymenoptera: Braconidae) hoverfly mimics (Diptera: Syrphidae)? *Proceedings and Transactions of the British Entomology and Natural History Society*, **17**, 60.

Quicke DLJ (1986a) Preliminary notes on homeochromatic associations within and between the Afrotropical Braconinae (Hym., Braconidae) and Lamiinae (Col., Cerambycidae). *Entomologist's Monthly Magazine*, **122**, 97–109.

Quicke DLJ (1986b) Warning coloration and mimicry. In C Betts (ed.), *Hymenopterist's Handbook*. AES, Brentwood, UK, pp. 73–80.

Quicke DLJ (1997) *Parasitic Wasps*. Chapman & Hall, London.

Quicke DLJ (2015) *The Braconid and Ichneumonid Parasitic Wasps: Biology, Systematics, Evolution and Ecology*. Wiley Blackwell, Oxford.

Quicke DLJ, Ingram SN, Proctor J, Huddleston T (1992) Batesian and Müllerian mimicry between species with connected life histories, with a new example involving braconid wasp parasites of *Phoracantha* beetles. *Journal of Natural History*, **26**, 1013–34.

Racey PA, Skinner JD (1979) Endocrine aspects of sexual mimicry in Spotted hyaenas *Crocuta crocuta*. *Journal of Zoology*, **187**, 315–26.

Raguso RA, Roy BA (1998) 'Floral' scent production by *Puccinia* rust fungi that mimic flowers. *Molecular Ecology*, **7**, 1127–36.

Raguso RA, Kelber A, Pfaff M, Levin RA, McDade LA (2007) Floral biology of North American *Oenothera* Sect. *Lavauxia* (Onagraceae): advertisements, rewards, and extreme variation in floral depth. *Annals of the Missouri Botanical Garden*, **94**, 236–57.

Rainey MR, Grether GF (2007) Competitive mimicry: synthesis of a neglected class of mimetic relationships. *Ecology*, **88**, 2440–48.

Ralphs MH (1997) Persistence of aversions to larkspur in naive and native cattle. *Journal of Range Management*, **50**, 367–70.

Ramachandran VS, Tyler CW, Gregory RL, Rogers-Ramachandran D, Duensing S, Pillsbury C, Ramachandran C (1996) Rapid adaptive camouflage in tropical flounders. *Nature*, **379**, 815–8.

Ramesh A, Vijayan S, Sreedharan S, Somanathan H, Uma DB (2016) Similar yet different: differential response of a praying mantis to ant-mimicking spiders. *Biological Journal of the Linnean Society*, doi:10.1111/bij.12793

Rand AS (1967) Predator-prey interactions and the evolution of aspect diversity. *Atlas de Simpioso Sobra a Biota Amazonica*, **5**, 73–83.

Randall JE (2005) A review of mimicry in marine fishes. *Zoological Studies*, **44**, 299–328.

Randall JE, Emery AR (1971) On the resemblance of the young of the fishes *Platax pinnatus* and *Plectorhynchus chaetodontoides* to flatworms and nudibranchs. *Zoologica, New York*, **56**, 115–19.

Randall JE, Guézé P (1980) The goatfish *Mulloidichthys mimicus* n. sp. (Pisces, Mullidae) from Oceania, a mimic of the snapper *Lutjanus kasmira* (Pisces, Lutjanidae). *Bulletin du Museum National d'Histoire Naturelle Ser. 4: Section A: Zoologie Biologie et Ecologie Animales*, **2**, 603–9.

Randall JE, Hartman WD (1968) Sponge-feeding fishes of the West Indies. *Marine Biology*, **1**, 216–26.

Randall JE, Kuiter RH (1989) The juvenile Indo-Pacific grouper *Anyperodon leucogrammicus*, a mimic of the wrasse *Halichoeres purpurascens* and allied species, with a review of the recent literature on mimicry in fishes. *Revue Française d'Aquariologie et Herpetologie*, **16**, 51–6.

Randall JE, Randall HA (1960) Examples of mimicry and protective resemblance in tropical marine fishes. *Bulletin of Marine Science of the Gulf and Caribbean*, **10**, 444–80.

Rangaiah K, Rao SP, Raju AJS [2004] (2006) Bird-pollination and fruiting phenology in *Spathodea campanulata* Beauv. (Bignoniaceae). *Beiträge zur Biologie der Pflanzen*, **73**, 395–408.

Rao KR, Fingerman M, Bartell CK (1967) Physiology of the white chromatophores in the fiddlercrab, *Uca pugilator*. *Biological Bulletin*, **133**, 606–17.

Raske AG (1967) Morphological and behavioral mimicry among beetles of the genus *Moneilema*. *The Pan-Pacific Entomologist*, **43**, 239–44.

Rasekh A, Michaud JP, Kharazi-Pakdel A, Allahyari H (2010) Ant mimicry by an aphid parasitoid, *Lysiphlebus fabarum*. *Journal of Insect Science*, **10**, 126.

Rashed A, Sherratt TN (2006) Mimicry in hoverflies (Diptera: Syrphidae): a field test of the competitive mimicry hypothesis. *Behavioral Ecology*, **18**, 337–44.

Rathcke BJ, Price PW (1976) Anomalous diversity of tropical ichneumonoid parasitoids: a predation hypothesis. *The American Naturalist*, **110**, 889–93.

Rausher MD (1979) Egg recognition: its advantage to a butterfly. *Animal Behaviour*, **27**, 1034–40.

Rätti O (1994) Female reactions to male absence after pairing in the pied flycatcher. *Behavioral Ecology and Sociobiology*, **35**, 201–3.

Raymond DL (1984) Wild birds prefer the familiar of striped and unstriped artificial prey. *Biological Journal of the Linnean Society*, **23**, 229–35.

Rechten C (1978) Interspecific mimicry in birdsong: does the Beau Geste hypothesis apply? *Animal Behaviour*, **26**, 305–6.

Redondo T, Arias de Reyna L (1988) Vocal mimicry of hosts by greater spotted cuckoo *Clamator glandarius*: further evidence. *The Condor*, **30**, 540–4.

Reid PJ, Shettleworth SJ (1992) Detection of cryptic prey – search image or search rate. *Journal of Experimental Psychology*, **18**, 273–86.

Reimchen TE (1979) Substratum heterogeneity, crypsis and colour polymorphism in an intertidal snail (*Littorina mariae*). *Canadian Journal of Zoology*, **57**, 1070–85.

Reimchen TE (1989) Shell colour ontogeny and tubeworm mimicry in a marine gastropod *Littorina mariae*. *Biological Journal of the Linnean Society*, **36**, 97–109.

Reiserer RS, Schuett GW (2008) Aggressive mimicry in neonates of the sidewinder rattlesnake *Crotalus cerastes* (Serpentes: Viperidae): stimulus control and visual perception of prey luring. *Biological Journal of the Linnean Society*, **95**, 81–91.

Reiskind J (1965) Behaviour of an avian predator in an experiment simulating Batesian mimicry. *Animal Behaviour*, **13**, 466–9.

Reiskind J (1977) Ant-mimicry in Panamanian clubionid and salticid spiders (Araneae: Clubionidae, Salticidae). *Biotropica*, **9**, 1–8.

Reiskind J, Levi HW (1967) *Anatea*, an ant-mimicking theridiid spider from New Caledonia (Araneae: Theridiidae). *Psyche*, **74**, 20–3.

Reit S (1970) *Masquerade: The Amazing Camouflage Deceptions of World War II*. Robert Hale, London.

Remmel T, Tammaru T (2011) Evidence for the higher importance of signal size over body size in aposematic signaling in insects. *Journal of Insect Science*, **11**: 4.

Ren Z-X, Li D-Z, Bernhardt P, Wang H (2011) Flowers of *Cypripedium fargesii* (Orchidaceae) fool flat-footed flies (Platypezidae) by faking fungus-infected foliage. *Proceedings of the National Academy of Sciences of the United States of America*, **108**, 7478–80.

Ren Z-X, Li D-Z, Bernhardt P, Wang H (2012) Correction for "Flowers of *Cypripedium fargesii* (Orchidaceae) fool flat-footed flies (Platypezidae) by faking fungus-infected foliage," by Zong-Xin Ren, De-Zhu Li, Peter Bernhardt, and Hong Wang, which appeared in issue 18, May 3, 2011, of Proc Natl Acad Sci USA. *Proceedings of the National Academy of Sciences of the United States of America*, **109**, 20776.

Ren Z-X, Wang H, Bernhardt P, Camilo G, Li D-Z (2014) Which food-mimic floral traits and environmental factors influence fecundity in a rare orchid, Calanthe yaoshanensis? Botanical Journal of the Linnean Society, **176**, 421–33.

Renner SS (2006) Rewardless flowers in the angiosperms and the role of insect cognition in their evolution. In: NM Waser and J Ollerton (eds), Plant–Pollinator Interactions: From Specialization to Generalization. University of Chicago Press, Chicago, pp. 123–44.

Renoult JP, Schaefer HM, Sallé B, Charpentier MJE (2011) The evolution of the multicoloured face of mandrills: insights from the perceptual space of colour vision. PLoS ONE, **6(12)**: e29117.

Rescorla RA, Wagner ARA (1972) A theory of Pavlovian conditioning. Variations in the effectiveness of reinforcement and nonreinforcement. In AH Black and WF Prokasy (eds), Classical Conditioning II: Current Research and Theory. Appleton-Century-Crofts, New York, pp. 64–99.

Resetarits WJ Jr, Binckley CA (2013) Is the pirate really a ghost? Evidence for generalized chemical camouflage in an aquatic predator, pirate perch Aphredoderus sayanus. The American Naturalist, **181**, 690–9.

Rettenmeyer CW (1970) Insect mimicry. Annual Review of Entomology, **15**, 43–74.

Ricklefs RE (2009) Aspect diversity in moths revisited. The American Naturalist, **173**, 411–6.

Ricklefs RE, O'Rourke KE (1975) Aspect diversity in moths: a temperate-tropical comparison. Evolution, **29**, 313–324.

Ridley HN (1930) The Dispersal of Plants Throughout the World. Reeve, Ashford, UK.

Ries L, Mullen SP (2008) A rare model limits the distribution of its more common mimic: a twist on frequency-dependent Batesian mimicry. Evolution, **62**, 1798–1803.

Riipi M, Alatalo RV, Lindström L, Mappes J (2001) Multiple benefits of gregariousness cover detectability costs in aposematic aggregations. Nature, **413**, 512–4.

Rios-Cardenas O, Darrah A, Morris MR (2010) Female mimicry and an enhanced sexually selected trait: what does it take to fool a male? Behaviour, **147**, 1443–60.

Ristau CA (1991) Before mindreading: attention, purposes and deception in birds? In A Whiten (ed), Natural Theories of Mind. Blackwell, Oxford, pp. 209–222.

Ritland DB (1991) Revising a classic butterfly mimicry scenario: demonstration of Mullerian mimicry between Florida viceroys (Limenitis archippus floridensis) and queens (Danaus gilippus berenice). Evolution, **45**, 918–34.

Ritland DB (1994) Variations in the palatability of queen butterflies (Danaus gilippus) and implications regarding mimicry. Ecology, **75**, 732–46.

Ritland DB, Brower LP (1991) The viceroy butterfly is not a Batesian mimic. Nature, **350**, 497–8.

Rivers TJ, Morin JG (2009) Plasticity of male mating behavior in a marine bioluminescent ostracode in both time and space. Animal Behaviour, **78**, 723–34.

Robbins RK (1980) The lycaenid "false head" hypothesis: historical review and quantitative analysis. Journal of the Lepidopterists' Society, **34**, 194–208.

Robbins RK (1981) The "false head" hypothesis: predation and wing pattern variation in lycaenid butterflies. The American Naturalist, **118**, 770–5.

Robbins RK (1985) Independent evolution of "false head" behavior in Riodinidae. Journal of the Lepidopterists' Society, **39**, 224–5.

Roberts TR (1982) A revision of the South and Southeast Asian angler-catfishes (Chacidae). Copeia, **1982**, 895–901.

Robertson DR (2013) Who resembles whom? Mimetic and coincidental look-alikes among tropical reef fishes. PLoS ONE, **8(1)**: e54939.

Robinson EJ Jr (1947) Notes on the life history of Leucochloridium fuscostriatum n. sp. provis. (Trematoda: Brachylaemidae). The Journal of Parasitology, **33**, 467–75.

Robinson MH (1969) Defences against visually hunting predators. Evolutionary Biology, **3**, 225–59.

Robinson MH (1981) A stick is a stick and not worth eating: on the definition of mimicry. Biological Journal of the Linnean Society, **16**, 15–20.

Robinson MH, Robinson B (1970) The stabilimentum of the orb web spider, Argiope argentata: an improbable defense against predators. Canadian Entomologist, **102**, 641–55.

Robinson MW, Donnelly S, Hutchinson AT, To J, Taylor NL, Norton RS, Perugini MA, Dalton JP (2011) A family of helminth molecules that modulate innate cell responses via molecular mimicry of host antimicrobial peptides. PLoS Pathogens, **7**: e1002042.

Rödel M-O, Brede C, Hirschfeld M, Schmitt T, Favreau P, Stöklin R, Wunder C, Mebs D (2013) Chemical camouflage – A frog's strategy to co-exist with aggressive ants. PLoS ONE, **8(12)**: e81950.

Rodriguez J, Pitts JP, von Dohlen CD, Wilson JS (2014) Müllerian mimicry as a result of codivergence between velvet ants and spider wasps. PLoS ONE, **9(11)**: e112942.

Rodríguez-Gironés MA, Santamaría L (2004) Why are so many bird flowers red? PLoS Biology, **2(10)**: e350.

Rohwer S (1975) The social significance of avian winter plumage variability. Evolution, **29**, 593–610.

Rohwer S (1977) Status signaling in Harris sparrows: some experiments in deception. Behaviour, **61**, 107–29.

Rohwer S (1978) Passerine subadult plumages and the deceptive acquisition of resources: test of critical assumption. The Condor, **80**, 173–9.

Rohwer S (1983) Testing the female mimicry hypothesis of delayed plumage maturation: a comment on Procter-Gray and Holmes. Evolution, **17**, 421–3.

Rohwer S, Niles DM (1979) The subadult plumage of male purple martins: variability, female mimicry and a century of evolution. Zeitschrift für Tierpsychologie, **51**, 282–300.

Rohwer S, Rohwer FC (1978) Status signaling in Harris sparrows: experimental deceptions achieved. Animal Behaviour, **26**, 1012–22.

Rohwer S, Fretwell SD, Niles DM (1980) Delayed maturation in passerine plumages and the deceptive acquisition of resources. The American Naturalist, **115**, 400–37.

Ronel M, Khateeb S, Lev-Yadun S (2009) Protective spiny modules in thistles of the Asteraceae in Israel. The Journal of the Torrey Botanical Society, **136**, 46–56.

Roosevelt T (1911) Revealing and concealing coloration in birds and mammals. Bulletin of the American Museum of Natural History, **30**, 218.

Roper TJ (1990) Responses of domestic chicks to artificially coloured insect prey: effects of previous experience and background colour. Animal Behaviour, **39**, 466–73.

Roper TJ (1993) Effects of novelty on taste-avoidance learning in chicks. Behaviour, **125**, 265–81.

Roper TJ (1994) Conspicuousness of prey retards reversal of learned avoidance. *Oikos*, **69**, 115–8.

Roper TJ, Cook SE (1989) Responses of chicks to brightly coloured insect prey. *Behaviour*, **110**, 276–93.

Roper TJ, Redston S (1987) Conspicuousness of distasteful prey affects the strength and durability of one-trial avoidance learning. *Animal Behaviour*, **35**, 739–47.

Roper TJ, Wistow R (1986) Aposematic colouration and avoidance learning in chicks. *Quarterly Journal of Experimental Psychology, Section B, Comparative Physiology & Psychology*, **38**, 141–9.

Rose MJ, Barthlott W (1994) Colored pollen in Cactaceae – a mimetic adaptation to hummingbird pollination. *Botanica Acta*, **107**, 402–6.

Rosenberg G (1989) Aposematism evolves by individual selection: evidence from marine gastropods with pelagic larvae. *Evolution*, **43**, 1811–3.

Rosenberg G (1991) Aposematism and synergistic selection in marine gastropods. *Evolution*, **45**, 451–4.

Ross RM, Quetin LB (1990) Mating behavior and spawning in two neustonic nudibranchs in the family Glaucidae. *American Malacological Bulletin*, **8**, 61–6.

Rota J, Wagner DL (2006) Predator mimicry: metalmark moths mimic their jumping spider predators. *PLoS ONE*, **1(1)**: e45.

Rothrock JT (1888) Mimicry among plants. *Proceedings of the Academy of Natural Sciences of Philadelphia*, **40**, 1–3.

Rothschild M (1961) Defensive odours and Müllerian mimicry among insects. *Transactions of the Royal Entomological Society of London*, **113**, 101–2.

Rothschild M (1962) Development of paddling and other behaviours in young black-headed gulls. *British Birds*, **55**, 114–7.

Rothschild M (1963) Is the buff ermine (*Spilosoma lutea* (Hufn.)) a mimic of the white ermine (*Spilosoma lubricipeda* (L.))? *Proceedings of the Royal Entomological Society of London*, **38**, 159–64.

Rothschild M (1964a) A note on the evolution of defensive and repellant odours of insects. *Entomologist*, **92**, 276–80.

Rothschild M (1964b) An extension of Dr Lincoln Brower's theory on bird predation and food specificity, together with some observations on bird memory in relation to aposematic colour patterns. *Entomologist*, **97**, 73–8.

Rothschild M (1971) Speculations about mimicry with Henry Ford. In ER Creed (ed.), *Ecological Genetics and Evolution*. Blackwell Scientific, Oxford, pp. 202–23.

Rothschild M (1972) Secondary plant substances and warning coloration in insects. In HF Van Emden (ed.), *Insect/Plant Relationships*. Blackwell, Oxford, pp. 59–83.

Rothschild M (1974) Modified stipules of *Passiflora* which resemble horned caterpillars. *Proceedings of the Royal Entomological Society of London*, **39**, 16.

Rothschild M (1975) Remarks on carotenoids in the evolution of signals. In LE Gilbert and PH Raven (eds), *Coevolution of Animals and Plants*. University of Texas Press, Austin, pp. 20–51.

Rothschild M (1981) The mimicrats must move with the times. *Biological Journal of the Linnean Society*, **16**, 21–3.

Rothschild M (1984) *Aide Mémoire* mimicry. *Ecological Entomology*, **9**, 311–9.

Rothschild M (1985a) British aposematic Lepidoptera. In JR Heath and AM Emmet (eds), *The Moths and Butterflies of Great Britain and Ireland*. Vol. 2 *(Cossidae to Heliodinidae)*. Harley Books, Colchester, pp. 9–62.

Rothschild M (1985b) Inherited beak-wiping behaviour. *Ibis*, **127**, 563–4.

Rothschild M (1986) The red smell of danger. *New Scientist*, **111**, 34–6.

Rothschild M (1991) Is the viceroy a Batesian mimic? *Nature*, **355**, 611–12.

Rothschild M, Feltwell J (1972) Carotenoids in the body tissues of aposematic and cryptic moths. *Proceedings of the Royal Entomological Society of London, C*, **37**, 29.

Rothschild M, Ford B (1968) Warning signals from a starling *Sturnus vulgaris* observing a bird rejecting unpalatable prey. *Ibis*, **110**, 104–5.

Rothschild M, Kellett DN (1972) Reactions of various predators to insects storing heart poisons (cardiac glycosides) in their tissues. *Journal of Entomology (A)*, **46**, 103–10.

Rothschild M, Lane C (1960) Warning and alarm signals by birds seizing aposematic insects. *Ibis*, **102**, 328–30.

Rothschild M, Mummery R (1985) Carotenoids and bile pigments in danaid and swallowtail butterflies. *Biological Journal of the Linnean Society*, **24**, 1–14.

Rothschild M, Schoonhoven LM (1977) Assessment of egg load by *Pieris brassicae* (Lepidoptera: Pieridae). *Nature*, **266**, 352–5.

Rothschild M, von Euw J, Aplin R, Harman RRM (1970) Toxic Lepidoptera. *Toxicon*, **8**, 293–9.

Rothschild M, Gardiner B, Valadon G, Mummery R (1975) Lack of response to background colour in *Pieris brassicae* pupae reared on carotenoid-free diet. *Nature*, **254**, 592–4.

Rothschild M, Rowan MG, Fairbairn JW (1977) Storage of cannabinoids by *Arctia caja* and *Zonocerus elegans* fed on chemically distinct strains of *Cannabis sativa*. *Nature*, **266**, 650–1.

Rothschild M, Marsh NA, Gardiner B (1978) Cardioactive substances in the monarch butterfly and *Euploea core* reared on leaf-free artificial diet. *Nature*, **275**, 649–50.

Rothschild M, Keutmann H, Lane NJ, Parsons J, Prince PW, Swales LS (1979) A study on the mode of action and composition of a toxin from the female abdomen and eggs of *Arctia caja* (L.) (Lep. Arctiidae): an electrophysiological, ultrastructural and biochemical analysis. *Toxicon*, **17**, 285–306.

Rothschild M, Moore BP, Brown WV (1984a) Some peculiar aspects of danaid/plant relationships. *Entomologia Experimentalis et Applicata*, **24**, 437–50.

Rothschild M, Moore BP, Brown WV (1984b) Pyrazines as warning odour components in the monarch butterfly, *Danaus plexippus*, and in moths of the genus *Zygaena* and *Amata* (Lepidoptera). *Biological Journal of the Linnean Society*, **23**, 375–80.

Rothschild M, Mummery R, Farrell C (1986) Carotenoids of butterfly models and their mimics (Lep: Papilionidae and Nymphalidae). *Biological Journal of the Linnean Society*, **28**, 359–72.

Rothstein SI (1978) Gegraphic variation in the nestling coloration of parasitic cowbirds. *The Auk*, **95**, 152–60.

Rothstein SI (1982) Mechanisms of avian egg recognition: which egg parameters elicit responses by rejecter species? *Behavioral Ecology and Sociobiology*, **11**, 229–39.

Rovner JS (1976) Detritus stabilimenta on the webs of *Cyclosa turbinata* (Araneae, Araneidae). *Journal of Arachnology*, **4**, 215–6.

Rowe C, Guilford T (1996) Hidden colour aversions in domestic chicks triggered by pyrazine odours of insect warning displays. *Nature*, **383**, 520–2.

Rowe C, Guilford T (1999) The evolution of multimodal warning displays. *Evolutionary Ecology*, **13**, 655–71.

Rowe C, Skelhorn J (2005) Colour biases are a question of taste. *Animal Behaviour*, **69**, 587–94.

Rowe C, Lindström L, Lyytinen A (2004) The importance of pattern similarity between Müllerian mimics in predator avoidance learning. *Proceedings of the Royal Society B: Biological Sciences*, **271**, 407–13.

Rowe MP, Coss RG, Owings DH (1986) Rattlesnake rattles and burrowing owl hisses: a case of acoustic Batesian mimicry. *Ethology*, **72**, 53–71.

Rowell-Rahier M, Pasteels JM (1986) Economics of chemical defense in Chrysomelinae. *Journal of Chemical Ecology*, **12**, 1189–203.

Rowell-Rahier M, Pasteels JM, Alonso-Media A, Brower LP (1995) Relative unpalatability of leaf beetles with either biosynthesized or sequestered chemical defence. *Animal Behaviour*, **49**, 709–14.

Rowland HM (2009) From Abbott Thayer to the present day: what have we learned about the function of countershading? *Philosophical Transactions of the Royal Society, Series B, Biological Sciences*, **364**, 519–27.

Rowland HM, Speed MP, Ruxton GD, Edmunds M, Stevens M, Harvey IF (2007a) Countershading enhances cryptic protection: an experiment with wild birds and artificial prey. *Animal Behaviour*, **74**, 1249–58.

Rowland HM, Ihalainen E, Lindström L, Mappes J, Speed MP (2007b) Co-mimics have a mutualistic relationship despite unequal defences. *Nature*, **448**, 64–7.

Rowland HM, Cuthill IC, Harvey IF, Speed MP, Ruxton GD (2008) Can't tell the caterpillars from the trees: countershading enhances survival in a woodland. *Proceedings of the Royal Society of London, Series B: Biological Sciences*, **275**, 2539–45.

Rowland HM, Hoogesteger T, Ruxton GD, Speed MP, Mappes J (2010) A tale of 2 signals: signal mimicry between aposematic species enhances predator avoidance learning. *Behavioral Ecology*, **21**, 851–860.

Rowlands D (1959) A case of mimicry in plants – *Vicia sativa* L. in lentil crops. *Genetica*, **30**, 435–46.

Rowley D, Jenkin CR (1962) Antigenic cross-reaction between host and parasite as a possible cause of pathogenicity. *Nature*, **193**, 151–4.

Roy BA (1993) Floral mimicry by a plant pathogen. *Nature*, **362**, 56–8.

Roy BA (1994) The effects of pathogen-induced pseudoflowers and buttercups on each others insect visitation. *Ecology*, **75**, 352–8.

Roy BA (1996) A fungal plant pathogen influences pollinator behavior and may influence reproduction of non-hosts. *Ecology*, **77**, 2445–57.

Roy BA, Raguso RA (1997) Olfactory versus visual cues in a floral mimicry system. *Oecologia (Berlin)*, **109**, 414–26.

Roy BA, Widmer A (1999) Floral mimicry: a fascinating yet poorly understood phenomenon. *Trends in Plant Science*, **4**, 325–30.

Roy L, Guilbert E, Bourgoin T (2007) Phylogenetic patterns of mimicry strategies in Darnini (Hemiptera: Membracidae). *Annales de la Société entomologique de France* (N.S.), **43**, 273–88.

Rubinoff I, Kropach C (1970) Differential reactions of Atlantic and pacific predators to sea snakes. *Nature*, **228**, 1288–90.

Rubino DL, McCarthy BC (2004) Presence of aposematic (warning) coloration in vascular plants of southeastern Ohio. *Journal of the Torrey Botanical Society*, **131**, 252–6.

Ruchon F, Laugier T, Quignard JP (1995) Alternative male reproductive strategies in the peacock blenny. *Journal of Fish Biology*, **47**, 826–40.

Rudge DW (2003) The role of photographs and films in Kettlewell's popularizations of the phenomenon of industrial melanism. *Science & Education*, **12**, 261–87.

Rudge DW (2005) Did Kettlewell commit fraud? Re-examining the evidence. *Public Understanding of Science*, **14**, 249–68.

Russell BC, Allen GR, Lubbock HR (1976) New cases of mimicry in marine fishes. *Journal of Zoology*, **180**, 407–23.

Rust CC, Meyer RK (1969) Hair color, molt, and testis size in male, short-tailed weasels treated with melatonin. *Science*, **165**, 921–2.

Ruxton GD (2002) The possible fitness benefits of striped coat coloration for zebra. *Mammal Review*, **32**, 237–44.

Ruxton GD (2006) Grasshoppers don't play possum. *Nature*, **440**, 880.

Ruxton GD (2009) Non-visual crypsis: A review of the empirical evidence for camouflage to senses other than vision. *Philosophical Transactions of the Royal Society of London, Series B: Biological Sciences*, **264**, 549–57.

Ruxton GD, Hansell MH (2011) Fishing with a bait or lure: a brief review of the cognitive issues. *Ethology*, **117**, 1–9.

Ruxton GD, Kennedy MW (2006) Peppers and poisons: the evolutionary ecology of bad taste. *Journal of Animal Ecology*, **75**, 1224–6.

Ruxton GD, Schaefer HM (2011) Alternative explanations for apparent mimicry, *Journal of Ecology*, **99**, 899–904.

Ruxton GD, Speed MP (2006) How can automimicry persist when predators can preferentially consume undefended mimics? *Proceedings of the Royal Society of London, Series B: Biological Sciences*, **273**, 373–8.

Ruxton GD, Sherratt TN, Speed MP (2004a) *Avoiding Attack: The Evolutionary Ecology of Crypsis, Warning Signals & Mimicry*. Oxford University Press, New York.

Ruxton GD, Speed MP, Kelly DJ (2004b) What, if anything, is the adaptive function of countershading? *Animal Behaviour*, **68**, 445–51.

Ruxton GD, Speed MP, Broom M (2007) The importance of initial protection of conspicuous mutants for the coevolution of defense and aposematic signaling of the defense. *Evolution*, **61**, 2165–74.

Ryan PG, Wilson RP, Cooper J (1987) Intraspecific mimicry and status signals in juvenile African penguins. *Behavioral Ecology & Sociobiology*, **20**, 69–76.

Saccheri IJ, Rousset F, Watts PC, Brakefield PM, Cook LM (2008) Selection and gene flow on a diminishing cline of melanic peppered moths. *Proceedings of the National Academy of Sciences of the U.S.A.*, **105**, 16212–7.

Saidel WM (1978) Analysis of flatfish camouflage. *American Zoologist*, **18**, 579.

Sakai S (2002) *Aristolochia* spp. (Aristolochiaceae) pollinated by flies breeding on decomposing flowers in Panama. *American Journal of Botany*, **89**, 527–34.

Salazar A, Fürstenau B, Quero C, Pérez-Hidalgod N, Carazo P, Font E, Martínez-Torres D (2015) Aggressive mimicry coexists with mutualism in an aphid. *Proceedings of the National Academy of Sciences, U.S.A.*, **112**, 1101–6.

Salzburger W, Braasch I, Meyer A (2007) Adaptive sequence evolution in a color gene involved in the formation of the characteristic egg-dummies of male haplochromine cichlid fishes. *BMC Biology*, **5**: 51.

Salzet M, Capron A, Stefano GB (2000) Molecular crosstalk in host–parasite relationships: schistosome– and leech–host interactions. *Parasitology Today*, **16**, 536–40.

Sánchez JM, Corbacho C, Muñoz del Viejo A, Parejo D (2004) Colony-site tenacity and egg color crypsis in the gull-billed tern. *Waterbirds*, **27**, 21–30.

Sánchez-Gullén RA, Van Gossum H, Cordero Rivera A (2005) Hybridization and the inheritance of female colour polymorphism in two ischnurid damselflies (Odonata: Coenagrionidae). *Biological Journal of the Linnean Society*, **85**, 471–81.

Sandeen MI (1950) Chromatophorotropins in the central nervous system of *Uca pugilator* with special reference to their origins and actions. *Physiological Zoology*, **23**, 337–52.

Sandre S-L, Tammaru T, Mänd T (2007) Size-dependent coloration in larvae of *Orgyia antiqua* (Lepidoptera: Lymantriidae): a trade-off between warning effect and detectability? *European Journal of Entomology*, **104**, 745–52.

Sannolo M, Gatti F, Scali S (2014) First record of thanatosis behaviour in *Malpolon monspessulanus* (Squamata: Colubridae). *Herpetology Notes*, **7**: 323.

Santos ME, Braasch I, Boileau N, Meyer BS, Sauteur L, Böhne A, Belting H-G, Affolter M, Salzburger W (2014) The evolution of cichlid fish egg-spots is linked with a *cis*-regulatory change. *Nature Communications*, **5**: 5149.

Santos RS (1985) Parentais e satelites: tácticas alternativas de acasalamento nos machos de *Blennius sanguineolentas* (Actinopterygii: Blenniidae). *Arquipélago – Serie Ciências de Natureza* **6**, 119–46.

Saporito RA, Garraffo HM, Donnelly MA, Edwards AL, Longino JT (2004) Formicine ants: An arthropod source for the pumiliotoxin alkaloids of dendrobatid poison frogs. *Proceedings of the National Academy of Sciences of the U.S.A.*, **101**, 8045–50.

Saporito RA, Donnelly MA, Norton R, Garraffo HM, Spande T, Daly JW (2007) Oribatid mites as a major dietary source for alkaloids in poison frogs. *Proceedings of the National Academy of Sciences of the U.S.A.*, **104**, 8885–90.

Sargeant AB, Eberhardt LE (1975) Death feigning by ducks in response to predation by red foxes (*Vulpes fulva*). *American Midland Naturalist*, **94**, 108–19.

Sargent TD (1966) Background selections of geometrid and noctuid moths. *Science*, **154**, 1674–5.

Sargent TD (1968) Cryptic moths: effects on background selections of painting circumocular scales. *Science*, **159**, 100–1.

Sargent TD (1969) Background selections of the pale and melanic forms of the cryptic moth, *Phigalia titea* (Cramer). *Nature*, **222**, 585–6.

Sargent TD (1973) Studies on the *Catocala* (Noctuidae) of southern New England. IV. A preliminary analysis of beak-damaged specimens, with discussion of anomaly as a potential anti-predator function of hindwing diversity. *Journal of the Lepidopterists' Society*, **27**, 175–92.

Sargent TD (1977) Studies on the *Catocala* (Noctuidae) of southern New England. V. The records of Sydney A. Hessel from Washington, Connecticut, 1961–1973. *Journal of the Lepidopterists' Society*, **31**, 1–16.

Sargent TD (1978) On the maintenance of stability in hindwing diversity among moths of the genus *Catocala* (Lepidoptera: Noctuidae). *Evolution*, **32**, 424–34.

Sargent TD (1995) On the relative acceptabilities of local butterflies and moths to local birds. *Journal of the Lepidopterists' Society*, **49**, 148–62.

Sargent TD, Keiper RR (1969) Behavioural adaptations of cryptic moths. I. Preliminary studies on bark-like species. *Journal of the Lepidopterists' Society*, **23**, 1–9.

Sargent TD, Owen DF (1975) Apparent stability in hindwing diversity in samples of moths of varying species composition. *Oikos*, **26**, 205–10.

Sato (1979) Experimental studies on parasitization by *Apanteles glomeratus*. IV. Factors leading a female to the host. *Physiological Entomology*, **4**, 63–70.

Saul-Gershenz LS, Millar JG (2006) Phoretic nest parasites use sexual deception to obtain transport to their host's nest. *Proceedings of the National Academy of Sciences, U.S.A.*, **103**, 14039–44.

Savage JM, Slowinski JB (1992) The colouration of the venomous coral snakes (family Elapidae) and their mimics (families Aniliidae and Colubridae). *Biological Journal of the Linnean Society*, **45**, 235–54.

Savage JM, Slowinski JB (1996) Evolution of coloration, urotomy and coral snake mimicry in the snake genus *Scaphiodontophis* (Serpentes: Colubridae). *Biological Journal of the Linnean Society*, **57**, 129–94.

Savile D (1976) Evolution of rust fungi (Uredinales) as reflected by their ecological problems. In T Dobzhansky, M Hecht and W Steere (eds), *Evolutionary Biology*. Plenum Press, New York, pp. 137–207.

Savile D (1980) Ecology, convergent evolution, and classification of the Uredinales. *Reports of the Tottori Mycological Institute (Japan)*, **18**, 275–81.

Sazima I (1991) Caudal luring in two neotropical pitvipers, *Bothrops jararaca* and *B. jararacussu*. *Copeia*, **1991**, 245–48.

Sazima I (2002a) Juvenile snooks (Centropomidae) as mimics of mojarras (Gerreidae), with a review of aggressive mimicry in fishes. *Environmental Biology of Fishes*, **65**, 37–45.

Sazima I (2002b) Juvenile grunt (Haemulidae) mimicking a venomous leatherjacket (Carangidae), with a summary of Batesian mimicry in marine fishes. *Aqua: Journal of Ichthyology and Aquatic Biology*, **6**, 61–8.

Sazima I (2010) Five instances of bird mimicry suggested for Neotropical birds: a brief reappraisal. *Revista Brasileira de Ornitologia*, **18**, 328–35.

Sazima I, Puorto G (1993) Feeding technique of juvenile *Tropidodryas striaticeps*: probable caudal luring in a colubrid snake. *Copeia*, **1993**, 222–6.

Sazima I, Krajewski JP, Bonaldo RM, Sazima C (2005) Wolf in a sheep's clothes: juvenile coney (*Cephalopholis fulva*) as an aggressive mimic of the brown chromis (*Chromis multilineata*). *Neotropical Ichthyology*, **3**, 315–8.

Sbordoni V, Bullini L (1971) Further observations on mimicry in *Zygaena ephialtes* (Lepidoptera Zygaenidae). *Fragmenta Entomologica*, **8**, 49–56.

Sbordoni V, Bullini L, Scarpelli G, Forestiero S, Rampini M (1979) Mimicry in the burnet moth *Zygaena ephialtes*: population studies and evidence of a Batesian Müllerian situation. *Ecological Entomology*, **4**, 83–93.

Scaife M (1976a) The response to eye-like shapes by birds. I. The effect of context: a predator and a strange bird. *Animal Behaviour*, **24**, 195–9.

Scaife M (1976b) The response to eye-like shapes by birds. II. The importance of staring, pairedness and shape. *Animal Behaviour*, **24**, 200–6.

Schaefer HM (2011) Why fruits go to the dark side. *Acta Oecologica*, **37**, 604–10.

Schaefer HM, Ruxton GD (2008) Fatal attraction: carnivorous plants roll out the red carpet to lure insects. *Biology Letters*, **4**, 153–5.

Schaefer HM, Ruxton GD (2009) Deception in plants: mimicry or perceptual exploitation? *Trends in Ecology and Evolution*, **24**, 676–85.

Schaefer HM, Ruxton GD (2014) Fenestration: a window of opportunity for carnivorous plants. *Biology Letters*, **10**: 20140134.

Schaefer HM, Stobbe N (2006) Disruptive coloration provides camouflage independent of background matching. *Proceedings of the Royal Society of London, Series B: Biological Sciences*, **273**, 2427–32.

Schaefer HM, Schaefer V, Vorobyev M (2007) Are fruit colors adapted to consumer vision and birds equally efficient in detecting colorful signals? *The American Naturalist*, **169**, 159–69.

Schaefer HM, Valido A, Jordano P (2014) Birds see the true colours of fruits to live off the fat of the land. *Proceedings of the Royal Society of London, Series B: Biological Sciences*, **281**: 20132516.

Schall JJ, Pianka ER (1980) Evolution of escape behaviour diversity. *The American Naturalist*, **115**, 551–66.

Scharff N, Hormiga G (2012) First evidence of aggressive chemical mimicry in the Malagasy orb weaving spider *Exechocentrus lancearius* Simon, 1889 (Arachnida: Araneae: Araneidae) and description of a second species in the genus. *Arthropod Systematics & Phylogeny*, **70**, 107–18.

Schelly R, Takahashi T, Bills R, Hori M (2007) The first case of aggressive mimicry among lamprologines in a new species of *Lepidiolamprologus* (Perciformes: Cichlidae) from Lake Tanganyika. *Zootaxa*, **1638**, 39–49.

Schemske DW (1981) Floral convergence and pollinator sharing in two bee-pollinated tropical herbs. *Ecology*, **62**, 946–54.

Schemske DW, Ågren J (1995) Deceit pollination and selection on female flower size in *Begonia involucrata*: an experimental approach. *Evolution*, **49**, 207–14.

Schemske DW, Ågren J, Le Corff J (1996) Deceit pollination in the monoecious, neotropical herb *Begonia oaxacana*. In DG Lloyd and SCH Barrett (eds.), *Floral Biology*. Chapman and Hall, London, pp. 292–318.

Schiestl FP (2005) On the success of a swindle: pollination by deception in orchids. *Naturwissenschaften*, **92**, 255–64.

Schiestl FP, Ayasse M (2002) Do changes in floral odor cause speciation in sexually deceptive orchids? *Plant Systematics and Evolution*, **234**, 111–9.

Schiestl FP, Peakall R (2005) Two orchids attract different pollinators with the same floral odour compound: ecological and evolutionary implications. *Functional Ecology*, **19**, 674–80.

Schiestl FP, Ayasse M, Paulus HF, Löfstedt C, Hansson BS, Ibarra F, Francke W (1999) Orchid pollination by sexual swindle. *Nature*, **399**, 421–2.

Schiestl FP, Ayasse M, Paulus HF, Löfstedt C, Hansson BS, Ibarra F, Francke W (2000) Sex pheromone mimicry in the early spider orchid (*Ophrys sphegodes*): patterns of hydrocarbons as the key mechanism for pollination by sexual deception. *Journal of Comparative Physiology A – Neuroethology, Sensory, Neural and Behavioral Physiology*, **186**, 567–74.

Schiestl FP, Peakall R, Mant JG, Ibarra F, Schulz C, Francke S, Francke W (2003) The chemistry of sexual deception in an orchid-wasp pollination system. *Science*, **302**, 437–8.

Schiestl FP, Peakall R, Mant JG (2004) Chemical communication in the sexually deceptive orchid genus *Cryptostylis*. *Botanical Journal of the Linnean Society*, **144**, 199–205.

Schlenoff DH (1985) The startle response of blue jays to *Catocala* (Lepidoptera: Noctuidae) prey models. *Animal Behaviour*, **33**, 1057–67.

Schlichter D (1976) Macromolecular mimicry: substances released by sea anemones and their role in the protection of anemone fishes. In GO Mackie (ed.), *Coelenterate Ecology and Behaviour*. Plenum Press, New York, pp. 433–41.

Schmalhofer VR (2000) Diet-induced and morphological color changes in juvenile crab spiders (Araneae, Thomisidae). *Journal of Arachnology*, **28**, 56–60.

Schmidt JO (1990) Hymenoptera venoms: striving toward the ultimate defense against vertebrates. In DL Evans and JO Schmidt (eds), *Insect Defenses: Adaptive Mechanisms and Strategies of Prey and Predators*. State University of New York Press, Albany, New York, pp. 387–419.

Schmidt JO, Blum MS, Overal WL (1983) Hemolytic activities of stinging insect venoms. *Archives of Insect Biochemistry and Physiology*, **1**, 155–60.

Schmidt RS (1960) Predator behaviour and the perfection of incipient mimetic resemblances. *Behaviour*, **16**, 149–58.

Schmidt V, Schaefer HM (2004) Unlearned preference for red may facilitate recognition of palatable food in young omnivorous birds. *Evolutionary Ecology Research*, **6**, 919–25.

Schmidt V, Schaefer HM, Winkler H (2004) Conspicuousness, not colour as foraging cue in plant–animal signalling. *Oikos*, **106**, 551–7.

Schneider D, Boppré M, Schneider H, Thompson WR, Boriack CJ, Petty RL, Meinwald J (1975) A pheromone precursor and its uptake in male *Danaus* butterflies. *Journal of Comparative Physiology*, **97**, 245–56.

Schröder I (1993) Concealed ovulation and clandestine copulation: a female contribution to human evolution. *Ethology and Sociobiology*, **14**, 381–9.

Schuett GW, Clark DL, Kraus F (1984) Feeding mimicry in the rattlesnake *Sistrurus catenatus*, with comments on the evolution of the rattle. *Animal Behaviour*, **32**, 625–6.

Schuler W, Hesse E (1985) On the function of warning coloration: a black and yellow pattern inhibits prey-attack by naive domestic chicks. *Behavioral Ecology and Sociobiology*, **116**, 249–55.

Schultz TD (1986) Role of structural colors in predator avoidance by tiger beetles of the genus *Cicindela* (Coleoptera: Cicindelidae). *Bulletin of the Entomological Society of America*, **32**, 142–6.

Schultz TD, Fincke OM (2013) Lost in the crowd or hidden in the grass: signal apparency of female polymorphic damselflies in alternative habitats. *Animal Behaviour*, **86**, 923–31.

Scopece G, Cozzolino S, Johnson SD, Schiestl FP (2010) Pollination efficiency and the evolution of specialized deceptive pollination systems. *The American Naturalist*, **175**, 98–105.

Scott D (1986) Sexual mimicry regulates the attractiveness of mated *Drosophila melanogaster* females. *Proceedings of the National Academy of Sciences, U.S.A.*, **83**, 8429–33.

Scott JA (1973) Lifespan of butterflies. *Journal of Research on the Lepidoptera*, **12**, 225–30.

Scott-Samuel NE, Baddeley R, Palmer CE, Cuthill IC (2011) Dazzle camouflage affects speed perception. *PLoS One*, **6**: 6.

Scriber JM, Deering MD, Francke LN, Wehling WF, Lederhouse RC (1996a) Notes on swallowtail population dynamics of three *Papilio* species in South-Central Florida (Lepidoptera: Papilionidae). *Holarctic Lepidoptera*, **5**, 53–62.

Scriber JM, Hagen RH, Lederhouse RC (1996b) Genetics of mimicry in the tiger swallowtail butterflies, *Papilio glaucus* and *P. canadensis* (Lepidoptera: Papilionidae). *Evolution*, **50**, 222–36.

Sedeek KE, Whittle E, Guthörl D, Grossniklaus U, Shanklin J, Schlüter PM (2016) Amino acid change in an orchid desaturase enables mimicry of the pollinator's sex pheromone. *Current Biology*, **26**, 1505–11.

Seevers CH (1965) The systematics, evolution and zoogeography of staphylinid beetles associated with army ants (Coleoptera: Staphylonidae). *Fieldiana, Zoology*, **47**, 137–351.

Seiber JN, Tuskes PM, Brower LP, Nelson CJ (1980) Pharmacodynamics of some individual milkweed cardenolides fed to larvae of the monarch butterfly (*Danaus plexippus*). *Journal of Chemical Ecology*, **6**, 321–39.

Seigel JA, Adamson TA (1983) Batesian mimicry between a cardinalfish (Apogonidae) and a venomous scorpionfish (Scorpaenidae) from the Philippine Islands. *Pacific Science*, **37**, 75–9.

Sekimura T, Fujihashi Y, Takeuchi Y (2014) A model for population dynamics of the mimetic butterfly *Papilio polytes* in the Sakishima Islands, Japan. *Journal of Theoretical Biology*, **361**, 133–40.

Selander RB, Miller JL, Mathieu JM (1963) Mimetic associations of lycid and cerambycid beetles (Coleoptera) in Coahuila, Mexico. *Journal of the Kansas Entomological Society*, **36**, 45–52.

Servedio MR (2000) The effects of predator learning, forgetting, and recognition errors on the evolution of warning coloration. *Evolution*, **54**, 751–63.

Servedio MR, Lande R (2003) Coevolution of an avian host and its parasitic cuckoo. *Evolution*, **57**, 1164–75.

Sexton OJ (1960) Experimental studies of artificial Batesian mimics. *Behaviour*, **15**, 244–52.

Shapiro AM (1980) Egg-load assessment and carryover diapause in *Anthocharis* (Pieridae). *Journal of the Lepidopterists' Society*, **34**, 307–15.

Shapiro AM (1981a) The pierid red-egg syndrome. *The American Naturalist*, **117**, 276–94.

Shapiro AM (1981b) Egg-mimics of *Streptanthus* (Cruciferae) deter oviposition by *Pieris sisymbrii* (Lepidoptera: Pieridae). *Oecologia (Berlin)*, **48**, 142–3.

Shcherbakov DE (2007) Mesozoic spider mimics – Cretaceous Mimarachnidae fam.n. (Homoptera: Fulgoroidea). *Russian Entomological Journal*, **16**, 259–64.

Shearer MK, Katz LS (2006) Female–female mounting among goats stimulates sexual performance in males. *Hormones and Behavior*, **50**, 33–7.

Shelford R (1902) Observations on some mimetic insects and spiders from Borneo and Singapore, with appendices containing descriptions of new species. *Proceedings of the General Meetings for Scientific Business of the Zoological Society of London*, **2**, 230–84, pls. XIX–XXIII.

Shelomi M (2011) Phasmid eggs do not survive digestion by quails and chickens. *Journal of Orthoptera Research*, **20**, 159–62.

Sheppard PM (1959) The evolution of mimicry: a problem in ecology and genetics. Cold Springs Harbour Symposium on Quantitative Biology, **24**, 131–40.

Sheppard PM, Turner JRG (1977) The existence of Müllerian mimicry. *Evolution*, **31**, 452–3.

Sheppard PM, Turner JRG, Brown KS Jr, Benson WW, Singer MC (1985) Genetics and the evolution of Muellerian mimicry in *Heliconius* butterflies. *Philosophical Transactions of the Royal Society of London, Series B: Biological Sciences*, **308**, 433–613.

Sherman AR (1910) At the sign of the northern flicker. *The Wilson Bulletin*, **22**, 135–71.

Sherratt TN (2001) The evolution of female-limited polymorphisms in damselflies: a signal detection model. *Ecology Letters*, **4**, 22–9.

Sherratt TN (2002a) The evolution of imperfect mimicry. *Behavioral Ecology*, **13**, 821–6.

Sherratt TN (2002b) The coevolution of warning signals. *Proceedings of the Royal Society of London, Series B: Biological Sciences*, **269**, 741–6.

Sherratt TN (2006) Spatial mosaic formation through frequency-dependent selection in Müllerian mimicry complexes. *Journal of Theoretical Biology*, **240**, 165–74.

Sherratt TN, Forbes MR (2001) Sexual differences in coloration of coenagrionid damselflies (Odonata): a case of intraspecific aposematism? *Animal Behaviour*, **62**, 653–60.

Sherratt TN, Franks DW (2005) Do unprofitable prey evolve traits that profitable prey find difficult to exploit? *Proceedings of the Royal Society of London, Series B: Biological Sciences*, **272**, 2441–7.

Sherratt TN, Wilkinson DM, Bain RS (2005) Explaining Dioscorides' "Double Difference": why are some mushrooms poisonous, and do they signal their unprofitability? *The American Naturalist*, **166**, 767–75.

Sherzer WH (1896) Pebble mimicry in Philippine Island beans. *Botanical Gazette*, **21**, 235–7.

Shettleworth SJ (1972) The role of novelty in learned avoidance of unpalatable 'prey' by domestic chicks (*Gallus gallus*). *Animal Behaviour*, **20**, 29–35.

Shi J, Cheng J, Luo D, Shangguan FZ, Luo YB (2007) Pollination syndromes predict brood-site deceptive pollination by female hoverflies in *Paphiopedilum dianthum* (Orchidaceae). *Acta Phytotaxonomica Sinica*, **45**, 551–60.

Shi J, Luo YB, Bernhardt P, Ran JC, Liu ZJ, Zhou Q (2008) Pollination by deceit in *Paphiopedilum barbigerum*: a staminode exploits the innate colour preferences of hoverflies (Syrphidae). *Plant Biology*, **11**, 17–28.

Shideler RT (1973) The importance of mimic pattern and position in an artificial mimicry situation. *Behaviour*, **47**, 268–80.

Shine R (1980) Ecology of the Australian death adder *Acanthophis antarcticus* (Elapidae): evidence for convergence with the Viperidae. *Herpetologica*, **36**, 281–9.

Shine R, Harlow P, Lemaster MP, Moore IT, Mason RT (2000) The transvestite serpent: why do male garter snakes court (some) other males? *Animal Behaviour*, **59**, 349–59.

Shine R, Langkilde T, Mason RT (2012) Facultative pheromonal mimicry in snakes: "she-males" attract courtship only when it is useful. *Behavioral Ecology and Sociobiology*, **66**, 691–5.

Shuster SM, Wade MJ (1991) Equal mating success among male reproductive strategies in a marine isopod. *Nature*, **350**, 608–10.

Shutt DA (1976) The effects of plant oestrogens on animal reproduction. *Endeavour*, **35**, 110–13.

Shuttleworth A (2016) Smells like debauchery: the chemical composition of semen-like, sweat-like and faintly foetid floral odours in *Xysmalobium* (Apocynaceae: Asclepiadoideae). *Biochemical Systematics and Ecology*, **66**, 63–75.

Shuttleworth A, Johnson SD (2010) The missing stink: sulphur compounds can mediate a shift between fly and wasp pollination systems. *Proceedings of the Royal Society of London, Series B: Biological Sciences*, **277**, 2811–9.

Sibley CG (1955) Behavioural mimicry in the titmice (Paridae) and certain other birds. *The Wilson Bulletin*, **67**, 128–32.

Sick H (1997) *Ornitologia brasileira, uma introdução*. Editora Nova Fronteira, Rio de Janeiro.

Sikkel PC, Hardison PD (1992) Interspecific feeding association between the goatfish *Mulloides martinicus* (Mullidae) and a possible aggressive mimic, the snapper *Ocyurus chrysurus* (Lutjanidae). *Copeia*, **1992**, 914–17.

Silberglied RE, Eisner T (1969) Mimicry of Hymenoptera by beetles with unconventional flight. *Science*, **163**, 486–8.

Silberglied RE, Aiello A, Windsor DM (1980) Disruptive coloration in butterflies: lack of support in *Anartia fatima*. *Science*, **209**, 617–19.

Sillén-Tullberg B[1] (1985a) Higher survival of an aposematic than of a cryptic form of a distasteful bug. *Oecologia (Berlin)*, **67**, 411–15.

Sillén-Tullberg B (1985b) The significance of coloration *per se*, independent of background, for predator avoidance of aposematic prey. *Animal Behaviour*, **33**, 1382–4.

Sillén-Tullberg B (1988) Evolution of gregariousness in aposematic butterfly larvae: a phylogenetic analysis. *Evolution*, **42**, 293–305.

Sillén-Tullberg B (1990) Do predators avoid groups of aposematic prey? An experimental test. *Animal Behaviour*, **40**, 856–60.

Sillén-Tullberg B (1992) Does gregariousness reduce attacks on aposematic prey? A reply to Cooper. *Animal Behaviour*, **43**, 165–7.

Sillén-Tullberg B, Bryant EH (1983) The evolution of aposematic coloration in distasteful prey: an individual selection model. *Evolution*, **37**, 993–1000.

Sillén-Tullberg B, Leimar O (1988) The evolution of gregariousness in distasteful insects as a defense against predators. *The American Naturalist*, **132**, 723–34.

Sillén-Tullberg B, Møller AP (1993) The relationship between "concealed ovulation" and mating systems in anthropoid primates. A phylogenetic analysis. *The American Naturalist*, **141**, 1–25.

Sillén-Tullberg B, Wiklund C, Järvi T (1982) Aposematic coloration in adults and larvae of *Lygaeus equestris* and its bearing on Müllerian mimicry: an experimental study on predation on living bugs by the great tit *Parus major*. *Oikos*, **39**, 131–6.

Simmons KEL (1972) Some adaptive features of seabird plumage types. *British Birds*, **65**, 465–79, 510–21.

Simmons RB, Weller SJ (2002) What kind of signals do mimetic tiger moths send? A phylogenetic test of wasp mimicry systems (Lepidoptera: Arctiidae: Euchromiini). *Proceedings of the Royal Society of London, Series B: Biological Sciences*, **269**, 983–90.

Simmons RE (1988) Food and the deceptive acquisition of mates by polygynous male harriers. *Behavioral Ecology and Sociobiology*, **23**, 83–92.

Sims SR (1983) The genetic and environmental basis of pupal colour dimorphism in *Papilio zelicaon* (Lepidoptera: Papilionidae). *Heredity*, **50**, 159–68.

Singer RB (2002) The pollination mechanism in *Trigonidium obtusum* Lindl. (Orchidaceae: Maxillariinae): sexual mimicry and trap-flowers. *Annals of Botany*, **89**, 157–63.

Singh I, Yadav AS, Mohanty KK, Katoch K, Sharma P, Mishra B, Bisht D, Gupta UD, Sengupta U (2015) Molecular mimicry between *Mycobacterium leprae* proteins (50S ribosomal protein L2 and Lysyl-tRNA synthetase) and myelin basic protein: a possible mechanism of nerve damage in leprosy. *Microbes and Infection*, **17**, 247–57.

Sivinski J (1981) Arthropods attracted to luminous fungi. *Psyche*, **88**, 383–90.

Sivinski J, Marshall S, Petersson E (1999) Kleptoparasitism and phoresy in the Diptera. *Florida Entomologist*, **82**, 179–97.

Skelhorn J, Rowe C (2005) Tasting the difference: do multiple defence chemicals interact in Müllerian mimicry? *Proceedings of the Royal Society of London, Series B: Biological Sciences*, **272**, 339–45.

Skelhorn J, Rowe C (2006a) Avian predators taste-reject aposematic prey on the basis of their chemical defence. *Biology Letters*, **2**, 348–50.

Skelhorn J, Rowe C (2006b) Taste-rejection by predators and the evolution of unpalatability in prey. *Behavioral Ecology & Sociobiology*, **60**, 550–5.

Skelhorn J, Rowe C (2007a) Predators' toxin burdens influence their strategic decisions to eat toxic prey. *Current Biology*, **17**, 1479–83.

Skelhorn J, Rowe C (2007b) Automimic frequency influences the foraging decisions of avian predators on aposematic prey. *Animal Behaviour*, **74**, 1563–72.

Skelhorn J, Rowe C (2009) Distastefulness as an antipredator defence strategy. *Animal Behaviour*, **78**, 761–6.

Skelhorn J, Ruxton GD (2010) Predators are less likely to misclassify masquerading prey when their models are present. *Biology Letters*, **6**, 597–9.

Skelhorn J, Ruxton GD (2011) Mimicking multiple models: polyphenetic masqueraders gain additional benefits from crypsis. *Behavioral Ecology*, **22**, 60–5.

Skelhorn J, Ruxton GD (2013) Size-dependent microhabitat selection by masquerading prey. *Behavioral Ecology*, **24**, 89–97.

Skelhorn J, Griksaitis D, Rowe C (2008) Colour biases are more than a question of taste. *Animal Behaviour*, **75**, 827–35.

Skelhorn J, Rowland HM, Speed MP, De Wert L, Quinn L, Jon Delf J, Ruxton GD (2010a) Size-dependent misclassification of masquerading prey. *Behavioral Ecology*, **21**, 1344–8.

Skelhorn J, Rowland HM, Ruxton GD (2010b) The evolution and ecology of masquerade. *Biological Journal of the Linnean Society*, **99**, 1–8.

Skelhorn J, Rowland HM, Speed MP, Ruxton GD (2010c) Masquerade: camouflage without crypsis. *Science* **327**, 51.

Skelhorn J, Rowland HM, Delf J, Speed MP, Ruxton GD (2011) Density-dependent predation influences the evolution and behaviour of masquerading prey. *Proceedings of the National Academy of Sciences, U.S.A.*, **108**, 6532–6.

Skrade PDB, Dinsmore SJ (2013) Egg crypsis in a ground-nesting shorebird influences nest survival. *Ecosphere*, **4**, 151

Slagsvold T, Dale S (1994) Why do female pied flycatchers mate with already mated males: deception or restricted mate sampling? *Behavioral Ecology and Sociobiology*, **34**, 239–50.

Slagsvold T, Sætre G-P (1991) Evolution of plumage color in male pied flycatchers (*Ficedula hypoleuca*): evidence for female mimicry. *Evolution*, **45**, 910–17.

Sláma K (1993) Ecdysteroids: insect hormones, plant defensive factors, or human medicine? *Phytoparasitica*, **21**, 3–8.

Sláma K, Williams CM (1965) The juvenile hormone. V. The sensitivity of the bug, *Pyrrhocoris apterus*, to a hormonally active factor in American paper-pulp. *Biological Bulletin*, **130**, 235–46.

Sledge MF, Dani FR, Cervo R, Dapporto L, Turillazzi S (2001) Recognition of social parasites as nest-mates: adoption of colony-specific host cuticular odours by the paper wasp parasite *Polistes sulcifer*. *Proceedings of the Royal Society of London, Series B: Biological Sciences*, **268**, 2253–60.

Slobodchikoff CN (1987) Aversive conditioning in a model-mimic system. *Animal Behaviour*, **35**, 75–80.

Smale L, Lee TM, Nelson R, Zucker I (1990) Prolactin counteracts effects of short day lengths on pelage growth in the meadow vole, *Microtus pennsylvanicus*. *Journal of Experimental Zoology*, **253**, 186–8.

Smilanich AM, Dyer LA, Chambers JQ, Bowers MD (2009) Immunological cost of chemical defence and the evolution of herbivore diet breadth. *Ecology Letters*, **12**, 612–21.

Smith AD, Wilson JS, Cognato AI (2015) The evolution of Batesian mimicry within the North American Asidini (Coleoptera: Tenebrionidae). *Cladistics*, **31**, 441–454.

Smith AG (1980) Environmental factors influencing pupal colour determination in Lepidoptera. II. Experiments with *Pieris rapae*, *Pieris napi* and *Pieris brassicae*. *Proceedings of the Royal Society of London, Series B: Biological Sciences*, **207**, 163–86.

Smith AP (1986) Ecology of a leaf color polymorphism in a tropical forest species: habitat segregation and herbivory. *Oecologia (Berlin)*, **69**, 283–7.

Smith BN, Meeuse BJD (1966) Production of volatile amines and skatole at anthesis in some arum lily species. *Plant Physiology*, **41**, 343–7.

Smith DAS (1973a) Batesian mimicry between *Danaus chrysippus* and *Hypolimnas misippus* (Lepidoptera) in Tanzania. *Nature*, **242**, 129–31.

Smith DAS (1975a) All-female broods in the polymorphic butterfly *Danaus chrysippus* L. and their ecological significance. *Heredity*, **34**, 363–71.

Smith DAS (1975b) Genetics of some polymorphic forms of the African butterfly *Danaus chrysippus* L. (Lepidoptera: Danaidae). *Entomologia Scandinavica*, **6**, 134–44.

Smith DAS (1976) Phenotypic diversity, mimicry and natural selection in the African butterfly *Hypolimnas misippus* L. (Lepidoptera: Nymphalidae). *Biological Journal of the Linnean Society*, **8**, 183–204.

Smith DAS (1978) The effect of cardiac glycoside storage on growth rate and adult size in the butterfly *Danaus chrysippus* (L.). *Experientia*, **34**, 845–6.

Smith DAS (1979) The significance of beak marks on the wings of an aposematic butterfly. *Nature*, **281**, 215–16.

Smith DAS (1981) Heterozygous advantage expressed through sexual selection in a polymorphic African butterfly. *Nature*, **289**, 174–5.

Smith DAS, Gordon IJ (1987) The genetics of the butterfly *Hypolimnas misippus* (L.): the classification of forms *misippus* and *inaria*. *Heredity*, **59**, 467–75.

Smith DAS, Shoesmith EA, Smith AG (1988) Pupal polymorphism in the butterfly *Danaus chrysippus* (L.): environmental, seasonal and genetic influences. *Biological Journal of the Linnean Society*, **33**, 17–50.

Smith DAS, Owen DF, Gordon IJ, Owiny AM (1993) Polymorphism and evolution in the butterfly *Danaus chrysippus* L. (Lepidoptera: Danainae). *Heredity*, **71**, 242–51.

Smith J, Kronforst MR (2013) Do *Heliconius* butterfly species exchange mimicry alleles? *Biology Letters*, **9**: 20130503.

Smith KGV (1956) On the Diptera associated with the stinkhorn (*Phallus impudicus* Pers.) with notes on other insects and invertebrates found on this fungus. *Proceedings of the Royal Entomological Society of London. Series A, General Entomology*, **31**, 49–55.

Smith RH (1974) Is the slow worm a Batesian mimic? *Nature*, **247**, 571–2.

Smith SM (1975) Innate recognition of coral snake pattern by a possible avian predator. *Science*, **187**, 759–60.

Smith SM (1977) Coral-snake pattern recognition and stimulus generalization by naive great kiskadees (Aves: Tyrannidae). *Nature*, **265**, 535–6.

Smith SM (1980) Responses of naive temperate birds to warning coloration. *American Midland Naturalist*, **103**, 344–52.

Smithers SR (1972) Recent advances in the immunology of schistosomiasis. *British Medical Bulletin*, **28**, 49–54.

Smithson A, Macnair MR (1997) Negative frequency-dependent selection by pollinators on artificial flowers without rewards. *Evolution*, **51**, 715–23.

Smith Trail DR (1980) Behavioural interactions between parasites and hosts: host suicide and the evolution of complex life cycles. *The American Naturalist*, **116**, 77–91.

Soane ID, Clarke B (1973) Evidence for apostatic selection by predators using olfactory cues. *Nature*, **241**, 62–3.

Solano-Ugalde A (2011) Notes on the roosting site, foraging behaviour, and plumage crypsis of the rufous potoo (*Nyctibius bracteatus*) from the Ecuadorian Amazon. *Boletín SAO*, **20**, 39–42.

Soler JJ, Avilés JM, Møller AP, Moreno J (2012) Attractive blue-green egg coloration and cuckoo-host coevolution. *Biological Journal of the Linnean Society*, **106**, 154–68.

Solis JC, de Lope F (1995) Nest and egg crypsis in the ground-nesting stone curlew *Burhinus oedicnemus*. *Journal of Avian Biology*, **26**, 135–8.

Solis MA, Yen S-H, Goolsby JH, Wright T, Pemberton R, Winotal A, Chattrukul U, Thagong A, Rimbut S (2005) *Siamusotima aranea*, a new stem-boring Musotimine (Lepidoptera: Crambidae) from Thailand feeding on *Lygodium flexuosum* (Schizaeaceae). *Annals of the Entomological Society of America*, **98**, 887–95.

Soltau U, Döttorl S, Liede-Schumann S (2009) Leaf variegation in *Caladium steudneriifolium* (Araceae): a case of mimicry? *Evolutionary Ecology*, **23**, 503–12.

Sommer V, Vasey PL (2006) *Homosexual Behaviour in Animals, an Evolutionary Perspective*. Cambridge University Press, Cambridge.

Sordahl TA (1988) The American avocet (*Recurvirostra americana*) as a paradigm for adult automimicry. *Evolutionary Ecology*, **2**, 189–96.

Sorenson MD, Payne RB (2005) A molecular genetic analysis of the cuckoo phylogeny. In RB Payne (ed.), *The Cuckoos*. Oxford University Press, Oxford, UK, pp. 68–94.

Sorley L (ed.) (2010) *The Vietnam War: An Assessment by South Vietnam's Generals*. Texas Tech University Press, Lubbock.

Sourakov A (2013a) Two heads are better than one: false head allows *Calycopis cecrops* (Lycaenidae) to escape predation by a jumping spider, *Phidippus pulcherrimus* (Salticidae). *Journal of Natural History*, **47**, 1047–54.

Sourakov A (2013b) Pierce faces in the Florida tigers: moths mimicking spiders. *News of the Lepidopterists' Society*, **55**, 60–1.

Spaethe J, Moser WH, Paulus HF (1997) Increase of pollinator attraction by means of a visual signal in the sexually deceptive orchid, *Ophrys heldreichii* (Orchidaceae). *Plant Systematics and Evolution*, **264**, 31–40.

Spaethe J, Streinzer M, Paulus HF (2010) Why sexually deceptive orchids have colored flowers. *Communicative & Integrative Biology*, **3**, 139–41.

Spear LB, Ainsley DG (1993) Kleptoparasitism by kermadec petrels, jaegers, and skuas in the eastern tropical Pacific: evidence of mimicry by two species of *Pterodroma*. *The Auk*, **110**, 222–33.

Speed MP (1993a) Muellerian mimicry and the psychology of predation. *Animal Behaviour*, **45**, 571–80.

Speed MP (1993b) When is mimicry 'good' for predators? *Animal Behaviour*, **46**, 1246–8.

Speed MP (1999a) Robot predators in virtual ecologies: the importance of memory in mimicry studies. *Animal Behaviour*, **57**, 203–13.

Speed MP (1999b) Batesian, quasi-Batesian or Müllerian mimicry? Theory and data in mimicry research. *Evolutionary Ecology*, **13**, 755–76.

Speed MP (2000) Warning signals, receiver psychology and predator memory. *Animal Behaviour*, **60**, 269–78.

Speed MP (2014) Mimicry. in Encyclopedia of Life Sciences (*eLS*). John Wiley & Sons, Ltd, Chichester.

Speed MP, Ruxton GD (2005a) Aposematism: what should our starting point be? *Proceedings of the Royal Society of London, Series B: Biological Sciences*, **272**, 431–8.

Speed MP, Ruxton GD (2005b) Warning displays in spiny animals: one (more) evolutionary route to aposematism. *Evolution*, **59**, 2499–508.

Speed MP, Ruxton GD (2007) How bright and how nasty: explaining diversity in warning signal strength. *Evolution*, **61**, 623–35.

Speed MP, Ruxton GD (2010) Imperfect Batesian mimicry and the conspicuousness costs of mimetic resemblance. *The American Naturalist*, **176**, E1–E14.

Speed MP, Turner JRG (1999) Learning and memory in mimicry: II. Do we understand the mimicry spectrum? *Biological Journal of the Linnean Society*, **67**, 281–312.

Speed MP, Alderson NJ, Hardman C, Ruxton GD (2000) Testing Müllerian mimicry: an experiment with wild birds. *Proceedings of the Royal Society of London, Series B: Biological Sciences*, **265**, 725–31.

Speed MP, Kelly DJ, Davidson AM, Ruxton GD (2005) Countershading enhances crypsis with some bird species but not others. *Behavioral Ecology*, **16**, 327–34.

Spinner Hansen L, Fernández González S, Toft S, Bilde T (2008) Thanatosis as an adaptive male mating strategy in the nuptial gift-giving spider *Pisaura mirabilis*. *Behavioral Ecology*, **19**, 546–51.

Spottiswoode CN (2010) The evolution of host-specific variation in cuckoo eggshell strength. *Journal of Evolutionary Biology*, **23**, 1792–9.

Spottiswoode CN (2013) A brood parasite selects for its own eggs traits. *Biology Letters*, **9**: 20130573.

Spottiswoode CN, Stevens M (2010) Visual modeling shows that avian host parents use multiple visual cues in rejecting parasitic eggs. *Proceedings of the National Academy of Sciences, U.S.A.*, **107**, 8672–6.

Spottiswoode CN, Stevens M (2011) How to evade a coevolving brood parasite: egg discrimination versus egg variability as host defences. *Proceedings of the Royal Society of London, Series B: Biological Sciences*, **278**, 3566–73.

Spottiswoode CN, Stevens M (2012) Host-parasite arms races and rapid changes in bird egg appearance. *The American Naturalist*, **179**, 633–48.

Spottiswoode CN, Stryjewski KF, Quader S, Colebrook-Robjent JFR, Sorenson MD (2011) Ancient host specificity within a single species of brood parasitic bird. *Proceedings of the National Academy of Sciences, U.S.A.*, **108**, 17738–42.

Sprengel CK (1793) *Das entdeckte Geheimnis der Natur in Bau und in der Befruchtung der Blumen*. Viewig, Berlin.

Springer VGW, Smith-Vaniz F (1972) Mimetic relationships involving the fishes of the family Blenniidae. *Smithsonian Contributions to Zoology*, **112**, 1–36.

Srinivasan MV, Davey M (1995) Strategies for active camouflage of motion. *Proceedings of the Royal Society of London, Series B: Biological Sciences*, **259**, 19–25.

Srivastava A, Borries C, Sommer V (1991) Homosexual mounting in free-ranging female Hanuman langurs (*Presbytis entellas*). *Archives of Sexual Behaviour*, **20**, 487–512.

Srygley RB (1994) Locomotor mimicry in butterflies? The associations of positions of centres of mass among groups of mimetic, unprofitable prey. *Philosophical Transactions of the Royal Society of London, Series B: Biological Sciences*, **343**, 145–55.

Srygley RB (2004) The aerodynamic costs of warning signals in palatable mimetic butterflies and their distasteful models. *Proceedings of the Royal Society of London, Series B: Biological Sciences*, **271**, 589–94.

Srygley RB, Chai P (1990a) Flight morphology of neotropical butterflies – palatability and distribution of mass to the thorax and abdomen. *Oecologia (Berlin)*, **84**, 491–9.

Srygley RB, Chai P (1990b) Predation and the elevation of thoracic temperature in brightly colored neotropical butterflies. *The American Naturalist*, **135**, 766–87.

Stachowicz JJ, Hay ME (1999) Reducing predation through chemically mediated camouflage: indirect effects of plant defenses on herbivores. *Ecology*, **80**, 495–509.

Stachowicz JJ, Hay ME (2000) Geographic variation in camouflage specialization by a decorator crab. *The American Naturalist*, **156**, 59–71.

Staddon JER, Gendron RP (1983) Optimal detection of cryptic prey may lead to predator switching. *The American Naturalist*, **122**, 843 8.

Stamp NE (1981) Behaviour of parasitized aposematic caterpillars: advantageous to the parasitoid or the host? *The American Naturalist*, **118**, 715–25.

Stamps JA, Gon SM III (1983) Sex-biased pattern variation in the prey of birds. *Annual Review of Ecology and Systematics*, **14**, 231–53.

Stankowich T, Caro T, Cox M (2011) Bold coloration and the evolution of aposematism in terrestrial carnivores. *Evolution*, **65**, 3090–9.

Starling M, Heinsohn R, Cockburn A, Langmore NE (2006) Cryptic gentes revealed in pallid cuckoos *Cuculus pallidus* using reflectance spectrophotometry. *Proceedings of the Royal Society of London, Series B: Biological Sciences*, **273**, 1929–34.

Starnecker G, Hazel WN (1999) Convergent evolution of neuroendocrine control of phenotypic plasticity in pupal colour in butterflies. *Proceedings of the Royal Society of London, Series B: Biological Sciences*, **266**, 2409–12.

Starrett A (1993) Adaptive resemblance: a unifying concept for mimicry and crypsis. *Biological Journal of the Linnean Society*, **48**, 299–317.

Stearn WT (1981) Henry Walter Bates (1812–1892), discoverer of Batesian mimicry. *Biological Journal of the Linnean Society*, **6**, 5–7.

Stebbins CE, Galán JE (2001) Structural mimicry in bacterial virulence. *Nature*, **412**, 701–5.

Stefanescu C (2004) Seasonal change in pupation behaviour and pupal mortality in a swallowtail butterfly. *Animal Biodiversity and Conservation*, **27**, 25–36.

Stehr G (1959) Hemolymph polymorphism in a moth and the nature of sex-controlled inheritance. *Evolution*, **13**, 537–60.

Steidle JLM, Dettner K (1993) Chemistry and morphology of the tergal gland of free-living adult aleocharinae (Coleoptera: Staphylinidae) and its phylogenetic significance. *Systematic Entomology*, **18**, 149–68.

Steiner KE, Whitehead VB, Johnson SD (1994) Floral and pollinator divergence in two sexually deceptive South African orchids. *American Journal of Botany*, **81**, 185–94.

Stensmyr MC, Urru I, Collu I, Celander M, Hansson BS, Angioy A-M (2002) Pollination: rotting smell of deadhorse arum florets. *Nature (London)*, **420**, 625–6.

Sterkel AK, Lorenzini JL, Fites JS, Vignesh KS, Sullivan TD, Wuthrich M, Brandhorst T, Hernandez-Santos N, Deepe DS, Jr., Klein BS (2016) Fungal mimicry of a mammalian aminopeptidase disables innate immunity and promotes pathogenicity. *Cell Host & Microbe*, **19**, 361–74.

Stermitz FR, Gardner DR, McFarland N (1988) Iridoid glycoside sequestration by two aposematic *Penstemon*-feeding geometrid larvae. *Journal of Chemical Ecology*, **14**, 435–41.

Sternburg JG, Waldbauer GP, Jeffords MR (1977) Batesian mimicry: selective advantage of colour pattern. *Science*, **195**, 681–3.

Sternfeld R (1910) Zur Schlangenfauna Deutsch-Südwestafrikas. Mehrere Fälle von Mimikry bei afrikanischen Schlangen. *Mit Mitteilungen aus dem Zoologischen Museum in Berlin*, **5**, 51–60.

Stevens M (2005) The role of eyespots as anti-predator mechanisms, principally demonstrated in the Lepidoptera. *Biological Reviews*, **80**, 573–88.

Stevens M (2007) Predator perception and the interrelation between different forms of protective coloration. *Proceedings of the Royal Society of London, Series B: Biological Sciences*, **274**, 1457–64.

Stevens M (2013a) Evolutionary ecology: knowing how to hide your eggs. *Current Biology*, **23**, R106–8.

Stevens M (2013b) Bird brood parasitism. *Current Biology*, **23**, R909–13.

Stevens M (2015) *Cheats and Deceits: How Animals and Plants Exploit and Mislead*. Oxford University Press, Oxford.

Stevens M, Merilaita S (2009a) Animal camouflage: current issues and new perspectives. *Philosophical Transactions of the Royal Society of London, Series B: Biological Sciences*, **364**, 423–7.

Stevens M, Merilaita S (2009b) Defining disruptive coloration and distinguishing its functions. *Philosophical Transactions of the Royal Society of London, Series B: Biological Sciences*, **364**, 481–8.

Stevens M, Merilaita S (eds) (2011) *Animal Camouflage: Mechanisms and Functions*. Cambridge University Press, Cambridge, UK.

Stevens M, Ruxton GD (2014) Do animal eyespots really mimic eyes? *Current Zoology*, **60**, 26–36.

Stevens M, Hopkins E, Hinde W, Adcock A, Connolly Y, Troscianko T, Cuthill IC (2007) Field experiments on the effectiveness of 'eyespots' as predator deterrents. *Animal Behaviour*, **74**, 1215–27.

Stevens M, Hardman CJ, Stubbins CL (2008a) Conspicuousness, not eye mimicry, makes 'eyespots' effective antipredator signals. *Behavioral Ecology*, **19**, 525–31.

Stevens M, Stubbins CL, Hardman CJ (2008b) The anti-predator function of 'eyespots' on camouflaged and conspicuous prey. *Behavioral Ecology and Sociobiology*, **62**, 1787–93.

Stevens M, Yule DH, Ruxton GD (2008c) Dazzle coloration and prey movement. *Proceedings of the Royal Society of London, Series B: Biological Sciences*, **275**, 2639–43.

Stevens M, Graham J, Winney IS, Cantor A (2008d) Testing Thayer's hypothesis: can camouflage work by distraction? *Biology Letters*, **4**, 648–50.

Stevens M, Cantor A, Graham J, Winney IS (2009a) The function of animal 'eyespots': conspicuousness but not eye mimicry is key. *Current Zoology*, **55**, 319–26.

Stevens M, Winney IS, Cantor A, Graham J (2009b) Outline and surface disruption in animal camouflage. *Proceedings of the Royal Society of London, Series B: Biological Sciences*, **276**, 781–6.

Stevens M, Castor-Perry SA, Price JRF (2009c) The protective value of conspicuous signals is not impaired by shape, size, or position asymmetry. *Behavioral Ecology*, **20**, 96–102.

Stevens M, Searle WT, Seymour JE, Marshall KLE, Ruxton GD (2011) Motion dazzle and camouflage as distinct anti-predator defenses. *BMC Biology*, **9**: 81.

Stevens M, Rong CP, Todd PA (2013a) Colour change and camouflage in the horned ghost crab *Ocypode ceratophthalmus*. *Biological Journal of the Linnean Society*, **109**, 257–70.

Stevens M, Marshall KLA, Troscianko J, Finlay S, Burnand D, Chadwick SL (2013b) Revealed by conspicuousness: distractive markings reduce camouflage. *Behavioral Ecology*, **24**, 213–22.

Stevens M, Troscianko J, Marshall KLA, Finlay S (2013c) What is camouflage through distractive markings? A reply to Merilaita et al. (2013). *Behavioral Ecology*, **24**, e1272–3.

Stevens M, Lown AE, Denton AM (2014a) Rockpool gobies change colour for camouflage. *PLoS ONE*, **9(10):** e110325.

Stevens M, Lown AE, Wood LE (2014b) Colour change and camouflage in juvenile shore crabs, *Carcinus maenas*. *Frontiers in Ecology and Evolution*, **2**: 14.

Steward RC (1977a) Melanism and selective predation in three species of moths. *Journal of Animal Ecology*, **46** 483–96.

Steward RC (1977b) Genetic control of the melanic forms of the moths *Diurnea flagella* and *Allophyes oxyacanthae*. *Heredity*, **39**, 235–41.

Steward RC (1985) Evolution of resting behaviour in polymorphic 'industrial melanic' moth species. *Biological Journal of the Linnean Society*, **24**, 285–93.

Stewart AJA, Lees DR (1987) Localized industrial melanism in the spittlebug *Philaenus spumarius* (L.) (Homoptera: Aphrophoridae) in Cardiff docks, south Wales. *Biological Journal of the Linnean Society*, **31**, 333–45.

Stiles EW (1979) Evolution of colour pattern and pubescence characteristics in male bumblebees: automimicry vs. thermoregulation. *Evolution*, **33**, 941–57.

Stobbe N, Schaefer HM (2008) Enhancement of chromatic contrast increases predation risk for striped butterflies. *Proceedings of the Royal Society B – Biological Sciences*, **275**, 1535–41.

Stoddard MC (2012) Mimicry and masquerade from the avian visual perspective. *Current Zoology*, **58**, 630–48.

Stoddard MC, Stevens M (2010) Pattern mimicry of host eggs by the common cuckoo, as seen through a bird's eye. *Proceedings of the Royal Society of London, Series B: Biological Sciences*, **277**, 1387–93.

Stoddard MC, Stevens M (2011) Avian vision and the evolution of egg colour mimicry in the common cuckoo. *Evolution*, **65**, 2004–13.

Stoddard PK, Markham MR (2008) Signal cloaking by electric fish. *Bioscience*, **58**, 415–25.

Stoeffler M, Maier TS, Tolasch T, Steidle JLM (2007) Foreign-language skills in rove-beetles? Evidence for chemical mimicry of ant alarm pheromones in myrmecophilous *Pella* beetles (Coleoptera: Staphylinidae). *Journal of Chemical Ecology*, **33**, 1382–92.

Stoeffler M, Boettinger L, Tolasch T, Steidle JLM (2013) The tergal gland secretion of the two rare myrmecophilous species *Zyras collaris* and *Z. haworthi* (Coleoptera: Staphylinidae) and the effect on *Lasius fuliginosus*. *Psyche*, **2013**: 601073.

Stokke BG, Moksnes A, Røskaft E (2002) Brood parasites as selective agents for evolution in passerine birds. *Evolution*, **56**, 199–205.

Stökl J, Paulus HF, Dafni A, Schulz C, Francke W, Ayasse M (2005) Pollinator attracting odour signals in sexually deceptive orchids of the *Ophrys fusca* group. *Plant Systematics and Evolution*, **254**, 105–20.

Stökl J, Twele R, Erdmann DH, Francke W, Ayasse M (2007) Comparison of the flower scent of the sexually deceptive orchid *Ophrys iricolor* and the female sex pheromone of its pollinator *Andrena morio*. *Chemoecology*, **17**, 231–3.

Stökl J, Schlüter PM, Stuessy TF, Paulus HF, Fraberger R, Erdmann DH, Schulz C, Francke W, Assum G, Ayasse M (2009) Speciation in sexually deceptive orchids: pollinator-driven selection maintains discrete odour phenotypes in hybridizing species. *Biological Journal of the Linnean Society*, **98**, 439–51.

Stökl J, Strutz A, Dafni A, Svatos A, Doubsky J, Knaden M, Sachse S, Hansson BS, Stensmyr MC (2010) A deceptive pollination system targeting drosophilids through olfactory mimicry of yeast. *Current Biology*, **20**, 1846–52.

Stökl J, Brodmann J, Dafni A, Ayasse M, Hansson BS (2011) Smells like aphids: orchid flowers mimic aphid alarm pheromones to attract hoverflies for pollination. *Proceedings of the Royal Society of London, Series B: Biological Sciences*, **278**, 1216–22.

Stoner CJ, Bininda-Emonds ORP, Caro T (2003) The adaptive significance of coloration in lagomorphs. *Biological Journal of the Linnean Society*, **79**, 309–28.

Stournaras KE, Lo E, Böhning-Gaese K, Cazetta E, Dehling DM, Schleuning M, Stoddard MC, Donoghue MJ, Prum RO, Schaefer HM (2013) How colorful are fruits? Limited color diversity in fleshy fruits on local and global scales. *New Phytologist*, **198**, 617–29.

Stowe MK, Tumlinson JH, Heath RR (1987) Chemical mimicry: bolas spiders emit components of moth prey species sex pheromones. *Science*, **236** 964–7.

Stradling DJ (1976) The nature of the mimetic patterns of the brassolid genera *Caligo* and *Eryphanis*. *Ecological Entomology*, **1**, 135–8.

Streinzer M, Paulus HF, Spaethe J (2009) Floral colour signal increases short-range detectability of a sexually deceptive orchid to its bee pollinator. *Journal of Experimental Biology*, **212**, 1365–70.

Strohm E, Kroiss J, Herzner G, Laurien-Kehnen C, Boland W, Schreier P, Schmitt T (2008) A cuckoo in wolves' clothing? Chemical mimicry in a specialized cuckoo wasp of the European beewolf (Hymenoptera, Chrysididae and Crabronidae). *Frontiers in Zoology*, **5**: 2.

Stuart-Fox D, Moussalli A (2008) Selection for social signalling drives the evolution of chameleon colour change. *PLoS Biology*, **6**: e25.

Stuart-Fox D, Moussalli A (2009) Camouflage, communication and thermoregulation: lessons from colour changing organisms. *Philosophical Transactions of the Royal Society of London, Series B, Biological Sciences*, **364**, 463–70.

Stuart-Fox D, Whiting MJ, Moussalli A (2006) Camouflage and colour change: antipredator responses to bird and snake predators across multiple populations in a dwarf chameleon. *Biological Journal of the Linnean Society*, **88**, 437–46.

Stuart-Fox D, Moussalli A, Whiting MJ (2008) Predator-specific camouflage in chameleons. *Biology Letters*, **4**, 326–9.

Stuckert AMM, Vanegas PJ, Summers K (2014) Experimental evidence for predator learning and Müllerian mimicry in Peruvian poison frogs (*Ranitomeya*, Dendrobatidae). *Evolutionary Ecology*, **28**, 413–26.

Sugiura N, Goubara M, Kitamura K, Inoue K (2002) Bumblebee pollination of *Cypripedium macranthos* var. *rebunense* (Orchidaceae): a possible case of floral mimicry of *Pedicularis schistostegia* (Orobanchiaceae). *Plant Systematics and Evolution*, **235**, 189–95.

Sulak KJ (1975) Cleaning behavior in the centrarchid fishes, *Lepomis macrochirus* and *Micropterus salmoides*. *Animal Behaviour*, **23**, 331–4.

Summers K, Clough ME (2001) The evolution of coloration and toxicity in the poison frog family (Dendrobatidae). *Proceedings of the National Academy of Sciences, U.S.A.*, **98**, 6227–32.

Sumner FB (1911) The adjustment of flatfishes to various backgrounds: a study of adaptive colour change. *Journal of Experimental Zoology*, **10**, 409–505.

Sun J, Bhushan B, Tong J (2013) Structural coloration in nature. *RSC Advances*, **3**, 14862–89.

Sundberg P (1979) *Tubulanus annulatus*, an aposematic nemertean? *Biological Journal of the Linnean Society*, **12**, 177–9.

Sundberg P (1987) A possible mechanism for the evolution of aposematic coloration in solitary nemerteans (phylum Nemertea). *Oikos*, **48**, 289–96.

Suomalainen E, Cook LM, Turner JRG (1973) Achiasmatic oogenesis in the heliconiine butterflies. *Hereditas*, **74**, 302–4.

Supple MA, Hines HM, Dasmahapatra KK, Lewis JJ, Nielsen DM, Lavoie C, Ray DA, Salazar C, McMillan WO, Counterman BA (2013) Genomic architecture of adaptive color pattern divergence and convergence in *Heliconius* butterflies. *Genome Research*, **23**, 1248–57.

Surlykke A, Miller LA (1985) The influence of arctiid moth clicks on bat echolocation: jamming or warning? *Journal of Comparative Physiology A*, **156**, 831–43.

Suzuki TK (2013) Modularity of a leaf moth-wing pattern and a versatile characteristic of the wing-pattern ground plan. *BMC Evolutionary Biology*, **13**: 158.

Suzuki TK, Tomita S, Sezutsu H (2014) Gradual and contingent evolutionary emergence of leaf mimicry in butterfly wing patterns. *BMC Evolutionary Biology*, **14**: 229.

Svádová K, Exnerová A, Štys P, Landová E, Valenta J, Fučíková, Socha R (2009) Role of different colours of aposematic insects in learning, memory and generalization of naïve bird predators. *Animal Behaviour*, **77**, 327–36.

Svennungsen TO, Holen ØH (2007) The evolutionary stability of automimicry. *Proceedings of the Royal Society of London, Series B: Biological Sciences*, **274**, 2055–62.

Sweet SS (1985) Geographic variation, convergent crypsis and mimicry in gopher snakes (*Pituophis melanoleucus*) and western rattlesnakes (*Crotalus viridis*). *Journal of Herpetology*, **19**, 55–67.

Swets JA (1964) *Signal Detection and Recognition by Human Observers*. Wiley, New York.

Sword GA (1999) Density-dependent warning coloration. *Nature*, **397**, 217.

Swynnerton CFM (1915a) A brief preliminary statement of a few of the results of five year's special testing of the theories on mimicry. *Proceedings of the Entomological Society of London*, **I**, 32–44.

Swynnerton CFM (1915b) Birds in relation to their prey: experiments on wood hoopoes, small hornbills, and a babbler. *Journal of the South African Ornithogical Union*, **11**, 32–108.

Swynnerton CFM (1926) An investigation into the defenses of butterflies of the genus *Charaxes*. *Proceedings of the Third Entomological Congress, Zurich (1925)*, **2**, 478–506.

Symula R, Schulte R, Summers K (2001) Molecular phylogenetic evidence for a mimetic radiation of Peruvian poison frogs supports a Müllerian mimicry hypothesis. *Proceedings of the Royal Society of London, Series B: Biological Sciences*, **268**, 2415–21.

Szidat L (1932) Ueber cysticerke Riesencercarien, insbesondere *Cercaria mirabilis* Braun und *Cercaria splendens* n. sp., und ihre Entwicklung im Magen von Raubfischen zu Trematoden der Gattung *Azygia* Looss. *Zeitschrift fur Parasitenkunde*, **4**, 477–505.

Taborsky M, Hudde B, Wirtz P (1987) Reproductive behaviour and ecology of *Symphodus* (*Crenilabrus*) *ocellatus*, a European wrasse with four types of male behaviour. *Behaviour*, **102**, 82–117.

Takada H, Hashimoto Y (1985) Association of the root aphid parasitoids *Aclitus sappaphis* and *Paralipsis eikoae* (Hymenoptera, Aphidiidae) with the aphid-attending ants *Pheidole fervida* and *Lasius niger* (Hymenoptera, Formicidae). *Kontyû, Tokyo*, **53**, 150–60.

Takahashi Y, Watanabe M (2010) Morph-specific fecundity and egg size in the female-dimorphic damselfly *Ischnura senegalensis*. *Zoological Science*, **27**, 325–29.

Takahashi Y, Yoshimura J, Morita S, Watanabe M (2010) Negative frequency dependent selection in female color polymorphism in a damselfly. *Evolution*, **64**, 3620–8.

Takasu F, Moskát C, Munoz R, Imanishi S, Nakamura H (2009) Adaptations in the common cuckoo (*Cuculus canorus*) to host eggs in a multiple-hosts system of brood parasitism. *Biological Journal of the Linnean Society*, **98**, 291–300.

Tan EJ, Li DQ (2009) Detritus decorations of an orb-weaving spider, *Cyclosa mulmeinensis* (Thorell): for food or camouflage? *Journal of Experimental Biology*, **212**, 1832–9.

Tanaka KD, Morimoto G, Stevens M, Ueda K (2011) Rethinking visual supernormal stimuli in cuckoos: visual modeling of host and parasite signals. *Behavioral Ecology*, **22**, 1012–9.

Taniguchi K, Maruyama M, Ichikawa T, Ito F (2005) A case of Batesian mimicry between a myrmecophilous staphylinid beetle, *Pella comes*, and its host ant, *Lasius* (*Dendrolasius*) *spathepus*: an experiment using the Japanese treefrog *Hyla japonica* as a real predator. *Insect Socieaux*, **52**, 320–2.

Tate DP (1994) Observations on nesting behavior of the common potoo in Venezuela. *Journal of Field Ornithology*, **65**, 447–52.

Taylor CH, Reader T, Gilbert F (2016) Hoverflies are imperfect mimics of wasp colouration. *Evolutionary Ecology*, **30**, 567–81.

Teichert H, Dötterl S, Frame D, Kirejtshuk A, Gottsberger G (2012) A novel pollination mode, saprocantharophily, in *Duguetia cadaverica* (Annonaceae): a stinkhorn (*Phallales*) flower mimic. *Flora*, **207**, 522–9.

Temrin H, Stenius S (1994) How reliable are behavioral cues for assessment of male mating status in polyterritorial wood warblers, *Phylloscopus sibilatrix*? *Behavioral Ecology and Sociobiology*, **35**, 147–52.

Tengö J, Bergström G (1976) Odor correspondence between *Mellita* females and males of their nest parasite *Nomada flavopicta* K. (Hymenoptera: Apoidea). *Journal of Chemical Ecology*, **2**, 57–65.

Tengö J, Bergström G (1977) Cleptoparasitism and odor mimetism in bees: do *Nomada* males imitate the odor of *Andrena* females? *Science*, **196**, 1117–19.

Terrick TD, Mumme RL, Burghardt GM (1995) Aposematic coloration enhances chemosensory recognition of noxious prey in the garter snake *Thamnophis radix*. *Animal Behaviour*, **49**, 857–66.

Tewksbury JJ, Nabhan GP (2001) Seed dispersal – directed deterrence by capsaicin in chillies. *Nature*, **412**, 403–4.

Teyssier T, Saenko SV, van der Marel D, Milinkovitch MC (2015) Photonic crystals cause active colour change in chameleons. *Nature Communications*, **6**: 6368.

Thayer AH (1896) The law which underlies protective coloration. *The Auk*, **13**, 124–9.

Thayer GH (1909) *Concealing-Colouration in the Animal Kingdom: An Exposition of the Laws of Disguise Tthrough Colour and Pattern: Being a Summary of Abbot H. Thayer's Discoveries*. The Macmillan Co., New York.

The Heliconius Genome Consortium (2012) Butterfly genome reveals promiscuous exchange of mimicry adaptations among species. *Nature*, **487**, 94–98.

Theis A, Salzburger W, Egger B (2012) The function of anal fin eggspots in the cichlid fish *Astatotilapia burtoni*. *PLoS One*, **7**: e29878.

Théry M (2007) Colours of background reflected light and of the prey's eye affect adaptive coloration in female crab spiders. *Animal Behaviour*, **73**, 797–804.

Théry M, Casas J (2002) Predator and prey views of spider camouflage. *Nature*, **415**, 133.

Théry M, Debut M, Gomez D, Casas J (2005) Specific color sensitivities of prey and predator explain camouflage in different visual systems. *Behavioral Ecology*, **16**, 25–9.

Thomas RJ, Marples NM, Cuthill IC, Takahashi M, Gibson EA (2003) Dietary conservatism may facilitate the initial evolution of aposematism. *Oikos*, **101**, 458–66.

Thompson MJ, Timmermans MJTN (2014) Characterising the phenotypic diversity of *Papilio dardanus* wing patterns using an extensive museum collection. *PLoS One*, **9(5)**: e96815.

Thompson MJ, Timmermans MJTN, Jiggins CD, Vogler AP (2014) The evolutionary genetics of highly divergent alleles of the mimicry locus in *Papilio dardanus*. *BMC Evolutionary Biology*, **14**: 140.

Thompson P, Mikellidou K (2011) Applying the Helmholtz illusion to fashion: horizontal stripes won't make you look fatter. *i-Perception*, **2**, 69–76.

Thompson RKR, Fotin RW, Boylan RJ, Sweet A, Graves CA, Lowitz CE (1981) Tonic immobility in Japanese quail can reduce the probability of sustained attack by cats. *Animal Learning & Behavior*, **9**, 145–9.

Thompson TE, Bennett I (1969) *Physalia* nematocysts: utilised by mollusks for defense. *Science*, **166**, 1532–3.

Thompson TE, Bennett I (1970) Observations on Australian Glaucidae (Mollusca: Opisthobranchia). *Zoological Journal of the Linnean Society*, **49**, 187–97.

Thompson V (1973) Spittlebug polymorphic for warning coloration. *Nature*, **242**, 126–8.

Thompson V (1974) Reply to: spittlebug morph mimics avian excrement. *Nature*, **250**, 352–3.

Thompson V (1984) Polymorphism under apostatic and aposematic selection. *Heredity*, **53**, 677–86.

Thomson JD (1980) Skewed flowering distributions and pollinator attraction. *Ecology*, **61**, 572–9.

Thornhill R (1979) Adaptive female-mimicking behaviour in a scorpionfly. *Science*, **205**, 412–4.

Thorogood R, Davies NB (2013) Hawk mimicry and the evolution of polymorphic cuckoos. *Chinese Birds*, **4**, 39–50.

Thresher RE (1978) Polymorphism, mimicry, and the evolution of the hamlets. *Bulletin of Marine Science*, **28**, 345–53.

Thurman CL (1988) Rhythmic physiological color change in Crustacea: a review. *Comparative Biochemistry and Physiology Part C Comparative Pharmacology*, **91**, 171–185.

Thurman TJ, Seymoure BM (2015) A bird's eye view of two mimetic tropical butterflies: coloration matches predator's sensitivity. *Journal of Zoology*, **298**, 159–68.

Till I (1987) Variability of expression of cyanogenesis in white clover (*Trifolium repens* L.). *Heredity*, **59**, 265–71.

Till-Bottraud I, Gouyon PH (1992) Intra-plant versus inter-plant Batesian mimicry – a model on cyanogenesis and herbivory in clonal plants. *The American Naturalist*, **139**, 509–20.

Timmermans MJTN, Baxter SW, Clark R, Heckel DG, Vogel H, Collins S, Papanicolaou A, Fukova I, Joron M, Thompson MJ, Jiggins CD, ffrench-Constant RH, Vogler AP (2014) Comparative genomics of the mimicry switch in *Papilio dardanus*. *Proceedings of the Royal Society of London, Series B: Biological Sciences*, **281**: 20140465.

Tinbergen L (1960) The natural control of insects in pine woods I. Factors influencing the intensity of predation by songbirds. *Archives Néerlandaises de Zoologie*, **13**, 265–343.

Tinbergen N (1951) *Study of Instinct*. Oxford University Press, Oxford.

Tinbergen N (1962) Foot-paddling in gulls. *British Birds*, **55**, 117–9.

Tinbergen N, Inpekoven M, France D (1967) An experiment on spacing-out as a defence against predation. *Behaviour*, **28**, 307–21.

Tinbergen N, Broekhuysen GJ, Feekes F, Houghton JCW, Kruuk H, Szulc E (1962) Egg shell removal by the black-headed gull, *Larus ridibundus* L.; a behaviour component of camouflage. *Behaviour*, **19**, 74–116.

Titcomb TC, Kikuchi DW, Pfennig DW (2014) More than mimicry? Evaluating scope for flicker-fusion as a defensive strategy in coral snake mimics. *Current Zoology*, **60**, 123–30.

Tobler M (2005) Feigning death in the Central American cichlid *Parachromis friedrichsthalii*. *Journal of Fish Biology*, **66**, 877–81.

Toledo LF, Haddad CFB (2009) Colors and some morphological traits as defensive mechanisms in anurans. *International Journal of Zoology*, **2009**: 910892.

Toledo LF, Sazima I, Haddad CFB (2010) Is it all death feigning? Case in anurans. *Journal of Natural History*, **44**, 1979–88.

Tollrian R, Harvell CD (1999) The evolution of inducible defences: current ideas. In R Tollrian and CD Harvell (eds), *The Ecology and Evolution of Inducible Defences*. Princeton University Press, Princeton, NJ, pp. 306–322.

Tooke W, Camire L (1991) Patterns of deception in intersexual and intrasexual mating strategies. *Ethology and Sociobiology*, **12**, 345–64.

Townes H (1972) The function of eye stalks in Diopsidae (Diptera). *Proceedings of the Entomological Society of Washington*, **74**, 85–6.

Tramer EJ (1994) Feeder access: deceptive use of alarm calls by a white-breasted nuthatch. *The Wilson Bulletin*, **106**, 573.

Treiber M (1979) Composites as host plants and crypts for *Synchlora aerata* (Geometridae). *Journal of the Lepidopterists' Society*, **33**, 239–44.

Tremblay RL, Ackerman JD, Zimmerman JK, Calvo RN (2005) Variation in sexual reproduction in orchids and its evolutionary consequences: a spasmodic journey to diversification. *Biological Journal of the Linnean Society*, **84**, 1–54.

Trimen R (1869) On some remarkable mimetic analogies among African butterflies. *Transactions of the Linnean Society of London*, **26**, 497–522.

Trnka A, Prokop P (2012) The effectiveness of hawk mimicry in protecting cuckoos from aggressive hosts. *Animal Behaviour*, **83**, 263–8.

Troscianko J, Lown AE, Hughes AE, Stevens M (2013) Defeating crypsis: detection and learning of camouflage strategies. *PLoS ONE*, **8(9)**: e73733.

Troscianko J, Wilson-Aggarwal J, Stevens M, & Spottiswoode CN (2016) Camouflage predicts survival in ground-nesting birds. *Scientific Reports*, **6**: 19966.

True JR (2003) Insect melanism: the molecules matter. *TRENDS in Ecology and Evolution*, **18**, 640–7.

Tsacas L, Desmier de Chenon R, Coutin R (1970) Observations sur le parasitisme larvaire d'*Hyperechia bomboides* (Dipt., Asilidae). *Annales de la Société Entomologique de France* (n.s.), **6**, 493–512.

Tsiakalos A, Routsias JG, Kordossis T, Moutsopoulos HM, Tzioufas AG, Sipsas NV (2011) Fine epitope specificity of anti-erythropoi-etin antibodies reveals molecular mimicry with HIV-1 p17 protein: a pathogenetic mechanism for HIV-1–related anemia. *Journal of Infectious Diseases*, **204**, 902–11.

Tso I-M, Huang J-P, Liao C-P (2007) Nocturnal hunting of a brightly coloured sit-and-wait predator. *Animal Behaviour*, **74**, 787–93.

Tsoularis A, Wallace J (2005) Reinforcement learning for a stochastic automaton modeling predation in stationary model-mimic environments. *Mathematical Biosciences*, **195**, 76–91.

Tsurui K, Honma A, Nishida T (2010) Camouflage effects of various colour-marking morphs against different microhabitat backgrounds in a polymorphic pygmy grasshopper *Tetrix japonica*. *PLoS ONE*, **5**, 1–7.

Tucker GM, Allen JA (1988) Apostatic selection by humans searching for computer-generated images on a colour monitor. *Heredity*, **60**, 329–34.

Tullberg BS, Hunter AS (1996) Evolution of larval gregariousness in relation to repellent defences and warning coloration in tree-feeding macrolepidoptera: a phylogenetic analysis based on independent contrasts. *Biological Journal of the Linnean Society*, **57**, 253–76.

Tullberg BS, Leimar O, Gamberale-Stille G (2000) Did aggregation favour the initial evolution of warning coloration? A novel world revisited. *Animal Behaviour*, **59**, 281–7.

Tullberg BS, Merilaita S, Wiklund C (2005) Aposematism and crypsis combined as a result of distance depence: functional versatility of the colour pattern in the swallowtail butterfly larva. *Proceedings of the Royal Society of London, Series B: Biological Sciences*, **272**, 1315–21.

Tullrot A (1994) The evolution of unpalatability and warning coloration in soft-bodied marine invertebrates. *Evolution*, **48**, 925–8.

Tullrot A, Sundberg P (1991) The conspicuous nudibranch *Polycera quadrilineata*: aposematic coloration and individual selection. *Animal Behaviour*, **41**, 175–6.

Tuno N (1998) Spore dispersal of *Dictyophora* fungi (Phallaceae) by flies. *Ecological Research*, **13**, 7–15.

Tuomi J, Augner M (1993) Synergistic selection of unpalatability in plants. *Evolution*, **47**, 668–72.

Turner ERA (1961) Survival values of different methods of camouflage as shown in a model population. *Proceedings of the Zoological Society of London*, **136**, 273–84.

Turner JRG (1970) Mimicry: a study in behaviour genetics, ecology and biochemistry. *Science Progress*, **58**, 219–35.

Turner JRG (1971a) Studies of Müllerian mimicry and its evolution in burnet moths and heliconid butterflies. In ER Creed (ed.), *Ecological Genetics and Evolution*. Blackwell, Oxford, pp. 224–60.

Turner JRG (1971b) Two thousand generations of hybridisation in a *Heliconius* butterfly. *Evolution*, **25**, 471–82.

Turner JRG (1975) Communal roosting in relation to warning colour in two heliconiine butterflies (Nymphalidae). *Journal of the Lepidopterists' Society*, **29**, 221–6.

Turner JRG (1977) Butterfly mimicry – genetical evolution of an adaptation. *Evolutionary Biology*, **10**, 163–206.

Turner JRG (1978) Why male butterflies are non-mimetic: natural selection, sexual selection, group selection, modification and sieving. *Biological Journal of the Linnean Society*, **10**, 385–432.

Turner JRG (1979) Oscillation of gene frequencies in Batesian mimics: a correction. *Biological Journal of the Linnean Society*, **11**, 397–8.

Turner JRG (1980) Oscillations of frequency in Batesian mimics, hawks and doves, and other simple frequency dependent polymorphisms. *Heredity*, **45**, 113–26.

Turner JRG (1981) Adaptation and evolution in *Heliconius*: a defense of neo-Darwinism. *Annual Review of Ecology and Systematics*, **12**, 99–121.

Turner JRG (1982) How do refuges produce biological diversity? Allopatry and parapatry, extinction and gene flow in mimetic butterflies. In GT Prance (ed.), *Biological Diversification in the Tropics*. Columbia University Press, New York, pp. 309–35.

Turner JRG (1983) Mimetic butterflies and punctuated equilibria: some old light on a new paradigm. *Biological Journal of the Linnean Society*, **20**, 277–300.

Turner JRG (1984) Mimicry: the palatability spectrum and its consequences. In RI Vane-Wright and PR. Ackery (eds), *The Biology of Butterflies*. Academic Press, London, pp. 141–61.

Turner JRG (1987) The evolutionary dynamics of Batesian and Müllerian mimicry: similarities and differences. *Ecological Entomology*, **12**, 81–95.

Turner JRG (1988) The evolution of mimicry: a solution to the problem of punctuated equilibrium. *The American Naturalist*, **131**, S42–66.

Turner JRG, Crane J (1962) The genetics of some polymorphic forms of the butterflies *Heliconius melpomene* Linnaeus and *Heliconius erato* Linnaeus. I. Major genes. *Zoologica, New York*, **47**, 141–52.

Turner JRG, Johnson MS (1980) Sex-limited mimicry: sexual selection, or gene dosage, or both? *Heredity* **45**, 137.

Turner JRG, Mallet JB (1996) Did forest islands drive the diversity of warningly coloured butterflies? Biotic drift and the shifting balance. *Philosophical Transactions of the Royal Society*, **B351**, 835–45.

Turner JRG, Sheppard PM (1975) Absence of crossing-over in female butterflies (*Heliconius*). *Heredity*, **34**, 265–9.

Turner JRG, Speed MP (1996) Learning and memory in mimicry. I. Simulations of laboratory experiments. *Philosophical Transactions of the Royal Society of London, Series B: Biological Sciences*, **351**, 1157–70.

Turner JRG, Kearny EP, Exton LS (1984) Mimicry and the Monte Carlo predator: the palatability spectrum and the origins of mimicry. *Biological Journal of the Linnean Society*, **23**, 247–68.

Tursch B, Daloze D, Dupont M, Hootele C, Kaisin M, Pasteels JM, Zimmermann D (1971) Coccinellin, the defensive alkaloid of the beetle *Coccinella septempunctata*. *Chemia*, **25**, 307.

Tutt JW (1896) *British Moths*. Routledge, London.

Tyler J (2001a) Are glow-worms *Lampyris noctiluca* (Linneaus) (Lampyridae) distasteful? *The Coleopterist*, **9**, 148.

Tyler J (2001b) A previously undescribed defence mechanism in the larval glow-worm *Lampyris noctiluca* (Linneaus) (Lampyridae)? *The Coleopterist*, **10**, 38.

Tyler J. McKinnon W, Lord GA, Hilton PJ (2008) A defensive steroidal pyrone in the glow-worm *Lampyris noctiluca* L. (Coleoptera: Lampyridae). *Physiological Entomology*, **33**, 167–70.

Tyrie EK, Hanlon RT, Siemann LA, Uyarra MC (2015) Coral reef flounders, *Bothus lunatus*, choose substrates on which they can achieve camouflage with their limited body pattern repertoire. *Biological Journal of the Linnean Society*, **114**, 629–38.

Uboni A, Bagnères A C, Christides J P, Lorenzi MC (2012) Cleptoparasites, social parasites and a common host: Chemical insignificance for visiting host nests, chemical mimicry for living in. *Journal of Insect Physiology*, **58**, 1259–64.

Uesugi K (1996) The adaptive significance of Batesian mimicry in the swallowtail butterfly, *Papilio polytes* (Insecta: Papilionidae) – associative learning in a predator. *Ethology*, **102**, 762–75.

Uglem I, Rosenqvist G, Wasslavik HS (2000) Phenotypic variation between dimorphic males in corkwing wrasse. *Journal of Fish Biology*, **57**, 1–14.

Uma D, Durkee CA, Herzner G, Weiss MR (2013) Double deception: ant-mimicking spiders elude both visually-and chemically-oriented predators. *PLoS ONE*, **8(11)**: e79660.

Umbers KDL, Fabricant SA, Gawryszewski FM, Seago AE, Herberstein ME (2014) Reversible colour change in Arthropoda. *Biological Reviews*, **89**, 820–48.

Ureña O, Hanson P (2010) A fly larva (Syrphidae: *Ocyptamus*) that preys on adult flies. *Revista de Biología Tropical*, **58**, 1157–63.

Urru I, Stensmyr MC, Hansson BS (2011) Pollination by brood-site deception. *Phytochemistry*, **72**, 1655–66.

Uyenoyama M, Feldman MW (1980) Theories of kin and group selection: a population genetics perspective. *Theoretical Population Biology*, **17**, 380–415.

Vakkari P (2009) Polymorphism in *Simyra albovenosa* (Lepidoptera, Noctuidae). I. Genetic control of the melanic forms. *Hereditas*, **93**, 181–4.

Vale A, Navarro L, Rojas D, Alvarez JC (2011) Breeding system and pollination by mimicry of the orchid *Tolumnia guibertiana* in Western Cuba. *Plant Species Biology*, **26**, 163–73.

Valkonen JK, Mappes J (2014) Resembling a viper: implications of mimicry for conservation of the endangered smooth snake. *Conservation Biology*, **28**, 1568–74.

Valkonen JK, Niskanen M, Björklund M, Mappes J (2011a) Disruption or aposematism? Significance of dorsal zig-zag pattern of European vipers. *Evolutionary Ecology*, **25**, 1047–63.

Valkonen JK, Nokelainen O, Mappes J (2011b) Antipredatory function of head shape for vipers and their mimics. *PLoS ONE*, **6**: e22272.

Valkonen JK, Nokelainen O, Niskanen M, Kilpimaa J, Björklund M, Mappes J (2012) Variation in predator species abundance can cause variable selection pressure on warning signaling prey. *Ecology and Evolution*, **2**, 1971–6.

Valkonen JK, Nokelainen O, Jokimöki M, Kuusinen E, Paloranta M, Peura M, Mappes J (2014) From deception to frankness: benefits of ontogenetic shift in the anti-predator strategy of alder moth *Acronicta alni* larvae. *Current Zoology*, **60**, 114–22.

Vallin A, Jakobsson S, Lind J, Wiklund C (2005) Prey survival by predator intimidation: an experimental study of peacock butterfly defence against blue tits. *Proceedings of the Royal Society of London, Series B: Biological Sciences*, **272**, 1203–7.

Vallin A, Jakobsson S, Lind J, Wiklund C (2006) Crypsis versus intimidation – anti-predation defence in three closely related butterflies. *Behavioral Ecology & Sociobiology* **59**, 455–9.

Vallin A, Jakobsson S, Wiklund C (2007) "An eye for an eye" – on the generality of the intimidating quality of eyespots in a butterfly and a hawkmoth. *Behavioral Ecology & Sociobiology*, **61**, 1419–24.

van Achterberg C (1989) *Pheloura* Gen. Nov., a neotropical genus with an extremely long pseudo ovipositor (Hymenoptera: Braconidae). *Entomologische Berichten*, **49**, 105–8.

Van der Cingel NA (2001) *An Atlas of Orchid Pollination, America, Africa, Asia and Australia*. A.A. Balkema, Rotterdam.

Vander Meer RK, Wojcik DP (1982) Chemical mimicry in the myrmecophilous beetle *Myrmecaphodius excavaticollis*. *Science*, **218**, 806–8.

Vander Meer RK, Jouvenaz DP, Wojcik DP (1989) Chemical mimicry in a parasitoid (Hymenoptera: Eucharitidae) of fire ants (Hymenoptera: Formicidae). *Journal of Chemical Ecology*, **15**, 2247–61.

van der Niet T, Hansen D, Johnson SD (2011) Carrion mimicry in a South African orchid: flowers attract a narrow subset of the fly assemblage on animal carcasses. *Annals of Botany*, **107**, 981–92.

Van der Meijden E, Crawley MJ, Nisbet RM (1998) The dynamics of a herbivore-plant interaction, the cinnabar moth and ragwort. In JP Dempster and IFG. McLean (eds), *Insect Populations*. Chapman & Hall, London, pp. 291–308.

van der Pijl L, Dodson CH (1966) *Orchid Flowers: Their Pollination and Evolution*. University of Miami Press, Coral Gables, FL.

van Die I, Cummings RD (2010) Glycan gimmickry by parasitic helminths: A strategy for modulating the host immune response? *Glycobiology*, **20**, 2–12.

Vane-Wright RI (1971) The systematics of *Drusillopsis* Oberthur (Satyrinae) and the supposed amathusiid *Bigaena* van Eecke (Lepidoptera: Nymphalidae), with some observations on Batesian mimicry. *Transactions of the Royal Entomological Society, London*, **123**, 97–123.

Vane-Wright RI (1975) An integrated classification for polymorphism and sexual dimorphism in butterflies. *Journal of Zoology*, **177**, 329–37.

Vane-Wright RI (1976) A unified classification of mimetic resemblances. *Biological Journal of the Linnean Society*, **8**, 25–56.

Vane-Wright RI (1979) Towards a theory of the evolution of butterfly colour patterns under directional and disruptive selection. *Biological Journal of the Linnean Society*, **11**, 141–52.

Vane-Wright RI (1980) On the definition of mimicry. *Biological Journal of the Linnean Society*, **13**, 1–6.

Vane-Wright RI (1981) Only connect. *Biological Journal of the Linnean Society*, **16**, 33–40.

Vane-Wright RI (1984) The role of pseudosexual selection in the evolution of butterfly colour patterns. In RI Vane-Wright and PR Ackery (eds), *The Biology of Butterflies*. Symposia of the Royal Entomological Society of London, no. I, London, pp. 251–3.

Vane-Wright RI (2009) An integrated classification for polymorphism and sexual dimorphism in butterflies. *Journal of Zoology*, **177**, 329–37.

Vane-Wright RI (2010) Mimicry affecting British Diptera. In PJ Chandler (ed.), *A Dipterist's Handbook* (2nd edn). The Amateur Entomologist, vol. 15. AES, Brentwood, UK, pp. 495–500.

Vane-Wright RI, Boppré M (1993) Visual and chemical signalling in butterflies: functional and phylogenetic perspectives. *Philosophical Transactions of the Royal Society B, Biological Sciences*, **340**, 197–205.

Vane-Wright RI, Raheem DC, Cieslak A, Vogler AP (1999) Evolution of the mimetic African swallowtail butterfly *Papilio dardanus*: molecular data confirm relationships with *P. phorcas* and *P. constantinus*. *Biological Journal of the Linnean Society*, **66**, 215–29.

Van Gossum H, Stoks R, de Bruyn L (2005) Lifetime fitness components in female colour morphs of a damselfly: density-or frequency-dependent selection? *Biological Journal of the Linnean Society*, **86**, 515–23.

van Sommeren VGL, Jackson THE (1959) Some comments on protective resemblance among African Lepidoptera (Rhopalocera). *Journal of the Lepidopterists' Society*, **13**, 121–50.

Van't Hof AE, Edmonds N, Daliková M, Marec F, Saccheri IJ (2011) Industrial melanism in British peppered moths has a singular and recent mutational origin. *Science*, **332**, 958–60.

Vasconcellos-Neto J, Lewinsohn TM (1984) Discrimination and release of unpalatable butterflies by *Nephila clavipes*, a neotropical orb-weaving spider. *Ecological Entomology*, **9**, 337–44.

Vasey PL, Duckworth N (2006) Sexual reward via vulvar, perineal, and anal stimulation: a proximate mechanism for female homosexual mounting in Japanese macaques. *Archives of Sexual Behavior*, **25**, 323–32.

Vaughan FA (1983) Startle responses of blue jays to visual stimuli presented during feeding. *Animal Behaviour*, **31**, 385–96.

Vavilov NI (1922) The law of homologous series in variation. *Journal of Genetics*, **12**, 47–89.

Vereecken NJ, Mahé G (2007) Larval aggregations of the blister beetle *Stenoria analis* (Schaum) (Coleoptera: Meloidae) sexually deceive patrolling males of their host, the solitary bee *Colletes hederae* Schmidt & Westrich (Hymenoptera: Colletidae). *Annales de la Société Entomologique de France*, **43**, 493–6.

Vereecken NJ, McNeil JN (2010) Cheaters and liars: chemical mimicry at its finest. *Canadian Journal of Zoology*, **88**, 725–52.

Vereecken NJ, Schiestl FP (2008) The evolution of imperfect floral mimicry. *Proceedings of the National Academy of Sciences, U.S.A.*, **105**, 7484–8.

Vereecken NJ, Wilson CA, Hötling S, Schulz S, Banketov SA, Mardulyn P (2012) Pre-adaptations and the evolution of pollination by sexual deception: Cope's rule of specialization revisited. *Proceedings of the Royal Society of London, Series B, Biological Sciences*, **279**, 4786–94.

Vereecken NJ, Dorchin A, Dafni A, Hötling S, Schulz S, Watts S (2013) A pollinator's eye view of a shelter mimicry system. *Annals of Botany*, **111**, 1155–65.

Vergara P, Fargallo JA (2007) Delayed plumage maturation in Eurasian kestrels: female mimicry, subordination signalling or both? *Animal Behaviour*, **74**, 1505–13.

Veselý P, Veselá S, Fuchs R (2013) The responses of Central European avian predators to an allopatric aposematic true bug. *Ethology, Ecology & Evolution*, **25**, 275–88.

Vidal Cordero JM, Moreno-Rueda G, López-Orta A, Marfil-Daza C, Ros-Santaella JL, Ortiz-Sánchez FJ (2012) Brighter-colored paper wasps (*Polistes dominula*) have larger poison glands. *Frontiers in Zoology*, **9**, 20.

Vitt L (1992) Lizard mimics millipede. *National Geographic Research and Exploration*, **8**, 76–95.

Vlasáková B, Jarau S (2011) Dioecious *Clusia nemorosa* achieves pollination by combining specialized and generalized floral rewards. *Plant Ecology*, **212**, 1327–37.

Vlieger L, Brakefield PM (2007) The deflection hypothesis: eyespots on the margins of butterfly wings do not influence predation by lizards. *Biological Journal of the Linnean Society*, **92**, 661–7.

Völkl W, Mackauer M (1993) Interactions between ants attending *Aphis fabae* ssp. *cirsiiacanthoidis* on thistles and foraging parasitoid wasps. *Journal of Insect Behavior*, **6**, 301–12.

Vogel S (1978) Pilzmuckenblumen als Pilzmimeten. I and II. *Flora (Jena)*, **167**, 329–66, 367–98.

Vogel S, Martens J (2000) A survey of the function of the lethal kettle traps of *Arisaema* (Araceae), with records of pollinating fungus gnats from Nepal. *Botanical Journal of the Linnean Society*, **133**, 61–100.

von Beeren C, Pohl S, Witte V (2012) On the use of adaptive resemblance terms in chemical ecology. *Psyche*, **2012**: 635761.

von Brockhusen F, Curio E (1975) Die innerartliche Variabilitat der Beutewahl beuteerfahrungsloser *Anolis*. *Experientia*, **31**, 45–6.

von Helversen B, Schooler LJ, Czienskowski U (2013) Are stripes beneficial? Dazzle camouflage influences perceived speed and hit rates. *PLoS ONE*, **8(4)**: e61173.

von Uexküll JJ [1934] *Streifzüge durch die Umwelten von Tieren und Menschen: Ein Bilderbuch unsichtbarer Welten*, J. Springer, Berlin, 102 pp; (translated 1992) A stroll through the world of animals and men: a picture book of invisible worlds. *Semiotica*, **89**, 319–91.

Vršanský P, Chorvát D (2013) Luminescent system of *Lucihormetica luckae* supported by fluorescence lifetime imaging. *Naturwissenschaften*, **100**, 1099–101.

Vršanský P, Chorvát D, Fritzsche I, Hain M, Ševčík R (2012) Light-mimicking cockroaches indicate Tertiary origin of recent terrestrial luminescence. *Naturwissenschaften*, **99**, 739–49.

Vukusic P, Chittka L (2013) Visual signals: color and light production. In SJ Simpson and AE Douglas (eds), *The Insects: Structure and Function* (5th edn). Cambridge University Press, Cambridge, UK, pp. 793–823.

Waage JK (1981) How the zebra got its stripes – biting flies as selective agents in the evolution of zebra colouration. *Journal of the Society of Southern Africa*, **44**, 351–8.

Waldbauer GP (1970) Mimicry of Hymenopteran antennae by Syrphidae. *Psyche*, **1970**, 45–9.

Waldbauer GP (1988a) Aposematism and Batesian mimicry: measuring mimetic advantage in natural habitats. *Evolutionary Biology*, **22**, 227–59.

Waldbauer GP (1988b) Asynchrony between Batesian mimics and their models. *The American Naturalist*, **131**, S103–21.

Waldbauer GP, LaBerge WE (1985) Phenological relationships of wasps, bumblebees, their mimics and insectivorous birds in northern Michigan. *Ecological Entomology*, **10**, 99–110.

Waldbauer GP, Sheldon JK (1971) Phenological relationships of some aculeate Hymenoptera, their dipteran mimics, and insectivorous birds. *Evolution*, **25**, 371–82.

Waldbauer GP, Sternburg JG (1975) Saturniid moths as mimics: an alternative interpretation of attempts to demonstrate mimetic advantage in nature. *Evolution*, **29**, 650–58.

Waldbauer GP, Sternburg JG (1983) A pitfall in using painted insects in studies of protective coloration. *Evolution*, **37**, 1085–6.

Waldbauer GP, Sternburg JG, Maier CT (1977) Phenological relationships of wasps, bumblebees, their mimics and insectivorous birds in an Illinois sand area. *Ecology*, **58**, 583–91.

Waldman B (1982) Sibling association among schooling tadpoles: field evidence and implications. *Animal Behaviour*, **30**, 700–13.

Waldman B, Adler K (1979) Toad tadpoles associate preferentially with siblings. *Nature*, **282**, 611–3.

Wallace AR (1863) List of birds collected in the island of Bouru (one of the Moluccas), with descriptions of new species. *Proceedings of the Zoological Society of London*, **1863**, 18–28.

Wallace AR (1865) On the phenomena of variation and geographical distribution as illustrated by the Papilionidae of the Malayan Region. *Transactions of the Linnean Society of London*, **25**, 1–71, + 8 plates.

Wallace AR (1867) Mimicry and other protective resemblances among animals. *Westminster and Foreign Quarterly Review*, **32**, 1–43. [Originally printed anonymously in the 1 July 1867 number of Volume 88 of the *Westminster Review (London edition)*: reproduced online 2010. Alfred Russel Wallace Classic Writings. Paper 8. http://digitalcommons.wku.edu/dlps_fac_arw/8].

Wallace AR (1869) *The Malay Archipelago*. Dover Publication, New York.

Wallace AR (1870) *Contributions to the Theory of Natural Selection*. Macmillan, London.

Wallace AR (1877) The colours of animals and plants, II. The colours of plants. *Macmillan's Magazine*, **36**, 464–71.

Wallace AR (1878) The colours of plants and the origin of the colour-sense. *Tropical Nature and Other Essays*. Macmillan, London, pp. 395–415. (preproduced online 2013. Cambridge Library Collection – Darwin, Evolution and Genetics. Cambridge: Cambridge University Press, pp. 221–48).

Wallace AR (1889) *Darwinism: An Exposition of the Theory of Natural Selection with Some of its Applications*. Macmillan and Company, London.

Wallace AR (1891) Mimicry, and other protective resemblances among animals. In AR Wallace, *Natural Selection and Tropical Nature*. Macmillan, London, pp. 34–90.

Waloff N (1961) Observations on the biology of *Perilitus dubius* (Wesmael) (Hymenoptera: Braconidae), a parasite of the chrysomelid beetle *Phytodecta olivacea* (Forster). *Proceedings of the Royal Entomological Society of London, Series A, General Entomology*, **36**, 96–102.

Wang B, Xia F, Engel MS, Perrichot V, Shi G, Zhang H, Chen J, Jarzembowski EA, Wappler T, Rust J (2016) Debris-carrying camouflage among diverse lineages of Cretaceous insects. *Scientific Advances*, **2**: e1501918.

Wang IJ (2008) Rapid color evolution in an aposematic species: a phylogenetic analysis of color variation in the strikingly polymorphic strawberry poison-dart frog. *Evolution*, **62**, 2742–59.

Wang IJ (2011) Inversely related aposematic traits: reduced conspicuousness evolves with increased toxicity in a polymorphic poison-dart frog. *Evolution*, **65**, 1637–49.

Wang Z, Schaefer HM (2012) Resting orientation enhances prey survival on strongly structured background. *Ecology Resources*, **27**, 107–13.

Wang Z, Schaefer HM (2014) Limits of selection against cheaters: birds prioritise visual fruit advertisement over taste. *Oecologia*, **174**, 1293–300.

Ward JM, Ruxton GD, Houston DC, McCafferty DJ (2007) Thermal consequences of turning white in winter: a comparative study of red grouse *Lagopus lagopus scoticus* and Scandinavian willow grouse *L. 1. lagopus*. *Wildlife Biology*, **13**, 120–9.

Warner JA, Latz MI, Case JF (1979) Cryptic bioluminescence in a midwater shrimp. *Science*, **203**, 1109–10.

Warren AD (2014) Deception on a web: chemical mimicry of saturniid sex pheromones by *Argiope* spiders (Araneidae). In [Program and Abstracts for] Southern Lepidopterists' Society and Association for Tropical Lepidoptera 2014 Annual Meeting,

McGuire Center for Lepidoptera and Biodiversity, Florida Museum of Natural History, University of Florida, Gainesville, 26–28 September, 2014, p. 22.

Wasmann E (1925) Die Ameisenmimikry. *Abhandlungen zur theoretischen Biologie, Berlin*, **19**, 1–164.

Watanabe H, Yano E (2013) Behavioral response of mantid *Tenodera aridifolia* (Mantodea: Mantidae) to windy conditions as a cryptic approach strategy for approaching prey. *Entomological Science*, **16**, 40–6.

Watson AC, Siemann LA, Hanlon RT (2014) Dynamic camouflage by Nassau groupers *Epinephelus striatus* on a Caribbean coral reef. *Journal of Fish Biology*, **85**, 1634–49.

Way MJ (1963) Mutualism between ants and honeydew producing Homoptera. *Annual Review of Entomology*, **8**, 307–44.

Webster RJ, Callahan A, Godin JGJ, Sherratt TN (2009) Behaviourally mediated crypsis in two nocturnal moths with contrasting appearance. *Philosophical Transactions of the Royal Society B, Biological Sciences*, **364**, 503–10.

Webster RJ, Hassall C, Herdman CM, Sherratt TN (2013) Disruptive camouflage impairs object recognition. *Biology Letters*, **9**: 20130501.

Weibel AC, Moore WS (2005) Plumage convergence in picoides woodpeckers based on a molecular phylogeny, with emphasis on the convergence in downy and hairy woodpeckers. *The Condor*, **107**, 797–809.

Weinstein AM, Davis BJ, Menz MNM, Dixon KW, Phillips RD (2016) Behaviour of sexually deceived ichneumonid wasps and its implications for pollination in *Cryptostylis* (Orchidaceae). *Biological Journal of the Linnean Society*, in press

Weldon PJ, Burghardt GM (1984) Deception divergence and sexual selection. *Zeitschrift für Tierpsychologie*, **65**, 89–102.

Weldon PJ, Rappole JH (1997) A survey of birds odorous or unpalatable to humans: possible indications of chemical defense. *Journal of Chemical Ecology*, **23**, 2609–33.

Welsh HH Jr, Lind AJ (2000) Evidence of lingual-luring by an aquatic snake. *Journal of Herpetology*, **34**, 67–74.

Wennersten L, Forsman A (2009) Does colour polymorphism enhance survival of prey populations? *Proceedings of the Royal Society of London, Series B: Biological Sciences*, **276**, 2187–94.

Werner Y (1983) Behavioural triangulation of the head in three boigine snakes: possible cases of mimicry. *Israel Journal of Zoology*, **32**, 205–28.

Werner Y (1985) Similarities of the colubrid snakes *Spalerosophis* and *Pythonodipsas* to vipers: an additional hypothesis. *Copeia*, **1985**, 266–8.

Wesołowska W, Wesołowski T (2014) Do *Leucochloridum* sporocysts manipulate the behaviour of their snail hosts? *Journal of Zoology*, **292**, 151–5.

West DA (1976) Aposematic coloration and mutualism in sponge-dwelling tropical zoanthids. In GO Mackie (ed.), *Coelenterate Ecology and Behaviour*. Plenum Press, New York, pp. 443–52.

West DA, Hazel WN (1979) Natural pupation sites of two swallowtail butterflies (Lepidoptera: Papilioninae): *Papilio polyxenes* Fabr., *P. glaucus* L. and *Battus philenor* (L.). *Ecological Entomology*, **4**, 387–92.

West DA, Hazel WN (1985) Pupal colour dimorphism in swallowtail butterflies: timing of the sensitive period and environmental control. *Physiological Entomology*, **10**, 113–9.

Westmoreland D, Kiltie RA (1996) Egg crypsis and clutch survival in three species of blackbirds (Icteridae). *Biological Journal of the Linnean Society*, **58**, 159–72.

Westoby M (1978) What are the biological bases of varied diets? *The American Naturalist*, **112**, 627–31.

Wheeler BC (2009) Monkeys crying wolf? Tufted capuchin monkeys use anti-predator calls to usurp resources from conspecifics. *Proceedings of the Royal Society of London, Series B: Biological Sciences*, **276**, 3013–8.

Wheeler DE, Holldobler B (1985) Cryptic phragmosis: the structural modifications. *Psyche*, **92**, 337–53.

Wheeler GC (1983) A mutillid mimic of an ant (Hymenoptera: Mutillidae and Formicidae). *Entomological News*, **94**, 143–4.

Wheelwright NT, Janson CH (1985) Colors of fruit displays of bird-dispersed plants in two tropical forests. *The American Naturalist*, **126**, 777–99.

White TE, Kemp DJ (2015) Technicolour deceit: a sensory basis for the study of colour-based lures. *Animal Behaviour*, **105**, 231–43.

Whitehead DR, Shelley RM (1992) Mimicry among aposematic Appalachian xystodesmid millipeds (Polydesmida: Chelodesmidea). *Proceedings of the Entomological Society of Washington*, **94**, 177–88.

Whitehead MR, Peakall R (2014) Pollinator specificity drives strong prepollination reproductive isolation in sympatric sexually deceptive orchids. *Evolution*, **68**, 1561–75.

Whitehouse MEA (1986) The foraging behaviour of *Argyrodes antipodiana* (Theridiidae), a kleptoparasite spider from New Zealand. *New Zealand Journal of Zoology*, **13**, 151–68.

Whiteley DAA, Owen DF, Smith DAS (1997) Massive polymorphism and natural selection in *Donacilla cornea* (Poli, 1791) (Bivalvia: Mesodesmatidae). *Biological Journal of the Linnean Society*, **62**, 475–94.

Whiting MJ, Webb JK, Keogh JS (2009) Flat lizard female mimics use sexual deception in visual but not chemical signals. *Proceedings of the Royal Society of London, Series B: Biological Sciences*, **276**, 1585–91.

Whitley GP (1935) Fishes from Princess Charlotte Bay, North Queensland. *Records of the South Australia Museum*, **5**, 345–63.

Whitman DW, Orsak L, Greene E (1988) Spider mimicry in fruit flies (Diptera: Tephritidae): further experiments on the deterrence of jumping spiders (Araneae: Salticidae) by *Zonosemata vittigera* (Coquillett). *Annals of the Entomological Society of America*, **81**, 532–6.

Wickler W (1962a) Ei-Attrappen und Maulbrüten bei afrikanischen Cichliden. *Zeitschrift für Tierpsychologie*, **19**, 129–64.

Wickler W (1962b) 'Egg-dummies' as natural releasers in mouth-breeding cichlids. *Nature*, **194**, 1092–3.

Wickler W (1965) Mimicry and the evolution of animal communication. *Nature*, **208**, 519–21.

Wickler W (1967) Socio-sexual signals and their intraspecific imitation among primates. In D Morris (ed.), *Primate Ethology*. Aldine, London & Chicago, pp. 69–147.

Wickler W (1968) *Mimicry in Plants and Animals*. World University Library, London.

Widder EA (1998) A predatory use of counterillumination by the squaloid shark, *Isistius brasiliensis*. *Environmental Biology of Fish*, **53**, 267–73.

Wiens D (1978) Mimicry in plants. *Evolutionary Biology*, **11**, 365–403.

Wiersma P, Muñoz-Garcia A, Walker A, Williams JB (2007) Tropical birds have a slow pace of life. *Proceedings of the National Academy of Sciences, U.S.A.*, **104**, 9340–5.

Wignall AE, Taylor PW (2011) Assassin bug uses aggressive mimicry to lure spider prey. *Proceedings of the Royal Society of London, Series B: Biological Sciences*, **278**, 1427–33.

Wijesinghe MR, Dayawansa PN (1998) A new anti-predator behavioural strategy in stilts suggestive of adaptive evolution. *Journal of South Asian Natural History*, **3**, 193–4.

Wiklund C (1974) Pupal colour polymorphism in *Papilio machaon* L. and the survival in the field of cryptic versus non-cryptic pupae. *Transactions of the Royal Entomological Society of London*, **127**, 73–84.

Wiklund C, Åhrberg C (1978) Host plants, nectar source plants, and habitat selection of males and females of *Anthocharis cardamines*. *Oikos*, **31**, 169–83.

Wiklund C, Järvi T (1982) Survival of distasteful insects after attack by naive birds: a reappraisal of the theory of aposematic coloration evolving through individual selection. *Evolution*, **36**, 998–1002.

Wiklund C, Sillén-Tullberg B (1985) Why distasteful butterflies have aposematic larvae and adults, but cryptic pupae: evidence from predation experiments on the monarch and the European swallowtail. *Evolution*, **39**, 1155–8.

Wilson RS, Jackson RR, Gentile K (1990) Spider web smokescreens: spider tricksters use background noise to mask stalking movements. *Animal Behaviour*, **51**, 313–26.

Williams CM (1970) Hormonal interactions between plants and insects. In E Sondheim and JB Simeone (eds), *Chemical Ecology*. Academic Press, New York, pp. 103–32.

Williams CR, Brodie Jr. ED, Tyler MJ, Walker SJ (2000) Antipredator mechanisms of Australian frogs. *Journal of Herpetology*, **34**, 431–43.

Williams D (2001) *Naval Camouflage 1914–1945*. Pen and Sword Books, Barnsley, UK.

Williams KS, Gilbert LE (1981) Insects as selective agents on plant vegetation morphology: egg mimicry reduces egg laying by butterflies. *Science*, **212**, 467–9.

Williams PH (1991) Phylogenetic relationships among bumble bees (*Bombus* Latr.): a reappraisal of morphological evidence. *Systematic Entomology*, **19**, 327–44.

Williams PH (2007) The distribution of bumblebee colour patterns worldwide: possible significance for thermoregulation, crypsis, and warning mimicry. *Biological Journal of the Linnean Society*, **92**, 97–118.

Williamson GB (1982) Plant mimicry: evolutionary constraints. *Biological Journal of the Linnean Society*, **18**, 49–58.

Willis EO (1963) Is the zone-tailed hawk a mimic of the turkey vulture? *Condor*, **65**, 313–7.

Willis EO (1976a) A possible reason for mimicry of a bird-eating hawk by an insect-eating kite. *The Auk*, **9**, 841–2.

Willis EO (1976b) Similarity of a tanager (*Orchesticus abeillei*) and an ovenbird (*Philydor rufus*): a possible case of mimicry. *Ciência e Cultura*, **28**, 1492–3.

Willis EO (1989) Mimicry in bird flocks of cloud forests in Southeastern Brazil. *Revista Brasileira de Biologia*, **49**, 615–9.

Willis RE, White CR, Merritt DJ (2011) Using light as a lure is an efficient predatory strategy in *Arachnocampa flava*, an Australian glowworm. *Journal of Comparative Physiology B*, **181**, 477–86.

Willmott HE, Foster SA (1995) The effects of rival male interaction on courtship and parental care in the fourspine stickleback, *Apeltes quadracus*. *Behaviour*, **132**, 997–1010.

Willmott KR, Mallet J (2004) Correlations between adult mimicry and larval host plants in ithomiine butterflies. *Proceedings of the Royal Society of London, Series B: Biological Sciences*, **271**, S266–9.

Willmott KR, Elias M, Sourakov A (2011) Two possible caterpillar mimicry complexes in neotropical danaine butterflies (Lepidoptera: Nymphalidae). *Annals of the Entomological Society of America*, **104**, 1108–18.

Willson MF, Ågren J (1989) Differential floral rewards and pollination by deceit in unisexual flowers. *Oikos*, **55**, 23–9.

Willson MF, Hoppes WG (1986) Foliar "flags" for avian frugivores: signal or serendipity? In A Estrada and TH Fleming (eds), *Frugivores and Seed Dispersal*. Junk, Dordrecht, pp. 55–69.

Willson MF, Melampy MN (1983) The effect of bicolored fruit displays on fruit removal by avian frugivores. *Oikos*, **41**, 27–31.

Willson MF, Thompson JN (1982) Phenology and ecology of color in bird-dispersed fruits, or why some fruits are red when they are 'green'? *Canadian Journal of Botany*, **60**, 701–13.

Willson MF, Whelan CJ (1990) The evolution of fruit color in fleshy-fruited plants. *The American Naturalist*, **136**, 790–809.

Willughby F (1678) *The Ornithology of Francis Willughby*. John Martyn, London.

Wilson D, Heinsohn R, Endler JA (2007) The adaptive significance of ontogenetic colour change in a tropical python. *Biology Letters*, **3**, 40–3.

Wilson JS, Williams KA, Forister ML, von Dohlen CD, Pitts JP (2012) Repeated evolution in overlapping mimicry rings among North American velvet ants. *Nature Communications*, **3**: 1272.

Wilson JS, Jahner JP, Forister ML, Sheehan ES, Williams KA, Pitts JP (2015) North American velvet ants form one of the world's largest known Müllerian mimicry complexes. *Current Biology*, **25**, R704–6.

Wilson RP, Ryan PG, James A, Wilson M-PT (1987) Conspicuous coloration may enhance prey capture in some piscivores. *Animal Behaviour*, **35**, 1558–60.

Wilts BD, Pirih P, Arikawa K, Stavenga DG (2013) Shiny wing scales cause spec(tac)ular camouflage of the angled sunbeam butterfly, *Curetis acuta*. *Biological Journal of the Linnean Society*, **109**, 279–89.

Winemiller KO (1990) Caudal eyespots as deterrents against fin predation in the Neotropical cichlid *Astronotus ocellatus*. *Copeia*, **1990**, 665–73.

Wing K (1983) *Tutelina similis* (Araneae: Salticidae): an ant mimic that feeds on ants. *Journal of the Kansas Entomological Society*, **56**, 55–8.

Wirtz P (1978) The behaviour of the Mediterranean *Tripterygion* species (Pisces, Blennoidei). *Zeitschrift für Tierpsychologie*, **48**, 142–74.

Witte V, Foitzik S, Hashim R, Maschwitz U, Schulz S (2009) Fine tuning of social integration by two myrmecophiles of the ponerine army ant, *Leptogenys distinguenda*. *Journal of Chemical Ecology*, **35**, 355–67.

Wittenberger JF (1981) *Animal Social Behavior*. Duxbury Press, Boston.

Wohlfahrt T (1954) Beobachtungen über Färbung und Zeichnung an Raupen und Puppen des Segelfalters *Iphiclides podalirius* (L.) and über die Ursache des Auftretens seiner Sommergeneration in Mitteleuropa. *Entomologische Zeitschrift*, **64**, 161–75.

Wohlfahrt T (1957) Über den Einfluss von Licht, Futterqualität und Temperatur auf Puppenruhe und Diapause des mitteleuropäischen Segelfalters *Iphiclides podalirius* (L.). *Wanderversammlung Deutscher Entomologen*, **8**, 6–14.

Wolda H (1967) Genetics of polymorphism in the land snail, *Cepaea nemoralis*. *Genetica*, **40**, 475–502.

Wollenberg KC, Veith M, Noonan BP, Lötters S (2006) Polymorphism versus species richness – systematics of large *Dendrobates* from the Eastern Guiana Shield (Amphibia: Dendrobatidae). *Copeia*, **2006**, 623–9.

Wong R[B]BM, Schiestl FP (2002) How an orchid harms its pollinator. *Proceedings of the Royal Society of London, Series B: Biological Sciences* **269**, 1329–32.

Wong R[B]BM, Salzmann C, Schiestl FP (2004) Pollinator attractiveness increases with distance from flowering orchids. *Proceedings of the Royal Society of London, Series B: Biological Sciences*, **271**, S212–14.

Woodcock G (1969) *Henry Walter Bates, Naturalist of the Amazons*. Faber and Faber, New York.

Woodcock TS, Larson BMH, Kevan PG, Inouye DW, Lunau K (2014) Flowers and flies II: floral attractants and rewards. *Journal of Pollination Ecology*, **12**, 63–94.

Woodward CL, Berry PE, Maas-van de Kamer H, Swing K (2007) *Tiputinia foetida*, a new mycoheterotrophic genus of Thismiaceae from Amazonian Ecuador, and a likely case of deceit pollination. *Taxon*, **56**, 157–62.

Wourms MK, Wasserman FE (1985a) Bird predation on Lepidoptera and the reliability of beak-marks in determining predation pressure. *Journal of the Lepidopterists' Society*, **39**, 239–61.

Wourms MK, Wasserman FE (1985b) Butterfly wing markings are more advantageous during handling than during the initial strike of an avian predator. *Evolution*, **39**, 845–51.

Wulff JL (1994) Sponge feeding by Caribbean angelfishes, trunkfishes, and filefishes. In RWM van Soest, TMG van Kempen and JC Braekman (eds), *Sponges in Time and Space*. Balkema, Rotterdam, pp. 265–71.

Wüster W, Allum CSE, Bjargardóttir IB, Bailey KL, Dawson KJ, Guenioui J, Lewis J, McGurk J, Moore AG, Niskanen M, Pollard CP (2004) Do aposematism and Batesian mimicry require bright colours? A test, using European viper markings. *Proceedings of the Royal Society of London, Series B: Biological Sciences*, **271**, 2495–9.

Wyman RL, Ward JA (1972) A cleaning symbiosis between the cichlid fishes *Etroplus maculatus* and *Etroplus suratensis* I. Description and possible evolution. *Copeia*, **1972**, 834–8.

Xu SQ, Schlüter PM, Schiestl P (2012) Pollinator-driven speciation in sexually deceptive orchids. *International Journal of Ecology*, **2012**: 285081.

Yachi S, Higashi M (1998) The evolution of warning signals. *Nature*, **394**, 882–4.

Yamauchi A (1993) A population dynamic model of Batesian mimicry. *Researches on Population Ecology*, **35**, 295–315.

Yamazaki K (2010) Leaf mines as visual defensive signals to herbivores. *Oikos*, **119**, 796–801.

Yamazaki K (2016) Caterpillar mimicry by plant galls as a visual defense against herbivores. *Journal of Theoretical Biology*, **404**, 10–14.

Yamazaki K, Lev-Yadun S (2015) Dense white trichome production by plants as possible mimicry of arthropod silk or fungal hyphae that deter herbivory. *Journal of Theoretical Biology*, **364**, 1–6.

Yanega D (1994) Arboreal, ant-mimicking mutillid wasps, *Pappognatha*; parasites of Neotropical *Euglossa* (Hymenoptera: Mutillidae and Apidae). *Biotropica*, **26**, 465–8.

Yang C, Antonov A, Cai Y, Stokke BG, Moksnes A, Roskaft E, Liang W (2012) Large hawk-cuckoo *Hierococcyx sparverioides* parasitism on the Chinese babax *Babax lanceolatus* may be an evolutionarily recent host-parasite system. *Ibis*, **154**, 202–4.

Yasukawa K, Searcy WA (1985) Song repertoires and density assessment in red winged blackbirds: further tests of the Beau Geste hypothesis. *Behavioral Ecology and Sociobiology*, **16**, 171–5.

Yeager CP (1992) Proboscis monkey (*Nasalis larvatus*) social organization: nature and possible functions of intergroup patterns of association. *American Journal of Primatology*, **26**, 133–7.

Yeargan KV (1994) Biology of bolas spiders. *Annual Review of Entomology*, **39**, 81–99.

Yeargan KV, Quate LW (1996) Juvenile bolas spiders attract psychodid flies. *Oecologia (Berlin)*, **106**, 266–71.

Yen S-H, Robinson GS, Quicke DLJ (2005) Phylogeny, systematics and evolution of mimetic wing patterns of *Eterusia* moths (Lepidoptera, Zygaenidae, Chalcosiinae). *Systematic Entomology*, **30**, 358–97.

Yeo P (1968) The evolutionary significance of the speciation of *Euphrasia* in Europe. *Evolution*, **22**, 736–47.

Yokoi T, Fujisaki K (2008) Hesitation behaviour of hoverflies *Sphaerophoria* spp. to avoid ambush by crab spiders. *Naturwissenschaften*, **96**, 195–200.

Yoshida A, Motoyama M, Kosaku A, Miyamoto K (1997) Antireflective nanoprotuberance array in the transparent wing of a hawkmoth, *Cephonodes hylas*. *Zoological Science*, **14**, 737–41.

Yoshino TP, Bayne CJ (1983) Mimicry of snail host antigens by miracidia and primary sporocysts of *Schistosoma mansoni*. *Parasite Immunology*, **5**, 317–28.

Yoshino TP, Wu X-J, Liu H, Gonzalez LA, Deelder AM, Hokke CH (2012) Glycotope sharing between snail hemolymph and larval schistosomes: larval transformation products alter shared glycan patterns of plasma proteins. *PLoS Neglected Tropical Diseases*, **6(3)**: e1569.

Young AM (1971) Wing coloration and reflectance in *Morpho* butterflies as related to reproductive behaviour and escape from predators. *Oecologia (Berlin)*, **7**, 209–22.

Young AM (1979) The evolution of eyespots in tropical butterflies in response to feeding on rotting fruit: an hypothesis. *Journal of the New York Entomological Society*, **87**, 66–77.

Young FN (1957) Notes on the habits of *Plusiotis gloriosa* Le Conte (Scarabaeidae). *The Coleopterist's Bulletin*, **11**, 67–70.

Young RE (1983) Oceanic bioluminescence: an overview of general functions. *Bulletin of Marine Science*, **33**, 829–45.

Young RE, Mencher FM (1980) Bioluminescence in mesopelagic squid: diel color change during counterillumination. *Science*, **208**, 1286–8.

Young RE, Roper CFE (1977) Intensity regulation of bioluminescence during countershading in living midwater animals. *Fishery Bulletin. United States Fish and Wildlife Service*, **75**, 239–52.

Young RT (1916) Some experiments on protective coloration. *Journal of Experimental Zoology*, **20**, 457–507.

Young TP, Stanton ML, Christian CE (2003) Effects of natural and simulated herbivory on spine lengths of *Acacia drepanolobium* in Kenya. *Oikos*, **101**, 171–9.

Yurtsever S (2000) On the polymorphic meadow spittlebug, *Philaenus spumarius* (L.) (Homoptera: Cercopidae). *Turkish Journal of Zoology*, **24**, 447–59.

Zabka H, Tembrock G (1986) Mimicry and crypsis – A behavioral approach to classification. *Behavioural Processes*, **13**, 159–76.

Zagrobelny M, Bak S, Rasmussen AV, Jørgensen B, Naumann CM, Møller BL (2004) Cyanogenic glucosides and plant–insect interactions. *Phytochemistry*, **65**, 293–306.

Zagrobelny M, Bak S, Olsen CE, Møller BL (2007a) Intimate roles for cyanogenic glucosides in the life cycle of *Zygaena filipendulae* (Lepidoptera, Zygaenidae). *Insect Biochemistry and Molecular Biology*, **37**, 1189–97.

Zagrobelny M, Bak S, Ekstrøm CT, Olsen CE, Møller BL (2007b) The cyanogenic glucoside composition of *Zygaena filipendulae* (Lepidoptera: Zygaenidae) as effected by feeding on wild-type and transgenic lotus populations with variable cyanogenic glucoside profiles. *Insect Biochemistry and Molecular Biology*, **37**, 10–8.

Zagrobelny M, Bak S, Møller BL (2008) Cyanogenesis in plants and arthropods. *Phytochemistry* **69**, 1457–68.

Zahavi A (1977) Reliability in communication systems and the evolution of altruism. In B Stonehouse and CM Perrins (eds), *Evolutionary Ecology*. Macmillan, London, pp. 253–9.

Zahavi A (1993) The fallacy of conventional signaling. *Philosophical Transactions of the Royal Society Society of London, Series B: Biological Sciences*, **338**, 227–30.

Zahiri R, Kitching IJ, Lafontaine JD, Mutanen M, Kaila L, Holloway JD, Wahlberg N (2011) A new molecular phylogeny offers hope for a stable family-level classification of the Noctuoidea (Lepidoptera). *Zoologica Scripta*, **40**, 158–73.

Zahiri R, Holloway JD, Kitching IJ, Lafontaine JD, Mutanen M, Wahlberg N (2012) Molecular phylogenetics of Erebidae (Lepidoptera, Noctuoidea). *Systematic Entomology*, **37**, 102–24.

Zaret TM (1977) Inhibition of cannibalism in *Cichla ocellaris* and hypothesis of predator mimicry among South American fishes. *Evolution*, **31**, 421–37.

Zhang S, Chen H-L, Chen K-Y, Huang J-J, Chang C-C, Piorkowski D, Liao C-P, Tso I-M (2015) A nocturnal cursorial predator attracts flying prey with a visual lure. *Animal Behaviour*, **102**, 119–25.

Zhao Z-S, Granucci F, Yeh L, Schaffer PA, Cantor H (1998) Molecular mimicry by Herpes Simplex Virus-Type 1: autoimmune disease after viral infection. *Science*, **279**, 1344–7.

Zimma B01, Ayasse M, Tengö J, Ibarra F, Schulz C, Francke W (2003) Do social parasitic bumblebees use chemical weapons? (Hymenoptera, Apidae). *Journal of Comparative Physiology A*, **189**, 769–75.

Zimmerman DA (1976) Comments on feeding habits and vulture-mimicry in the zone-tailed hawk. *The Condor*, **78**, 420–1.

Zimova M, Mills LS, Lukacs PM, Mitchell MS (2014) Snowshoe hares display limited phenotypic plasticity to mismatch in seasonal camouflage. *Proceedings of the Royal Society of London, Series B: Biological Sciences*, **281**: 20140029.

Zug GR, Vitt LJ, Caldwell JP (2001) *Herpetology. An Introductory Biology of Amphibians and Reptiles*. Academic Press, San Diego.

Zolnerowich GA (1992) Unique *Amycle* nymph (Homoptera: Fulgoridae) that mimics jumping spiders (Araneae: Salticidae). *Journal of the New York Entomological Society*, **100**, 498–502.

Zompro O, Fritzsche I (1999) *Lucihormetica* n. gen. n. sp., the first record of luminescence in an orthopteroid insect (Dictyoptera: Blaberidae: Blaberinae: Brachycolini). *Amazoniana*, **15**, 211–9.

Zylinski S, Johnsen S (2011) Mesopelagic cephalopods switch between transparency and pigmentation to optimise camouflage in the deep. *Current Biology*, **21**, 1937–41.

Zylinski S, Osorio D, Shohet A (2009a) Cuttlefish camouflage: context dependent body pattern use during motion. *Proceedings of the Royal Society of London, Series B: Biological Sciences*, **276**, 3963–9.

Zylinski S, Osorio D, Shohet A (2009b) Perception of edges and visual texture in the camouflage of the common cuttlefish, *Sepia officinalis*. *Philosophical Transactions of the Royal Society of London, Series B: Biological Sciences*, **364**, 439–48.

Zylinski S, Osorio D, Shohet, A (2009c) Edge detection and texture classification by cuttlefish. *Journal of Vision*, **9**, 1–10.

AUTHOR INDEX

Index of senior authors, and second authors in papers with two or more authors, and also last authors in papers with three or more authors because of the common practice of senior authors taking the last place in author lists. With the growing trend towards papers with many authors, to list all would be rather space-wasting.

Considerable effort has been put in to assuring that authors are listed only once even if they appear with different numbers of given initials in their publications. In a few cases where an author has married and amended their name accordingly, they may be cited twice. Only in the one case of a very prolific author have I made relevant reference to the alternative name used; in most other cases the identity should be apparent.

Abbott CE, 99
Abramowitz AA, 54
Acheampong A, 99
Ackerman JD, 372–3, 382
Adam A, 31
Adamo JM, 214
Adamson TA, 239
Adler K, 132
Ågren J, 374, 393–4
Ahl JSB, 354
Åhrberg C, 154
Aiello A, 194, 199
Ainsley DG, 306
Akcali CK, 242
Åkesson S, 89
Akino T, 51, 53, 73, 233, 237, 313, 346
Akkaynak D, 63
Akoury TS, 99
Akre RD, 347, 349
Alatalo RV, 14, 73, 128–30, 132–4, 139, 152–3, 204, 207, 210, 246
Alcami A, 425
Alcock J, 3, 128, 143, 182, 194, 198, 210, 248, 318, 354–5, 376
Alderson NJ, 126, 204–5
Aldrich JR, 132
Alexander HM, 418
Alexander T, 173
Alexandersson R, 385
Alexandrou MA, 239

Allahyari H, 350
Allain D, 37
Allan RA, 233, 236
Allen AA, 99
Allen GR, 323
Allen JA, 20, 31, 73, 75, 77, 204, 206, 441
Allen JJ, 56, 58, 63
Allen WL, 58, 61, 71–2
Allum CSE, 245
Almada VC, 355
Alonso-Mejía A, 145
Alpert G, 347
Alpert T, 347
Alvarez F, 336, 339
Alvarez JC, 272
Alvarez-Lopez H, 35
Alves RJV, 392
Amaoka K, 315
Amcoff M, 11, 362–3
Amoore JE, 390
Anderson BC, 384–5, 387
Anderson CM, 368, 404
Anderson KP, 204
Anderson MG, 11, 339–40
Anderson SC, 317
Andersson M, 157
Andreadis PT, 317
Andrén C, 37
Ángel-Giraldo P, 215
Angers B, 218

Angersbach D, 35
Angiov A-M, 390
Antoniazzi MM, 147
Antonov A, 335
Antonovics J, 418
Aplin RT, 139
Arauco R, 301
Arca M, 225
Arenas LM, 104
Ari C, 56
Arias M, 134
Arias de Reyna L, 339
Arieli R, 368
Arillo A, 99
Armstrong EA, 277
Arndt E, 347
Arnold EN, 279
Arnold SJ, 190, 355, 357
Arnqvist G, 11, 362–3
Aronsson M, 104, 109
Askew RR, 41, 89
Atkins D, 54
Atkins EL, 249
Atkinson EC, 324
Atkinson WD, 74
Atwood JT, 388
Aubin-Horth N, 423
Aubret F, 157
Augner M, 165, 295
Auko TH, 89

Mimicry, Crypsis, Masquerade and other Adaptive Resemblances, First Edition. Donald L. J. Quicke.
© 2017 John Wiley & Sons Ltd. Published 2017 by John Wiley & Sons Ltd.

Averill CK, 309
Aviezera I, 165
Avila JL, 424
Avila-Pires TCS, 279
Avilés JM, 334, 336–7
Avise JC, 354
Ayasse M, 344, 373–5, 377, 379, 381, 389

Bachmann T, 307
Backwell PRY, 54, 408
Baddeley R, 58, 61, 69, 71–2
Badruddine H, 122
Badura LL, 37
Bagnara JT, 49
Bagnères AG, 344
Bailey VA, 253
Bailey W, 354
Bain RS, 166–7, 209
Bak S, 145–6
Baker ECS, 336
Baker HG, 393, 431
Baker RR, 151
Balgooyen TG, 137
Ballantyne LA, 325
Balleto E, 99
Balogh ACV, 172, 180–1
Balph DF, 442
Balph MH, 442
Bantock CR, 204
Bänziger H, 382, 392
Baptista LF, 402
Barber JR, 275
Barbosa A, 58, 151
Barboza G, 373, 376
Barlow BA, 293
Barlow GW, 61, 355
Barnard CJ, 286
Barnard KH, 320
Barnett CA, 205–6
Barrett JA, 183, 201
Barrett KH, 117
Barrion AT, 233, 319
Barrow RA, 379
Barta Z, 337
Bartell CK, 54
Barthlott W, 290, 397
Barton NH, 217
Batchelor S, 204
Bateman RM, 375
Bates HW, 2, 4, 135, 214–5, 271
Bateson M, 205–6, 262
Bateson W, 83
Batra LR, 418
Batra S, 418
Bawa KS, 393
Baxter SW, 219–20, 222
Baylis JR, 401–2

Bayne CJ, 423
Beach SA, 365
Bear A, 308
Beardsell DV, 382, 384–5
Beattie AJ, 372
Beatty CD, 196
Beccaloni GW, 162, 214, 216
Beck J, 138
Beckers GJL, 242
Beddington JR, 333
Beebe W, 216
Beecher MD, 402
Beehler BM, 124
Beharry S, 248
Beirinckx K, 196
Belk MC, 37
Bell GRR, 58
Belt T, 201
Benitez-Vieyra S, 386
Benjamin DR, 166
Bennett ATD, 209
Bennett I, 147
Benson WW, 81, 162, 175, 204,
    216, 299
Bent AC, 277
Bequaert J, 419
Berenbaum MR, 224
Berger D, 53
Bergström G, 346, 348, 379
Bergstrom PW, 277
Bermingham E, 325
Bernays E, 145, 295
Bernhardt P, 388
Berry PE, 392
Berry RJ, 37
Bessière J-M, 351
Bezemer TM, 233
Bezzerides AL, 119
Bhushan B, 47
Bickel CT, 368, 404
Biedermann PHW, 300
Bierzychudek P, 382, 385
Bilde T, 99
Billen J, 231
Billiard S, 132
Binckley CA, 309, 312
Bininda-Emonds ORP, 29, 36–7
Bionda R, 37
Blahó M, 68–9
Blakiston T, 173
Blamires SJ, 23, 93
Blanco MA, 138, 373, 376
Blest AD, 120, 138, 257, 262
Blomquist GJ, 349
Blount JD, 118
Blow NC, 204
Blum MS, 132, 146, 227, 349, 351

Blut C, 260, 267–8
Boal JG, 354
Boardman M, 41, 89
Boatwright JS, 163
Bobisud LE, 180, 228, 381
Bocak L, 153
Boettinger L, 349
Boggs CL, 121
Bohl E, 46
Bohman B, 379, 440
Boileau N, 322
Boland W, 147
Bolin JF, 392
Bologna MA, 147
Bonaldo RM, 286
Bonavita-Cougourdan A, 348
Bond AB, 2–3, 73–5, 97, 171
Bond JE, 242
Bonser RHC, 61, 437
Booth CL, 82
Boppré M, 144–5, 202, 251
Borg-Karlson A-K, 379
Borgia G, 288
Borries C, 367
Bostanchi H, 317
Bots J, 358
Bourgoin T, 23
Bowden SR, 35
Bowers MD, 32, 120, 136, 144, 150, 152,
    192–3, 201, 224
Boyd RS, 418
Boyden TC, 136, 139, 373–4, 382, 385
Boyer D, 317
Braasch I, 362
Braby MF, 239
Brach V, 227
Bradley BJ, 24
Bradshaw SD, 317
Brady J, 67
Brady PC, 49
Brakefield PM, 148
Brain CK, 246
Brakefield PM, 37, 153, 266–7
Brandao CRF, 233
Brandon RA, 239, 240, 246
Brandt A, 207
Brassington RJ, 319
Brattstrom BH, 240
Brede C, 342
Breed MD, 349
Bretagnolle V, 309
Brilot BO, 262
Brim-Box J, 417
Brinton BA, 47
Bristowe WS, 319
Britton G, 50
Bro-Jørgensen J, 365–6

Broadhead E, 40
Broadley RA, 312
Brockmann HJ, 320, 366
Brodie III ED, 69, 119, 240, 242–4
Brodmann J, 373, 389
Broekhuysen GJ, 60–1
Bronner GN, 419
Brooke M de L, 13, 332, 334, 336–7, 339
Brooker AMH, 338
Brooker LC, 338
Brooker MG, 338
Broom M, 108–9, 116
Brower AVZ, 146, 162, 214, 224–5
Brower JVZ, 3, 124, 137, 172, 182,
    191–4, 198, 201, 248–9, 253, 329,
    429–30
Brower LP, 3, 120, 124, 136, 140, 142–6,
    172, 182, 188–90, 192–4, 198, 201,
    248–9, 329, 429–30
Brown C, 354–5
Brown CP, 318
Brown IL, 136, 201
Brown JH, 285, 395, 397
Brown JL, 76
Brown KS, Jr, 72, 140–1, 130, 118,
    144–6, 149, 162, 199, 204, 216, 299,
    395, 397
Brown RM, 244
Brown SM, 222
Brown VK, 290
Brown WV, 120, 140, 142, 146–7
Bruyns PV, 142
Bryant EH, 112, 114–15
Bshary R, 321
Bucher KE, 302
Bückmann D, 35
Budelmann BU, 58
Budenberg WJ, 382
Bull JJ, 109, 112–13
Bullini L, 196, 228
Bunk E, 344
Bunkley-Williams L, 323
Burd M, 72
Burg JS, 425
Burger BV, 419
Burger J, 136
Burgess KS, 388
Burghardt GM, 99, 125, 190, 250,
    354, 408
Burkhardt EL, 290–1
Burns JM, 203, 262, 284
Burns KC, 290, 293–4, 411
Burns-Balogh P, 386
Burtt EH, 286, 309
Butler RS, 417
Buttery RG, 390
Buxton PA, 145

Byrkjedal I, 277
Byrne R, 367
Bytebier B, 374

Cabrera A, 351
Cahn MG, 297
Cain AJ, 37, 74, 77, 80
Cairns DK, 309
Calder DM, 386
Caldwell GS, 130
Caldwell JP, 277
Caley MJ, 195, 197–8
Callahan A, 44
Calvo RN, 374
Camara MD, 150
Camargo MGG, 411
Camazine SM, 108, 166
Cambefort JP, 243
Cameron KM, 374
Camire L, 367
Campbell HW, 71
Campbell JA, 242, 317
Campbell SE, 402
Campitelli BE, 290, 297
Canfield MR, 44
Cannatella DC, 119, 153
Cannon L, 122
Cantor A, 63, 260, 263–5
Cantor H, 425
Cantrell MA, 285
Canyon DV, 293
Capon RJ, 233, 236
Capron A, 423
Carbutt C, 385, 387
Card A, 235
Cardoso GC, 24
Cardoso MZ, 111, 140
Carey JR, 139
Cariou M-L, 373
Carlberg U, 99
Carlisle TR, 360
Carlson SP, 54
Carlsson B-G, 364
Carmona-Díaz G, 385
Carneiro LA, 355
Caro T, 29, 36–7, 63, 157–8, 368
Carpenter CC, 317
Carpenter GC, 317
Carpenter GDH, 2, 120, 131, 136–7,
    139, 199
Carrasco-Urra F, 294
Carrick R, 20, 229
Carroll J, 121
Carruthers JM, 344
Casas J, 54
Case JF, 29
Cassey P, 61

Castner JL, 23
Castor-Perry SA, 260
Castoriades N, 122
Castro RMC, 315
Catania KC, 328
Catarino MF, 307
Catchpole CK, 402
Cazetta E, 411
Ceccarelli FS, 236
Cehláriková P, 109
Chadwick SL, 63–4, 66
Chai P, 135, 138, 214–15
Chance MRA, 429
Chandler S, 56
Chandra HS, 332
Chang S, 85
Chanter DO, 199
Charleston M, 162, 216, 218
Charlesworth B, 172, 183–4
Charlesworth D, 172, 183–4
Charpentier MJE, 406
Chase MW, 394
Chazot N, 72
Chemsak JA, 99
Chen Q, 291, 390
Chen H-L, 312
Cheney KL, 56, 119, 149, 321–2
Cheng J, 382
Chiao C-C, 31, 56, 58
China WE, 255
Chiou T-H, 49
Chiszar D, 317
Chittka L, 42, 47, 55–6, 211
Chopard L, 349
Chorvát D, 155–6
Chouteau M, 218
Christen H, 424
Christian CE, 165
Christiansen SS, 100
Christy JH, 408
Chu M, 288
Chubb C, 58
Churchi MA, 99
Ciotek L, 376
Cipresso Pereira PH, 285
Claridge MF, 250
Clark DL, 306, 317, 434
Clark L, 125
Clark R, 222
Clarke B, 73, 78, 204
Clarke CA, 32, 34, 39, 198–9, 203,
    220–3, 248
Clarke FMM, 203, 221
Clarke GL, 29
Clarke JM, 56
Claus R, 418
Claussen D, 100

Clee C, 227
Clément J-L, 344, 348
Clements MA, 382, 384–5
Cloudsley-Thompson JL, 6, 23, 42, 156, 284, 433
Clough ME, 119, 239
Clucas B, 140
Clutton-Brock T, 405
Cnaani J, 181
Cobb NA, 419
Cobben RH, 231
Coccuci AA, 386
Cockrell BJ, 189
Cockrum PA, 144
Cocucci AA, 376, 388
Cody ML, 240
Coe RL, 270, 402
Cohen D, 373
Cohen JA, 150, 189
Cohen JH, 31
Coleman E, 376, 380
Collenette CL, 136–7
Collett TS, 257
Collinge WE, 135
Collins CT, 124
Collins SL, 82
Colombo JM, 185–6
Compton SG, 415
Congdon JD, 246
Conner WE, 142
Connor OR, 61
Cook LM, 37–41, 89, 184, 194
Cook SE, 130, 358
Cooney LJ, 413
Cooper J, 408
Cooper JM, 20, 31, 73
Cooper SM, 165
Cooper WE Jr, 132, 250, 279
Coppinger RP, 127, 248
Corbacho C, 61
Cordero A, 358
Cordero Rivera A, 358
Cori R, 183
Cornwell CJ, 58
Cortes F, 149
Côrtes Figuera JE, 30
Cortesi F, 56, 119, 322
Corvino JM, 430
Coscia EM, 404
Coss RG, 156, 270
Cossart P, 421
Côté IM, 322
Cott HB, 2–3, 6, 14, 20, 29, 61, 88, 124, 129, 438, 442
Cotterill JV, 303
Cotton PA, 204
Couldridge VCK, 361

Counterman BA, 217, 219
Cournoyer BL, 31
Coutin R, 253
Cowan DP, 303
Cowen RK, 47
Cowie RJ, 153
Cox AHM, 156
Cox CL, 242, 244
Cox GW, 277
Cox M, 158
Cox S, 56
Coyne JA, 14, 38
Cozzolino S, 375–6
Craig CL, 311
Crane J, 217
Cranfield MR, 85
Cranston F, 201
Cratsley CK, 325
Crawley MJ, xvi, 152
Creed ER, 40
Cremer S, 355, 357
Crewe RM, 346
Crews D, 355–6
Cronin TW, 46, 49
Croston R, 339
Croze H, 75
Croze HJ, 194
Crozier RH, 236
Cruz A, 338
Cuervo H, 424
Culvenor CCJ, 145, 152
Cummings ME, 49, 119, 153
Cummings RD, 423
Cunha GR, 404
Cunningham MW, 424
Curio E, 190
Curran MC, 47
Currey JD, 37
Cushing PE, 234, 236
Cusick ME, 424
Cuthill IC, 3, 14, 51, 88, 61–3, 66, 69, 71–2, 122, 155, 165, 209, 260, 262, 269
Cuthill JH, 162, 216, 218
Cutler B, 234–5
Cygan JP, 277
Czajka M, 319
Czienskowski U, 69, 71

D'Arcy Ward R, 139
D'Ettorre P, 348
D'Rozario V, 233, 237
DaCosta JM, 340
Dafni A, 350, 372–5, 378, 382, 385–6, 437
Dafni J, 285, 441
Dahlgren U, 28
Dale S, 364–5

Dalen JL, 411
Dallas JF, 405
Daloze D, 144, 146
Dalton JP, 423
Daly JW, 119, 124, 147
Dalziell AH, 3, 402
Damian RT, 422, 424
Damman H, 120
Dani FR, 344
Dare J, 47
Darlington PJ, 153
Darnell MZ, 54
Darrah A, 356
Darst C, 119, 153
Darwin C, 1, 201, 297, 372
Davey M, 308
Davidson DW, 415
Davidson E, 317
Davies NB, 13, 332, 334, 336–7, 339, 341–2
Davis A, 389
Davis JRA, 260
Davis Rabosky AR, 242, 244
Davis RH, 145–6
Dawkins MS, 75–6, 104, 116–17, 119, 178, 216, 441
Dawkins R, 255, 339, 360, 436
Dawson SM, 408
Dayawansa PN, 277
De Bona S, 264
de Bruyn L, 358
De Cock R, 155
de Gunst JH, 227
de Ibarra NH, 386
de Jong PW, 146, 429
de Jonge J, 355
de Lope F, 58
de Oliveira Calleia F, 324
de Ruiter AJH, 355
de Ruiter L, 3, 6, 23, 94
de Vlieger J, 401
de Vos H, 146
de Wert L, 88, 159
de Wet JMJ, 415
Debut M, 54
Deegan JF II, 163
Deeleman-Reinhold CL, 233, 237
Deering MD, 201
Del-Claro K, 231, 233
del Rio MG, 227
Delay LS, 414
DeLoache JS, 130
Demottoni P, 185
Dentinger BTM, 389
Denton AM, 29, 47, 54
Denton EJ, 29, 47
Depczynski M, 354
Deppe C, 276

DeRiemer K, 149
Deroe C, 144
Desmier de Chenon R, 253
Dettner K, 120, 146–7, 233, 346–7,
    349–50, 381
Detto T, 54
DeVries PJ, 137, 216
Dewhirst RA, 27–8
Di Giusto B, 351
Diamant A, 285, 441
Diamond AR Jr, 418
Diamond J, 2–3, 97, 171
Diamond JM, 306, 401
Didham RK, 381
Dieckhoff C, 350
Dill LM, 188
Dimitrova M, 31, 63–4, 66–8
Dinan L, 439
Dinsmore SJ, 60
Dinter K, 347
Dipper FA, 354–5
Ditmars RL, 244
Dittrich WH, 200–209
Dixey FA, 439
Dixson AF, 368, 403
Dodson CH, 13, 373, 376, 394
Dolenská M, 131, 153
Dominey WJ, 354–5
Donnelly MA, 147, 153
Donnelly S, 423
Doostdar H, 297
Dorchin A, 392
dos Santos MB, 97
Dötterl S, 290, 293, 297–8, 373, 391–2
Doyle RJ, 424
Drea CM, 404
Dressler R, 387
Drijfhout FP, 344, 347
Drumm KE, 37
Drummond BA, 234, 443
Duckworth N, 366
Duckworth WD, 45
Dudgeon CL, 239
Dudley R, 99
Duff HC, 294
Duffey SS, 297
Duffy K, 429
Dukas R, 13, 313, 355, 393
Dumbacher JP, 124, 147
Duncan CJ, 207
Duncan MJ, 37
Dunlap-Pianka H, 121
Dunn CW, 316
Dunn ER, 240
Dunning DC, 250, 256–7
Dupuy D, 382
Durkee CA, 235, 236
Dussourd DE, 145

Duval C, 61
Dvořáková R, 368
Dyer AG, 30
Dyer LA, 150

Eagle JV, 323
East ML, 355, 404
Eaton MD, 30
Eaton RL, 240
Eberhard WG, 93, 307, 326, 431, 442
Eberhardt LE, 97
Eberhardt MJW, 329
Ebert D, 374, 379
Ebert K, 311
Edelstam C, 320
Edgar JA, 144–5, 152
Edgehouse M, 318
Edmonds N, 40
Edmunds J, 93, 126
Edmunds M, xiv, 5–6, 11, 27–8, 50–1,
    104, 118, 122, 126, 137, 142, 147,
    154, 188, 190, 210–11, 220, 227,
    229, 235, 249, 260, 433, 437, 441
Edmund-Hill 417
Edwards DP, 301
Edwards GB, 227, 231
Eeley H, 217
Egger B, 360–1
Egri Á, 68–9
Ehrlén J, 386
Ehrlich AH, 135, 285
Ehrlich PR, 135, 285
Eichlin TD, 45
Eisenmann E, 277
Eisikowitch D, 299
Eisner M, 145, 147, 170
Elsner T, 85, 108, 139, 143–5, 147, 153,
    170, 249–51, 260, 325
El Mayas H, 418
Elgar MA, 233–4, 236, 319
Elias M, 72, 224
Eliyahu D, 319
Ellis A, 373, 391
Elmes GW, 237
Emery AR, 5
Emlen JM, 172, 176, 181
Emmel TC, 23, 89, 153
Emsley MG, 162, 216, 240, 434
Endara L, 389
Endler JA, 3–6, 8–9, 13, 20, 24, 41, 72,
    75, 82, 104, 109, 117, 133, 150, 169,
    161, 175, 430, 437, 439
Endress ME, 142
Engen S, 112, 114
Ennos AR, xiv, 249
Enquist M, 115–16, 122, 188
Epstein WW, 415
Eriksson S, 284

Erlanson C, 419
Errard C, 348
Escobar-Lasso S, 97
Estabrook GF, 180
Estrada C, 219
Eterovick PC, 30
Evans DL, 31, 99, 113, 122, 229,
    283–4
Evans HE, 199, 329
Evans SM, 285
Ewao H, 355
Ewer RF, 97
Ewert J-P, 61
Exnerová A, 47, 106, 109, 132
Exton LS, 109, 177, 198

Fabricant SA, 47, 50, 54, 72
Fairbairn JW, 140
Fargallo JA, 400
Farrell C, 140, 150
Farrell TM, 317
Feeney WE, 56, 338, 342
Feitosa RM, 231
Feldman MW, 108
Feller KD, 46
Feltwell J, 150
Fenton A, 161
Fenton MB, 257
Ferguson GP, 58
Fernandes WG, 302
Fernández González S, 99
Ferreira WC Jr., 175
ffrench-Constant RH, 185, 223
Fiedler K, 138
Field SA, 355, 357
Fields PG, 37
Finckh LM, 333, 338
Fingerman M, 54
Fink LS, 53
Finlay S, 63
Finn J, 151, 354
Fiore G, 58
Fisher RA, 3, 112, 172, 183,
    204, 211
Fisher RM, 157
Fishlyn DA, 53, 73
Fitzpatrick BM, 74
Fitzpatrick JL, 270
Fitzsimons VFM, 246
Flach A, 379
Flanagan NS, 219
Flegr J, 368
Fleishman LJ, 308
Floren A, 251
Florey E, 58
Flower TP, 402
Foitzik S, 347
Foot G, 350

Forbes MR, 358
Forbes P, 2
Ford B, 112
Ford EB, 2, 31, 81, 172, 183
Ford HA, 207, 209
Forrest AD, 375
Forrester LJ, 390
Forsman A, 41, 63, 69, 76–9, 109, 132, 239
Forsyth A, 318, 354–5
Forsythe JW, 58, 151
Forti LC, 350
Foster MA, 400
Foster MS, 414
Foster RG, 54
Foster SA, 355
Fotin RW, 100
Fowler HG, 350
Frahn JL, 144
France D, 81
Francini RB, 145
Francke W, 344, 374, 379, 381
Francq EN, 97
Frank LG, 404–5
Franks DW, 116, 120, 172, 185
Frazer JFD, 228
Free CA, 333
Freeland WJ, 139
Freitas AVL, 151
French V, 266
Fresno M, 424
Fretwell SD, 400
Frezal L, 185, 223
Frick C, 142
Fricke HW, 255, 432
Frisch AJ, 195
Frith CB, 288
Frith HW, 288
Fritzsche I, 155
Froesch D, 58
Frolova EN, 416
Frymire D, 13, 373
Fuchs R, 129, 131, 153
Fuentes M, 101, 413–14
Fuhrman FA, 149
Fuhrman GJ, 149
Fujihashi Y, 203
Fujii R, 50
Fujii Y, 146
Fujinami RS, 424
Fujisaki K, 313
Fujiwara H, 203
Fulgione D, 54
Fullard JH, 257
Fürstenau B, 349
Fürtbauer I, 368
Furth DG, 150

Gagliano M, 270, 354
Gagliardo A, 132–3
Gagné RJ, 302
Gahan CJ, 153
Galán JE, 424
Gallego-Ropero MC, 231
Gamberale-Stille G, 82, 104, 108–11, 119, 123, 130, 132–4, 172, 180–1, 190
Gamble FW, 54, 56
Gans C, 242–3, 254
García DA, 246
Garcia J, 108
Garcia JE, 30
Garcia KC, 425
Garcia TS, 56
Gardiner B, 35, 189
Gardner DR, 224
Garman S, 155
Garnett WB, 347
Garraffo HM, 147
Garrett K, 99
Garwood MP, 354–5
Gaskett AC, 376, 380
Gatti F, 97
Gatz Jr AJ, 286
Gaul AT, 157
Gaume L, 351
Gavrilets S, 185
Geburtig MA, 73
Gehlbach FR, 276
Gehring C, 425
Geiselhardt SF, 346
Gelis RA, 277
Gemeno C, 326
Gendron RP, 75–6, 80–1
Gentile K, 308
Gentile L, 209
Gerald GW, 100
Gerhardt U, 319
Getty T, 76, 181, 186, 207
Ghanem E, 294
Giard A, 100
Gibb JA, 75
Gibbs HL, 332
Gibernau M, 389–90
Gibran FZ, 315
Gibson DO, 144–5, 150, 188, 224
Gibson EA, 128
Gibson G, 67
Gibson RW, 300
Gigord LDB, 372, 387
Giguère LA, 46
Gilbert F, 207–9, 229
Gilbert LE, 89, 121, 161, 216–17, 294, 299
Gilerson AA, 49

Gillingham JC, 317
Gillis JE, 42
Gilpin-Brown JB, 47
Gingerich PD, 240
Giorgis P, 376
Gironès N, 424
Gittleman JL, 109–10, 124
Givnish TJ, 290–1, 374
Glazier SC, 120, 142, 144, 190
Glickman SE, 404–5
Gloag R, 338
Gluckman T-L, 24, 341
Gnezdilov VM, 199, 231
Gochfeld M, 136, 277
Goddard GHR, 122
Godden DH, 99
Godfray HCJ, 152
Godfrey D, 69
Goetz MA, 325
Goffinet B, 419
Gohli J, 85, 127
Goldblatt P, 388
Golding YC, xiv, 249
Goldman BD, 37
Goldschmidt RB, 3
Goldschmidt T, 362
Gomez D, xiv
Gómez JM, 165
Gon SM III, 199
Gonçalves EJ, 355
Gonçalves Rodrigues LR, 77
González A, 120, 145, 250, 260
González F, 392
González-Duran GA, 97
Goodale E, 287–8, 402
Goodale MA, 126–7
Goodman JD, 244
Goodman JM, 244
Gordo M, 324
Gordon IJ, 220
Gordon L, 334
Goris RC, 250
Gosler AG, 61
Götmark F, 124, 129
Gotthard K, 53
Gottsberger G, 392
Goubara M, 385
Gould JL, 89
Gould KS, 413
Gould SJ, 87
Gouyon PH, 296
Gower BA, 37
Gower D, 99
Grafen A, 117, 315
Graham J, 63
Graham RR, 307
Grant BS, 39, 50, 52

Grant GG, 277
Grant JB, 82
Grant KG, 395
Grant RP, 108, 120
Granucci F, 425
Gray M, 92
Grayson J, 50–1
Green PR, 209
Greene E, 50, 251–2
Greene HW, 2, 99, 163, 240, 242, 275, 317
Greeney HF, 277
Greenwood JJD, 73, 76, 81, 181, 204
Greenwood PJ, 112, 124
Greenwood RG, 149
Greer AT, 47
Greig-Smith PW, 413
Grenot C, 97
Grether GF, 323–4, 432
Grewcock D, 208–9
Gribble M, 402
Griksaitis D, 131
Grim L, xiv, 3, 332, 339
Grimaldi DA, 346, 389
Grobecker DB, 307, 315
Grober MS, 155, 355
Groom PK, 294
Gross MR, 354, 356
Groth I, 379
Grubb PJ, 290
Grüneberg H, 77
Grutter AS, 321–2
Gudger EW, 315
Guézé P, 285
Guilbert F, 23
Guilford T, 5, 75–6, 104, 108, 112, 116–17, 119, 122–3, 129, 132–3, 140, 155, 157, 165, 179, 190, 339, 358, 435–6, 440–1
Gumbert A, 386
Guthrie RD, 41, 408, 444
Gutman M, 302

Haag WR, 417
Haas F, 432
Hacker SD, 97
Haddad CFB, 32, 88, 97, 99, 163, 276
Haddock SHD, 316
Hadeler KP, 185
Hadley ME, 49
Hafernik J, 255
Hagen RH, 202, 223
Hagman M, 109, 239, 317
Hahn TP, 199
Hailman JP, 20, 88, 271, 320
Hakkarainen H, 404
Hallwachs W, 262, 271, 284

Halpern M, 165–6
Halpin CG, 120, 205, 207
Hammerschmidt K, 423
Hamner WM, 47
Handel SN, 381
Hanks JP, 41
Hanlon RT, 31, 54, 58, 63, 151, 354–5
Hansell MH, 85, 315
Hansen D, 391
Hansen M, 147
Hansen T, 224
Hanski I, 166
Hansknecht KA, 317
Hanson P, 315
Hanson TC, 250
Hansson BS, 388–9
Haraguchi D, 100
Harari AR, 366
Harcombe WR, 242
Harder LD, 374
Hardison PD, 286
Hardman CJ, 260, 263
Hare JF, 145
Harlan JR, 415
Harland DP, 308
Harlow P, 357
Harman RRM, 140, 145
Harper GR, 162, 173
Harper JL, 297
Harper RD, 29
Harris AC, 253
Harris LG, 149
Harris RHTP, 67
Hart NS, 31
Hartfield PD, 417
Hartman WD, 74
Harvell CD, 426
Harvey IF, 28
Harvey JA, 233
Harvey PH, 104, 109–10, 112–13, 124, 204
Harvis CA, 144
Hasegawa E, 348
Hashim R, 231
Hashimoto Y, 233
Hass M, 35
Hassall C, 63, 66, 210
Hasson O, 308
Håstad O, 31
Hastings A, 185
Hatle JD, 139
Hauber ME, 11, 336–40
Haugsten Holen Ø, 224
Hauser MD, 367
Havenhand JN, 354–5
Havlíček J, 368
Hawkins BA, 302

Hawkins CE, 405
Hawryshyn CW, 49
Hay ME, 83, 85, 147
Hayashi M, 85
Haynes KF, 326
Hazel WN, 32, 34–5, 203
Heal J, 249
Healy SD, 270, 288, 402
Heath RR, 326
Heatwole H, 317
Hebert FO, 423
Hebert PDN, 114, 151
Hecht MK, 240
Hefetz A, 348
Hegna RH, 153, 161, 201
Heiling AM, 42
Heimpel GE, 350
Heinrich B, 82, 88, 375, 386–7
Heinsohn R, 82, 334
Heinze J, 348, 355, 357
Heistermann M, 368
Heller R, 190
Helmholtz H v, 67, 367
Hemmi JM, 54
Henderson G, 182
Henning F, 362
Henning SF, 239
Hensel JL, 119
Henson R, 351
Herberstein ME, 42, 50, 54, 72, 307, 313, 376
Heredia MD, 35
Herrel J, 203
Herrera C, 104
Herrström J, 63
Hert E, 360, 362
Hetherington TE, 150–1, 201, 203
Hesse E, 130
Hetterscheide WLA, 390
Hetz M, 121, 193
Hewitson WC, 278
Hicks K, 85
Higashi M, 108
Higginson AD, 88
Hilker M, 147
Hill CJ, 293
Hill KBR, 326, 431
Hill RI, 216, 277
Hilton DFJ, 358, 429–30, 435–6
Hilton PJ, 155
Hines HM, 217, 219
Hinman KE, 242
Hinton HE, 284
Hobbhahn N, 374
Hochberg FG, 151
Hockey PAR, 60
Hocking B, 40

Hodges JK, 368
Hoekstra HE, 37
Hofer H, 355, 404
Högstedt G, 85, 127
Hokke CH, 423
Holen ØH, 120–1, 190, 225
Holldobler B, 439
Holliday JC, 419
Holling CS, 80–1, 126, 172–3, 181, 430
Hollingsworth ML, 375
Holloway GJ, 146, 207, 266, 429
Holloway JD, 45
Holmes JS, 97, 99
Holmes RT, 400
Holmgren NMA, 188
Holt D, 276
Holwell GI, 307, 313
Holz C, 147
Holzkamp G, 145
Honma A, 31, 99, 438
Honza M, 335
Hoogesteger T, 196
Hoogmoed MS, 279
Hooper J, 37
Hopkins E, 260, 262, 269
Hopkins GHE, 145
Hoppen HO, 418
Hoppes WG, 414
Hori M, 323
Horiger N, 146
Hormiga G, 326
Horsley DT, 204
Horton CC, 326
Horváth G, 68–9
Hossie TJ, 269, 271
Hotta M, 397
Hounsome MV, 151
Houraannoet MW, 216
Houston AI, 31
How MJ, 69
Howard RD, 401–2, 408
Howard RR, 240
Howard RW, 239, 347, 349
Howarth B, 227, 229
Hower AS, 193–4
Howse PE, 206, 260, 441
Hristov NI, 142
Hsiao C, 146
Hsiao TH, 146
Huang J-P, 312–13
Hudde B, 354–5
Huddleston T, 11, 253, 436
Hughes KA, 74
Hughes L, 415
Huheey JE, 172–6, 180, 185, 192, 198,
    215, 228–9, 239–40
Huheey RB, 246

Hultgren KM, 53, 83, 85
Hunt DM, 31
Hunt S, 339
Hunter AS, 104, 134–5
Huson LW, 262
Husseini J, 119
Huttunen H, 109, 169
Huxley J, 258

Ichiishi S, 350
Igic B, 287
Ihalainen E, 32, 205, 210, 248
Iijima T, 203
Ikin M, 76, 78
Imafuku I, 214
Imperio S, 37
Inbar M, 157, 297, 300–2
Inglis IR, 262
Ingram III W, 71
Ingram JR, 425
Ingram SN, 11, 253, 436
Inoue K, 385
Inpekoven M, 81
Iserbyt A, 358
Itino T, 397
Ito F, 231, 349
Ito S, 355
Ivri Y, 373, 385
Iwasa Y, 220–1
Iyengar VK, 144–5
Izally R, 74
Izhaki I, 302, 414

Jablonski P, 277
Jablonski PG, 43
Jackson JF, 71, 234, 443
Jackson RR, 231, 272, 296–7, 300, 319,
    328, 432
Jackson THE, 285, 430
Jacobs CH, 163
Jacobson RL, 121
Jaffé K, 350–1
Jahner JP, 227
Jakobsson S, 283
Janson CH, 411, 413
Janzen DH, 139, 243, 262, 284
Janzen FJ, 242
Jarau S, 394
Jared C, 147
Järvi T, 104, 112, 114, 121, 210
Jeffords MR, 194–5
Jefson M, 120
Jenkin CR, 424
Jenkins PF, 408
Jennings DT, 326
Jersáková J, 372, 379, 385–6
Jespersen DC, 180

Jetz W, 123
Jiggins CD, 14, 162, 216–17, 219,
    219–20
Jin X-H, 388–9
Jiroš P, 236
Jocque R, 326
Jödicke R, 355, 358
Joel DM, 350–1
Johnsen S, 45
Johnson KM, 326
Johnson MS, 201
Johnson SD, 372–4, 379, 381, 384–7,
    389, 391, 419, 429
Johnson TW, 277
Johnston JP, 204
Johnston SD, 386
Johnstone RA, 225
Jones DA, 145, 295
Jones FM, 131
Jones GP, 322–3
Jones GR, 342
Jones RB, 262–3
Jones RS, 161
Jones TC, 99
Joneschild DE, 58, 151
Jordan TM, 47
Jordano P, 410–11
Jormalainen V, 31
Joron M, 132, 134, 161–2, 173, 185,
    201, 206, 217, 218–21, 223, 430
Jouvenaz DP, 349
Jouventin P, 243
Joyce FJ, 229
Jukema J, 354
June-Wells M, 214
Juniper BE, 350
Jürgens A, 372, 379, 385–6, 389, 391, 419

Kacelnik A, 334
Kafatos FC, 147
Kaiser R, 389
Kaitala V, 116
Kalshoven LGE, 237
Kamagata S, 54
Kamil AC, 43, 73–4
Kamilar JM, 24
Kang C-K, 43, 214
Kano Y, 355
Kapan DD, 196, 219
Kardong KV, 246
Karg H, 418
Karlsson M, 41
Karpestam E, 41, 76–7
Kasai A, 54
Kassarov L, 137–8
Katayamay K, 99
Kato M, 397

Kats LB, 54
Katz LS, 366
Kauert G, 147
Kaufmann JH, 315
Kawaharab AY, 275
Kawakita A, 342
Kayser H, 35
Kazemi B, 109–11
Kearny EP, 109, 177, 198
Keeble FW, 54, 56
Keeton WT, 89
Keiper RR, 43
Keiser ED Jr, 318
Kelber A, 390
Keller L-A, 338
Keller MA, 355, 357
Kellett DN, 120, 142, 248
Kelley JL, 270
Kelley LA, 270, 288, 402
Kelly C, 332
Kelly CA, 144–5
Kelly DJ, 3, 24, 28, 108, 124
Kemp DJ, 310
Kennedy MW, 126
Kenward B, 147
Keogh JS, 357
Kessell SR, 189
Kettlewell HBD, 37–8, 40–1, 429
Keutmann H, 145–6
Kevan PG, 388
Khateeb S, 165
Khera S, 3
Kikuchi DW, 69, 198–9, 207
Kikuchi Q, 211
Kilner RM, 329, 337–9
Kiltie RA, 27, 41, 60
Kimball BA, 302
Kindlmann, 372
King DH, 99
King JMB, 77
King WW, 246
Kingsland S, 3, 86
Kirby WE, 436
Kirchner WH, 346
Kiritani K, 136
Kirkpatrick TW, 100, 229
Kirvan CA, 424
Kisbenedek T, 337
Kistner DH, 349
Kitamura T, 214
Kitching IJ, 45
Kite GC, 373, 389–90
Kjellsson G, 382
Kjernsmo K, 270
Klän S, 307
Klein BS, 424
Klein NK, 340

Kleinholz LH, 54
Kloft W, 233, 349
Kloock CT, 319
Klots AB, 153
Knapp JJ, 237
Knapp RA, 360
Knill R, 77
Knipper H, 255
Kobak CJ, 363
Kobayashi TM, 140
Kocher T, 325
Kodandaramaiah U, 260, 266–7
Kodric-Brown A, 395, 397
Koeniger G, 326
Koeniger N, 326
Kofman S, 266
Kokko H, 188, 441
Kokpol U, 351
Koller G, 56
Kolm N, 11, 362–3
Kolyer JM, 138
Komárek S, 2–3, 37, 61, 438, 440
Kondou K, 355
Köpf A, 147
Kontur S, 23
Korpimaki E, 404
Korshikov E, 121
Korsnes L, 334
Kotagama SW, 287–8, 402
Kozak KM, 216
Kraft GT, 302
Krajewski JP, 286, 322
Král J, 236
Kraus F, 306, 317, 434
Krebs JR, 339, 399, 408
Krebs RA, 201–2
Kreuter K, 344
Kreuzwieser J, 350–1
Kroiss J, 344–5
Kronforst MR, 203, 216, 220
Kropach C, 130
Kropf M, 387
Kruse PR, 141
Kugler H, 389
Kuiter RH, 323
Kullenberg B, 379
Kumano N, 100
Kunin WE, 387–8
Kunte K, 201, 203
Kunze A, 146–7
Kunze J, 386
Kuriwada T, 100
Kurtz J, 423
Kyalangalilwa B, 163

Laaksonen T, 283
Labanick GM, 239

LaBerge WE, 229–30
Lachaud J-P, 239
Lack D, 60, 336, 400
Lamar WW, 242
Lamberts SWJ, 37
Lamborn WA, 137
Lamont BB, 294
Lande R, 172, 216, 334
Landeau L, 286–7
Landová E, 109, 132
Lane C, 112, 157
Langham GM, 218
Langkilde T, 356
Langmore NE, 329, 332, 334, 337–9
Lanteri AA, 227
Larin Z, 32
Lariviere S, 159
Larsen TB, 99, 139, 266
Larson BMH, 373, 389
Latimer J, 255
Latz MI, 29
Laufer H, 354
Laugier T, 356
Launchbaugh KL, 302
Laverty TM, 388
Lawrence LB, 76
Lawton JH, 290, 333
Le Conte Y, 348
Le Corff J, 393
Le Poul Y, 185
Lea RG, 438
Leaich HR, 99
Leal M, 317
Leão Feitosa JL, 285
Leavell BC, 275
Lecuit M, 421
Lederhouse RC, 82, 201–2, 223
Lee TL, 116–17
Lee TM, 37
Leenders TAAM, 242
Lees DR, 40–1
Lehtonen J, 394–5
Lehtonen TK, 362
Lei H-Y, 425
Leimar O, 104, 109–11, 113–16, 122, 132–3, 172, 175, 180–1
Lemoli F, 348
Lenko K, 231, 233
Lenoir A, 348
Leonard ML, 340
Leong TM, 233, 237
Leroy A, 326
Leroy J-M, 326
Leroy Y, 161, 315
Lev-Yadun S, 41, 85, 157, 163, 165–6, 290–1, 295, 297, 299–302, 413–14, 426

Levi HW, 199, 234
Levin DA, 140, 387
Lewinsohn TM, 124–5, 136
Lewis SM, 147, 325
Lhomme P, 344
Li CC, 78
Li D, 313, 319
Li D-Z, 382, 388
Li DQ, 93
Li Z-Y, 388–9
Liang W, 335, 342
Liao C-P, 312–13
Liao Q, 346
Libbey JE, 424
Lichter-Marck IH, xiii, 104
Liede S, 373
Liede-Schumann S, 290, 293, 297–8
Liepert C, 120, 233, 346–7, 350, 381
Lilley GM, 307–8
Lillo MJ, 153
Lima NRW, 363
Limbaugh C, 322
Lin Y-S, 425
Lind AJ, 317
Lindell LE, 69
Linder HP, 374, 385
Lindroth CH, 150
Lindstedt C, 32, 53, 109, 169
Lindström L, 14, 123, 126, 128–30, 132, 173, 188, 204–5, 210, 441
Linsley EG, 99, 147, 151, 153, 253
Liu M-H, 23, 93
Llaurens V, 132, 218
Lloyd FE, 350
Lloyd H, 348
Lloyd JE, 325, 363, 431
Lloyd Morgan C, 130
Lo E, 410
LoBue V, 130
Löfqvist J, 348
Lohman DL, 216
Lomáscolo SB, 411
Londoño GA, 246
Long EC, 199
Longino JT, 147
Longman HA, 326
Lorenzi MC, 344
Lorenzini JL, 424
Lorimer N, 40
Losey GS, 320, 323
Lotem A, 336–7
Lötters S, 148
Lovell PG, 58, 60
Lowe CH, 56
Lowenstine M, 29
Lowitz CE, 100

Lown AE, 29, 54, 56, 104
Lubbock HR, 323
Lubbock J, 295
Lucanus O, 318
Ludin P, 424
Luedeman JK, 181
Lumsden RD, 423
Lunau K, 260, 267–8, 373, 389
Luo C, 357
Luo YB, 382
Lynch BM, 204
Lynn SK, 181–2
Lyon BE, 151, 400
Lythgoe JN, 54, 69
Lyytinen A, 31, 129, 139, 152, 205, 266–7

Maan ME, 119
Maass E, 392
MacDougal JM, 299
MacDougall A, 178, 216, 441
Macías-Garcia C, 345–5
Mackauer M, 350
Mackie GO, 155
Mackintosh NJ, 75
Macleay C, 354
Macleod RM, 37
Macnair MR, 372, 387
Macquart D, 389–90
Madden D, 165
Maddison WP, 167
Madin LP, 97
Madsen J, 124
Magnus DBE, 201, 255
Magoolagan L, 161
Magrath RD, 3, 287, 402
Magurran AE, 270
Mahic G, 255
Mahon K, 159
Maier CT, 227, 229
Maier TS, 349
Maisch A, 35
Majerus MEN, 37
Malcicka M, 233
Malcolm SB, 3, 8, 112, 188–9, 435
Malcolm WM, 41
Mallet J[LB], 39, 117, 161–2, 173, 175, 186, 188, 202, 206, 216–17, 219–20, 430, 442
Malosse C, 348
Mänd T, 109
Mangin A, 156
Mani GS, 144–5, 224
Mank JE, 354
Manley DG, 227
Manning JC, 388
Mant JG, 378–9

Mappes J, 31, 53, 92, 109, 126, 128–30, 132–4, 139, 150, 152–3, 161–2, 169, 173, 175, 188, 196, 201, 204–5, 207, 210, 224, 244, 246, 248, 250, 264, 266–7, 441
Maran T, 3, 14, 15
Marchetti K, 329, 332, 337
Marcon D, 175
Mardan M, 326
Marden JH, 215
Mardulyn P, 392
Marek PE, 155, 242
Marien D, 240
Marino P, 419
Mariscal RN, 149
Markham MR, 73
Marples NM, 109, 123–4, 128, 150, 153
Marques Tozetti AM, 97
Marsaioli AJ, 379
Marsh DC, 248
Marsh NA, 167, 189, 221
Marshall DC, 326, 431
Marshall GAK, 2, 135, 153, 250, 433
Marshall JN, 322
Marshall KLA, 24, 63–4, 66
Marshall NJ, 45, 149
Marshall S, 392
Martens J, 388
Martin C, 348
Martin DJ, 155
Martin JW, 258
Martin SH, 216
Martin SJ, 344, 347
Martinet L, 37
Martínez MAS, 246
Martínez-Torres D, 349
Marthus de Oliveira MCL, 97
Martos F, 373
Maruyama M, 349
Mascia Loos FE, 368
Maser I, 424
Mason JR, 125
Mason P, 60, 334
Mason RT, 355–7
Massei G, 303
Masterman AT, 239
Masters AR, 440
Masters WM, 250
Masumoto T, 154
Matessi C, 183
Mather MH, 251
Mathew AP, 234
Mäthger LM, 54, 58
Mathieu JM, 153
Matsuura K, 417
Matthysen E, 155
Mattiello T, 58

Mawdsley JR, 227
May PG, 317
May RM, 333
Mayer RT, 297
Maynard Smith J, 436
McAtee WL, 237
McCafferty DJ, 36
McCallum ML, 248
McCarthy BC, 163
McClintock JB, 45
McCosker JE, 250
McDade LA, 390
McDaniel CA, 349
McDiarmid RW, 2, 163, 240, 242
McFarland N, 224
McGraw KJ, 119
McGuire JA, 242, 244
McGuire L, 181–2
McIver JD, 199, 231, 234, 443, 438
McKaye KR, 318, 325
McKinnon W, 155
McKone MJ, 242
McLachlan J, 287
McLean KA, 439
McMillan WO, 217, 219
McMorris FR, 181
McNamara JC, 54
McNeil JM, 37
McNeil JN, 372, 418
McRae SB, 339
Mebs D, 147, 342
Meck HR, 189
Meeuse BJD, 390
Meinwald J, 120, 144, 146, 166, 325
Meinwald YC, 146
Melampy MN, 413
Mellencamp K, 35
Mencher FM, 29
Menz MHM, 381
Merilaita S, 2, 5, 23, 29, 31, 41, 53, 61,
    63–4, 67, 72, 76–7, 82, 109, 132,
    155, 270, 283
Merritt DJ, 155, 312
Mertens R, 88, 97, 240, 250, 276
Messenger JB, 58
Messier F, 159
Meunier M, 37
Meve U, 373, 391
Meyer A, 362
Meyer K, 145
Meyer RK, 37
Meyer-Rochow VB, 45
Michaels AA, 45
Michaud JP, 350
Michelangeli F, 350
Michida K, 153–4
Midgley JJ, 163, 374, 419

Mikellidou K, 367
Miles DH, 351
Milewski AV, 165
Milinkovitch MC, 56
Milius S, 97
Millar JG, 255
Miller JL, 153
Miller LA, 257
Miller WE, 277
Millett J, 350
Millot J, 227, 435
Mills LS, 37
Milne AA, 19
Milne K, 347
Minkiewicz R, 83
Minno MC, 23, 89
Mirsky N, 426
Mirza ZA, 97
Mitchell BK, 99
Mitchell MS, 37
Miyamoto K, 45
Miyatake M, 99
Miyatake T, 99
Mizumoto M, 99
Mizuno T, 313
Mochizuki R, 346
Mody NV, 351
Moermond TC, 277
Moffitt CM, 142, 188, 190
Moksnes A, 334, 337
Moland E, 322–3
Molleman F, 139
Møller AP, 100, 334, 337, 342, 365,
    368, 402
Møller BL, 145
Mollon JD, 31
Moment GB, 77–8, 441
Mönnig HO, 415
Montevecchi WA, 60
Montgomerie RD, 151, 400
Montgomery JC, 47
Moody S, 140
Moon HP, xiii
Moon J-Y, 43
Moore AJ, 242
Moore BP, 120, 123, 140, 142, 146–7, 440
Moore BR, 367
Moore PD, 415
Moore W, 155
Moore WS, 401
Moran AP, 425
Moré M, 388
Morellato LPC, 411
Moreno J, 336
Moreno-Rueda G, 119
Morgan MJ, 31
Morgan RA, 76

Morhart JA, 349
Mori A, 250, 342, 348
Morimoto G, 339
Morin JG, 364
Morita S, 385–6
Moritz RFA, 346
Morrell GM, 207, 438
Morrell R, 271
Morris D, 355, 364, 405
Morris MR, 356
Morse DH, 313
Morton ES, 288
Moser WH, 380
Moskát C, 334, 336–7
Mosley S, 204
Mostler G, 207–8
Motoyama M, 45
Motta PC, 140
Motta PC, 246
Mouritsen KN, 124
Moussalli A, 30, 49–50, 57, 434
Mousseau TA, 100
Moynihan MH, 286
Mueller C, 121
Mueller TC, 206, 320
Mühlmann H, 207
Müller F, 172–3, 214–5
Mukherjee R, 266
Mullen SP, 131, 172
Mumme RL, 125
Mummery R, 35, 140, 150
Mundy NI, 341
Mundy PJ, 339
Munn C, 406
Muñoz-Garcia A, 400
Murphy JB, 317
Murphy PM, 125
Murray BA, 317
Murray D, 216
Musselman LJ, 392
Muyshondt A, 135
Muyshondt A Jr., 135
Myczko Ł, 100
Myers CW, 119

Nabhan GP, 426
Nachman MW, 37
Nadav Shashar N, 49
Nadeau NJ, 14, 216–17, 219
Nagamitsu T, 350
Nagasaka K, 123
Nagel P, 346
Nagy TR, 37
Nahrstedt A, 145
Naisbit RE, 219
Nakamura H, 332, 334, 336–7
Nakamura K, 51, 53, 73

Nakamuta K, 85
Nakata E, 54
Nakata K, 93
Nambara T, 145
Nams VO, 76
Napier J, 403
Napier P, 403
Nardi G, 145
Nation JL, 347
Naud M-J, 354–5
Navarro L, 373
Ne'eman G, 41, 163, 301, 414
Nedvĕd O, 131, 153
Negro JJ, 320
Neiland MR, 381
Neill WT, 317
Neitz J, 163
Nelson CJ, 189
Nelson JW, 250
Nelson XJ, 231, 233, 235–6, 307, 328, 432
Nembaware V, 425
Nentwig W, 23, 227
Neudecker S, 61, 270
Neuhaus RJ, 381
Neville PA, 262
Newman E, 384
Newman L, 122
Ngugi HK, 418–19
Nicholson AJ, 172, 253
Nickle DA, 23
Nicol C, 123, 140, 440
Nicolai J, 438
Nicolaus LK, 154
Niemelä P, 297
Nierenberg L, 373
Nieves-Aldrey JL, 227
Nijhout HF, 89, 222
Niles DM, 400
Nilson G, 37
Nilsson D, 424
Nilsson LA, 372, 382, 384–7
Nilsson AA, 100
Nisbet RM, 152
Nishida R, 140, 145–6, 189, 440–1
Nishida T, 31, 99, 417, 438
Nishikawa H, 203
Nishino H, 99
Niskanen M, 244
Niu Y, 291
Noble DG, 339
Noble J, 116
Nojima S, 319
Nokelainen O, 53, 92, 153, 161–2, 201, 246, 250
Nomakuchi S, 154
Nomura M, 85
Nonacs P, 117

Noor MAF, 50, 52
Norbutus AJ, 144–5
Norma-Rashid Y, 313
Norman MD, 151, 354
Normandale CL, 262
Norn S, 141
Norris JN, 302
Norris KS, 56
Northcote TG, 46
Noy-Meir L, 373
Nunez-Farfan J, 136
Nur U, 183
Nuss M, 392
Nylin S, 82

O'Brien CA, 417
O'Day WT, 28
O'Donald P, 126, 183, 210
O'Hanlon JC, 307, 313
O'Rourke KE, 283
O'Donnell RP, 315
O'Hanlon JC, 313
Oaten A, 181
Oberprieler C, 295
Ödeen A, 31
Odenaal FJ, 136
Oelschlägel B, 392
Ohara Y, 123
Ohguchi O, 286
Ohsaki N, 123, 151, 201, 211
Ojala K, 205
Oku S, 99
Okuda N, 355
Oldstone MBA, 425
Olendorf R, 74
Oliveira C, 239
Oliveira FFR, 30
Oliveira PO, 438
Oliveira RF, 355
Oliver JC, 167
Oliver M, 99
Ololsson M, 65, 66, 68, 267
Olofsson O, 284
Oniki Y, 60
Ono A, 315
Ormond RFG, 323
Orr LP, 14, 240
Orsak LJ, 251–2
Ortiz-Sánchez FJ, 119
Oshima N, 54
Osorio D, 58, 63, 71, 211
Ostner J, 368
Otto S, 251
Oudeman JTh., 90, 438
Outomuro D, 215
Overal WL, 227
Owada M, 225

Owen ARG, 175, 178, 180, 429
Owen DF, 2–3, 32, 39, 77, 78, 199, 220, 225, 282, 441
Owen RE, 175, 178, 180, 225, 429, 441
Owen-Smith N, 165
Owings DH, 140, 155
Owiny AM, 3, 220

Paarmann W, 347
Pabón-Mora N, 392
Padovani Ferreira B, 285
Page LM, 360
Pagnucco K, 340
Palmerino CC, 108
Palokangas P, 404
Palumbo A, 58
Panchen AL, 188
Pangle WM, 365–6
Papa R, 162, 217, 219–20
Papadopulos AST, 372, 382, 385
Papageorgis C, 72
Papaj D, 155
Papaj DR, 181
Papavero N, 231
Papenfuss TJ, 317
Pardo-Diaz C, 220
Parejo D, 61
Parellada X, 317
Parker GA, 151, 365–6
Parker J, 346
Parker WS, 246
Parnell RS, 50, 52
Parsons J, 145
Partidge JC, 47
Passmore NI, 408
Pasteels JM, 144–6, 150
Pasteur G, 2, 4–5, 8–9, 11–12, 20, 83, 88, 97, 100, 156, 240, 243, 251, 255, 277, 299, 367, 376, 382, 402–3, 405, 422, 429, 431–2, 434–9, 441, 443–4
Patrick K, 429
Paulus HF, 379–81
Pauw A, 385–6
Pawlowski B, 368
Paxton CGM, 270
Paxton RJ, 104, 109, 112–13
Payne LL, 340
Payne RB, 332, 340–1
Peacor CR, 394
Peakall R, 372, 374, 376, 378–9, 381, 440
Pearce CEM, 181
Pearson RG, 365–6
Peckham EG, 439
Peckham GW, 439
Pedersen HC, 337

Pegram KV, 153
Pekár S, 7–8, 236
Pellissier L, 388
Pellmyr O, 386
Penney HD, 210
Penz C, 151
Peres CA, 414
Pérez-Lachaud G, 239
Perfecto I, 349
Perrard A, 225
Peschke K, 346, 404
Peter CI, 374, 385–6
Petersen WGB, 419
Petersson E, 392
Petocz RG, 408, 444
Pfennig DW, 69, 131, 161, 173, 198–9, 207, 211, 242
Pfennig KS, 242
Pfrender M, 357
Phelps L, 423
Phillips BL, 317
Phillips DW, 53, 73
Phillips G, 309–10
Phillips RA, 376, 379, 381
Pianka ER, 80, 246, 279
Pickett JA, 300
Pie MR, 181–2, 231
Piclowski Z, 286
Pierce NE, 32, 85, 239, 346
Piersma T, 354
Pietrewicz AT, 43
Pietsch TW, 307, 315
Pilecki C, 126, 210
Pinheiro CEG, 136–8, 151, 195
Pinto AC, 392
Pirih P, 47, 49
Pitts JP, 227
Pizarro-Cerdá J, 421
Place NJ, 404
Plaisted KC, 75
Platt AP, 198, 248
Pliske TE, 120, 145
Plowright RC, 225, 441
Pocock RI, 157–8, 231
Pohl S, 4
Polacicová L, 335
Policha T, 389
Polidori C, 227
Pollard CP, 245
Polte S, 299
Ponce de León P, 424
Poole RW, 224
Popescu C, 40
Portugal AHA, 73
Potratz CJ, 180
Pough FH, 61, 71, 162, 189, 240, 242, 246, 248, 284, 317

Poulton EB, 2, 24, 32, 50–1, 83, 122, 137, 153, 221, 233, 237, 253, 430, 439–40
Pouyanne A, 439
Powch I, 404–5
Powell MP, 372, 382, 385
Powell RA, 276
Powell S, 233
Powell W, 382
Prates I, 147
Prendergast MM, 425
Price EG, 415
Price JRF, 260
Price PW, 214, 302
Procházka P, 335
Procter-Gray E, 400
Proctor HC, 374
Proctor MCF, 382
Prohammer LA, 99
Prokop P, 341
Proksch P, 146, 147
Provenza FD, 302
Provenzale A, 37
Prudic KL, 3, 167
Pruett-Jones S, 124
Prum RO, 401
Przeczek K, 121
Przetak G, 389–90
Puebla O, 325
Punnett RC, 3, 14
Puorto G, 317
Puurtinen M, 116
Pye JD, 257

Quate LW, 326
Quetin LB, 149
Quicke DLJ, 2, 11, 72, 100, 152, 225, 249–50, 253, 424, 436
Quignard JP, 356

Raats D, 165
Racey PA, 355, 404–5
Radcliffe CW, 317
Raguso RA, 388, 390, 418–19
Raheem DC, 202
Raia P, 54
Rainey MR, 323–4, 432
Raju AJS, 397
Ralphs MH, 108
Ramachandran C, 56
Ramachandran VS, 56
Ramesh A, 236
Rampini M, 196, 228
Rand AS, 74, 77
Randall HA, 239
Randall JE, 5, 74, 239, 285, 320, 323
Rangaiah K, 397

Rao KR, 54
Rao SP, 397
Rappole JH, 124
Rasekh A, 350
Rashed A, 209, 323, 432
Raske AG, 193
Rasmont P, 344
Rasmussen FN, 382
Rathcke BJ, 214
Ratnayake CP, 287, 402
Rätti O, 365
Rausher MD, 136, 154, 299
Raymond DL, 73
Read AF, 339
Reader T, 207, 209
Realpe E, 215
Rechten C, 402
Redondo T, 339
Redston S, 124
Reed CM, 58
Reid PJ, 76
Reimchen TE, 31, 94
Reinhold K, 299
Reiserer RS, 317
Reiskind J, 123–6, 199, 204, 234
Reit S, 86
Reitsma N, 267
Relethford JH, 368
Remmel T, 109
Ren Z-X, 382, 388–9
Renfree MB, 405
Rennenberg H, 350–1
Renner SS, 372, 387, 395
Renoult JP, 406
Resch JF, 166
Rescorla RA, 180
Resetarits WJ Jr, 309, 312
Rettenmeyer CW, 147, 229, 329, 349
Rezende CM, 392
Rice SP, 350
Ricklefs RE, 283
Ridley AR, 402
Ridley HN, 414
Ries L, 172
Riipi M, 133–4
Rimbut S, 251
Rios-Cardenas O, 356
Ristau CA, 277
Ritland DB, 3, 188
Rivers TJ, 364
Riviére G, 348
Roba J, 144
Robbins RK, 271, 274
Roberts NW, 47
Roberts TR, 315
Robertson DR, 323
Robinson B, 93

Robinson EJ Jr, 416
Robinson GS, 2
Robinson MH, 5–6, 94, 434, 441
Robinson MW, 423
Robinson SK, 333
Robson DS, 85
Rodd FH, 74
Rödel M-O, 342
Rodriguez J, 227
Rodríguez-Gironés MA, 395
Roeder KD, 256–7
Rohe F, 324
Rohr D, 30
Rohwer FC, 401
Rohwer S, 400–1, 442
Roitberg BD, 251
Rojas B, 117
Romesburg HC, 442
Ronel M, 165
Rong CP, 54, 63
Roosevelt T, 3
Roper CFE, 29
Roper TJ, 123–4, 129–30, 410, 431
Rose MJ, 397
Rosenberg G, 122
Rosenqvist G, 354
Roskaft E, 334, 337
Ross HA, 11, 339–40
Ross RM, 149
Rossini C, 120, 145, 250, 260
Rota J, 251–2
Rotheray GE, 227
Rothrock JT, 290
Rothschild M, xiii, 3, 5–6, 11, 35, 72,
    108–9, 112, 120, 131, 135–7,
    139–40, 142, 145–6, 150, 152, 154,
    157, 167, 189, 224, 228, 250, 295,
    299, 315, 429, 440
Rothstein SI, 60, 334, 336, 339
Roubik DW, 342
Rousset F, 39–40
Roulsias JG, 425
Rovner JS, 93
Rowan MG, 140
Rowe C, 120, 123, 126–7, 129, 131, 157,
    190, 205–7
Rowe MP, 139, 155
Rowell-Rahier M, 145–6, 150
Rowland HM, 20, 24, 28, 88, 94–7, 118,
    196, 210, 248
Rowlands D, 415
Rowley D, 424
Roy BA, 382, 389, 417–18
Roy L, 23
Ruano F, 348
Rubino DL, 163
Rubinoff I, 130

Rubinoff RW, 130
Ruchon F, 356
Rudge DW, 37
Ruiz MC, 350
Rumball DA, 69
Rusiniak KW, 108
Russell BC, 323
Russell WMS, 429
Rust CC, 37
Rust J, 83
Rutila J, 337
Rutowski RL, 153
Ruxton GD, 2–3, 5, 8, 20, 24, 26, 28, 36,
    47, 50–1, 58, 67, 69–70, 73, 78, 82,
    88, 94–7, 99, 104, 108–9, 116,
    118–19, 126, 128, 146, 160, 163,
    172–3, 180–1, 185, 190, 204, 211,
    260, 274, 276, 280, 290, 315, 350–1,
    426, 430, 432, 438, 441
Ryan PG, 408
Ryerson WN, 120, 142

Saccheri IJ, 39–40
Saenko SV, 56
Sætre G-P, 404
Saidel WM, 56
Saillant PA, 257
Sakai M, 99
Sakai S, 392
Salazar A, 349
Salazar BA, 139
Salvy M, 348
Salzburger W, 322, 360–2
Salzet M, 423
Salzmann C, 381
Samia DSM, 342
Samuelson L, 401
Sanap DW, 97
Sánchez JM, 61
Sánchez-Guillén RA, 358
Sandoval ML, 51
Sandre S-L, 109
Sanford MT, 347
Sannasi A, 146
Sannolo M, 97
Santamaría L, 395
Santos AJ, 355
Santos ME, 362
Santos RS, 355
Santos X, 317
Saporito RA, 147, 153
Sargeant AB, 97
Sargent RC, 360
Sargent TD, 38, 41, 43, 206, 282–3
Sato, 82
Saul-Gershenz LS, 255
Savage JM, 240, 243, 279

Savile D, 418
Savolainen V, 372, 382, 385
Sazima C, 322
Sazima I, 97, 99, 286, 317, 320, 322–3,
    406, 438
Sbordoni V, 196, 228
Scaccabarozzi D, 376
Scaife M, 265
Scali S, 97
Schaefer HM, 8, 41, 45, 61, 63–4, 290,
    350–1, 406, 410–11, 413–14
Schaefer V, 410, 414
Schal C, 319
Schall JJ, 80
Scharff N, 326
Scheerer U, 350–1
Schelly R, 323
Schemske DW, 382, 393–4
Scherm H, 418–19
Schiestl FP, 372, 374, 376, 378–80, 382
Schiestl P, 379
Schlenoff DH, 284
Schlichter D, 422
Schluter D, 56, 195, 197, 198
Schlüter PM, 375, 379
Schmalhofer VR, 56
Schmidt JO, 227
Schmidt RS, 207
Schmidt V, 411
Schmitt T, 344–5
Schmuff N, 348
Schneider D, 144
Schnitzler CE, 316
Schooler LJ, 69, 71
Schoonhoven LM, 154
Schröder I, 368
Schroeder FC, 120, 146, 147, 170
Schuett GW, 300, 317, 434
Schuler W, 130
Schulte R, 163, 239
Schulte TD, 71, 111, 118
Schulz S, 347
Schupp EW, 101
Scopece G, 376
Scott D, 114, 355
Scott JA, 121
Scott-Samuel NE, 69
Scriber JM, 201–2, 223
Searcy WA, 408
Searle WT, 69–70
Sedeek KE, 375
Seevers CH, 346
Segura M, 339
Sehlke G, 347
Seiber JN, 189
Seidel JL, 415
Seigel JA, 239

Sekimura T, 203
Selander RB, 153
Sengupta U, 424
Senou H, 315
Seoighe C, 425
Servedio MR, 116, 334
Ševčík R, 155
Sexton OJ, 194–6, 207
Seymoure BM, 218
Sezutsu H, 23, 89
Shand J, 54
Shapiro AM, 154, 199, 300
Shcherbakov DE, 253
Shcherbina TV, 416
Shearer MK, 366
Sheldon JK, 229
Shelford R, 237
Shelley RM, 242
Shelomi M, 415
Sheppard PM, 2, 32, 34, 74, 77, 80, 162,
    175, 184, 198–9, 207, 216, 219,
    221–23
Cheront W, 67
Sherman AR, 156
Sherman PW, 158
Sherratt TN, 2, 5, 20, 24, 26, 44, 47, 63,
    66, 77, 97, 104, 116, 119–20, 128,
    146, 163, 166–7, 172–3, 180–2,
    185, 196, 209–11, 219, 269, 271,
    274, 276, 280, 323, 358, 430,
    432, 441
Sherzer WH, 77
Shettleworth SJ, 76, 124
Shi J, 382
Shideler RT, 248
Shimada K, 146
Shimizu Y, 333
Shine R, 317, 356, 357
Shoesmith EA, 32, 35
Shohet A, 58, 71
Shook K, 74
Shorrocks B, 40
Short RV, 405
Shuster SM, 354–5
Shutt DA, 427
Shuttleworth A, 373, 389–90
Sibley CG, 156
Sick H, 406
Siemann LA, 56
Sih A, 56
Sikkel PC, 286
Silberglied RE, 194, 249
Sillén-Tullberg B (see also Tullberg BS),
    32–3, 104, 112–15, 121, 128, 132,
    134, 210, 368
Silva Absalão R, 77
Silvestre R, 89

Sime KR, 224–5
Simmonds MSJ, 390
Simmons JA, 257
Simmons KEL, 309
Simmons RB, 206, 227–8
Simmons RE, 364
Sims SR, 34
Singer MC, 117, 162, 216–17, 219
Singer MS, xiii, 104
Singer RB, 381
Singfield J, 388
Singh I, 424
Sipsas NV, 425
Sivinski J, 166, 392
Skelhorn J, 20, 50–1, 88, 94–7, 120,
    126–7, 131, 190, 205, 207
Skinner JD, 355, 404
Skórka P, 100
Skrade PDB, 60
Slagsvold T, 364–5, 404
Sláma K, 427
Slatkin M, 172
Slobodchikoff CN, 121, 193
Slowinski JB, 240, 243, 279
Smale L, 37
Smilanich AM, 150
Smith AG, 32, 35, 131
Smith AP, 165, 291, 297, 413
Smith BN, 390
Smith DAS, 3, 32, 35, 77, 136, 189,
    202, 220
Smith J, 216, 220
Smith KGV, 419
Smith MH, 37
Smith RH, 246
Smith SM, 111, 246, 413
Smith Trail DR, 152
Smith-Vaniz F, 323
Smithers SR, 422
Smithson A, 372, 387
Smyth MEB, 181
Sneddon I, 126–7
Snyder LE, 349
Soane ID, 73
Socha R, 106
Solano-Ugalde A, 23, 88
Soler JJ, 336
Solis JC, 58
Solis MA, 251
Soltau U, 290, 293, 297–8
Sommer V, 365, 367
Sordahl TA, 151
Sorenson MD, 332, 340–1
Sorger T, 368
Sorley L, 140
Sota T, 342

Soule N, 419
Sourakov A, 136, 210, 224, 275
Spaethe J, 380–1, 386
Spalink D, 374
Spear LB, 306
Speed MP, 2–5, 20, 24, 26, 28, 47, 78, 82,
    97, 104, 108–9, 116–19, 121, 126,
    128–9, 146, 153, 159–61, 163,
    172–3, 175, 177–81, 190, 204–5,
    210–11, 215–16, 274, 276, 280,
    430–2, 438, 440–1
Spence W, 436
Spencer KA, 58, 60, 161
Sperling FAH, 167
Spinner Hansen L, 99
Spottiswoode CN, 58, 332, 334–6,
    338, 342
Sprengel CK, 372
Springer VGW, 323
Srinivasan MV, 63, 308
Srivastava A, 367
Srygley RB, 135, 214–15, 437
Stachowicz JJ, 53, 83, 85, 147
Staddon JER, 75–6, 80–1
Stafford D, 99
Stamp NE, 152
Stamps JA, 199
Stankowich T, 158
Stanton ML, 165
Starling M, 334
Starnecker G, 35
Starrett A, 5, 8, 23, 89, 422, 430, 434,
    439–40
Stavenga DG, 47, 49
Stawarczyk T, 277
Stearn WT, xiii, 2
Stebbing GL, 422
Stefanescu C, 34–5
Stefano GB, 423
Steglik I, 290, 297
Stehr G, 201
Steidle JLM, 349
Steiner KE, 374, 381, 385–6
Stenius S, 365
Stensmyr MC, 388, 390
Stephens JA, 423
Stephens PA, 118
Sterkel AK, 424
Stermitz FR, 224
Sternburg JG, 194–5, 229
Sternfeld R, 244
Stetson MH, 37
Stevens M, 2, 5, 23, 29, 31, 54, 56, 58,
    61–4, 65, 66, 69–70, 104, 246, 260,
    262–5, 269, 332, 334–6, 339
Steward RC, 39–42
Stewart AJA, 40

Stiles EW, 37
Stinchcombe JR, 290, 297
Stobbe N, 61, 63–4, 67
Stoddard MC, 334–5, 338, 341
Stoddard PK, 73, 402
Stoeffler M, 349
Stokke BG, 334, 342
Stökl J, 375, 377, 379, 389–90
Stoks R, 358
Stonedahl G, 231, 234, 443, 438
Stoner CJ, 29, 36–7
Stournaras KE, 410
Stout MJ, 297
Stowe MK, 326
Stradling DJ, 260, 267
Streil G, 147
Streinzer M, 380–1
Stringer IA, 312
Strohm E, 344–5
Strutz A, 390
Stryjewski KF, 332
Stuart-Fox D, 30, 49–50, 57, 434
Stubbins CL, 260, 263
Stuckert AMM, 218
Štys P, 47, 132
Sugiura N, 385
Sulák KJ, 320
Summers K, 119, 162, 218, 239
Sumner FB, 56
Sun H, 291
Sun J, 47
Sundberg P, 122, 154
Suomalainen E, 184
Supple MA, 219
Surlykke A, 257
Suzuki N, 154
Suzuki TK, 23, 89
Švádová K, 106
Svennungsen TO, 190
Swales LS, 145–6
Swedo SE, 434
Sweet JJ, 140
Swenson DP, 45
Swets JA, 181
Swing K, 392
Swofford DC, 360
Sword GA, 112
Swynnerton CFM, 2, 112, 120, 151
Symula R, 163, 239
Székely T, 337
Szidat L, 416
Szlachetko DL, 386
Szulc E, 60–1

Taborsky M, 354–5
Tagashira N, 350
Takada H, 233

Takahashi J-I, 347
Takahashi T, 323
Takahashi Y, 358
Takakura K-i, 438
Takasu F, 334, 337
Takeuchi Y, 203
Tammaru T, 109
Tan EJ, 93
Tanaka C, 417
Tanaka KD, 339
Taniguchi K, 349
Tate DP, 23
Tattersall GJ, 30
Taylor CH, 207, 209
Taylor HH, 54
Taylor MI, 239
Taylor PW, 319
Teichert H, 392
Tembrock G, 13, 15, 307, 354
Temrin H, 365
Tengö J, 346, 379
Terborgh J, 286–7
Terrick TD, 125
Tesei A, 185
Tewksbury JJ, 426
Teyssier T, 56
Thayer AH, 24
Thayer GH, 24, 61, 63
The Heliconius Genome Consortium, 220
Theis A, 360–1
Théry M, xiv, 54, 56, 218
Thinh TH, 225
Thomas GE, 204
Thomas JA, 237
Thomas R, 317
Thomas RJ, 108, 128
Thompson JN, 414
Thompson MJ, 111–2
Thompson P, 367
Thompson RKR, 100
Thompson TE, 147
Thompson V, 151, 181
Thomson JD, 392
Thornhill R, 357
Thornton IWB, 223
Thorogood R, 341
Thresher RE, 323
Throop HL, 242
Thurman CL, 54
Thurman TJ, 218
Till I, 295
Till-Bottraud I, 296
Timmermann BN, 3
Timmermans MJTN, 222–3
Tinaut A, 348
Tinbergen L, 74–5
Tinbergen N, 60–1, 81, 315, 339

Titcomb TC, 69
Tobler M, 318
Todd PA, 54, 63
Toft S, 99
Toledo LF, 32, 88, 97, 99, 163, 276
Tollrian R, 426
Tomita S, 23, 89
Tong J, 47
Tonra C, 339
Tooke W, 367
Townes H, 228
Trad BM, 89
Tramer EJ, 406
Trapanese M, 54
Trauth SE, 248
Travassos MA, 239
Tregenza T, 151, 354
Treiber M, 85
Tremblay RL, 374
Trigo JR, 73, 140, 145, 149
Trimen R, 203
Trnka A, 341
Troscianko J, 58, 63, 104, 342
Troscianko TS, 61
True JR, 38
Tryjanowski P, 100
Tsacas L, 253
Tsiakalos A, 425
Tso I-M, 23, 93, 312–13
Tsoularis A, 172
Tsurui K, 31
Tucker GM, 73
Tuckerman RD, 157
Tullberg BS (*see also* Sillén-Tullberg B), 72, 82, 104, 109, 119, 123, 130, 132–3, 135, 155, 175
Tullrot A, 122
Tomlinson JH, 326
Tuni C, 99
Tuno N, 119
Tuomi J, 31, 104, 109, 165, 297
Turillazzi S, 344
Turner ERA, 27
Turner JRG, 3–4, 6, 39, 76, 78, 109, 162, 172, 175–7, 179, 183–6, 198, 201–4, 207, 215–17, 219, 223, 438
Tursch B, 146
Tuskes PM, 189
Tutt JW, 37
Twele R, 373, 377, 389
Tyler CW, 56
Tyler J, 155
Tyrie EK, 56

Uboni A, 344
Ueda K, 339
Uesugi K, 224

Uglem I, 354
Uma DB, 235–6
Umbers KDL, 50, 54
Ureña O, 315
Urru I, 388, 390
Uyarra MC, 56
Uyenoyama M, 108

Vaca JF, 277
Vakkari P, 40
Vale A, 373
Valero A, 345–5
Valido A, 410–11
Valkonen JK, 92, 153, 162, 244, 246, 250, 264
Vallin A, 267, 283
Valverde J, 424
Vamosi SM, 121
van Achterberg C, 231
van den Hurk R, 355
van der Bank M, 163
Van der Cingel NA, 373
Van der Meijden E, 152
van der Niet T, 391
van der Pijl L, 376, 394
van Die I, 423
van Dragt RG, 54
Van Gossum H, 181–2, 358
van Roosmalen MGM, 414
van Sommeren VGL, 285, 430
Van't Hof AE, 40
Vander Meer RK, 349
Vandermeer JH, 349
Vane-Wright RI, xiv 4–6, 8, 10–12, 14, 199, 202–3, 227, 249, 251, 253, 284–5, 295, 307, 323–4, 402
Vanegas PJ, 218
Vasconcellos-Neto J, 124–5, 136, 204
Vasconcelos HL, 350
Vasey PL, 365–6
Vaughan FA, 284
Vavilov NI, 414–15
Vaze VV, 97
Veith M, 148
Vereecken NJ, 255, 372, 379–80, 392, 418
Vergara P, 400
Vernon JG, 358
Vertainen L, 14
Veselá S, 129
Veselý P, 129
Victorsson J, 31
Vidal-Cordero JM, 119
Videler JJ, 355
Vijayan S, 236
Vikan JR, 334
Villalba JJ, 302
Villemant C, 225

Viraktamath CA, 199, 231
Visicchio R, 342
Vitt L, 242
Vitt LJ, 246, 277, 279
Vittoz P, 388
Vlasáková B, 394
Vlieger L, 267
Vogel S, 373, 388–9
Vogler AP, 202, 222–3
Vogt R, 295
Völkl W, 350
von Beeren C, 4
von Brockhusen F, 190
von Euw J, 140, 145
von Helversen B, 69, 71
von Uexküll JJ, 74
Vorobyev M, 410, 414
Vos H, 146, 429
Vrijenhoek RC, 363
Vršanský P, 155
Vukusic P, 47

Waage JK, 67
Waas JR, 339
Wade MJ, 99, 334–5
Wagner ARA, 180
Wagner DL, 251–2
Wagner H, 307
Wahlberg N, 45, 216
Wakamura S, 51, 53, 73
Wako A, 147
Waldbauer GP, 20, 194–5, 228–30, 284
Waldman B, 132
Walker HJ, 61–2, 65
Walker SJ, 250
Wallace AR, 2, 135, 153, 201, 214, 248, 251, 290, 342, 384–5, 401
Wallace J, 172
Waloff N, 326
Wang B, 83
Wang H, 382, 388
Wang IJ, 119, 163
Wang Y, 404
Wang Z, 41, 45, 413
Wanke S, 392
Ward JA, 322
Ward JM, 36
Wardlaw JC, 237
Ware AB, 415
Waring GL, 302
Warner DD, 181
Warner JA, 29
Warren AD, 326
Warwick T, 74
Wasmann E, 443
Wasserman FE, 136, 274–5
Wasslavik HS, 354

Watanabe H, 308
Watanabe K, 323
Watanabe M, 140, 358
Watson AC, 56, 151
Watts S, 392
Way MJ, 233, 349
Webb JK, 357
Weber NO, 99
Webster RJ, 44, 63, 66
Wee S-L, 389
Wei C, 357
Weibel AC, 401
Weiner C, 37
Weintraub JD, 290–1
Weiss MR, 235–6
Welbergen JA, 3, 341–2, 402
Weldele ML, 404
Weldon PJ, 124, 354, 408
Weller SJ, 206, 227–8
Welsh HH Jr, 317
Welsh JH, 54
Wennersten L, 78–9
Werner Y, 244, 250
Wesolowska W, 416
Wesolowski T, 416
West DA, 32, 34, 160–1, 201–2
Westcott PW, 329
Westmoreland D, 60
Westoby M, 139, 415
Weston PH, 378–9
Wheeler BC, 406
Wheeler DE, 439
Wheeler GC, 227
Wheelwright NT, 411, 413
Whelan CJ, 410
Wheye D, 136, 201
Whibley A, 163, 217
Whitaker MR, 139
White CR, 312
White JDM, 419
White R, 277
White TE, 310
White WT, 239
Whitehead DR, 242
Whitehead MR, 379, 394–5
Whitehead VB, 381
Whitehouse MEA, 319
Whiteley DAA, 77–8
Whiteman E, 325
Whiten A, 367
Whiting MJ, 30, 49, 357
Whitley GP, 239
Whitman DW, 139, 251–2
Whittle E, 375
Wickiser KJ, 31, 56
Wickler W, xiv, 2–5, 9–11, 13–14, 16, 93, 121, 153, 240, 242, 253, 255, 260,

274, 276, 295, 312, 319–20, 326,
338, 350, 355, 360, 362, 394, 403–4,
416–17, 419, 434–5, 439, 444
Widder EA, 315
Widmer A, 375, 382
Wiemer DF, 146
Wiens D, 3, 5, 23, 290, 293, 297, 392,
419, 433, 439, 441
Wiernasz DC, 192
Wiersma P, 400
Wignall AE, 319
Wijesinghe MR, 277
Wiklund C, 32–3, 63, 66, 68, 72,
104, 112, 114, 121, 154, 210, 267,
283–4
Wilbrandt J, 260, 267–8
Wilcock C, 381
Wilcox RS, 308, 319
Wiley JW, 338
Wilkinson DM, 166–7
Williams CM, 427
Williams CR, 250
Williams D, 86
Williams EG, 381, 384–5
Williams EH, 144, 323
Williams JB, 400
Williams KA, 227
Williams KS, 299
Williams PH, 37, 342
Williamson GB, 294, 351, 414
Williamson JE, 354–5
Willis EO, 286, 320
Willis RE, 312
Willmott HE, 355
Willmott KR, 72, 216, 224
Willson MF, 393, 410, 413–14
Willughby F, 328
Wilson C, 351
Wilson CA, 392
Wilson D, 89
Wilson DM, 204
Wilson JS, 227
Wilson RP, 9, 286, 309, 408
Wilson-Aggarwal J, 58
Wilts BD, 47, 49
Windsor DM, 194
Winemiller KO, 270
Wing K, 236

Wink M, 142
Winkler H, 411
Winney IS, 63, 260, 263–5
Winnick CG, 376
Wirtz P, 354–5
Wistow R, 124, 431
Witte L, 146
Witte V, 4, 347
Wittenberger JF, 306
Wohlfahrt T, 34
Wojcik DP, 349
Wolda H, 77
Wollenberg KC, 148
Wong R[B]BM, 378, 381
Wood EM, 204
Wood K, 276
Wood LE, 54
Woodcock G, 2
Woodcock TS, 373, 389
Woodell SRJ, 372, 374
Woodson CD, 47
Woodward CL, 392
Work K, 56
Wourms MK, 136, 274–5
Wray GA, 89
Wright PG, 47
Wu X-J, 423
Wulff JL, 74
Wüster W, 245
Wylde M, xiii
Wyman RL, 322

Xia C, 390
Xia F, 83
Xu SQ, 379

Yachi S, 108
Yadav AS, 424
Yagi T, 153
Yamaguchi S, 212
Yamamura K, 136
Yamaoka R, 233, 346, 348
Yamashita H, 136
Yamauchi A, 185
Yamazaki K, 297, 299
Yanega D, 227
Yang C, 335
Yano E, 308

Yasukawa K, 408
Yeager CP, 406
Yeargan KV, 326
Yeh T-M, 425
Yek SH, 349
Yen S-H, 2, 251
Yeo P, 382, 385
Yokoi T, 313
Yoshida A, 45
Yoshimura J, 358
Yoshino TP, 423
Young AM, 151, 262
Young FN, 47
Young RE, 28–9
Young RT, 2
Young TP, 165
Yu DW, 301
Yule DH, 69, 260
Yurtsever S, 151

Zabka H, 13, 15, 307, 354
Zagrobelny M, 145–6
Zahavi A, 117, 214,
336–7, 435
Zahiri R, 45
Zamora R, 165
Zanette L, 340
Zanker JM, 69
Zaret TM, 439
Zhang S, 312
Zhang W, 203
Zhao Z-S, 425
Zhou Q, 382
Ziegler R, 146–7
Ziegler T, 368
Zimma BO1, 344
Zimmerman DA, 320
Zimmermann D, 146
Zimova M, 37
Zivanov D, 146
Zöld A, 337
Zölnerowich GA, 251
Zompro O, 155
Zschokke S, 270
Zuanon J, 307
Zucker I, 37
Zug GR, 277
Zylinski S, 45, 58, 71

GENERAL INDEX

Abbot H. Thayer, 3, 24, 27, 86
adult mimicry by juveniles, 151, 408
aide mémoire mimicry, 97, 127, 155, 159, 206, 210, 228, 245, 250, 257, 260, 429, 440
aerial crypsis, 38, 429
aggressive mimicry, xiii–xiv, 6, 9, 12–16, 85–6, 92, 231, 233–4, 236, 286, 255, 305–51, 363, 366, 406, 410, 416, 419, 429–31, 439, 444
alarm calls, 9, 12, 128, 287, 394–406
  mimicry of, to gain food, 8, 401–2, 408
  mimicry of, to protect paternity, 365–6
  mimicry of, for protection, 287–8
alarm pheromones, 231, 233, 300, 349, 389, 391,
alluring prey, 269, 324, 390
ambush predators, 54, 55, 71, 139, 306, 313
anal gland spray, 158–9
androchromatism, 358–9
angling, 269, 324, 390
anthocyanins, 410, 413, 430
antiapostatic selection, 73, 74, 78, 182–3, 204, 430
aposematism
  and longevity, 121
  and symmetry, 63
  evolution of, 2–3, 104–29
  in plants, 163–6
  with non-chemical defence, 149–50
apostatic selection, 73–4, 183, 204, 430
appeasement pheromones, 342, 344, 404
appetite suppression, 426
Area de Conservación Guanacaste, 141–2, 245
aristolochic acid, 147, 149
Aristotelian mimicry, 277, 280, 430
arithmetic mimicry, 10, 12, 260, 430

artificial sweeteners, 425–6
aspartame, 425–6
aspect diversity, 74, 77, 150, 283, 430
asymmetry, 62–3, 65, 85, 131
attack deflection, 259–61, 263, 267, 269–71, 274–5, 277, 279
autoimmune disease, xiv, 424, 425, 430
automimicry, 12, 97, 120, 151, 164, 188–90, 216, 393, 405, 430–1
  see also weapon automimicry
autotomy, 80, 277, 430, 442, 443
aversion induction, 123

background matching, 4, 24, 29, 37, 42, 47, 53, 58, 60–1, 64, 90, 110, 290, 422
background selection, 30, 41–2, 56, 58
bacterial pathogens, 424–5
Baden-Powell, 254
Bakerian mimicry, 13, 393, 431, 433
Bates, Henry, xiii, 1–2, 7
Batesian load see mimetic load
Batesian mimicry, 6 7, 11, 13–14, 16, 20, 73, 81, 104, 119, 121–2, 142, 150, 155–6, 172, 174, 179–81, 183, 185, 187 8, 190, 192 3, 198, 201, 203–6, 214–16, 239–40, 246–51, 253, 320, 323, 430–1
  in plants, 293, 381
Batesian–Müllerian spectrum, 4
Batesian–Poultonian mimicry, 431
Batesian–Wallacian mimicry, 431
beak marks on butterfly wings, 135–7, 243
beak-wiping, 112, 284
Beau Geste hypothesis, 399, 402, 408, 431
bioluminescence, xiii, 45, 435
  aggressive mimicry, 307, 312, 315–16, 325, 431

and sex, 364
aposematism, 155–6, 166, 430
counterillumination, 28–9
bird-dropping mimicry, 23, 88–9, 91–3, 95, 150, 318
birds
  broken wing display see Aristotelian mimicry
  edibility of, 124
  eggs, 60–1
  owl feathers, 307
  pecking order, 404–1
  pollination by, 394–7
  song, 8, 324, 340, 401–2, 408
  UV vision in, 30
  wing tips, 309
bluff, 86, 253–4, 271, 284
brood-site deception, 388–9
Browerian mimicry, 188, 431
bulling see female–female mounting

calcium oxalate, 165, 394
cantharidine, 145, 146, 431
capsaicin, 163, 426
cardenolides see cardiac glycosides
cardiac glycosides, 120, 140, 141, 143, 144, 188, 190, 432
carotenoids, 35, 50, 139, 147, 149, 153
carrion mimicry, 16, 112, 120, 302, 389, 391–2, 419, 441
caterpillars, 3, 5–6, 23–4, 26–8, 30, 32, 34–5, 50–3, 61, 72–5, 78, 81–3, 85, 88–9, 91, 94–7, 100, 104, 107, 109, 112–14, 120, 122–3, 131–4, 140–3, 145, 151–2, 160–1, 163, 165, 167, 209, 216, 224, 225, 227, 237–9, 243–4, 246, 248, 255, 260, 269, 271–2, 294, 299–300, 307, 416, 435, 437

Mimicry, Crypsis, Masquerade and other Adaptive Resemblances, First Edition. Donald L. J. Quicke.
© 2017 John Wiley & Sons Ltd. Published 2017 by John Wiley & Sons Ltd.

Chagas' disease, 424
chemical mimicry, 346–9
chromatophores, 45, 49–50, 54, 57–9, 432
chromosomes, 40, 184, 201, 203, 220, 222–3, 332, 436, 441
circadian rhythms, 54
cleaner fish mimicry, 320–2
climate change, 37
clitoris, 403–5, 439
  of spotted hyaena, 404, 439
cognition, 8, 76, 104, 116, 127, 159, 162, 178, 201, 210–11, 224, 237, 258, 262, 270, 277, 285, 290, 323, 329, 334, 336, 339–40, 344, 347–9, 354, 362, 367, 410, 423, 433, 441
colour change, 49–58
  effect of diet, 51, 85
  medium paced, 54–6, 285, 325
  rapid, 56–7, 354
  slow, 50–3
compound mimicry, 235, 432
computer 'games', 43, 69, 77, 196
coral snake problem see Emsleyan mimicry
corpse mimicry see thanatosis
countershading, 3–4, 7, 20, 24, 26–8, 47, 58–9, 94, 157–8, 269, 432, 434, 438, 441
cryptic oestrus, 368
cuckoldry, 354, 401, 433
  in birds, 329–42
  in insects, 329, 342, 344–5
cuticular hydrocarbons, 50, 85, 233, 236, 239, 344, 346, 349–50, 376
cyanogenesis, 296
cyanogenic glycosides, 145, 303, 436–7, 441

dazzle, 3, 10
  in plants, 291
  military, 86
  motion, 67, 69, 71, 161, 197
decoys
  duck, 328
  military, 86
deimatic displays, 182, 260, 283–4, 433
dermal light sensitivity, 54
de-stinging behaviour, 216
detectability dietary conservatism, 109
disc equation, 80–1, 173, 188
disguise, 7
dispersal
  of individuals, 117–18
  of propagules, 410–20
disruptive patterns, 3, 7, 27, 58, 60–5, 67, 72, 90, 97, 157, 239, 244, 315, 433

disruptive selection, 76–7, 184
distractive markings, 3, 63–4, 66–8, 161, 194, 277, 328
dodecyl acetate, 344
Dodsonian mimicry, 382
domestic chicks, 75, 94–5, 104, 106, 108–9, 120, 122–6, 130–3, 154, 157, 190, 204, 262, 265–6
dual signals, xiv, 11, 45, 72–3, 433

echolocation, 255, 257
edge detection, 63
edge-intercepting patches, 46, 61–6
egg dummies/spots, on fish, 299, 360–2, 433–4
egg load assessment, 154
egg mimics, on plants, 299–300
eggs
  cuckoo, 332–8
  ground nesting birds, 58, 60
elaiosomes, 415, 433
electrosensory crypsis, 73
empathic learning see observational learning
Emsleyan mimicry, 16, 121, 173, 240–6, 254, 433, 437
EPB see exploitation of perceptual biases
epitopes, 342, 423–4
escape mimicry, 129, 150, 434
ESS see evolutionary stable strategies
eumelanin, 437
evolutionary chase, 162, 185, 192, 338
evolutionary stable strategies (ESSs), 15, 44, 100, 115–17, 131, 296, 305, 350, 358, 394–5, 402, 434
exogenous camouflage, 8, 53, 82–3, 434
exploitation of perceptual biases, 8, 207, 290
eyes
  concealment, 62, 270
  in transparent animals, 45
  stripes, 62, 270
eyespots, 15, 62, 89, 137, 192, 250, 260–77, 282–4, 433
  blinking/winking, 271–2

false heads, 271–7
family selection, 113, 134, 152
feeding mimicry see aggressive mimicry
female–female mounting, 365
female mimicry by males, 16, 354–7, 404
femmes fatales, 325
flags to attract attention, 396, 411–14
flash colouration, 260–1, 281–2, 442
flirting, 328–9
flocking see schooling
foetid odour, 392

food dummies for sex, 11, 362–3
fruit colour and fruit dispersal, 410–14
functional response, 173, 186, 434
fungal pathogens, 424
fungi, xiii, 108, 166–7, 410, 419, 432, 440, 443
  as models, 388–90, 417–18
  pathogenic, 424

galls, 300, 302
gastroenteritis, 425
genetics, xiv, 2, 34, 37, 41, 60, 77, 153, 182, 203, 216–20, 223, 266, 358
  population, 338
gentes, 332, 334, 337, 434
Georges Pasteur, 11
ghillie suits, 85–6
glucosides, 3, 122, 145
glucosinolates, 435
glycosides, 120, 140–4, 146, 149, 188–90, 224, 303, 426, 432, 435
green beard selection, 112
gregariousness, 109, 114, 131–4, 139, 152, 162, 169, 236, 386
guanine, 45, 54, 56
Guillain-Barré syndrome, 424–5

Hairy-Downy game, 401
handicaps, 117–19, 214, 414, 435
handling time, 80, 94, 96, 122, 155, 173
Helmholtz illusion, 67, 367
herding, 434
hierarchies, 9, 363, 400–1, 403
hissing
  by snakes, 156–7, 254
  mimicry of by birds, 156
homosexual behaviour see female–female mounting
honest signals, 85, 112, 117, 119, 159, 163, 284, 408
honest mimicry, 122, 127
horns, automimicry of, 407–8
hunger, 173, 177, 188, 190

illicium, 315, 434, 436
immunity, 243, 424
imperfect mimicry, 78, 196, 196, 206–11, 441
individual selection, 5, 104, 112–14, 121–2, 134, 165
innate
  colour preference, 382, 410–11
  recognition, 110, 127–31, 155, 231, 236, 242, 250, 260, 262, 283–4, 290, 303
insectivorous plants, 350–1
interspecific social dominance mimicry

iridescence, 45, 47, 151–2, 344
iridoid glycosides, 140–1, 149, 432
iridophores, 49, 54, 56–7, 362, 436

jamming, 255–7
juvenile mimicry by adults, 400, 404, 408

kin selection, 112, 121–2, 131, 133–4, 366, 435–6, 443
kleptocnidae, 147
Kirbyan mimicry, 11, 436

lauryl acetate see dodecyl acetate
leaf damage, 82–3
lichen mimicry, 7, 26, 29, 38
linamarin, 145, 436
linkage maps, 219, 222
lithium chloride, 125, 302, 436
locomotor mimicry, 214–15, 233, 235, 437
longevity, 121, 149, 166, 242, 290
lotaustralin, 145–6, 437
lures, 306–7
    angler fish and relatives, 315–16
    bolas spiders, 9, 13, 316
    caudal, 317–18
    dragonfly, 318
    pretending to be dead, 318
    snake tails, 317
    snake tongues, 317–18

magnet effect
    in pollination, 374, 386
    in seed dispersal, 411
male alternative reproductive tactics, 354–7
male mimicry by females, 355, 358–9
masquerade 4–6, 8–10, 20, 22–3, 25, 50–1, 76, 81, 86, 88–9, 91–7, 104, 109, 290, 307, 422, 431, 436–7, 439, 442
melanin, 35, 53–4, 309, 437
melanism, 37, 58, 69
    fire, 40–1
    industrial, 37–40, 436
melatonin, 37
memory, 117–18, 126, 154, 173–5, 177–80, 185, 189, 337, 387, 429
memory jogging see aide mémoire mimicry
Mertensian mimicry see Emsleyan mimicry
metapopulations, 116, 131, 211
metallic colouration, 106, 108, 150, 196, 203, 344
methyl salicylate, 123
military camouflage, masquerade and bluff, 3, 10, 85–6, 139, 255
military uniforms, 12, 307
mimetic load, 162, 203–4, 220

mimic and model relative abundance, 95, 172–3, 175, 186, 188, 192, 196, 198, 204, 206, 242, 322
mimicry definitions, 2–5, 7–10, 13
miracidia, 423
mobbing, 287, 320
mobbing call mimicry, 288, 402
monocrotaline, 143
Monte Carlo simulations, 177–9, 216
moss mimicry, 26, 29
motion camouflage/dazzle, 24, 69–71, 86, 161, 437
Müllerian mimicry, 4, 8, 10–11, 14–15, 131, 145, 151, 156, 162, 165, 172–5, 179, 181, 185–6, 188, 192, 195, 198, 204–6, 215–16, 219, 224, 227, 239, 242, 251, 285, 286, 430, 433, 436–7
    in plants, 295, 382
mummy-berry, 417–19
muscimol, 108, 166
musk, 368, 418
mustard oils, 138
myrmecomorphy, 331–6, 346, 349, 438

nanocrystals, 56
native Americans, 11, 328
nectarless flowers see unrewarding flowers
negative frequency-dependent selection see apostatic selection
nematocysts, 177, 432
neophobia, 109, 116, 123, 128–9, 438
neopeptide, 35, 58
nestlings, 58, 60–1, 141–2, 229, 244, 246, 248, 259, 280–1, 332, 338–9, 364, 430, 436
nests
    bird, xv, 60, 05, 153, 157, 259–60, 277, 288, 320, 329, 332–3
    social and other insects, 89, 104, 225, 227, 231, 233, 235, 237, 239, 251, 255–6, 301
neurohormone, 54, 99
neuropeptide, 35, 58
nitrophenathrene see aristolochic acid
novel world, 128–9, 132–3, 196, 438
novelty, 124, 127–30, 154
nuchal glands, 250
nuptial gifts, 99, 357

object mimicry see masquerade
observational learning, 242, 255
odours
    defensive, 108, 120, 158; see also pyrazine
    floral, 375–6, 379, 392, 418
oestrogens (phyto-), 247
oestrus, 365–6, 368, 403–4, 418
olfactory crypsis, 344, 350

ontogenetic changes, 32–4, 37, 53, 81–2, 89, 97, 231, 234
osmeteria, 120
Oudeman's phenomenon, 90, 438
outline concealment/disruption, 265
ovulation, 368
Owen–Owen effect, 178–9, 438

padded bras, 368
paddling for worms, 315
parasitic 'worms', 8, 107, 122, 159, 302, 415–16, 422–3, 430
parasitoid wasps, 82, 100, 120, 149, 151–2, 186, 227, 233, 237, 239, 253–5, 297, 302, 326, 333, 347, 350, 424, 426, 433, 438
Pavlovian predators, 206, 438
pebble mimicry, 80, 88, 290, 433, 439
pelt colour, 35, 61, 140, 198, 240–4, 308, 355, 386
penis, 13, 400, 403–6, 439–40
petiole clipping, 82
phallic mimicry 405
phaneric mimesis 88, 439
phenology 133, 220, 384
pheromones, 13, 143–5, 194, 201, 231, 233, 255, 256, 300–1, 307, 313, 319, 326, 342, 348–9, 356, 372–3, 375–6, 378–9, 381–2, 389, 391, 404, 431, 433, 439, 440, 442
phylogenetics, xiii, 6, 8, 26, 57, 72, 83, 89, 119, 121, 124, 133–4, 147, 158, 167–8, 203, 214–16, 218, 227–8, 231, 236, 242, 270, 302, 309, 311, 329, 332, 341–2, 348, 354, 368, 374, 376, 378–9, 381–2, 391, 411, 441
phytoecdysteroids, 427
plant galls, 300, 302
pleiotropy, 360, 439
PMRF see pupal melanisation-reducing factor
poison glands, 119, 147
    see also venom
polarisation, of light, 49, 54, 68
pollen, 121, 323, 372, 381–3, 386–9, 392, 394, 397, 410, 418–19
pollination by birds, 394–7
polyamines, 390
polymorphism, 32, 34–5, 51–2, 73–4, 76–80, 129, 150, 161–2, 175, 184–5, 190, 196, 199–201, 203–4, 211, 216–17, 219–23, 229, 267, 291, 308, 320, 333, 341, 349, 354, 357–8, 375, 386–7, 393, 430
    and sex-limited mimicry, 184, 198–203, 211, 223, 358, 442

population genetics, 338
    see also genetics
predator mimicry, 9, 12, 89, 251–3
predator–prey systems, 253–4
prickles see thorns of plants
prolactin, 37, 437
pseudonectaries, 394, 440
pseudoovipositor, 231
pseudopenis, 404–5, 440
pseudosting, 230, 440
puddling, 135, 260
pupal colour, 32–5
pupal melanisation-reducing
    factor, 35
putrid odours see odours
pyrazine, 120, 123–4, 139, 157, 206,
    379, 440
pyrrolizidine alkaloids (PAs), 120, 140–2,
    144–5, 440

quasi-Batesian mimicry, 178, 188, 206,
    323, 440
quasi-Müllerian mimicry, 12
quinine, 77, 109–10, 124, 126–7,
    130–2, 143, 204–7, 209,
    243, 441

rain mimicry, 315
raphids, 165, 441
receiver psychology, 104, 172, 441
red egg syndrome, 154, 300
reflectance, 27, 31, 37, 41, 42, 45, 47,
    54, 374, 482
    spectra, 55–7, 66, 334, 337,
        383–4, 393
    UV, 152, 313, 335, 410–11
reflex bleeding, 120, 122, 142, 147
reflexive selection, 77, 79, 150,
    430, 441
reverse mimicry, 271, 273–7
rewardless flowers see unrewarding
    flowers
ringed patterns, 107, 141, 242

salicin, 149, 441
satiric mimicry see imperfect mimicry
Scheinsaftblumen, 374
schistosomiasis, 423
schools and schooling, 9–10, 12,
    204, 260, 285, 286, 311, 322,
    434, 441
sea birds
    as mimics, 306–7
    plumage colouration, 309
    taste, 124

seaweed, 23, 61, 84–5, 88, 94,
    97, 429
search image, 74–8, 80–1, 188, 204,
    286, 430, 441
seasonal colour polymorphism, 32–7
seedless fruit, 100–1
seeds, xiv–xvi, 23, 41, 100–1, 106,
    122, 126, 138, 150, 163, 165,
    300, 368, 374, 387, 409–15,
    433, 438
selection experiments, 17, 34, 99, 362
self shadow concealment, 432, 441
semiotics, 3
sequestering toxins, 144, 167, 426
    from animals, 145–7, 431
    from plants, 3, 72, 109, 120, 140,
        142–5, 147, 149, 151, 169,
        188–90, 214, 224, 248, 255, 432,
        435, 440–1
sex-limited mimicry see polymorphism
sex pheromones, 13, 143–5, 201, 255–6,
    307, 319, 326, 372–3, 376, 378–9,
    381–2, 404, 431
sex skin, 368, 403
sexual deception in pollination, 372,
    374–6, 392, 394
sexual mimicry in animals, 353–69
shadowing, 308–9
she-males, 356–7
shoaling see schooling
simulation modelling see Monte Carlo
    simulation
sluggish behaviour, 139
sneaky mating, 400–1
Snell's window, 47, 309–10, 442
social mimicry, 400–1, 404, 442
social parasites, 342, 344, 346–8, 433
Spanish fly, 108, 431
sparkles, 260, 267–9
spatially explicit models, 116–18, 172,
    217–20
spectral sensitivity see visual spectra
spicules, 147
spines in animals, 11, 82, 159–60, 228,
    360, 441
    see also thorns of plants
sporocysts, 16, 416, 423, 437, 442
stabilimenta, 92, 93, 442
status signalling, 401
stealth, 307
stick-like caterpillars, 6, 50–2, 94–7
studliness, 400
subadult plumage, 400, 403
sulphur compounds (odourous), 158,
    303, 389, 391–2, 419, 435

supernormal stimuli, 336, 339–40,
    360, 404
survivorship plots, 28, 64, 65,
    264, 265, 268
swaying movement, 308
switching between prey, 73–75,
    176, 442
symmetry deception see reverse
    mimicry
synergistic selection, 104, 165–6

tailed moths and reverse mimicry in
    flight, 275
tail-shedding, 277, 279–80
tannins, 50
taste avoidance learning, 124, 154
termite balls, 417
testosterone, 401, 405
thanatosis, 12, 15, 97–100, 182, 283,
    318, 432, 441, 443
thermoregulation, 36–8, 67
thorns of plants (also prickles and spines)
    advertisement of, 163–5
    mimicry of, 7, 23, 93, 161
toxin load, 138, 173, 186, 205
trade-offs, 30–1, 36–7, 61, 75, 85, 100
    137–8, 147, 149, 201, 206–7, 210,
    215, 279, 291, 308, 322, 335, 337,
    358, 393, 395, 404, 443
transformational mimicry see ontogenetic
    changes
transparency, 20, 45–7, 72, 106
transvestism, 203, 356
    see also female mimicry by males; male
        mimicry by females
trophallaxis, 346, 349, 443
trypanosome, 424
two-step models of evolution
    of aposematism, 114, 195
    of mimicry, 3, 14, 172, 198

unfamiliarity see novelty
unrewarding flowers, 374–5, 387, 393–4
urotomy see autotomy; tail-shedding
UV (ultraviolet), 384, 392, 410,
    411, 418
    absorbance, 350
    aposematism, 152
    fluorescence, 155–6
    paint, 32
    reflectance, 45, 54–6, 152–3, 267–8,
        310–11, 313, 335, 374, 384, 392,
        410–11, 418
    vision, 3, 30–1, 54, 106, 209,
        267, 384

Vavilovian mimicry, 414, 415, 443, 444
venom, 8, 72, 107, 121, 129, 149,
    153, 157, 199, 213, 214, 225,
    227, 230, 239, 240, 241–4, 246,
    248, 250, 346, 422, 424, 433,
    438–9
vertebrate mimicry by invertebrates, 246
viper head shape, 244, 248
viruses, 425
vision models, 56, 69, 104, 157, 207,
    218, 334–6, 338, 341, 386, 392,
    410–11
visual spectra, 30–1, 56, 57, 165, 339

war see military...
warships, 86
warning colouration, 63, 109, 112,
    115–16, 119–20, 129, 131, 134,
    149, 155, 157–9, 165, 185, 240, 439
  in mammals, 157–9
  in plants, 163–5
warning sounds, 155–7
wasp mimicry, 152, 208–9, 226–9,
    249, 251
  by orchids, 301
Wasmannian mimicry, 13, 231, 233,
    346, 443

weapon
    advertisement, 158–60
    automimicry, 163, 165, 407–8, 430, 443
Wicklerian mimicry, 403–5, 444
Wicklerian–Barlowian mimicry, 13
Wicklerian–Guthrian mimicry, 405–6
wild birds, 3, 27–8, 31, 43, 63–4, 66, 73,
    76–8, 152, 204, 206, 208–9, 246,
    250, 262, 266, 268, 414
winter plumage and pelts, 35–7
'wolf in sheep's clothing', 322

zig–zag patterns in snakes, 244, 247–8

TAXONOMIC INDEX

Because of the broad taxonomic scope of the book, I have added the major group of organism to which each example belongs, as well as a finer division (usually a family). However, to keep things relatively simple I have used different levels for the higher groups, some being given as kingdoms (Plantae for green plants and Fungi for fungi), phyla (e.g. Cnidaria for sea anemones, hydroids, etc., Platyhelminthes for flatworms and flukes), some as classes (e.g. Actinopterygii, the bony fish, Aves, the birds, Amphibia for frogs (Anura) and salamanders (Urodela) and Reptilia for snakes and lizards (Squamata), etc.), whereas for the insects which display so many mimetic resemblances, I have used the orders, the main ones being the Lepidoptera (butterflies and moths), Hymenoptera (wasps, bees and ants), Coleoptera (beetles), Hemiptera (bugs), Orthoptera (crickets, katydids and grasshoppers), Phasmatodea (leaf and stick insects) and Odonata (dragonflies and damselflies). Crustaceans are in the Crustacea and spiders (as well as mites) are in the Arachnida.

Common names are nearly always referred to the species that is specifically dealt with, but in a few cases some non-specific references are given directly.

aardwolf *see Proteles cristata*
Abelmoschus moschatus (Plantae: Malvaceae), 368
Abelmosk *see Abelmoschus moschatus*
Abies balsamea (Plantae: Pinaceae), 427
Abisara neophron (Lepidoptera: Riodinidae), 273
Abraxas (Lepidoptera: Geometridae), 145
Abrus precatorius (Plantae: Fabaceae), 412, 411
Acacia (Plantae: Fabaceae), 163–4, 418
A. drepanolobium see Vachellia drepanolobium
Acanthiza pusilla (Aves: Acanthizidae), 287
Acanthurus pyroferus (Actinopterygii: Acanthuridae), 323–4
Accipiter (Aves: Accipitridae), 100, 129, 286, 320, 341–3
A. badius (Aves: Accipitridae), 341
A. bicolor (Aves: Accipitridae), 320
A. gentilis (Aves: Accipitridae), 100, 286
A. nisus (Aves: Accipitridae), 129, 262, 341–3
Acer saccharum (Plantae: Sapindaceae), 44
Achaea janata (Lepidoptera: Noctuidae), 281
Achaeus spinosus (Crustacea: Malacostraca: Inachidae), 84

Acherontia atropos (Lepidoptera: Sphingidae), 346
Acherontia styx (Lepidoptera: Sphingidae), 27
Acheta domesticus (Orthoptera: Gryllidae), 193
Acinonyx jubatus (Mammalia: Felidae), 240
Acipenseridae (Actinopterygii), 97
Acmaeodera (Coleoptera: Buprestidae), 249
Acraea eponina (Lepidoptera: Nymphalidae), 139
A. insignis (Lepidoptera: Nymphalidae), 137
A. johnstoni (Lepidoptera: Nymphalidae), 137
Acraeini (Lepidoptera: Nymphalidae), 124–5, 137, 145
Acrocephalus arundinaceus (Aves: Acrocephalidae), 335–7, 341
Acronicta alni (Lepidoptera: Noctuidae), 89
Actias luna (Leipodoptera: Saturniidae), 275
Adalia bipunctata (Coleoptera: Coccinelidae), 40, 153, 175, 429
Adanson jumping spider *see Hasarius adansoni*
adder *see Vipera berus*

Aegithalos caudatus (Aves: Paridae), 85
Aegolius funereus (Aves: Strigidae), 364
Aeolidae (Mollusca: Gastropoda), 122, 147
Aepyrocerus melampus (Mammalia: Bovidae), 165
Aeropetes tulbaghia (Lepidoptera: Nymphalidae), 384–5
Aeshna (Odonata: Aeshnidae), 318, 358
A. palmata (Odonata: Aeshnidae), 318
Aethusa cynapium (Plantae: Apiaceae), 295
African civet *see Civetticus civetta*
African grey parrot *see Psittacus erithacus*
African oystercatcher *see Haematopus moquini*
African queen butterfly *see Danaus chrysippus*
African tawny-flanked prinia *see Prinia subflava*
Afzelia rhomboidea (Plantae: Fabaceae), 411
Agalaia panamensis (Hymenoptera: Vespidae), 226
Aganacris (Orthoptera: Tettigoniidae), 228, 251
Agaricus muscari (Fungi: Basidiomycota: Agaricaceae), 166
Agave (Plantae: Agavaceae), 163–5
A. victoriareginae (Plantae: Agavaceae), 164

Agelaius phoeniceus (Aves: Icteridae), 127–8, 131, 145, 229, 400

Aglais urticae (Lepidoptera: Nymphalidae), 283

Agonopteryx pulvipennella (Lepidoptera: Oecophoridae), 41

Agraulis vanillae (Lepidoptera: Nymphalidae), 124

Agrilus (Coleoptera: Buprestidae), 150

Ajuga chamaepitys (Plantae: Lamiaceae), 295

A. ophrydis (Plantae: Lamiaceae), 384, 385

Alcathoe (Lepidoptera: Sesiidae), 230

Alcea setosa (Plantaceae: Malvaceae), 300

alder moth *see Acronicta alni*

Aleiodes alternator (Hymenoptera: Braconidae), 100

Aleochara curtula (Coleoptera, Staphylinidae), 404

alligator snapping turtle *see Macroclemys temminckii*

Allocyclosa (Arachnidae: Araneae: Araneidae), 93

Allograpta javana (Diptera: Syrphidae), 382

Alloteuthis subulata (Mollusca: Cephalopoda: Loliginidae), 59

Allophyes oxyacanthae (Lepidoptera: Noctuidae), 12

Alpinia (Plantae: Zingiberaceae), 412

Altica bicarinata (Coleoptera: Chrysomelidae), 150

Alyssum see *Hormathophylla*

Amanita (Fungi: Basidiomycota: Amanitaceae), 166

Amata (Lepidoptera: Erebidae: Arctiinae), 106, 196, 440

A. polymita (Lepidoptera: Erebidae: Arctiinae), 106

Amathuxidia amythaon (Lepidoptera: Nymphalidae), 160

Amauris albimaculata (Lepidoptera: Nymphalidae: Danainae), 137

A. niavius (Lepidoptera: Nymphalidae: Danainae), 137, 222

Amblyglyphidodon curacao (Actinopterygii: Pomacentridae), 322

Ambystoma barbouri (Amphibia, Urodela), 56

A. maculatum (Amphibia: Urodela: Ambystomatidae), 357

A. texanum (Amphibia: Urodela: Ambystomatidae), 56

A. tigrinum (Amphibia: Urodela: Ambystomatidae), 357

American avocet *see Recurvirostra americana*

American kestrels *see Falco sparverius*

American peppered moth *see Biston betularia cognataria*

American redstart *see Setophaga ruticilla*

Amischotolype (Plantae: Commelinaceae), 412

Amorphophallus titanum (Plantae: Araceae), 390

Amphidromus (Mollusca: Camaenidae), 77

Amphiprion (Actinopterygii: Pomacentridae), 149, 161, 422

A. clarkii (Actinopterygii: Pomacentridae), 422

Amyclaea (Arachnida: Araneae: Thomisidae), 233

Amycle (Hemiptera: Fulgoridae), 251, 253

Amyema biniflora (Plantae: Loranthaceae), 293

Anacamptis israelitica (Plantae: Orchidaceae), 385

Anaea (Lepidoptera: Nymphalidae), 89

Anartia amalthea (Lepidoptera: Nymphalidae), 127, 210

A. fatima (Lepidoptera: Nymphalidae), 194

A. jatrophae (Lepidoptera: Nymphalidae), 127

Andrena (Hymenoptera: Apidae), 346, 373, 379–81, 418

A. aliciae (Hymenoptera: Apidae), 418

A. combinata (Hymenoptera: Apidae), 379

anemone fish *see Amphiprion*

Anemone patens see Pulsatilla patens

angled sunbeam butterfly *see Curetis acuta*

angler catfish *see Chacidae*, 315

angler fish *see Lophiidae, Ceratiidae, Melanocetidae*, 307, 315

Anguis fragilis (Reptilia: Squamata: Anguidae), 246, 277

Anisakis (Nematoda: Anisakidae), 122, 130

anoles *see Anolis*

Anolis (Reptilia: Squamata: Polychrotidae), 136, 190, 194–6, 266

A. carolinensis (Reptilia: Squamata: Polychrotidae), 194, 196, 266–7

A. linearopus (Reptilia: Squamata: Polychrotidae), 190

Anomalopidae (Actinopterygii), 315

Anomalospiza imberbis (Aves: Viduidae), 331, 334, 336, 342

antbird *see Thamnophilidae*

Anthemis nobilis (Plantae: Asteraceae), 295

Anthistathmophtera (Lepidoptera: Saturniidae), 275

Antilocapra americana (Mammalia: Antilocapridae), 407–8

ants, 13, 50, 73, 85, 89, 120, 139, 147, 214, 225, 227, 231–9, 301, 328–9, 342, 344, 346–51, 362–3, 376, 415, 429, 432–3, 436, 438–9, 442–3

Antennarius maculatus (Actinopterygii: Antennariidae), 316

A. striatus (Aves: Viduidae), 316

anther smut fungus *see Microbotryum violaceum*

Anthia (Coleoptera: Carabidae), 246

Anthus pratensis (Aves: Motacillidae), 334–5

ants *see Formicidae*

Aphantochilus rogersi (Araneae: Thomisidae), 236

Aphredoderus sayanus (Actinopterygii: Aphredoderidae), 309, 312

Aphis fabae (Insecta: Hemiptera), 350

A. nerii (Insecta: Hemiptera), 112

Apis cerana (Hymenoptera: Apidae), 313, 389

A. mellifera (Hymenoptera: Apidae), 55, 192, 227, 249, 326, 344–5, 347, 389

Apocynaceae (Plantae), 82, 142, 144, 224, 373, 383, 389, 391–2, 426, 440

Apocynum cannabinum (Plantae: Apocynaceae), 82

Arabis holboellii (Plantae: Brassicaceae), 418

Arachnocampa (Diptera: Keroplatidae), 312

Araneidae (Arachnida), 6, 91–2, 99, 124, 234

Araschnia levana (Lepidoptera: Nymphalidae), 32

Arawacus sino (Lepidoptera: Lycaenidae), 273

Archaeoprepona (Lepidoptera: Nymphalidae), 151, 272

Archidendron ramiflorum (Plantae: Fabaceae), 413

Arctia caja (Lepidoptera: Erebidae), 110, 110, 297

arctic fox *see Vulpes lagopus*

arctic hare *see Lepus arcticus*

Arctiinae (Lepidoptera: Erebidae), 45, 72, 82, 120, 140–2, 144–6, 151, 162, 196, 210, 226–7, 255

Arctium tomentosum (Plantae: Asteraceae), 299

Ardea ibis (Aves: Ardeidae), 229

Ardeidae (Aves), 315

Argema (Lepidoptera: Saturniidae), 275

Argia vivida (Odonata: Coenagrionidae), 318

Argiope (Arachnida: Araneae: Araneidae), 93, 311, 326

A. argentata (Arachnida: Araneae: Araneidae), 311

A. trifasciata (Arachnida: Araneae: Araneidae), 326

Argogorytes (Hymenoptera: Sphecidae), 379

Argyrodes (Arachnida: Araneae: Theridiidae), 319

Arisarum vulgarum (Plantae: Araceae), 301

Aristolochia (Plantae: Aristolochiaceae), 150, 200, 388, 392

Arizona bird–dropping moth *see Ponometia elegantula*

Arum palaestinum (Plantae: Araceae), 390

Asarum (Plantae: Aristolochiaceae), 373

Ascaris lumbricoides (Nematoda: Ascarididae), 424

Asclepias curassavica (Plantae: Apocynaceae), 142–3, 224, 382–3, 385

Asilidae (Diptera), 11, 308, 329

Asota monycha (Lepidoptera: Noctuidae), 255

A. plana (Lepidoptera: Noctuidae), 107

Aspidomorpha (Coleoptera: Chrysomelidae), 46–7

Aspidontus taeniatus (Actinopterygii: Blenniidae), 320–1

Aspisoma (Coleoptera: Lampyridae), 325

assassin bugs *see Stenolemus bituberus*

Astatotilapia burtoni (Actinopterygii: Cichlidae), 360, 362

A. elegans (Actinopterygii: Cichlidae), 358, 362

A. latifasciata (Actinopterygii: Cichlidae), 361

Asteraceae (Plantae), 55, 72, 142, 144, 152, 165, 232, 291, 295, 299, 301, 368, 374, 418, 426, 440

Asterorhombus fijiensis (Actinopterygii: Bothidae), 315

Astragalus bisulcatum (Plantae: Fabaceae), 303

Astronotus ocellatus (Actinopterygii: Cichlidae), 270

Astropyga (Echinoidea: Diadematidae), 11

Athalia rosae (Hymenoptera: Tenthredinidae), 122

Athene cunicularia (Aves: Strigidae), 155

Atta cephalotis (Hymenoptera: Formicidae), 233

Augrabies flat lizard *see Platysaurus broadleyi*

Australian leafwing butterfly *see Doleschallia bisaltide*

Autographa bactrea (Lepidoptera: Noctuidae), 63

A. gamma (Lepidoptera: Noctuidae), 63

Babax lanceolatus (Aves: Leiothrichidae), 355

Baccharis (Plantae: Asteraceae), 306

badger *see Meles meles*

balsam fir *see Abies balsamea*

Baltimore checkerspot butterfly *see Euphydryas phaeton*

banana *see Musa ?vellutina*

banded sea snake/krait *see Laticauda colubrina*

banded snake eel *see Myrichthys colubrinus*

banded sunfish *see Enneacanthus obesus*

banded treebrown butterfly *see Lethe confuse*

barn owl *see Tyto alba*

barn swallow *see Hirundo rustica*

bats, 142, 255–7, 275, 397

batfish *see Chaca chaca*

Battus philenor (Lepidoptera: Papilionidae), 32, 34, 136–8, 194–5, 198, 203, 223

B. polydama (Lepidoptera: Papilionidae), 194

beaked coralfish *see Chelmon rostratus*

bee orchids *see Ophrys* spp

bees, 13, 16, 54–5, 157, 194, 227, 229, 249, 255, 256, 313, 326, 342, 344, 345, 346–7, 373, 375, 379–82, 384, 385, 386, 389, 391, 418

bee hawkmoth *see Hemaris tityus*

beewolf *see Philanthus triangulum*

Begonia involucrata (Plantae: Begoniaceae), 393–4

Bellevalia flexuosa (Plantae: Liliaceae), 385

Belonogaster (Hymenoptera: Vespidae), 227

Berberis vulgaris (Plantae: Berberidaceae), 418

Biblis hyperia (Lepidoptera: Nymphalidae), 125, 210

bicoloured hawk *see Accipiter bicolor*

Bicyclus anyana (Lepidoptera: Nymphalidae: Satyrinae), 266

B. safitza (Lepidoptera: Nymphalidae: Satyrinae), 137

bigeye snapper *see Lutjanus lutjanus*

Biomphalaria glabrata (Mollusca: Planorbidae), 423

Bipalium rauchi (Platyhelminthes: Turbellaria), 107

bird's-foot trefoil *see Lotus corniculatus*

Biston betularia (Lepidoptera: Geometridae), 37–42, 50–53

B. betularia cognataria (Lepidoptera: Geometridae), 50, 52

B. hirtaria (Lepidoptera: Geometridae), 94

B. robustum (Lepidoptera: Geometridae), 50–1, 53, 73

blackbird, European *see Turdus merula* American *see Agelaius phoeniceus*

blackcap *see Sylvia atricapilla*

black-capped chickadee *see Parus atricapillus*

blackfly *see Simuliidae*

black-headed gull *see Chroicocephalus ridibundus*

black-winged stilt *see Himantopus hymatopus*

Blastomyces dermatitidis (Fungi: Ascomycota: Ajellomycetaceae), 424

Blattella germanica (Blattodea: Blattellidae), 319

Blenina chrysochlora (Lepidoptera: Nolidae), 281

Blennius pholis (Actinopterygii: Benniidae), 94

blister beetle *see Meloidae*

blueberry *see Vaccinium*

blue butterflies *see Lycaenidae*

bluegill sunfish *see Lepomis macrochirus*

blue-grey gnatcatcher *see Polioptila caerulea*

blue jay *see Cyanocitta cristata*

bluestreak cleaner wrasse *see Labroides dimidiatus*

blue-striped fangblenny *see Plagiotremus rhinorhynchus*

blue-stripe snapper *see Lutjanus kasmira*

blue tit *see Cyanistes caeruleus*

Bokermannohyla luctuosa (Amphibia: Anura: Hylidae), 98

bolas spiders *see Mastophora cornigera*

Bolitophila see Arachnocampe

Bombus (Hymenoptera: Apidae), 15, 157, 192, 229, 342, 344, 373, 385–6, 391

Bombus ('*Psithyrus*') (Hymenoptera: Apidae), 15, 157, 342, 344

Bombus americanorum (Hymenoptera: Apidae), 192

B. pennsylvanicus (Hymenoptera: Apidae), 229

B. terrestris (Hymenoptera: Apidae), 344

Bombyx mori (Lepidoptera: Bombycidae), 223

boomslang *see Dispholidus typus*

Boquila trifoliolata (Plantae: Lardizabalaceae), 294

Boraginaceae (Plantae), 144, 440

Bosellia (Mollusca: Boselliidae), 22

Bothus ocellatus (Actinopterygii: Bothidae), 56

Bracca (Lepidoptera: Geometridae), 244

Braconidae (Hymenoptera), 350

Bradypodion (Reptilia: Squamata: Chamaeleonidae), 57

brambling *see Fringilla montrifringilla*

Brassica (Plantae: Brassicaceae), 139, 154, 165, 290, 301, 387, 415, 418, 432, 435
flower colour
B. hirta (Plantae: Brassicaceae), 94, 387
B. kaber (Plantae: Brassicaceae), 387
Brentia (Lepidoptera: Choreutidae), 251–2
brimstone moth *see Opisthograptis luteolata*
bristly cutworm *see Lacinipolia renigera*
bronze cuckoos *see Chalcites*
Broussonetia (Plantae: Moraceae), 297
brown-eared bulbuls *see Hypsipetes amaurotis*
brown-headed cowbird *see Molothrus ater*
brown lemming *see L. trimucronatus*
brown snake *see Storeria dekayi*
brown tanager *see Orchesticus abeille*
buff-fronted foliage-gleaner *see Philydor rufus*
bufftip moth *see Phalera bucephala*
Bufo terrestris (Amphibia: Anura: Bufonidae), 192, 249
bullhorn wattle *see Vachellia cornigera*
Bungarus (Reptilia: Squamata: Elapidae), 69, 107, 244, 276
Buprestidae (Coleoptera), 108, 150, 201, 249
burrowing owl *see Athene cunicularia*
Buteo albonotatus (Aves: Accipitridae), 319
Buteo buteo (Aves: Accipitridae), 246
buzzard *see Buteo buteo*
by-the-wind sailor, *see Velella*

cacti (Plantae: Cactaceae), 163, 397
Caenocoris nerii (Hemiptera: Lygaeidae), 122
Caladenia barbarossa (Plantae: Orchidaceae), 379
Caladium steudneriifolium (Plantae: Araceae), 397–9
Calamus (Plantae: Arecaceae), 160, 164
Calicotome villosa (Plantae: Fabaceae), 165
Caligo (Lepidoptera: Nymphalidae), 260, 264
Calliophis intestinalis (Reptilia: Squamata: Elapidae), 275
Calloplesiops altivelis (Actinopterygii: Plesiopidae), 250
Callosamia promethea (Lepidoptera: Saturniidae), 194–5
Calluna (Plantae: Ericaceae), 31, 385
Calycopis cecrops (Lepidoptera: Lycaenidae), 275
Calypso bulbosa (Plantae: Orchidaceae), 373–4
Camelina sativa (Plantae: Brassicaceae), 415
Camellia japonica (Plantae: Theaceae), 51, 53

Campaea perlata (Lepidoptera: Geometridae), 42
Camponotus (Hymenoptera: Formicidae), 233, 235–6, 347
Camponotus nearcticus (Hymenoptera: Formicidae), 235
C. sericeiventris (Hymenoptera: Formicidae), 233
C. sericeus (Hymenoptera: Formicidae), 236
Campanula (Plantae: Campanulaceae), 382–5
C. persicifolia (Plantae: Campanulaceae), 382, 384–5
C. rotundifolia (Plantae: Campanulaceae), 382, 385
Campylobacter jejuni (Bacteria: Campylobacteraceae), 425
candy-shouldered thorn *see Ennomos alniaria*
Cantharidae (Coleoptera), 146
Canthigaster valentini (Actinopterygii: Tetraodontidae), 195, 197–8
Caprimulgidae (Aves), 277
Cardiocondyla obscurior (Hymenoptera: Formicidae), 357
Carica papaya (Plantae: Caricaceae), 393
carrion crow *see Corvus corone*
Carthamus (Plantae: Asteraceae), 165, 299
Castanopsis cuspidata (Plantae: Fagaceae), 51, 53
castor oil plant *see Ricinus communis*
Catephia alchymista (Lepidoptera: Erebidae), 281
Cathartes aura (Aves: Cathartidae), 319
Catocala (Lepidoptera: Noctuidae), 42–44, 73, 74, 281–2, 284
C. antinympha (Lepidoptera: Noctuidae), 42
C. cerogama (Lepidoptera: Noctuidae), 43–4
C. ilia (Lepidoptera: Noctuidae), 281
catsnake *see Telescopus fallax*
Celaenia (Arachnida: Araneae: Araneidae), 92
Cemophora coccinea (Reptilia: Squamata: Colubridae), 240–1
Centaurea ibirica (Plantae: Asteraceae), 165
Centris (Hymenoptera: Apidae), 13, 373
Centropyge vrolikii (Actinopterygii: Pomacanthidae), 323
Cepaea nemoralis (Mollusca: Gastropoda: Helicidae), 77, 80, 204
Cephalanthera rubra (Plantae: Orchidaceae), 382, 384–5, 391
Cephalopholis fulva (Actinopterygii: Serranidae), 323

cephalopods
Cephalotes specularis (Hymenoptera: Formicidae), 233
Cephonodes hylas (Lepidoptera: Sphingidae), 45
Cercaspis carinatus see Lycodon carinatus
Cercopithecus pygerythrus (Mammalia: Cercopithecidae), 403
Cercotrichas galactotes (Aves: Muscicapidae), 336
Cercyonis pegala (Lepidoptera: Nymphalidae), 192–3
Ceriagrion (Odonata: Coenagrionidae), 358
Ceriana (Diptera: Syrphidae), 226
Ceropegia sandersonii (Plantae: Apocynaceae), 389
Cerura vinula (Lepidoptera: Notodontidae), 61
Chaca chaca (Actinopterygii: Chacidae), 315
Chacidae (Actinopterygii), 315
Chaetodipus intermedius (Mammalia: Heteromyidae), 37
Chaetodon (Actinopterygii: Chaetodontidae), 270
Chaetodontidae (Actinopterygii), 62, 270
Chaetognatha, 45
chaffinch *see Fringilla coelebs*
Chalcites see Chrysococcyx
Chalcites basalis see Chrysococcyx basalis
Chalcotropis gulosus (Arachnida: Araneae: Salticidae), 235
chameleons *see Bradypodion*
chameleon prawn *see Hippolyte varians*
Chaoborus (Diptera: Chaoboridae), 46, 285, 328
Charadrius hiaticula (Aves: Charadriidae), 280
C. melodus (Aves: Charadriidae), 60
C. vociferus (Aves: Charadriidae), 280
Chaunus icterus see Rhinella icterica
Charaxes (Lepidoptera: Nymphalidae), 191
cheetah *see Acinonyx jubatus*
Cheilosia illustrata (Diptera: Syrphidae), 250
Cheilosia lucida (Diptera: Syrphidae), 388
Cheirophyllum temulum (Plantae: Apiaceae), 55
Chelmon rostratus (Actinopterygii: Chaetodontidae), 62
Chelostoma (Hymenoptera: Apidae), 382, 384–5
C. campanularum (Hymenoptera: Apidae), 382
C. fuliginosum (Hymenoptera: Apidae), 382
chestnut-crowned becard *see Pachyramphus castaneus*
chicks and chickens *see Gallus gallus*

chilli peppers *see Solanum fructicosa*

Chiloglottis reflexa (Plantae: Orchidaceae), 378, 381

C. trapeziformis (Plantae: Orchidaceae), 378–9

C. valida (Plantae: Orchidaceae), 378

Chinese character moth *see Cilix glaucata*

Chionactis palarostris (Reptilia: Squamata: Colubridae), 240

Chlamydera maculata (Aves: Ptilonorhynchidae), 287–8

Chlamydosaurus kingii (Reptilia: squamata: Agamindae)., 254

Chlorobalius leucoviridis (Orthoptera: Tettigoniidae), 325, 327

Chlosyne harrisii (Lepidoptera: Nymphalidae), 192–3

chocolate surgeonfih *see Acanthurus pyroferos*

chocolate tip star *see Protoreaster nodosus*

Chondrodactylus angulifer (Reptilia: Squamata: Gekkonidae), 246

Chondrohierax uncinatus (Aves: Accipitridae), 320

Christmas tree worm *see Spirobranchus giganteus*

Chroicocephalus ridibundus (Aves: Laridae), 315

Chromis multilineata (Actinopterygii: Pomacentridae), 322

Chromodorididae (Mollusca: Gastropoda), 149

Chromodoris magnifica (Mollusca: Gastropoda; Chromodorididae), 149

C. quadricolor (Mollusca: Gastropoda: Chromodorididae), 149

Chrysanthemum balsamita (Plantae: Asteraceae), 42

Chrysididae (Hymenoptera), 342, 344

Chrysina (Coleoptera: Scarabaeidae), 47

Chrysobothris humilis (Coleoptera: Buprestidae), 201, 203

Chrysococcyx basalis (Aves: Cuculidae), 337

C. caprius (Aves: Cuculidae), 338

C. lucidus (Aves: Cuculidae), 339–40

Chrysomelidae (Coleoptera), 46–7, 108, 144–7, 150, 201, 326

Chrysopa slossonae (Neuroptera: Chrysopidae), 85, 346

Chrysoplectrum (Lepidoptera: Hesperiidae), 141

Chrysotoxum (Diptera: Syrphidae), 208, 226

Cicadetta calliope (Hemiptera: Cicadidae), 327

C. viridis (Hemiptera: Cicadidae), 327

Cilix glaucata (Lepidoptera: Drepanidae), 89

Cimifuga (Plantae: Ranunculaceae), 386

cinnabar moth *see Tyria jacobaeae*

Circotettix rabula (Orthoptera: Acrididae), 42

Circus (Aves: Accipitridae), 320

C. cyaneus (Aves: Accipitridae), 364

Cirsium arvense (Plantae: Asteraceae), 418

Citellus osgoodi (Mammalia: Sciuridae), 41

Cithaerias menander (Lepidoptera: Nymphalidae), 46

citrus canker *see Xanthomonas axonopodis*

Civetticus civetta (Mammalia: Viverridae), 368

Cladomelea (Arachnida: Araneae: Araneidae), 326

Cladosporium (Fungi: Ascomycota: Davidiellaceae), 388

Clamator glandarius (Aves: Cuculidae), 339

Clathrus archeri (Fungi: Basidiomycota: Phallaceae), 419

cleaner fish *see Labroides*

Clerodendrum (Plantae: Lamiaceae) 411–12

cliff-flycatcher *see Hirundinea ferruginea*

clover *see Trifolium repens*

Clusia nemorosa (Plantae: Clusiaceae), 393

Clytus (Coleoptera: Cerambycidae), 228

Cnemidophorus (Reptilia: Squamata: Teiidae), 80

Cnidaria, 5, 11, 20, 45, 83, 122, 147, 155, 159, 161, 255, 316, 422, 432

Coccinella septempunctata (Coleoptera: Coccinellidae), 153

Coccinellidae (Coleoptera), 104, 119–20, 122, 146, 163, 429, 432

Cochleosoma lipscombiae (Reptilia: Orchidaceae), 373

C. lipscombiae (Plantae: Orchidaceae), 373

cocklebur *see Xanthium strumarium*

Coendou (Mammalia: Erethizontidae), 157

Coenophlebia (Lepidoptera: Nymphalidae), 89

Colaptes auratus (Aves: Picidae), 156

Coleonyx variegatus (Reptilia: Squamata: Gekkonidae), 246, 279

Coleosoma floridanum (Arachnida: Araneae: Therediidae), 203

Colletes cunicularius (Hymenoptera: Apidae), 375, 380

C. hederae (Hymenoptera: Apidae), 256

Colobura (Lepidoptera: Nymphalidae), 151

Colobus verus (Mammalia: Cercopithecidae), 404

Colorado potato beetle *see Leptinotarsa decemlineata*

Colotis (Lepidoptera: Pieridae), 137

coltsfoot *see Tussilago farfara*

Columba livia (Aves: Columbidae), 73, 76, 208, 286

Combretum constrictum (Plantae: Combretaceae), 396

comet fish *see Calloplesiops altivelis*

comma butterfly *see Polygonia c-album*

common buckeye butterfly *see Junonia coenia*

common earl butterfly *see Tanaecia julii*

common grackle *see Quiscalus quiscula*

common wood nymph *see Cercyonis pegala*

coney *see Cephalopholis fulva*

Conium maculatum (Plantae: Apiaceae), 295

convolvulus hawkmoth *see Herse convolvuli*

Copiopteryx (Lepidoptera: Saturniidae), 275

Coptosperma littorale (Plantae: Rubiaceae), 314

coquina *see Donax variabilis*

coral trout *see Plectropomus*

corkwing wrasse *see Crenilabrus melos*

Coronella austriaca (Reptilia: Squamata: Colubridae), 246–7

Corvus corone (Aves: Corvidae), 112

Corybas (Plantae: Orchidaceae), 373

Corydalis benecincta (Plantae: Papaveraceae), 290

Corydoras (Actinopterygii: Callichthyidae), 239

Corynopoma riisei (Actinopterygii: Characidae), 16, 363

Cosmophasis bitaeniata (Arachnida: Araneae: Salticidae), 236

Costus (Plantae: Costaceae), 382

Cosymbia or *Idaea* (Lepidoptera: Geometridae), 42

Cotesia euphydridis (Hymenoptera: Braconidae), 152

C. glomerata (Hymenoptera: Braconidae), 82

C. popularis (Hymenoptera: Braconidae), 152

C. rubecula (Hymenoptera: Braconidae), 355, 357

Coturnix japonica (Aves: Phasianidae), 33, 58

cowbirds *see Molothrus*

crabs

    decorator crabs *see Hyastenus elatus, Maja varrucosa, M. squinado, Pisa tetraodon, Pugettia producta*

    kelp crab *see Pugettia producta*

    Samurai *see Heikeopsis japonica*

Crangon (Crustacea: Malacostraca: Crangonidae), 56

Crematogaster (Hymenoptera: Formicidae), 233, 236, 328, 349, 415

C. ampla (Hymenoptera: Formicidae), 233

Crenicichla (Actinopterygii: Cichlidae), 270

Crenilabrus see Symphodus

Creobroter gemmatus (Matodea: Hymenopodidae), 314

Creophilus maxillosus (Coleoptera: Staphylinidae), 120

crested porcupine *see Hystrix cristata*

Crinoidea (Echinodermata), 79

crocodile (toys), 130–1

Crocuta crocuta (Mammalia: Hyaenidae), 355, 404–5

Crotalaria (Plantae: Fabaceae), 144–5

Crotalus viridis (Reptilia: Squamata: Viperidae), 155

crows (Aves: Corvidae), 124, 416

Cryptoprocta ferox (Mammalia: Viverridae), 405

Cryptostylis (Plantae: Orchidaceae), 376, 377, 379–81

C. erecta (Plantae: Orchidaceae), 377

Ctenodactylidae, 97

Ctenophora, 45, 47, 147

Ctenophorus fordi (Squamata: Agamidae), 30

cuckoo (bird),
common *see Cuculus canorus*
Dieldrick *see Chrysococcyx caprius*
hawk *see Hierococcyx varius*

cuckoo bumble bee *see Bombus* ('*Psithyrus*')

cuckoo finch *see Anomalospiza imberbis*

cuckoo wasp *see Chrysididae*

Cuculus canorus (Aves: Cuculidae), 329, 332, 338–9, 343

C. fugax (Aves: Cuculidae), 339

C. pallidus (Aves: Cuculidae), 334

Curetis acuta (Lepidoptera: Lycaenidae), 47–9

C. bulis (Lepidoptera: Lycaenidae), 49

Cyanistes caeruleus (Aves: Paridae), 28, 31, 32, 63, 66, 68, 109, 110, 129, 204, 267

Cyanocitta cristata (Aves: Corvidae), 6, 73, 127, 131, 192, 274, 284

Cyclosa (Arachnida: Araneae: Araneidae), 93

C. argentoalba (Arachnida: Araneae: Araneidae), 93

C. mulmeinensis (Arachnida: Araneae: Araneidae), 93

Cydonia oblonga (Plantae: Rosaceae), 414

Cylas formicarius (Coleoptera: Curculionidae), 100

Cymothoe herminia (Lepidoptera: Nymphalidae), 137

Cynops pyrrhogaster (Amphibia: Urodela: Salamandridae), 153

Cynoscion nebulosus (Actinopterygii: Scianidae), 31

Cyphoma gibbosum, (Mollusca: Ovulidae), 122

Cyphotilapia gibberosa (Actinopterygii: Cichlidae), 322

Cyprepidium (Plantae: Orchidaceae), 373

C. fargesii (Plantae: Orchidaceae), 388

Cyrtocara eucinostomus (Actinopterygii: Cichlidae), 325

C. orthognathus (Actinopterygii: Cichlidae), 325

C. pleurotaenia (Actinopterygii: Cichlidae), 325

Cyrtodactylus australotitiwangsaensis (Reptilia: Squamata: Gckkonidae), 244

Cystophora (Phaeophyceae [brown algae]), 22

Cytomegalovirus (Viruses: Herpesviridae), 425

Dactyloplus coccus (Hemiptera: Dactylopiidae), 146–7

Dactylorhiza sambucina (Plantae: Orchidaceae), 373, 387

Damaliscus lunatus (Mammalia: Bovidae), 365

damselfish *see Pomatocentrus amboinensis*

Danainae (Lepidoptera: Nymphalidae), 72, 107, 112, 124–5, 135, 137–40, 142–4, 162, 248

Danaus chrysippus (Lepidoptera: Nymphalidae), 35, 139, 142, 144, 175, 188–90, 202, 220–2, 224

D. gilippus (Lepidoptera: Nymphalidae), 125, 188, 201

D. plexippus (Lepidoptera: Nymphalidae), 3, 33, 121, 138, 142–3, 145, 188, 194, 224, 440

Daphnia (Crustacea: Cladocera), 46, 285–6

Dasycaris (Crustaceae: Decapoda: Pontaniidae), 21

Dasymutilla (Hymenoptera: Mutillidae), 227

Dasypeltis scabra (Reptilia: Squamata: Colubridae), 244

Daucus carota (Plantae: Apiaceae), 297

dead-horse arum *see Helicodiceros muscivorus*

death's head hawkmoth *see Acherontia atropos*

Deilephila elpenor (Lepidoptera: Sphingidae), 53, 271

Dendrobates (Amphibia: Anura: Dendrobatidae), 119, 148, 153, 163

D. auratus (Amphibia: Anura: Dendrobatidae), 148

D. imitator (Amphibia: Anura: Dendrobatidae), 162

D. leucomelas (Amphibia: Anura: Dendrobatidae), 148

D. pumilio see Oophaga pumilio

D. tinctorius (Amphibia: Anura: Dendrobatidae), 148

Dendrobium infundibulum (Plantae: Orchidaceae), 382

D. sinense (Plantae: Orchidaceae), 373, 389

Dendrophidion dendrophis (Reptilia: Squamata: Colubridae),

Desmometopa (Diptera: Milichidae), 389

destroying angel *see Amanita*

Diacamma (Hymenoptera: Formicidae), 233

Diamphidia (Coleoptera: Chrysomelidae), 150

Diaprepes abbreviatus (Coleoptera: Curculionidae), 366

dice snake *see Natrix tesselata*

Dicrostichus see Ordgarius

Dicrurus (Aves: Dicruridae), 287, 342–3, 402

D. adsimilis (Aves: Dicruridae), 402

D. paradiseus (Aves: Dicruridae), 287

Dictyota (Phaeophyceae [brown algae]: Dictyotaceae), 83, 147

Didelphis virginiana (Mammalia: Didelphidae), 108, 166

Digitalis (Plantae: Plantaginaceae), 142, 432

Dinia eagrus (Lepidoptera: Erebidae: Arctiinae), 45

Diodontidae (Actinopterygii), 253

Dionaea muscipula (Plantae: Droseraceae), 350–1

Dioscoreophyllum volkensii (Plantae: Menispermaceae), 426

Disa (Plantae: Orchidaceae), 374, 381, 384–7

D. atricapilla (Plantae: Orchidaceae), 374, 381

D. bivalvata (Plantae: Orchidaceae), 374, 381

D. cephalotes (Plantae: Orchidaceae), 385

D. draconis (Plantae: Orchidaceae), 384–5

D. ferruginea (Plantae: Orchidaceae), 384–5

D. filicornis (Plantae: Orchidaceae), 374

D. harveiana (Plantae: Orchidaceae), 374, 385

D. karooica (Plantae: Orchidaceae), 385–6

D. nivea (Plantae: Orchidaceae), 387

D. oreophila (Plantae: Orchidaceae), 385

D. pulchra (Plantae: Orchidaceae), 384–5

D. tenuifolia (Plantae: Orchidaceae), 374, 386

Dismorphia theucharila (Lepidoptera: Pieridae), 2

Disonycha (Coleoptera: Chrysomelidae), 150

Dispholidus typus (Reptilia: Squamata: Colubridae), 57

Diurus maculata (Plantae: Orchidaceae), 384–5

Dolabrifera dolabrifera (Mollusca: Aplysiidae), 22

Doleschallia bisaltide (Lepidoptera: Nymphalidae), 89

Dolichovespula arenaria (Hymenoptera: Vespidae), 157, 344–5

Donacilla cornea (Mollusca: Donacidae), 77

Donax variabilis (Mollusca: Donacidae), 77

Douglas ground squirrel *see Tamiasciurus douglasii*

downy woodpecker *see Picoides pubescens*

Doxocopa (Lepidoptera: Nymphalidae), 151

Dracula (Plantae: Orchidaceae), 373, 389–90

Drakaea (Plantae: Orchidaceae), 376–9, 381, 440

D. glyptodon (Plantae: Orchidaceae), 378

D. livida (Plantae: Orchidaceae), 379

D. micrantha (Plantae: Orchidaceae), 377

Dryas iulia (Lepidoptera: Nymphalidae), 124, 163

dronefly *see Eristalis tenax*

drongo (also *see Dicrurus*, 124

drongo cuckoo *see Surniculus lugubris*

Drosera rotundifolia (Plantae: Droseraceae), 350–1

Drosicha (Diptera: Monophlebidae), 105

Drosophila (Diptera: Drosophilidae), 56, 351, 355, 362–3

Drusilla (Coleoptera: Staphylinidae), 349

dunnock *see Prunella modularis*

dusky dottyback *see Pseudochromis fuscus*

dwarf boa *see Tropidophiidae*

dwarf chameleons *see Brapypodion*

Dysdercus cingulatus (Hemiptera: Pyrrhocoridae), 129

early thorn moth *see Selenia dentaria*

earthworms (Annelida: Oligochaeta), 125, 315

eastern groundsel *see Senecio vernalis*

eastern milksnake *see Lampropeltis triangulum*

eastern tiger swallowtail *see Papilio glaucus*

Ecballium elaterium (Plantae: Cucurbitaceae), 13, 393

Echinochloa crus-galli (Plantae: Poaceae), 415

Echinodermata, 77, 79, 155, 160

Eciton (Hymenoptera: Formicidae), 346

Ecsenius gravieri (Actinopterygii: Blenniidae), 323

Ectatomma ruidum (Hymenoptera: Formicidae), 239, 349

elder-flowered orchid *see Dactylorhiza sambucina*

Eleodes (Coleoptera: Tenebrionidae), 8, 121, 193

E. longicollis (Coleoptera: Tenebrionidae), 193

E. obscura (Coleoptera: Tenebrionidae), 121

Elysiella pusilla (Mollusca: Opisthobranchia: Plakobranchidae), 22

Elytroleptus (Coleoptera: Cerambycidae), 147

Emballonuridae (Mammalia: Chiroptera), 257

emperor moths *see Saturniidae*

Enallagma (Odonata: Coenagrionidae), 358–9

E. civile (Odonata: Coenagrionidae), 359

E. geminatum (Odonata: Coenagrionidae), 359

E. hageni (Odonata: Coenagrionidae), 359

Enneacanthus obesus (Actinopterygii: Centrarchidae), 309, 312

Ennomos alniaria (Lepidoptera: Geometridae), 94

Epidendrum ibaguense (Plantae: Orchidaceae), 385

E. radicans (Plantae: Orchidaceae), 382–3, 385

Epipactis consimilis (Plantae: Orchidaceae), 373

E. veratrifolia (Plantae: Orchidaceae), 391

Episyrphus balteatus (Diptera: Syrphidae), 209, 382

Eptesicus fuscus (Mammalia: Chiroptera: Vespertilionidae), 142

Equetus lanceolatus (Actinopterygii: Sciaenidae), 61

Equus burchelli (Mammalia: Equidae), 67

E. grevyi (Mammalia: Equidae), 67

E. zebra (Mammalia: Equidae), 67

Erebidae (Lepidoptera), 45, 53, 72, 82, 106, 140–2, 144, 146, 152, 162, 196, 210, 226–7, 238, 255, 281

Eremias lugubris (Squamata: Lacertidae), 246

Eriovixia laglaisei (Arachnida: Araneae: Araneidae), 91

Eristalis tenax (Diptera: Syrphidae), 199, 208, 249

Erithacus rubecula (Aves: Muscicapidae), 27, 28, 128, 150, 332

ermine *see Mustella erminea*

Ero furcata (Arachnida: Araneae: Mimetidae), 319

Erpeton tentacularis (Reptilia: Squamata: Colubridae), 328

Erythrina (Plantae: Fabaceae), 164, 414

Erythrolamprus guentheri (Reptilia: Squamata: Colubridae), 163

Etheostoma (Actinopterygii: Percidae), 360

E. flabellare (Actinopterygii: Percidae), 360

Euantissa pulchra (Mantodea: Hymenopodidae), 236

Eucalyptus (Plantae: Myrtaceae), 293

Eucera berlandi (Hymenoptera: Apidae), 380

E. clypeata (Hymenoptera: Apidae), 385

Eucharitidae (Hymenoptera), 239

Eucheuma arnoldii (Rhodophyta [red algae]), 302

Euchloe ausonides (Lepidoptera: Pieridae), 154

Eudaemonia (Lepidoptera: Saturniidae), 275

Eudryas unio (Lepidoptera: Noctuidae), 89

Euglyphis jessiehillae (Lepidoptera: Lasiocampidae), 141

E. lankesteri (Lepidoptera: Lasiocampidae), 141

Eumaeus atala (Lepidoptera: Lycaenidae), 32

Eumomota superciliosa (Aves: Momotidae), 131, 242

Eumorpha (Lepidoptera: Sphingidae), 53, 271–2

E. fasciata (Lepidoptera: Sphingidae), 53

Eupemphix nattereri see Physalaemus nattereri

Euphorbia cyparissias (Plantae: Euphorbiaceae), 295, 418

Euphrasia (Plantae: Scrophulariaceae), 385

Euphydryas phaeton (Lepidoptera: Nymphalidae), 152, 192–3

Euphyia intermediata (Lepidoptera: Geometridae), 43–4

Euphysa (Cnidaria: Corymorphidae), 155

Eupithecia (Lepidoptera: Geometridae), 307

Euploea (Lepidoptera: Nymphalidae), 137, 189, 199–200, 248

E. mulciber (Lepidoptera: Nymphalidae), 199–200

Eurasian starlings *see Sturnus vulgaris*

Euryattus (Arachnida: Araneae: Salticidae), 319

Eurybrachyidae (Hemiptera), 274

Euschistus conspersus (Hemiptera: Pentatomidae), 128

Eutamias quadrimaculatus (Mammalia: Sciuridae), 117

E. sibiricus (Mammalia: Sciuridae), 140

Exechocentrus (Arachnida: Araneae: Araneidae), 326

Exochomus (Coleoptera: Coccinelidae), 153

eyebright *see Euphrasia*

Fabaceae (Plantae), 23, 144–5, 163–5, 231, 295, 384–5, 411–15, 418, 427, 440

Fagopyrum esculentum (Plantae: Polygonaceae), 387

Falco sparverius (Aves: Falconidae), 286

F. tinnunculus (Aves: Falconidae), 400

fallen bark looper *see Gastrophora henricaria*

false cobra *see see Rhagerhis moilensis*

false wanderer butterfly *see Pseudacraea eurytus*

fangblenny *see Aspidontus taeniatus, Plagiotremus rhinorhynchos*

Fasciola hepatica (Platyhelminthes: Trematoda: Fasciolidae), 423

fawn-breasted tanagers *see Pipraeidea melanonota*

Feniseca tarquinius (Lepidoptera: Lycaenidae), 346

fennel *see Foeniculum vulgare*

Ficedula hypoleuca (Aves: Muscicapdae), 130, 139, 266, 283, 364, 400

Ficus (Plantae: Moraceae), 200, 414

figs *see Ficus*

firebug *see Pyrrhocoris apterus*

fireflies *see Lampyridae, Aspisoma, Photinus, Photuris*

fiscal shrike *see Lanius collaris*

fish-eating spider *see Dolomedes*

flamingo tongue snail *see Cyphoma gibbosum*

flashlight fish *see Anomalopidae*

Flatidae (Hemiptera), 255

flea beetles *see Altica*

flicker *see Colaptes auratus*

Foeniculum vulgare (Plantae: Apiaceae), 224–5, 284

fool's parsley *see Aethusa cynapium*

fork-tailed drongo *see Dicrurus adsimilis*

Formicidae (Hymenoptera), 51, 349, 357

Formiscurra indicus (Hemiptera: Caliscelidae), 231–2

fossa *see Cryptoprocta ferox*

Fountainea nobilis (Lepidoptera: Nymphalidae), 89

foxglove *see Digitalis*

Freesia (Plantae: Iridaceae), 123

friarbirds *see Philemon*

frill-necked lizard, *see Chlamydosaurus kingii*

Fringilla coelebs (Aves, Frigillidae), 27, 100, 153

F. montifringilla (Aves, Frigillidae), 335

Fucus serratus (Phaeophyceae [brown algae]), 94

Fusarium semitectum (Fungi: Ascomycota: Nectriaceae), 418

Gadus morhua (Actinopterygii: Gadidae), 122

Galbula ruficauda (Aves: Galbulidae), 138, 215, 218

Gallus gallus (Aves: Phasianidae), 2, 75, 82, 94–7, 104, 106, 108–10, 118–20, 122–4, 126–7, 129–34, 157, 190, 204, 262–3, 265–6

Gamasomorpha maschwitzi (Arachnida: Araneae: Oonpidae), 347

garden tiger moth *see Arctia caja*

garden warbler *see Acrocephalus arundinaceus*

garter snakes *see Thamnophis elegans, T. radix, T. sirtalis parietalis*

Gasterosteus aculeatus (Actinopterygii: Gasterosteidae), 423

Gastrodia similis (Plantae: Orchidaceae), 373

Gastrophora henricaria (Lepidoptera: Geometridae), 261

Gaylussacia (Plantae: Ericaceae), 418

Geometridae (Lepidoptera), 2, 25, 37, 42–4, 50, 52–3, 85, 88, 92, 94, 135, 149, 224, 227, 244, 261, 309

German cockroach *see Blattella germanica*

Germerium junin (Aves: Acanthizidae), 339, 340

G. magnirostris (Aves: Acanthizidae), 338

giant leaf insect *see Phyllium giganteum*

giant purple grasshopper *see Titanacris albipes*

Girardinichthys multiradiatus (Actinopterygii: Goodeiidae), 354

glass catfish *see Kryptopterus bicirrhis*

glasswing butterfy *see Greta oto*

Glaucidium californicum (Aves: Strigidae), 276–7, also see 264

G. gnomum (Aves: Strigidae), 277

Glaucus atlanticus (Mollusca: Gastropoda: Glaucidae), 147

globe orchid *see Traunsteinera globosa*

Glossina (Diptera: Glossinidae), 67

glow-worm *see Lampyris noctiluca*

Gnathodentex aureolineatus (Actinopterygii: Lethrinidae), 285

goatfish *see Mulloidichthys*

Gobiosoma (Actinopterygii: Gobiidae), 320

Gobius paganellus (Actinopterygii: Gobiidae), 56

goldenrod *see Senecio vernalis*

goldenrod spider *see Misumena vatia*

gold-lined sea-bream *see Gnathodentex aurolineatus*

gold spangle moth *see Autographa bactrea*

Goniobranchus kuniei (Mollusca: Gastropoda: Chromodorididae), 149

Goniurellia tridens (Diptera: Tephritidae), 92

Gorteria diffusa (Plantae: Asteraceae), 374

goshawk *see Accipiter gentilis*

Grant's gazelle *see Nanger granti*

Graphosoma (Hemiptera: Pentatomidae), 105

greater racket-tailed drongo *see Dicrurus paradiseus*

great kiskadee *see Pitangus*

great orange tip butterfly *see Hebomoia glaucippe*

great reed warbler *see Acrocephalus arundinaceus*

great tit *see Parus major*

green-banded broodsac *see Leucochloridium paradoxum*

green-brindled crescent moth *see Allophyes oxyacanthae*

Greta oto (Lepidoptera: Nymphalidae), 46

grey-banded kingsnake *see Lampropeltis alterna*

gulf flounder *see Paralichthys albigutta*

gulls *see Laridae*

gundis *see Ctenodactylidae*

guppy *see Poecilia reticulata*

Gymnothorax meleagris (Actinopterygii: Muraenidae), 250

Haekelia rubra (Ctenophora), 147

Haematopus moquini (Aves: Haematopodidae), 60

Habropoda pallida (Hymenoptera: Apidae), 255

Haemulon chrysargyreum (Actinopterygii: Haemulidae), 286

hairy woodpecker *see Picoides villosus*

Hakea trifurcata (Plantae: Proteaceae), 294

Halimeda discoidea (Plantae: Chlorophyta [green alga]: Halimedaceae), 22

hamadryas baboon *see Papio hamadryas*

Haplochromis wingatii (Actinopterygii: Cichlidae), 360

Harmonia axyridis (Coleoptera: Coccinellidae), 105, 119, 153

Harmonia dimidiata (Coleoptera: Coccinellidae), 105
Harpagus diodon (Aves: Accipitridae), 320
Harris's checkerspot butterfly *see Chlosyne harrisii*
Harris sparrow *see Zonotrichia querula*
harvestmen *see Parampheres ronae*
Hasarius adansoni (Arachnida: Araneae: Salticidae), 99
'hawks'
  as mimics (see also *Buteo*), 319–20
  as models (see also *Cathartes*), 7, 26, 88, 192, 211, 227, 239, 295
Hebomoia glaucippe (Lepidoptera: Pieridae), 260
Hedychrum rutilans (Hymenoptera: Chrysididae), 344–5
Heikea see Heikeopsis
Heikeopsis japonica (Crustacea: Malacostraca: Dorippidae), 257
Heliamphora heterodoxa (Plantae: Sarraceniaceae), 351
H. tatei (Plantae: Sarraceniaceae), 351
Helianthus annuus (Plantae: Asteraceae), 125
Helicobacter pylori (Bacteria: Helicobacteraceae), 425
Helicodiceros muscivorus (Plantae: Araceae), 390
Heliconius clysonymus (Lepidoptera: Nymphalidae), 300
H. cydno (Lepidoptera: Nymphalidae), 196, 219
H. dorsis (Lepidoptera: Nymphalidae), 163
H. eleuchia (Lepidoptera: Nymphalidae), 196
H. erato (Lepidoptera: Nymphalidae), 124, 161–2, 204, 216–20, 223
H. cthilla (Lepidoptera: Nymphalidae), 125, 163
H. melpomene (Lepidoptera: Nymphalidae), 124, 162–3, 216–20, 222–3
H. numata (Lepidoptera: Nymphalidae), 124, 162, 216–20, 223
H. sapho (Lepidoptera: Nymphalidae), 196, 219
H. sara (Lepidoptera: Nymphalidae), 124, 163
Heliotropium (Plantae: Boraginaceae), 144
Hemaris tityus (Lepidoptera: Sphingidae), 46
Hemeroplanes (Lepidoptera: Sphingidae), 271, 272
Hemibungarus calligaster (Reptilia: Squamata: Elapidae), 244
Hemipepsis hilaris (Hymenoptera: Pompilidae), 381
hemlock *see Conium maculatum*

hemp dogbane *see Apocynum cannabinum*
heron *see Ardeidae*
Herse convolvuli (Lepidoptera: Sphingidae), 27
Heterochroma sarepta (Lepidoptera: Erebidae), 141
Heterodon platyrhinos (Reptilia: Squamata: Colubridae), 99
Heteropoda venatoria (Arachnida: Araneae: Sparassidae), 312
Heterorhabditis bacteriophora (Nematoda: Heterorhabditidae), 161
Hierococcyx varius (Aves: Cuculidae), 335, 341
Himacerus mirmicoides (Hemiptera: Nabidae), 232
Himantolophus groenlandicus (Actinopterygii: Himantolophidae), 315
Himantopus himantopus (Aves: Recurvirostridae), 277
Hippocampus (Actinopterygii: Syngnathidae), 21
Hippolyte varians (Crustacea: Malacostraca: Hippolytidae), 56
Hippopodius (Cnidaria: Hippopodiidae), 155
Hippotragus niger (Mammalia: Bovidae), 408
Hirundinea ferruginea (Aves: Tyrannidae), 196
Hirundo rustica (Aves: Hirundinidae), 342, 365
Historis (Lepidoptera: Nymphalidae), 89
hoatzin *see Opisthocomus hoazin*
hognosed snake *see Heterodon platyrhinos*
Homo erectus† (Mammalia: Hominidae), 406
Homodes (Lepidoptera: Erebidae), 237–8
honey badger *see Mellivora capensis*
honey bee *see Apis mellifera*
honeyguides *see Indicator indicator*
Hoodia gordonii (Plantae: Apocynaceae), 426
Hormathophylla spinosa (Plantae: Brassicaceae), 165
Horsfield's bronze-cuckoo *see Chalcites basalis*
Horsfield's hawk-cuckoo *see Cuculus fugax*
house cricket *see Acheta domesticus*
house sparrow *see Passer domesticus*
hoverflies (Diptera: Syrphidae), 16, 207–11, 227, 229, 251, 313, 323, 346–7, 373, 382, 389, 391, 436
huckleberry *see Gaylussacia*
huntsman spider *see Heteropoda venatoria*
Hyaena hyaena (Mammalia: Hyaenidae), 240

Hyalophora promethea see Callosamia promethean
Hyastenus elatus (Crustacea: Malocostraca: Epialtidae), 83
Hybognathus nuchalis (Actinopterygii: Cyprinidae), 286–7
Hydrophyllum virginianum (Plantae: Boraginaceae), 297
Hyla (Amphibia: Anura: Hylidae), 192, 309
H. cinerea (Amphibia: Anura: Hylidae), 192
Hylobittacus apicalis (Mecoptera: Bittacidae), 357
Hymenopus coronatus (Mantodea: Hymenopodidae), 312, 314
Hyperaspis trifurcata (Coleoptera: Coccinellidae), 147
Hyperechia (Diptera: Asilidae), 253
Hypna (Lepidoptera: Nymphalidae), 151
Hypolimnas bolina (Lepidoptera: Nymphalidae), 220–1, 248
H. misippus (Lepidoptera: Nymphalidae), 220–1
Hypomecis roboraria (Lepidoptera: Geometridae), 43
Hypoplectrus (Actinopterygii: Serranidae), 323, 325
H. indigo (Actinopterygii: Serranidae), 323–4, 329–31, 338–40
H. unicolor (Actinopterygii: Serranidae), 323
Hypsa monycha see Asota monycha
Hypsiglena torquata (Reptilia: Squamata: Colubridae), 279
Hypsipetes amaurotis (Aves: Pycnonotidae), 224
Hystrix cristata (Mammalia: Hystricidae), 158

Idolum see Idolomantis
Idolomantis diabolicum (Mantodea: Empusidae), 7, 312
Idotea balthica (Crustacea: Isopoda), 61, 77
impala *see Aepyrocerus melampus*
Inachis io (Lepidoptera: Nymphalidae), 35, 261, 283
Indian leaf butterfly *see Kallima inachus*
Indicator indicator (Aves: Indicatoridae), 330, 332, 338
indigobird *see Vidua*
Iphiclides podalirius (Lepidoptera: Papilionidae), 34–5
Ipomoea rosea (Plantae: Convolvulaceae), 27
Iris atropurpurea (Plantae: Iridaceae), 392
Ischnochiton striolatus (Mollusca: Neoloricata: Ischnochitonidae), 77

Ischnocnema guentheri (Amphibia: anura: Brachycephalidae), 98

Ischnura (Odonata: Coenagrionidae), 358

Isopoda (Crustaceae: Malacostraca), 46, 61, 77, 312, 320, 354–5

Ithomia (Lepidoptera: Nymphalidae), 121, 125

Ithomiini (Lepidoptera: Nymphalidae: Danainae), 45, 72, 124–5, 135, 138–9, 144, 162, 199, 204, 215–16

Ithyphallus coralloides (Fungi: Phallaceae), 419

Ityraea gregoryi (Hemiptera: Flatidae), 255

jacamar, 138, 177, 215, 218

Jacksonoides queenslandicus (Arachnidae: Araneae: Salticidae), 308

Jankowskia fuscaria (Lepidoptera: Geometridae), 43

Japanese quail *see Coturnix japonica*

Japetella heathi (Mollusca: Cephalopoda: Bolitaenidae), 45

jewel beetles *see* Buprestidae

Jorunna (Mollusca: Discodorididae), 22

Juglans regia (Plantae: Juglandaceae), 414

junglefowl *see Gallus gallus*

Junonia almana (Lepidoptera: Nymphalidae: Nymphalinae), 138, 261, 267

J. coenia (Lepidoptera: Nymphalidae: Nymphalinae), 150

J. terea (Lepidoptera: Nymphalidae: Nymphalinae), 137

Jynx (Aves: Picidae), 88, 156

Kallima inachus (Nymphalidae), 89–90

Kapala sulcifacies (Hymenoptera: Eucharitidae), 239

katydids, 25–6, 89–90, 327

kelp crabs *see Pugettia* and *Mimulus*

Kermadec petrel *see Pterodroma neglecta*

kestrel *see Falco tinnunculus*

Kiefferia pericarpiicola (Diptera: Cecidomyiidae), 299

Kikihia scutellaris (Hemiptera: Cicadidae), 327

killdeer *see Charadrius vociferus*

kingbirds *see Tyrannus melancholicus*

kingfishers (Aves: Alcedinidae), 124

Kinyongia (Reptilia: Squamata: Chamaeleonidae), 25

Knautia (Plantae: Dipsacaceae), 386

Kniphofia uvaria (Plantae: Xanthorrhoeaceae), 384–5

Koh-i-Noor butterfly *see Amathuxidia amythaon*

koho salmon *see Oncorhynchus kisutch*

Korthasella salicornioides (Plantae: Viscaceae), 293

Kryptopterus bicirrhis (Actinopterygii: Siluridae), 45

kudu *see Tragelephus*

Labroides dimidiatus (Actinopterygii: Labridae), 320–2

Lacinipolia renigera (Lepidoptera: Noctuidae), 326

ladybirds *see* Coccinellidae

Lagopus muta (Aves: Phasianidae), 36

Lamium album (Plantae: Lamiaceae), 295

Lamprologus lemairii (Actinopterygii: Cichlidae), 318

Lampropeltis alterna (Reptilia: Squamata: Colubridae), 198

L. elapsoides (Reptilia: Squamata: Colubridae), 240, 242

L. triangulum (Reptilia: Squamata: Colubridae), 198, 240–3

Lampsilis ovata (Mollusca: Unionidae), 416

Lampyris noctiluca (Coleoptera: Lampyridae), 155

Lanio versicolor (Aves: Thraupidae), 406

Laniocera hypopyrra (Aves: Tityridae), 246

Lanius collaris (Aves: Laniidae), 57

L. collurio (Aves: Laniidae), 335

L. excubitor (Aves: Laniidae), 324

Lantana camara (Plantae: Verbenaceae), 382–3, 385

Laothoe populi (Lepidoptera: Sphingidae), 50, 51

largemouth bass *see Micropterus salmoides*

Laridae (Aves), 315

Larinioides cornutus (Arachnida: Araneae: Araneidae), 99

Larus ridibundus see Chroicocephalus ridibundus

Lasius fuliginosus (Hymenoptera: Formicidae), 349

L. niger (Hymenoptera: Formicidae), 233, 350

L. sakagamii (Hymenoptera: Formicidae), 233

Lathyrus vernus (Plantae: Fabaceae), 386

Laticauda colubrina, (Reptilia: Squamata: Elapidae), 239

leaf beetles *see* Chrysomelidae

leafcutter ant *see Atta cephalotis*

leaffish *see Monocirrhus polyacanthus*

leafy sea dragon *see Phyllopteryx eques*

Lebia (Coleoptera: Carabidae), 150

Lebistes (Coleoptera: Carabidae), 150

Leistotrophus versicolor (Coleoptera: Staphylinidae), 318, 354–5

lemmings *see Lemmus* spp

Lemmus lemmus (Mammalia: Cricetidae), 157

L. trimucronatus (Mammalia: Cricetidae), 157

Lens esculenta (Plantae: Fabaceae), 415

Lentinellus ursinus (Fungi: Basidiomycota: Auriscalpiaceae), 166

leopard frog *see Rana pipiens*

Lepanthes (Plantae: Orchidaceae), 373

Lepomis macrochirus (Actinopterygii: Centrarchidae), 354

Leptailurus serval (Mammalia: Felidae), 276

Leptasterias (Echinodermata: Asteroidea: Asteriidae), 73

Leptinotarsa (Coleoptera: Chrysomelidae), 146

Leptogenys (Hymenoptera: Formicidae), 347

Leptopelis rufus (Amphibia: Anura: Arthroleptidae), 99

Leptospermum scoparium (Plantae: Myrtaceae), 293

Lepus americanus (Mammalia: Leporidae), 37

Lepus arctias (Mammalia: Leporidae), 35

Lethe confusa (Lepidoptera: Nymphalidae), 136

Leucauge magnifica (Arachnida: Araneae: Theriidae), 311–13

Leucanella ?hosmera (Lepidoptera: Saturniidae), 159

Leucochloridium macrostomum (Platyhelminthes: Trematoda: Leucochloridiidae), 16

Leucochloridium paradoxum (Platyhelminthes: Trematoda: Leucochloridiidae), 416

Leucorampha see Hemeroplanes

Libinia dubia (Crustacea: Malacostraca: Pisidae), 83, 147

Lichenomorphus ?carlosmendesi (Orthoptera: Tettigoniidae), 28

Limenitis archippus (Lepidoptera: Nymphalidae), 3, 150

Linum usitatissimum (Plantae: Linaceae), 415

lionfish *see Pterois*

Lissodelphis borealis (Mammalia: Delphinidae), 311

Lissopimpla (Hymenoptera: Ichneumonidae), 376–7, 380–1

Lithops (Plantae: Aizoaceae), 15, 290

Littorina mariae (Mollusca: Gastropoda: Littorinidae), 94

liver fluke *see Fasciola hepatica*

lizards, mimicry by *see Eremias*

Lobelia cardinalis (Plantae: Campanulaceae), 397

lobster moth *see Stauropus*

Lobobunaea acetes (Lepidoptera:
  Saturniidae), 261
Loligo pealeii (Mollusca: Cephalopoda:
  Loliginidae), 54
L. vulgaris (Mollusca: Cephalopoda:
  Loliginidae), 59
long-tailed tit *see Aegithalos caudatus*
long-tongued flies *see* Nemestrinidae,
  Tabanidae
Lopinga achine (Lepidoptera:
  Nymphalidae), 267
Lotus corniculatus (Plantae: Fabaceae),
  145, 295
lubber grasshopper *see Romalea guttata*
Lucihormetica (Blattodea: Blaberidae), 156
luna moth *see Actias luna*
lupin *see Lupinus*
Lupinus (Plantae: Fabaceae), 231
Lutjanus kasmira (Actinopterygii:
  Lutjanidae), 285
L. lutjanus (Actinopterygii:
  Lutjanidae), 285
Lycaenidae (Lepidoptera), 32, 49, 239,
  271, 273–5, 284, 346, 447
Lycodon carinatus (Reptilia: Squamata:
  Colubridae), 244
L. ophiophagus (Reptilia: Squamata:
  Colubridae), 107
Lycus (Coleoptera: Lycidae), 120
Lygaeus (Hemiptera: Lygaeidae), 128, 210
L. equestris (Hemiptera: Lygaeidae), 128
Lysiphlebus cardui (Hymenoptera:
  Braconidae), 350
Lytta vesicatoria (Coleoptera: Meloidae),
  108, 146

Macrobdella diplotertia (Annelida:
  Hirudinea), 248
Macrocilix maia (Lepidoptera:
  Geometridae), 92
Macroclemys temminckii (Reptilia:
  Chelydridae), 315, 317
Macrocyclops albidus (Crustaceae:
  Copepoda: Cyclopidae), 423
Macrocystis pyrifera (Phaeophyceae
  [brown algae]: Laminariaceae), 53
Macropisthodon plumbicolor (Reptilia:
  Squamata: Colubridae), 250
Macrotritopus defilippi (Mollusca:
  Cephalopoda), 151
Macroxiphus (Orthoptera:
  Tettigoniidae), 232
Maculinea ribeli see Phengaris ribeli
magpie moths *see Abraxas*
Maja squinado (Crustaceae: Malacostraca:
  Majidae), 83
M. varrucosa (Crustaceae: Malacostraca:
  Majidae), 83

Majidea zangueberica (Plantae:
  Sapindaceae), 413
Malacostraca (Crustacea), 54
Malayan krait *see Bungarus candidus*
Malayatelura ponerophila (Hexapoda:
  Thysanura), 347
Mallada desjardinsi (Neuroptera:
  Chrysopidae), 85
Mallee dragon lizard *see Ctenophorus fordi*
Mallophora bomboides (Diptera:
  Asilidae), 192
Malpolon moilensis see Rhagerhis moilensis
Malurus cyaneus (Aves: Maluridae), 339
Malvaceae (Plantae), 296, 300,
  368, 386
mandrill *see Mandrillus sphinx*
Mandrillus sphinx (Mammalia:
  Cercopithecidae), 243, 406, 444
Manduca florestan (Lepidoptera:
  Sphingidae), 142
mangrove saltmarsh watersnake *see
  Nerodia clarkii compressicauda*
Mantellidae (Amphibia: Anura), 147
Mantispidae (Neuroptera), 220
Maoricicada campbelli (Hemiptera:
  Cicadidae), 327
Markia hystrix (Orthoptera:
  Tettigoniidae), 26
massasauga *see Sistrurus catenatus*
Mastophora cornigera (Arachnida:
  Araneae: Araneidae), 326
M. hutchinsoni (Arachnida: Araneae:
  Araneidae), 326
Matricaria chamomilla (Plantae:
  Asteraceae), 295
mayapple *see Podophyllum peltatum*
meadow pipit *see Anthus pratensis*
Mecaenichthys immaculatus
  (Actinopterygii: Pomacentridae), 62
Mechanitis polymnia (Lepidoptera:
  Nymphalidae), 73, 125
Medicago sativa (Plantae: Fabaceae), 427
meerkat *see Suricata suricata*
Megacyllene robiniae (Coleoptera:
  Cerambycidae), 228
Megapalpus nitidus (Diptera:
  Bombyliidae), 374
Meiacanthus atrodorsalis (Actinopterygii:
  Blenniidae), 323
M. nigrolineatus (Actinopterygii:
  Blenniidae), 323
Meles meles (Mammalia: Mustelidae), 158
Melinaea (Lepidoptera: Ithomidae), 139,
  162, 217–18
Mellivora capensis (Mammalia:
  Mustelidae), 158, 240
Meloidae (Coleoptera), 105, 146–7, 256,
  431, 443

Meloe franciscanus (Coleoptera:
  Meloidae), 255
Melospiza melodia (Aves: Melospizidae),
  339, 402
Memphis (Lepidoptera: Nymphalidae), 89
Meneris tulbagha see Aeropetes tulbagha
Menura novaehollandiae (Aves:
  Menuridae), 402
Mephitis mephitis (Mammalia:
  Mephitidae), 76, 158, 160
Meris alticola (Lepidoptera:
  Geometridae), 224
Metallyticus splendidus (Mantodea:
  Metallyticidae), 151
Metriorrhynchus rhipidius (Coleoptera:
  Lycidae), 120
Meris alticola (Lepidoptera:
  Geometridae), 224
mgambo tree *see Majidea zangueberica*
Mickey-mouse plant *see Ochna integerrima*
Microbotryum violaceum (Fungi:
  Basidiomycetes: Microbotryaceae), 418
Microdon albicomatus (Diptera:
  Syrphidae), 347
M. piperi (Diptera: Syrphidae), 347
Microhylidae (Amphibia: Anura), 342
Micropterus salmoides (Actinopterygii:
  Centrarchidae), 286–7, 417
M. dolomieu (Actinopterygii:
  Centrarchidae), 417
Micruroides euryxanthus (Reptilia:
  Squamata: Elapidae), 240
Micrurus alleni (Reptilia: Squamata:
  Elapidae), 243
Micrurus diastema (Reptilia: Squamata:
  Elapidae), 241
Micrurus frontalis (Reptilia: Squamata:
  Elapidae), 242
Micrurus fulvius (Reptilia: Squamata:
  Elapidae), 198, 240–2
Micrurus multifasciatus (Reptilia:
  Squamata: Elapidae), 243
Micrurus nigrocinctus (Reptilia: Squamata:
  Elapidae), 243
Micrurus pyrrhocryptus (Reptilia:
  Squamata: Elapidae), 241
Milichidae (Diptera), 389
milk thistle *see Silybum marianum*
milkweed butterflies *see* Daniinae,
  Danaus spp.
millipedes, 105, 155, 242, 244
  see also *Orthomorpha*
Mimetica (Orthoptera: Tettigoniidae), 89
Mimetus (Arachnida: Araneae:
  Mimetidae), 319
Mimidae (Aves), 367
Mimoides pausanias (Lepidoptera:
  Papilionidae), 218

Mimophis (Reptilia: Squamata: Lamprophiidae), 97

Mimulus foliata (Crustacea: Malacostraca: Epialtidae), 85

Mimulus moschatus (Plantae: Phrymaceae), 368

Mimus polyglottos (Aves: Mimidae), 402

Mischocyttarus mastigophorus (Hymenoptera: Vespidae), 229

Missulena occatoria (Arachnida: Araneae: Theraphosodae), 8

Misumena vatia (Arachnida: Araneae: Thomisidae), 54–5

Misumenoides formosipes (Arachnida: Araneae: Thomisidae), 56

Misumenops asperatus (Arachnida: Araneae: Thomisidae), 56

mocker swallowtail *see Papilio dardanus*

mockingbird *see Mimus polyglottos*

mole *see Talpa europaea*

Molossidae (Mammalia: Chiroptera), 257

Molothrus ater (Aves: Icteridae), 336, 339

M. bonariensis (Aves: Icteridae), 334, 338

monarch butterflies *see Daniinae, Danaus* spp

Moneilema appressum (Coleoptera: Cerambycidae), 193

Monilinia (Fungi: Sclerotiniaceae), 417–19

monkey orange tree *see Strychnos madagascariensis*

monkeys, 2, 12, 130–1, 307, 403–4, 406

Monocirrhus polyacanthus (Actinopterygii: Polycentridae), 307

Moorish gecko *see Tarentola mauritanica*

Moraceae (Plantae), 297

moray eel *see Gymnothorax meleagris*

Morelia viridis (Reptilia: Squamata: Pythonidae), 82

Mormolyca ringens (Plantae: Orchidaceae), 379

Morpho (Lepidoptera: Nymphalidae: Morphinae), 35, 151

Morus (Plantae: Moraceae), 297

moss *see Splachnaceae*

Motacilla alba (Aves: Motacillidae), 335

motmots *see Eumomota superciliosa*

Mulloidichthys martinicus (Actinopterygii: Mullidae), 286

M. vanicolensis (Actinopterygii: Mullidae), 285

mummy-berry *see Monilinia*

Muraenidae (Actinopterygii), 250

Musa (Plantae: Musaceae), 396–7

Musca domestica (Diptera: Muscidae), 252

muskflower *see Mimulus moschatus*

muskwood *see Olearia argophylla*

Mustela erminea (Mammalia: Mustelidae), 36, 276

Mutilla europaea (Hymenoptera: Mutillidae), 344

Mutillidae (Hymenoptera), 250, 344

Mydas heros (Diptera: Mydaidae), 250

Myja longicornis (Mollusca: Nudibranchia: Tergipedidae), 22

Mylabris (Coleoptera: Meloidae), 105

Myobatrachidae (Amphibia: Anura), 147

Myrichthys colubrinus (Actinopterygii: Ophichthidae), 239

Myrmarachne (Araneae: Salticidae), 211, 233–7, 308, 328

M. melanotarsa (Araneae: Salticidae), 328

M. plataleoides (Araneae: Salticidae), 234

Myrmecaphodius excavaticollis (Coleoptera: Aphodidae), 349

Myrmecopsis (Lepidoptera: Erebidae: Arctiinae), 226–8

Myrmica incompleta (Hymenoptera: Formicidae), 347

Mystus leucophasis (Actinopterygii: Bagridae), 24

Myzinum quinquecinctum (Hymenoptera: Tiphiidae), 230

Myzus persicae (Hemiptera: Aphidae), 300

Nanger granti (Mammalia: Bovidae), 407–8

Nasalis larvatus (Mammalia: Cercopithecidae), 406

Natrix natrix (Reptilia: Squamata: Colubridae), 97

N. sipedon (Reptilia: Squamata: Colubridae), 71

N. tessellata (Reptilia: Squamata: Colubridae), 97–8

Nectariniidae (Aves), 394

Nehalennia irene (Odonata: Coenagrionidae), 358

Nemertea, 122

Nemestrinidae (Diptera), 374, 385, 387

Nemoria arizonica (Lepidoptera: Geometridae), 50

Neochmia (Bathilda) ruficauda (Aves: Estrilidae), 150

Neodiprion sertifer (Hymenoptera: Tenthredinidae), 132, 167

Neogea (Araneae: Araneidae), 93

Neolamprologus sexfasciatus (Actinopterygii: Cichlidae), 322

Neoterpes graefiaria (Lepidoptera: Geometridae), 224

Neozeloboria (Hymenoptera: Tiphiidae), 378–9, 381

Nepenthes rafflesiana (Plantae: Nepenthidae), 351

Nephila clavipes (Arachnida: Araneae: Nephilidae), 124–5, *see also* 311

Neritina communis (Mollusca: Neritidae), 77

Nerium (Plantae: Apocynaceae), 142, 200

Nerodia clarkii compressicauda (Reptilia: Squamata: Colubridae), 317–18

Neurergus kaiseri (Amphibia: Urodela: Salamandridae), 98

Nicrophorus (Coleoptera: Silphidae), 105, 157, 250

N. investigator (Coleoptera: Silphidae), 157

Nomada (Hymenoptera: Apidae), 55, 346

northern harrier *see Circus cyaneus*

northern mountain swordtail *see Xiphophorus nezahualcoyotl*

northern shrike *see Lanius excubitor*

Norwegian lemming *see Lemmus lemmus*

Notoacmea paleacea (Mollusca: Gastropoda: Lottiidae), 73

Notophthalmus viridescens (Amphibia: Urodela), 248

Notothlaspi rosulatum (Plantae: Brassicaceae), 290

Nucella lamellosa (Mollusca: Muricidae), 77

Nyctibius (Aves: Nyctibiidae), 23, 88

N. jamaicensis (Aves: Nyctibiidae), 23

nymphalid/Nymphalidae (Lepidoptera), 2–3, 32, 35, 45–6, 53, 68, 73, 89–90, 124, 127–8, 135–8, 150–1, 160, 192, 199–202, 220, 248, 260–2, 266–7, 277, 279, 283–4, 384–5

Ochna integerrima (Plantae: Ochnaceae), 412

Octomeria (Plantae: Orchidaceae), 389

octopus stinkhorn *see Clathrus archeri*

Ocypode ceratophthalmus (Crustacea: Malacostraca: Ocypodidae), 54

Ocyptamus (Diptera: Syrphidae), 313, 315

Odontonema strictum (Plantae: Acanthaceae), 396

Oecophylla (Hymenoptera: Formicidae), 237

O. smaragdina (Hymenoptera: Formicidae), 231, 233–4, 236

Oecodoma see Atta

Oedemera femorata (Coleoptera: Oedemeridae), 147

Oedemeridae (Coleoptera), 108, 146–7, 431

Ogcocephalus vespertilio (Actinopterygii: Ogcocephalidae), 315

Olea europaea (Plantae: Oleaceae), 414

Olearia argophylla (Plantae: Asteraceae), 368

olive *see Olea europaea*

Ommatoptera pictifolia (Orthoptera: Tettigoniidae), 282

Onchocerca lienalis (Nematoda: Onchocercidae), 423–4

Onchocerca volvulus (Nematoda: Onchocercidae), 422, 424

Oncidiinae (Plantae: Orchidaceae), 382

Oncidium (Plantae: Orchidaceae), 13, 373, 382, 385

Oncidium cosymbephorum (Plantae: Orchidaceae), 385

Oncidium lucayanum (Plantae: Orchidaceae), 373

Oncidium nebulosum (Plantae: Orchidaceae), 382

Oncorhynchus kisutch (Actinopterygii: Salmonidae), 45

O. masou (Actinopterygii: Salmonidae), 356

Onopordum (Plantae: Asteraceae), 299

Oophaga pumilio (Amphibia: Anura: Dendrobatidae), 119, 148, 153, 163

Ophiopsila riisei (Echinodermata: Ophiurida: Ophiocomidae), 155

Ophrys (Plantae: Orchidaceae), 16, 375–7, 379–81, 384, 385

O. apifera (Plantae: Orchidaceae), 376–7

O. bilunulata (Plantae: Orchidaceae), 375

O. bombyliflora (Plantae: Orchidaceae), 376

O. cephalonica (Plantae: Orchidaceae), 380

O. exaltata (Plantae: Orchidaceae) 377, 380

O. fabrella (Plantae: Orchidaceae), 375

O. fusca (Plantae: Orchidaceae), 379

O. fusca lutea (Plantae: Orchidaceae), 379

O. heldreichii (Plantae: Orchidaceae), 380

O. insectifera (Plantae: Orchidaceae), 376, 377, 379

O. iricolor (Plantae: Orchidaceae), 379

O. lupercalis (Plantae: Orchidaceae), 375

O. lutea (Plantae: Orchidaceae), 379

O. speculum (Plantae: Orchidaceae), 379

O. sphegodes (Plantae: Orchidaceae), 376, 379, 381

Opisthocomus hoazin (Aves: Opisthocomidae), 124

Opisthograptis luteolata (Lepidoptera: Geometridae), 94–5

orange-spotted tiger clearwing *see Mechanitis polymnia*

Orasema (Hymenoptera: Eucharitidae), 239

Orchesticus abeille (Avies: Thraupidae), 287

Orchidaceae, 372–85, 387–9, 391, 394, 440

orchid mantis *see Hymenopus coronatus*

orchid spider *see Leucage magnifica*

orchids *see* Orchidaceae

Orchis (Plantae: Orchidaceae), 372, 385

O. israelitica see Anacamptis israelitica

O. morio (Plantae: Orchidaceae), 372

Ordgarius (Arachnida: Araneae: Araneidae), 326

Oreina (Coleoptera: Chrysomelidae), 145

Oregonia gracilis (Crustacea: Oregoniidae), 84

Oreochromis (Actinopterygii: Cichlidae), 360–1

oriole *see Oriolus*

Oriolus (Aves: Oreolidae), 401

Ormosia (Plantae: Fabaceae), 414

O. isthamensis (Plantae: Fabaceae), 414

O. macrocalyx (Plantae: Fabaceae), 414

Ornithoscatoides see Phrynarachne

Orophus tesselatus (Orthoptera: Tettigoniidae), 89

Oroplema plagifera (Lepidoptera: Uraniidae), 92

Orthomorpha (Myriapodae: Paradoxosomatidae), 105

oryx and *Oryx* (Mammalia: Bovidae), 407–8

Oryza punctata (Plantae: Poaceae), 415

Oryza rufipogon (Plantae: Poaceae), 415

Osmunda japonica (Plantae: Osmundaceae), 299

owl *see Glaucidium, Tyto*

owl butterfly *see Caligo*

Ozark Highlands leech *see Macrobdella diplotertia*

Pachliopta aristolochiae (Lepidoptera: Papilionidae), 200, 203, 214, 224

Pachylomerus (Coleoptera: Scarabeidae), 419

Pachyramphus castaneus (Aves: Tityridae), 287

Paeonia clusii (Plantae: Paeoniaceae), 411

Palaemonetes pugio (Crustacea: Malacostraca: Palaemonidae), 47

pallid cuckoo *see Cuculus pallidus*

Paltothyreus tarsatus (Hymenoptera: Formicidae), 342

panic moth *see Saucrobotys futilalis*

Panthera (Mammalia: Felidae), 323

papaya *see Carica papaya*

paper wasp *see Polistes, Sulcopolistes*

Paphiopedilum barbigerum (Plantae: Orchidaceae), 382

P. dianthum (Plantae: Orchidaceae), 382

Papilio aegeus (Lepidoptera: Papilionidae), 168, 202

P. canadensis (Lepidoptera: Papilionidae), 168, 223, 269

P. cresphontes (Lepidoptera: Papilionidae), 89, 91, 168

P. dardanus (Lepidoptera: Papilionidae), 168, 184–5, 199, 202–3, 220–3, 358

P. demodocus (Lepidoptera: Papilionidae), 33, 168

P. glaucus (Lepidoptera: Papilionidae), 82, 168, 201–2, 223

P. machaon (Lepidoptera: Papilionidae), 32–3, 72, 122, 132, 168, 224–5, 284

P. memnon (Lepidoptera: Papilionidae), 168, 223

P. palimedes (Lepidoptera: Papilionidae), 89

P. paradoxa (Lepidoptera: Papilionidae), 199–200

P. polymnestor (Lepidoptera: Papilionidae), 89

P. polytes (Lepidoptera: Papilionidae), 32, 34, 168, 203, 215, 223

P. polyxenes (Lepidoptera: Papilionidae), 224–5

P. troilus (Lepidoptera: Papilionidae), 168

P. xuthus (Lepidoptera: Papilionidae), 215

P. zelicaon (Lepidoptera: Papilionidae), 34, 168

Papio hamadryas (Mammalia: Cercopithecidae), 403, 444

Parablennius sanguinolentus (Actinopterygii: Blenniidae), 356

Paracerceis sculpta (Crustacea: Isopoda), 157

Parachartergus (Hymenoptera: Vespidae), 227–8

Paracheirodon innesi (Actinopterygii: Characidae), 54

Parachromis friedrichsthalii (Actinopterygii: Cichlidae), 318

Paracletus cimiciformis (Hemiptera: Aphididae), 349

Parage xiphia (Lepidoptera: Nymphalidae), 53

Paralichthys albigutta (Actinopterygii: Paralichthyidae), 56

Paraluteres prionurus (Actinopterygii: Monacanthidae), 197

Parampheres ronae (Arachnida: Opiliones), 250

Paraplectana duodecimmaculata (Araneae: Araneidae), 153

Parasemia plantaginis (Lepidoptera: Erebidae: Arctiinae), 53, 162, 201

Parazoanthus (Cnidaria: Anthozoa: Parazoantidae), 159

Parides anchises (Lepidoptera: Papilionidae), 194

P. arcas (Lepidoptera: Papilionidae), 194

P. erithalion (Lepidoptera: Papilionidae), 194

P. sesostris (Lepidoptera: Papilionidae), 194

Parnassia palustris (Plantae: Celastraceae), 394

Parnassius (Lepidoptera: Papilionidae), 291

Parus atricapillus (Aves: Paridae), 125–6

P. major (Aves: Paridae), 27, 31, 106, 109, 112, 128–9, 132–3, 153, 196, 205, 210, 224, 248, 267, 284, 402, 408

Paspalum paspaloides (Plantae: Poaceae), 300

Passer domesticus (Aves: Passeridae), 126, 131, 135

Passiflora (Plantae: Passifloraceae), 294, 299–300, 434

P. biflora (Plantae: Passifloraceae), 300

P. candollei (Plantae: Passifloraceae), 299

P. laurifoliae (Plantae: Passifloraceae), 300

P. oerstedii (Plantae: Passifloraceae), 299

Pauropsalta (Hemiptera: Cicadidae), 327

pea *see Pisum sativum*

peach blossom moth *see Thyatira batis*

peacock butterfly *see Inachis io*

peanut tree *see Sterculia quadrifida*

Peckhamia picata (Arachnida: Araneae: Salticidae), 235

Pedicularis (Plantae: Orobanchaceae), 385

Pelamis platura (Reptilia: Squamata: Elapidae), 130

Pelargonium stipulaceum (Plantae: Geraniaceae), 385–6

Pella (Coleoptera: Staphylinidae), 349

Pellionia repens (Plantae: Urticaceae), 397

penguins *see Spheniscus*

Pennaria disticha (Cnidaria: Hydrozoa), 22

Penstemon (Plantae: Plantaginaceae), 224

Pentadiplandra brazzeana (Plantae: Pentadiplandraceae), 425

peppered moth *see Biston betularia*

Pepsis (Hymenoptera: Pompilidae), 225, 228, 250

Perca flavescens (Actinopterygii: Percidae), 417

Perigonia lusca (Lepidoptera: Sphingidae), 142

Perilitus dubius (Hymenoptera: Braconidae), 326

Peromyscus floridanus (Mammalia: Cricetidae), 250

P. leucopus (Mammalia: Cricetidae), 137

P. polionotus (Mammalia: Cricetidae), 37

Perrhybris pyrrha (Lepidoptera: Pieridae), 199

Pertyia sericea (Coleoptera: Cerambycidae), 233

Peucetia viridans (Arachnida: Araneae: Oxyopidae), 306

Phacochoerus aethiopicus (Mammalia: Suidae), 408

Phalera bucephala (Lepidoptera: Notodontidae), 167

Phallus impudicus (Fungi: Phallaceae), 419

P. indusiatus (Fungi: Phallaceae), 419

Phasmatodea (Insecta), 25, 282, 415

Phauda (Lepidoptera: Zygaenidae), 91

Pheloura (Hymenoptera: Braconidae), 231

Phengaris ribeli (Lepidoptera: Lycaenidae), 237, 239

Phiale formosa (Arachnida: Araneae: Salticidae), 252

Phidippus appacheanus (Arachnida: Araneae: Salticidae), 252

P. pulcherrimus (Arachnida: Araneae: Salticidae), 275

Phigalia pilosaria (Lepidoptera: Geometridae), 42

Phigalia titea (Lepidoptera: Geometridae), 38

Philaenus spumarius (Hemiptera: Cercopidae), 40, 151

Philanthus triangulum (Hymenoptera: Sphecidae), 344–5

Philemon (Aves: Meliphagidae), 401

Philoliche aethiopica (Diptera: Tabanidae), 384–5

P. gulosa (Diptera: Tabanidae), 385–6

Philydor rufus (Aves: Fumariidae), 286

Phoebis sennae (Lepidoptera: Pieridae), 141

Phoenicurus phoenicurus (Aves: Muscicapidae), 337

Pholcus phalangioides (Arachnida: Araneae: Pholcidae), 319

Phoridae (Diptera), 318

Photeros annecohenae (Crustacea: Ostracoda), 364

Photinus (Coleoptera: Lampyridae), 194–6, 325, 363, 438

P. pyralis (Coleoptera: Lampyridae), 194, 196

Photuris (Coleoptera: Lampyridae), 16, 325, 363, 438

P. trivittata (Coleoptera: Lampyridae), 325

Phrynarachne (Arachnida: Araneae: Thomisidae), 91–2

P. ceylonica (Arachnida: Araneae: Thomisidae), 91–2

P. rothschildi (Arachnida: Araneae: Thomisidae), 92

Phrynomantis microps (Amphibia: Anura: Microhylidae), 342

Phyllium giganteum (Phasmatodea: Phyllidae), 25

Phyllobates (Amphibia: Anura: Dendrobatidae), 124, 147

P. terribilis (Amphibia: Anura: Dendrobatidae), 147

Phyllopteryx eques (Actinopterygii: Syngnathidae), 23

Phylloscopus humei (Aves: Phylloscopidae), 329

P. sibilatrix (Aves: Phylloscopidae), 156, 365

Phyllospadix (Plantae: Zosteraceae), 73

Phymateus saxosus (Orthoptera: Pyrgomorphidae), 282

Physalaemus nattereri (Amphibia: Anura: Leiuperidae), 276

Physalia physalis (Cnidaria: Hydrozoa: Physalidae), 45, 147

Phytolacca americana (Plantae: Phytolaccaceae), 413

Picoides pubescens (Aves Picidae), 401

P. villosus (Aves: Picidae), 401

pied flycatcher *see Ficedula hypoleuca*

pied wagtail *see Motacilla alba*

Pierella astyoche (Lepidoptera: Nymphalidae), xx

pierid/Pieridae (Lepidoptera), 2, 32, 35, 45, 122, 125, 135, 137, 139, 141, 154, 199, 260, 299, 300–1

Pieris brassicae (Lepidoptera: Pieridae), 35, 82, 122, 154

P. protodice (Lepidoptera: Pieridae), 154

P. rapae (Lepidoptera: Pieridae), 154

P. sisymbrii (Lepidoptera: Pieridae), 299, 300

pigeon *see Columba livia*

Pinus halepensis (Plantae: Pinaceae), 41

P. sylvestris (Plantae: Pinaceae), 100

Pipraeidea melanonota (Aves: Thraupidae), 204

piping plover *see Charadrius melodius*

Pipistrellus pipistrellus (Mammalia: Vespertilionidae), 257

pipistrelle bats *see Pipistrellus pipistrellus*

pirate perch *see Aphredoderus sayanus*

Pisa tetraodon (Crustacea: Malacostraca: Epialtidae), 83

Pisaura mirabilis (Arachnida: Araneae: Pisauridae), 99

Pison xanthopus (Hymenoptera: Sphecidae), 235

Pistacia terebinthus (Plantae: Anacardiaceae), 413

Pisum elatius (Plantae: Fabaceae), 165

P. humile (Plantae: Fabaceae), 165

P. sativum (Plantae: Fabaceae), 165

Pitangus sulphuratus (Aves: Tyrannidae), 131, 242

pitcher plant *see Nepenthes rafflesiana*

Pitohui (Aves), 124

pitviper *see Trimeresurus popeiorum*

Plagiotremus laudandus (Actinopterygii: Blenniidae), 322, 323

P. rhinorhynchos (Actinopterygii: Blenniidae), 321–2

P. tapeinosoma (Actinopterygii: Blenniidae), 322

P. townsendi (Actinopterygii: Blenniidae), 323

plain nawab butterfly *see Polyura hebe*

Platypleura (Hemiptera: Cicadidae), 282

Platysaurus broadleyi (Reptila: Squamata: Cordylidae), 355, 357

Plebeia droryana (Hymenoptera: Apidae), 381

Plecodus straeleni (Actinopterygii: Cichlidae), 322

Plectroglyphidodon lacrymatus (Actinopterygii: Pomacentridae), 323

Plectropomus (Actinopterygii: Serranidae), 195

Plesiopidae (Actinopterygii), 250

Plethodon jordani (Amphibia: Urodela: Plethodontidae), 119, 240, 357

Pleurosoma (Lepidoptera: Erebidae: Arctiinae), 227–8

Pleurothallis (Plantae: Orchidaceae), 389

Plexippus paykulli (Arachnida: Araneae: Salticidae), 308

Ploceus cucullatus (Aves: Ploceidae), 338

Plotosus anguillaris (Actinopterygii: Plotosidae), 255

plover *see Charadrius*

Pluvialis dominica (Aves: Charadriidae), 60

pocket clam *see Lampsilis*

Podalonia canescens (Hymenoptera: Sphecidae), 381

Podargus strigoides (Aves: Podargidae), 88

Podophyllum peltatum (Plantae: Berberidaceae), 386

Poecilia reticulata (Actinopterygii: Poeciliidae), 270, 318

Poeciliopsis monacha-lucida (Actinopterygii: Poeciliidae), 363

Pogonogaster sp. (Mantodea: Thespidae), 26

poison arrow/dart frogs *see Dendrobates*

poison-fang blenny *see Meiacanthus atrodorsalis*

Poliopastea laciades (Lepidoptera: Erebidae), 141

Polioptila caerulea (Aves: Polioptilidae), 85

Polistes atrimandibularis see Sulcopolistes atrimandibularis

P. buglumis (Hymenoptera: Vespidae), 344

P. dominula (Hymenoptera: Vespidae), 344

P. sulcifer (Hymenoptera: Vespidae), 344

Polybia (Hymenoptera: Vespidae), 226–8

P. dimidiata (Hymenoptera: Vespidae), 105, 226

Polycentrus schomburgkii (Actinopterygii: Polycentridae), 356

Polyergus rufescens (Hymenoptera: Formicidae), 348

Polygonia c-album (Lepidoptera: Nymphalidae), 63, 68, 283

Polyura hebe (Lepidoptera: Nymphalidae: Charaxinae), 23

Pomacanthus (Actinopterygii: Pomacanthidae), 324

Pomatocentrus amboinensis (Actinopterygii: Pomatocentridae), 270

Ponometia elegantula (Lepidoptera: Noctuidae), 89

Populus (Plantae: Salicaceae), 50, 51, 269

P. tremuloides (Plantae: Salicaceae), 269

porcupine fish *see Diodontidae*

Portuguese man-o-war *see Physalia physalis*

Portia (Arachnida: Araneae: Salticidae), 235, 308, 319

P. fimbriata (Arachnida: Araneae: Salticidae), 308, 319

possum *see Didelphis virginiana*

potoo *see Nyctibius jamaicensis*

Pranburia mahannopi (Arachnida: Araneae: Corinnidae), 237

Prepona (Lepidoptera: Nymphalidae), 151

Prinia subflava (Aves: Cisticolidae), 331, 334, 336

Probopyrus pandalicola (Crustacea: Malocostraca: Isopoda), 47

Prociphilus tessellatus (Hemiptera: Aphididae), 85, 346

Procoeca ganglbaueri (Diptera: Nemestrinidae), 387

pronghorn *see Antilocapra Americana*

Protaeolidiella juliae (Mollusca: Nudibrancia: Pleurolidiidae), 22

Proteles cristata (Mammalia: Hyannidae), 240

Protogonius hippon (Lepidoptera: Nymphalidae), 128

Protoreaster nodosus (Echinoidea: Asteroidea: Oreasteridae), 159

Prunella modularis (Aves: Prunellidae), 27, 335, 337

Prunus serotina (Plantae: Rosaceae), 413

P. yedoensis (Plantae: Rosaceae), 51, 53

Pseudacraea eurytus (Lepidoptera: Nymphalidae), 199

P. lucretia (Lepidoptera: Nymphalidae), 137

Pseudamphithoides incurvaria (Crustacea: Malacostraca: Amphipoda), 147

Pseudocerastes urarachnoides (Reptilia: Squamata: Viperidae), 317–18

Pseudochromis fuscus (Actinopterygii: Pseudochromidae), 56

Pseudocreobotra wahlbergi (Mantodea: Hymemopodidae), 314

Pseudosphinx tetrio (Lepidoptera: Sphingidae), 141, 243

Pseudotropheus aurora (Actinopterygii: Cichlidae), 360

Psithyrus see Bombus ('*Psithyrus*')

Psittacus erithacus (Aves: Psittacidae), 402

Psocoptera (Hexapoda), 40

Pterobrycon (Actinopterygii: Characidae), 315, 362

Pterochroza ocellata (Orthoptera: Tettigoniidae), 89

Pterocymbium diversifolium (Plantae: Malvaceae), 296

Pterodroma arminjoniana (Aves: Procellariidae), 306

P. neglecta (Aves: Procellariidae), 306

Pterois (Actinopterygii: Scorpaenidae), 151

Ptilonorhynchus maculatus see Chlamydera maculata

Puccinia (Fungi: Pucciniaceae), 417, 418

Puccinia arrhenatheri (Fungi: Pucciniaceae), 418

P. monoica (Fungi: Pucciniaceae), 418

P. punctiformis (Fungi: Pucciniaceae), 418

pufferfish *see Tetraodontidae*, 195, 197, 198

Pugettia producta (Crustacea: Malacostraca: Epialtidae), 53, 84

P. richii (Crustacea: Malacostraca: Epialtidae), 85

Pulsatilla patens (Plantae: Ranunculaceae), 418

Pultenaea scabra (Plantae: Fabaceae), 385

Pungitius pungitus (Actinopterygii: Gasterosteidae), 356

puss moth *see Cerura vinula*

pygmy grasshopper *see Tetrix subulata*

pygmy owl *see Glaucidium californicum*

Pygoscelis adelae (Aves: Spheniscidae), 311

pyjama nudibranch *see Chromodoris quadricolor*

Pyractonema (Coleoptera: Lampyridae), 325

Pyrophorus (Coleoptera: Elateridae), 155–6

Pyrops astarte (Hemiptera: Fulgoridae), 282

P. pyrorhyncha (Hemiptera: Fulgoridae), 282

Pyrrhocoris apterus (Hemiptera: Pyrrhocoridae), 106, 112, 129, 132, 427

Pythonodipsas (Reptilia: Squamata: Colubridae), 244
Pytilia (Aves: Estrildidae), 330, 340

Quercus (Plantae: Fagaceae), 47, 50, 62
queen butterfly *see Danaus gilippus*
queen butterfly *see Danaus gilippus*
quince *see Cydonia oblonga*
Quiscalus quiscula (Aves: Icteridae), 127, 229

Rafflesia kerri (Plantae: Rafflesiaceae/ Euphorbiaceae), 392
ragwort *see Senecio jacobaea*
Rana pipiens (Amphibia: Anura: Ranidae), 139
Rangifer tarandus (Mammlia: Cervidae), 36
Ranitomeya imitator (Amphibia: Anura: Dendrobatidae), 218
Ranunculus inamoenus (Plantae: Ranunculaceae), 418
ratel *see Mellivora capensis*
rattan palm *see Calamus*
rattlesnakes *see Crotalus viridis, Sistrurus*
Ravenelia (Fungi: Basidiomycota: Raveneliaceae), 418
Recurvirostra americana (Aves: Recurvirostridae), 151
red-backed shrike *see Lanius collurio*
red drum *see Sciaenops ocellatus*
red-flanked bluetail *see Tarsiger cyanurus*
red flour beetle *see Tribolium castaneum*
red-headed mouse spider *see Missulena occatoria*
redstart *see Phoenicurus phoenicurus*
red-winged blackbird *see Agelaius phoeniceus*
reindeer *see Rangifer tarandus*
Reticulitermes flavipes (Isoptera: Rhinotermitidae), 349
R. okinawensis (Isoptera: Rhinotermitidae), 412
Rhagerhis moilensis (Reptilia: Squamata: Colubridae), 250
Rhamnaceae (Plantae), 392
Ramphocelus carbo (Aves: Thraupidae), 124
Rhinella icterica (Amphibia: Anura: Bufonidae), 98
Rhinoclemmys pulcherimma (Reptilia: Testudines: Geoemydidae), 243, 245
Rhodnius prolixus (Hemiptera: Reduviidae), 424
Rhodogorgon (Rhodophyta [red algae]), 302
Ribbon worm *see Nemertea*
rice *see Oryza*
Ricinus communis (Plantae: Euphorbiaceae), 415

robberflies *see Asilidae*
robin
    American *see Turdus migratorius*
    European *see Erithacus rubecula*
rock pocket mice *see Chaetodipus intermedius*
Rogas geniculatus see Aleiodes alternator
Romalea guttata (=*microptera*) (Orthoptera: Romaleidae), 139
Rosa (Plantae: Rosaceae), 164, 233, 243
Rudbeckia (Plantae: Asteraceae), 418
rufous bush chat *see Cercotrichas galactotes*
rufous-collared sparrow *see Zonotrichia capensis*
rufous-tailed jacamar *see Galbula ruficauda*
rufous-tailed scrub robin *see Cercotrichas galactotes*
rufous-thighed kite *see Harpagus diodon*

sable antelope *see Hippotragus niger*
Sagra (Coleoptera: Chrysomelidae), 99, 108
samurai crab *see Heikeopsis japonica*
Salix (Plantae: Salicaceae), 3, 50–1, 441
salps *see Urochordata*
Salticidae (Arachnida: Araneae), 6, 99, 227, 234, 236, 251–2, 319, 328
Salaria pavo (Actinopterygii: Blenniidae), 356
Sambucus nigra (Plantae: Caprifoliaceae), 413
Sander vitreus (Actinopterygii: Percidae), 417
Sapphirina (Crustacea: Copepodidae), 45
Saracenia flava (Plantae: Sarracenaceae), 351
Sarcopoterium spinosum (Plantae: Rosaceae), 164
Sargassum (Chromalveolata: Sargassaceae), 22, 97
sawfly *see Diprion or Neodiprion*
Sathrophyllia (Orthoptera: Tettigoniidae), 89
Saturnia pavonia (Lepidoptera: Saturniidae), 31, 261
S. pavoniella (Lepidoptera: Saturniidae), 31, 159
Satyrium pumilum (Plantae: Orchidaceae), 373, 391
Saucrobotys futilalis (Lepidoptera: Pyralidae), 82
Saxinis deserticola (Coleoptera: Chrysomelidae), 201
Scabiosa (Plantae: Dipsacaceae), 385, 386
Scalenus (Coleoptera: Cerambycidae), 226
Scaphiodontophis (Reptilia: Squamata: Colubridae), 279

Scaphura see Aganacris
Scaptodrosophila bangi (Diptera: Drosophilidae), 373
scarlet bean *see Archidendron ramiflorum*
scarlet kingsnake *see Lampropeltis elapsoides*
Sceliphron caementarium (Hymenoptera: Sphecidae), 236
Schistocerca emarginata (Orthoptera: Acrididae), 112
S. lineata see Schistocerca emarginata
Schistocephalus solidus (Platyhelminthes: Cestoda), 423
Schistosoma mansoni (Platyhelminthes: Trematoda), 423
schistosomes *see Schistosoma mansoni*
Schizotus pectinicornis (Coleoptera: Pyrochroidae), 147
Sciaenops ocellatus (Actinopterygii: Sciaenidae), 31
Scinax fuscomarginatus (Amphibia: Anura: Hylidae), 98
Sciurus carolinensis (Mammalia: Sciuridae), 76
S. niger (Mammalia: Sciuridae), 41
Sclerochiton ilicifolius (Plantae: Acanthaceae), 425
Scorpaenidae (Actinopterygii), 239
Scorpaenodes (Actinopterygii: Scorpaenidae), 239
scorpion fish *see Scorpaenidae*
scorpionflies *see Hylobittacus apicalis*, 16
Scymnus frontalis (Coleoptera: Coccinelidae), 132
Scytodidae (Arachnida: Araneae), 6, 319
seahorse *see Hippocampus*
sea sapphires *see Sapphirina*
sea-slugs, 11, 22, 149
sedge sprite *see Nehalennia irene*
Velum donkei la Haplophryne
Ceratiidae), 54–5
Semioptila (Lepidoptera: Himantoperidae), 275
Senecio jacobaea (Plantae: Asteraceae), 72, 152
S. vernalis (Plantae: Asteraceae), 55–6
Sepia apama (Mollusca: Cephalopoda: Sepiidae), 354–5
S. officinalis (Mollusca: Cephalopoda: Sepiidae), 59–60
S. plangon (Mollusca: Cephalopoda: Sepiidae), 354–55
Serapias vomeracea (Plantae: Orchidaceae), 373
serval *see Leptailurus serval*
Setophaga petechia (Aves: Parulidae), 339
S. ruticilla (Aves: Parulidae), 400

seven-spot ladybird *see Coccinella septempunctata*

sexton beetle *see Nicrophorus*

sharp-angled carpet moth *see Euphyia intermediata*

sheep, as herbivores

shikra falcon *see Accipiter badius*

shrikes *see Lanius* spp.

Siamusotima aranea (Lepidoptera: Crambidae), 251

Siberian chipmunk *see Eutamias sibiricus*

Siderone (Lepidoptera: Nymphalidae), 89

Silphidae (Coleoptera), 250

silver-beaked tanager *see Ramphocelus carbo*

silver-Y moth *see Autographa gamma*

Silybum marianum (Plantae: Asteraceae), 165, 291

Simophis rhinostoma (Reptilia: Squamata: Colubridae), 242

Simulidae (Diptera), 422, 424

Simyra albovenosa (Lepidoptera: Noctuidae), 40

Siphamia (Actinopterygii: Apogonidae), 11, 233

Siraitia grosvenorii (Plantae: Cucurbitaceae), 426

Sistrurus catenatus (Reptilia: Squamata: Viperidae), 317

S. miliarius (Reptilia: Squamata: Viperidae), 317

Sitta carolinensis (Aves: Sittidae), 406

skipper butterfly *see Chrysoplectrum*

slow worm *see Anguis fragilis*

smallmouth bass *see M. dolomieu*

smallmouth grunt *see Haemulon chrysargyreum*

small tortoiseshell butterfly *see Aglais urticae*

Smerinthus ocellatus (Lepidoptera: Sphingidae), 50, 260, 283

smooth snake *see Coronella austriaca*

snail-eating hook-billed kite *see Chondrohierax uncinatus*

snakes, 2, 13, 16, 30, 57, 61–2, 69, 71, 72, 97–100, 107, 121, 124–5, 130–1, 140, 151–2, 155–7, 163, 173, 190, 198–9, 213–14, 225, 239–48, 250, 254, 267, 271–2, 275, 276, 279, 284, 308, 317–18, 320, 328, 355–7, 433, 436, 438, 443

snapper (fish) *see Lutjanus*

snowshoe hare *see Lepus americanus*

Solanderia (Cnidaria: Hydrozoa), 22

Solanaceae (Plantae), 388

Solanum berthaultii (Plantae: Solanaceae), 300

S. fructicosa (Plantae: Solanaceae), 426

Solea vulgaris (Actinopterygii: Soleidae), 239

Solenopsis (Hymenoptera: Formicidae), 239, 349, 439

Solenostomus paradoxus (Actinopterygii: Solenostomidae), 21

song sparrow *see Melospiza melodia*

song thrush *see Turdus philomelos*

sorghum (Plantae: Graminaceae), 415

southern Titiwangsa bent-toed gecko *see Cyrtodactylus australotitiwangsaensis*

Spalerosophis (Reptilia: Squamata: Colubridae), 244

Spanish fly *see Lytta vesicatoria*

sparrowhawk, *see Accipiter nisus*

Spathodea campanulata (Plantae: Bignoniaceae), 395, 396

Sphaerophoria (Diptera: Syrphidae), 313

Sphecidae (Hymenoptera), 225, 227, 229, 236, 429

Spermophilus

Sphecosoma (Lepidoptera: Erebidae: Arctiinae), 227–8

Spheniscus (Aves: Spheniscidae), 8, 286, 309, 311, 408

S. demersus (Aves: Spheniscidae), 408

spiders, 7, 8, 10, 13, 54–56, 85, 89, 92–3, 99, 112, 121, 124–5, 130, 136, 153, 199, 203, 210, 211, 225, 227, 231–8, 251–3, 275, 299, 306–8, 310–13, 315, 317–19, 323, 326, 328, 347, 376, 392, 429, 431, 433–4, 438–40, 442–3

spider crab *see Achaeus spinosus*

Spilomyia hamifera (Diptera: Syrphidae), 157

Spindasis lohita (Lepidoptera: Lycaenidae), 273

Spirobranchus giganteus (Annelida: Polychaeta: Serpulidae), 79

Spirorbis (Annelida: Spirorbidae), 94, 97

Splachnaceae (Plantae), 419

sponge decorator crab *see Hyastenus elatus*

sponge zoanthid *see Parazoanthus*

spotted bowerbird *see Chlamydera maculata*

squirrels *see Sciurus*, *Xerus* and *Citellus*

staghorn damselfish *see Amblyglyphidodon curacao*

Stapelia (Plantae: Apocynaceae), 373, 389, 391

Staphylinidae (Coleoptera), 318

starfinch *see Bathilda ruficauda*

starling *see Sturnus vulgaris*

Stauropus (Lepidoptera: Notodontidae), 237–8

Stegaspis (Hemiptera: Membracidae), 233

Stegostoma fasciatum (Chondrichthyes: Stegostomatidae), 239

Stelis (Plantae: Orchidaceae), 394

Stenella coeruleoalba (Mammalia: Delphinidae), 311

Stenolemus bituberus (Hemiptera: Reduviidae), 319

Stenomorpha marginata (Coleoptera: Tenebrionidae), 121, 193

Stenoria analis (Coleoptera: Meloidae), 255–6

Sterculia quadrifida (Plantae: Sapindaceae), 412–13

Stevardia riisei see Corynopoma riisei

Stevia rebaudiana (Plantae: Asteraceae), 426

stick caterpillar *see Geometridae*

stick insect *see Phasmatodea*

stinging nettle *see Urtica dioica*

stoat *see Mustella ermine*

stonefish (Actinopterygii: Synanceiidae), 239

stone plants *see Lithops*

Storeria dekayi (Reptilia: Squamata: Colubridae), 100

Strangalia (Coleoptera: Cerambycidae), 228

Stratiomyidae (Diptera), 150, 228

Streptanthus breweri (Plantae: Brassicaceae), 299

S. glandulosus (Plantae: Brassicaceae), 300–1

Strophanthus (Plantae: Apocynaceae), 142

striped skunk *see Mephitis mephitis*

Strychnos madagascariensis (Plantae: Loganiaceae), 419

sturgeon *see Acipenseridae*

Sturnus vulgaris (Aves: Sturnidae), 27, 112, 126, 205, 262, 402

Subpsaltria yangi (Hemiptera: Cicadidae), 357

Sulcopolistes atrimandibularis (Hymenoptera: Vespidae), 344

Succinea putris (Mollusca: Gastropoda: Succineidae), 416

sunbird *see Necteriniidae*

sundew *see Drosera*

superb fairy-wren *see Malurus cyaneus*

superb lyrebird *see Menura novaehollandiae*

Suricata suricata (Mammalia: Herpestidae), 403

Surniculus lugubris (Aves: Cuculidae), 343

swallow *see Hirundo*

swallowtail butterflies *see Papilio*, Papilionidae

sweetpotato weevil *see Cylas formicarius*

swellfish *see Tetraodontidae*

sword-tailed characin *see Corynopoma riisei*

Sylvia atricapilla (Aves: Sylviidae), 410, 411, 413
Symphodus (*Crenilabrus*) *melos* (Actinopterygii, Labriidae), 354, 356
S. (*Crenilabrus*) *ocellatus* (Actinopterygii, Labriidae), 354, 356
Synageles occidentalis (Arachnida: Araneae: Salticidae), 235
Synchlora aerata (Lepidoptera: Geometridae), 85
S. frondaria (Lepidoptera: Geometridae), 85
Synodontus nigriventris (Actinopterygii: Mochokidae), 24
Syntomoides imaon (Lepidoptera: Erebidae), 106
Syrphus corollae (Diptera: Syrphidae), 229
S. ribesii (Diptera: Syrphidae), 209
Systella rafflesii (Orthoptera: Tettigoniidae), 89

Tabanidae (Diptera), 69, 208, 374, 385
Tachinidae (Diptera), 150, 208
Taeniopygia guttata (Aves: Estrildidae), 124, 128
Tagesoidea nigrofasciata (Phasmatodea: Diapheromeridae), 282
Talima beckeri (Lepidoptera: Limacodidae), 159
Talpa europaea (Mammalia: Talpidae), 315
Tamiasciurus douglasii (Mammalia: Sciuridae), 156
Tanaecia julii (Lepidoptera: Nymphalidae), 279
tanager see *Lanio versicolor*
Tanaorhinus viridiluteatus (Lepidoptera: Geometridae), 25
Tarentola mauritanica (Reptilia: Squamata: Gekkonidae), 51
Tarsiger cyanurus (Aves: Muscicapidae), 339
tawny-flanked prinia see *Prinia subflava*
tawny frogmouth see *Podargus strigoides*
Telescopus dhara (Reptilia: Squamata: Colubridae), 244, 250
T. fallax (Reptilia: Squamata: Colubridae), 244, 250
Tenebrio molitor (Coleoptera: Tenebrionidae), 126, 130, 150, 194, 196, 206–8, 256–7, 262
Tengmalm's owl see *Aegolius funereus*
Tenodera aridifolia (Mantodea: Mantidae), 308
tentacle snake see *Erpeton tentacularis*
Tetramorium (Hymenoptera: Formicidae), 85, 349
Tetanolita mynesalis (Lepidoptera: Erebidae), 326
Tetraodontidae (Actinopterygii), 197, 253
Tetraplodon (Plantae: Splachnaceae), 419

Tetrix subulata (Orthoptera: Tetrigidae), 41, 76
Thamnomanes schistogynus (Aves: Thamnophilidae), 406
Thamnophilidae (Aves), 277
Thamnophis elegans (Reptilia: Squamata: Colubridae), 190
T. ordinoides (Reptilia: Squamata: Colubridae), 69
T. radix (Reptilia: Squamata: Colubridae), 125
T. sirtalis (Reptilia: Squamata: Colubridae), 30, 190, 356
Thaumatococcus daniellii (Plantae: Marantaceae), 425
Thaumoctopus mimicus (Mollusca: Cephalopoda), 151
Theretra oldenlandiae (Lepidoptera: Sphingidae), 52
Thereus pedusa (Lepidoptera: Lycaenidae), 273
Thomisus labefactus (Arachnida: Araneae: Thomisidae), 306, 313
T. onustus (Arachnida: Araneae: Thomisidae), 56
T. spectabilis (Arachnida: Araneae: Thomisidae), 42
Thyatira batis (Lepidoptera: Thyatiridae), 61, 63–4
Ticherra acte (Lepidoptera: Lycaenidae), 273
tiger see *Panthera tigris*
tiger moths see Arctiinae
Tilapia see *Oreochromis*
Tinolius (Lepidoptera: Noctuidae), 107
Tirumala cf. *gautama* (Lepidoptera: Nymphalidae: Danainae), 107
Titanacris albipes (Orthoptera: Romaleidae), 282
toad see *Bufo terrestris*
Tomaiuris (Aves: Ibla ‘Thinocoridae’), 88
topi antelope see *Damaliscus lunatus*
Toxocarpus (Plantae: Apocynaceae), 200
Tozeuma carolinense (Crustacea: Hippolytidae), 31
Trachinus draco (Actinopterygii: Trachinidae), 239
T. vipera (Actinopterygii: Trachinidae), 239
Tragelephus (Mammalia: Bovidae), 165
Traunsteinera globosa (Plantae: Orchidaceae), 386
tree porcupine see *Coendou*
Triatoma infestans (Hemiptera: Reduviidae), 424
Tribolium castaneum (Coleoptera: Tenebrionidae), 99

Trichocentrum ascendens (Plantae: Orchidaceae), 382
Trichopsenius frosti (Coleoptera: Staphylinidae), 349
Trichura (Lepidoptera: Erebidae: Arctiinae), 230
Trifolium pratense (Plantae: Fabaceae), 303
T. repens (Plantae: Fabaceae), 303
T. subterraneum (Plantae: Fabaceae), 427
Trigonidium obtusum (Plantae: Orchidaceae), 381
Trimeresurus popeiorum (Reptilia: Squamata: Viperidae), 62
Trioxys angelicae (Hymenoptera: Braconidae), 350
Tripterygion (Actinopterygii: Tripterygiidae), 356
Tritoniopsis triticea (Plantae: Iridaceae), 384–5
Trogon melanocephalus (Aves: Trogonidae), 142
Tropidophiidae (Reptilia: Squamata), 99
Tropidothorax leucopterus (Hemiptera: Lygaeidae), 123
truffles see *Tuber*
Trypanosoma cruzi (Excavata: Kinetoplastida), 124
tsetse fly see *Glossina*
Tuber (Fungi: Actinomycetes: Tuberaceae), 418
Turbellaria see *Bipalium rauchi*
Turdus merula (Aves: Turdidae), 27–8, 75, 129, 210
T. migratorius (Aves: Turdidae), 336
T. philomelos (Aves: Turdidae), 27, 204, 210, 335
turkey moray eel see *Gymnothorax meleagris*
turkey vulture see *Cathartes aura*
Turnera sidoides (Plantae: Turneraceae), 306
turquoise-browed motmot see *Eumomota superciliosa*
Tutelina similis (Arachnida: Araneae: Salticidae), 236
two-spot ladybird see *Adalia bipunctata*
Typophyllum (Orthoptera: Tettigoniidae), 89
Tyrannus melancholicus (Aves: Tyrannidae), 138, 196
Tyria jacobaeae (Lepidoptera: Erebidae: Arctiinae), 72, 152
Tyto alba (Aves: Tytonidae), 307

Uca annulipes (Crustacea: Malacostraca: Ocypodidae), 408
Uca capricornis (Crustacea: Malacostraca: Ocypodidae), 54

underwing moths *see Catocala*
upside-down catfish *see Synodontus nigriventris*
Urabunana marshalli (Hemiptera: Cicadidae), 327
Urochordata, 45
Urtica dioica (Plantae: Urticaceae), 295
Ustilago violacea see Microbotryum violaceum
Utetheisa jacobaeae see Tyria jacobaeae
U. ornatrix (Lepidoptera: Erebidae: Arctiinae), 145

Vaccinium (Plantae: Ericaceae), 417–19
Vachellia (Plantae: Fabaceae), 164
V. cornigera (Plantae: Fabaceae), 164, 326
V. drepanolobium (Plantae: Fabaceae), 165
Valeriana (Plantae: Caprifoliaceae), 386, 436
Varroa jacobsoni (Arachnida: Mesostigmata), 347
Velella (Cnidaria: Hydrozoa: Porpitidae), 45
velvet ant *see* Mutillidae
Venus flytrap *see Dionaea muscipula*
vervet monkey *see Cercopithecus pygerythrus*
Vespa affinis (Hymenoptera: Vespidae), 326
V. bicolor (Hymenoptera: Vespidae), 373, 389
V. crabro (Hymenoptera: Vespidae), 347
V. dybalskii (Hymenoptera: Vespidae), 347
V. simillima (Hymenoptera: Vespidae), 347
V. velutina (Hymenoptera: Vespidae), 225
Vespertilionidae (Mammalia), 257
Vespula (Hymenoptera: Vespidae), 208, 249, 344, 373
viceroy butterfly *see Limenitis archippus*
Vicia sativa (Plantae: Fabaceae), 415
Vidua (Aves Viduidae), 329–31, 338–40, 342
V. purpurascens (Aves Viduidae), 323, 330
Vipera berus (Reptilia: Squamata: Viperidae), 69, 246–7

V. palaestinae (Reptilia: Squamata: Viperidae), 244
Vulpes lagopus (Mammalia: Canidae), 36

walnut *see Juglans regia*
walleye *see Sander vitreus*
warthog *see Phacochoerus aethiopicus*
Watsonia lepida (Plantae: Iridaceae), 384–5
wavy-lined emerald moth *see Synchlora aerata*
weaver ants *see Oecophylla*
whistling acacia *see Acacia drepanolobium*
white-breasted nuthatch *see Sitta carolinensis*
white butterflies *see Colotis, Pieris,* Pieridae
white clover *see Trifolium repens*
white dead nettle *see Lamium album*
white-footed mice *see Peromyscus leucopus.*
white-nosed coati *see Nasua narica*
wolf snake *see Lycodon ophiophagus*
woodand brown butterfly *see Lopinga achine*
wood nymph moth *see Eudryas unio*
wood tiger moth *see Parasemia plantaginis*
wood warbler *see Phylloscopus*
woolly alder aphid *see Prociphilus tessellatus*
wrangler grasshopper *see Circotettix rabula*
wryneck *see Jynx*

Xanthium strumarium (Plantae: Asteraceae), 232, 301
Xanthomonas axonopodis (Bacteria: Proteobacteria), 425
Xenocarcinus (Crustacea: Decapoda: Epialtidae), 21
Xerus erythropus (Mammalia; Sciuridae) 97
Xiphophorus nezahualcoyotl (Actinopterygii: Poeciliidae), 356

yellow-bellied sea snake *see Pelamis platura*
yellow-browed leaf warbler *see Phylloscopus humei*

yellowfin goatfish *see Mulloidichthys vanicolensis*
yellow goatfish *see M. martinicus*
yellowjacket cichlid *see Parachromis friedrichsthalii*
yellow perch *see Perca flavescens*
yellow umbrella stick insect *see Tagesoidea nigrofasciata*
yellow warbler *see Setophaga petechia*

Zacryptocerus (Hymenoptera: Formicidae), 236
Zaluzianskya microsiphon (Plantae: Scrophulariaceae), 387
Zaretis itys (Lepidoptera: Nymphalidae), 89
Zaspilothynnus trilobatus (Hymenoptera: Typhiidae), 377–8
zebra *see Equus burchelli, grevyi & zebra*
zebra finch *see Taeniopygia guttata*
zebra shark *see Stegostoma fasciatum*
Zelurus (Hemiptera: Reduviidae), 228
Zicrona coerulea (Hemiptera: Pentatomidae), 150
Zodarion (Arachnida: Araneae: Zodariidae), 236
zone-tailed hawk *see Buteo albonotatus*
Zonosemata vittigera (Diptera: Tephritidae), 252
Zonotrichia capensis (Aves: Emberizidae), 334
Z. leucophrys (Aves: Emberizidae), 402
Z. querula (Aves: Emberizidae), 401
Zygaena (Lepidoptera: Zygaenidae), 100, 145, 196, 440
Z. ephialtes (Lepidoptera: Zygaenidae), 196
Zygaenidae (Lepidoptera), 91, 116
Zygiella x-notata (Arachnida: Araneae: Araneaeidae), 112
Zygothrica (Diptera: Drosophilidae), 373, 389
Zyras (Coleoptera: Staphylinidae), 349